THE POWER ELECTRONICS HANDBOOK

INDUSTRIAL ELECTRONICS SERIES

Series Editor
J. David Irwin, *Auburn University*

TITLES INCLUDED IN THE SERIES

Supervised and Unsupervised Pattern Recognition: Feature Extraction and Computational Intelligence
Evangelia Micheli-Tzanakou, *Rutgers University*

Switched Reluctance Motor Drives: Modeling, Simulation, Analysis, Design, and Applications
R. Krishnan, *Virginia Tech*

The Power Electronics Handbook
Timothy L. Skvarenina, *Purdue University*

The Handbook of Applied Computational Intelligence
Mary Lou Padgett, *Auburn University*
Nicolaos B. Karayiannis, *University of Houston*
Lofti A. Zadeh, *University of California, Berkeley*

The Handbook of Applied Neurocontrols
Mary Lou Padgett, *Auburn University*
Charles C. Jorgensen, *NASA Ames Research Center*
Paul Werbos, *National Science Foundation*

THE POWER ELECTRONICS HANDBOOK

INDUSTRIAL ELECTRONICS SERIES

Edited by
TIMOTHY L. SKVARENINA
Purdue University
West Lafayette, Indiana

CRC PRESS

Boca Raton   London   New York   Washington, D.C.

Library of Congress Cataloging-in-Publication Data

The power electronics handbook / edited by Timothy L. Skvarenina.
    p. cm. — (Industrial electronics series)
    Includes bibliographical references and index.
    ISBN 0-8493-7336-0 (alk. paper)
    1. Power electronics. I. Skvarenina, Timothy L. II. Series.

TK7881.15 .P673 2001
621.31′7—dc21
                                                                                                                  2001043047

This book contains information obtained from authentic and highly regarded sources. Reprinted material is quoted with permission, and sources are indicated. A wide variety of references are listed. Reasonable efforts have been made to publish reliable data and information, but the authors and the publisher cannot assume responsibility for the validity of all materials or for the consequences of their use.

Neither this book nor any part may be reproduced or transmitted in any form or by any means, electronic or mechanical, including photocopying, microfilming, and recording, or by any information storage or retrieval system, without prior permission in writing from the publisher.

All rights reserved. Authorization to photocopy items for internal or personal use, or the personal or internal use of specific clients, may be granted by CRC Press LLC, provided that $1.50 per page photocopied is paid directly to Copyright Clearance Center, 222 Rosewood Drive, Danvers, MA 01923 USA The fee code for users of the Transactional Reporting Service is ISBN 0-8493-7336-0/02/$0.00+$1.50. The fee is subject to change without notice. For organizations that have been granted a photocopy license by the CCC, a separate system of payment has been arranged.

The consent of CRC Press LLC does not extend to copying for general distribution, for promotion, for creating new works, or for resale. Specific permission must be obtained in writing from CRC Press LLC for such copying.

Direct all inquiries to CRC Press LLC, 2000 N.W. Corporate Blvd., Boca Raton, Florida 33431.

Trademark Notice: Product or corporate names may be trademarks or registered trademarks, and are used only for identification and explanation, without intent to infringe.

Visit the CRC Press Web site at www.crcpress.com

© 2002 by CRC Press LLC

No claim to original U.S. Government works
International Standard Book Number 0-8493-7336-0
Library of Congress Card Number 2001043047
Printed in the United States of America 1 2 3 4 5 6 7 8 9 0
Printed on acid-free paper

Preface

Introduction

The control of electric power with power electronic devices has become increasingly important over the last 20 years. Whole new classes of motors have been enabled by power electronics, and the future offers the possibility of more effective control of the electric power grid using power electronics. *The Power Electronics Handbook* is intended to provide a reference that is both concise and useful for individuals, ranging from students in engineering to experienced, practicing professionals. The Handbook covers the very wide range of topics that comprise the subject of power electronics blending many of the traditional topics with the new and innovative technologies that are at the leading edge of advances being made in this subject. Emphasis has been placed on the practical application of the technologies discussed to enhance the value of the book to the reader and to enable a clearer understanding of the material. The presentations are deliberately tutorial in nature, and examples of the practical use of the technology described have been included.

The contributors to this Handbook span the globe and include some of the leading authorities in their areas of expertise. They are from industry, government, and academia. All of them have been chosen because of their intimate knowledge of their subjects as well as their ability to present them in an easily understandable manner.

Organization

The book is organized into three parts. Part I presents an overview of the semiconductor devices that are used, or projected to be used, in power electronic devices. Part II explains the operation of circuits used in power electronic devices, and Part III describes a number of applications for power electronics, including motor drives, utility applications, and electric vehicles.

The Power Electronics Handbook is designed to provide both the young engineer and the experienced professional with answers to questions involving the wide spectrum of power electronics technology covered in this book. The hope is that the topical coverage, as well as the numerous avenues to its access, will effectively satisfy the reader's needs.

Acknowledgments

First and foremost, I wish to thank the authors of the individual sections and the editorial advisors for their assistance. Obviously, this handbook would not be possible without them. I would like to thank all the people who were involved in the preparation of this handbook at CRC Press, especially Nora Konopka and Christine Andreasen for their guidance and patience. Finally, my deepest appreciation goes to my wife Carol who graciously allows me to pursue activities such as this despite the time involved.

The Editor

Timothy L. Skvarenina received his B.S.E.E. and M.S.E.E. degrees from the Illinois Institute of Technology in 1969 and 1970, respectively, and his Ph.D. in electrical engineering from Purdue University in 1979. In 1970, he entered active duty with the U.S. Air Force, where he served 21 years, retiring as a lieutenant colonel in 1991. During his Air Force career, he spent 6 years designing, constructing, and inspecting electric power distribution projects for a variety of facilities. He also was assigned to the faculty of the Air Force Institute of Technology (AFIT) for 3 years, where he taught and researched conventional power systems and pulsed-power systems, including railguns, high-power switches, and magnetocumulative generators. Dr. Skvarenina received the Air Force Meritorious Service Medal for his contributions to the AFIT curriculum in 1984. He also spent 4 years with the Strategic Defense Initiative Office (SDIO), where he conducted and directed large-scale systems analysis studies. He received the Department of Defense Superior Service Medal in 1991 for his contributions to SDIO.

In 1991, Dr. Skvarenina joined the faculty of the School of Technology at Purdue University, where he currently teaches undergraduate courses in electrical machines and power systems, as well as a graduate course in facilities engineering. He is a senior member of the IEEE; a member of the American Society for Engineering Education (ASEE), Tau Beta Pi, and Eta Kappa Nu; and a registered professional engineer in the state of Colorado.

Dr. Skvarenina has been active in both IEEE and ASEE. He has held the offices of secretary, vice-chair, and chair of the Central Indiana chapter of the IEEE Power Engineering Society. At the national level he is a member of the Power Engineering Society Education Committee. He has also been active in the IEEE Education Society, serving as an associate editor of the *Transactions on Education* and co-program chair for the 1999 and 2003 Frontiers in Education Conferences. For his activity and contributions to the Education Society, he received the IEEE Third Millennium Medal in 2000.

Within ASEE, Dr. Skvarenina has been an active member of the Energy Conversion and Conservation Division, serving in a series of offices including division chair. In 1999, he was elected by the ASEE membership to the Board of Directors for a 2-year term as Chair, Professional Interest Council III. In June 2000, he was elected by the Board of Directors as Vice-President for Profession Interest Councils for the year 2000–2001.

Dr. Skvarenina is the principal author of a textbook, *Electric Power and Controls*, published in 2001. He has authored or co-authored more than 25 papers in the areas of power systems, power electronics, pulsed-power systems, and engineering education.

Editorial Advisors

Mariesa Crow
University of Missouri-Rolla
Rolla, Missouri

Farhad Nozari
Boeing Corporation
Seattle, Washington

Scott Sudhoff
Purdue University
West Lafayette, Indiana

Annette von Jouanne
Oregon State University
Corvallis, Oregon

Oleg Wasynczuk
Purdue University
West Lafayette, Indiana

Contributors

Ali Agah
Sharif University of Technology
Tehran, Iran

Ashish Agrawal
University of Alaska Fairbanks
Fairbanks, Alaska

Hirofumi Akagi
Tokyo Institute of Technology
Tokyo, Japan

Sohail Anwar
Pennsylvania State University
Altoona, Pennsylvania

Rajapandian Ayyanar
Arizona State University
Tempe, Arizona

Vrej Barkhordarian
International Rectifier
El Segundo, California

Ronald H. Brown
Marquette University
Milwaukee, Wisconsin

Patrick L. Chapman
University of Illinois
 at Urbana-Champaign
Urbana, Illinois

Badrul H. Chowdhury
University of Missouri-Rolla
Rolla, Missouri

Keith Corzine
University of Wisconsin-
 Milwaukee
Milwaukee, Wisconsin

Dariusz Czarkowski
Polytechnic University
Brooklyn, New York

Alexander Domijan, Jr.
University of Florida
Gainesville, Florida

Mehrdad Ehsani
Texas A&M University
College Station, Texas

Ali Emadi
Illinois Institute of Technology
Chicago, Illinois

Ali Feliachi
West Virginia University
Morgantown, West Virginia

Wayne Galli
Southwest Power Pool
Little Rock, Arkansas

Michael Giesselmann
Texas Tech University
Lubbock, Texas

Tilak Gopalarathnam
Texas A&M University
College Station, Texas

Sam Guccione
Eastern Illinois University
Charleston, Illinois

Sándor Halász
Budapest University
 of Technology
 and Economics
Budapest, Hungary

Azra Hasanovic
West Virginia University
Morgantown, West Virginia

John Hecklesmiller
Best Power Technology, Inc.
Nededah, Wisconsin

Alex Q. Huang
Virginia Polytechnic Institute
 and State University
Blacksburg, Virginia

Iqbal Husain
The University of Akron
Akron, Ohio

Amit Kumar Jain
University of Minnesota
Minneapolis, Minnesota

Attila Karpati
Budapest University
 of Technology
 and Economics
Budapest, Hungary

Philip T. Krein
University of Illinois
  at Urbana-Champaign
Urbana, Illinois

Dave Layden
Best Power Technology, Inc.
Nededah, Wisconsin

Daniel Logue
University of Illinois
  at Urbana-Champaign
Urbana, Illinois

Javad Mahdavi
Sharif University
  of Technology
Tehran, Iran

Paolo Mattavelli
University of Padova
Padova, Italy

Roger Messenger
Florida Atlantic University
Boca Raton, Florida

István Nagy
Budapest University
  of Technology
  and Economics
Budapest, Hungary

Tahmid Ur Rahman
Texas A&M University
College Station, Texas

Kaushik Rajashekara
Delphi Automotive Systems
Kokomo, Indiana

Michael E. Ropp
South Dakota State University
Brookings, South Dakota

Hossein Salehfar
University of North Dakota
Grand Forks, North Dakota

Bipin Satavalekar
University of Alaska Fairbanks
Fairbanks, Alaska

Karl Schoder
West Virginia University
Morgantown, West Virginia

Daniel Jeffrey Shortt
Cedarville University
Cedarville, Ohio

Timothy L. Skvarenina
Purdue University
West Lafayette, Indiana

Zhidong Song
University of Florida
Gainesville, Florida

Giorgio Spiazzi
University of Padova
Padova, Italy

Ana Stankovic
Cleveland State University
Cleveland, Ohio

Ralph Staus
Pennsylvania State University
Reading, Pennsylvania

Laura Steffek
Best Power Technology, Inc.
Nededah, Wisconsin

Roman Stemprok
University of North Texas
Denton, Texas

Mahesh M. Swamy
Yaskawa Electric America
Waukegan, Illinois

Hamid A. Toliyat
Texas A&M University
College Station, Texas

Eric Walters
P. C. Krause and Associates
West Lafayette, Indiana

Oleg Wasynczuk
Purdue University
West Lafayette, Indiana

Richard W. Wies
University of Alaska
  Fairbanks
Fairbanks, Alaska

Brian Young
Best Power Technology, Inc.
Nededah, Wisconsin

Contents

PART I Power Electronic Devices

1 Power Electronics
  1.1 Overview   *Kaushik Rajashekara* 1-2
  1.2 Diodes   *Sohail Anwar* 1-9
  1.3 Schottky Diodes   *Sohail Anwar* 1-15
  1.4 Thyristors   *Sohail Anwar* 1-18
  1.5 Power Bipolar Junction Transistors   *Sohail Anwar* 1-24
  1.6 MOSFETs   *Vrej Barkhordarian* 1-31
  1.7 General Power Semiconductor Switch Requirements   *Alex Q. Huang* 1-57
  1.8 Gate Turn-Off Thyristors   *Alex Q. Huang* 1-62
  1.9 Insulated Gate Bipolar Transistors   *Alex Q. Huang* 1-74
  1.10 Gate-Commutated Thyristors and Other Hard-Driven GTOs   *Alex Q. Huang* 1-79
  1.11 Comparison Testing of Switches   *Alex Q. Huang* 1-86

PART II Power Electronic Circuits and Controls

2 DC-DC Converters
  2.1 Overview   *Richard Wies, Bipin Satavalekar, and Ashish Agrawal* 2-1
  2.2 Choppers   *Javad Mahdavi, Ali Agah, and Ali Emadi* 2-3
  2.3 Buck Converters   *Richard Wies, Bipin Satavalekar, and Ashish Agrawal* 2-8
  2.4 Boost Converters   *Richard Wies, Bipin Satavalekar, and Ashish Agrawal* 2-12
  2.5 Cúk Converter   *Richard Wies, Bipin Satavalekar, and Ashish Agrawal* 2-14
  2.6 Buck–Boost Converters   *Daniel Jeffrey Shortt* 2-17

3 AC-AC Conversion   *Sándor Halász*
  3.1 Introduction 3-1
  3.2 Cycloconverters 3-1
  3.3 Matrix Converters 3-3

4 Rectifiers
  4.1 Uncontrolled Single-Phase Rectifiers   *Sam Guccione* 4-1
  4.2 Uncontrolled and Controlled Rectifiers   *Mahesh M. Swamy* 4-4
  4.3 Three-Phase Pulse-Width-Modulated Boost-Type Rectifiers   *Ana Stankovic* 4-33

5 Inverters

- 5.1 Overview *Michael Giesselmann* 5-1
- 5.2 DC-AC Conversion *Attila Karpati* 5-8
- 5.3 Resonant Converters *István Nagy* 5-25
- 5.4 Series-Resonant Inverters *Dariusz Czarkowski* 5-42
- 5.5 Resonant DC-Link Inverters *Michael B. Ropp* 5-56
- 5.6 Auxiliary Resonant Commutated Pole Inverters
  Eric Walters and Oleg Wasynczuk 5-67

6 Multilevel Converters *Keith Corzine*

- 6.1 Introduction 6-1
- 6.2 Multilevel Voltage Source Modulation 6-2
- 6.3 Fundamental Multilevel Converter Topologies 6-7
- 6.4 Cascaded Multilevel Converter Topologies 6-15
- 6.5 Multilevel Converter Laboratory Examples 6-17
- 6.6 Conclusion 6-21

7 Modulation Strategies

- 7.1 Introduction *Michael Giesselmann* 7-1
- 7.2 Six-Step Modulation *Michael Giesselmann* 7-2
- 7.3 Pulse Width Modulation *Michael Giesselmann* 7-2
- 7.4 Third Harmonic Injection for Voltage Boost of SPWM Signals
  Michael Giesselmann 7-9
- 7.5 Generation of PWM Signals Using Microcontrollers and DSPs
  Michael Giesselmann 7-11
- 7.6 Voltage-Source-Based Current Regulation *Michael Giesselmann* 7-12
- 7.7 Hysteresis Feedback Control *Hossein Salehfar* 7-14
- 7.8 Space-Vector Pulse Width Modulation
  Hamid A. Toliyat and Tahmid Ur Rahman 7-28

8 Sliding-Mode Control of Switched-Mode Power Supplies
Giorgio Spiazzi and Paolo Mattavelli

- 8.1 Introduction 8-2
- 8.2 Introduction to Sliding-Mode Control 8-2
- 8.3 Basics of Sliding-Mode Theory 8-5
- 8.4 Application of Sliding-Mode Control to DC-DC Converters—Basic Principle 8-8
- 8.5 Sliding-Mode Control of Buck DC-DC Converters 8-9
- 8.6 Extension to Boost and Buck–Boost DC-DC Converters 8-14
- 8.7 Extension to Cúk and SEPIC DC-DC Converters 8-18
- 8.8 General-Purpose Sliding-Mode Control Implementation 8-20
- 8.9 Conclusions 8-22

Part III Applications and Systems Considerations

9 DC Motor Drives *Ralph Staus*
9.1 DC Motor Basics 9-1
9.2 DC Speed Control 9-2
9.3 DC Drive Basics 9-3
9.4 Transistor PWM DC Drives 9-4
9.5 SCR DC Drives 9-5

10 AC Machines Controlled as DC Machines (Brushless DC Machines/Electronics) *Hamid A. Toliyat and Tilak Gopalarathnam*
10.1 Introduction 10-1
10.2 Machine Construction 10-2
10.3 Motor Characteristics 10-4
10.4 Power Electronic Converter 10-7
10.5 Position Sensing 10-9
10.6 Pulsating Torque Components 10-11
10.7 Torque-Speed Characteristics 10-11
10.8 Applications 10-15

11 Control of Induction Machine Drives *Daniel Logue and Philip T. Krein*
11.1 Introduction 11-1
11.2 Scalar Induction Machine Control 11-2
11.3 Vector Control of Induction Machines 11-4
11.4 Summary 11-17

12 Permanent-Magnet Synchronous Machine Drives *Patrick L. Chapman*
12.1 Introduction 12-1
12.2 Construction of PMSM Drive Systems 12-2
12.3 Simulation and Model 12-3
12.4 Controlling the PMSM 12-6
12.5 Advanced Topics in PMSM Drives 12-9

13 Switched Reluctance Machines *Iqbal Husain*
13.1 Introduction 13-1
13.2 SRM Configuration 13-2
13.3 Basic Principle of Operation 13-4
13.4 Design 13-9
13.5 Converter Topologies 13-11
13.6 Control Strategies 13-14
13.7 Sensorless Control 13-16
13.8 Applications 13-19

14 Step Motor Drives *Ronald H. Brown*
14.1 Introduction **14**-1
14.2 Types and Operation of Step Motors **14**-2
14.3 Step Motor Models **14**-8
14.4 Control of Step Motors **14**-12

15 Servo Drives *Sándor Halász*
15.1 DC Drives **15**-2
15.2 Induction Motor Drives **15**-3

16 Uninterruptible Power Supplies *Laura Steffek, John Hacklesmiller, Dave Layden, and Brian Young*
16.1 UPS Functions **16**-1
16.2 Static UPS Topologies **16**-3
16.3 Rotary UPSs **16**-6
16.4 Alternate AC and DC Sources **16**-7

17 Power Quality and Utility Interface Issues
17.1 Overview *Wayne Galli* **17**-1
17.2 Power Quality Considerations *Timothy L. Skvarenina* ... **17**-3
17.3 Passive Harmonic Filters *Badrul H. Chowdhury* **17**-20
17.4 Active Filters for Power Conditioning *Hirofumi Akagi* ... **17**-30
17.5 Unity Power Factor Rectification *Rajapandian Ayyanar and Amit Kumar Jain* **17**-49

18 Photovoltaic Cells and Systems *Roger Messenger*
18.1 Introduction **18**-1
18.2 Solar Cell Fundamentals **18**-1
18.3 Utility Interactive PV Applications **18**-4
18.4 Stand-Alone PV Systems **18**-7

19 Flexible, Reliable, and Intelligent Electrical Energy Delivery Systems *Alexander Domijan, Jr. and Zhidong Song*
19.1 Introduction **19**-1
19.2 The Concept of FRIENDS **19**-2
19.3 Development of FRIENDS **19**-5
19.4 The Advanced Power Electronic Technologies within QCCs ... **19**-7
19.5 Significance of FRIENDS **19**-9
19.6 Realization of FRIENDS **19**-11
19.7 Conclusions **19**-12

20 Unified Power Flow Controllers *Ali Feliachi, Azra Hasanovic, and Karl Schoder*
20.1 Introduction **20**-1
20.2 Power Flow on a Transmission Line **20**-2

20.3 UPFC Description and Operation 20-4
20.4 UPFC Modeling 20-9
20.5 Control Design 20-14
20.6 Case Study 20-20
20.7 Conclusion 20-24
Acknowledgment 20-25

21 More-Electric Vehicles *Ali Emadi and Mehrdad Ehsani*
21.1 Aircraft *Ali Emadi and Mehrdad Ehsani* 21-1
21.2 Terrestrial Vehicles *Ali Emadi and Mehrdad Ehsani* 21-6

22 Principles of Magnetics *Roman Stemprok*
22.1 Introduction 22-1
22.2 Nature of a Magnetic Field 22-1
22.3 Electromagnetism 22-2
22.4 Magnetic Flux Density 22-3
22.5 Magnetic Circuits 22-3
22.6 Magnetic Field Intensity 22-4
22.7 Maxwell's Equations 22-5
22.8 Inductance 22-7
22.9 Practical Considerations 22-7

23 Computer Simulation of Power Electronics *Michael Giesselmann*
23.1 Introduction 23-1
23.2 Code Qualification and Model Validation 23-2
23.3 Basic Concepts—Simulation of a Buck Converter 23-3
23.4 Advanced Techniques—Simulation of a Full-Bridge (H-Bridge) Converter ... 23-10
23.5 Conclusions 23-22

Index I-1

I

Power Electronic Devices

1 Power Electronics *Kaushik Rajashekara, Sohail Anwar, Vrej Barkhordarian, Alex Q. Huang*... 1-1

Overview • Diodes • Schottky Diodes • Thyristors • Power Bipolar Junction Transistors • MOSFETs • General Power Semiconductor Switch Requirements • Gate Turn-Off Thyristors • Insulated Gate Bipolar Transistors • Gate-Commutated Thyristors and Other Hard-Driven GTOs • Comparison Testing of Switches

1
Power Electronics

Kaushik Rajashekara
Delphi Automotive Systems

Sohail Anwar
Pennsylvania State University

Vrej Barkhordarian
International Rectifier

Alex Q. Huang
Virginia Polytechnic Institute and State University

1.1 Overview 1-2
   Thyristor and Triac • Gate Turn-Off Thyristor • Reverse-Conducting Thyristor (RCT) and Asymmetrical Silicon-Controlled Rectifier (ASCR) • Power Transistor • Power MOSFET • Insulated-Gate Bipolar Transistor (IGBT) • MOS-Controlled Thyristor (MCT)

1.2 Diodes 1-9
   Characteristics • Principal Ratings for Diodes • Rectifier Circuits • Testing a Power Diode • Protection of Power Diodes

1.3 Schottky Diodes 1-15
   Characteristics • Data Specifications • Testing of Schottky Diodes

1.4 Thyristors 1-18
   The Basics of Silicon-Controlled Rectifiers (SCR) • Characteristics • SCR Turn-Off Circuits • SCR Ratings • The DIAC • The Triac • The Silicon-Controlled Switch • The Gate Turn-Off Thyristor • Data Sheet for a Typical Thyristor

1.5 Power Bipolar Junction Transistors 1-28
   The Volt-Ampere Characteristics of a BJT • BJT Biasing • BJT Power Losses • BJT Testing • BJT Protection

1.6 MOSFETs 1-45
   Static Characteristics • Dynamic Characteristics • Applications

1.7 General Power Semiconductor Switch Requirements 1-60

1.8 Gate Turn-Off Thyristors 1-65
   GTO Forward Conduction • GTO Turn-Off and Forward Blocking • Practical GTO Turn-Off Operation • Dynamic Avalanche • Non-Uniform Turn-Off Process among GTO Cells • Summary

1.9 Insulated Gate Bipolar Transistors 1-77
   IGBT Structure and Operation

1.10 Gate-Commutated Thyristors and Other Hard-Driven GTOs 1-82
   Unity Gain Turn-Off Operation • Hard-Driven GTOs

1.11 Comparison Testing of Switches 1-89
   Pulse Tester Used for Characterization • Devices Used for Comparison • Unity Gain Verification • Gate Drive Circuits • Forward Conduction Loss Characterization • Switching Tests • Discussion • Comparison Conclusions

1.1 Overview

Kaushik Rajashekara

The modern age of power electronics began with the introduction of thyristors in the late 1950s. Now there are several types of power devices available for high-power and high-frequency applications. The most notable power devices are gate turn-off thyristors, power Darlington transistors, power MOSFETs, and insulated-gate bipolar transistors (IGBTs). Power semiconductor devices are the most important functional elements in all power conversion applications. The power devices are mainly used as switches to convert power from one form to another. They are used in motor control systems, uninterrupted power supplies, high-voltage DC transmission, power supplies, induction heating, and in many other power conversion applications. A review of the basic characteristics of these power devices is presented in this section.

Thyristor and Triac

The thyristor, also called a silicon-controlled rectifier (SCR), is basically a four-layer three-junction *pnpn* device. It has three terminals: anode, cathode, and gate. The device is turned on by applying a short pulse across the gate and cathode. Once the device turns on, the gate loses its control to turn off the device. The turn-off is achieved by applying a reverse voltage across the anode and cathode. The thyristor symbol and its volt–ampere characteristics are shown in Fig. 1.1. There are basically two classifications of thyristors: converter grade and inverter grade. The difference between a converter-grade and an inverter-grade thyristor is the low turn-off time (on the order of a few microseconds) for the latter. The converter-grade thyristors are slow type and are used in natural commutation (or phase-controlled) applications.

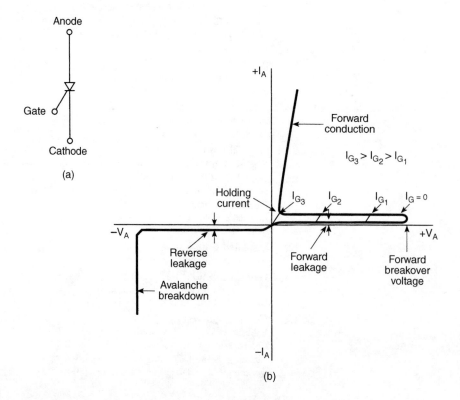

FIGURE 1.1 (a) Thyristor symbol and (b) volt–ampere characteristics. (From Bose, B.K., *Modern Power Electronics: Evaluation, Technology, and Applications*, p. 5. © 1992 IEEE. With permission.)

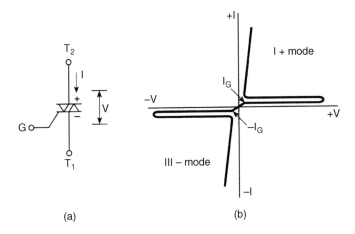

FIGURE 1.2 (a) Triac symbol and (b) volt–ampere characteristics. (From Bose, B.K., *Modern Power Electronics: Evaluation, Technology, and Applications*, p. 5. © 1992 IEEE. With permission.)

Inverter-grade thyristors are used in forced commutation applications such as DC-DC choppers and DC-AC inverters. The inverter-grade thyristors are turned off by forcing the current to zero using an external commutation circuit. This requires additional commutating components, thus resulting in additional losses in the inverter.

Thyristors are highly rugged devices in terms of transient currents, di/dt, and dv/dt capability. The forward voltage drop in thyristors is about 1.5 to 2 V, and even at higher currents of the order of 1000 A, it seldom exceeds 3 V. While the forward voltage determines the on-state power loss of the device at any given current, the switching power loss becomes a dominating factor affecting the device junction temperature at high operating frequencies. Because of this, the maximum switching frequencies possible using thyristors are limited in comparison with other power devices considered in this section.

Thyristors have $I^2 t$ withstand capability and can be protected by fuses. The nonrepetitive surge current capability for thyristors is about 10 times their rated root mean square (rms) current. They must be protected by snubber networks for dv/dt and di/dt effects. If the specified dv/dt is exceeded, thyristors may start conducting without applying a gate pulse. In DC-to-AC conversion applications, it is necessary to use an antiparallel diode of similar rating across each main thyristor. Thyristors are available up to 6000 V, 3500 A.

A triac is functionally a pair of converter-grade thyristors connected in antiparallel. The triac symbol and volt–ampere characteristics are shown in Fig. 1.2. Because of the integration, the triac has poor reapplied dv/dt, poor gate current sensitivity at turn-on, and longer turn-off time. Triacs are mainly used in phase control applications such as in AC regulators for lighting and fan control and in solid-state AC relays.

Gate Turn-Off Thyristor

The GTO is a power switching device that can be turned on by a short pulse of gate current and turned off by a reverse gate pulse. This reverse gate current amplitude is dependent on the anode current to be turned off. Hence there is no need for an external commutation circuit to turn it off. Because turn-off is provided by bypassing carriers directly to the gate circuit, its turn-off time is short, thus giving it more capability for high-frequency operation than thyristors. The GTO symbol and turn-off characteristics are shown in Fig. 1.3.

GTOs have the $I^2 t$ withstand capability and hence can be protected by semiconductor fuses. For reliable operation of GTOs, the critical aspects are proper design of the gate turn-off circuit and the snubber circuit. A GTO has a poor turn-off current gain of the order of 4 to 5. For example, a 2000-A peak current GTO may require as high as 500 A of reverse gate current. Also, a GTO has the tendency to latch at temperatures above 125°C. GTOs are available up to about 4500 V, 2500 A.

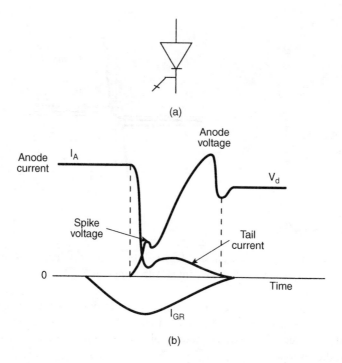

FIGURE 1.3 (a) GTO symbol and (b) turn-off characteristics. (From Bose, B.K., *Modern Power Electronics: Evaluation, Technology, and Applications*, p. 5. © 1992 IEEE. With permission.)

Reverse-Conducting Thyristor (RCT) and Asymmetrical Silicon-Controlled Rectifier (ASCR)

Normally in inverter applications, a diode in antiparallel is connected to the thyristor for commutation/freewheeling purposes. In RCTs, the diode is integrated with a fast switching thyristor in a single silicon chip. Thus, the number of power devices could be reduced. This integration brings forth a substantial improvement of the static and dynamic characteristics as well as its overall circuit performance.

The RCTs are designed mainly for specific applications such as traction drives. The antiparallel diode limits the reverse voltage across the thyristor to 1 to 2 V. Also, because of the reverse recovery behavior of the diodes, the thyristor may see very high reapplied dv/dt when the diode recovers from its reverse voltage. This necessitates use of large RC snubber networks to suppress voltage transients. As the range of application of thyristors and diodes extends into higher frequencies, their reverse recovery charge becomes increasingly important. High reverse recovery charge results in high power dissipation during switching.

The ASCR has similar forward blocking capability to an inverter-grade thyristor, but it has a limited reverse blocking (about 20 to 30 V) capability. It has an on-state voltage drop of about 25% less than an inverter-grade thyristor of a similar rating. The ASCR features a fast turn-off time; thus it can work at a higher frequency than an SCR. Since the turn-off time is down by a factor of nearly 2, the size of the commutating components can be halved. Because of this, the switching losses will also be low.

Gate-assisted turn-off techniques are used to even further reduce the turn-off time of an ASCR. The application of a negative voltage to the gate during turn-off helps to evacuate stored charge in the device and aids the recovery mechanisms. This will, in effect, reduce the turn-off time by a factor of up to 2 over the conventional device.

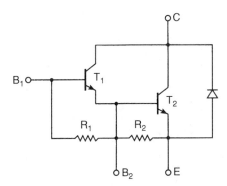

FIGURE 1.4 A two-stage Darlington transistor with bypass diode. (From Bose, B.K., *Modern Power Electronics: Evaluation, Technology, and Applications*, p. 6. © 1992 IEEE. With permission.)

Power Transistor

Power transistors are used in applications ranging from a few to several hundred kilowatts and switching frequencies up to about 10 kHz. Power transistors used in power conversion applications are generally *npn* type. The power transistor is turned on by supplying sufficient base current, and this base drive has to be maintained throughout its conduction period. It is turned off by removing the base drive and making the base voltage slightly negative (within $-V_{BE(max)}$). The saturation voltage of the device is normally 0.5 to 2.5 V and increases as the current increases. Hence, the on-state losses increase more than proportionately with current. The transistor off-state losses are much lower than the on-state losses because the leakage current of the device is of the order of a few milliamperes. Because of relatively larger switching times, the switching loss significantly increases with switching frequency. Power transistors can block only forward voltages. The reverse peak voltage rating of these devices is as low as 5 to 10 V.

Power transistors do not have I^2t withstand capability. In other words, they can absorb only very little energy before breakdown. Therefore, they cannot be protected by semiconductor fuses, and thus an electronic protection method has to be used.

To eliminate high base current requirements, Darlington configurations are commonly used. They are available in monolithic or in isolated packages. The basic Darlington configuration is shown schematically in Fig. 1.4. The Darlington configuration presents a specific advantage in that it can considerably increase the current switched by the transistor for a given base drive. The $V_{CE(sat)}$ for the Darlington is generally more than that of a single transistor of similar rating with corresponding increase in on-state power loss. During switching, the reverse-biased collector junction may show hot-spot breakdown effects that are specified by reverse-bias safe operating area (RBSOA) and forward-bias safe operating area (FBSOA). Modern devices with highly interdigited emitter base geometry force more uniform current distribution and therefore considerably improve secondary breakdown effects. Normally, a well-designed switching aid network constrains the device operation well within the SOAs.

Power MOSFET

Power MOSFETs are marketed by different manufacturers with differences in internal geometry and with different names such as MegaMOS, HEXFET, SIPMOS, and TMOS. They have unique features that make them potentially attractive for switching applications. They are essentially voltage-driven rather than current-driven devices, unlike bipolar transistors.

The gate of a MOSFET is isolated electrically from the source by a layer of silicon oxide. The gate draws only a minute leakage current on the order of nanoamperes. Hence, the gate drive circuit is simple and power loss in the gate control circuit is practically negligible. Although in steady state the gate draws virtually no current, this is not so under transient conditions. The gate-to-source and gate-to-drain

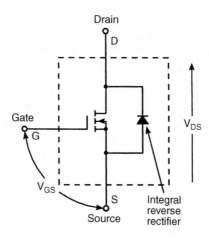

FIGURE 1.5 Power MOSFET circuit symbol. (From Bose, B.K., *Modern Power Electronics: Evaluation, Technology, and Applications*, p. 7. © 1992 IEEE. With permission.)

capacitances have to be charged and discharged appropriately to obtain the desired switching speed, and the drive circuit must have a sufficiently low output impedance to supply the required charging and discharging currents. The circuit symbol of a power MOSFET is shown in Fig. 1.5.

Power MOSFETs are majority carrier devices, and there is no minority carrier storage time. Hence, they have exceptionally fast rise and fall times. They are essentially resistive devices when turned on, while bipolar transistors present a more or less constant $V_{CE(sat)}$ over the normal operating range. Power dissipation in MOSFETs is $Id^2 R_{DS(on)}$, and in bipolars it is $I_C V_{CE(sat)}$. At low currents, therefore, a power MOSFET may have a lower conduction loss than a comparable bipolar device, but at higher currents, the conduction loss will exceed that of bipolars. Also, the $R_{DS(on)}$ increases with temperature.

An important feature of a power MOSFET is the absence of a secondary breakdown effect, which is present in a bipolar transistor, and as a result, it has an extremely rugged switching performance. In MOSFETs, $R_{DS(on)}$ increases with temperature, and thus the current is automatically diverted away from the hot spot. The drain body junction appears as an antiparallel diode between source and drain. Thus, power MOSFETs will not support voltage in the reverse direction. Although this inverse diode is relatively fast, it is slow by comparison with the MOSFET. Recent devices have the diode recovery time as low as 100 ns. Since MOSFETs cannot be protected by fuses, an electronic protection technique has to be used.

With the advancement in MOS technology, ruggedized MOSFETs are replacing the conventional MOSFETs. The need to ruggedize power MOSFETs is related to device reliability. If a MOSFET is operating within its specification range at all times, its chances for failing catastrophically are minimal. However, if its absolute maximum rating is exceeded, failure probability increases dramatically. Under actual operating conditions, a MOSFET may be subjected to transients—either externally from the power bus supplying the circuit or from the circuit itself due, for example, to inductive kicks going beyond the absolute maximum ratings. Such conditions are likely in almost every application, and in most cases are beyond a designer's control. Rugged devices are made to be more tolerant for overvoltage transients. Ruggedness is the ability of a MOSFET to operate in an environment of dynamic electrical stresses, without activating any of the parasitic bipolar junction transistors. The rugged device can withstand higher levels of diode recovery dv/dt and static dv/dt.

Insulated-Gate Bipolar Transistor (IGBT)

The IGBT has the high input impedance and high-speed characteristics of a MOSFET with the conductivity characteristic (low saturation voltage) of a bipolar transistor. The IGBT is turned on by applying a positive voltage between the gate and emitter and, as in the MOSFET, it is turned off by making the gate signal zero or slightly negative. The IGBT has a much lower voltage drop than a MOSFET of similar ratings.

Power Electronics

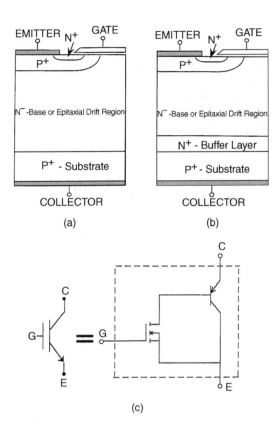

FIGURE 1.6 (a) Nonpunch-through IGBT, (b) punch-through IGBT, (c) IGBT equivalent circuit.

The structure of an IGBT is more like a thyristor and MOSFET. For a given IGBT, there is a critical value of collector current that will cause a large enough voltage drop to activate the thyristor. Hence, the device manufacturer specifies the peak allowable collector current that can flow without latch-up occurring. There is also a corresponding gate source voltage that permits this current to flow that should not be exceeded.

Like the power MOSFET, the IGBT does not exhibit the secondary breakdown phenomenon common to bipolar transistors. However, care should be taken not to exceed the maximum power dissipation and specified maximum junction temperature of the device under all conditions for guaranteed reliable operation. The on-state voltage of the IGBT is heavily dependent on the gate voltage. To obtain a low on-state voltage, a sufficiently high gate voltage must be applied.

In general, IGBTs can be classified as punch-through (PT) and nonpunch-through (NPT) structures, as shown in Fig. 1.6. In the PT IGBT, an N^+ buffer layer is normally introduced between the P^+ substrate and the N^- epitaxial layer, so that the whole N^- drift region is depleted when the device is blocking the off-state voltage, and the electrical field shape inside the N^- drift region is close to a rectangular shape. Because a shorter N^- region can be used in the punch-through IGBT, a better trade-off between the forward voltage drop and turn-off time can be achieved. PT IGBTs are available up to about 1200 V.

High-voltage IGBTs are realized through a nonpunch-through process. The devices are built on an N^- wafer substrate which serves as the N^- base drift region. Experimental NPT IGBTs of up to about 4 kV have been reported in the literature. NPT IGBTs are more robust than PT IGBTs, particularly under short circuit conditions. But NPT IGBTs have a higher forward voltage drop than the PT IGBTs.

The PT IGBTs cannot be as easily paralleled as MOSFETs. The factors that inhibit current sharing of parallel-connected IGBTs are (1) on-state current unbalance, caused by V_{CE}(sat) distribution and main circuit wiring resistance distribution, and (2) current unbalance at turn-on and turn-off, caused by the switching time difference of the parallel connected devices and circuit wiring inductance distribution. The NPT IGBTs can be paralleled because of their positive temperature coefficient property.

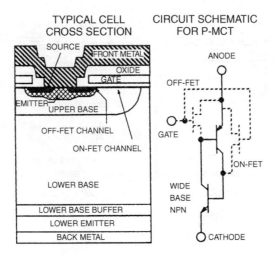

FIGURE 1.7 Typical cell cross section and circuit schematic for P-MCT. (From Harris Semiconductor, *User's Guide of MOS Controlled Thyristor*. With permission.)

MOS-Controlled Thyristor (MCT)

The MCT is a new type of power semiconductor device that combines the capabilities of thyristor voltage and current with MOS gated turn-on and turn-off. It is a high-power, high-frequency, low-conduction drop and a rugged device, which is more likely to be used in the future for medium and high power applications. A cross-sectional structure of a *p*-type MCT with its circuit schematic is shown in Fig. 1.7. The MCT has a thyristor type structure with three junctions and *pnpn* layers between the anode and cathode. In a practical MCT, about 100,000 cells similar to the one shown are paralleled to achieve the desired current rating. MCT is turned on by a negative voltage pulse at the gate with respect to the anode, and is turned off by a positive voltage pulse.

The MCT was announced by the General Electric R&D Center on November 30, 1988. Harris Semiconductor Corporation has developed two generations of *p*-MCTs. Gen-1 *p*-MCTs are available at 65 A/1000 V and 75 A/600 V with peak controllable current of 120 A. Gen-2 *p*-MCTs are being developed at similar current and voltage ratings, with much improved turn-on capability and switching speed. The reason for developing a *p*-MCT is the fact that the current density that can be turned off is two or three times higher than that of an *n*-MCT; but *n*-MCTs are the ones needed for many practical applications.

The advantage of an MCT over IGBT is its low forward voltage drop. *n*-type MCTs will be expected to have a similar forward voltage drop, but with an improved reverse bias safe operating area and switching speed. MCTs have relatively low switching times and storage time. The MCT is capable of high current densities and blocking voltages in both directions. Since the power gain of an MCT is extremely high, it could be driven directly from logic gates. An MCT has high *di/dt* (of the order of 2500 A/μs) and high *dv/dt* (of the order of 20,000 V/μs) capability.

The MCT, because of its superior characteristics, shows a tremendous possibility for applications such as motor drives, uninterrupted power supplies, static VAR compensators, and high power active power line conditioners.

The current and future power semiconductor devices developmental direction is shown in Fig. 1.8. High-temperature operation capability and low forward voltage drop operation can be obtained if silicon is replaced by silicon carbide material for producing power devices. The silicon carbide has a higher band gap than silicon. Hence, higher breakdown voltage devices could be developed. Silicon carbide devices have excellent switching characteristics and stable blocking voltages at higher temperatures. But the silicon carbide devices are still in the very early stages of development.

Power Electronics

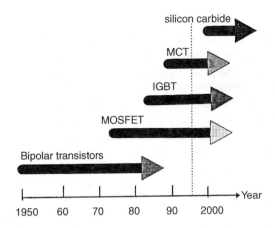

FIGURE 1.8 Current and future power semiconductor devices development direction. (From Huang, A.Q., Recent developments of power semiconductor devices, *VPEC Seminar Proceedings*, pp. 1–9. With permission.)

References

Bose, B.K., *Modern Power Electronics: Evaluation, Technology, and Applications*, IEEE Press, New York, 1992.
Harris Semiconductor, *User's Guide of MOS Controlled Thyristor*.
Huang, A.Q., Recent developments of power semiconductor devices, in *VPEC Seminar Proceedings*, September 1995, 1–9.
Mohan, N. and T. Undeland, *Power Electronics: Converters, Applications, and Design*, John Wiley & Sons, New York, 1995.
Wojslawowicz, J., Ruggedized transistors emerging as power MOSFET standard-bearers, *Power Technics Magazine*, January 1988, 29–32.

Further Information

Bird, B.M. and K.G. King, *An Introduction to Power Electronics*, Wiley-Interscience, New York, 1984.
Sittig, R. and P. Roggwiller, *Semiconductor Devices for Power Conditioning*, Plenum, New York, 1982.
Temple, V.A.K., Advances in MOS controlled thyristor technology and capability, *Power Conversion*, 544–554, Oct. 1989.
Williams, B.W., *Power Electronics, Devices, Drivers and Applications*, John Wiley, New York, 1987.

1.2 Diodes

Sohail Anwar

Power diodes play an important role in power electronics circuits. They are mainly used as uncontrolled rectifiers to convert single-phase or three-phase AC voltage to DC. They are also used to provide a path for the current flow in inductive loads. Typical types of semiconductor materials used to construct diodes are silicon and germanium. Power diodes are usually constructed using silicon because silicon diodes can operate at higher current and at higher junction temperatures than germanium diodes. The symbol for a semiconductor diode is given in Fig. 1.9. The terminal voltage and current are represented as V_d and I_d, respectively. Figure 1.10 shows the structure of a diode. It has an anode (A) terminal and a cathode (K) terminal. The diode is constructed by joining together two pieces of semiconductor material—a *p*-type and an *n*-type—to form a *pn*-junction. When the anode terminal is positive with respect to the cathode terminal, the *pn*-junction becomes forward-biased and the diode conducts current with a relatively low voltage drop. When the cathode terminal is positive with respect to the anode terminal, the *pn*-junction becomes reverse-biased and the current flow is blocked. The arrow on the diode symbol in Fig. 1.9 shows the direction of conventional current flow when the diode conducts.

Characteristics

The voltage-current characteristics of a diode are shown in Fig. 1.11. In the forward region, the diode starts conducting as the anode voltage is increased with respect to the cathode. The voltage where the current starts to increase rapidly is called the knee voltage of the diode. For a silicon diode, the knee voltage is approximately 0.7 V. Above the knee voltage, small increases in the diode voltage produce large increases in the diode current. If the diode current is too large, excessive heat will be generated, which can destroy the diode. When the diode is reverse-biased, diode current is very small for all values of reverse voltage less than the diode breakdown voltage. At breakdown, the diode current increases rapidly for small increases in diode voltage.

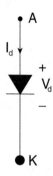

FIGURE 1.9 Diode symbol.

Principal Ratings for Diodes

Figures 1.12 and 1.13 show typical data sheets for power diodes.

Maximum Average Forward Current

The maximum average forward current ($I_{f(avg)max}$) is the current a diode can safely handle when forward biased. Power diodes are available in ratings from a few amperes to several hundred amperes. For example, the power diode D_6 described in the data specification sheet (Fig 1.12) can handle up to 6 A in the forward direction when used as a rectifier.

Peak Inverse Voltage

The peak inverse voltage (PIV) of a diode is the maximum reverse voltage that can be connected across a diode without breakdown. The peak inverse voltage is also called peak reverse voltage or reverse breakdown voltage. The PIV ratings of power diodes extend from a few volts to several thousand volts. For example, the power diode D_6 has a PIV rating of up to 1600 V, as shown in Fig. 1.12.

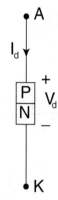

FIGURE 1.10 Diode structure.

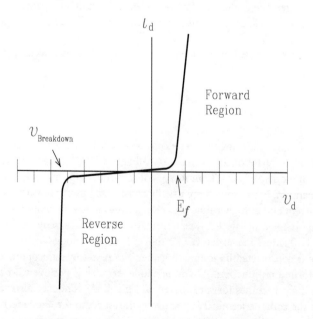

FIGURE 1.11 Diode voltage-current characteristic.

Rectifier Diode D6

Technical Data

Typical applications: All purpose high power rectifier diodes, Non-controllable and half controlled rectifiers. Free-wheeling diodes.

Type No.	V_{RRM} (Volts)	V_{RSM} (Volts)
D6/02	200	300
D6/04	400	500
D6/08	800	900
D6/12	1200	1300
D6/16	1600	1700

Features

- Reverse voltage upto 1600 V.
- Hermatic glass to metal seal
- C : Cathode to stud
- A : Anode to stud

Symbol	Conditions	Values
$I_{F(AV)}$	Sin 180; Tcase = 130 °C	6 A
I_{FSM}	Tvj = 25 °C; 10 ms	190 A
	Tvj = 180 °C; 10 ms	160 A
I^2t	Tvj = 25 °C	180 A^{2}s
	Tvj = 180 °C	130 A^{2}s
I_{RRM}	Tvj = 180 °C	2.2 mA max
V_F	Tvj = 25 °C; I_F = 15 A	1.25 V max
V_0	Tvj = 180 °C	0.85 V
R_0	Tvj = 180 °C	25 mΩ
$R_{th(j-c)}$		3.8 °C/W
$R_{th(c-h)}$		1.0 °C/W
T_{vj}		180 °C
T_{stg}		-40.....+ 180 °C
Mounting torque	SI units	2 Nm
Weight	Approx	20 g
Case outline		C/P

FIGURE 1.12 Diode data sheet—ratings. (From USHA, India. With permission.)

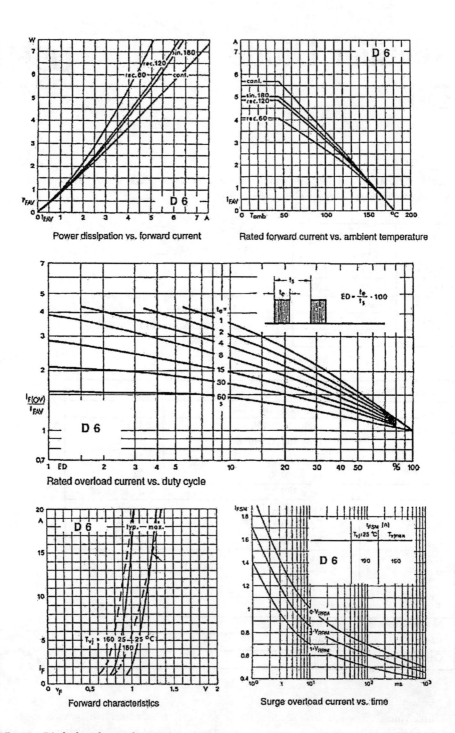

FIGURE 1.13 Diode data sheet—characteristic curves.

Power Electronics

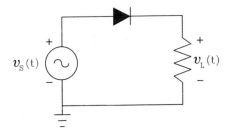

FIGURE 1.14 Basic circuit for half-wave rectifier.

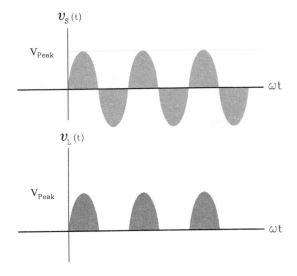

FIGURE 1.15 Input and output voltage waveforms for the circuit in Fig. 1.14.

Maximum Surge Current

The I_{FSM} (forward surge maximum) rating is the maximum current that the diode can handle as an occasional transient or from a circuit fault. The I_{FSM} rating for the power diode D_6 is up to 190 A, as shown in Fig 1.12.

Maximum Junction Temperature

This parameter defines the maximum junction temperature that a diode can withstand without failure. The maximum junction temperature for the power diode D_6 is 180°C.

Rectifier Circuits

Rectifier circuits produce a DC voltage or current from an AC source. The diode is an essential component of these circuits. Figure 1.14 shows a half-wave rectifier circuit using a diode. During the positive half cycle of the source voltage, the diode is forward-biased and conducts for $v_s(t) > E_f$. The value of E_f for germanium is 0.2 V and for silicon it is 0.7 V. During the negative half cycle of $v_s(t)$, the diode is reverse-biased and does not conduct. The voltage $v_L(t)$ across the load R_L is shown in Fig. 1.15.

The half-wave rectifier circuit produces a pulsating direct current that uses only the positive half cycle of the source voltage. The full-wave rectifier shown in Fig. 1.16 uses both half cycles of source voltage. During the positive half cycle of $v_s(t)$, diodes D_1 and D_2 are forward-biased and conduct. Diodes D_3 and D_4 are reverse-biased and do not conduct. During the negative half cycle of $v_s(t)$, diodes D_1 and D_2 are reverse-biased and do not conduct, whereas diodes D_3 and D_4 are forward-biased and conduct. The voltage $v_L(t)$ across the load R_L is shown in Fig. 1.17.

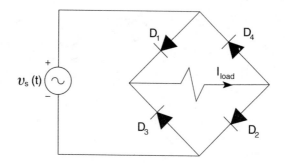

FIGURE 1.16 Basic circuit for full-wave rectifier.

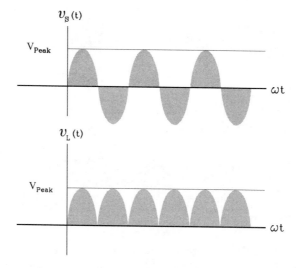

FIGURE 1.17 Input and output voltage waveforms for the circuit in Fig. 1.16.

Testing a Power Diode

An ohmmeter can be used to test power diodes. The ohmmeter is connected so that the diode is forward-biased. This should give a low resistance reading. Reversing the ohmmeter leads should give a very high resistance or even an infinite reading. A very low resistance reading in both directions indicates a shorted diode. A high resistance reading in both directions indicates an open diode.

Protection of Power Diodes

A power diode must be protected against over current, over voltage, and transients.

When a diode is reverse-biased, it acts like an open circuit. If the reverse bias voltage exceeds the breakdown voltage, a large current flow results. With this high voltage and large current, power dissipation at the diode junction may exceed its maximum value, destroying the diode. For the diode protection, it is a usual practice to choose a diode with a peak reverse voltage rating that is 1.2 times higher than the expected voltage during normal operating conditions.

Current ratings for diodes are based on the maximum junction temperatures. As a safety precaution, it is recommended that the diode current be kept below this rated value. Electrical transients can cause higher-than-normal voltages across a diode. To protect a diode from the transients, an RC series circuit may be connected across the diode to reduce the rate of change of voltage.

1.3 Schottky Diodes

Sohail Anwar

Bonding a metal, such as aluminum or platinum, to *n*-type silicon forms a Schottky diode. The Schottky diode is often used in integrated circuits for high-speed switching applications. An example of a high-speed switching application is a detector at microwave frequencies. The Schottky diode has a voltage-current characteristic similar to that of a silicon *pn*-junction diode. The Schottky is a subgroup of the TTL family and is designed to reduce the propagation delay time of the standard TTL IC chips. The construction of the Schottky diode is shown in Fig. 1.18a, and its symbol is shown in Fig. 1.18b.

Characteristics

The low-noise characteristics of the Schottky diode make it ideal for application in power monitors of low-level radio frequency, detectors for high frequency, and Doppler radar mixers. One of the main advantages of the Schottky barrier diode is its low forward voltage drop compared with that of a silicon diode. In the reverse direction, both the breakdown voltage and the capacitance of a Schottky barrier diode behave very much like those of a one-sided step junction. In the one-sided step junction, the doping level of the semiconductor determines the breakdown voltage. Because of the finite radius at the edges of the diode and because of its sensitivity to surface cleanliness, the breakdown voltage is always somewhat lower than theoretical predictions.

Data Specifications

The data specification sheet for a DSS 20-0015B power Schottky diode is provided as an example in Figs. 1.19 and 1.20. Specifications will vary depending on the application and model of Schottky diode.

Testing of Schottky Diodes

Two ways of testing the diodes use either a voltmeter or a digital multimeter. The voltmeter should be set to the low resistance scale. A single diode or rectifier should read a low resistance, typically, 2/3 scale from the resistance in the forward direction. In the reverse direction, the resistance should be nearly infinite. It should not read near 0 Ω in the shorted or open directions. The diode will result in a higher

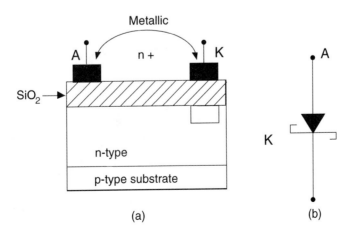

FIGURE 1.18 Diagram (a) and symbol (b) of the Schottky diode.

Power Schottky Rectifier

$I_{FAV} = 20$ A
$V_{RRM} = 15$ V
$V_F = 0.33$ V

Preliminary Data

V_{RSM} V	V_{RRM} V	Type
15	15	DSS 20-0015B

TO-220 AC

A = Anode, C = Cathode , TAB = Cathode

Symbol	Conditions	Maximum Ratings	
I_{FRMS}		35	A
I_{FAVM}	$T_C = 135$ C; rectangular, d = 0.5	20	A
I_{FSM}	$T_{VJ} = 45$ ¡C; $t_p = 10$ ms (50 Hz), sine	350	A
E_{AS}	I_{AS} = tbd A; L = 180 H; $T_{VJ} = 25$ ¡C; non repetitive	tbd	mJ
I_{AR}	$V_A = 1.5$ V_{RRM} typ.; f=10 kHz; repetitive	tbd	A
$(dv/dt)_{cr}$		tbd	V/ s
T_{VJ}		-55...+150	C
T_{VJM}		150	C
T_{stg}		-55...+150	C
P_{tot}	$T_C = 25$ C 9	0	W
M_d	mounting torque	0.4...0.6	Nm
Weight	typical	2	g

Symbol	Conditions	Characteristic Values	
		typ.	max.
I_R	$T_{VJ} = 25$ ¡C $V_R = V_{RRM}$		10 mA
	$T_{VJ} = 100$ ¡C $V_R = V_{RRM}$		200 mA
V_F	$I_F = 20$ A; $T_{VJ} = 125$ ¡C		0.33 V
	$I_F = 20$ A; $T_{VJ} = 25$ ¡C		0.45 V
	$I_F = 40$ A; $T_{VJ} = 125$ ¡C		0.43 V
R_{thJC}			1.4 K/W
R_{thCH}		0.5	K/W

Features
¥ International standard package
Very low V_F
Extremely low switching losses
Low I_{RM}-values
Epoxy meets UL 94V-0

Applications
Rectifiers in switch mode power supplies (SMPS)
Free wheeling diode in low voltage converters

Advantages
High reliability circuit operation
Low voltage peaks for reduced protection circuits
Low noise switching
Low losses

Dimensions see outlines.pdf

Pulse test: Pulse Width = 5 ms, Duty Cycle < 2.0 %
Data according to IEC 60747 and per diode unless otherwise specified

IXYS reserves the right to change limits, Conditions and dimensions.

FIGURE 1.19 Data specification sheet for a DSS 20-00105B power Schottky diode (front). (Courtesy of IXYS.)

Power Electronics

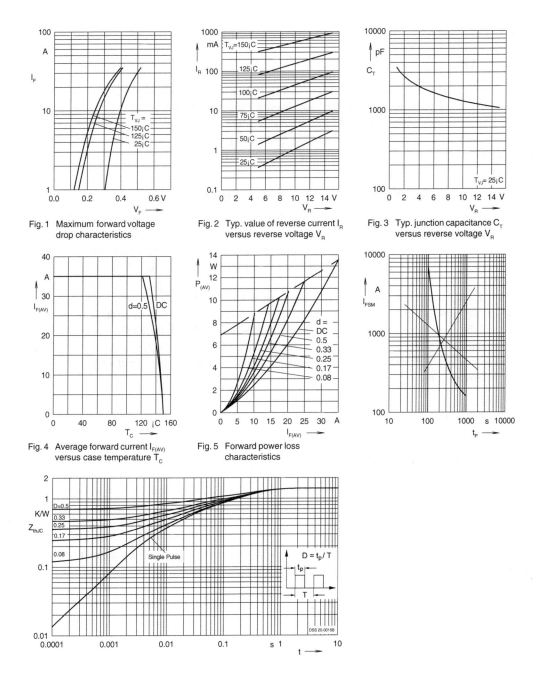

FIGURE 1.20 Data specification sheet for a DSS 20-00105B power Schottky diode (reverse).

scale reading of resistance as a result of its lower voltage drop. What is being measured is the resistance at a particular low current point; it is not the actual resistance in a power rectifier circuit.

The digital multimeter will usually have a diode test mode. When using this mode, a silicon diode should read between 0.5 to 0.8 V in the forward direction and open in the reverse direction. A germanium diode will be in the range of 0.2 to 0.4 V in the forward direction. By using the normal resistance range, these diodes will usually show open for any semiconductor junction since the voltmeter does not apply enough voltage to reach the value of the forward drop.

1.4 Thyristors

Sohail Anwar

Thyristors are four-layer *pnpn* power semiconductor devices. These devices switch between conducting and nonconducting states in response to a control signal. Thyristors are used in timing circuits, AC motor speed control, light dimmers, and switching circuits. Small thyristors are also used as pulse sources for large thyristors. The thyristor family includes the silicon-controlled rectifier (SCR), the DIAC, the Triac, the silicon-controlled switch (SCS), and the gate turn-off thyristor (GTO).

The Basics of Silicon-Controlled Rectifiers (SCR)

The SCR is the most commonly used electrical power controller. An SCR is sometimes called a *pnpn* diode because it conducts electrical current in only one direction. Figure 1.21a shows the SCR symbol. It has three terminals: the anode (A), the cathode (K), and the gate (G). The anode and the cathode are the power terminals and the gate is the control terminal. The structure of an SCR is shown in Fig. 1.21b.

When the SCR is forward-biased, that is, when the anode of an SCR is made more positive with respect to the cathode, the two outermost *pn*-junctions are forward-biased. The middle *pn*-junction is reverse-biased and the current cannot flow. If a small gate current is now applied, it forward-biases the middle *pn*-junction and allows a much larger current to flow through the device. The SCR stays ON even if the gate current is removed. SCR shutoff occurs only when the anode current becomes less than a level called the holding current (I_H).

Characteristics

The volt-ampere characteristic of an SCR is shown in Fig. 1.22. If the forward bias is increased to the forward breakover voltage, V_{FBO}, the SCR turns ON. The value of forward breakover voltage is controlled by the gate current I_G. If the gate-cathode *pn*-junction is forward-biased, the SCR is turned ON at a lower breakover voltage than with the gate open. As shown in Fig. 1.22, the breakover voltage decreases with an increase in the gate current. At a low gate current, the SCR turns ON at a lower forward anode voltage. At a higher gate current, the SCR turns ON at a still lower value of forward anode voltage.

When the SCR is reverse-biased, there is a small reverse leakage current (I_R). If the reverse bias is increased until the voltage reaches the reverse breakdown voltage ($V_{(BR)R}$), the reverse current will increase sharply. If the current is not limited to a safe value, the SCR may be destroyed.

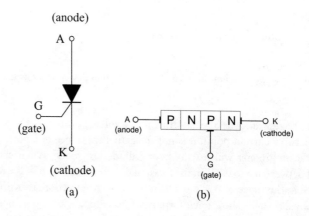

FIGURE 1.21 (a) The SCR symbol; (b) the SCR structure.

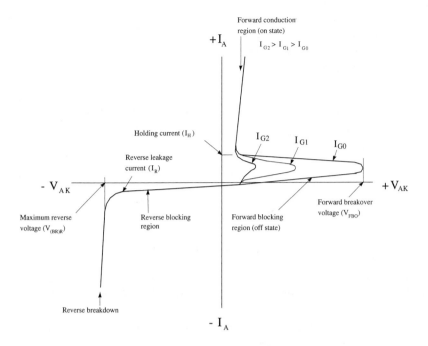

FIGURE 1.22 SCR characteristics.

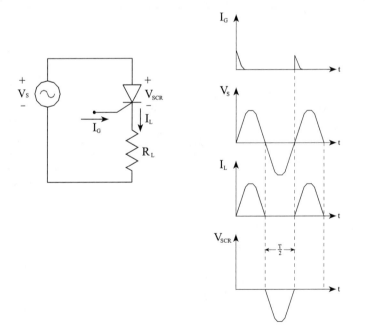

FIGURE 1.23 An SCR turn-off circuit.

SCR Turn-Off Circuits

If an SCR is forward-biased and a gate signal is applied, the device turns ON. Once the anode current is above I_H, the gate loses control. The only way to turn OFF the SCR is to make the anode terminal negative with respect to the cathode or to decrease the anode current below I_H. The process of SCR turnoff is called commutation. Figure 1.23 shows an SCR commutation circuit. This type of commutation method is called

AC line commutation. The load current I_L flows during the positive half cycle of the source voltage. The SCR is reverse-biased during the negative half cycle of the source voltage. With a zero gate current, the SCR will turn OFF if the turn-off time of the SCR is less than the duration of the half cycle.

SCR Ratings

A data sheet for a typical thyristor follows this section and includes the following information:

Surge Current Rating (I_{FM})—The surge current rating (I_{FM}) of an SCR is the peak anode current an SCR can handle for a short duration.

Latching Current (I_L)—A minimum anode current must flow through the SCR in order for it to stay ON initially after the gate signal is removed. This current is called the latching current (I_L).

Holding Current (I_H)—After the SCR is latched on, a certain minimum value of anode current is needed to maintain conduction. If the anode current is reduced below this minimum value, the SCR will turn OFF.

Peak Repetitive Reverse Voltage (V_{RRM})—The maximum instantaneous voltage that an SCR can withstand, without breakdown, in the reverse direction.

Peak Repetitive Forward Blocking Voltage (V_{DRM})—The maximum instantaneous voltage that the SCR can block in the forward direction. If the V_{DRM} rating is exceeded, the SCR will conduct without a gate voltage.

Nonrepetitive Peak Reverse Voltage (V_{RSM})—The maximum transient reverse voltage that the SCR can withstand.

Maximum Gate Trigger Current (I_{GTM})—The maximum DC gate current allowed to turn the SCR ON.

Minimum Gate Trigger Voltage (V_{GT})—The minimum DC gate-to-cathode voltage required to trigger the SCR.

Minimum Gate Trigger Current (I_{GT})—The minimum DC gate current necessary to turn the SCR ON.

The DIAC

A DIAC is a three-layer, low-voltage, low-current semiconductor switch. The DIAC symbol is shown in Fig. 1.24a. The DIAC structure is shown in Fig. 1.24b. The DIAC can be switched from the OFF to the ON state for either polarity of applied voltage.

The volt-ampere characteristic of a DIAC is shown in Fig. 1.25. When Anode 1 is made more positive than Anode 2, a small leakage current flows until the breakover voltage V_{BO} is reached. Beyond V_{BO}, the

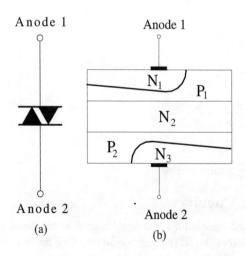

FIGURE 1.24 (a) The DIAC symbol; (b) the DIAC structure.

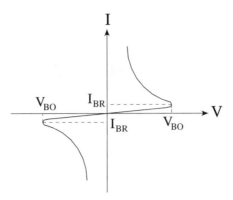

FIGURE 1.25 The DIAC characteristics.

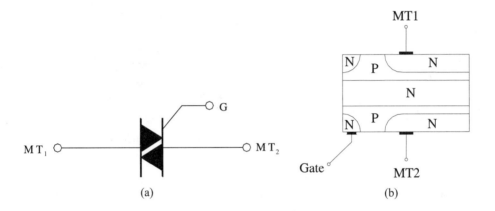

FIGURE 1.26 (a) The Triac symbol; (b) the Triac structure.

DIAC will conduct. When Anode 2 is made more positive relative to Anode 1, a similar phenomenon occurs. The breakover voltages for the DIAC are almost the same in magnitude in either direction. DIACs are commonly used to trigger larger thyristors such as SCRs and Triacs.

The Triac

The Triac is a three-terminal semiconductor switch. It is triggered into conduction in both the forward and the reverse directions by a gate signal in a manner similar to the action of an SCR. The Triac symbol is shown in Fig. 1.26a and the Triac structure is shown in Fig. 1.26b.

The volt-ampere characteristic of the Triac is shown in Fig. 1.27. The breakover voltage of the Triac can be controlled by the application of a positive or negative signal to the gate. As the magnitude of the gate signal increases, the breakover voltage decreases. Once the Triac is in the ON state, the gate signal can be removed and the Triac will remain ON until the main current falls below the holding current (I_H) value.

The Silicon-Controlled Switch

The SCS is a four-layer *pnpn* device. The SCS symbol is shown in Fig. 1.28a and the SCS structure is shown in Fig. 1.28b. The SCS has two gates labeled as the anode gate (AG) and the cathode gate (KG). An SCS can be turned ON by the application of a negative gate pulse at the anode gate. When the SCS is in the ON state, it can be turned OFF by the application of a positive pulse at the anode gate or a negative pulse at the cathode gate.

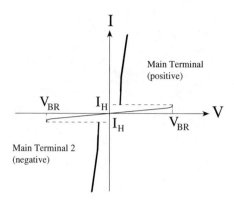

FIGURE 1.27 The Triac characteristics.

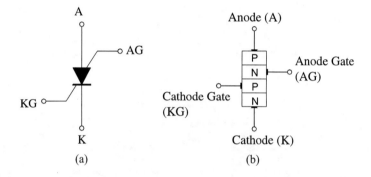

FIGURE 1.28 (a) The SCS symbol; (b) the SCS structure.

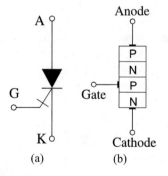

FIGURE 1.29 (a) The GTO symbol; (b) the GTO structure.

The Gate Turn-Off Thyristor

The GTO is a power semiconductor switch that turns ON by a positive gate signal. It can be turned OFF by a negative gate signal. The GTO symbol is shown in Fig. 1.29a and the GTO structure is shown in Fig. 1.29b. The GTO voltage and current ratings are lower than those of SCRs. The GTO turn-off time is lower than that of SCR. The turn-on time is the same as that of an SCR.

Data Sheet for a Typical Thyristor

Figures 1.30 to 1.35 are the data sheets for a typical thyristor.

Power Electronics   1-23

Philips Semiconductors   **Product specification**

Thyristors   BT151S series
                                                                                          BT151M series

GENERAL DESCRIPTION

Passivated thyristors in a plastic envelope, suitable for surface mounting, intended for use in applications requiring high bidirectional blocking voltage capability and high thermal cycling performance. Typical applications include motor control, industrial and domestic lighting, heating and static switching.

QUICK REFERENCE DATA

SYMBOL	PARAMETER	MAX.	MAX.	MAX.	UNIT
	BT151S (or BT151M)-	500R	650R	800R	
V_{DRM}, V_{RRM}	Repetitive peak off-state voltages	500	650	800	V
$I_{T(AV)}$	Average on-state current	7.5	7.5	7.5	A
$I_{T(RMS)}$	RMS on-state current	12	12	12	A
I_{TSM}	Non-repetitive peak on-state current	100	100	100	A

PINNING - SOT428

PIN NUMBER	Standard S	Alternative M
1	cathode	gate
2	anode	anode
3	gate	cathode
tab	anode	anode

PIN CONFIGURATION

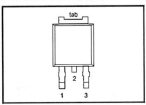

SYMBOL

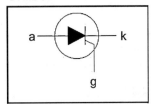

LIMITING VALUES
Limiting values in accordance with the Absolute Maximum System (IEC 134).

SYMBOL	PARAMETER	CONDITIONS	MIN.	MAX.			UNIT
				-500R	-650R	-800R	
V_{DRM}, V_{RRM}	Repetitive peak off-state voltages		-	500[1]	650[1]	800	V
$I_{T(AV)}$	Average on-state current	half sine wave; $T_{mb} \leq 103$ °C	-	7.5			A
$I_{T(RMS)}$	RMS on-state current	all conduction angles	-	12			A
I_{TSM}	Non-repetitive peak on-state current	half sine wave; $T_j = 25$ °C prior to surge					
		t = 10 ms	-	100			A
		t = 8.3 ms	-	110			A
I^2t	I^2t for fusing	t = 10 ms	-	50			A^2s
dI_T/dt	Repetitive rate of rise of on-state current after triggering	$I_{TM} = 20$ A; $I_G = 50$ mA; $dI_G/dt = 50$ mA/μs	-	50			A/μs
I_{GM}	Peak gate current		-	2			A
V_{GM}	Peak gate voltage		-	5			V
V_{RGM}	Peak reverse gate voltage		-	5			V
P_{GM}	Peak gate power		-	5			W
$P_{G(AV)}$	Average gate power	over any 20 ms period	-	0.5			W
T_{stg}	Storage temperature		-40	150			°C
T_j	Operating junction temperature		-	125			°C

[1] Although not recommended, off-state voltages up to 800V may be applied without damage, but the thyristor may switch to the on-state. The rate of rise of current should not exceed 15 A/μs.

June 1999   1   Rev 1.200

FIGURE 1.30 Page 1 of a data sheet for a typical thyristor. (From Philips Semiconductors. With permission.)

Philips Semiconductors | Product specification

Thyristors

BT151S series
BT151M series

THERMAL RESISTANCES

SYMBOL	PARAMETER	CONDITIONS	MIN.	TYP.	MAX.	UNIT
$R_{th\,j\text{-}mb}$	Thermal resistance junction to mounting base		-	-	1.8	K/W
$R_{th\,j\text{-}a}$	Thermal resistance junction to ambient	pcb (FR4) mounted; footprint as in Fig.14	-	75	-	K/W

STATIC CHARACTERISTICS

$T_j = 25\,°C$ unless otherwise stated

SYMBOL	PARAMETER	CONDITIONS	MIN.	TYP.	MAX.	UNIT
I_{GT}	Gate trigger current	$V_D = 12\,V$; $I_T = 0.1\,A$	-	2	15	mA
I_L	Latching current	$V_D = 12\,V$; $I_{GT} = 0.1\,A$	-	10	40	mA
I_H	Holding current	$V_D = 12\,V$; $I_{GT} = 0.1\,A$	-	7	20	mA
V_T	On-state voltage	$I_T = 23\,A$	-	1.4	1.75	V
V_{GT}	Gate trigger voltage	$V_D = 12\,V$; $I_T = 0.1\,A$	-	0.6	1.5	V
		$V_D = V_{DRM(max)}$; $I_T = 0.1\,A$; $T_j = 125\,°C$	0.25	0.4	-	V
I_D, I_R	Off-state leakage current	$V_D = V_{DRM(max)}$; $V_R = V_{RRM(max)}$; $T_j = 125\,°C$	-	0.1	0.5	mA

DYNAMIC CHARACTERISTICS

$T_j = 25\,°C$ unless otherwise stated

SYMBOL	PARAMETER	CONDITIONS	MIN.	TYP.	MAX.	UNIT
dV_D/dt	Critical rate of rise of off-state voltage	$V_{DM} = 67\%\,V_{DRM(max)}$; $T_j = 125\,°C$; exponential waveform;				
		Gate open circuit	50	130	-	V/µs
		$R_{GK} = 100\,\Omega$	200	1000	-	V/µs
t_{gt}	Gate controlled turn-on time	$I_{TM} = 40\,A$; $V_D = V_{DRM(max)}$; $I_G = 0.1\,A$; $dI_G/dt = 5\,A/µs$	-	2	-	µs
t_q	Circuit commutated turn-off time	$V_D = 67\%\,V_{DRM(max)}$; $T_j = 125\,°C$; $I_{TM} = 20\,A$; $V_R = 25\,V$; $dI_{TM}/dt = 30\,A/µs$; $dV_D/dt = 50\,V/µs$; $R_{GK} = 100\,\Omega$	-	70	-	µs

June 1999   Rev 1.200

FIGURE 1.31 Page 2 of a data sheet for a typical thyristor. (From Philips Semiconductors. With permission.)

Philips Semiconductors — Product specification

Thyristors

BT151S series
BT151M series

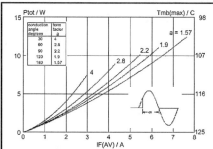

Fig.1. Maximum on-state dissipation, P_{tot}, versus average on-state current, $I_{T(AV)}$, where a = form factor = $I_{T(RMS)}/I_{T(AV)}$.

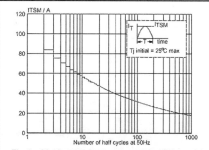

Fig.4. Maximum permissible non-repetitive peak on-state current I_{TSM}, versus number of cycles, for sinusoidal currents, $f = 50$ Hz.

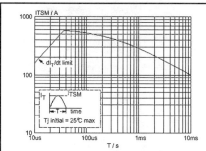

Fig.2. Maximum permissible non-repetitive peak on-state current I_{TSM}, versus pulse width t_p, for sinusoidal currents, $t_p \le 10$ms.

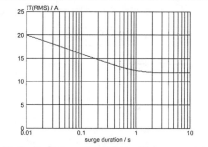

Fig.5. Maximum permissible repetitive rms on-state current $I_{T(RMS)}$, versus surge duration, for sinusoidal currents, $f = 50$ Hz; $T_{mb} \le 103°C$.

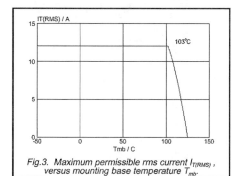

Fig.3. Maximum permissible rms current $I_{T(RMS)}$, versus mounting base temperature T_{mb}.

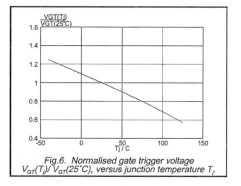

Fig.6. Normalised gate trigger voltage $V_{GT}(T_j)/V_{GT}(25°C)$, versus junction temperature T_j.

June 1999 — Rev 1.200

FIGURE 1.32 Page 3 of a data sheet for a typical thyristor. (From Philips Semiconductors. With permission.)

Philips Semiconductors — Product specification

Thyristors

BT151S series
BT151M series

Fig. 7. Normalised gate trigger current $I_{GT}(T_j)/I_{GT}(25°C)$, versus junction temperature T_j.

Fig. 10. Typical and maximum on-state characteristic.

Fig. 8. Normalised latching current $I_L(T_j)/I_L(25°C)$, versus junction temperature T_j.

Fig. 11. Transient thermal impedance $Z_{th\,j\text{-}mb}$, versus pulse width t_p.

Fig. 9. Normalised holding current $I_H(T_j)/I_H(25°C)$, versus junction temperature T_j.

Fig. 12. Typical, critical rate of rise of off-state voltage, dV_D/dt versus junction temperature T_j.

June 1999 — Rev 1.200

FIGURE 1.33 Page 4 of a data sheet for a typical thyristor. (From Philips Semiconductors. With permission.)

Power Electronics   1-27

Philips Semiconductors   Product specification

Thyristors   BT151S series
                                                          BT151M series

MECHANICAL DATA

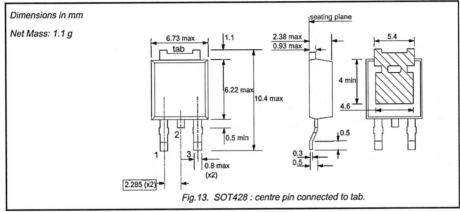

Fig.13. SOT428 : centre pin connected to tab.

MOUNTING INSTRUCTIONS

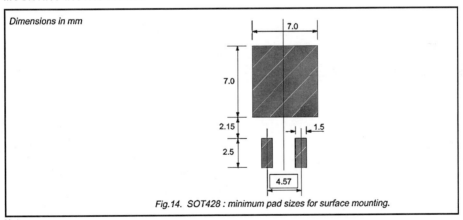

Fig.14. SOT428 : minimum pad sizes for surface mounting.

Notes
1. Plastic meets UL94 V0 at 1/8".

June 1999   5   Rev 1.200

FIGURE 1.34 Page 5 of a data sheet for a typical thyristor. (From Philips Semiconductors. With permission.)

Philips Semiconductors	Product specification
Thyristors	BT151S series BT151M series

DEFINITIONS

Data sheet status	
Objective specification	This data sheet contains target or goal specifications for product development.
Preliminary specification	This data sheet contains preliminary data; supplementary data may be published later.
Product specification	This data sheet contains final product specifications.
Limiting values	
Limiting values are given in accordance with the Absolute Maximum Rating System (IEC 134). Stress above one or more of the limiting values may cause permanent damage to the device. These are stress ratings only and operation of the device at these or at any other conditions above those given in the Characteristics sections of this specification is not implied. Exposure to limiting values for extended periods may affect device reliability.	
Application information	
Where application information is given, it is advisory and does not form part of the specification.	
© Philips Electronics N.V. 1999	
All rights are reserved. Reproduction in whole or in part is prohibited without the prior written consent of the copyright owner.	
The information presented in this document does not form part of any quotation or contract, it is believed to be accurate and reliable and may be changed without notice. No liability will be accepted by the publisher for any consequence of its use. Publication thereof does not convey nor imply any license under patent or other industrial or intellectual property rights.	

LIFE SUPPORT APPLICATIONS

These products are not designed for use in life support appliances, devices or systems where malfunction of these products can be reasonably expected to result in personal injury. Philips customers using or selling these products for use in such applications do so at their own risk and agree to fully indemnify Philips for any damages resulting from such improper use or sale.

June 1999   6   Rev 1.200

FIGURE 1.35 Page 6 of a data sheet for a typical thyristor. (From Philips Semiconductors. With permission.)

1.5 Power Bipolar Junction Transistors

Sohail Anwar

Power bipolar junction transistors (BJTs) play a vital role in power circuits. Like most other power devices, power transistors are generally constructed using silicon. The use of silicon allows operation of a BJT at higher currents and junction temperatures, which leads to the use of power transistors in AC applications where ranges of up to several hundred kilowatts are essential.

The power transistor is part of a family of three-layer devices. The three layers or terminals of a transistor are the base, the collector, and the emitter. Effectively, the transistor is equivalent to having two *pn*-diode junctions stacked in opposite directions to each other. The two types of a transistor are termed *npn* and *pnp*. The *npn*-type transistor has a higher current-to-voltage rating than the *pnp* and is preferred for most power conversion applications. The easiest way to distinguish an *npn*-type transistor from a *pnp*-type is by virtue of the schematic or circuit symbol. The *pnp* type has an arrowhead on the emitter that points toward the base. Figure 1.36 shows the structure and the symbol of a *pnp*-type transistor. The *npn*-type transistor has an arrowhead pointing away from the base. Figure 1.37 shows the structure and the symbol of an *npn*-type transistor.

When used as a switch, the transistor controls the power from the source to the load by supplying sufficient base current. This small current from the driving circuit through the base–emitter, which must be maintained, turns on the collector—emitter path. Removing the current from the base–emitter path and making the base voltage slightly negative turns off the switch. Even though the base–emitter path may only utilize a small amount of current, the collector–emitter path is capable of carrying a much higher current.

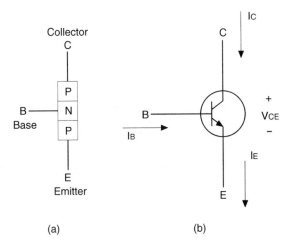

FIGURE 1.36 *pnp* transistor structure (a) and circuit symbol (b).

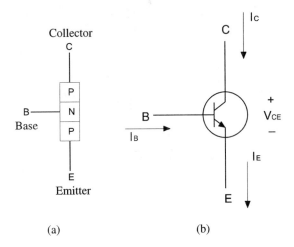

FIGURE 1.37 *npn* transistor structure (a) and circuit symbol (b).

The Volt-Ampere Characteristics of a BJT

The volt-ampere characteristics of a BJT are shown in Fig. 1.38. Power transistors have exceptional characteristics as an ideal switch and they are primarily used as switches. In this type of application, they make use of the common emitter connection shown in Fig. 1.39. The three regions of operation for a transistor that must be taken into consideration are the cutoff, saturation, and the active region. When the base current (I_B) is zero, the collector current (I_C) is insignificant and the transistor is driven into the cutoff region. The transistor is now in the OFF state. The collector–base and base–emitter junctions are reverse-biased in the cutoff region or OFF state, and the transistor behaves as an open switch. The base current (I_B) determines the saturation current. This occurs when the base current is sufficient to drive the transistor into saturation. During saturation, both junctions are forward-biased and the transistor acts like a closed switch. The saturation voltage increases with an increase in current and is normally between 0.5 to 2.5 V. The active region of the transistor is mainly used for amplifier applications and should be avoided for switching operation. In the active region, the collector–base junction is reversed-biased and the base–emitter junction is forward-biased.

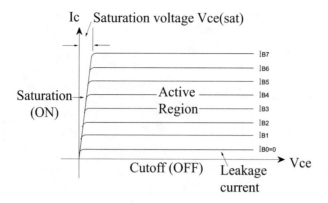

FIGURE 1.38 BJT V-I characteristic.

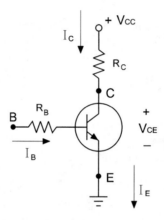

FIGURE 1.39 Biasing of a transistor.

BJT Biasing

When a transistor is used as a switch, the control circuit provides the necessary base current. The current of the base determines the ON or OFF state of the transistor switch. The collector and the emitter of the transistor form the power terminals of the switch.

The DC load line represents all of the possible operating points of a transistor and is shown in Fig. 1.40. The operating point is where the load line and the base current intersect and is determined by the values of V_{CC} and R_C.

In the ON state, the ideal operating point occurs when the collector current I_C is equal to V_{CC}/R_C and V_{CE} is zero. The actual operating point occurs when the load line intersects the base current at the saturation point. This occurs when the base current equals the saturation current or $I_B = I_{B(sat)}$. At this point, the collector current is maximum and the transistor has a small voltage drop across the collector–emitter terminals called the saturation voltage $V_{CE(sat)}$.

In the OFF state, or cutoff point, the ideal operating point occurs when the collector current I_C is zero and the collector–emitter voltage V_{CE} is equal to the supply voltage V_{CC}. The actual operating point, in the OFF state, occurs when the load line intersects the base current ($I_B = 0$). At the cutoff point, the collector current is the leakage current. By applying Kirchoff's voltage law around the output loop, the collector–emitter voltage (V_{CE}) can be found.

The operating points between the saturation and cutoff constitute the active region. When operating in the active region, high power dissipation occurs due to the relatively high values of collector current

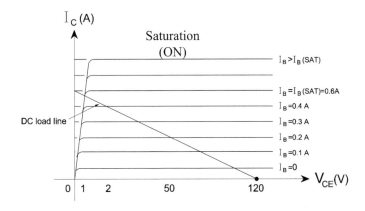

FIGURE 1.40 DC load line.

I_C and collector–emitter voltage V_{CE}. For satisfactory operation, a slightly higher than minimum base current will ensure a saturated ON state and will result in reduced turn-on time and power dissipation.

BJT Power Losses

The four types of transistor power losses are the ON-state and OFF-state losses and turn-ON and turn-OFF switching loss. OFF-state transistor losses are much lower than ON-state losses since the leakage current of the device is within a few milliamps. Essentially, when a transistor is in the off state, whatever the value of collector–emitter voltage, there is no collector current. Switching losses depend on switching frequency. The highest possible switching frequency of the transistor is limited by the losses due to the rate of switching. In other words, the higher the switching frequency, the more power loss in the transistor.

BJT Testing

Testing of the state of a transistors can be done with a multimeter. When a transistor is forward-biased, the base–collector and base–emitter regions should have a low resistance. When reverse-biased, the base–collector and base–emitter regions should have a high resistance. When testing the resistance between the collector and the emitter, the resistance reading should result in a much higher than forward bias base–collector and base–emitter resistance. However, faulty power transistors can appear shorted when measuring resistance across the collector and emitter, but still pass both junction tests.

BJT Protection

Transistors must be protected against high currents and voltages to prevent damage to the device. Since they are able to absorb very little energy before breakdown, semiconductor fuses cannot protect them. Thermal conditions are vitally important and can occur during high-frequency switching. Some of the most common types of BJT protection are overcurrent and overvoltage protection. Electronic protection techniques are also frequently used to provide needed protection for transistors.

Overcurrent protection turns the transistor OFF when the collector–emitter voltage and collector current reach a preset value. When the transistor is in the ON state, an increase in collector–emitter voltage causes an increase in the collector current and therefore an increase in junction temperature. Since the BJT has a negative temperature coefficient, the increase in temperature causes a decrease in resistance and results in an even higher collector current. This condition, called positive feedback, could eventually lead to thermal runaway and destroy the transistor. One such method of overcurrent protection limits the base current during an external fault. With the base current limited, the device current will be limited at the saturation point, with respect to the base current, and the device will hold some value of the voltage. This feature turns the transistor off without being damaged and is used for providing

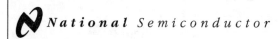

July 2000

LM195/LM395
Ultra Reliable Power Transistors

General Description

The LM195/LM395 are fast, monolithic power integrated circuits with complete overload protection. These devices, which act as high gain power transistors, have included on the chip, current limiting, power limiting, and thermal overload protection making them virtually impossible to destroy from any type of overload. In the standard TO-3 transistor power package, the LM195 will deliver load currents in excess of 1.0A and can switch 40V in 500 ns.

The inclusion of thermal limiting, a feature not easily available in discrete designs, provides virtually absolute protection against overload. Excessive power dissipation or inadequate heat sinking causes the thermal limiting circuitry to turn off the device preventing excessive heating.

The LM195 offers a significant increase in reliability as well as simplifying power circuitry. In some applications, where protection is unusually difficult, such as switching regulators, lamp or solenoid drivers where normal power dissipation is low, the LM195 is especially advantageous.

The LM195 is easy to use and only a few precautions need be observed. Excessive collector to emitter voltage can destroy the LM195 as with any power transistor. When the device is used as an emitter follower with low source impedance, it is necessary to insert a 5.0k resistor in series with the base lead to prevent possible emitter follower oscillations. Although the device is usually stable as an emitter follower, the resistor eliminates the possibility of trouble without degrading performance. Finally, since it has good high frequency response, supply bypassing is recommended.

For low-power applications (under 100 mA), refer to the LP395 Ultra Reliable Power Transistor.

The LM195/LM395 are available in the standard TO-3, Kovar TO-5, and TO-220 packages. The LM195 is rated for operation from −55 C to +150 C and the LM395 from 0 C to +125 C.

Features

- Internal thermal limiting
- Greater than 1.0A output current
- 3.0 A typical base current
- 500 ns switching time
- 2.0V saturation
- Base can be driven up to 40V without damage
- Directly interfaces with CMOS or TTL
- 100% electrical burn-in

Simplified Circuit

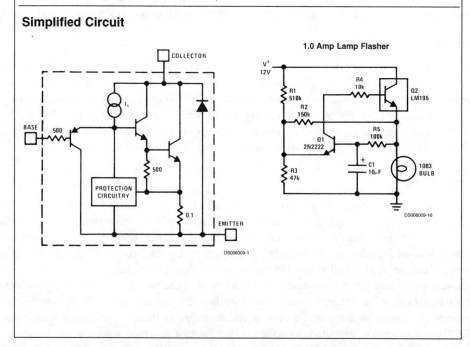

FIGURE 1.41 Typical data sheet for a power transistor (page 1). (From National Semiconductor. With permission.)

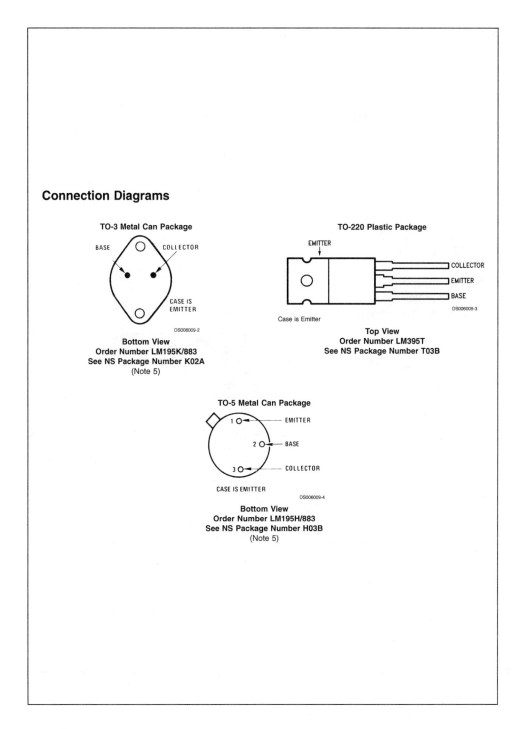

FIGURE 1.42 Typical data sheet for a power transistor (page 2). (From National Semiconductor. With permission.)

Absolute Maximum Ratings (Note 1)

If Military/Aerospace specified devices are required, please contact the National Semiconductor Sales Office/Distributors for availability and specifications.

Collector to Emitter Voltage	
LM195	42V
LM395	36V
Collector to Base Voltage	
LM195	42V
LM395	36V
Base to Emitter Voltage (Forward)	
LM195	42V
LM395	36V
Base to Emitter Voltage (Reverse)	20V
Collector Current	Internally Limited
Power Dissipation	Internally Limited
Operating Temperature Range	
LM195	55 C to +150 C
LM395	0 C to +125 C
Storage Temperature Range	65 C to +150 C
Lead Temperature	
(Soldering, 10 sec.)	260 C

Preconditioning
100% Burn-In In Thermal Limit

Electrical Characteristics
(Note 2)

Parameter	Conditions	LM195 Min	LM195 Typ	LM195 Max	LM395 Min	LM395 Typ	LM395 Max	Units
Collector-Emitter Operating Voltage (Note 4)	$I_Q \leq I_C \leq I_{MAX}$			42			36	V
Base to Emitter Breakdown Voltage	$0 \leq V_{CE} \leq V_{CEMAX}$	42			36	60		V
Collector Current								
TO-3, TO-220	$V_{CE} \leq 15V$	1.2	2.2		1.0	2.2		A
TO-5	$V_{CE} \leq 7.0V$	1.2	1.8		1.0	1.8		A
Saturation Voltage	$I_C \leq 1.0A, T_A = 25\,C$		1.8	2.0		1.8	2.2	V
Base Current	$0 \leq I_C \leq I_{MAX}$, $0 \leq V_{CE} \leq V_{CEMAX}$		3.0	5.0		3.0	10	A
Quiescent Current (I_Q)	$V_{be} = 0$, $0 \leq V_{CE} \leq V_{CEMAX}$		2.0	5.0		2.0	10	mA
Base to Emitter Voltage	$I_C = 1.0A, T_A = +25\,C$		0.9			0.9		V
Switching Time	$V_{CE} = 36V, R_L = 36\Omega, T_A = 25\,C$		500			500		ns
Thermal Resistance Junction to Case (Note 3)	TO-3 Package (K)		2.3	3.0		2.3	3.0	C/W
	TO-5 Package (H)		12	15		12	15	C/W
	TO-220 Package (T)					4	6	C/W

Note 1: »Absolute Maximum Ratings..indicate limits beyond which damage to the device may occur. Operating Ratings indicate conditions for which the device is functional, but do not guarantee specific performance limits.
Note 2: Unless otherwise specified, these specifications apply for $55\,C \leq T_j \leq +150\,C$ for the LM195 and $0\,C \leq \ldots \leq +125\,C$ for the LM395.
Note 3: Without a heat sink, the thermal resistance of the TO-5 package is about +150 C/W, while that of the TO-3 package is +35 C/W.
Note 4: Selected devices with higher breakdown available.
Note 5: Refer to RETS195H and RETS195K drawings of military LM195H and LM195K versions for specifications.

FIGURE 1.43 Typical data sheet for a power transistor (page 3). (From National Semiconductor. With permission.)

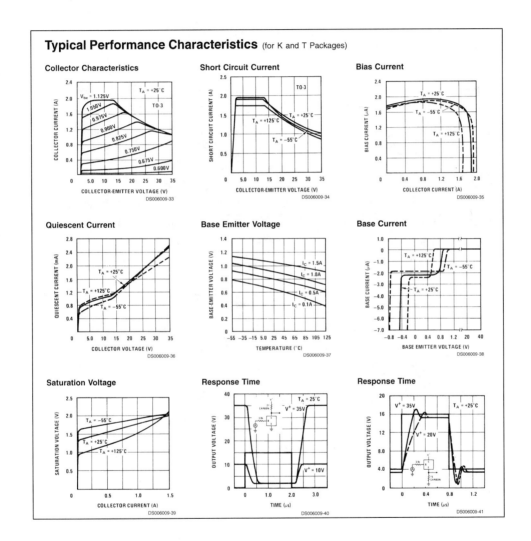

FIGURE 1.44 Typical data sheet for a power transistor (page 4). (From National Semiconductor. With permission.)

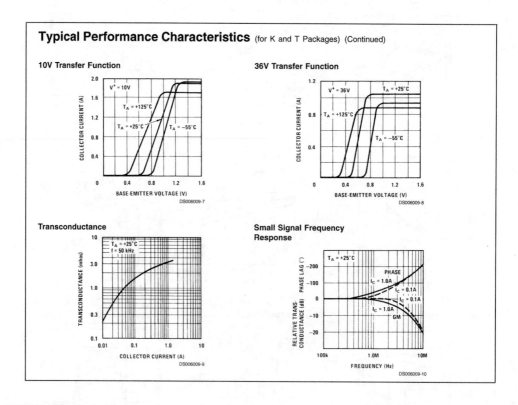

FIGURE 1.45 Typical data sheet for a power transistor (page 5). (From National Semiconductor. With permission.)

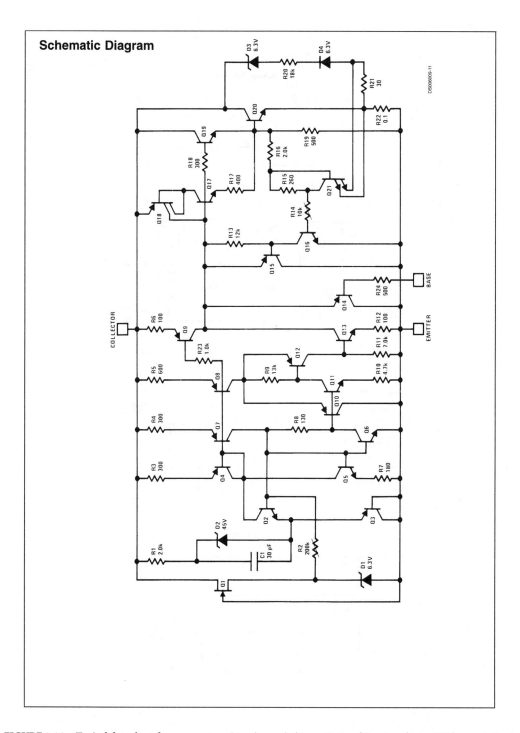

FIGURE 1.46 Typical data sheet for a power transistor (page 6). (From National Semiconductor. With permission.)

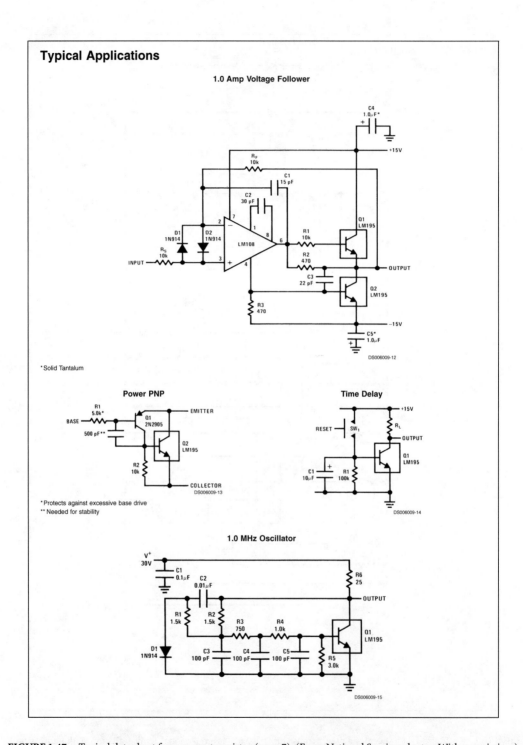

FIGURE 1.47 Typical data sheet for a power transistor (page 7). (From National Semiconductor. With permission.)

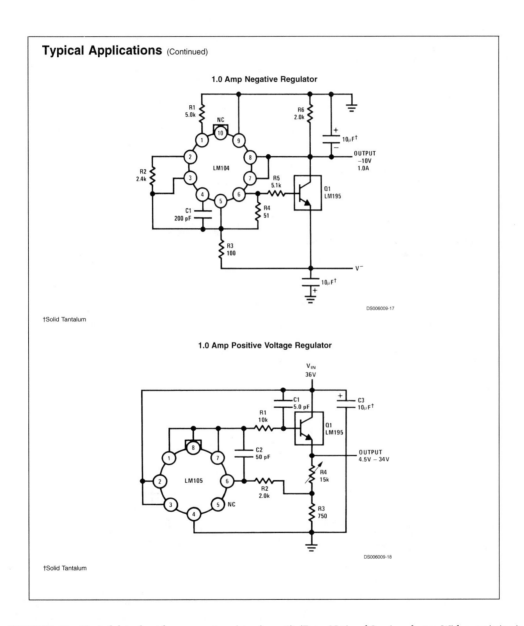

FIGURE 1.48 Typical data sheet for a power transistor (page 8). (From National Semiconductor. With permission.)

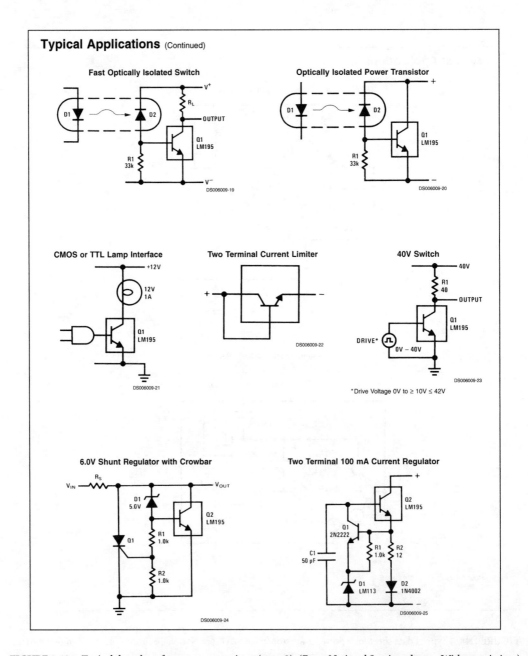

FIGURE 1.49 Typical data sheet for a power transistor (page 9). (From National Semiconductor. With permission.)

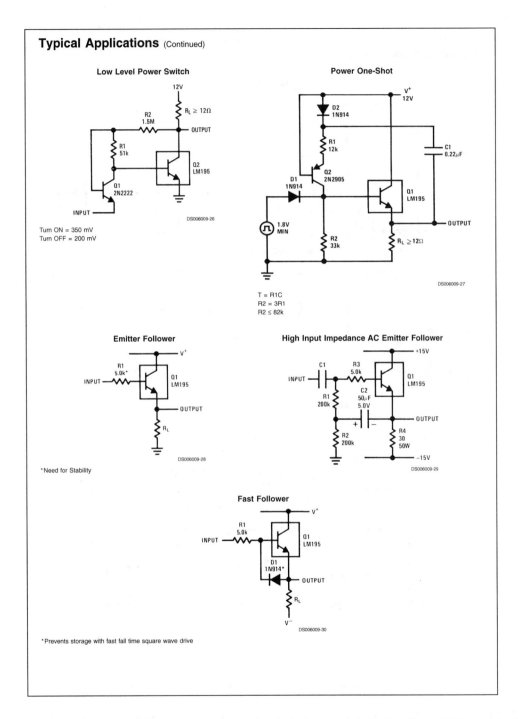

FIGURE 1.50 Typical data sheet for a power transistor (page 10). (From National Semiconductor. With permission.)

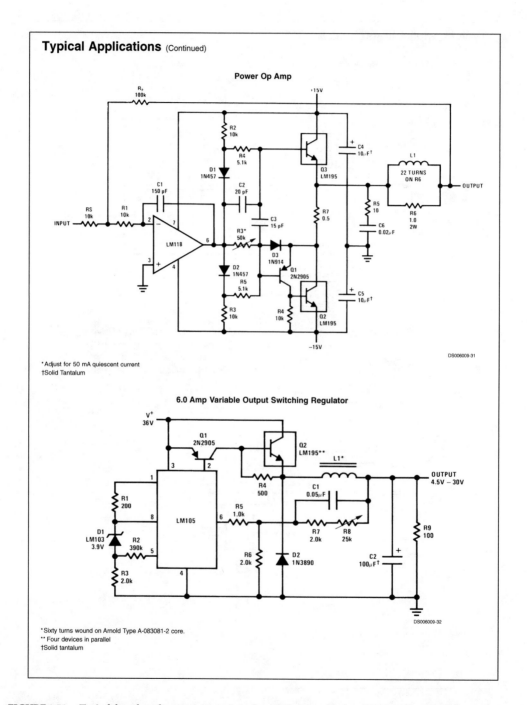

FIGURE 1.51 Typical data sheet for a power transistor (page 11). (From National Semiconductor. With permission.)

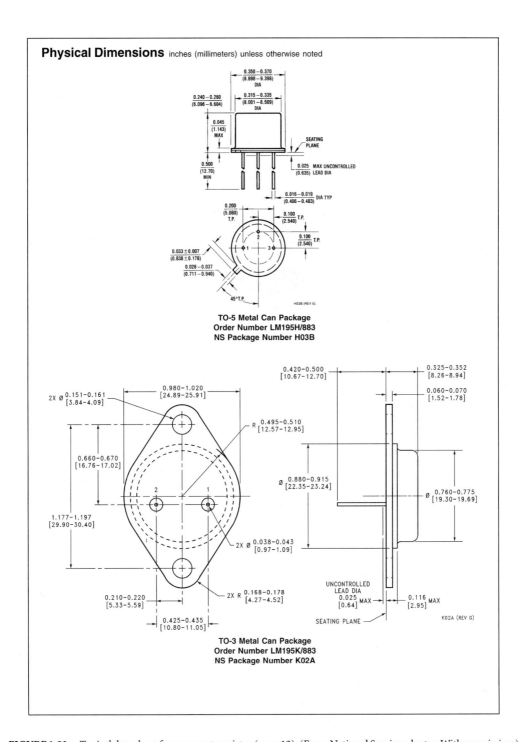

FIGURE 1.52 Typical data sheet for a power transistor (page 12). (From National Semiconductor. With permission.)

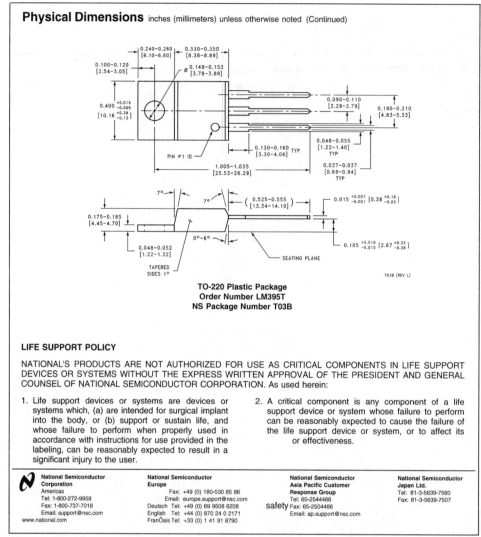

FIGURE 1.53 Typical data sheet for a power transistor (page 13). (From National Semiconductor. With permission.)

protection in low power converters by limiting the current during an external fault. Other methods of overcurrent protection for more severe faults use a shorting switch, or shunt switch, in parallel with the transistor. When a fault is detected, an external circuit activates the parallel shorting switch, providing an alternate path for the fault current.

Overvoltage protection is used to protect a transistor from high voltages. When a transistor is in the OFF state, high collector-base reverse–bias voltages can cause avalanche breakdown. Avalanche breakdown occurs when the reverse voltage exceeds the reverse voltage limit of the collector–base region. High collector–base reverse-bias voltages can easily damage the transistor. One simple method to ensure overcurrent protection of a transistor is to connect an antiparallel diode across the transistor.

Most power transistors are unable to block reverse voltages in excess of 20 V. Reverse voltages can easily damage the transistor and therefore they should not be used in AC control applications without a reverse shunting diode connected between the emitter and the collector.

A typical data sheet for a power transistor is provided in Figs. 1.41 through 1.53.

1.6 MOSFETs

Vrej Barkhordarian

The metal-oxide-semiconductor field-effect transistor (MOSFET) is the most commonly used active device in very large scale integrated (VLSI) circuits. Figure 1.54 shows the device schematic, current-voltage characteristics, transfer characteristics and device symbol for a MOSFET. It is a lateral device and though very suitable for integration into integrated circuits, it has severe limitations at high power levels. The power MOSFET design is based on the original field-effect transistor and, since its invention in the early 1970s, has gone through several evolutionary steps. The processing of power MOSFETs is very similar to that of today's VLSI circuits although the device geometry is significantly different from the

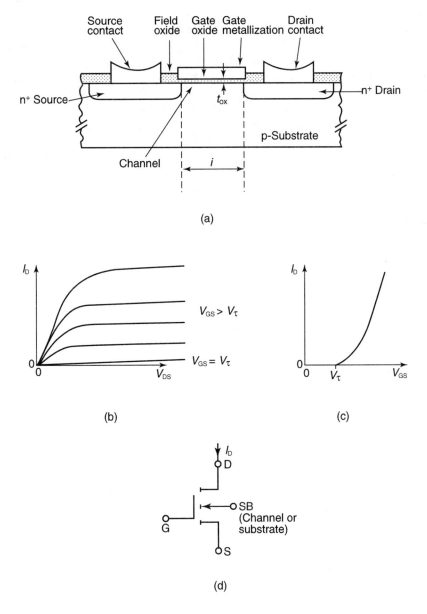

FIGURE 1.54 (a) Schematic diagram, (b) current-voltage characteristics, (c) transfer characteristics, and (d) device symbol for an *n*-channel enhancement mode MOSFET.

design used in these circuits. Power MOSFETs are commonly used as switches in power electronic applications.

The invention of the power MOSFET was partly driven by the limitations of bipolar power transistors which, until recently, were the devices of choice in power electronics applications. Although it is not possible to define absolutely the operating boundaries of a power device, we will loosely refer to the power device as any device that is capable of switching at least 1A. The bipolar power transistor is a current-controlled device and a large base drive current as high as one fifth of the collector current is required to keep the device in the on state. Also, higher reverse base drive currents are required to obtain fast turn-off. Despite the very advanced state of manufacturability and lower costs of bipolar power transistors, these limitations have made the base drive circuit design more complicated and hence more expensive. There are two further limitations to the bipolar power transistor. First, both electrons and holes contribute to conduction in BJTs. Presence of holes with their higher carrier lifetime causes the switching speed to be several orders of magnitude slower than for a power MOSFET of similar size and voltage rating. Secondly, the BJTs suffer from thermal runaway. The forward voltage drop of a BJT decreases with increasing temperature causing diversion of current to a single device when several devices are paralleled. Power MOSFETs, on the other hand, are majority carrier devices with no minority carrier injection. They are superior to the BJTs in high-frequency applications where switching power losses are important and can withstand simultaneous application of high current and voltage without undergoing destructive failure due to second breakdown. Power MOSFETs can also be paralleled easily since the forward voltage drop increases with increasing temperature, ensuring an even distribution of current among all components. However, at high breakdown voltages (>~200V) the on-state voltage drop of the power MOSFET becomes higher than that of a similar size bipolar device with a similar voltage rating, making it more attractive to use the bipolar power transistor at the expense of worse high-frequency performance. Figure 1.55 shows the present current-voltage limitations of power MOSFETs and BJTs. New materials, structures and processing techniques are expected to push these limits out over time. A relatively new device which combines the high-frequency advantages of the MOSFET with the low on-state voltage drop of high voltage BJTs is the insulated-gate-bipolar transistor (IGBT).

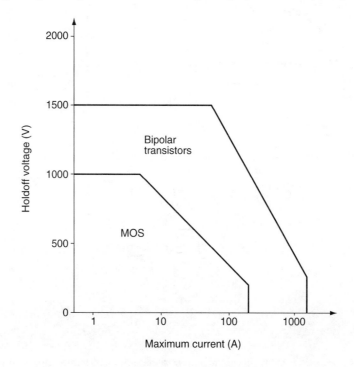

FIGURE 1.55 Current-voltage limitations of MOSFETs and BJTs.

Power Electronics

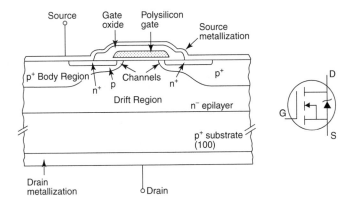

FIGURE 1.56 Schematic diagram for an *n*-channel power MOSFET and the device symbol.

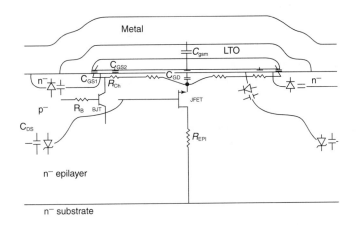

FIGURE 1.57 The origin of parasitic components for a power MOSFET.

MOSFETs used in integrated circuits are lateral devices with gate, source and drain all on the top of the device and with current flow taking place in a path parallel to the surface. Although this design lends itself to integration, it is not suitable for discrete power device applications due to large distances required between source and drain in order to maintain isolation. Having all three terminals as the upper surface makes the metallization and isolation of terminals more complicated from the processing point of view. The vertical double diffused MOSFET solves this problem by using the substrate of the device as the drain terminal. Figure 1.56 shows the schematic diagram and the circuit symbol for an *n*-channel power MOSFET. When a positive bias greater than the threshold voltage is applied to the gate, the silicon surface in the channel region is inverted and a current starts to flow between the source and drain. For gate voltages of less than V_t no surface inversion occurs in the channel and the device remains in the off-state. The current in this device flows horizontally along the inverted channel first and then vertically between the drain and source. The term "double-diffused" refers to the two consecutive ion implantation steps using the poly as a mask. For an *n*-channel device, the regions formed by double implant and subsequent diffusion are first *p*-type to define the channel and then *n*-type to define the source. The *p*-body implant is performed in a separate step. The terms "body drift" and "body-drain" diodes are used interchangeably to denote the *p-n* junction formed by this *p*-body implant and the drift region.

Figure 1.57 shows the physical origin of the parasitic components in an *n*-channel power MOSFET. The parasitic JFET appearing between the two body implants restricts current flow when the depletion

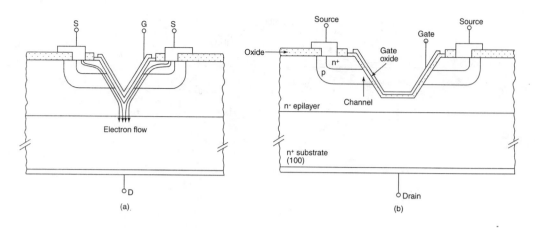

FIGURE 1.58 Schematic diagram of (a) V-groove trench MOSFET showing the current crowding at the apex and (b) truncated V-groove design.

widths of the two adjacent body diodes extend into the drift region with increasing drain voltage. Poly line-width and the epi layer resistivity under the poly are two important design parameters for minimizing the JFET effect. The parasitic BJT can make the device susceptible to unwanted device turn-on and premature breakdown. The base resistance R_B has to be minimized through careful design of the doping and distance under the source region. These two components and the parasitic resistances are discussed further in the next sections. There are several parasitic capacitances associated with the power MOSFET as shown in Fig. 1.57. C_{GS} is the capacitance due to the overlap of the source and the channel regions by the polysilicon gate and is independent of applied voltage. G_{GD} is made up of two parts. The first part is the capacitance associated with the overlap of the polysilicon gate and the silicon underneath in the JFET region. The second part is the capacitance associated with the depletion region immediately under the gate. C_{GD} is a nonlinear function of voltage and is discussed further in the "Dynamic Characteristics" section. Finally, C_{DS} is the capacitance associated with the body-drift diode and varies inversely with the square root of the drain-source bias.

There are currently two designs of power MOSFETs. These are usually referred to as the planar and the trench designs. The planar design has already been introduced in the schematics of Figs. 1.56 and 1.57. Two variations of the trench power MOSFET are shown in Fig. 1.58. The V-groove device is fabricated by etching a groove in the silicon after the double diffusion step. The use of an anisotropic etch results in the sides of the groove to be at an angle of 54.7° to the surface of the wafer. Etching stops when the groove sides, which are planes, reach each other. The gate oxide and gate poly or metallization are then grown in the groove followed by the source metallization. Current crowding at the apex of the V-groove reduces current handling capability. In a truncated V-groove design, the anisotropic etch is stopped before this point is reached. The trench technology has the advantage of higher cell density but is more difficult to manufacture compared with the planar device.

Static Characteristics

One of the important features of the power MOSFET is the very high input impedance which simplifies the gate drive circuitry and reduces cost. It is a voltage-controlled device with to gate current flow during operation. Figure 1.59 shows I–V characteristics of an enhancement mode (normally off) power MOSFET. Data sheets contain typical graphs which can be used to determine if the device is in the fully on state or in the constant-current region for a given value of gate bias and drain current. Temperature effect on threshold voltage (about 6 mV/C reduction) and the difference between typical values of parameters and the maximums should be taken into account.

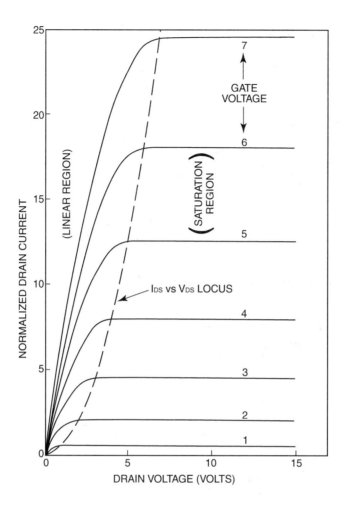

FIGURE 1.59 Current-voltage characteristics of a power MOSFET.

Breakdown Voltage

This is the drain voltage at which the reverse-biased body-drift diode breaks down and a significant current starts to flow between the source and drain by the avalanche multiplication process, while the gate and source are shorted together. Breakdown voltage, BV_{DSS}, is normally measured at a drain current of 250 μA. For drain voltages below BV_{DSS} and with no bias on the gate, no channel is formed under the gate at the surface and the drain voltage is entirely supported by the reverse-biased body-drift pn junction. There are two related phenomena which can occur in poorly designed and processed devices. These are punch-through and reach-through.

Punch-through is observed when the depletion region on the source side of the body-drift pn-junction reaches the source region at drain voltages below the rated avalanche voltage of the device. This provides a current path between source and drain and causes a soft breakdown characteristic as shown in Fig. 1.60. The leakage current flowing between source and drain is denoted by I_{DSS}. Careful selection and optimization of the doping profile used in the fabrication of a power MOSFET is therefore very important. Figure 1.61 shows a typical diffusion profile for a power MOSFET. The surface concentration of the body diffusion and the channel length (distance between the two pn-junctions formed by the source diffusion and the channel diffusion) will determine whether punch-through will occur or not. There are trade-offs to be made between on-resistance R_{dson} which requires shorter channel lengths and punch-through avoidance which requires longer channel lengths. An approximate equation giving the depletion region width as a function of silicon

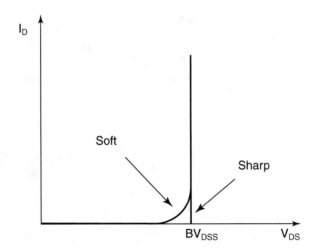

FIGURE 1.60 Breakdown characteristics of a power MOSFET showing the ideal (sharp) and nonideal (soft) behaviors.

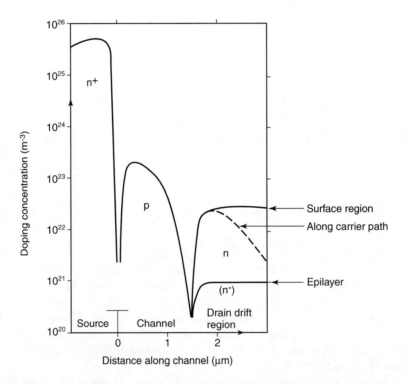

FIGURE 1.61 Typical doping profile of a power MOSFET, in a direction parallel to the device surface. Threshold voltage is determined by the peak carrier concentration in the channel region.

background doping is given by:

$$W \approx \sqrt{\frac{4\epsilon_s KT}{q^2 N_A} \ln\left[\frac{N_A}{n_i}\right]} \tag{1.1}$$

where ϵ_s is semiconductor permittivity, K is Boltzmann's constant, T is temperature in K, q is electronic charge, N_A is background doping, and n_i is the intrinsic carrier density.

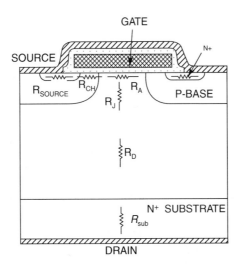

FIGURE 1.62 The origin of the internal resistances in a power MOSFET.

Also, higher channel implant dose is beneficial from the punch-through point of view since depletion width will be smaller, but the R_{dson} will suffer through reduced carrier mobility. The design of the doping profile involves choosing channel and source implant doses, diffusion times and temperatures that give a designed threshold voltage while simultaneously minimizing R_{dson} and I_{DSS}. Optimizing these performance parameters with manufacturability in mind is one of the challenges of power MOSFET design.

The reach-through phenomenon, on the other hand, occurs when the depletion region on the drift side of the body-drift *pn*-junction reaches the epilayer-substrate interface before avalanching takes place in the epi. Once the depletion edge enters the high carrier concentration substrate, a further increase in drain voltage will cause the electric field to quickly reach the critical value of 2×10^5 V/cm at which avalanching begins.

Other factors that affect the breakdown voltage of power MOSFETs for a given epitaxial layer include termination design, cell spacing (poly line width) and curvature of the body diode depletion region in the epi which is a function of diffusion depth. Power MOSFETs are designed such that avalanche breakdown occurs in the active area first.

On-Resistance

The on-state resistance of a power MOSFET is made up of several components as shown in Fig. 1.62.

$$R_{dson} = R_{source} + R_{ch} + R_A + R_J + R_D + R_{sub} + R_{wcml} \qquad (1.2)$$

where

R_{source} = source diffusion resistance
R_{ch} = channel resistance
R_A = accumulation resistance
R_J = the "JFET" component-resistance of the region between the two body regions
R_D = drift region resistance
R_{sub} = the substrate resistance; wafers with resistivities of up to 20 mΩ-cm are used for high-voltage devices and less than 5 mΩ-cm for low-voltage devices
R_{wcml} = sum of bond wire resistance, contact resistance between the source and drain metallization and the silicon, metallization resistance, and leadframe contributions; these are normally negligible in high-voltage devices but can become significant in low-voltage devices

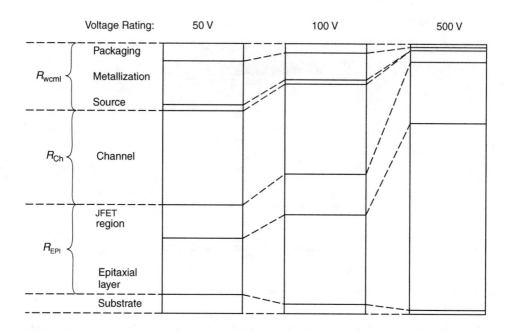

FIGURE 1.63 Relative contributions to R_{dson} in devices with different voltage ratings.

Figure 1.63 shows the relative importance of each of the components to R_{dson} over the voltage spectrum. As can be seen, at high voltages the R_{dson} is dominated by epi resistance and the JFET component. This component is higher in high-voltage devices due to the higher resistivity or lower background carrier concentration in the epi. At lower voltages, the R_{dson} is dominated by the channel resistance and the contributions from the metal to semiconductor contact, metallization, bond wires, and leadframe. The substrate contribution becomes more significant for lower breakdown voltage devices.

Transconductance

This parameter is a measure of the sensitivity of drain current to changes in gate-source bias and is defined as:

$$g_{fs} = \left(\frac{\Delta I_D}{\Delta V_{gs}}\right) \quad V_{ds} \text{ constant} \tag{1.3}$$

i.e., the gradient of the I_d vs. V_{gs} graph. In the saturation region, g_{fs} is given by:

$$g_{fs} = \mu C_{ox} \frac{W}{L}(V_{gs} - V_{th}) \tag{1.4}$$

This parameter is normally quoted for a V_{gs} that gives a drain current equal to about one half of the maximum current rating value and for a V_{DS} that ensures operation in the constant current region. With mobility μ fixed for a given semiconductor, the design parameters influencing transconductance of a MOSFET are gate width W, channel length L, and gate oxide thickness t_{ox} and hence C_{ox}. Gate width is the total polysilicon gate perimeter of the cellular structure and increases in proportion to the active area as the cell density increases. The cell density has increased over the years from around half a million per square inch in 1980 to around 8 million for planar MOSFETs and around 12 million for the trench technology at the present time. The limiting factor for even higher cell densities is the photolithography process control and resolution which allows contacts to be made to the source metallization in the center of the cells.

Reduced channel length is beneficial to both g_{fs} and on-resistance, with punch-through as a trade-off. The lower limit of this length is set by the ability to control the double-diffusion process and is around 1 to 2 μm today. Finally, reductions in gate oxide thickness give higher C_{ox} and higher g_{fs}. The reduction in oxide thickness will reduce V_{th} unless channel implant dose is increased which in turn will cause a higher R_{dson}. Ultimately, the lower limit of t_{ox} is set by the maximum gate-source voltage rating. This is ±30 V for high-voltage devices and ±20 V for lower-voltage logic-level devices used in portable electronic applications.

Threshold Voltage

This is defined as the minimum gate electrode bias required to strongly invert the surface under the poly and form a conducting channel between the source and the drain regions. V_{th} is usually measured at a drain-source current of 250 μA. A value of 2 to 4 V for high-voltage devices with thicker gate oxides and logic-compatible values of 1 to 2 V for lower-voltage devices with thinner gate oxides are common. With power MOSFETs finding increasing use in portable electronics and wireless communications where battery power is at a premium, the trend is toward lower values of R_{dson} and V_{th}. Gate oxide quality and integrity become major issues as the gate oxide thickness is reduced to achieve lower V_{th}. An approximate expression for V_{th} is given by:

$$V_{th} \approx \frac{\sqrt{4\epsilon_s KTN_A \ln(N_A/n_i)}}{(\epsilon_{ox}/t_{ox})} + \frac{2KT}{q} \ln(N_A/n_i) \tag{1.5}$$

where ϵ_{ox} and t_{ox} are oxide permittivity and thickness and the other parameters are defined in Eq. (1.1).

Processing methods used and their influence on the chemistry of the silicon surface have pronounced effects on V_{th}. Fixed and mobile surface and interface charges as well as charges in the gate oxide act to change the value of V_{th} from the intended value. Therefore, control of these charges in the process is necessary for obtaining consistent V_{th} values in production. Also, the presence of mobile charges away from the gate oxide and oxide/silicon interface may find their way to the device surface over the lifetime of the device and cause a gradual shift in V_{th}. For example, sodium ions in the low-temperature oxide (LTO) or in the metallization can cause a shift in V_{th} by changing the charge distribution at the interface. Accelerated life-tests are used by manufacturers to evaluate new processes and also to monitor V_{th} shift in production. Monitoring and control of contamination in the clean room equipment are routinely carried out by capacitance-voltage measurements of test diodes.

In real devices, V_{th} is altered by the unequal metal and semiconductor work functions. Denoting the barrier height between the metal and silicon oxide as ϕ_B, the work function difference is given by:

$$q\phi_{ms} = q\phi_B + q\chi_o - (q\chi + E_g/2 + q\psi_B) \tag{1.6}$$

where ψ_B is the potential difference between the intrinsic and Fermi levels in the semiconductor; χ and χ_o are the semiconductor and oxide electron affinities and E_g is the semiconductor band-gap energy.

Taking into account this effect and also the various fixed and mobile charges that may alter the value of V_{th} from that given above, the expression for V_{th} becomes:

$$V_{th} = \phi_{ms} + 2\psi_B - \left(\frac{Q_s + Q_{ss} + Q_I + Q_{FC}}{C_{ax}}\right) \tag{1.7}$$

where

Q_s = surface charge, which is a function of surface potential and determines channel conductivity
Q_{ss} = interface state charge (typically 10^{10} to 10^{12} cm^{-2}); caused by dangling bonds at the semiconductor surface, these can charge and discharge with changes in the surface potential
Q_I = charge due to mobile ions in the oxide
Q_{FC} = fixed surface charge at the silicon–oxide interface

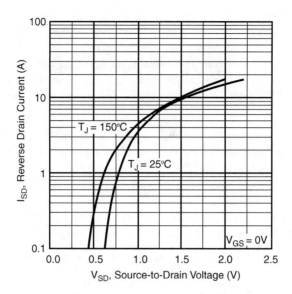

FIGURE 1.64 Typical source-drain (body) diode forward voltage characteristics.

It is worth mentioning that the success of silicon devices lies partly in the low density of these interface states which is due to the existence of native oxide in silicon as opposed to other semiconductors such as GaAs where such a native oxide does not exist and oxide layers have to be deposited with several orders of magnitude higher interface state densities.

Diode Forward Voltage (V_F or V_{SD})

This is the guaranteed maximum forward drop of the body-drain diode at a specified value of source current. Figure 1.64 shows a typical I–V characteristic for this diode at two temperatures. p-Channel devices have higher values of V_F due to the higher contact resistance between metal and p-silicon compared with n-type silicon. Maximum values of 1.6 V for high-voltage devices (>100 V) and values of 1.0 V for low-voltage devices (<100 V) are common.

Power Dissipation

The maximum allowable power dissipation which will raise the die temperature to the maximum allowable when the case temperature is held at 25°C is an important parameter and is given by:

$$P_d = \left(\frac{T_{jmax} - 25}{R_{thJC}}\right) \qquad (1.8)$$

where T_{jmax} is the maximum allowable temperature of the pn junction in the device (normally 150 or 175°C) and R_{thJC} is the junction to cause thermal impedance of the device.

Dynamic Characteristics

Switching and Transient Response

When the MOSFET is used as a switch, its basic function is to control the drain current by the gate voltage. Figure 1.65 shows the transfer characteristics and an equivalent circuit model often used for the analysis of MOSFET switching performance. For a detailed discussion of this topic see Chapter 4 in Grant and Gower (1989). The following is a summary of the important points.

The switching performance of a device is determined by the time required to establish voltage changes across capacitances and current changes in inductances. R_G is the distributed resistance of the gate and is approximately inversely proportional to active area. Values of around 20 Ω-mm^2 are common for the

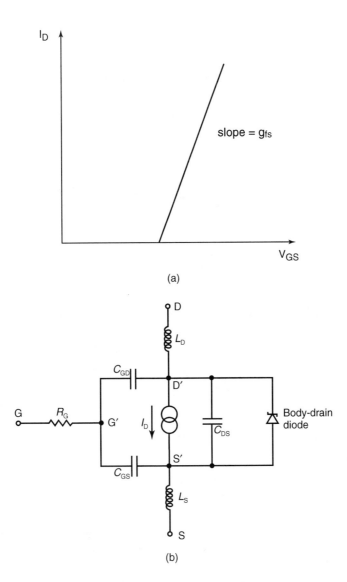

FIGURE 1.65 (a) Transfer characteristics and (b) an equivalent circuit diagram showing the MOSFET parasitic components that have the greatest effect on switching speed.

product of R_G and active area for polysilicon gates. L_S and L_D are source and drain lead inductances and are around a few tens of nH. The physical origin of the capacitances C_{GS}, C_{GD}, and C_{DS} were discussed in the introduction of this chapter regarding the device schematic shown in Fig. 1.57. The typical values of input (C_{iss}), output (C_{oss}), and reverse transfer (C_{rss}) capacitances given in the data sheets are used by circuit designers as a starting point in determining circuit component values. The data sheet capacitances are defined in terms of the equivalent circuit capacitances as:

$$C_{iss} = C_{GS} + C_{GD}, \quad C_{DS} \text{ shorted}$$
$$C_{rss} = C_{GD}$$
$$C_{oss} = C_{DS} + C_{GD}$$

The gate-to-drain capacitance C_{GD} is a nonlinear function of voltage and is the most important parameter since it provides a feedback loop between the output and the input of the circuit. C_{GD} is also called the

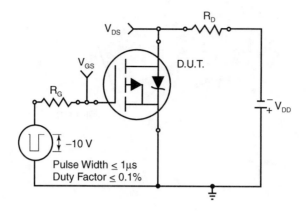

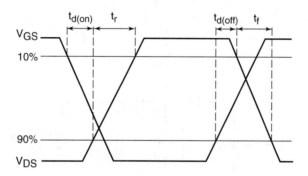

FIGURE 1.66 Switching time test circuit and resulting V_{GS} and V_{DS} waveforms.

Miller capacitance since it causes the total dynamic input capacitance to become greater than the sum of the static capacitances.

Figure 1.66 shows a typical switching time test circuit. Also shown are the components of the rise and fall times with reference to the V_{GS} and V_{DS} waveforms. Turn-on delay, $t_{d(on)}$, is the time taken to charge the input capacitance of the device before drain current conduction can start. Similarly, turn-off delay $t_{d(off)}$ is the time taken to discharge the capacitance after the gate is switched off.

Gate Charge

Although input capacitance values are useful, they do not provide accurate results when comparing the switching performances of two devices from different manufacturers. Effects of device size and transconductance make such comparisons more difficult. A more useful parameter from the circuit design point of view is the gate charge rather than capacitance. Most manufacturers include both parameters on their data sheets. Figure 1.67 shows a typical gate charge waveform and the test circuit. When the gate is connected to the supply voltage, V_{GS} starts to increase until it reaches V_{th}, at which point the drain current starts to flow and the C_{GS} starts to charge. During the period t_1 to t_2, C_{GS} continues to charge, the gate voltage continues to rise and the drain current rises proportionally. At time t_2, C_{GS} is completely charged and the drain current reaches the predetermined current I_D and stays constant while the drain voltage starts to fall. With reference to the equivalent circuit model of the MOSFET shown in Fig. 1.67, it can be seen that with C_{GS} fully charged at t_2, V_{GS} becomes constant and the drive current starts to charge the Miller capacitance C_{GD}. This continues until time t_3. Note that the charge time for the Miller capacitance is larger than that for the gate to source capacitance C_{GS}, due to the rapidly changing drain voltage between t_2 ant t_3 (current = $C\, dV/dt$). Once both of the capacitances C_{GS} and C_{GD} are fully charged, the gate voltage

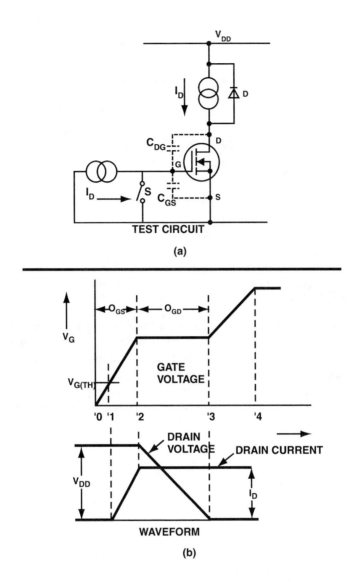

FIGURE 1.67 (a) Gate charge test circuit and (b) resulting gate and drain waveforms.

V_{GS} starts increasing again until it reaches the supply voltage at time t_4. The gate charge ($Q_{GS} + Q_{GD}$) corresponding to time t_3 is the bare minimum charge required to switch the device on. Good circuit design practice dictates the use of a higher gate voltage than the bare minimum required for switching and therefore the gate charge used in the calculations is Q_G corresponding to t_4.

The advantage of using gate charge is that the designer can easily calculate the amount of current required from the drive circuit to switch the device on in a desired length of time; since $Q = CV$ and $I = C\ dV/dt$ then Q = time × current. For example, a device with a gate charge of 20 nC can be turned on in 20 μs if a current of 1 mA is supplied to the gate or it can turn on in 20 ns if the gate current is increased to 1 A. These simple calculations would not have been possible with input capacitance values.

dV/dt Capability

This is also called the peak diode recovery and is defined as the maximum rate of rise of drain-source voltage allowed. If this rate is exceeded then the voltage across the gate-source terminals may become higher than the threshold voltage of the device, forcing the device into the current conduction mode and

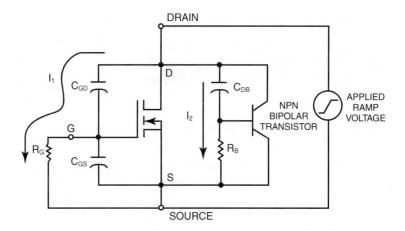

FIGURE 1.68 Equivalent circuit model of a power MOSFET showing the two possible mechanisms for dV/dt-induced turn-on. (From Baliga, B. J., *Modern Power Devices*, © 1987 John Wiley & Sons, Inc., New York. Reprinted by permission of John Wiley & Sons, Inc.)

under certain conditions a catastrophic failure may occur. There are two possible mechanisms by which a dV/dt induced turn-on may take place. Figure 1.68 shows the equivalent circuit model of a power MOSFET, including the parasitic BJT. The first mechanism of dv/dt induced turn-on becomes active through the feedback action of the gate-drain capacitance C_{GD}. When a voltage ramp appears across the drain and source terminals of the device, a current I_1 flows through the gate resistance R_G by means of the gate-drain capacitance C_{GD}. R_G is the total gate resistance in the circuit and the voltage drop across it is given by:

$$V_{GS} = I_1 R_G$$
$$= R_G C_{GD}\left(\frac{dV}{dt}\right) \qquad (1.9)$$

When the gate voltage V_{GS} exceeds the threshold voltage of the device V_{th}, the device is forced into conduction. The dV/dt capability for this mechanism is thus set by:

$$\left(\frac{dV}{dt}\right) = \frac{V_{th}}{R_G C_{GD}} \qquad (1.10)$$

It is clear that low V_{th} devices are more prone to dV/dt turn-on. The negative temperature coefficient of V_{th} is of special importance in applications where high-temperature environments are present. Also, gate circuit impedance has to be chosen carefully in order to avoid this effect. C_{GD} is an internal device parameter and is determined by the overlap area between poly gate and silicon and gate oxide thickness. Higher gate oxide thicknesses reduce C_{GD} and also increase V_{th}, both advantageous to dV/dt rating, as long as the higher V_{th} is acceptable in the application.

The second mechanism for the dV/dt turn-on in MOSFETs is through the parasitic BJT as shown in Fig. 1.69. The capacitance associated with the depletion region of the body diode extending into the drift region is denoted as C_{DB} and appears between the base of the BJT and the drain of the MOSFET. This capacitance gives rise to a current I_2 which flows through the base resistance R_B when a voltage ramp appears across the drain-source terminals. With analogy to the first mechanism, the dV/dt capability of this mechanism is given by:

$$\left(\frac{dV}{dt}\right) = \frac{V_{BE}}{R_B C_{DB}} \qquad (1.11)$$

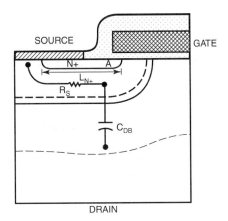

FIGURE 1.69 Physical origin of the parasitic BJT components that may cause dV/dt-induced turn-on in power MOSFET. (From Baliga, B. J., *Modern Power Devices,* © 1987 John Wiley & Sons, Inc., New York. Reprinted by permission of John Wiley & Sons, Inc.)

If the voltage that develops across R_B is greater than about 0.7 V, then the base–emitter junction is forward-biased and the parasitic BJT is turned on. Under the conditions of high dV/dt and large values of R_B, the breakdown voltage of the MOSFET will be limited to that of the open-base breakdown voltage of the BJT. If the applied drain voltage is greater than the open-base breakdown voltage, then the MOSFET will enter avalanche and may be destroyed if the current is not limited externally.

Increasing dV/dt capability therefore requires reducing the base resistance R_B by increasing the body region doping and reducing the distance the current I_2 has to flow laterally before it is collected by the source metallization. As in the first mode, the BJT related dV/dt capability becomes worse at higher temperatures since R_B increases and V_{BE} decreases with increasing temperature.

Applications

The following are two of the major markets where power MOSFETs are finding increasing applications as either logic-controlled or analog switches.

Portable Electronics and Wireless Communication

With the recent advances in the portable electronic products, low R_{dson}, logic level surface mount power MOSFET are experiencing explosive demand. A portable computer, for example, uses power MOSFETs in the AC-DC converters, the DC-DC converters and voltage regulators, load management switches, battery charger circuitry, and reverse battery protection. Required features of MOSFETs in these applications are small size, low power dissipation, and low on-resistance for extended battery life. Reduction of both conduction and switching losses are important considerations in the design of MOSFETs aimed at this market.

Automotive

Mechanical contact breakers have mostly been replaced by semiconductor devices in ignition circuits in modern cars. A suitable semiconductor device must be capable of blocking high voltages in a severe environment where line voltage surges are common due to the opening and closing of switches and the connection and disconnection of inductive loads during maintenance and loose connections. Bipolar transistors with their susceptibility to secondary breakdown are not suited whereas power MOSFETs with avalanche capability are ideally suited. Voltage transients are clamped by the avalanching of the MOSFET without the need to use any external protection circuits.

In 12-V battery vehicles the most commonly used MOSFETs are rated at 50 or 60 V breakdown voltages. The significant guard-banding is necessary in order to avoid device failure due to the alternator producing high voltages after shedding a heavy load.

The other features of power MOSFETs which make them suitable for the automotive applications are high dV/dt ratings, high-temperature performance, ruggedness and high reliability. Logic level, surface mount devices with low R_{dson} have recently found application in this field. The smaller footprint of surface mounts offers space savings and the lower R_{dson} does away with the need to parallel devices to reduce on-resistance. This in turn translates into fewer device counts and heat-sinks which lowers the overall cost.

In addition to ignition control, power MOSFETs are used in anti-lock brake (ABS) systems, electronic power steering (EPS) systems, air bags, electronic suspension, and numerous motor control applications such as power windows, power seats, radiator fan, wipers, fuel pump, etc.

References

Baliga, B.J. 1987. *Modern Power Devices,* John Wiley & Sons, New York.
Grant, D.A. and Gower, I. 1989. *Power MOSFETs—Theory and Applications,* John Wiley & Sons, New York.
International Rectifier, 1995, *HEXFET Power MOSFET Designer's Manual—Application Notes and Reliability Data,* International Rectifier, El Segundo, CA.
Oxner, E.S. 1982. *Power FETs and Their Applications,* Prentice-Hall, Englewood Cliffs, NJ.
Sze, S.M. 1981. *Physics of Semiconductor Devices,* John Wiley & Sons, New York.

1.7 General Power Semiconductor Switch Requirements

Alex Q. Huang

A power semiconductor switch is a component that can either conduct a current when it is commanded ON or block a voltage when it is commanded OFF through a control. This change of conductivity is made possible in a semiconductor by specially arranged device structures that control the carrier transportation. The time that it takes to change the conductivity is also reduced to the microsecond level compared with the millisecond level of a mechanical switch. By employing this kind of switch, a properly designed electrical system can control the flow of electric energy, shaping the electricity into desired forms.

Parameters describing the performance of a power conversion system include reliability, efficiency, size, and cost. The power switch plays an important role in determining these system-level performances [1]. To facilitate the analysis, a simple buck converter shown in Fig. 1.70a (buck converter) and 1.70b (its switching waveforms) is used as an example. There are two switches SW and D_F in the circuit. The purpose of this circuit is to deliver energy from a power source with a higher voltage V_{CC} to the load with a lower voltage V_O requirement. When the power switch SW is on, the energy is delivered from the source V_{CC} through switch SW, inductor L to the load. When the output voltage is high enough, this energy link will be shut down by turning off SW. Energies stored in L and C_O will maintain the load voltage. The typical circuit waveforms are depicted in Figs. 1.70a and b its switching waveforms. The circuit has four different operating modes: (1) (t_0–t_1) SW off and D_F on; (2) (t_1–t_3) SW turn-on and D_F turn-off; (3) (t_3–t_4) SW on and D_F off; (4) (t_4–t_6) SW turn-off and D_F turn-on.

Generally, the following parameters are important for a semiconductor switch designed for power conversion applications:

1. Maximum current carrying capability
2. Maximum voltage blocking capability
3. Forward voltage drop during ON and its temperature dependency
4. Leakage current during OFF
5. Thermal capability

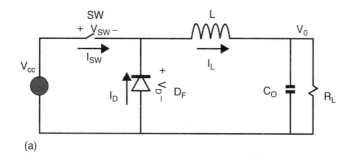
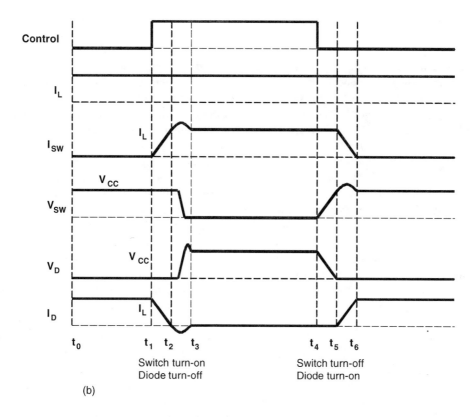

FIGURE 1.70 (a) Buck converter and (b) its switching waveforms.

6. Switching transition times during both turn-on and turn-off
7. Capability to stand dV/dt when the switch is OFF or during turn-off
8. Capability to stand dI/dt when the switch is ON or during turn-on
9. Controllable dI/dt or dV/dt capability during switching transition
10. Ability to withstand both high current and voltage simultaneously
11. Switching losses
12. Control power requirement and control circuit complexity

The above items can be further divided into three categories: static, dynamic, and control parameters. Items 1 to 5 relate to the static performance of a switch. Both current and voltage ratings describe the power handling capability of a switch. For a certain application, devices with higher current and voltage ratings are more robust to transient overcurrent and voltage due to switching transitions or circuit faults, increasing the system level reliability. For the buck converter, the nominal current of SW when it is on is equal to the current of the output inductor. However, SW will experience higher peak current during the turn-on period between t_2 and t_3 due to diode D_F reverse recovery. When the load R_L is shorted or D_F is fail shorted, SW will observe a much higher fault current.

Lower forward voltage drop and leakage current lead to a lower power loss, which is good from the energy efficiency and the thermal management point of view. Between t_0 and t_1, SW is on and its power dissipation is ($I_L V_F$), where V_F is the forward voltage drop of SW. Between t_3 and t_4, SW is off and its power dissipation is ($V_{CC} I_{LKG}$), where I_{LKG} is the leakage current of SW. Good thermal capability, which refers to the thermal resistance from the device to ambient and the maximum temperature the device can withstand, allows the device to operate at its full power rating instead of being limited by the thermal management.

Items 6 to 11 are related to the dynamic performance of a switch. Short transition times are required to increase the switching frequency and reduce the switching loss. The latter is caused by the overlap of current and voltage on the switch. For the buck converter, the turn-on transition time of SW is ($t_3 - t_1$) and the turn-off transition time is ($t_6 - t_4$). The current/voltage of the switch overlaps; hence, its switching losses are approximately proportional to the switching times. Item 7 describes the external dV/dt immunity of the device. In a system, the switch is generally exposed to a complex electromagnetic environment. However, the state and the operation of the switch should only be controlled by its control command instead of the environment. When the switch is in the OFF state or during turn-off operation, the switch should stay OFF or continue its turn-off process no matter what the external dV/dt across its anode and cathode (or collector/emitter) is. Similarly, there is a dI/dt requirement when the switch is ON or during the turn-on transition. Devices with a large cell size such as the gate turn-off (GTO) thyristor have lower dI/dt limitations because of the longer time required for uniform current distribution.

While a good switch should be able to withstand severe dynamic voltage and current changes, it should also be able to provide the system with an acceptable electromagnetic noise. This requires the controllable dI/dt and dV/dt capabilities from the switch [2]. A typical *turn-on* operation of a switch in a power conversion system is associated with a *turn-off* process of another switch (or diode). The dI/dt is generally determined by the *turn-on switch* and shared by the *turn-off switch*, which may not be able to withstand the high dI/dt. For example, a diode has a turn-off problem and high turn-off dI/dt may overstress it. In the buck converter, the *turn-off* of the diode D_F is accompanied with the turn-on of SW starting from t_1. The falling dI/dt of the D_F of the is equal to that of rising dI/dt of the SW. After t_2, D_F enters its reverse recovery process, experiencing its highest instant power before its current finally goes to zero. To protect these associated devices effectively, the maximum turn-on dI/dt should be limited. Similarly, a typical turn-off operation of a switch in a power conversion circuit is associated with a turn-on process of another switch (or diode). The dV/dt is generally determined by the turn-off switch and shared by the turn-on switch, which may not be able to withstand the high dV/dt. The maximum dV/dt of the active switch should be limited to protect the associated switches. Both dV/dt and dI/dt controls normally require a device to possess a forward-biased safe operation area (FBSOA) [3]. The FBSOA defines a maximum V–I region in which the device can be commanded to operate with simultaneous high voltage and current. The device current can be controlled through its gate (or base) and the length of the operation is only limited by its thermal limitation. Devices with FBSOA normally have an active region in which the device current is determined by the control signal level, as is shown in Fig. 1.71. It should be noted, however, that dI/dt control in practice means slowing down the transient process and increasing the turn-on loss.

During a typical inductive turn-off process, the voltage of a switch will rise and its current will decrease. During the transition, the device observes both high voltage and high current simultaneously. Figure 1.72 depicts the typical voltage–current trajectory of an inductive turn-off process as is the case in the buck circuit shown in Figs. 1.70a and b, between t_4 and t_6 in time domain. The current of the device stays constant while its voltage rises. Its current begins to decrease once its voltage reaches its nominal value. The voltage spike is caused by the dI/dt and stray inductance in the current commutation loop. On the I–V plane of the device, the curve that defines the maximum voltage and current boundary within which the device can turn off safely, is referred to as the reverse-biased safe operation area (RBSOA) [4] of the device. Obviously, the RBSOA of a device should be larger than all its possible turn-off I–V trajectories. Devices without a large enough RBSOA need an external circuit (such as an auxiliary soft-switching circuit

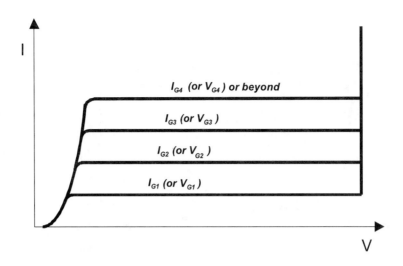

FIGURE 1.71 Forward *I–V* characteristic of a device and its FBSOA (shaded area) definition. The control of the device may be current or voltage.

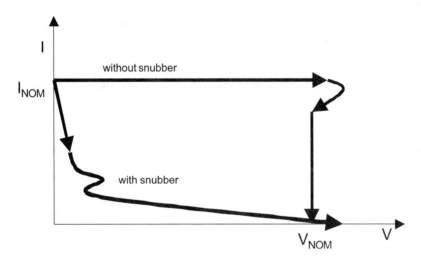

FIGURE 1.72 Turn-off *I–V* trajectories of a device under typical inductive load condition with and without a turn-off snubber.

or a *dV/dt* snubber) to shape their turn-off *I–V* trajectories to a smaller one to ensure safe turn-off operation. Devices with turn-off snubbers can therefore survive with a much smaller RBSOA. However, a *dV/dt* snubber increases the component count of the system, hence the system's size and cost. The turn-off operation conducted without the help of a snubber is called snubberless turn-off or hard turn-off, whereas a process with the help of a snubber is called snubbered turn-off.

During the turn-on transition, a switch will also observe both high voltage and high current simultaneously. Figure 1.73 depicts the typical voltage–current trajectory of an inductive turn-on process as is the case in the buck circuit shown in Fig. 1.70 between t_1 and t_3 in time domain. The voltage of the device stays constant while its current increases until it hits the nominal current level of the device. The current overshoot is due to the reverse recovery of an associated diode (or a switch). A device without a large enough FBSOA needs an external snubber circuit to help its *I–V* trajectory, as is shown in Fig. 1.73. The stress on the device can be significantly reduced with the turn-on snubber. However, a turn-on snubber circuit also increases the component count, size, and cost of a system.

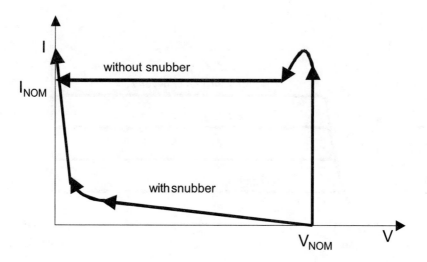

FIGURE 1.73 Turn-on I–V trajectories of a device under a typical inductive load condition.

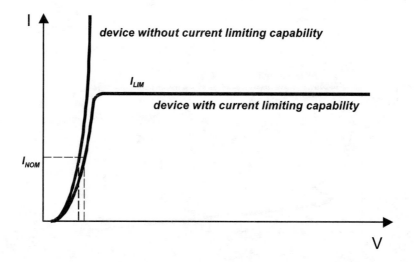

FIGURE 1.74 Forward I–V characteristics of two types of devices with and without self-current limiting capability.

Item 10 defines the capability of a switch to withstand high instant power. However, this capability during turn-on and turn-off will be different for a semiconductor device because of the difference in free carrier distribution. RBSOA is mostly used to describe the turn-off capability of a device, while FBSOA is used to measure its turn-on capability. FBSOA, as implied by its name, is also used to measure the ability of a device to withstand high voltage and high current under DC and short-circuit conditions.

A load short circuit is a threat to the device that is ON or is turning on in a typical circuit. A temporary load short can introduce an extremely high current that generates high instant power dissipation, leading to the failure of the switch. To effectively protect the switch under a short-circuit condition, the ability to limit its maximum current at a given DC voltage is required. In this case, the peak instant power is ($V_{CC}I_{LIM}$), whereas for a device without this ability it is (V_{CC}^2/r), where V_{CC} is the DC voltage, I_{LIM} is the maximum current limitation of the device, and r is the effective resistance of a device while it is ON. Since r is normally low in a practical device, the instant power of a device under a load short circuit without the maximum current limitation is much higher. Figure 1.74 shows the I–V characteristics during the ON state for devices with or without the self-current limitation capability.

The ability of a switch to limit its maximum current regardless of the voltage applied is an effective method to limit its instant power. A device with FBSOA capability normally has self-current limiting capability and, hence, can survive a short-circuit fault for a short time as determined by its thermal limitation [5].

References

1. M. Nishihara, Power electronics diversity, presented at IPEC '90, 1990, 21–28.
2. R. Chokhawala et al., Gate drive considerations for IGBT modules, presented at IAS '92, 1992, 1186–1195.
3. B.J. Baliga, Trends in power semiconductor devices, *Electron Devices IEEE Trans.*, 43(10) 1717–1731, 1996.
4. D.Y. Chen, G. Carpenter, and F.C. Lee, RBSOA characterization of GTO devices, in *PESC '93 Record*, 24th Annual IEEE 1993, 489–495.
5. H.C. Eckel et al., Optimization of the turn-off performance of IGBT at over current and short circuit current, in *5th European Conference on Power Electronics and Application Rec.*, 1993, 317–322.

1.8 Gate Turn-Off Thyristors

Alex Q. Huang

The first power semiconductor switch that was put in use was the silicon controllable rectifier (SCR) [1] invented in 1950s. The SCR is a latch-up device with only two stable states: ON and OFF. It does not have FBSOA. It can be switched from OFF to ON by issuing a command in the form of a small gate-triggering current. This will initiate a positive-feedback process that will eventually turn the device on. The SCR has a good trade-off between its forward voltage drop and blocking voltage because of the strong conductivity modulation provided by the injections of both electrons and holes. Moreover, the structure of an SCR is very simple from a manufacturing point of view because its gate can be placed at one small region. The size of a single SCR can therefore be easily expanded to increase the current capability of the device without too many processing problems. There are 8.0 kA/10.0 kV SCRs commercially available that use a 6-in. silicon wafer for current conduction. However, SCRs cannot be turned off through their gate controls.

Because of the limitation of the turn-off controllability of the SCR, the gate turn-off (GTO) thyristor [2] was subsequently developed. As its name denotes, a GTO is a device that can be turned off through its gate control. Its basic structure is very similar to that of an SCR. However, many gate fingers are placed in the GTO surrounding its cathode. During a turn-off operation, the latch-up mechanism can be broken through the gate control. A GTO is thus a device with full gate control and similar high current–voltage rating of an SCR. To date, the GTO has the highest power rating and the best trade-off between the blocking voltage and the conduction loss of any fully controllable switch. However, the dynamic performance of GTOs is poor. A GTO is slow in both turn-on and turn-off. It lacks FBSOA and has poor RBSOA so it requires snubbers to control dV/dt during the turn-off transition and dI/dt during turn-on transition.

The GTO thyristor was one of the very first power semiconductor switches with full gate control. It has served many power applications ranging from low power (below 100 W) in its early years to high power up to hundreds of megawatts. A state-of-the-art GTO can be fabricated on a silicon wafer as big as 6 in. and can be rated up to 6.0 kA and 6.0 kV [3]. This rating is much higher than the ratings of any other fully controllable devices.

The GTO static parameters are excellent: low conduction loss due to its double-sided minority carrier injection, high blocking voltage, and low cost due to its fabrication on a large single wafer. However, its dynamic performance is poor. The requirements of a dV/dt snubber during turn-off operation, a dI/dt

snubber during turn-on operation, and minimum on and off times make the GTO difficult to use. To improve the dynamic performance of the GTO while keeping its good static performance, a better understanding of the mechanism of the GTO is necessary. In this section, the basic operating principle of the GTO, its advantages and disadvantages, and the mechanism that determines its performance are summarized and discussed. A new gate-driving concept, namely, unity-gain turn-off, is then introduced. The advantages of this special driving method are analyzed and discussed. Finally, all known approaches that make use of this special driving technique are summarized.

GTO Forward Conduction

Figure 1.75a illustrates the cell structure and the doping profile of a typical high power GTO. Figure 1.75b shows the two-transistor GTO model; and Fig. 1.75c is a photograph of a 4-in. GTO along with its gate lead. The structure is a three-terminal, four-layer *pnpn* structure with a lightly doped n^- voltage-blocking layer in the center [4]. The electrode on the external p^+ layer is called the anode where the current normally flows into the device. The electrode on the external n^+ layer is called the cathode from where the current normally flows out. The electrode on the internal p layer (p-base) is called the gate, which is used for control.

The operating principle of a GTO can be understood through its equivalent circuit model shown in Fig. 1.75b. The *pnp* transistor represents the top three layers of the GTO, whereas the *npn* transistor

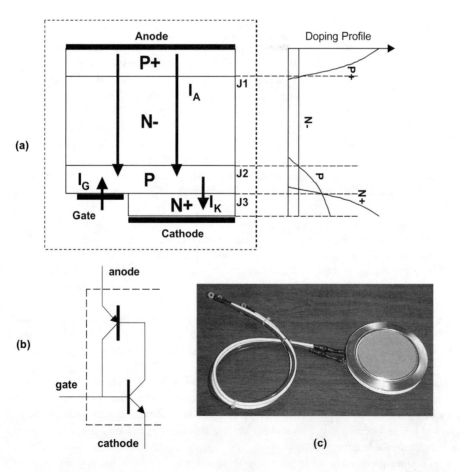

FIGURE 1.75 (a) GTO cell structure and its doping profile; (b) The two-transistor GTO model; (c) a photograph of a 4-in. GTO along with its gate lead.

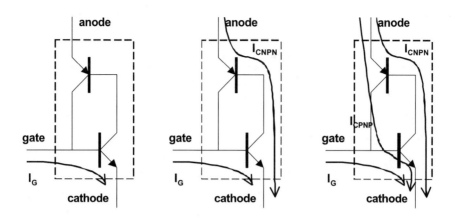

FIGURE 1.76 Turn-on and current-sustaining process in a GTO.

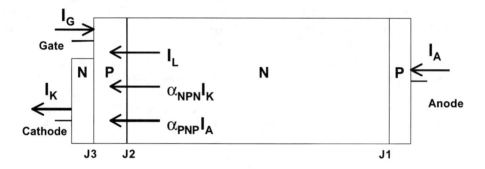

FIGURE 1.77 Current flow in a GTO with gate drive current.

represents the bottom three layers of the GTO. Since the n^- layer serves as the base of the *pnp* and the collector of the *npn*, and the internal *p* layer serves as the base of the *npn* and the collector of the *pnp*, the two transistors are cross-coupled. This structure has two stable states: ON and OFF, which are determined by its gate control. When a current is injected into the GTO from its gate to its cathode, the *npn* structure is turned on and its collector current flows from the anode of the GTO through J_1 junction. Since J_1 is the emitter junction of the *pnp* structure, the collector current of the *pnp* is then the base current of the *npn*. The two transistors therefore provide base currents to each other, forming a positive feedback among them until they reach a self-sustaining state commonly known as latch-up or latched. Under the latched condition, high-level minority carrier injections are available from the anode to the cathode, with all three *pn* junctions forward-biased. A high conductivity therefore exists from anode to cathode, allowing high current to flow from the anode to the cathode. Figure 1.76 illustrates this turn-on process.

At the silicon level, the turn-on of junction J_3 results in the injection of electrons into the *p*-base region. These electrons diffuse across the *p*-base and are mostly collected by the reverse biased junction J_2. To maintain the continuity of the current, junction J_1 will supply a current by injecting holes into the n^- region. Part of these holes will diffuse across the n^- region and are collected by junction J_2, resulting in more electron injection from junction J_3. When both transistors operate at sufficient current gain, a positive feedback mechanism is sufficient to result in latch-up.

Let the common base current gain of the *pnp* and *npn* be α_{pnp} and α_{npn}, respectively. Normally, α_{pnp} is lower than α_{npn} since the *pnp* is a wide-base structure. The current flow inside a GTO is illustrated in Fig. 1.77. At junction J_2, the current due to cathode side injection is $\alpha_{npn}I_K$; the current due to anode side injection is $\alpha_{pnp}I_A$; and the leakage current is I_L. According to Kirchhoff's law,

$$I_A = \alpha_{pnp}I_A + \alpha_{npn}I_K + I_L \tag{1.12}$$

and

$$I_A = I_K - I_G \tag{1.13}$$

Combining these equations,

$$I_A = (\alpha_{pnp}I_G + I_L)/(1 - \alpha_{pnp} - \alpha_{npn}) \tag{1.14}$$

This equation shows that the thyristor structure can sustain its anode current by itself once the sum of the common base current gain ($\alpha_{pnp} + \alpha_{npn}$) of both transistors is approaching unity. For a GTO, α_{npn} is designed low and is normally depending on I_G to ensure its gate turn-off capability. This will be discussed later. With this self-sustaining capability, the gate of a GTO does not need to supply a lot of current and does not need to be very close to its cathode as is necessary in a bipolar junction transistor (BJT) design. The dimension of a typical GTO cell shown in Fig. 1.75 is 100 to 150 μm wide. This is very large compared with the micron and/or even submicron process used for modern MOSFETs and insulated gate bipolar transistors (IGBTs). The large cell size design is cost-effective and makes it possible to fabricate large single-die devices to boost their current capability. A state-of-the-art GTO die is as large as 6-in. in diameter with a turn-off current capability of up to 6.0 kA [3]. Figure 1.75c shows a large GTO fabricated by ABB. The GTO shown is fabricated on a 4-in. silicon wafer consisting of thousands of cells like the one shown in Fig. 1.75 and packaged in a so-called press-pack or hockey-puck package.

The large cell structure in the GTO introduces a current spreading problem during the turn-on transition of a GTO. When a gate current is injected, the turn-on occurs first in the vicinity of the gate contact. The conduction area then spreads across the rest of the cathode area. This can be characterized by a propagation velocity called the *spreading velocity* [5]. Experimental measurements [6] have shown a typical spreading velocity of 5000 cm/s. This velocity also depends on the GTO design parameters, the gate turn-on injection current, and its dI_G/dt.

Because of this spreading velocity, it takes time for the whole GTO cell to turn on. To avoid overstressing the part of the cell that is turned on first, the increasing rate of the anode current should be limited. This sets the maximum turn-on dI/dt limitation for a GTO.

The major advantages of the GTO are its low forward voltage drop and high-voltage blocking capability. These can be understood as the major benefits of its double-side minority carrier injection mechanism. For high-voltage GTO, a thick and lightly doped n-base is needed (see Fig. 1.75). The forward voltage drop in this case is mainly determined by the resistive voltage drop in the voltage-blocking region where minority carriers play an important role.

Figure 1.78a shows the minority carrier distribution in the n^- region of a GTO and Fig. 1.78b shows the case of an IGBT (see Section 1.9). For the same blocking voltage design, their n^- regions should have similar thickness and doping. Since there is only one transistor in the IGBT structure, minority carriers can only be injected from one side; therefore, the conductivity modulation in the n^- region is weaker than that of the GTO. In the GTO, since there are two transistors, minority carriers can be injected from both ends, making a more uniform plasma distribution in the whole area. For a 4.5-kV state-of-the-art GTO, its forward voltage drop at a current density of 50 A/cm^2 can be as low as 2.0 V [7] if a constant gate current injection presents. Figure 1.79 shows the on-state characteristics of a state-of-the-art GTO manufactured by ABB [7]. The forward voltage drop at 2000 A is only about 1.5 V for this 4.5-kV GTO. This result is typical of a low conduction loss GTO.

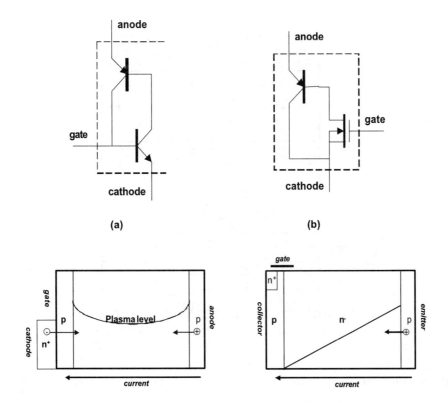

FIGURE 1.78 On-state minority carrier distribution in the voltage blocking region for (a) GTO and (b) IGBT.

GTO Turn-Off and Forward Blocking

If the GTO gate is pulling out current from the GTO, the current injection into the *npn* base will be reduced. Once this is reduced below a certain level, the collector current of the *npn*, and hence the base current of the *pnp*, will also decrease, leading to the reduced *pnp* collector current. This will further reduce the base current of the *npn* since it is the difference between the collector current of the *pnp* and the gate pullout current. This positive-feedback process will eventually turn off the GTO.

Figure 1.80 shows the current flow inside the GTO when its gate is pulling out current to turn off the device. The base drive current required to maintain current conduction in the *npn* transistor is $(1 - \alpha_{npn})I_K$. The base drive current available to the *npn* transistor in this case is $(\alpha_{pnp}I_A - I_G)$. Thus, the condition to turn off the GTO through the gate control is given by:

$$\alpha_{pnp}I_A - I_G < (1 - \alpha_{npn})I_K \tag{1.15}$$

Since

$$I_K = I_A - I_G \tag{1.16}$$

the condition to turn off the GTO is

$$I_G > \frac{(\alpha_{pnp} + \alpha_{npn} - 1)}{\alpha_{npn}} I_A \tag{1.17}$$

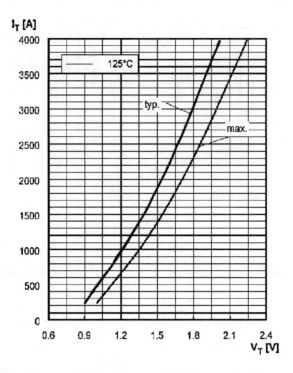

FIGURE 1.79 On-state characteristics of 5SGT 40L4502, a 4-kA, 4.5-kV GTO from ABB.

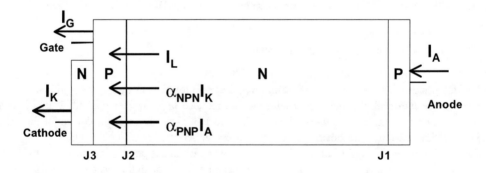

FIGURE 1.80 Current flow inside the GTO when its gate is pulling out current.

The ratio of the anode current to the gate current at which level a GTO is turned off is defined as the turn-off gain. From Eq. (1.17), the maximum turn-off gain [4] can be expressed as:

$$\beta_m \equiv \frac{I_A}{I_G} = \frac{\alpha_{npn}}{\alpha_{pnp} + \alpha_{npn} - 1} \qquad (1.18)$$

A large turn-off gain is normally desirable to reduce the current requirements of the gate driver. A lower ($\alpha_{pnp} + \alpha_{npn}$) value is necessary to ensure a reasonable turn-off gain. It is also important to point out that α_{npn} in Eq. 1.18 is not a constant; normally it decreases when gate current I_G increases.

When a GTO is OFF, its junction J_2 is reverse-biased and can support a high voltage applied between its anode and cathode, as shown in Fig. 1.81a. If the junction J_3 is reverse-biased or shorted by the gate

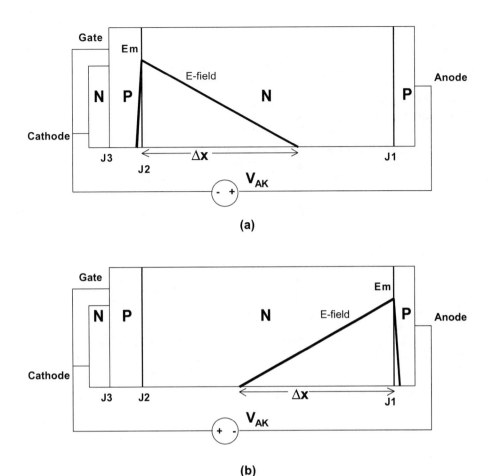

FIGURE 1.81 Electric field profile when a GTO is blocking forward (a) and reverse voltage (b).

driver, the maximum forward blocking voltage BV_{AK} of the GTO is determined by the avalanche breakdown capability of the *pnp* transistor under the open-base condition [8]. This voltage can be expressed as:

$$BV_{AK} = (1 - \alpha_{pnp})^{1/n} BV_{j2} \qquad (1.19)$$

where α_{pnp} is the common base current gain of the *pnp* structure at low current levels; n is an empirical constant, and BV_{j2} is the avalanche breakdown voltage of the *pn*-junction J_2. Since this *pnp* has a wide base structure, its common base current gain α_{pnp} is low compared with a normal bipolar transistor. Thus, the forward voltage blocking capability BV_{AK} of a GTO is very close to the breakdown voltage of junction J_2.

A GTO can also block a reverse voltage by its junction J_1, as shown in Fig. 1.81b. When the junction J_3 is gated off, the reverse voltage blocking capability is similarly determined by the avalanche breakdown of the *pnp* structure under the open-base condition. A GTO with both forward and reverse blocking capability is called symmetric blocking GTO. Most GTOs manufactured today, however, are asymmetric GTOs because the reverse blocking capability is not utilized (J_1 junction not designed to support high reverse voltage) or cannot be utilized because of other design requirements, such as the need to introduce anode-short at junction J_1 to speed the turn-off.

Practical GTO Turn-Off Operation

The turn-off capability of a GTO is limited dominantly by a non-uniform current distribution (also called current filamentation) problem during turn-off transient. This causes current to concentrate to a few GTO cells and destroy the device with the high power stress. Furthermore, the current filamentation is believed to be initiated by the dynamic avalanche (see next section) in an inhomogeneous, large-area GTO.

A GTO normally requires a dV/dt snubber circuit to conduct actual turn-off operation under high voltage and high current condition. This is because a large current GTO turn-off will fail without such a dV/dt snubber as a result of its small RBSOA. This small RBSOA is caused by a non-uniform current distribution or current filamentation problem in the GTO.

Figure 1.82 shows a practical setup in which a typical dV/dt snubber formed by D_S, R_S, and C_S is used, and Fig. 1.83 shows a typical GTO turn-off characteristic under snubbered condition. Before t_0, the GTO is ON, so a current is built up in the load inductor L_L and the device under test (DUT). The anode current I_A

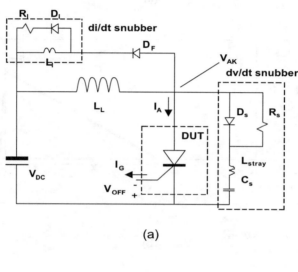

(a)

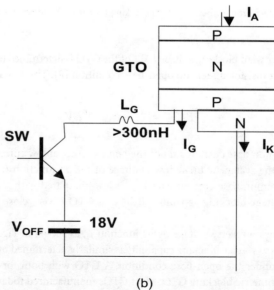

(b)

FIGURE 1.82 (a) The turn-off circuit of a GTO with a typical RCD dV/dt snubber. (b) Typical GTO gate drive circuit with a large gate inductance L_G.

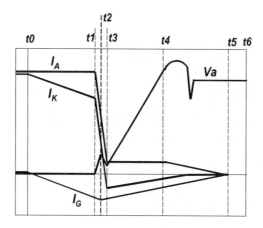

FIGURE 1.83 Typical GTO turn-off characteristics under dV/dt snubber condition.

is approximately equal to the cathode current I_K because the gate current I_G is negligible. Starting from time t_0, a negative voltage V_{OFF} is applied to the gate of the GTO. The gate current I_G then decreases linearly at a rate determined by the applied negative turn-off gate voltage V_{OFF} and the gate lead stray inductance L_G. At t_1, the device could not maintain the latch any longer so the anode current begins to decay. The current from the load inductor is diverted to the dV/dt snubber path. At t_2, when the anode current observes its maximum dI/dt, the anode voltage shows a spike due to the stray inductance L_S in the dV/dt snubber path. At t_3, the anode current enters its tail stage. At t_4, the anode voltage reaches the DC link voltage so the freewheeling diode D_F will be conducting. The energy in the stray inductance in the loop of power supply, freewheeling diode, and the dV/dt snubber is released to the snubber capacitor, causing another voltage peak. The anode voltage dip between t_4 and t_5 is due to the reverse recovery of the dV/dt snubber diode D_S. The turn-off trajectory of a GTO with a dV/dt snubber is significantly reduced, as shown in Fig. 1.72 (see Section 1.7 on General Power Semiconductor Switch Requirements).

Dynamic Avalanche

Under a high electric field, an avalanche process occurs within the silicon. The static critical electric field is a function of the doping profile. The lower the doping, the lower the critical avalanche electric field. The static avalanche voltage of a single side abrupt pn-junction is determined by both the critical electric field and the depletion region width.

While the reverse-biased junction conducts high current, as is the case of a GTO turn-off with or without a dV/dt snubber, the avalanche voltage decreases significantly because of the existence of carriers in the depletion region. This process is called dynamic avalanche [9]. Figure 1.84 shows the cross section of a pnp transistor under both current and voltage stress. A GTO turn-off with a dV/dt snubber enters the pnp conduction mode between t_2 and t_3, as shown in Fig. 1.82. Assuming carriers in the depletion region are moving at their saturation speed, then both the anode current density and the anode–cathode voltage can be expressed as:

$$J_A = qpv_s \tag{1.20}$$

$$V_{AK} = E_m W_E/2 \tag{1.21}$$

$$\approx (\varepsilon_s E_C^2)/2qp = (\varepsilon_s E_C^2)/(J_A/v_s) \tag{1.22}$$

where p is hole density in the depletion region, E_C is the critical electric field causing avalanche breakdown, and v_S is the saturation velocity of holes. In the depletion region, holes are the only carriers. In the presence of holes in the depletion region, the charge density in the depletion region is higher compared with the case without the current, so the peak electric field is also higher at the same width of the depletion region.

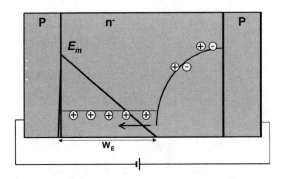

FIGURE 1.84 Dynamic avalanche in the blocking junction of a *pnp* structure.

At the point when dynamic avalanche happens, the power density of the device, which is the product of both the current and the voltage applied on the device, can thus be expressed as:

$$J_A V_{AK} = \varepsilon_s v_s E_C^2 / 2 \qquad (1.23)$$

which is about 200 to 300 kW/cm^2 for silicon.

The onset of dynamic avalanche itself is not a stable condition because the generated carrier is not sufficient to maintain the current. Hence it is not a failure condition from the point of view of the physics of the device. The dynamic avalanche is, however, widely regarded as the failure mechanism of the GTO because it will initiate a non-uniform current distribution among large-wafer-size GTOs. The current crowding or current filament formed after the onset of dynamic avalanche is enough to destroy the device at one location in the form of a melted spot [10].

Non-Uniform Turn-Off Process among GTO Cells

For a high-power GTO, the experimentally obtained instant turn-off power it can withstand is far below the value set by the dynamic avalanche breakdown shown in Eq. (1.21). So a GTO needs help from a *dV/dt* snubber to shape its turn-off *I–V* trajectory, as is shown in Fig. 1.72, and to lower the maximum average instant power the external circuit can apply. Non-uniform current distribution or current filament [10] among GTO cells during the turn-off operation accounts for this limitation. The current filament can be formed at the beginning of the turn-off due to differences in storage times or caused by the onset of the dynamic avalanche during the turn-off when the voltage and current are both high [11].

Current Filamentation Caused by Storage Time Difference

The non-uniform turn-off process can be understood by considering two GTO cells in parallel, as is shown in Figs. 1.85 and 1.87. The two cells are identical except for their storage time. This storage time difference is considered unavoidable in high-current GTOs because of differences in carrier lifetime, wafer thickness, and doping. Although only two cells are shown, GTO1 can represent a group of slower cells whereas GTO2 represents a group of faster cells. The turn-off process starts from t_0. Since it has a shorter storage time, GTO2 turns off earlier at t_1. The current originally shared by GTO2 is now transferred to GTO1. At t_2, GTO1 is turned off at twice its previous current. Turn-off failure can happen if its current at t_2 exceeds the maximum turn-off capability of GTO1. This can easily be the case when the number of faster cells is much larger than the number of slower cells. This type of failure typically occurs at the very beginning of the GTO turn-off before voltage rises and is caused by a rapid formation of current filament due to storage time difference.

What makes this type of failure likely is that there is also a positive feedback mechanism that will further increase the storage time difference, as shown in Fig. 1.86. At higher current density, the common-base

Power Electronics

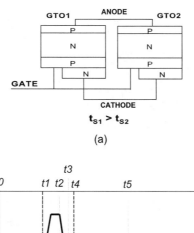

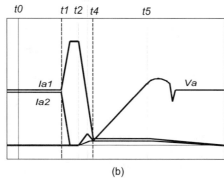

FIGURE 1.85 Current crowding among two GTO cells as a result of their storage time difference.

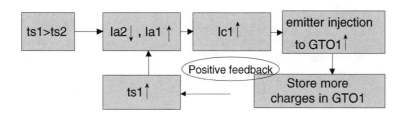

FIGURE 1.86 Positive feedback mechanism enhances the storage time difference and pushes the current filament into the slowest cell.

current gains of both transistors in GTO1 increase. Thus, its turn-off gain becomes even lower according to Eq. (1.16), requiring more gate current for turn-off, hence increasing its storage time. Considering that the typical storage time of a high-current GTO is in the range of 20 μs, there is enough time for the dangerous current filament to form.

Current Filamentation Caused by Onset of Dynamic Avalanche

Even if the above-discussed failure did not occur because GTO1 is beginning to turn off before the current filament density is too high, another failure mechanism can exist. At t_2, where both current and voltage are high, GTO1 is subject to much higher instant power stress than that of GTO2. The dynamic avalanche could then occur first at GTO1 and initiate another positive feedback that will further increase the localized current density (hence the name current filament) and enable the relatch of GTO1. Dynamic avalanche in a few cells can be viewed as an effective increase in the conductivity of those cells. If the number of slower cells is much smaller than that of faster cells, the current density in GTO1 can then become extremely high. This process can occur very quickly around t_2 with the area of the current filament smaller and smaller and the current density higher and higher (due to positive feedback). The excessive energy

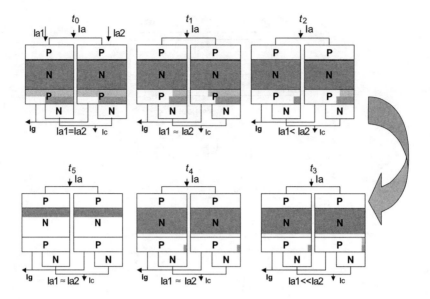

FIGURE 1.87 Semiconductor level analysis of the non-uniform turn-off process. The shaded region represents stored charge (plasma) in the GTO.

dissipated on the stressed cells can cause permanent failure because the temperature can be very high. After the device failure, a GTO device loses its blocking capability and behaves resistively.

It can therefore be concluded that the combination of storage time differences and the onset of possible localized dynamic avalanche makes the turn-off capability of a GTO small. Practical RBSOAs of high-power GTOs are below the 50 kW/cm^2 power constant line. This low limitation mandates the use of a dV/dt snubber. Even with a dV/dt snubber, GTO failure can still occur if the instant power is too high at t_2 and exceeds the RBSOA limit of the GTO at that time instance. The voltage spike at t_2 can be reduced by minimizing the stray inductance of the dV/dt snubber. The size of the snubber (C_s value) is typically between 3 to 6 μF. The disadvantages of using a dV/dt snubber are the increased count and size of the components, its high-energy loss, and its increased thermal management requirement to cool resistor R_s.

The concerns of non-uniform current distribution also mandate a minimum on-time rating for GTOs to ensure that conduction current is uniformly distributed in the ON state before a turn-off can be performed. Minimum off-time is also a commonly used rating for the GTO to guarantee the tail current of the GTO is completely gone and the GTO cells are all in the OFF state.

Summary

Advantages of the GTO include:

1. High current–voltage capability
2. Low conduction loss
3. Low cost

Disadvantages

1. Non-uniform turn-off—poor RBSOA and dV/dt snubber required
2. Non-uniform turn-on—dI/dt snubber required
3. Current control—high gating power
4. Long switching time—long storage time, minimum on-time and off-time requirements
5. No current limitation capability (FBSOA)

References

1. S.K. Gandhi, *Semiconductor Power Devices*, Wiley, New York, 1977.
2. E.D. Wolley, Gate Turn-Off in P-N-P-N devices, *IEEE Trans. Electron Devices*, ED-13, 590–597, 1966.
3. Mitsubishi GTO FG6000AU-120D data sheet.
4. B.J. Baliga, *Power Semiconductor Devices*, PWS Publishing Company, Boston, 1996.
5. W.H. Dodson and R.L. Longini, Probed determination of turn-on spread of large area thyristors, *IEEE Trans. Electron Devices*, ED-13, 478–484, 1966.
6. H.J. Ruhl, Spreading velocity of the active area boundary in a thyristor, *IEEE Trans. Electron Devices*, ED-17, 672–680, 1970.
7. ABB GTO 5SGT 40L4502 data sheet.
8. P.L. Hower and V.G.K. Reddi, Avalanche breakdown in transistors, *IEEE Trans. Electron Devices*, ED-17, 320–335, 1970.
9. I. Takata, T. Hikichi, and M. Inoue, High voltage bipolar transistor with new concepts, presented at *IEEE IAS '92*, 1992, 1126–1134.
10. Y. Shimizu, S. Kimura, H. Kozaka, N. Matsuura, T. Tanaka, and N. Monma, A study on maximum turn-off current of a high-power GTO, *Electron Devices, IEEE Trans.*, 462, 413–419, 1999.
11. K. Lilja and H. Gruning, Onset of current filamentation in GTO devices, in *PESC '90 Record, 21st Annual IEEE*, 398–406, 1990.

1.9 Insulated Gate Bipolar Transistors

Alex Q. Huang

When the development of power MOSFETs encountered difficulty in increasing their current-handling capability, the idea of a MOS-controlled bipolar device was developed to overcome the problem. This effort led to today's insulated gate bipolar transistor (IGBT) [1]. The IGBT fundamentally changes the BJT current control into voltage control while maintaining the advantages of the BJT. In addition, the use of a wide-base *pnp* transistor in the IGBT structure results in a much improved conductivity modulation effect than a conventional BJT, pushing the voltage rating of the IGBT toward the level of GTOs. The internal *pnp* structure also does not have the second breakdown problem as a conventional *npn* structure because the high voltage is supported by the base region of the *pnp* transistor instead of by the collector region as is the case for a conventional *npn* transistor. IGBTs also have excellent RBSOA and FRSOA. Having undergone several years' development, IGBTs have become the best device for applications in the range of 600 to 3000 V.

Although there are a number of other devices that have been developed or are being developed, the workhorse power semiconductor devices today are SCRs, GTOs, MOSFETs, and IGBTs. Each of these devices dominates a specialized power arena. The MOSFET has excellent dynamic and static performance. It dominates low voltage applications below 600 V. The IGBT is slower than the MOSFET but has better forward voltage drop above 600 V. It dominates applications from 600 to 3000 V. At an even higher voltage level, the GTO becomes the dominant device with better current-carrying capability but much slower dynamic response. Without turn-off capability, the SCR has an even better current conduction capability, so it is suitable for even higher power AC applications where gate-controlled turn-off capability is not necessary.

For a typical application, the switching frequency is an important index in determining system performance. Generally, the higher the switching frequency, the better the dynamic performance of the system, the smaller the size of the system due to reduced passive components, and the lower the cost of the system due to savings on passive components. The practical switching frequency of an application system is a trade-off of many issues including maximum device switching frequency, maximum magnetic switching frequency, switching losses of the power switches, overall system efficiency, etc. In the low power field where the MOSFET plays the major role, the switching frequency is normally subject to system efficiency and/or magnetic considerations instead of device limitations. In the medium power field, where the IGBT plays the major role, the situation changes. At the lower end, the limitation of the device does not dominate since the lower-rating IGBT is normally fast enough. However, when the power rating is higher, the IGBT switching speed decreases and the switching losses increase significantly. The practical switching frequency is thus subject to the limitation of the device. When the power level moves even higher, the GTO is the only available device. Since it has several tens of microseconds switching time, significant turn-off, and dV/dt snubber loss, the GTO is traditionally the limitation of the switching frequency of the system.

The above trend shows that when the power level moves higher, power semiconductor devices limit the maximum system switching frequency, hence the performance of the system, especially at the GTO level. To meet the increasing demand for better performance in high-power systems, many efforts have been made to improve the performance of high-power semiconductor devices. Among them, one effort is to push the IGBT toward higher power ratings based on the module concept. With its good dynamic performance, high-power systems equipped with IGBTs can operate at a much higher switching frequency and have many benefits compared with a conventional GTO system. The state-of-the-art IGBT rating is currently 3.3 kV/1.2 kA [2], which is at the low end of that of the GTO.

IGBT Structure and Operation

The name insulated gate bipolar transistor stems from its operation based on an internal interaction between an insulated-gate FET (IGFET) and a bipolar transistor. It has previously been called an IGT (insulated-gate transistor), an IGR (insulated-gate rectifier), a COMFET (conductivity-modulated field-effect transistor), a GEMFET (gain-enhanced MOSFET), a BiFET (bipolar FET), and an injector FET. IGBTs have been successfully used since they were first demonstrated in 1982 and are currently the most widely used power semiconductor switches with applications from several kilowatts to a few megawatts.

A cross section of the planar junction–based IGBT structure introduced in the 1980s is shown in Fig. 1.88a. The IGBT structure is similar to that of a planar power MOSFET except the difference in the substrate doping type. The fabrication of the IGBT therefore is almost the same as a power MOSFET. This has made its manufacture relatively easy immediately after conception, and its ratings have grown at a rapid pace as a result of the ability to scale up both the current and the blocking voltage ratings. Today, the largest single-chip IGBT can carry about 100 A and block more than 3000 V. Larger current IGBTs are also introduced by paralleling more IGBT chips in a single package. These IGBTs are also called IGBT modules. Figure 1.89 shows a photograph of a 1200-A, 3300-V IGBT module fabricated by Mitsubishi.

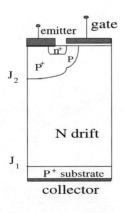
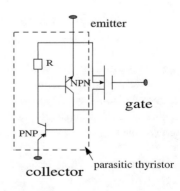

FIGURE 1.88 (a) Cross section of the IGBT structure and (b) equivalent circuit for the IGBT.

FIGURE 1.89 Photograph of a 1200-A, 3300-V IGBT module in which 24 1-cm^2 IGBT dies are paralleled together by wire bonds.

The equivalent circuit for the IGBT, shown in Fig. 1.88b, consists of a wide-base *pnp* bipolar transistor driven by a short-channel MOSFET. Notice the main current path for the IGBT is not through the *pnp* transistor but through the indicated path. In the IGBT structure, when a positive bias voltage larger than the threshold voltage of the DMOS channel is applied to the gate electrode, an inversion layer is formed along the *p*-base surface of the DMOS, and the DMOS channel is turned ON. Also an accumulation layer of electrons is formed at the surface of the *n* region below the gate. When a positive bias is applied to the collector, electrons flow from the n^+ emitter contact via the DMOS channel and the accumulation layer into the n^- drift region. This provides the base drive current for the vertical *pnp* transistor in the IGBT structure. Since the emitter junction (J_1) for this bipolar transistor is forward-biased, the p^+ region injects holes into the n^- base region. When the positive bias on the collector terminal of the IGBT is increased, the injected hole concentration increases and reduces the resistance of the n^- drift region. Consequently, the IGBT can operate at much higher current densities than the VDMOS even when it is designed to support high blocking voltages.

As long as the gate bias is sufficiently large to produce a strong inversion layer and an accumulation layer of electrons at the n^- base region surface, the IGBT forward conduction characteristic resembles that of a *pin* diode. Therefore, the IGBT can also be considered a *pin* diode in series with a MOSFET. Electron injections are provided by the accumulation layer electrons beneath the gate and between the adjacent *p*-body regions. However, not all injected holes recombine with these electrons; instead, some of the holes are collected by the *p*-body region, which acts as the collector region of the parasitic *pnp* transistor. IGBT design for low conduction drop requires minimizing the parasitic *pnp* transistor current and maximizing the *pin* current that maximizes the conductivity modulation.

However, if the DMOS channel becomes pinched off and the electron current saturates, the hole current also saturates because of the saturation of the base drive current for the *pnp* transistor. Consequently, the device operates with current saturation in its active region with a gate-controlled output current. This current saturation characteristic is useful for applications in which the device is required to sustain a short-circuit condition.

When the gate voltage is lower than the threshold voltage of the DMOS, the inversion layer cannot sustain and the electron current via the DMOS channel is terminated. The IGBT then operates in the

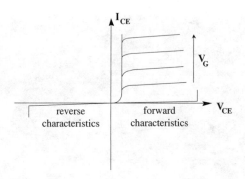

FIGURE 1.90 Output characteristics of the IGBT.

forward blocking mode. A large voltage can then be supported by the reverse-biased p-base/n-drift junction (J_2). Figure 1.90 shows a typical output characteristic of the IGBT.

The IGBT was the first commercially successful device based upon combining the physics of MOS-gate control with bipolar current conduction. Due to the injection of a high concentration of holes from the p^+ substrate into the n^- drift, the conductivity of the long n^- region is modulated and the IGBT exhibits *pin* diodelike on-state characteristics with a low forward voltage drop. Thus, the IGBT exhibits excellent current-carrying characteristics with forward conduction current densities 20 times higher than that of a power MOSFET and five times greater than that of a bipolar transistor operating at a current gain of 10. Since the input signal for the IGBT is a voltage applied to the MOS-gate, the IGBT has the high input impedance of the power MOSFET and can be classified as a voltage-controlled device. However, unlike the power MOSFET, the switching speed of the IGBT is limited by the time taken to remove the stored charges in the drift region due to the injection of holes during on-state current conduction. The turn-off time for the IGBT is dictated by the conduction modulation of the drift region and the minority carrier lifetime. The frontier is specified dominantly by the current gain of the wide-base *pnp* transistor, and the latter can be controlled by a lifetime control process, such as electron irradiation. Although the lifetime control process can be successful in reducing the turn-off time, it was found that there was a trade-off between the on-state voltage drop (conduction loss) and the turn-off time (switching loss). A shorter minority carrier lifetime makes the switching loss of the IGBT lower, but the shorter minority carrier lifetime also results in a higher conduction loss.

One of the problems encountered when operating the IGBT at high current levels has been the latch-up of the parasitic *pnpn* thyristor structure inherent in the device structure. Latch-up of this thyristor can occur, causing losses of gate-controlled current conduction. Since the current gains of the *npn* and *pnp* transistors increase with increasing temperature, the latching current decreases with increasing temperature. This effect is also aggravated by an increase in the resistance of the p-base with temperature due to a decrease in the mobility of holes. Many methods have been explored to suppress the latch-up of the parasitic thyristor, such as the use of a deep p^+ diffusion, a shallow p^+ diffusion, or a self-aligned sidewall diffusion of n^+ emitter. State-of-the-art IGBTs have basically solved this problem, and latch-up does not occur for all gate voltages applied. These IGBTs therefore exhibit close to square FBSOA.

Traditionally, IGBTs are fabricated on a lightly doped epitaxial substrate, such as the one shown in Fig. 1.88a. Because of the difficulty of growing the lightly doped epitaxial layer, the breakdown voltage of this type of IGBT is limited to below 1000 V. To benefit from such a design, an n buffer layer is normally introduced between the p^+ substrate and the n^- epitaxial layer, so that the whole n^- drift region is depleted when the device is blocking the off-state voltage, and the electric field shape within the n^- drift region is close to rectangular. This type of design is referred to as Punch-Through IGBT (PT IGBT), as shown in Fig. 1.91a. The PT structure allows it to support the same forward blocking voltage with about half the thickness of the n^- base region of the *pnp* transistor, resulting in a greatly improved trade-off relationship between the forward voltage drop and the turn-off time. Thus, the PT structure together with lifetime control is preferred for IGBTs with forward blocking capabilities of up to 1200 V.

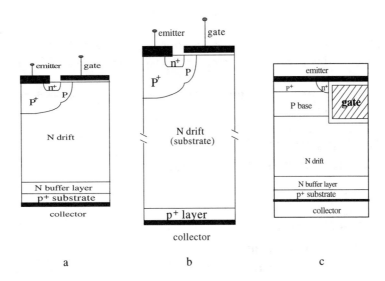

FIGURE 1.91 (a) PT IGBT structure, (b) NPT IGBT structure, and (c) UMOS gate PT IGBT structure.

For higher blocking voltages, the thickness of the drift region becomes too large for cost-effective epitaxial growth. Another type of design, the *Non-Punch-Through IGBT* (NPT IGBT, as shown in Fig. 1.91b), is gaining popularity. In the NPT IGBTs, devices are built on an n^- wafer substrate that serves as the n^- base drift region. The collector is implanted from the backside of the wafer after proper wafer thinning, and no field stopping n buffer layer is applied to the NPT IGBT. In this concept, the shape of the electric field is triangular in the forward blocking state, which makes a longer n^- base region necessary to achieve the same breakdown voltage as compared with the PT IGBT. However, the NPT IGBT offers some advantages over the PT IGBT. For example, the injection efficiency from the collector side can be controlled (due to the use of implanted p^+ region) and devices with voltage ratings as high as 4 kV can be realized. Further, by optimizing the emitter efficiency of carriers from the p^+ collector layer and the transport factor of carriers in the n^- base, the trade-off between the forward voltage drop and the turn-off time for the NPT IGBT can be improved to become similar to that of the PT type IGBT without lifetime control.

Generally speaking, the current tail in the NPT IGBT is longer than the PT IGBT, but the NPT IGBT is more robust than the PT IGBT, particularly under a short-circuit condition. The trench gate IGBT (UMOS-gate IGBT) structure is shown in Fig 1.91c. With the UMOS structure in place of the DMOS gate structure in the IGBT, the channel density is greatly increased and the JFET region is eliminated. In addition, the electron-hole concentration is enhanced at the bottom of the trench because an accumulation layer forms. This creates a catenary-type carrier distribution profile (see Fig. 1.91) in the IGBT, which resembles that obtained in a thyristor or *pin* diode. These improvements lead to a large reduction in the on-state voltage drop until it approaches that of a *pin* diode, hence approaching the theoretical limit of a silicon device. The latching current density of the UMOS IGBT structure is superior to that of the DMOS structure. This is attributed to the improved hole current flow path in the UMOS structure. As shown in Fig. 1.90c, the hole current flow can take place along a vertical trajectory in the UMOS structure, whereas in the DMOS structure hole current flow occurs below the n^+ emitter in the lateral direction. The resistance for the hole current that causes the latch-up is determined only by the depth of the n^+ emitter region. A shallow p^+ layer can be used, as shown in the figure, to reduce this resistance. As a consequence, the RBSOA of the UMOS IGBT structure is superior to that of the DMOS IGBT structure. Further, because of a very strong percentage of electron current flow in the trench gate IGBT, the turn-off speed of the trench-based IGBT is generally faster than the DMOS-based IGBT. It can be anticipated that trench gate IGBTs will replace the DMOS IGBT structures in the future.

References

1. Baliga, B.J., Adler, M.S., Love, R.P., Gray, P.V., and Zommer, N., The insulated gate transistor: a new three terminal MOS controlled bipolar power device, *IEEE Trans. Electron Devices*, ED-31, 821–828, 1984.
2. Brunner, H., Hier, T., Porst, A., and Spanke, R., 3300V IGBT module for traction application, in *EPE Conf. Rec.*, 1056–1059, 1995.

1.10 Gate-Commutated Thyristors and Other Hard-Driven GTOs

Alex Q. Huang

Unity Gain Turn-Off Operation

Traditional GTO Gate Drive Circuit

Traditional GTOs are generally designed with a turn-off gain of 3 to 5. This is the result of trade-offs between the performances of the GTO and the current (hence power) requirement its gate drive circuit. Figure 1.92b shows the typical turn-off gate drive circuitry for a traditional GTO. A negative turn-off voltage source V_{OFF} is connected to the GTO gate–cathode junction J_3 through the turn-off control switch SW. Since both sides of the junction J_3 are highly doped, its breakdown voltage BV_{GC} is practically about 20 V and can hardly be increased. The turn-off voltage V_{OFF} is selected below the junction J_3 breakdown voltage to avoid constant breakdown of this junction when the GTO is in the off state. To turn off the GTO, switch SW is turned on so the negative turn-off voltage V_{OFF} is applied on the GTO gate–cathode junction. The current originally flowing through the cathode is then diverted to the gate, causing cathode current I_K to decrease and the gate current to increase. Because of the existence of the GTO gate lead stray inductance L_G, which is practically on the order of several hundreds of nanohenry determined by the lead structure and length, the cathode current will decrease linearly and the gate current will increase linearly. This current commutation rate is thus given by:

$$dI_G/dt = V_{OFF}/L_G \qquad (1.24)$$

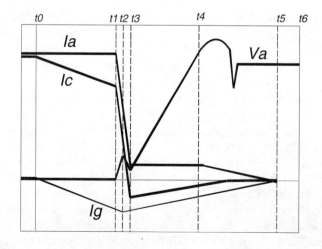

FIGURE 1.92 Typical turn-off characteristics of a GTO.

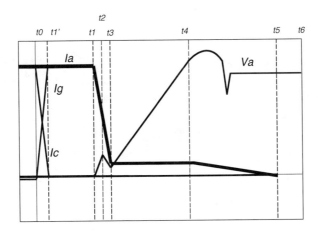

FIGURE 1.93 GTO turn-off waveform under unity gain.

The higher the turn-off gate current slew rate, the shorter the storage time. To obtain the shortest storage time, the turn-off voltage is normally selected very close to BV_{GC} to realize highest turn-off gate dI_G/dt. The typical turn-off gate dI_G/dt is on the order of several tens of amperes per microsecond, and the typical storage time of a high current GTO is about 20 μs. Figure 1.93 shows the typical current and voltage waveforms of a GTO turning off with a turn-off gain higher than 1. After the GTO is turned off, its gate current will drop back to 0 slowly by breaking down the GTO gate–cathode junction due to the energy stored in L_G. The energy required from the gate driver during this turn-off transition is the integration of the gate current times the turn-off voltage V_{OFF}. This energy is significant because the gate current lasts for a long period.

Because of the long transient process, storage time differences among GTO cells become bigger and the non-uniform current redistribution after t_1 is significant. The practical RBSOA of a GTO is normally much lower than the 200 kW/cm^2 limit set by the dynamic avalanche because of the non-uniform current turn-off (storage time differences and localized dynamic avalanche).

Unity Turn-Off Gain of the GTO

If the gate driver of a GTO is very fast so the gate current can increase rapidly to the anode current level and the cathode current decreases to zero before the anode current begins to decay, then the current and voltage waveforms of the device are as shown in Fig. 1.93. According to the definition above, the turn-off gain in this case is unity.

The internal turn-off process of the GTO changes significantly under the unity turn-off gain condition. Most important is that the GTO turn-off is now conducted in the *pnp* transistor mode after the unity gain is established. Figure 1.94 shows minority carrier distribution during the turn-off transition. Inside the *p*-base, there are two functioning parts of minority carriers (electrons). The first part is the electrons related to the bias of the gate–cathode *pn*-junction; the second part is the electrons related to the forward bias of junction J_2. Before the turn-off process at point t_0, minority carriers have been accumulated in the *p*-base and n^- region. Starting from t_0, the cathode current decreases rapidly and the gate current increases rapidly in the reverse direction. By t'_1, the cathode current comes to zero so minority carriers associated with the gate–cathode junction are removed. Zero cathode current cuts minority carrier injection from the n^+ side into the *p*-base. From this moment, the GTO is like an open-base *pnp*-transistor instead of a *pnpn* latch-up structure. This difference makes the GTO more rugged during turn-off transition. Negative gate current continues the extraction of minority carriers out from the *p*-base until t_1 when they are totally removed.

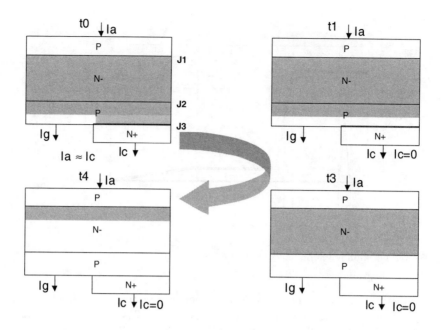

FIGURE 1.94 Internal process of a GTO under unity gain turn-off.

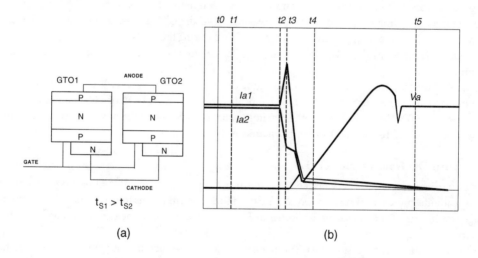

FIGURE 1.95 Turn-off waveforms of GTO cells under unity gain turn-off condition.

Advantages of Unity Gain Turn-Off

With unity turn-off gain, the storage time of a GTO is significantly reduced. The storage time in this case is the time required to remove minority carriers in the p-base. In the normal GTO case, the gate current is much less than the anode current so the removal speed is slow. Furthermore, the cathode current is not reduced to zero so minority carrier injection continues during the whole storage phase. With unity turn-off gain, the gate current is as high as the anode current, leading to a rapid carrier removing speed. Also, the cathode current is reduced to zero, hence instantly stopping the minority carrier injection into the p-base. Generally, the storage time of a GTO under unity turn-off gain is about 1 μs compared with that of about 20 μs in a normal GTO case with high turn-off gain.

Another important performance improvement with unity gain turn-off is in the RBSOA. As is analyzed above, the GTO current tends to crowd toward the cell with a longer storage time. This process

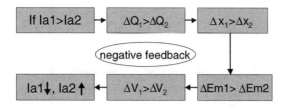

FIGURE 1.96 Negative feedback mechanism in the current sharing of two paralleled GTO cells during voltage increasing phase.

significantly limits the average instant power a GTO can withstand so a dV/dt snubber circuitry is normally required to limit the voltage level, hence the instant power stress, during turn-off transition.

GTO cells under unity gain turn-off have a tendency toward uniform current sharing, hence large RBSOA. First, the current filamentation due to the difference of storage time is greatly reduced because the absolute storage time is reduced to less than 1 μs. During the voltage rising phase after the storage time, if one cell still shares more current, that cell will have a faster carrier extraction rate and hence will turn off that cell faster. There is therefore a negative feedback process with current sharing instead of a positive one. This negative process is shown in Fig. 1.96.

With this uniform current distribution tendency provided by the unity turn-off gain, a GTO as a whole can be assumed to be more uniform in current sharing and hence can withstand much higher average instant power during turn-off transition. The RBSOA should now be pushed toward a power constant of 200 kW/cm^2 as predicted by Eq. (1.23) (in Section 1.8). This RBSOA is sufficiently large that a GTO should be able to perform turn-off operation even without the help of a dV/dt snubber. It should also be pointed out again that the onset of dynamic avalanche may not be the actual RBSOA boundary because if it does not initiate a runaway current filament, it is not a destructive one. Experimental results [1] on IGCT turn-off, however, suggest that the dynamic avalanche is not uniform and that it does lead to failure of a device. Unity gain turn-off is therefore effective in removing any current filament problem associated with storage-time differences and the dynamic avalanche soon after the current filament is formed.

Hard-Driven GTOs

Unity turn-off gain can significantly improve the performance of a GTO in several aspects, including RBSOA and turn-off storage time. Several innovative approaches have been proposed to realize unity turn-off gain. In all approaches, achieving unity turn-off gain is critical. This would require that cathode current be commuted to the gate path very fast. To turn off a 4-kA GTO with unity gain, the commutation rate should be higher than 6 kA/μs. This high commutation rate requirement distinguishes the performance of each of the devices discussed below. According to their realizations, they can be classified into two different categories: hard-driven type and MOS-controlled type. Hard-driven type approaches use a powerful gate driver to realize unity turn-off gain. The gate driver supplies the gate current and the gating power. Falling in this category is the integrated gate commutated thyristor (IGCT) [2]. The MOS-controlled approaches use MOSFETs to aid the turn-off process of the GTO. Other than the unity turn-off gain, these approaches also save control power for the turn-off process. Falling in this category are the emitter turn-off (ETO) [3] thyristor and the MOS turn-off (MTO) [4] thyristor.

IGCT

The key to achieving a hard-driven or unity-gain turn-off condition lies in the gate current commutation rate. A rate as high as 6 kA/μs is required for 4-kA turn-off. Two methods have been demonstrated for the implementation of a hard-driven GTO. The first is to hold the gate loop inductance low enough (3 nH) that a DC gate voltage less than the breakdown voltage of the gate–cathode junction (18 to 22 V) can generate a slew rate of 6 kA/μs. This approach is used in the IGCT/GCT [2, 5] (IGCT is an ABB product, GCT is by Mitsubishi, but the concept is the same), where a special low-inductance GTO housing and a

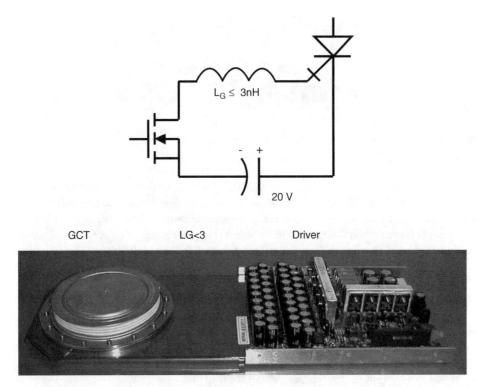

FIGURE 1.97 GCT operation principle and two GCTs developed by Mitsubishi and ABB. (Photographs courtesy of Mitsubishi (top) and ABB (bottom).)

carefully designed gate driver meet this requirement. The power consumption by the GCT driver is greatly reduced compared with that of a conventional GTO driver, since the gate current is present for a much shorter period of time [6]. Figure 1.97 shows the external view of the two commercially available GCTs.

The key disadvantage of the GCT approach is the high cost associated with the low-inductance housing design for the GTO and the low inductance and high current design for the gate driver.

MTO

Figure 1.98a shows the turn-off principle of a MTO™ [4, 7] developed by Silicon Power Corporation. The MTO device packages a number of low-voltage MOSFETs within a normal GTO device housing to form a current path that is in parallel with the emitter junction of the GTO. Therefore, the MTO looks

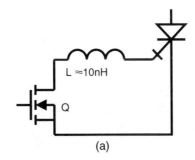

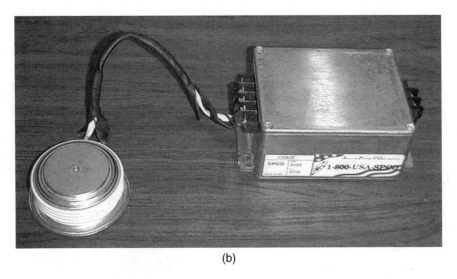

FIGURE 1.98 MTO equivalent circuit (a) and 500-A/4.5-kV MTO with gate driver (b) developed by Silicon Power Corp. (Photograph courtesy of Silicon Power Corp.)

just like a conventional GTO from the outside. The turn-off is initiated by turning on the MOSFET that shorts the GTO emitter junction. MTO, like the ETO, is therefore a MOS turn-off device requiring very little turn-off gate power. To achieve a high gate current commutation rate, very low gate inductance (<0.1 nH) is required.

Because of the use of the hybrid approach, a prototype 500-A, 4500-V device is available from SPCO. The major problem for the MTO, however, is still the limitation of the RBSOA [7]. This is because the gate current commutation rate is determined by the packaged gate inductance, which has to be reduced to below 0.1 nH. There are three reasons for this. First, in the MTO the commutation rate is determined by

$$\left(\frac{dI_G}{dt}\right)_{max} \leq \frac{0.7}{L_G} \qquad (1.25)$$

Second, the resistive voltage in the GTO p-base region and the MOSFET determines the peak gate current that can be commutated:

$$I_{G_{max}} \leq \frac{0.7}{R_{MOS} + R_{p\text{-base}}} \qquad (1.26)$$

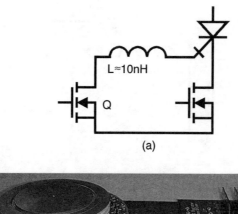

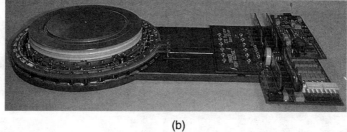

FIGURE 1.99 ETO equivalent circuit (a) and a 4-kA/6-kV ETO (b). An ETO gate driver is also shown.

Third, since there is no reverse bias voltage applied to the GTO emitter junction in the MTO, it is very easy to become latched again. Snubberless turn-off capability of the MTO is therefore lower than the GCT and ETO.

ETO

The method to achieve unity gain in the ETO thyristor is to insert an additional switch in series with the cathode of the GTO. The cathode of the GTO is the emitter of the internal *npn* transistor, so the series switch is referred to as the emitter switch and the new device is termed ETO. Turning off the emitter switch generates a high transient voltage long enough to commutate the emitter current to the gate path even with a higher parasitic inductance present. Because of this higher tolerance for parasitic inductance, conventional GTOs can be used in the ETO. An additional switch is connected to the gate of the GTO, and is complementary to the emitter switch. These switches are implemented with many paralleled low-voltage, high-current MOSFETs to minimize the additional conduction loss due to the emitter switch. The typical value for the conduction loss due to the series switch is 0.2 V at the average GTO current rating. The turn-off driving power for the ETO is negligible, since the turn-off is purely due to the removal of a MOSFET gate signal. The ETO in many aspects is similar to the IGBT. For example, the turn-off mechanism used in IGBT is also an emitter turn-off, and the IGBT always turns off in the rugged *pnp* transistor mode.

Figure 1.99 shows the equivalent circuit and hardware photograph of the developed 4-kA, 6-kV ETO by Virginia Tech. Other lower current rating ETOs have also been demonstrated by Virginia Tech. Because of the use of hybrid approach based on conventional GTO, ETO devices have clear advantages in terms of cost and gate drive power requirement over GCTs. ETO devices also have two other advantages when compared with the GCT. One is its feasibility of having a FBSOA [3, 8], and the other is its simplicity in overcurrent protection [8].

Conclusions

These newly developed GTOs (IGCT, MTO, and ETO) all utilize the unity gain turn-off concept and have dramatically improved performance compared with conventional GTOs. Quantitative comparisons of these devices are provided in a separate section on high-power IGBTs (Section 1.9).

References

1. I. Takata, M. Bessho, K. Koyanagi, M. Akamatsu, K. Satoh, K. Kurachi, and T. Nakagawa, Snubberless turn-off capability of four-inch 4.5k V GCT thyristor, presented at IEEE International Symposium on Power Semiconductor Devices and Ics, 1998, 177–180.
2. P.K. Steimer, H.E. Gruning, J. Werninger, E. Carrol, S. Klaka, and S. Linder, IGCT—a new emerging technology for high power, low cost inverters, presented at IEEE Industry Applications Society Annual Meeting, New Orleans, Louisiana, Oct. 5–9, 1997, 1592–1599.
3. Y. Li, A.Q. Huang, and F.C. Lee, Introducing the emitter turn-off thyristor, presented at 1998 IEEE Industry Applications Society 33rd Annual Meeting, 1998, 860–864.
4. D.E. Piccone, R.W. De Doncker, J.A. Barrow, and W.H. Tobin, The MTO thyristor—a new high power bipolar MOS thyristor, presented at IEEE Industry Applications Society 31st Annual Meeting, Oct. 6–10, 1996, 1472–1473.
5. E.R. Motto and M. Yamamoto, New High Power Semiconductors: High Voltage IGBTs and GCTs, in *PCIM'98 Power Electronics Conference Proceedings*, 1998, 296–302.
6. Mitsubishi GCT FGC4000BX-90DS data sheet.
7. A.Q. Huang, Y. Li, K. Motto, and B. Zhang, MTO thyristor—an efficient replacement for standard GTO, presented at IEEE Industry Applications Society 34th Annual Meeting, 1990, 364–372.
8. Y. Li, A.Q. Huang, and K. Motto, Experimental and numerical study of the emitter turn-off thyristor (ETO), *IEEE Trans. Power Electron.*, May 2000.

1.11 Comparison Testing of Switches

Alex Q. Huang

Pulse Tester Used for Characterization

In a typical power device dynamic test, the device under test (DUT) is initially off, and the high-voltage capacitor bank is charged to set the voltage the DUT will experience during switching. A typical pulse tester is shown in Fig. 1.100 and a typical waveform of the test is shown in Fig. 1.101. The so-called double pulse testing will capture one device turn-on event and one device turn-off event. The double-pulse test consists of the following complete events:

t_0–t_1: At time t_0, the control system initiates a pulse to the gate driver for the DUT. The DUT turns on and the high voltage capacitor bank charges the load inductor. After the current reaches the desired value at t_1, the DUT gate driver is commanded to turn off.

t_1–t_2: From time t_1 to t_2, no changes to the device are seen. During this time, referred to as the storage time, internal processes in the device initiate the turn-off process.

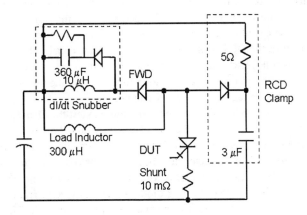

FIGURE 1.100 Pulse tester schematic diagram.

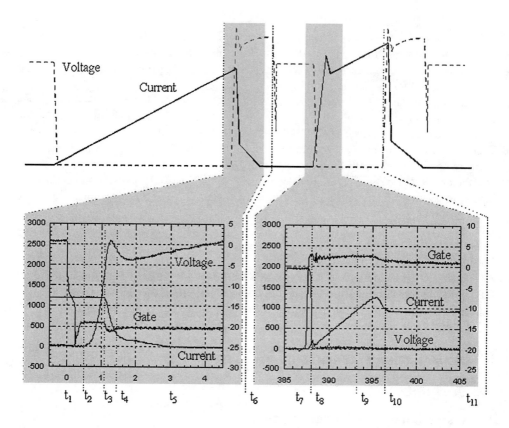

FIGURE 1.101 Double-pulse tester waveforms.

t_2–t_3: At time t_2, the anode voltage begins to rise, as the turn-off process has begun. The freewheeling diode is still reverse-biased so the current cannot yet begin to fall.

t_3–t_4: At time t_3, the anode voltage reaches the bus voltage and the main device current begins to fall. The current that had been flowing through the DUT is commutated into the freewheeling diode. This is the highest stress interval of the turn-off transition, as the current and voltage are simultaneously high during this interval

t_4–t_5: At time t_4, the main current fall is completed and the current tail phase begins. The current tail continues until t_5. At this point the device can be said to have completed the turn-off process.

t_5–t_6: During this time, the dI/dt snubber resistor carries the current, inducing additional voltage stress on the main DUT. The snubber inductor is charging during this time, and becomes charged at t_6. The snubber diode then goes through a reverse-recovery process.

t_6–t_7: During this time, the DUT is off and blocking a voltage equal to the input capacitor voltage. The current is still freewheeling through the load inductor and the freewheeling diode. This current will continue to circulate for a long time because the only energy dissipation is due to the conduction voltage of the freewheeling diode.

t_7–t_2: At this time, the controller initiates the second pulse to test the turn-on of the device. Nothing external occurs until t_8, which is the end of the turn-on delay time.

t_8–t_9: During this time, the load current begins to commutate into the DUT from the freewheeling diode. The dI/dt snubber inductor determines the rate of current transfer.

t_9–t_{10}: At time t_9, the load inductor current is completely commutated into the DUT and out of the freewheeling diode. The freewheeling diode undergoes reverse recovery during this period and releases a significant amount of reverse current into the DUT. It is important that the DUT have fully switched on by now or the diode recovery current will induce large power loss.

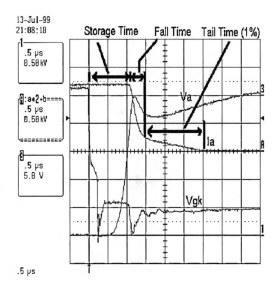

FIGURE 1.102 Switching time definition waveform.

t_{10}–t_{11}: During this time, the device is on and the current is rising because of the input voltage divided by the load inductance. This is equivalent to the interval t_0–t_1 from the first pulse. The same sequence will continue for the turn-off of the second pulse as that for the first pulse.

The current through the device under test is measured with a precision current shunt in series with the cathode (or emitter for an IGBT). All delay times are defined with respect to the actual gate of the device, so gate driver internal delays are not included. Conventionally, fall time is defined as when the current decreases from 90% of its initial value to 10%, but a different definition is used here. For the high-voltage devices, the current tail value can be greater than 10% of the initial current value, so it is unreasonable to include this time in the fall time. Therefore, the definition used here is that the fall time ends and the tail time begins when the current slope visibly changes. This is physically justified because for all three devices the current tail means that the main turn-off process is complete and the open-base *pnp* transistor is removing the remaining carriers. A sample waveform is shown in Fig. 1.102. Current tail time is defined from the end of the current fall time until the anode/collector current decreases to 1% of the initial current.

Devices Used for Comparison

To compare these various semiconductor technologies, two IGBTs, an IGCT, a GCT, and three ETOs were used [1]. One IGBT and the GCT are made by Mitsubishi, and the ETOs have been developed by researchers at Virginia Tech. The other IGBT is made by EUPEC, and the IGCT is from ABB. The IGBTs, CM1200HA-66H and FZ1200R33KF2, are rated for 1200 A (DC) and 3300 V, and are packaged in plastic modules 14 by 19 cm in size. The IGCT and the GCT are both 4500-V devices, which are rated for 4000 A maximum controllable current. The first ETO used, ETO4060s, is rated for 6000 V and 4000 A controllable current, and is based on a Toshiba GTO. The IGCT, the GCT, and the ETO4060s are packaged in 93-mm press-packs and, with gate drivers, have a maximum width of around 20 cm. The second ETO used, ETO1045s, is a small (53-mm) device rated for 4500 V and 1000 A. This ETO is based on a Westcode GTO. The ETO1045s is obviously of a lower rating than the GCT and IGCT, but it uses a fast conventional GTO, whereas the ETO4060s is based on a GTO designed for about 300 Hz. One final device used is a newly designed ETO, the ETO4045A, which is based on an ABB GTO similar to the thyristor used in the IGCT. The average current ratings for the IGCT, GCT, ETO4045A, and ETO4060s are 1200 A, whereas the ETO1045 is suitable for about 450 A average. When the switching losses of the IGBT and a safe

FIGURE 1.103 Clockwise from top left: ETO4060, EUPEC HVIGBT, Mitsubishi HVIGBT, ABB IGCT, Mitsubishi GCT. (Photograph courtesy of Mitsubishi.)

temperature margin are considered, the average operating current for this device should be between 600 and 800 A. Figure 1.103 shows most of the devices tested.

One significant difficulty in comparing this type of device is that the ratings, and even the ratings system, are different for the different devices. For GTO-based devices, the current ratings are the peak controllable current, whereas IGBTs use a DC current rating. The IGBTs tested have a controllable current rating of twice the DC rating, which translates to a 2400 A rating in the GTO system. These IGBTs consist of many small dies in parallel, giving a net current density much smaller than that of the GTO-based devicess. The rms current for the IGCT, the GCT, and the ETO4045A is about 1800 A, and the RMS current rating of the ETO 4060 is about 1600 A, although the devices have the same average rating (1200 A) from the manufacturers.

Unity Gain Verification

Because of the strict requirements on the gate loop stray inductance for the IGCT and the ETO, it is very difficult to insert a current probe directly to monitor the gate current. Fortunately, the unity gain of the IGCT and the ETO can be verified by observing easily probed voltage signals. It is critical for the performance of these devices that unity gain has been achieved, so some effort is made to verify unity gain and predict the maximum current that can be turned off while maintaining the hard-driven condition.

In the case of the IGCT, monitoring the gate-to-cathode voltage at the terminals of the IGCT thyristor can show the unity gain. When the gate voltage becomes −20 V, which is equal to the power supply in

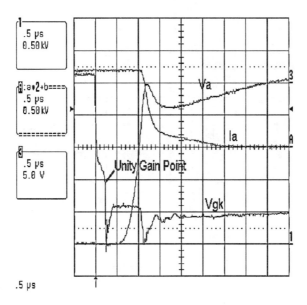

FIGURE 1.104 GCT unity gain.

FIGURE 1.105 Mitsubishi GCT gate driver. (Photograph courtesy of Mitsubishi.)

the gate driver, then clearly no voltage drop is occurring across the parasitic gate inductance. This in turn implies that dI_G/dt is zero, so the gate current has completed commutation. A typical GCT waveform showing the gate voltage is shown in Fig. 1.104. The inside of the GCT driver box is shown in Fig. 1.105.

Unity gain of the ETO can be verified by observing the drain-to-source voltage of the series switch. When the current is commutating, the voltage across this switch quickly rises to the breakdown voltage of the MOSFETs (60 V). When the voltage across this switch begins to fall, then the net cathode current of the GTO is negative, which discharges the output capacitors of the MOSFETs. Therefore, the ETO unity gain corresponds to the falling edge of the emitter switch voltage. A turn-off waveform showing the ETO emitter switch voltage is shown in Fig. 1.106.

Based on the unity gain observation, the rate of current commutation for the devices can be estimated by dividing the anode current by the time required for unity gain. This method yields a lower result than truly occurs because the total current commutated is slightly greater than the anode current due to a reverse recovery effect of the gate to cathode *pn*-junction. Even with this conservative estimation

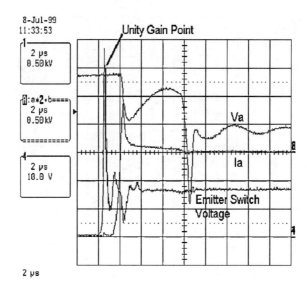

FIGURE 1.106 ETO4060 unity gain.

of the dI/dt of the gate current, the GCT and the ETO are both capable of approximately 6000 A/μs commutation rate.

Gate Drive Circuits

The performance of all semiconductor switches depends on the gate driver circuit. This is especially true for the GCT, where the device will be unable to operate in the snubberless mode if the gate driver is not drawing the gate current out fast enough to achieve unity gain. The drivers for the ETO and IGBT are less difficult to implement since the driver is not required to provide high current.

From a schematic point of view, the GCT driver is very simple, consisting primarily of a capacitor bank and a switch made from many parallel MOSFETs. The PCB layout and component selection is critical because of the very strict stray inductance requirement imposed on the switching loop. Additionally, there is a portion of the driver devoted to turning on the GCT. This is done by injecting a high-current (200-A) pulse into the gate for 5 μs and then injecting 10 A into the gate throughout the on time. This part of the driver dissipates significant power because of the linear transistors controlling the exact current level, but the implementation of this part of the gate driver is simple. The GCT driver contains minimum on-time and off-time protection to allow the device to be always in a uniform state prior to switching. No overcurrent protection is used for the GCT at the driver level. Although the total gating power is still very small compared with the main power, all the gating power must be supplied by an external isolated supply that must have an isolation capability and dV/dt rejection to match that of the GCT.

Because of the different thyristor design used by ABB in the IGCT, the driving power for this device has been greatly reduced. This is accomplished by increasing the current gain of the thyristor so less gate current is required to maintain the on state. This leads to a DC injection current of only 2 A. In addition, the IGCT driver uses a switching rather than linear circuit for pulse injection, which reduces losses as well.

For the ETO driver, three gates have to be controlled—the GTO current injection, the emitter switch, and the gate switch. Fortunately, the emitter switch and gate switch are easily controlled by using one inverting driver and one non-inverting driver controlled by the same input. The only function of the GTO gate is to inject the turn-on current just as in the case of the GCT. The ETO driver developed at the Center for Power Electronic Systems (CPES) also contains minimum on-time and off-time protection. In addition, the emitter switch MOSFET can be used as a linear resistor to approximate the anode current, which can be used for on-driver overcurrent protection. Like the GCT driver, the ETO driver requires an external isolated power supply, although the power consumption is much lower.

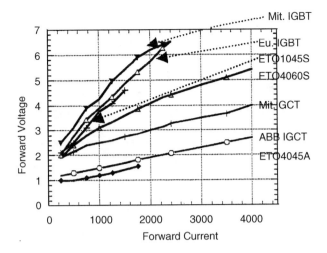

FIGURE 1.107 Forward conduction voltage.

The IGBT driver is very easy to implement, since it has only a single MOS-gate to control. The peak gate current for the tested IGBT is about 10 A, which flows for about 2 µs at every switching event. The IGBT driver can be used to control actively the dI/dt and dV/dt of the collector, but this feature was not implemented for this test. Information about active driver techniques can be found in many papers such as Lee et al. [2]. The IGBT driver implements an overcurrent protection by means of desaturation detection. IGBT drivers consume so little power that commercial DC-DC converter modules can be used to provide the isolation internally for the high-side switch.

Forward Conduction Loss Characterization

The forward current vs. voltage characteristics for all of these devices can be found easily. As can be seen from Fig. 1.107, the thyristors have a clear advantage in conduction loss over the IGBT, even though their active die area is less than that of the IGBT. If the relationship between breakdown voltage and conduction loss is found, the advantage of the latching devices becomes even greater. The 4.5-kV thyristors have the lowest conduction loss, followed by the 6-kV thyristor, and then the IGBT are the worst even if the loss is not normalized to die area. The ABB transparent anode and punch-through base design show an advantage in the forward conduction test, as the higher gain allows the device to latch into an extremely low loss conduction mode. This holds true for the ABB IGCT as well as for the ETO4045A, which is based on an ABB GTO with the same transparent anode and punch-through base design.

Switching Tests

Switching performance of high-power devices has been greatly enhanced by the hard-driven GTOs and the HVIGBTs appearing to challenge the slow GTO technology. Typical operation frequencies of the high-power GTOs range from line frequency (50/60 Hz) to a high of about 500 Hz. In contrast, the HVIGBT can be operated at up to 1500 Hz, and the hard-driven GTOs can operate at 1 kHz or more. This increase in frequency leads to dramatically reduced filters and lower distortion in the typical inverter applications.

To evaluate the performance of these devices, they were operated with DC voltages of 1.5 and 2 kV on the pulse tester without any turn-off snubbers. The limiting factor in the amount of current that could be switched off safely was the clamping diode used to limit the voltage spike on the switch. During reverse recovery, the voltage across this diode approaches its breakdown (4.5 kV) at the same time the anode (or collector for IGBTs) voltage of the device under test approaches zero, as circled in Fig. 1.108. For the GCT and the ETOs, no reverse voltage was acceptable because of the lack of either reverse

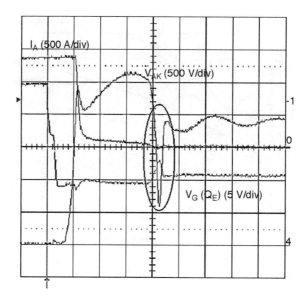

FIGURE 1.108 Typical turn-off waveform.

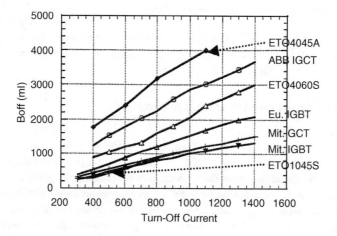

FIGURE 1.109 A 2-kV snubberless switching loss.

conduction capability (such as an antiparallel diode) or reverse voltage blocking capability. GTO-based devices can achieve reverse voltage blocking easily, but these tested GTOs are anode-shorted types, which trade away the reverse blocking capability for better switching performance, especially in the current tail phase. The ABB design uses a transparent anode rather than anode shorts, which also eliminates the reverse blocking capability. The transparent anode technology makes the current gain of the device change as a function of the current flowing so that it will have a high gain at low current and a lower gain at high current. The switching losses for each device were calculated by first multiplying the voltage across the device by the current being conducted, and then integrating this instantaneous power during the switching time to find the switching loss. The results of the switching loss tests were compared for the IGBT, the GCT, and the ETOs. These results are shown for a 2-kV bus in Fig. 1.109.

As expected, the IGBT holds the advantage in this test with the lowest turn-off loss overall. Surprisingly, the loss of the GCT and the ETO1045 is only marginally higher than the IGBT loss. The primary advantage of the IGBT in switching loss is in the initial voltage rising phase, which occurs much faster than in the thyristors. This is because the MOSFET channel in the IGBT can turn off faster than the *npn* transistor in the GTOs, and the channel is better distributed through the IGBT than are the gates of the GTOs. The amount of carrier stored in the GTOs is also higher than in the IGBT, resulting in slower dV/dt. It is not surprising that the ultrahigh-voltage ETO4060 has significantly more switching loss than the lower-voltage devices. The probable reasons for the high switching loss of this device are a high carrier lifetime in the GTO, a strong *pnp* transistor, which can maintain the current longer with the base open, and a GTO design optimized for low-frequency, high-power operation. The theory of hard-driven GTOs predicts no improvement in turn-off loss when compared with traditionally driven GTOs, only an improved safe operating area and higher speed. This shows that the GCT is very well optimized for performance as well as for a low internal inductance. The transparent anode of the ABB IGCT proved a disadvantage in this test, as switching times and switching losses were noticeably worse than with the anode-shorted devices.

As can be seen in Fig. 1.110, the switching times for all of these devices are short and very consistent. The ETOs and the GCT have long storage times at very low current levels, but the storage time is very consistent at 600 A and beyond. The current fall times for all devices characterized except the IGCT are around 250 ns and are essentially independent of the current being switched, as shown in Fig. 1.111. The IGCT has a very long current fall time at low current levels, although the speed improves at higher currents. The IGCT tail has a very large magnitude, which again shows that the anode-shorted structure of the GCT and the ETOs offers advantages in this area.

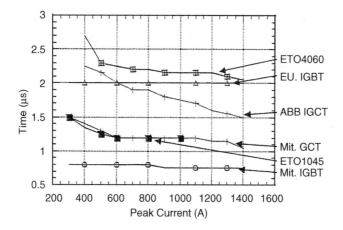

FIGURE 1.110 Storage (or delay) time comparison.

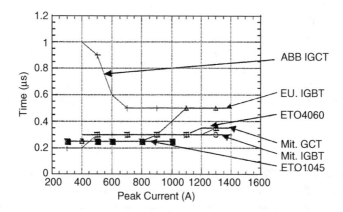

FIGURE 1.111 Fall time comparison.

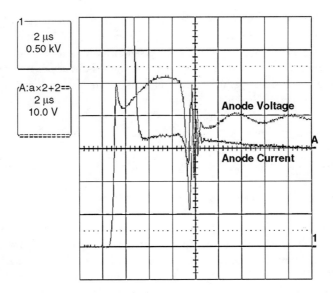

FIGURE 1.112 GCT current tail detail.

Because of the large (10 μH) turn-on inductor, the turn-on loss for all devices is negligible. All the thyristors hold a very slight advantage over the IGBT in terms of voltage fall time at turn-on, but the current is so low during this time that there is no significant difference in loss. It must be noted that the IGBT can be operated without the turn-on snubber at the expense of significantly increased switching loss, but doing so requires a more complex gate driver design. This is due to the ability of the IGBT to control the exact collector current by operating in the linear region. The GCT completely lacks this operating mode. Theoretical analysis predicts the existence of this forward-biased Safe Operating Area for the ETO [3], but no experimental verification has been performed except at low current [4]. For current tail comparison, the tail current was examined on a very high resolution (10 A/div) to see all the effects. Immediately after the main current fall, the tail current decreases rapidly for all the devices tested. However, the current tail can take a long time to finish decreasing to zero after this initial fast fall. The detail of the GCT current tail is shown in Fig. 1.112 after turning off 1200 A. The current tail can indicate the strength of the *pnp* transistor within an IGBT or a GTO. The long tail observed for the ETO4060 indicates a stronger *pnp*, which helps reduce the conduction losses. The GCT demonstrates the shortest current tail of all of the devices tested, which is further evidence of the very good internal design. The drawback of this performance is that the effective current gain of the GCT is reduced, thus requiring more DC gate current injection during conduction. The IGBT and the ETO1045 have only slightly worse current tails than the GCT.

Traditionally, the GTO switching frequencies were limited by the times required for the GTO to complete the switching transitions. In particular, a very long minimum off-time had to be observed due to some parts of the GTO remaining latched for more than 100 μs. The devices tested here all have very fast switching times, but the switching loss is rather high because of the very large currents and voltages considered. Therefore, the switching frequency is thermally limited by the switching loss. Soft switching techniques may allow these devices to achieve much higher operating frequencies (~10 kHz) if the switching loss can be reduced.

Discussion

Packaging technology is very different for the IGBT modules compared with GTO packaging. The IGBT modules use many parallel dies, which are wire-bonded and housed in a plastic module. Since a GTO can be fabricated on a single wafer, press-pack ("hockey-puck") housings are utilized. The reliability record for the press-pack devices is much higher than wire-bond modules, largely due to a better tolerance for thermal cycling. Additionally, the press-pack allows double-sided cooling to lower the thermal impedance. However, the IGBT achieves similar thermal impedance overall because of the much larger die area and the consequently large baseplate. The IGBT baseplate is electrically isolated from the heat sink, but the press-pack heat sinks are directly connected to the anode and cathode terminals. As a result, liquid-cooled systems with press-pack devices must rely on oil or deionized water to prevent the coolant from conducting current. The main advantage of the IGBT module is its ease of use, with the isolated baseplate leading to easy heat-sinking. The collector and emitter terminals are conveniently located for connection to a laminated busbar to reduce the parasitic inductance and hence the voltage spike. Additionally, the IGBT module does not require any external mechanical clamp for mounting, as the press-pack housing requires. The reliability of the press-pack is a key issue, and this package is preferred for many applications where long life is necessary.

Although failures are obviously unwanted, the characteristics of the device after a failure should be considered. This can make a big difference in how much damage is done to the rest of a system and how difficult repair will be. After a failure, any of these devices will become short-circuited. The current will then increase until either all the energy available has been consumed or an external circuit acts to stop the fault current. For the wire-bond IGBT, all the current will concentrate into the die that broke down. This will usually destroy the wire bonds for that die as a result of the huge current flowing. After failure, the IGBT can become an open circuit. This is a very dangerous condition for series-connected devices or multilevel converters, as the voltage will no longer be shared, thus exposing the other devices in the chain to the risk of overvoltage [5]. The press-pack devices will remain shorted since the die is directly connected with the metal contacts. There is some concern about the wire-bond MOSFETs in the ETO emitter and gate switches, although no failure of these MOSFETs has yet been seen even after destruction of the GTO. Another issue related to the packaging is explosion damage. The press-pack is very strong, and as a result explosions are very unlikely in this type of package. Plastic modules can easily shatter the housing, which leads to damage to nearby components.

As previously mentioned, an IGBT can actively control the collector voltage and current during the switching events. This feature of the device can lead to reduced EMI as well as elimination of the dI/dt (turn-on) snubber. However, elimination of this snubber in high-power, quasi-zero impedance source (voltage-fed) converters may not be desirable because of the other benefits the snubber offers. These include elimination of damage due to cross-conduction of bridge switches ("shoot-through"), or load short-circuiting, and improved fault management. If the rate of rise of current in a fault condition is controlled, a fast device such as the (I)GCT, ETO, or IGBT can respond in time to turn off the fault current with the semiconductor switches. For GTO systems, the GTO could not respond in time to interrupt a fault current, so the protection commonly used was to turn all the bridge switches on and wait for fuses to open. The ability of the ETO and IGBT to automatically detect and respond to overcurrents enhances the safe operation of high-power systems. In addition, the IGBT can self-limit the current that will be conducted, so operation within the switching capability of the device's can be ensured. Thyristor devices will conduct an extremely high surge current that is much higher than their interrupting capability, which requires control logic to prevent the devices from switching off during this time.

Comparison Conclusions

As can be seen from the switching times, all of the devices tested here offer very fast switching times relative to their power ratings. In addition, even the worst conduction loss from the IGBT is still acceptable when compared with the blocking voltage. For very high power systems, the IGCT, the GCT, the ETO4045A, and the ETO4060s are capable of handling extremely high power levels. The GCT is very fast for its high rating, and the only drawback is the difficult to construct gate driver and its power consumption. The ABB IGCT and the ETO4045A trade away switching loss to reduce driver power and conduction loss, so these devices are particularly suited to advanced topologies that reduce the necessary switching frequency or to soft-switching applications that can reduce the switching loss. The ETO4060 offers very high ratings with minimal driving power, even though the switching is not quite as good as the GCT; however, it is better than the IGCT. The IGBT offers the best switching speed and loss of any of the devices tested and the simplest drive. However, the GCT and small ETO are amazingly close to the IGBT in switching loss considering their latching nature and nearly 50% higher voltage rating. The performance of all devices tested here is very good, especially compared with the conventional GTO applications.

References

1. K. Motto, Y. Li, and A.Q. Huang, Comparison of the state-of-the-art in high power IGBTs, IGCTs, and ETOs, in *Conf. Rec. IEEE-APEC,* 2000, 1129–1136.
2. H.-G. Lee, Y.-H. Lee, B.-S. Suh, and D.-S. Hyun, An improved gate control scheme for snubberless operation of high power IGBTs, in *Conf. Rec. IEEE-IAS,* 1997, 975–982.
3. Y. Li, A.Q. Huang, and K. Motto, Experimental and numerical study of the Emitter Turn-Off thyristor (ETO), *IEEE Trans. Power Electron.,* 15(3), 2000, 561–574.
4. Z. Xu, Y. Bai, Y. Li, and A.Q. Huang, Experimental demonstration of the forward biased safe operation area of the emitter turn-off thyristor, in *Proc. CPES-VT Seminar,* 2000, 448–455.
5. S. Bernet, R. Teichmann, A. Zuckerberger, and P. Steimer, Comparison of high power IGBTs and hard driven GTOs for high power inverters, in *Conf. Rec. IEEE-APEC,* 1998, 711–718.

II

Power Electronic Circuits and Controls

2 **DC-DC Converters** *Richard Wies, Bipin Satavalekar, Ashish Agrawal, Javad Mahdavi, Ali Agah, Ali Emadi, Daniel Jeffrey Shortt* 2-1
   Overview • Choppers • Buck Converters • Boost Converters • Cúk Converter • Buck–Boost Converters

3 **AC-AC Conversion** *Sándor Halász* 3-1
   Introduction • Cycloconverters • Matrix Converters

4 **Rectifiers** *Sam Guccione, Mahesh M. Swamy, Ana Stankovic* 4-1
   Uncontrolled Single-Phase Rectifiers • Uncontrolled and Controlled Rectifiers • Three-Phase Pulse-Width-Modulated Boost-Type Rectifiers

5 **Inverters** *Michael Giesselmann, Attila Karpati, István Nagy, Dariusz Czarkowski, Michael E. Ropp* 5-1
   Overview • DC-AC Conversion • Resonant Converters • Series-Resonant Inverters • Resonant DC-Link Inverters • Auxiliary Resonant Commutated Pole Inverters

6 **Multilevel Converters** *Keith Corzine* 6-1
   Introduction • Multilevel Voltage Source Modulation • Fundamental Multilevel Converter Topologies • Cascaded Multilevel Converter Topologies • Multilevel Converter Laboratory Examples • Conclusions

7 **Modulation Strategies** *Michael Giesselmann, Hossein Salehfar, Hamid A. Toliyat, Tahmid Ur Rahman* 7-1
   Introduction • Six-Step Modulation • Pulse Width Modulation • Third Harmonic Injection for Voltage Boost of SPWM Signals • Generation of PWM Signals Using Microcontrollers and DSPs • Voltage Source–Based Current Regulation • Hysteresis Feedback Control • Space-Vector Pulse Width Modulation

8 **Sliding-Mode Control of Switched-Model Power Supplies** *Giorgio Spiazzi, Paolo Mattavelli* 8-1
   Introduction • Introduction to Sliding-Mode Control • Basics of Sliding-Mode Theory • Application of Sliding-Mode Control to DC-DC Converters—Basic Principle • Sliding-Mode Control of Buck DC-DC Converters • Extension to Boost and Buck–Boost DC-DC Converters • Extension to Cúk and SEPIC DC-DC Converters • General-Purpose Sliding-Mode Control Implementation • Conclusions

2
DC-DC Converters

Richard Wies
University of Alaska Fairbanks

Bipin Satavalekar
University of Alaska Fairbanks

Ashish Agrawal
University of Alaska Fairbanks

Javad Mahdavi
Sharif University of Technology

Ali Agah
Sharif University of Technology

Ali Emadi
Illinois Institute of Technology

Daniel Jeffrey Shortt
Cedarville University

2.1 Overview 2-1
    References
2.2 Choppers 2-3
    One-Quadrant Choppers • Two-Quadrant Choppers • Four-Quadrant Choppers
2.3 Buck Converters 2-8
    Ideal Buck Circuit • Continuous-Conduction Mode • Discontinuous-Conduction Mode • References
2.4 Boost Converters 2-12
    Ideal Boost Circuit • Continuous-Conduction Mode • Discontinuous-Conduction Mode • References
2.5 Cúk Converter 2-14
    Nonisolated Operation • Practical Cúk Converter • References
2.6 Buck–Boost Converters 2-17
    Circuit-Analysis • Small Signal Transfer Functions • Component Selection • Flyback Power Stage • Summary • References

2.1 Overview

Richard Wies, Bipin Satavalekar, and Ashish Agrawal

The purpose of a DC-DC converter is to supply a regulated DC output voltage to a variable-load resistance from a fluctuating DC input voltage. In many cases the DC input voltage is obtained by rectifying a line voltage that is changing in magnitude. DC-DC converters are commonly used in applications requiring regulated DC power, such as computers, medical instrumentation, communication devices, television receivers, and battery chargers [1, 2]. DC-DC converters are also used to provide a regulated variable DC voltage for DC motor speed control applications.

The output voltage in DC-DC converters is generally controlled using a switching concept, as illustrated by the basic DC-DC converter shown in Fig. 2.1. Early DC-DC converters were known as choppers with silicon-controlled rectifiers (SCRs) used as the switching mechanisms. Modern DC-DC converters classified as switch mode power supplies (SMPS) employ insulated gate bipolar transistors (IGBTs) and metal oxide silicon field effect transistors (MOSFETs).

The switch mode power supply has several functions [3]:

1. Step down an unregulated DC input voltage to produce a regulated DC output voltage using a buck or step-down converter.
2. Step up an unregulated DC input voltage to produce a regulated DC output voltage using a boost or step-up converter.

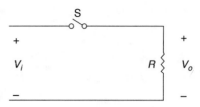

FIGURE 2.1 Basic DC-DC converter.

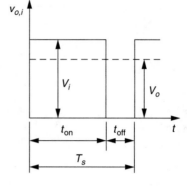

FIGURE 2.2 DC-DC converter voltage waveforms. (From Mohan, N., Undeland, T. M., and Robbins, W. P., *Power Electronics: Converters, Applications, and Design*, 2nd ed., John Wiley & Sons, New York, 1995. With permission from John Wiley & Sons.)

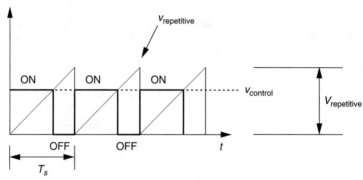

FIGURE 2.3 Pulsewidth modulation concept. (From Mohan, N., Undeland, T. M., and Robbins, W. P., *Power Electronics: Converters, Applications, and Design*, 2nd ed., John Wiley & Sons, New York, 1995. With permission from John Wiley & Sons.)

3. Step down and then step up an unregulated DC input voltage to produce a regulated DC output voltage using a buck–boost converter.
4. Invert the DC input voltage using a Cúk converter.
5. Produce multiple DC outputs using a combination of SMPS topologies.

The regulation of the average output voltage in a DC-DC converter is a function of the on-time t_{on} of the switch, the pulse width, and the switching frequency f_s as illustrated in Fig. 2.2. Pulse width modulation (PWM) is the most widely used method of controlling the output voltage. The PWM concept is illustrated in Fig. 2.3. The output voltage control depends on the duty ratio D. The duty ratio is defined as

$$D = \frac{t_{on}}{T_s} = \frac{v_{control}}{V_{repetitive}} \qquad (2.1)$$

based on the on-time t_{on} of the switch and the switching period T_s. PWM switching involves comparing the level of a control voltage $v_{control}$ to the level of a repetitive waveform as illustrated in Fig. 2.3 [2]. The on-time of the switch is defined as the portion of the switching period where the value of the repetitive waveform is less than the control voltage. The switching period (switching frequency) remains constant while the control voltage level is adjusted to change the on-time and therefore the duty ratio of the switch. The switching frequency is usually chosen above 20 kHz so the noise is outside the audio range [2, 3].

DC-DC converters operate in one of two modes depending on the characteristics of the output current [1, 2]:

1. Continuous conduction
2. Discontinuous conduction

The continuous-conduction mode is defined by continuous output current (greater than zero) over the entire switching period, whereas the discontinuous conduction mode is defined by discontinuous output current (equal to zero) during any portion of the switching period. Each mode is discussed in relationship to the buck and boost converters in subsequent sections.

References

1. Agrawal, J. P., *Power Electronics Systems: Theory and Design*, Prentice-Hall, Upper Saddle River, NJ, 2001, chap. 6.
2. Mohan, N., Undeland, T. M., and Robbins, W. P., *Power Electronics: Converters, Applications, and Design*, 2nd ed., John Wiley & Sons, New York, 1995, chap. 7.
3. Venkat, R., *Switch Mode Power Supply*, University of Technology, Sydney, Australia, March 1, 2001, available at http://www.ee.uts.edu.au/~venkat/pe_html/pe07_nc8.htm.

2.2 Choppers

Javad Mahdavi, Ali Agah, and Ali Emadi

Choppers are DC-DC converters that are used for transferring electrical energy from a DC source into another DC source, which may be a passive load. These converters are widely used in regulated switching power supplies and DC motor drive applications.

DC-DC converters that are discussed in this section are one-quadrant, two-quadrant, and four-quadrant choppers. Step-down (buck) converter and step-up (boost) converters are basic one-quadrant converter topologies. The two-quadrant chopper, which, in fact, is a current reversible converter, is the combination of the two basic topologies. The full-bridge converter is derived from the step-down converter.

One-Quadrant Choppers

In one-quadrant choppers, the average DC output voltage is usually kept at a desired level, as there are fluctuations in input voltage and output load. These choppers operate only in first quadrant of v–i plane. In fact, output and input voltages and currents are always positive. Therefore, these converters are called one-quadrant choppers.

One method of controlling the output voltage employs switching at a constant frequency, i.e., a constant switching time period ($T = t_{on} + t_{off}$), and adjusting the on-duration of the switch to control the average output voltage. In this method, which is called pulse-width modulation (PWM), the switch duty ratio d is defined as the ratio of the on-duration to the switching time period.

$$d = \frac{t_{on}}{T} \tag{2.2}$$

In the other control method, both the switching frequency and the on-duration of the switch are varied. This method is mainly used in converters with force-commutated thyristors.

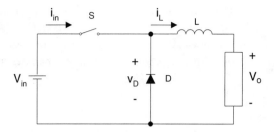

FIGURE 2.4 Step-down buck converter.

Choppers can have two distinct modes of operation, which have significantly different characteristics: continuous-conduction and discontinuous-conduction modes. In practice, a converter may operate in both modes. Therefore, converter control should be designed for both modes of operation.

Step-Down (Buck) Converter

A step-down converter produces an average output voltage, which is lower than the DC input voltage V_{in}. The basic circuit of a step-down converter is shown in Fig. 2.4.

In continuous-conduction mode of operation, assuming an ideal switch, when the switch is on for the time duration t_{on}, the inductor current passes through the switch, and the diode becomes reverse-biased. This results in a positive voltage $(V_{in} - V_o)$ across the inductor, which, in turn, causes a linear increase in the inductor current i_L. When the switch is turned off, because of the inductive energy storage, i_L continues to flow. This current flows through the diode and decreases. Average output voltage can be calculated in terms of the switch duty ratio as:

$$v_{o,\,ave.} = \frac{1}{T}\int_0^T v_o(t)\,dt = \frac{1}{T}\left(\int_0^{t_{on}} V_{in}\,dt + \int_{t_{on}}^T 0.0\right) = \frac{t_{on}}{T}V_{in} = dV_{in} \quad (2.3)$$

$v_{o,\,ave.}$ can be controlled by varying the duty ratio $(d = t_{on}/T)$ of the switch. Another important observation is that the average output voltage varies linearly with the control voltage. However, in the discontinuous-conduction mode of operation, the linear relation between input and output voltages is not valid. Figure 2.5 shows $(v_{o,\,ave.}/v_{in,\,ave.}) - i_{o,\,ave.}$ characteristic of a step-down converter in continuous and discontinuous conduction modes of operation.

Step-Up (Boost) Converter

Schematic diagram of a step-up boost converter is shown in Fig. 2.6. In this converter, the output voltage is always greater than the input voltage. When the switch is on, the diode is reversed-biased, thus isolating the output stage. The input voltage source supplies energy to the inductor. When the switch is off, the output stage receives energy from the inductor as well as the input source.

In the continuous-conduction mode of operation, considering d as the duty ratio, the input–output relation is as follows:

$$v_{o,\,ave.} = \frac{1}{1-d}V_{in} \quad (2.4)$$

If input voltage is not constant, V_{in} is the average of the input voltage. In this case, relation (2.3) is an approximation. In the discontinuous-conduction mode of operation, relation (2.3) is not valid. Figure 2.7 shows $(v_{in,\,ave.}/v_{o,\,ave.}) - i_{L,\,ave.}$ characteristic of a step-up converter in the continuous- and discontinuous-conduction modes of operation.

DC-DC Converters

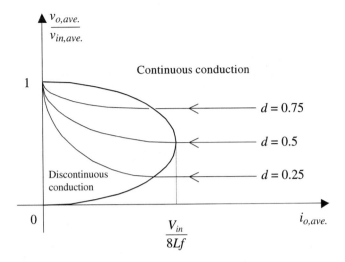

FIGURE 2.5 $(v_{o,ave.}/v_{in, ave.}) - i_{o,ave.}$ characteristic of a step-down converter.

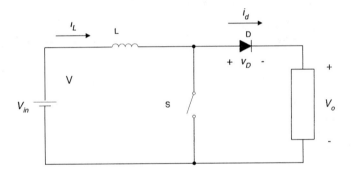

FIGURE 2.6 Step-up boost converter.

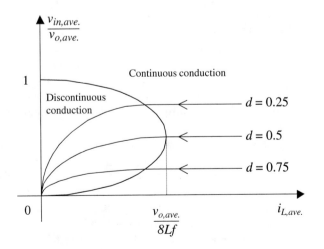

FIGURE 2.7 $(v_{in, ave.}/v_{o, ave.}) - i_{L, ave.}$ characteristic of a step-down converter.

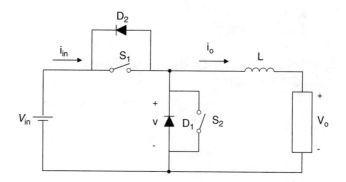

FIGURE 2.8 A current reversible chopper.

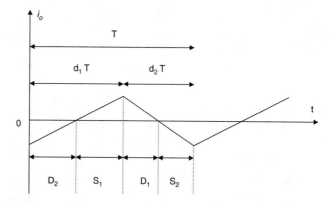

FIGURE 2.9 Output current of a two-quadrant chopper.

Two-Quadrant Choppers

A two-quadrant chopper has the ability to operate in two quadrants of the (v–i) plane. Therefore, input and output voltages are positive; however, input and output currents can be positive or negative. Thus, these converters are also named current reversible choppers. They are composed of two basic chopper circuits. In fact, a two-quadrant DC-DC converter is achieved by a combination of two basic chopper circuits, a step-down chopper and a step-up chopper, as is shown in Fig. 2.8.

The step-down chopper is composed of S_1 and D_1, and electric energy is supplied to the load. The step-up chopper is composed of S_2 and D_2; electric energy is fed back to the source. Reversible current choppers can transfer from operating in the power mode to operating in the regenerative mode very smoothly and quickly by changing only the control signals for S_1 and S_2, without using any mechanical contacts.

Figure 2.9 depicts the output current of a two-quadrant chopper. d_1 and $d_2 = 1 - d_1$ are the duty ratios of step-down and step-up converters, respectively. By changing d_1 and d_2, not only the amplitude of the average of the output current changes, but it can also be positive and negative, leading to two-quadrant operation.

For each of step-down and step-up operating mode, relations (2.3) and (2.4) are applicable for continuous currents. However, in discontinuous-conduction modes of operation, relations (2.3) and (2.4) are not valid. Figure 2.10 shows the ($v_{o,\,ave.}/v_{in,\,ave.}$) – $i_{o,\,ave.}$ characteristic of a two-quadrant converter in continuous- and discontinuous-conduction modes of operation. As is shown in Fig. 2.10, for changing the operating mode both from step-down to step-up operation and in the opposite direction,

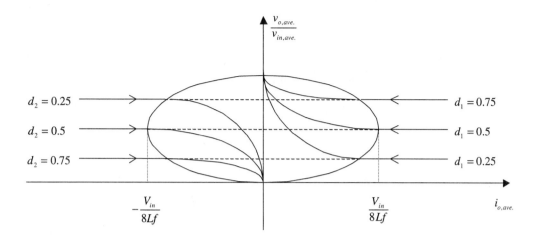

FIGURE 2.10 $(v_{o,\text{ave.}}/v_{\text{in, ave.}}) - i_{o,\text{ave.}}$ characteristic of a two-quadrant converter.

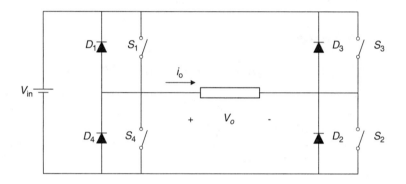

FIGURE 2.11 A full-bridge four-quadrant chopper.

the operating mode must move from the discontinuous-current region. However, by applying $d_2 = 1 - d_1$, the operating point will never move into the discontinuous-conduction region of the two basic converters. In Fig. 2.10, the broken lines indicate passage from step-down operation to step-up operation, and vice versa. In fact, because of this specific command—the relation between the two duty ratios—the converter operating point always stays in the continuous-conduction mode.

Four-Quadrant Choppers

In four-quadrant choppers, not only can the output current be positive and negative, but the output voltage also can be positive and negative. These choppers are full-bridge DC-DC converters, as is shown in Fig. 2.11. The main advantage of these converters is that the average of the output voltage can be controlled in magnitude as well as in polarity. A four-quadrant chopper is a combination of two two-quadrant choppers in order to achieve negative average output voltage and/or negative average output current.

The four-quadrant operation of the full-bridge DC-DC converter, as shown in Fig. 2.12, for the first two quadrants of the $(v-i)$ plane is achieved by switching S_1 and S_2 and considering D_1 and D_2 like a two-quadrant chopper. For the other two quadrants of the $(v-i)$ plane, the operation is achieved by switching S_3 and S_4 and considering D_3 and D_4 as another two-quadrant chopper, which is connected to the load in the opposite direction of the first two-quadrant chopper.

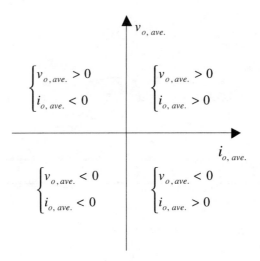

FIGURE 2.12 Four-quadrant operation of a full-bridge chopper.

2.3 Buck Converters

Richard Wies, Bipin Satavalekar, and Ashish Agrawal

The buck or step-down converter regulates the average DC output voltage at a level lower than the input or source voltage. This is accomplished through controlled switching where the DC input voltage is turned on and off periodically, resulting in a lower average output voltage [1]. The buck converter is commonly used in regulated DC power supplies like those in computers and instrumentation [1, 2]. The buck converter is also used to provide a variable DC voltage to the armature of a DC motor for variable speed drive applications [2].

Ideal Buck Circuit

The circuit that models the basic operation of the buck converter with an ideal single-pole double-throw switch and a purely resistive load is shown in Fig. 2.13. The output voltage equals the input voltage when the switch is in position 1 and the output voltage is zero when the switch is in position 2. The resulting output voltage is a rectangular voltage waveform with an average value as shown in Fig. 2.2 (in Section 2.1). The average output voltage level is varied by adjusting the time the switch is in position 1 and 2 or the duty ratio. The resulting average output voltage V_o is given in terms of the duty ratio and the input voltage V_i by Eq. (2.5) [2].

$$V_o = DV_i \qquad (2.5)$$

The square wave output voltage for the ideal circuit of the buck converter contains an undesirable amount of voltage ripple. The circuit is modified by adding an inductor L in series and a capacitor C in parallel with the load resistor as shown in Fig. 2.14. The inductor reduces the ripple in the current through

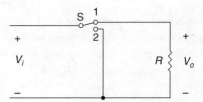

FIGURE 2.13 Ideal buck converter.

DC-DC Converters

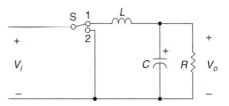

FIGURE 2.14 Modified buck converter with *LC* filter. (From Mohan, N., Undeland, T. M., and Robbins, W. P., *Power Electronics: Converters, Applications, and Design*, 2nd ed., John Wiley & Sons, New York, 1995. With permission from John Wiley & Sons.)

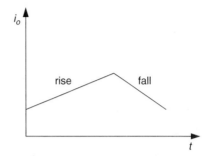

FIGURE 2.15 Rise and fall of load current in buck converter.

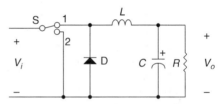

FIGURE 2.16 Buck converter with practical switch.

the load resistor, while the capacitor directly reduces the ripple in the output voltage. Since the current through the load resistor is the same as that of the inductor, the voltage across the load resistor (output voltage) contains less ripple.

The current through the inductor increases with the switch in position 1. As the current through the inductor increases, the energy stored in the inductor increases. When the switch changes to position 2, the current through the load resistor decreases as the energy stored in the inductor decreases. The rise and fall of current through the load resistor is linear if the time constant due to the *LR* combination is relatively large compared with the on- and off-time of the switch as shown in Fig. 2.15 [3]. A capacitor is added in parallel with the load resistor to reduce further the ripple content in the output voltage. The combination of the inductor and capacitor reduces the output voltage ripple to very low levels.

The circuit in Fig. 2.14 is designed assuming that the switch is ideal. A practical model of the switch is designed using a diode and power semiconductor switch as shown in Fig. 2.16. A freewheeling diode is used with the switch in position 2 since the inductor current freewheels through the switch. The switch is controlled by a scheme such as pulse width or frequency modulation.

Continuous-Conduction Mode

The continuous-conduction mode of operation occurs when the current through the inductor in the circuit of Fig. 2.14 is continuous. This means that the inductor current is always greater than zero. The average output voltage in the continuous-conduction mode is the same as that derived in Eq. (2.5) for the ideal circuit. As the conduction of current through the inductor occurs during the entire switching period, the average output voltage is the product of the duty ratio and the DC input voltage. The operation

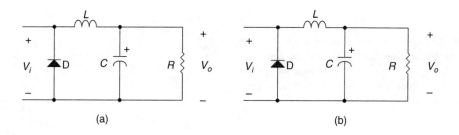

FIGURE 2.17 Buck converter switch states: (a) switch in position 1; (b) switch in position 2. (From Mohan, N., Undeland, T. M., and Robbins, W. P., *Power Electronics: Converters, Applications, and Design*, 2nd ed., John Wiley & Sons, New York, 1995. With permission from John Wiley & Sons.)

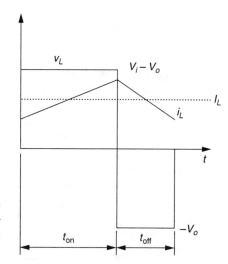

FIGURE 2.18 Inductor voltage and current for continuous mode of buck converter. (From Mohan, N., Undeland, T. M., and Robbins, W. P., *Power Electronics: Converters, Applications, and Design*, 2nd ed., John Wiley & Sons, New York, 1995. With permission from John Wiley & Sons.)

of this circuit resembles a DC transformer according to Eq. (2.6) based on the time-integral of the inductor voltage equal to zero over one switching period [2].

$$D = \frac{V_o}{V_i} = \frac{I_i}{I_o} \tag{2.6}$$

The operation of the circuit in steady state consists of two states as illustrated in Fig. 2.17 [2, 4]. The first state with the switch in position 1 has the diode reverse-biased and current flows through the inductor from the voltage source to the load. The switch changes to position 2 at the end of the on-time and the inductor current then freewheels through the diode. The process starts again at the end of the switching period with the switch returning to position 1. A representative set of inductor voltage and current waveforms for the continuous-conduction mode is shown in Fig. 2.18.

Discontinuous-Conduction Mode

The discontinuous mode of operation occurs when the value of the load current is less than or equal to zero at the end of a given switching period. Assuming a linear rise and fall of current through the inductor, the boundary point between continuous- and discontinuous-current conduction occurs when the average inductor current over one switching period is half of the peak value, as illustrated in Fig. 2.19. The average inductor current at the boundary point is calculated using Eq. (2.7) [2].

$$I_{LB} = \frac{1}{2} i_{L(\text{peak})} = \frac{DT_s}{2L}(V_i - V_o) \tag{2.7}$$

DC-DC Converters

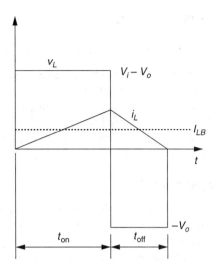

FIGURE 2.19 Inductor current at boundary point for discontinuous mode of buck converter. (From Mohan, N., Undeland, T. M., and Robbins, W. P., *Power Electronics: Converters, Applications, and Design*, 2nd ed., John Wiley & Sons, New York, 1995. With permission from John Wiley & Sons.)

The input voltage or output voltage is kept constant depending on the application. If the input voltage remains constant, then the average inductor current at the boundary is calculated by replacing the output voltage in Eq. (2.7) with Eq. (2.5), which yields the expression in Eq. (2.8) [2].

$$I_{LB} = \frac{DT_s}{2L}(V_i)(1-D) \tag{2.8}$$

The voltage ratio is now defined according to Eq. (2.9) [2]:

$$\frac{V_o}{V_i} = \frac{D^2}{D^2 + \frac{1}{4}\left(\frac{I_o}{I_{LB(max)}}\right)} \tag{2.9}$$

If the output voltage remains constant, then the average inductor current at the boundary is calculated by replacing the input voltage in Eq. (2.7) with Eq. (2.5), which yields the expression in Eq. (2.10) [2]:

$$I_{LB} = \frac{T_s}{2L}(V_o)(1-D) \tag{2.10}$$

The duty ratio is defined according to Eq. (2.11) by manipulating Eq. (2.9) [2]:

$$D = \frac{V_o}{V_i}\left(\frac{I_o/I_{LB(max)}}{1-\left(\frac{V_o}{V_i}\right)}\right)^{\frac{1}{2}} \tag{2.11}$$

References

1. Agrawal, J. P., *Power Electronics Systems: Theory and Design*, Prentice-Hall, Upper Saddle River, NJ, 2001, chap. 6.
2. Mohan, N., Undeland, T. M., and Robbins, W. P., *Power Electronics: Converters, Applications, and Design*, 2nd ed., John Wiley & Sons, New York, 1995, chap. 7.
3. Hoft, R. G., *Semiconductor Power Electronics*, Van Nostrand Reinhold, New York, 1986, chap. 5.
4. Venkat, R., *Switch Mode Power Supply*, University of Technology, Sydney, Australia, 01 March 2001, available at http://www.ee.uts.edu.au/~venkat/pe_html/pe07_nc8.htm.

2.4 Boost Converters

Richard Wies, Bipin Satavalekar, and Ashish Agrawal

A boost converter regulates the average output voltage at a level higher than the input or source voltage. For this reason the boost converter is often referred to as a step-up converter or regulator. The DC input voltage is in series with a large inductor acting as a current source. A switch in parallel with the current source and the output is turned off periodically, providing energy from the inductor and the source to increase the average output voltage. The boost converter is commonly used in regulated DC power supplies and regenerative braking of DC motors [1, 2].

Ideal Boost Circuit

The circuit that models the basic operation of the boost converter is shown in Fig. 2.20 [2, 3]. The ideal boost converter uses the same components as the buck converter with different placement. The input voltage in series with the inductor acts as a current source. The energy stored in the inductor builds up when the switch is closed. When the switch is opened, current continues to flow through the inductor to the load. Since the source and the discharging inductor are both providing energy with the switch open, the effect is to boost the voltage across the load. The load consists of a resistor in parallel with a filter capacitor. The capacitor voltage is larger than the input voltage. The capacitor is large to keep a constant output voltage and acts to reduce the ripple in the output voltage.

Continuous-Conduction Mode

The continuous-conduction mode of operation occurs when the current through the inductor in the circuit of Fig. 2.20 is continuous with the inductor current always greater than zero. The operation of the circuit in steady state consists of two states, as illustrated in Fig. 2.21 [2, 3]. The first state with the switch closed has current charging the inductor from the voltage source. The switch opens at the end of the on-time and the inductor discharges current to the load with the input voltage source still connected. This results in an output voltage across the capacitor larger than the input voltage. The output voltage remains constant if the RC time constant is significantly larger than the on-time of the switch.

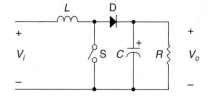

FIGURE 2.20 Basic boost converter. (From Mohan, N., Undeland, T. M., and Robbins, W. P., *Power Electronics: Converters, Applications, and Design*, 2nd ed., John Wiley & Sons, New York, 1995. With permission from John Wiley & Sons.)

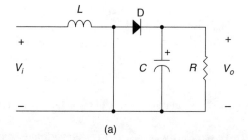

(a)

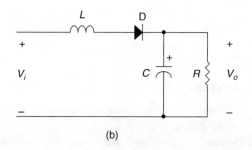

(b)

FIGURE 2.21 Basic boost converter switch states: (a) switch closed; (b) switch open. (From Mohan, N., Undeland, T. M., and Robbins, W. P., *Power Electronics: Converters, Applications, and Design*, 2nd ed., John Wiley & Sons, New York, 1995. With permission from John Wiley & Sons.)

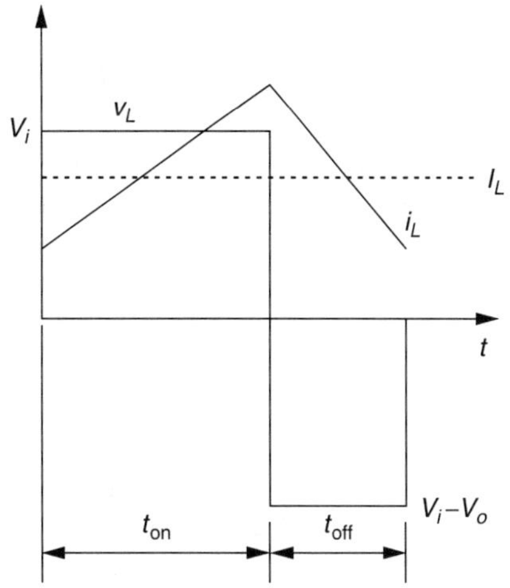

DC-DC Converters

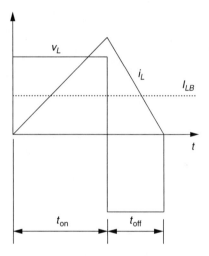

FIGURE 2.22 Inductor voltage and current waveforms for continuous mode of boost converter. (From Mohan, N., Undeland, T. M., and Robbins, W. P., *Power Electronics: Converters, Applications, and Design*, 2nd ed., John Wiley & Sons, New York, 1995. With permission from John Wiley & Sons.)

FIGURE 2.23 Inductor current at boundary point for discontinuous mode of boost converter. (From Mohan, N., Undeland, T. M., and Robbins, W. P., *Power Electronics: Converters, Applications, and Design*, 2nd ed., John Wiley & Sons, New York, 1995. With permission from John Wiley & Sons.)

A representative set of inductor voltage and current waveforms for the continuous conduction mode is shown in Fig. 2.22 [2].

The voltage ratio for a boost converter is derived based on the time-integral of the inductor voltage equal to zero over one switching period. The voltage ratio is equivalent to the ratio of the switching period to the off-time of the switch as illustrated by Eq. (2.12) [2].

$$\frac{V_o}{V_i} = \frac{I_i}{I_o} = \frac{T_s}{t_{\text{off}}} = \frac{T_s}{T_s - t_{\text{off}}} = \frac{T}{1-D} \qquad (2.12)$$

The current ratio is derived from the voltage ratio assuming that the input power is equal to the output power, as with ideal transformer analysis.

Discontinuous-Conduction Mode

The discontinuous mode of operation occurs when the value of the load current is less than or equal to zero at the end of a given switching period. Assuming a linear rise and fall of current through the inductor, the boundary point between continuous- and discontinuous-current conduction occurs when the average inductor current over one switching period is half the peak value, as illustrated in Fig. 2.23 [2]. The average inductor current at the boundary point is calculated using Eq. (2.13) [2].

$$I_{LB} = \frac{1}{2} i_{L(\text{peak})} = \frac{V_o T_s}{2L} D(1-D) \qquad (2.13)$$

The output current at the boundary condition is derived by using the current ratio of Eq. (2.12) in Eq. (2.13) with the inductor current equal to the input current. This results in Eq. (2.14) [2]:

$$I_{OB} = \frac{V_o T_s}{2L} D(1-D)^2 \qquad (2.14)$$

For the boost converter in discontinuous mode, the output voltage V_o is generally kept constant while the duty ratio D varies in response to changes in the input voltage V_i.

The duty ratio is defined as a function of the output current for various values of the voltage ratio according to Eq. (2.15) [2]:

$$D = \left[\frac{4}{27} \frac{V_o}{V_i} \left(\frac{V_o}{V_i} - 1 \right) \frac{I_o}{I_{oB(\max)}} \right]^{\frac{1}{2}} \qquad (2.15)$$

References

1. Agrawal, J. P., *Power Electronics Systems: Theory and Design*, Prentice-Hall, Upper Saddle River, NJ, 2001, chap. 6.
2. Mohan, N., Undeland, T. M., and Robbins, W. P., *Power Electronics: Converters, Applications, and Design*, 2nd ed., John Wiley & Sons, New York, 1995, chap. 7.
3. Venkat, R., *Switch Mode Power Supply*, University of Technology, Sydney, Australia, 01 March 2001, available at http://www.ee.uts.edu.au/~venkat/pe_html/pe07_nc8.htm.

2.5 Cúk Converter

Richard Wies, Bipin Satavalekar, and Ashish Agrawal

The Cúk converter is a switched-mode power supply named after the inventor Dr. Slobodan Cúk. The basic nonisolated Cúk converter shown in Fig. 2.24 is designed based on the principle of using two buck–boost converters to provide an inverted DC output voltage [1]. The advantage of the basic nonisolated Cúk converter over the standard buck–boost converter is to provide regulated DC output voltage at higher efficiency with identical components due to an integrated magnetic structure, reduced ripple currents, and reduced switching losses [2, 3]. The integrated magnetic structure of the isolated Cúk converter consists of the isolation transformer and the two inductors in a single core. As a result, the ripple currents in the inductors are driven into the primary and secondary windings of the isolation transformer. Also, the single core results in reduced flux paths, which improves the overall efficiency of the converter.

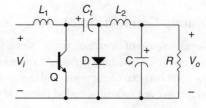

FIGURE 2.24 Nonisolated Cúk converter. (From Mohan, N., Undeland, T. M., and Robbins, W. P., *Power Electronics: Converters, Applications, and Design*, 2nd ed., John Wiley & Sons, New York, 1995. With permission from John Wiley & Sons.)

Nonisolated Operation

The basic nonisolated Cúk converter is a switching power supply with two inductors, two capacitors, a diode, and a transistor switch as illustrated in Fig. 2.24 [1, 2]. The transfer capacitor C_t stores and transfers energy from the input to the output. The average value of the inductor voltages for steady-state operation is zero. As a result, the voltage across the transfer capacitor is assumed to be the average value V_{C_t} in steady state and is the sum of the input and output voltages. The inductor currents are assumed to be continuous for steady-state operation.

The operation of the basic nonisolated Cúk converter in steady state consists of two transistor states, as illustrated in Fig. 2.25 [1, 2]. In the first state when the transistor is off, the inductor currents flow through the diode and energy is stored in the transfer capacitor from the input and the inductor L_1. The energy stored in the inductor L_2 is transferred to the output. As a result, both of the inductor currents are linearly decreasing in the off-state. In the second state when the transistor is on, the inductor currents flow through the transistor and the transfer capacitor discharges while energy is stored in the inductor L_1. As the transfer capacitor discharges through the transistor, energy is stored in the inductor L_2. Consequently, both of the inductor currents are linearly increasing in the on-state. A representative set of inductor voltage and current waveforms for the nonisolated Cúk converter are shown in Figs. 2.26 and 2.27 [1].

The voltage and current ratio for the nonisolated Cúk converter can be derived by assuming the inductor currents, which correspond to the input current and output current, are ripple-free [1]. This results

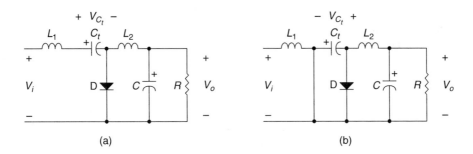

FIGURE 2.25 Cúk converter switch states: (a) switch open; (b) switch closed. (From Mohan, N., Undeland, T. M., and Robbins, W. P., *Power Electronics: Converters, Applications, and Design*, 2nd ed., John Wiley & Sons, New York, 1995. With permission from John Wiley & Sons.)

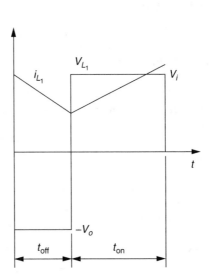

FIGURE 2.26 Inductor 1, voltage and current waveforms for Cúk converter. (From Mohan, N., Undeland, T. M., and Robbins, W. P., *Power Electronics: Converters, Applications, and Design*, 2nd ed., John Wiley & Sons, New York, 1995. With permission from John Wiley & Sons.)

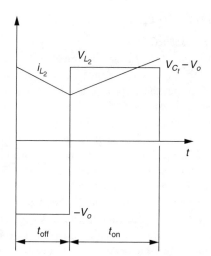

FIGURE 2.27 Inductor 2, voltage and current waveforms for Cúk converter. (From Mohan, N., Undeland, T. M., and Robbins, W. P., *Power Electronics: Converters, Applications, and Design*, 2nd ed., John Wiley & Sons, New York, 1995. With permission from John Wiley & Sons.)

in an equal charging and discharging of the transfer capacitor during the off-state and the on-state. The charging and discharging are defined in Eq. (2.16) in terms of the product of current and time [1].

$$I_{L_1} t_{\text{off}} = I_{L_2} t_{\text{on}} \qquad (2.16)$$

The resulting current ratio is expressed in Eq. (2.17) by substituting $I_{L_1} = I_i$, $I_{L_2} = I_o$, $t_{\text{off}} = (1 - D)T_s$, and $t_{\text{on}} = DT_s$ into Eq. (2.16) [1].

$$\frac{I_o}{I_i} = \frac{1-D}{D} \qquad (2.17)$$

If the input power is equal to the output power for the ideal case, the voltage ratio in Eq. (2.18) is determined as the inverse of the current ratio using the analysis of an ideal transformer [1].

$$\frac{V_o}{V_i} = \frac{D}{1-D} \qquad (2.18)$$

Practical Cúk Converter

The advantages of the practical isolated Cúk converter discussed earlier are the integrated magnetic structure, reduced ripple currents, and reduced switching losses. With the use of a single transformer to provide isolation and the two inductors required in the circuit, the ripple in the inductor currents is essentially reduced to zero. This reduces the amount of external filtering required, but the transfer capacitor carries the ripple from both inductors. This requires a transfer capacitor with a large ripple current capacity. For futher information and a more-detailed analysis of the practical Cúk converter, see Refs. 2 and 3.

References

1. Mohan, N., Undeland, T. M., and Robbins, W. P., *Power Electronics: Converters, Applications, and Design*, 2nd ed., John Wiley & Sons, New York, 1995, chap. 7.
2. TESLAco, CUKonverter Technology, 1996, 23 February 2001, available at http://www.teslaco.com/inverter.htm.
3. Cúk, S. and Middlebrook, R. D., *Advances in Switched-Mode Power Conversion*, Vol. 1 and 2, TESLAco, Pasadena, CA, 1981.

2.6 Buck–Boost Converters

Daniel Jeffrey Shortt

A schematic of the buck–boost converter circuit (in one of its simplest forms) is shown below in Fig. 2.28. The main power switch is shown to be a bipolar transistor, but it could be a power MOSFET, or any other device that could be turned on (and off) in a controlled fashion. This converter processes the power from a DC-biased source (high-voltage ripple) to a DC output (containing low-voltage ripple). The DC output voltage value can be chosen to be higher or lower than the input DC voltage. *Note*: The output load is represented by a resistor, R_L, but in real life can be something much more complicated. In a general sense, this circuit processes power from input to output with "square wave" technology, that is, the circuit produces waveforms that have sharp edges (such as those shown in Fig. 2.29). (There are converters that develop sine waves and semi-sine waves in the power process. They will not be discussed here.) The waveforms in Fig. 2.29 have a square-wave (or semi-square-wave) appearance and are indicative of current waveforms in a typical DC-DC converter. In fact, the i_L waveform is in a similar shape as the inductor (L) current in the buck–boost converter of Fig. 2.28, i_D can represent the diode current, and i_C, the capacitor current.

The operation of this converter is nonlinear and discrete; however, it can be represented by a cyclic change of power stage topologies. The three topologies for this converter, the equations for those topologies,

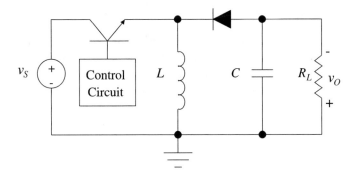

FIGURE 2.28 Buck–boost converter.

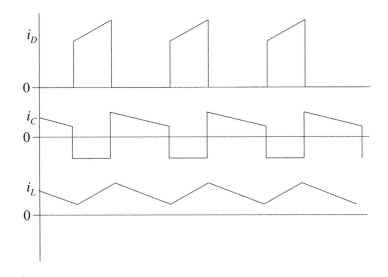

FIGURE 2.29 Typical current waveforms in a buck–boost converter.

and the small-signal transfer functions are presented in this section. For specific details of the derivation of each of these items, see the technical articles and papers listed in the References.

Circuit Analysis

The buck–boost converter has cyclic changes in topology due to the switching action of the semiconductor devices. During a cycle of operation, the main power switch is turned on and off; the diode responds to this by switching off and on.

Continuous-Current Mode

Figure 2.30 illustrates the topology where the main power switch is on and the diode is reverse-biased; thus, it is off. For the purpose of illustration the semiconductor devices are assumed to be ideal.

There are two independent state variables that contain the information describing the operation of this circuit: the inductor current, i_L, and the capacitor voltage, v_C. Two differential equations in terms of these variables, the output voltage, v_O, and the source voltage, v_S, for the designated Topology 1 are shown below.

$$\frac{di_L}{dt} = \frac{v_S}{L} \tag{2.19}$$

$$\frac{dv_C}{dt} = \frac{v_O}{R_L C} \tag{2.20}$$

Please note that the inductor is receiving energy from the source and being charged up, while the capacitor is being discharged into the output load, R_L, and the output voltage is falling.

Figure 2.31 shows the change in topology when the main power switch turns off. The inductor maintains current flow in the same direction so that the diode is forward-biased. The differential

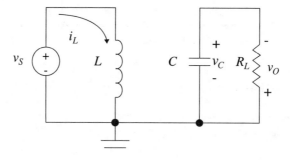

FIGURE 2.30 Topology 1 for the buck–boost converter.

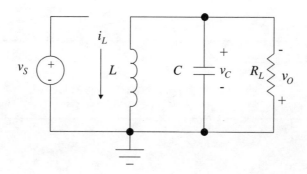

FIGURE 2.31 Topology 2 for the buck–boost converter.

equations for the designated Topology 2 are shown below. Please note that the inductor is transferring the energy it has obtained from the source into the capacitor; the capacitor is being charged up as the inductor is being discharged, and the output voltage is rising.

$$\frac{di_L}{dt} = -\frac{v_C}{L} \qquad (2.21)$$

$$\frac{dv_C}{dt} = \frac{i_L}{C} + \frac{v_O}{R_L C} \qquad (2.22)$$

Another topology change will occur if the inductor has transferred all of its energy out into the capacitor. In that case the inductor current will fall to zero. This will be examined later in the section. The inductor current is assumed to be nonzero.

These four linear time-invariant differential equations describe the state of the buck–boost converter. The power stage analysis is linear for each interval; however, for the complete operational cycle, it becomes a piecewise linear problem. The on-time or off-time of the main power switch may vary from cycle to cycle, further complicating the analysis.

Various modeling schemes have been proposed using nonlinear techniques that would in essence "combine" these equations. Basically there are two approaches: numerical (universal) and analytical (mathematical) techniques [1, 2]. In analytical techniques, a closed-form expression representing the operation of the converter is obtained, enabling a qualitative analysis to be performed [1]. The numerical techniques use various algorithms to produce an accurate quantitative result. However, simple relations among the system parameters are not easily obtainable. Numerical techniques are not to be considered at this time, because the desire at this point is to obtain a closed-form solution from which a considerable amount of design insight can be obtained.

Analytical techniques can be divided into two different system descriptions, discrete and continuous. The discrete system description makes no simplifying assumption on the basis of converter application. This description could be used in any application where the linearization of a periodically changing structure is sought. This method is accurate, but very complicated. The derived expressions are complex and cumbersome, which impedes its practical usefulness, and physical insight into the system operation is not easily obtainable.

An important continuous analytical technique is the averaging technique by Wester and Middlebrook [3]. It is easy to implement and gives physical insight into the operation of a buck–boost converter. Through circuit manipulation, analytical expressions were derived to determine the appropriate expressions. Middlebrook and Cúk [4, 5] modified the technique to average the state space descriptions (variables) over a complete cycle. Shortt and Lee [6–8] used a discrete sample of the average state space representation to develop a modeling technique that would enable a judicious control selection to be made. Vorpérian et al. [9] developed an equivalent circuit model for a pulse width modulation (PWM) switch that can be used in the analysis of this converter.

For the averaging technique each interval in the cycle is described by its state space representation (differential equation). Figure 2.32 shows the waveform of the continuous, instantaneous inductor current (that is,

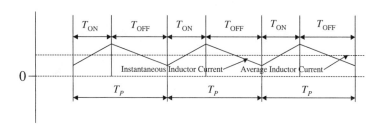

FIGURE 2.32 Continuous inductor current.

i_L does not equal zero at any point in time) and the average inductor current for the buck–boost converter (Fig. 2.30). The instantaneous current is cyclic with a time period equal to T_P s; the main power switch is on for T_{ON} s and off for T_{OFF} s. The equations are averaged to give a single period representation, as shown below:

$$\dot{i}_L = -d'\frac{v_C}{L} + d\frac{v_S}{L} \qquad (2.23)$$

$$\dot{v}_C = d'\frac{i_L}{C} + \frac{v_O}{R_L C} \qquad (2.24)$$

where i_L and v_C are average state variables, $d = T_{ON}/T_P$ and $d' = T_{OFF}/T_P$. Please note: $d + d' = 1$.

To study the small signal behavior, the time-varying system described in Eqs. (2.3) and (2.4) can be linearized using small signal perturbation techniques. By using these techniques, the inputs are assumed to vary around a steady-state operating point. Taking a first-order Fourier series approximation, the inputs are represented by the sum of a DC or steady-state term and an AC variation or sinusoidal term. Introducing variations in the line voltage and duty cycle by the following substitutions

$$v_S = V_S + \hat{v}_S, \qquad d = D + \hat{d}, \qquad d' = D' - \hat{d}$$

cause perturbations in the state and output, as shown below. In the above and following equations the variables in capital letters represent the DC or steady-state term; and the variables with the symbol "^" above them represent the AC variation or sinusoidal term.

$$\dot{i}_L = \dot{I}_L + \dot{\hat{i}}_L, \qquad i_L = I_L + \hat{i}_L, \qquad \dot{v}_C = \dot{V}_C + \dot{\hat{v}}_C, \qquad v_C = V_C + \hat{v}_C, \qquad v_O = V_O + \hat{v}_O$$

Figure 2.33 shows the type of change that is being modeled for an inductor current perturbation of Fig. 2.32. Note the T_{ON} and T_{OFF} slowly change from cycle to cycle, which produces a slight change in the inductor current from cycle to cycle.

The derivative of a DC term is zero, so the above equations can be simplified to the following:

$$\dot{i}_L = \dot{\hat{i}}_L, \qquad i_L = I_L + \hat{i}_L, \qquad \dot{v}_C = \dot{\hat{v}}_C, \qquad v_C = V_C + \hat{v}_C, \qquad v_O = V_O + \hat{v}_O$$

Substituting these equations into (2.23) and (2.24), separating the DC (steady-state) terms and the AC (sinusoidal) terms results in the following:

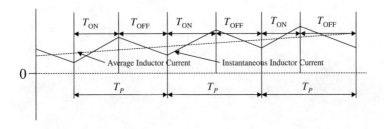

FIGURE 2.33 Inductor current perturbation.

DC-DC Converters

DC terms:

$$\frac{D'V_C}{L} + \frac{DV_S}{L} = 0 \tag{2.25}$$

$$\frac{D'I_L}{C} + \frac{V_O}{R_L C} = 0 \tag{2.26}$$

AC terms (neglecting the higher-order terms):

$$\dot{\hat{i}}_L = \frac{D'\hat{v}_C}{L} + \frac{D\hat{v}_S}{L} + \frac{V_S - V_C}{L}\hat{d} \tag{2.27}$$

$$\dot{\hat{v}}_C = \frac{D'}{C}\hat{i}_L + \frac{\hat{v}_O}{R_L C} + \frac{I_L}{C}\hat{d} \tag{2.28}$$

The equation

$$\frac{V_C}{V_S} = -\frac{D}{D'}$$

is derived from Eq. (2.25). Note that from Fig. 2.28, $v_C = -v_O$, giving $V_C = -V_O$ and $\hat{v}_C = -\hat{v}_O$; substituting this into the previous equation results in:

$$\frac{V_O}{V_S} = \frac{D}{D'} \tag{2.29}$$

Equation (2.29) states that the ratio of the DC output voltage to the DC input voltage is equal to the ratio of the power switch on-time to the power switch off-time. The expression for the DC inductor current term is

$$I_L = -\frac{V_O}{D'R_L} \tag{2.30}$$

Equations (2.27) and (2.28) constitute the small signal model of a buck–boost converter.

Another method that is utilized to extract the small signal model is to realize an equivalent circuit model from Eqs. (2.23) and (2.24). Figure 2.34 is the average circuit model of the buck–boost converter.

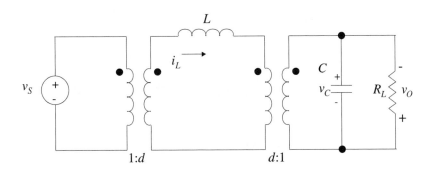

FIGURE 2.34 Average circuit model of the buck–boost converter.

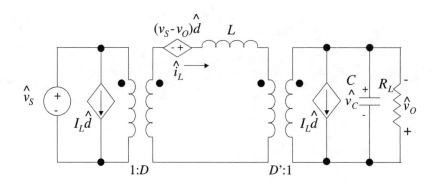

FIGURE 2.35 The small signal circuit model.

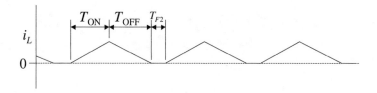

FIGURE 2.36 Discontinuous inductor current.

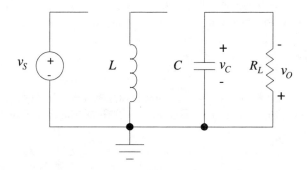

FIGURE 2.37 Topology 3 for the buck–boost converter (discontinuous inductor current).

(For a quantitative, numerical analysis, this circuit can be simulated with SPICE or an equivalent simulation package, as demonstrated in Ref. 10.)

Introducing perturbations into the state and output, removing the DC conditions, neglecting the small nonlinear terms, and simplifying the structure, results in Fig. 2.35.

Discontinuous-Current Mode

Figure 2.36 shows the waveform of the discontinuous inductor current for the buck–boost converter (Fig. 2.30). Note that the inductor current is equal to zero for $T_{F2}s$. This results in an additional (third) topology change, shown in Fig. 2.37.

Since the inductor current is zero for this portion of the switching cycle, there is only one state equation that can be determined.

$$\frac{dv_C}{dt} = \frac{v_O}{R_L C} \tag{2.31}$$

This equation indicates that the capacitor is now discharging its energy into the load resistor, R_L, and the output voltage is falling.

DC-DC Converters

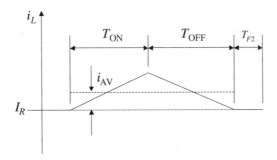

FIGURE 2.38 General form of discontinuous inductor current.

The modeling of this particular mode is presented in Refs. 5, 11, and 12. A general discussion is provided here as it applies to the model of the buck–boost converter in the discontinuous inductor current mode.

For this case, the inductor current does not behave as a true state variable, since $di_L/dt = 0$, thereby reducing the system order by one. Figure 2.38 illustrates the general form of the inductor current. The equations for the T_{on} time interval are the same as Eqs. (2.19) and (2.20), except $i_L = I_R + i_L^*$, where I_R represents the DC level at which the inductor current begins and i_L^*, the value of the time-varying inductor current. The equations for the T_{OFF} interval are the same as Eqs. (2.21) and (2.22) except $i_L = I_R + i_L^{**}$, where i_L^{**} represents the value of the time-varying inductor current. By combining these sets of equations with Eq. (2.31) by the averaging technique, the equations listed below are obtained.

$$\frac{di_L}{dt} = \frac{1}{T_P}\int_0^{T_{ON}} \frac{v_S}{L} dt + \frac{1}{T_P}\int_{T_{ON}}^{T_{ON}+T_{OFF}} \left(\frac{-v_C}{L}\right) dt = 0 \tag{2.32}$$

$$\frac{dv_C}{dt} = \frac{1}{T_P}\int_0^{T_{ON}} \left(-\frac{v_C}{R_L C}\right) dt + \frac{1}{T_P}\int_{T_{ON}}^{T_{ON}+T_{OFF}} \left(\frac{I_R + i_L^*}{C} - \frac{v_C}{R_L C}\right) dt + \frac{1}{T_P}\int_{T_{ON}+T_{OFF}}^{T_{ON}+T_{OFF}+T_{F2}} \left(-\frac{v_C}{R_L C}\right) dt \tag{2.33}$$

For the buck–boost converter case $I_R = 0$; also, from Fig. 2.38, note that

$$\int_{T_{ON}}^{T_{ON}+T_{OFF}} i_L^* \, dt = \left(\frac{1}{2}\frac{v_S}{L} T_{ON}\right) T_{OFF} = i_{AV} T_{OFF} \tag{2.34}$$

The variable i_{AV} is the average value of the inductor during the $T_{ON} + T_{OFF}$ time, not for the whole cycle. Substituting into Eqs. (2.32) and (2.33) results in the following:

$$\frac{dv_C}{dt} = -\frac{T_{ON}}{T_P}\frac{v_C}{R_L C} + \frac{T_{OFF}}{T_P}\frac{i_{AV}}{C} - \frac{T_{OFF}}{T_P}\frac{v_C}{R_L C} - \frac{T_{F2}}{T_P}\frac{v_C}{R_L C} \tag{2.35}$$

Let

$$d_1 = \frac{T_{ON}}{T_P}, \quad d_2 = \frac{T_{OFF}}{T_P}, \quad d_3 = \frac{T_{F2}}{T_P}$$

and substitute into the above equation.

$$\dot{v}_C = d_2\frac{i_{AV}}{C} - \frac{v_C}{R_L C} \tag{2.36}$$

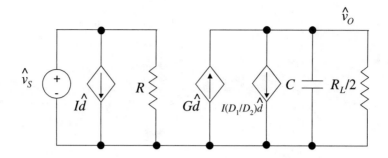

FIGURE 2.39 Buck–boost converter small signal model for the discontinuous mode.

Note that, $d_1 + d_2 + d_3 = 1$ and

$$i_{AV} = \frac{1}{2}\frac{v_s}{L}T_{ON} \tag{2.37}$$

At this point, the same perturbation techniques, as presented previously, are used to obtain the small signal model. Introducing variations in the line voltage and duty cycle

$$v_S = V_S + \hat{v}_S, \quad d_1 = D_1 + \hat{d}_1, \quad d_2 = D_2 + \hat{d}_2, \quad d_3 = D_3 + \hat{d}_3$$

produce perturbations in the state and output; separating the DC and AC terms and simplifying results in

$$\dot{v}_C = -\frac{\hat{v}_C}{R_L C} + \frac{V_S T_{OFF}}{2LC}\hat{d}_1 + \frac{V_S T_{ON}}{2LC}\hat{d}_2 + \frac{T_{ON} T_{OFF}}{2LCT_P}\hat{v}_S \tag{2.38}$$

where

$$T_{ON} = \frac{V}{V_S}\sqrt{\frac{2LT_P}{R_L}} \tag{2.39}$$

$$T_{OFF} = \sqrt{\frac{2LT_P}{R_L}} \tag{2.40}$$

$$\frac{V_C}{V_S} = \frac{D_1}{D_2} \tag{2.41}$$

A circuit model (Fig. 2.39) can be realized from the above equations. The process is not shown here; however, please see Ref. 5 for the details of the circuit derivation and presentation. This concludes the circuit analysis portion of this section. In the next section the appropriate transfer functions to be used in the design and implementation of the buck–boost converter are presented. The above small signal model is used to derive them. For more detail, please see Refs. 3, 5, and 9.

Small Signal Transfer Functions

The analysis done in the previous section enables the development of transfer functions that describe the buck–boost converter stability performance and input to output signal attenuation. The transfer functions are illustrated in Fig. 2.40. This figure assumes there is only one feedback (the output voltage) loop; for more complicated feedback schemes, please see Refs. 7 and 8.

The continuous-current mode transfer functions are derived by using the Laplace transform to solve for the output voltage and duty cycle variations in Eqs. (2.27) and (2.28). Equation (2.38) is used to derive the discontinuous-current mode transfer functions.

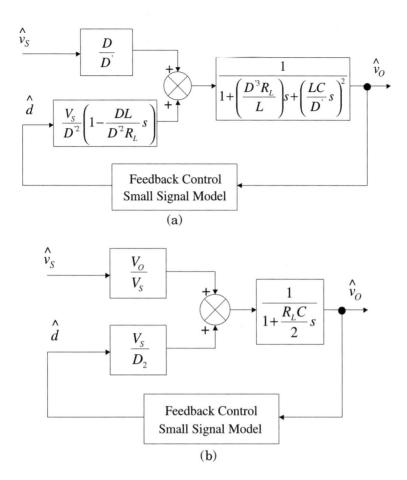

FIGURE 2.40 Block diagram of transfer functions for the buck–boost converter: (a) continuous-current mode; (b) discontinuous-current mode.

Component Selection

The component values can be chosen based on several constraints. The constraints that are to be discussed here are not exhaustive, but are only mentioned to provide an introduction into the selection process. Some component values can be based on an arbitrary selection. For example, the frequency of the converter is the designer's choice. As the frequency rises, the volume of the inductor (which is usually the biggest component in the converter) decreases and its temperature rises. The component value can be selected based on a given frequency value, which is assumed to be optimized based on the previously mentioned constraints. However, the frequency value can also be chosen based on experience. There is to be no discussion on the optimization of the switching frequency in this section; please see Ref. 13 for a detailed explanation of the process to optimize the converter switching frequency. Thus, an assumption made at this point is that the switching frequency has been selected.

Inductor Value

The inductor has to be large enough to handle the output power, according to the energy transfer equation shown below.

$$\frac{1}{2} L i_{peak}^2 = PT_P \qquad (2.42)$$

where, i_{peak} is the maximum value of the inductor current, P is the output power, and T_P is the time period of the switching cycle. If the desire for the current is to be continuous, then the inequality shown below must be satisfied.

$$\frac{2L}{R_L} > \frac{T_P}{\left(\frac{V}{V_S}+1\right)^2} \quad (2.43)$$

The inequality (2.43) is derived from Eqs. (2.39) and (2.40). The total of those equations has to be greater than the cycle time, T_P, for the converter to be in the continuous-current mode. If the designer desires the converter to be in the discontinuous mode, then the inequality sign in (2.43) is reversed so that $T_{ON} + T_{OFF}$ is less than T_P.

Satisfying the above two constraints, (2.42) and (2.43), should provide an inductor that is minimal, but probably not optimal. Using a circuit simulation package, such as PSpice, to simulate and check the converter action can help determine an optimal value.

Capacitor Value

The capacitor value is chosen based on the specified ripple voltage, V_{PP}, the switching frequency (actually the T_{OFF} for the buck–boost converter), and the allowable capacitor ripple current, $i_{allowable}$. The following inequality describes the relationship of the previously mentioned items:

$$C > \frac{i_{allowable}}{V_{PP}} T_{OFF} \quad (2.44)$$

As with the inductor value, this constraint provides a minimal capacitor value, but probably not an optimal one. Any value that is chosen should be used in a simulation to test the value for feasibility. The latent assumption made here is that the capacitor is an ideal one. In actuality, a practical capacitor can be modeled as a linear combination of resistors, inductors, and capacitors. This complicates the previously discussed models greatly. The equivalent series resistance (ESR) and the equivalent series inductance (ESL) (Fig. 2.41), probably have the biggest influence on the effective capacitance, because of their effect on the capacitor ripple voltage. Both, in general, tend to raise ripple voltage. This may require an iteration involving a simulation using the catalog or given values for the ESR and ESL in a more realistic model of the capacitor.

FIGURE 2.41 A practical capacitor model.

Main Power Switch and Output Power Diode

The main switching transistor and diode should be chosen based on the inductor current peak value. As with all of these components, the final component value selection should have appropriate design margins. These margins, however, do vary with the scope of the mission of the individual converter.

The main power switch function is to provide a path for the inductor to receive energy from the source; that is, the switch connects the source to the inductor at the appropriate time in the switching cycle. The switch can dissipate a significant amount of power if not chosen properly or not connected to an appropriately designed heat sink. So, in addition to ensuring that the switch can handle the peak current and voltage values, the power dissipation must be checked. For a bipolar transistor, assuming the efficiency

DC-DC Converters

of the converter is very high, the on-state power dissipation can be expressed as the following:

$$P_{DISS} = V_{CEsat}\left(\frac{D}{D'}I_O\right) \quad (2.45)$$

where $I_O = P/V_O$. If a MOSFET device is chosen, the on-state dissipation is the following:

$$P_{DISS} = \left(\frac{D}{D'}I_O\right)^2 R_{DS(on)} \quad (2.46)$$

The output power diode provides the path for the inductor to discharge its energy to the output; it connects the inductor to the output when the main power switch is off. Its voltage drop is primarily responsible for power dissipation. If V_d is the on-state voltage drop of the diode, then its power dissipation is expressed as

$$P_{DISS} = V_D I_O \quad (2.47)$$

A judicious selection for the diode can be made using the above calculated power value, the peak output current, and output voltage.

Flyback Power Stage

A popular version of the buck–boost converter, shown in Fig. 2.28, is the variation shown in Fig. 2.42, the flyback converter. The flyback converter provides isolation from input to output: note the output voltage is not inverted as in the simpler buck–boost converter version of Fig. 2.28. These things are accomplished because of the two-winding or coupled inductor. The inductor now serves a dual purpose: it transfers energy from the source to the output and provides input to output voltage. This is a popular power stage used in off-line (110 V_{AC} or 220 V_{AC}) applications, particularly with multiple output voltages. Power diodes, capacitors, and windings on the two-winding inductor (power transformer) are added in the appropriate fashion to provide additional outputs.

The process discussed previously can be used to determine the small signal model and DC operating point of this converter. The state variables for this converter are the capacitor voltage, v_C, and the flux density, ϕ, of the two-winding inductor:

$$N_P\frac{d\phi}{dt} = L_P\frac{di}{dt} \quad \text{and} \quad N_S\frac{d\phi}{dt} = L_S\frac{di}{dt}$$

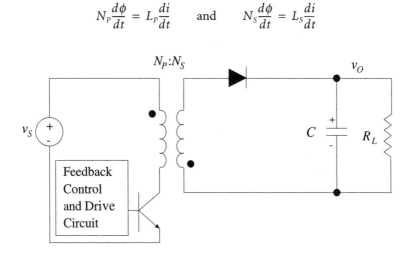

FIGURE 2.42 Flyback power converter.

The input voltage can be expressed as $(N_S/N_P)v_S$, instead of just v_s. Making these substitutions allows the development of the small signal model that is shown below.

The continuous-current mode small signal model is

$$\dot{\hat{i}}_L = \frac{D'\hat{v}_C}{L} + \frac{D\frac{N_S}{N_P}\hat{v}_S}{L} + \frac{\frac{N_S}{N_P}V_S - V_C}{L}\hat{d} \tag{2.48}$$

$$\dot{\hat{v}}_C = \frac{D'}{C}\frac{N_S}{L_S}\hat{\phi} + \frac{\hat{v}_O}{R_L C} + \frac{\frac{N_S}{L_S}\Phi}{C}\hat{d} \tag{2.49}$$

with

$$\frac{V_O}{V_S} = \frac{N_S}{N_P}\frac{D}{D'} \tag{2.50}$$

and

$$\Phi = -\frac{V_O}{D'R_L}\frac{L_S}{N_S} \tag{2.51}$$

The discontinuous-current mode small signal model is

$$\dot{\hat{v}}_C = -\frac{\hat{v}_C}{R_L C} + \frac{\frac{N_S}{N_P}V_S T_{\text{OFF}}}{2L_S C}\hat{d}_1 + \frac{\frac{N_S}{N_P}V_S T_{\text{ON}}}{2L_S C}\hat{d}_2 + \frac{T_{\text{ON}} T_{\text{OFF}}}{2L_S C T_P}\frac{N_S}{N_P}\hat{v}_S \tag{2.52}$$

with

$$T_{\text{ON}} = \frac{V}{\frac{N_S}{N_P}V_S}\sqrt{\frac{2L_S T_P}{R_L}} \tag{2.53}$$

$$T_{\text{OFF}} = \sqrt{\frac{2L_S T_P}{R_L}} \tag{2.54}$$

$$\frac{V_C}{V_S} = \frac{N_S}{N_P}\frac{D_1}{D_2} \tag{2.55}$$

Summary

This section has presented and analyzed a buck–boost converter (Fig. 2.28). The topological changes have been presented and a discussion of the state space has shown a modeling process (averaging), which can be used to design the converter. This process models a linear time-varying structure in a relatively simple way so that significant design insight can be obtained. The small signal model that was presented can be used to analyze the stability and the input-to-output signal attenuation of the buck–boost converter.

References

1. Middlebrook, R. D. and Cúk, S., Modeling and analysis methods for DC-to-DC switching converters, presented at IEEE Int. Semiconductor Power Converter Conference, 1977.
2. Owen, H. A., Capel, A., and Ferrante, J. G., Simulation and analysis methods for sampled power electronic systems, in *IEEE Power Electronics Specialists Conference Record*, 1976, 45–55.
3. Wester, G. W. and Middlebrook, R. D., Low-frequency characterization of switched DC-DC converters, *IEEE Trans. Aerospace Electron. Syst.*, AES-9(3), 376–385, 1973.
4. Middlebrook R. D. and Cúk, S., A general unified approach to modeling switching-converter power stages, in *IEEE Power Electronics Specialists Conference Record*, 1976, 18–34.
5. Cúk, S. and Middlebrook, R. D., A general unified approach to modeling switching DC-to-DC converters in discontinuous conduction mode, in *IEEE Power Electronics Specialists Conference Record*, 1977, 36–57.
6. Lee, F. C. and Shortt, D. J., Improved model for predicting the dynamic performance of high bandwidth and multiloop power converters, in *POWERCON 11 Record*, 1984, E-3, 1–14.
7. Shortt, D. J. and Lee, F. C., Extensions of the discrete-average models for converter power stages, in *PESC Record*, 1983, 23–37; *IEEE Trans. Aerospace Electron. Syst.*, AES-20(3), 279–289, 1984.
8. Shortt, D. J. and Lee, F. C., An improved switching converter model using discrete and average techniques, in *PESC Record*, 1982, 199–212; *IEEE Trans. Aerospace Electron. Syst.*, AES-19(2), 1983.
9. Vorpérian, V., Tymerski, R., and Lee, F. C., Equivalent circuit models for resonant and PWM switches, *IEEE Trans. Power Electron.*, 4(2), 1989.
10. Bello, V., Computer-aided analysis of switching regulators using SPICE2, in *IEEE Power Electronics Specialists Conference Record*, 1980, 3–11.
11. Chen, D. Y., Owen, H. A., and Wilson, T. G., Computer-aided design and graphics applied to the study of inductor energy storage DC-to-DC electronic converters, *IEEE Trans. Aerospace Electronic Syst.*, AES-9(4), 585–597, 1973.
12. Lee, F. C., Yu, Y., and Triner, J. E., Modeling of switching regulator power stages with and without zero-inductor current dwell time, in *IEEE Power Electronics Specialists Conference Record*, 1976, 62–72.
13. Rahman, S. and Lee, F. C., Nonlinear program based optimization of boost and buck–boost converter designs, in *IEEE Power Electronics Specialists Conference Record*, 1981, 180–191.

3
AC-AC Conversion

Sándor Halász
Budapest University of Technology and Economics

3.1 Introduction 3-1
3.2 Cycloconverters 3-1
3.3 Matrix Converters 3-3

3.1 Introduction

AC-AC converters as shown in Fig. 3.1 are frequency converters. They produce an AC voltage in which both the frequency and voltage can be varied directly from the AC line voltage, e.g., from a 60- or 50-Hz source. There are two major classes of AC-AC, or so-called direct static frequency converters, as shown in Fig. 3.1.

1. Cycloconverters, which are constructed using naturally commutated thyristors. The commutation voltage is ensured by the supply voltage. These are so-called line commutated converters.
2. Matrix converters, which are constructed using full-controlled static devices, such as transistors or GTOs (gate turn-off thyristors).

3.2 Cycloconverters

In Figs. 3.2 and 3.3, the two typical types of cycloconverters are presented. In the first case there are two three-phase midpoint controlled rectifiers connected back to back. The second case shows two three-phase bridge rectifier converters connected back to back. Both are used for three-phase to three-phase conversion. In Fig. 3.4 the single-phase output voltage and current waves are presented for the

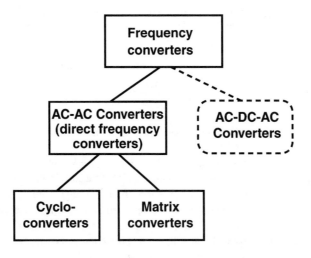

FIGURE 3.1 Classification of frequency converters.

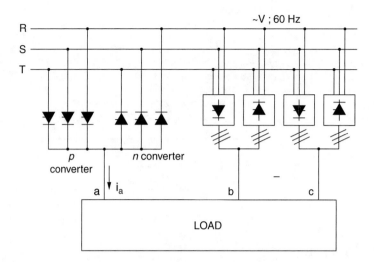

FIGURE 3.2 Cycloconverter scheme with three-phase midpoint controlled rectifier.

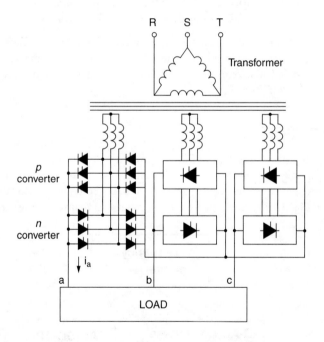

FIGURE 3.3 Cycloconverter scheme with three-phase bridge controlled rectifier.

bridge rectifier circuits. The output voltage V_a and current i_a have V_{a1} and i_{a1} fundamental components with ϕ_1 phase displacement and numerous harmonics. Because of the load inductance, the current harmonics will be significantly lower than the voltage harmonics. The firing angles are α_P and α_N for the p and n converters, respectively. In general, the controls are designed so that only the thyristors of either the p or n converter is firing, which produces a current in the desired direction. During this period the other converter is blocked. When the current changes direction, both converters must be blocked for a short time.

It is possible to operate without blocking the converters. In this case, their average voltage must be the same, and therefore the relation $\alpha_p = 180 - \alpha_n$ is valid. However, additional inductances are necessary to limit the circulating currents between two converters since the instantaneous voltages of the two converters differ from one another.

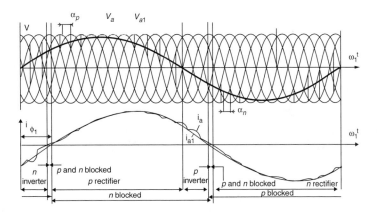

FIGURE 3.4 Voltage and current vs. time for cycloconverter with three-phase bridge converters.

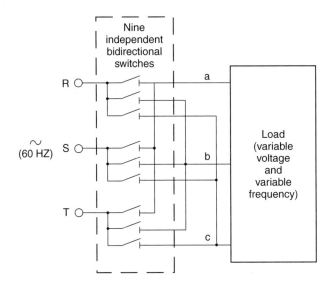

FIGURE 3.5 Three-phase to three-phase matrix converter.

The phase control of the p and n converters is modulated by a sine or trapezoidal wave. The content of the harmonics for sine modulation is lower; however, the maximum value of the output voltage is lower than that for trapezoidal modulation. During every cycle of the output voltage both of the converters must work as rectifiers and inverters.

The shape of the output voltage goes from bad to worse with an increase in the output voltage and the output frequency. If the frequency reaches the well-defined value the current harmonics become unacceptable. This frequency is usually 33% of supply frequency for three-phase midpoint (Fig. 3.2) and 50% for three-phase bridge (Fig. 3.3) converters.

The cycloconverter is usually used for three-phase, high-power, low-speed synchronous motor drives and rarely employed for induction motor drives.

3.3 Matrix Converters

The three-phase to three-phase matrix converter is presented in Fig. 3.5. Using the bidirectional switches, any phase of the load can be connected to any phase of the input voltage, e.g., the zero value of the load phase voltages is maintained by connecting all the load phases to the same input phase. Using pulse-width

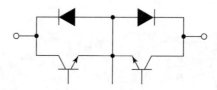

FIGURE 3.6 Bidirectional switch.

modulation techniques, the load voltage and the load frequency are controlled from zero to their maximum values. The maximum voltage is usually close to the input voltage, but the maximum frequency can be several times that of the input frequency and is only limited by practical considerations. The bidirectional switches must be capable of permitting current flow in either direction. In Fig. 3.6 one possible configuration of the bidirectional switch is shown.

Matrix converters require the use of numerous switches and well-established control methods. Some additional elements are necessary for the safe commutation of the bidirectional switches. These disadvantages of matrix converters prevent their use in industrial applications.

References

Guggi, L. and Pelly, B. R. 1976. *Static Power Frequency Changes*, John Wiley & Sons, New York.
Pelly, B. R. 1976. *Thyristor Phase-Controlled Converters*, John Wiley & Sons, New York.

4
Rectifiers

Sam Guccione
Eastern Illinois University

Mahesh M. Swamy
Yaskawa Electric America

Ana Stankovic
Cleveland State University

4.1 Uncontrolled Single-Phase Rectifiers 4-1
   Single-Phase Half-Wave Rectifiers • Single-Phase Full-Wave Rectifiers

4.2 Uncontrolled and Controlled Rectifiers 4-4
   Uncontrolled Rectifiers • Controlled Rectifiers • Conclusion

4.3 Three-Phase Pulse-Width-Modulated Boost-Type Rectifiers 4-33
   Introduction • Indirect Current Control of a Unity Power Factor Sinusoidal Current Boost-Type Rectifier • Appendix

4.1 Uncontrolled Single-Phase Rectifiers

Sam Guccione

Single-Phase Half-Wave Rectifiers

Operation

A single-phase half-wave rectifier consists of a single diode connected as shown in Fig. 4.1. This is the simplest of the rectifier circuits. It produces an output waveform that is half of the incoming AC voltage waveform. The positive pulse output waveform shown in Fig. 4.1 occurs because of the forward-bias condition of the diode. A diode experiences a forward-bias condition when its anode is at a higher potential than its cathode. Reverse bias occurs when its anode is lower than its cathode.

During the positive portion of the input waveform, the diode becomes forward biased, which allows current to pass through the diode from anode to cathode, such that it flows through the load to produce a positive output pulse waveform. Over the negative portion of the input waveform, the diode is reverse-biased ideally so no current flows. Thus, the output waveform is zero or nearly zero during this portion of the input waveform.

Because real diodes have real internal electrical characteristics, the peak output voltage in volts of a real diode operating in a half-wave rectifier circuit is

$$V_{P(out)} = V_{P(in)} - V_F \qquad (4.1)$$

where $V_{P(in)}$ is the peak value of the input voltage waveform and V_F is the forward-bias voltage drop across the diode. This output voltage is used to determine one of the specification values in the selection of a diode for use in a half-wave rectifier.

Other voltage and current values are important to the operation and selection of diodes in rectifier circuits.

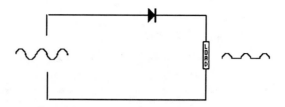

FIGURE 4.1 Single-phase half-wave rectifier.

Important Diode Current Characteristics

Peak Forward Current

The peak forward or rectified forward current, I_{FM}, in amperes is the current that flows through the diode as a result of the current demand of the load resistor. It is determined from the peak output voltage Eq. (4.1) as

$$I_{FM} = V_{P(out)}/R_L \tag{4.2}$$

where R_L is the load resistance in ohms. I_{FM} is also a specification value used to select a diode for use in a rectifier. Choose a diode with an I_{FM} that is equal to or greater than the I_{FM} calculated in Eq. (4.2).

rms Forward Current

Since rms values are useful, the rms value of forward current in amperes is determined from

$$I_{FRMS} = I_{FM} \times 0.707 \tag{4.3}$$

This value is sometimes called the maximum rms forward current.

Mean Forward Current

To find the continuous forward current that the diode in a half-wave rectifier circuit is subjected to, the mean or average rectified current, I_{FAV}, can be found from

$$I_{FAV} = I_{FM}/\pi \tag{4.4}$$

Because this average current is a continuous value, it is sometimes suggested that a diode be selected that has an I_{FAV} value of 1.25 times that determined from Eq. (4.4).

Single Cycle Surge Current

One additional current is important in rectifier circuits. That current is the single cycle surge current, I_{FSM}. This is the peak forward surge current that exists for one cycle or one half cycle for nonrepetitive conditions. This could be due to a power-on transient or other situations.

Important Diode Voltage Characteristics

Average Output Voltage

The average output voltage of a half-wave rectifier is determined from

$$V_{AVG\,(out)} = V_{P(in)}/\pi \tag{4.5}$$

Repetitive Peak Reverse Voltage

Another characteristic that is important to the operation of rectifier circuits is the voltage that the diode experiences during reverse bias. When the diode is reversed, it experiences a voltage that is equal to the value of the negative peak input voltage. For example, if the negative peak input voltage is 300 V, then the peak reverse voltage (prv) rating of the diode must be at least 300 V or higher. The prv rating is for

Rectifiers

a repetitive input waveform, thus producing a repetitive peak reverse voltage value. A nonrepetitive prv is also an important specification value, as will be described below.

The repetitive peak reverse voltage is given different names. It is called variously the peak reverse voltage, peak inverse voltage, maximum reverse voltage (V_{RM}), and maximum working peak reverse voltage (V_{RWM}). The most common name is the repetitive peak reverse voltage, V_{RRM}. The repetitive peak reverse voltage is one of the critical specification values that are important when selecting a diode for operation in half-wave rectifier circuits.

Forward Voltage Drop

The value of the maximum forward voltage, V_F, is the voltage value that occurs across a diode when it becomes forward biased. It is a small value usually in the range of 0.5 V to several volts. V_F is sometimes identified as the maximum forward voltage drop, V_{FM}. The threshold value of the forward voltage is sometimes listed in specifications as $V_{F(TO)}$.

Nonrepetitive Peak Reverse Voltage

Diodes used in rectifiers are also specified in terms of their characteristics to nonrepetitive conditions. This is usually identified as the voltage rating for a single transient wave. The symbol, V_{RSM}, is used. V_{RSM} is a specification value. This voltage is sometimes identified as the nonrepetitive transient peak reverse voltage.

Single-Phase Full-Wave Rectifiers

Operation

A single-phase full-wave rectifier consists of four diodes arranged as shown in Fig. 4.2 in what is called a bridge. This rectifier circuit produces an output waveform that is the positive half of the incoming AC voltage waveform and the inverted negative half. The bias path for the positive output pulse is through diode D_1, then the load, then D_4, and back to the other side of the power supply. The current flow through the load is in the down direction for the figure shown. Diodes D_2 and D_3 are reverse-biased during this part.

The bias path for the negative cycle of the input waveform is through diode D_3, then the load, then D_2, and back to the opposite side of the power supply. The current flow through the load resistor is once again down. That is, it is flowing through the load in the same direction as during the positive cycle of the input waveform. Diodes D_1 and D_4 are reverse-biased during this part. The resulting output waveform is a series of positive pulses without the "gaps" of the half-wave rectifier output.

As in the half-wave rectifier circuit description, real diodes have real characteristics, which affect the circuit voltages and currents. The peak output voltage in volts of a full-wave bridge rectifier with real diodes is

$$V_{P(\text{out})} = V_{P(\text{in})} - 2 \times V_F \tag{4.6}$$

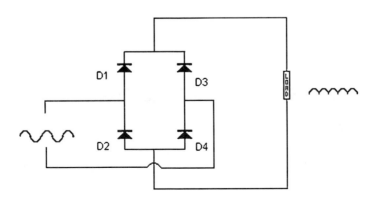

FIGURE 4.2 Single-phase full-wave bridge rectifier.

where V_F is the forward-bias voltage drop across one diode. Because there are two forward-biased diodes in the current path, the total drop would be twice the drop of one diode.

As in the half-wave rectifier, there are other voltages and currents that are important to the operation and selection of diodes in a full-wave rectifier. Only those values that are different from the half-wave circuit will be identified here. The other values are the same between a half-wave and a full-wave rectifier.

Important Diode Current Characteristics

Peak Rectified Forward Current

The peak rectified forward current, I_{FM}, in amperes has the same equation (4.1) as for the half-wave rectifier. The difference is that the value $V_{P(out)}$ is as shown in Eq. (4.6).

rms Forward Current

The rms value is computed using the same Eq. (4.2).

Average Forward Current

The mean or average forward current for a full-wave rectifier is twice the value for a half-wave rectifier. The equation is

$$I_{FAV} = 2 \times I_{FM}/\pi \tag{4.7}$$

Single-Cycle Surge Current

This current is the same for either type of rectifier.

Important Diode Voltage Characteristics

Average Output Voltage

The average output voltage of a full-wave rectifier is twice that of a half-wave rectifier. It is determined from

$$V_{AVG(out)} = 2 \times V_{P(in)}/\pi \tag{4.8}$$

Repetitive Peak Reverse Voltage

The repetitive peak reverse voltage, V_{RRM}, is slightly different for a full-wave bridge rectifier. It is determined by

$$V_{RRM} = V_{P(out)} - V_F \tag{4.9}$$

where $V_{P(out)}$ and V_F have been defined before in Eq. (4.1).

Forward Voltage Drop

This voltage is the same for either type of rectifier.

Nonrepetitive Peak Reverse Voltage

This voltage is the same for either type of rectifier.

4.2 Uncontrolled and Controlled Rectifiers

Mahesh M. Swamy

Rectifiers are electronic circuits that convert bidirectional voltage to unidirectional voltage. This process can be accomplished either by mechanical means like in the case of DC machines employing commutators or by static means employing semiconductor devices. Static rectification is more efficient and reliable compared to rotating commutators. This section covers rectification of electric power for industrial and commercial use. In other words, we will not be discussing small signal rectification that generally involves low power and low voltage signals. Static power rectifiers can be classified into two broad groups. They are (1) uncontrolled rectifiers and (2) controlled rectifiers. Uncontrolled rectifiers make use of power semiconductor diodes while controlled rectifiers make use of thyristors (SCRs), gate turn-off thyristors (GTOs), and MOSFET-controlled thyristors (MCTs).

Rectifiers, in general, are widely used in power electronics to rectify single-phase as well as three-phase voltages. DC power supplies used in computers, consumer electronics, and a host of other applications typically make use of single-phase rectifiers. Industrial applications include, but are not limited to, industrial drives, metal extraction processes, industrial heating, power generation and transmission, etc. Most industrial applications of large power rating typically employ three-phase rectification processes.

Uncontrolled rectifiers in single-phase as well as in three-phase circuits will be discussed, as will controlled rectifiers. Application issues regarding uncontrolled and controlled rectifiers will be briefly discussed within each section.

Uncontrolled Rectifiers

The simplest uncontrolled rectifier use can be found in single-phase circuits. There are two types of uncontrolled rectification. They are (1) half-wave rectification and (2) full-wave rectification. Half-wave and full-wave rectification techniques have been used in single-phase as well as in three-phase circuits. As mentioned earlier, uncontrolled rectifiers make use of diodes. Diodes are two-terminal semiconductor devices that allow flow of current in only one direction. The two terminals of a diode are known as the anode and the cathode.

Mechanics of Diode Conduction

The anode is formed when a pure semiconductor material, typically silicon, is doped with impurities that have fewer valence electrons than silicon. Silicon has an atomic number of 14, which according to Bohr's atomic model means that the K and L shells are completely filled by 10 electrons and the remaining 4 electrons occupy the M shell. The M shell can hold a maximum of 18 electrons. In a silicon crystal, every atom is bound to four other atoms, which are placed at the corners of a regular tetrahedron. The bonding, which involves sharing of a valence electron with a neighboring atom is known as covalent bonding. When a Group 3 element (typically boron, aluminum, gallium, and indium) is doped into the silicon lattice structure, three of the four covalent bonds are made. However, one bonding site is vacant in the silicon lattice structure. This creates vacancies or *holes* in the semiconductor. In the presence of either a thermal field or an electrical field, electrons from a neighboring lattice or from an external agency tend to migrate to fill this vacancy. The vacancy or *hole* can also be said to move toward the approaching electron, thereby creating a mobile hole and hence current flow. Such a semiconductor material is also known as lightly doped semiconductor material or p-type. Similarly, the cathode is formed when silicon is doped with impurities that have higher valence electrons than silicon. This would mean elements belonging to Group 5. Typical doping impurities of this group are phosphorus, arsenic, and antimony. When a Group 5 element is doped into the silicon lattice structure, it oversatisfies the covalent bonding sites available in the silicon lattice structure, creating excess or loose electrons in the valence shell. In the presence of either a thermal field or an electrical field, these loose electrons easily get detached from the lattice structure and are free to conduct electricity. Such a semiconductor material is also known as heavily doped semiconductor material or n-type.

The structure of the final doped crystal even after the addition of *acceptor* impurities (Group 3) or *donor* impurities (Group 5), remains electrically neutral. The available electrons balance the net positive charge and there is no charge imbalance.

When a p-type material is joined with an n-type material, a pn-junction is formed. Some loose electrons from the n-type material migrate to fill the holes in the p-type material and some holes in the p-type migrate to meet with the loose electrons in the n-type material. Such a movement causes the p-type structure to develop a slight negative charge and the n-type structure to develop some positive charge. These slight positive and negative charges in the n-type and p-type areas, respectively, prevent further migration of electrons from *n*-type to *p*-type and holes from *p*-type to *n*-type areas. In other words, an energy barrier is automatically created due to the movement of charges within the crystalline lattice structure. Keep in mind that the combined material is still electrically neutral and no charge imbalance exists.

When a positive potential greater than the barrier potential is applied across the *pn*-junction, then electrons from the *n*-type area migrate to combine with the holes in the *p*-type area, and vice versa. The *pn*-junction is said to be *forward-biased*. Movement of charge particles constitutes current flow. Current is said to flow from the anode to the cathode when the potential at the anode is higher than the potential at the cathode by a minimum threshold voltage also known as the junction barrier voltage. The magnitude of current flow is high when the externally applied positive potential across the *pn*-junction is high.

When the polarity of the applied voltage across the *pn*-junction is reversed compared to the case described above, then the flow of current ceases. The holes in the *p*-type area move away from the *n*-type area and the electrons in the *n*-type area move away from the *p*-type area. The *pn*-junction is said to be *reverse-biased*. In fact, the holes in the *p*-type area get attracted to the negative external potential and similarly the electrons in the *n*-type area get attracted to the positive external potential. This creates a depletion region at the *pn*-junction and there are almost no charge carriers flowing in the depletion region. This phenomenon brings us to the important observation that a *pn*-junction can be utilized to force current to flow only in one direction, depending on the polarity of the applied voltage across it. Such a semiconductor device is known as a *diode*. Electrical circuits employing diodes for the purpose of making the current flow in a unidirectional manner through a load are known as *rectifiers*. The voltage-current characteristic of a typical power semiconductor diode along with its symbol is shown in Fig. 4.3.

Single-Phase Half-Wave Rectifier Circuits

A single-phase half-wave rectifier circuit employs one diode. A typical circuit, which makes use of a half-wave rectifier, is shown in Fig. 4.4.

A single-phase AC source is applied across the primary windings of a transformer. The secondary of the transformer consists of a diode and a resistive load. This is typical since many consumer electronic items including computers utilize single-phase power.

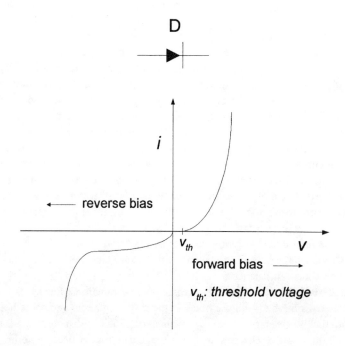

FIGURE 4.3 Typical *v–i* characteristic of a semiconductor diode and its symbol.

Rectifiers 4-7

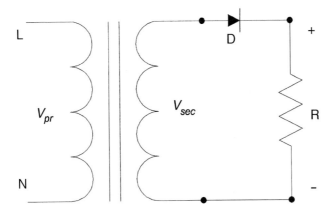

FIGURE 4.4 Electrical schematic of a single-phase half-wave rectifier circuit feeding a resistive load. Average output voltage is V_o.

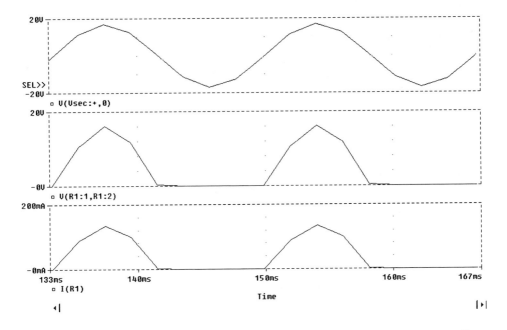

FIGURE 4.5 Typical waveforms at various points in the circuit of Fig. 4.4. For a purely resistive load, $V_o = \sqrt{2} \times V_{sec}/\pi$.

Typically, the primary side is connected to a single-phase AC source, which could be 120 V, 60 Hz, 100 V, 50 Hz, 220 V, 50 Hz, or any other utility source. The secondary side voltage is generally stepped down and rectified to achieve low DC voltage for consumer applications. The secondary voltage, the voltage across the load resistor, and the current through it is shown in Fig. 4.5.

As one can see, when the voltage across the anode-cathode of diode D_1 in Fig. 4.4 goes negative, the diode does not conduct and no voltage appears across the load resistor R. The current through R follows the voltage across it. The value of the secondary voltage is chosen to be 12 VAC and the value of R is chosen to be 120 Ω. Since, only one half of the input voltage waveform is allowed to pass onto the output, such a rectifier is known as a *half-wave* rectifier. The voltage ripple across the load resistor is rather large and, in typical power supplies, such ripples are unacceptable. The current through the load is discontinuous and the current through the secondary of the transformer is unidirectional. The AC component in the secondary of the transformer is balanced by a corresponding AC component in the primary winding.

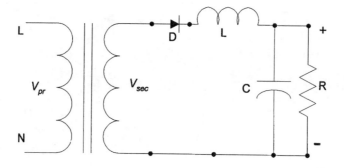

FIGURE 4.6 Modified circuit of Fig. 4.4 employing smoothing filters.

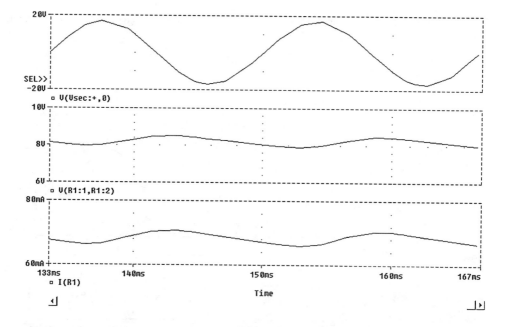

FIGURE 4.7 Voltage across load resistor R and current through it for the circuit in Fig. 4.6.

However, the DC component in the secondary does not induce any voltage on the primary side and hence is not compensated for. This DC current component through the transformer secondary can cause the transformer to saturate and is not advisable for large power applications. In order to smooth the output voltage across the load resistor R and to make the load current continuous, a smoothing filter circuit comprised of either a large DC capacitor or a combination of a series inductor and shunt DC capacitor is employed. Such a circuit is shown in Fig. 4.6.

The resulting waveforms are shown in Fig. 4.7. It is interesting to see that the voltage across the load resistor has very little ripple and the current through it is smooth. However, the value of the filter components employed is large and is generally not economically feasible. For example, in order to get a voltage waveform across the load resistor R, which has less than 6% peak-peak voltage ripple, the value of inductance that had to be used is 100 mH and the value of the capacitor is 1000 μF. In order to improve the performance without adding bulky filter components, it is a good practice to employ full-wave rectifiers. The circuit in Fig. 4.4 can be easily modified into a full-wave rectifier. The transformer is changed from a single secondary winding to a center-tapped secondary winding. Two diodes are now employed instead of one. The new circuit is shown in Fig. 4.8.

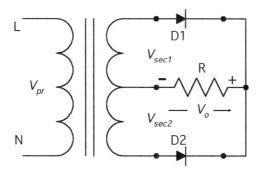

FIGURE 4.8 Electrical schematic of a single-phase full-wave rectifier circuit. Average output voltage is V_o.

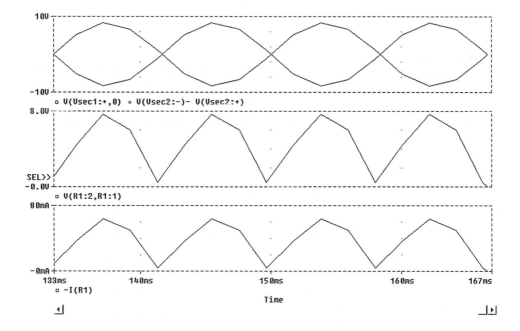

FIGURE 4.9 Typical waveforms at various points in the circuit of Fig. 4.8. For a purely resistive load, $V_o = 2 \times \sqrt{2} \times V_{sec}/\pi$.

Full-Wave Rectifiers

The waveforms for the circuit of Fig. 4.8 are shown in Fig. 4.9. The voltage across the load resistor is a full-wave rectified voltage. The current has subtle discontinuities but can be improved by employing smaller size filter components. A typical filter for the circuit of Fig. 4.8 may include only a capacitor. The waveforms obtained are shown in Fig. 4.10.

Yet another way of reducing the size of the filter components is to increase the frequency of the supply. In many power supply applications similar to the one used in computers, a high frequency AC supply is achieved by means of switching. The high frequency AC is then level translated via a ferrite core transformer with multiple secondary windings. The secondary voltages are then rectified employing a simple circuit as shown in Fig. 4.4 or Fig. 4.6 with much smaller filters. The resulting voltage across the load resistor is then maintained to have a peak-peak voltage ripple of less than 1%.

Full-wave rectification can be achieved without the use of center-tap transformers. Such circuits make use of four diodes in single-phase circuits and six diodes in three-phase circuits. The circuit configuration

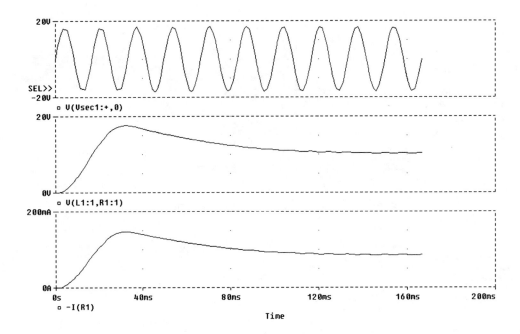

FIGURE 4.10 Voltage across the load resistor and current through it with the same filter components as in Fig. 4.6. Notice the conspicuous reduction in ripple across R.

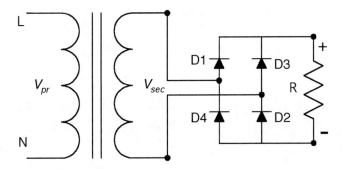

FIGURE 4.11 Schematic representation of a single-phase full-wave H-bridge rectifier.

is typically referred to as the *H-bridge* circuit. A single-phase full-wave H-bridge topology is shown in Fig. 4.11. The main difference between the circuit topology shown in Figs. 4.8 and 4.11 is that the H-bridge circuit employs four diodes while the topology of Fig. 4.8 utilizes only two diodes. However, a center-tap transformer of a higher power rating is needed for the circuit of Fig. 4.8. The voltage and current stresses in the diodes in Fig. 4.8 are also greater than that occurring in the diodes of Fig. 4.11.

In order to comprehend the basic difference in the two topologies, it is interesting to compare the component ratings for the same power output. To make the comparison easy, let both topologies employ very large filter inductors such that the current through R is constant and ripple-free. Let this current through R be denoted by I_{dc}. Let the power being supplied to the load be denoted by P_{dc}. The output power and the load current are then related by the following expression:

$$P_{dc} = I_{dc}^2 \times R$$

The rms current flowing through the first secondary winding in the topology in Fig. 4.8 will be $I_{dc}/\sqrt{2}$. This is because the current through a secondary winding flows only when the corresponding diode is forward-biased. This means that the current through the secondary winding will flow only for one half cycle. If the voltage at the secondary is assumed to be V, the VA rating of the secondary winding of the transformer in Fig. 4.8 will be given by:

$$VA = V \times I_{dc}/\sqrt{2}$$
$$VA_2 = V \times I_{dc}/\sqrt{2}$$
$$VA = VA_1 + VA_2 = \sqrt{2} \times V \times I_{dc}$$

This is the secondary-side VA rating for the transformer shown in Fig. 4.8.

For the isolation transformer shown in Fig. 4.11, let the secondary voltage be V and the load current be of a constant value I_{dc}. Since, in the topology of Fig. 4.11, the secondary winding carries the current I_{dc} when diodes D_1 and D_2 conduct and as well as when diodes D_3 and D_4 conduct, the rms value of the secondary winding current is I_{dc}. Hence, the VA rating of the secondary winding of the transformer shown in Fig. 4.11 is $V \times I_{dc}$, which is less than that needed in the topology of Fig. 4.8. Note that the primary VA rating for both cases remains the same since in both cases the power being transferred from the source to the load remains the same.

When diode D_2 in the circuit of Fig. 4.8 conducts, the secondary voltage of the second winding V_{sec2} (= V) appears at the cathode of diode D_1. The voltage being blocked by diode D_1 can thus reach two times the peak secondary voltage (= $2 \times V_{pk}$) (Fig. 4.9). In the topology of Fig. 4.11, when diodes D_1 and D_2 conduct, the voltage V_{sec} (= V), which is same as V_{sec2} appears across D_3 as well as across D_4. This means that the diodes have to withstand only one times the peak of the secondary voltage, V_{pk}. The rms value of the current flowing through the diodes in both topologies is the same. Hence, from the diode voltage rating as well as from the secondary VA rating points of view, the topology of Fig. 4.11 is better than that of Fig. 4.8. Further, the topology in Fig. 4.11 can be directly connected to a single-phase AC source and does not need a center-topped transformer. The voltage waveform across the load resistor is similar to that shown in Figs. 4.9 and 4.10.

In many industrial applications, the topology shown in Fig. 4.11 is used along with a DC filter capacitor to smooth the ripples across the load resistor. The load resistor is simply a representative of a load. It could be an inverter system or a high-frequency resonant link. In any case, the diode rectifier-bridge would see a representative load resistor. The DC filter capacitor will be large in size compared to an H-bridge configuration based on three-phase supply system. When the rectified power is large, it is advisable to add a DC-link inductor. This can reduce the size of the capacitor to some extent and reduce the current ripple through the load. When the rectifier is turned on initially with the capacitor at zero voltage, a large amplitude of charging current will flow into the filter capacitor through a pair of conducting diodes. The diodes $D_1 \sim D_4$ should be rated to handle this large surge current. In order to limit the high inrush current, it is a normal practice to add a charging resistor in series with the filter capacitor. The charging resistor limits the inrush current but creates a significant power loss if it is left in the circuit under normal operation. Typically, a contactor is used to short-circuit the charging resistor after the capacitor is charged to a desired level. The resistor is thus electrically nonfunctional during normal operating conditions. A typical arrangement showing a single-phase full-wave H-bridge rectifier system for an inverter application is shown in Fig. 4.12.

The charging current at time of turn-on is shown in a simulated waveform in Fig. 4.13. Note that the contacts across the soft-charge resistor are closed under normal operation. The contacts across the soft-charge resistor are initiated by various means. The coil for the contacts could be powered from the input AC supply and a timer or it could be powered on by a logic controller that senses the level of voltage across the DC bus capacitor or senses the rate of change in voltage across the DC bus capacitor. A simulated waveform depicting the inrush with and without a soft-charge resistor is shown in Fig. 4.13a and b, respectively.

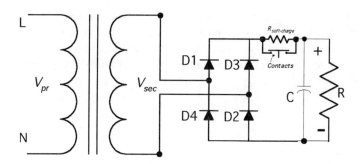

FIGURE 4.12 Single-phase H-bridge circuit for use with power electronic circuits.

For larger power applications, typically above 1.5 kW, it is advisable to use a higher power supply. In some applications, two of the three phases of a three-phase power system are used as the source powering the rectifier of Fig. 4.11 The line-line voltage could be either 240 or 480 VAC. Under those circumstances, one may go up to 10 kW of load power before adopting a full three-phase H-bridge configuration. Beyond 10 kW, the size of the capacitor becomes too large to achieve a peak-peak voltage ripple of less than 5%. Hence, it is advisable then to employ three-phase rectifier configurations.

Three-Phase Rectifiers (Half-Wave and Full-Wave)

Similar to the single-phase case, there exist half-wave and full-wave three-phase rectifier circuits. Again, similar to the single-phase case, the half-wave rectifier in the three-phase case also yields DC components in the source current. The source has to be large enough to handle this. Therefore, it is not advisable to use three-phase half-wave rectifier topology for large power applications. The three-phase half-wave rectifier employs three diodes while the full-wave H-bridge configuration employs six diodes. Typical three-phase half-wave and full-wave topologies are shown in Fig. 4.14.

In the half-wave rectifier shown in Fig. 4.14a, the shape of the output voltage and current through the resistive load is dictated by the instantaneous value of the source voltages, L1, L2, and L3. These source voltages are phase shifted in time by 120 electrical degrees, which corresponds to approximately 5.55 ms for a 60 Hz system. This means that if one considers the L1 phase to reach its peak value at time t_1, the L2 phase will achieve its peak 120 electrical degrees later (t_1 + 5.55 ms), and L3 will achieve its peak 120 electrical degrees later than L2 (t_1 + 5.55 ms + 5.55 ms). Since all three phases are connected to the same output resistor R, the phase that provides the highest instantaneous voltage is the phase that appears across R. In other words, the phase with the highest instantaneous voltage reverse biases the diodes of the other two phases and prevents them from conducting, which consequently prevents those phase voltages from appearing across R. Since a particular phase is connected to only one diode in Fig. 4.14a, only three pulses, each of 120° duration, appear across the load resistor, R. Typical output voltage across R for the circuit of Fig. 4.14a is shown in Fig. 4.15a.

A similar explanation can be provided to explain the voltage waveform across a purely resistive load in the case of the three-phase full-wave rectifier shown in Fig. 4.14b. The output voltage that appears across R is the highest instantaneous line-line voltage and not simply the phase voltage. Since there are six such intervals, each of 60 electrical degrees duration in a given cycle, the output voltage waveform will have six pulses in one cycle (Fig. 4.15b). Since a phase is connected to two diodes (diode pair), each phase conducts current out and into itself, thereby eliminating the DC component in one complete cycle.

The waveform for a three-phase full-wave rectifier with a purely resistive load is shown in Fig. 4.15b. Note that the number of humps in Fig. 4.15a is only three in one AC cycle, while the number of humps in Fig. 4.15b is six in one AC cycle.

In both the configurations shown in Fig. 4.14, the load current does not become discontinuous due to three-phase operation. Comparing this to the single-phase half-wave and full-wave rectifier, one can say that the output voltage ripple is much lower in three-phase rectifier systems compared to single-phase

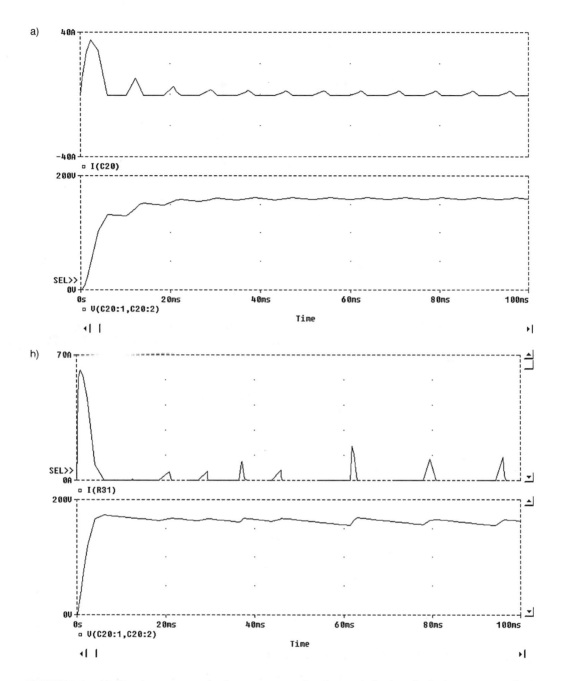

FIGURE 4.13 (a) Charging current and voltage across capacitor for a typical value of soft-charge resistor of 2 Ω. The DC bus capacitor is about 1000 μF. The load is approximately 200 Ω. (b) Charging current and voltage across capacitor for no soft charge resistor. The current is limited by the system impedance and by the diode forward resistance. The DC bus capacitor is about 1000 μF. The load is approximately 200 Ω.

rectifier systems. Hence, with the use of moderately sized filters, three-phase full-wave rectifiers can be operated at hundred to thousands of kilowatts. The only limitation would be the size of the diodes used and power system harmonics, which will be discussed next. Since there are six humps in the output voltage waveform per electrical cycle, the three-phase full-wave rectifier shown in Fig. 4.14b is also known as a six-pulse rectifier system.

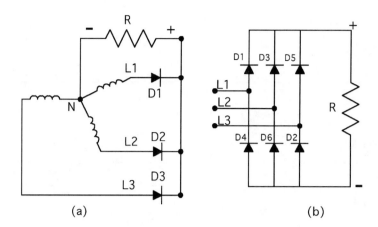

FIGURE 4.14 Schematic representation of three-phase rectifier configurations: (a) half-wave rectifier needing a neutral point, N; and (b) full-wave rectifier.

Average Output Voltage

In order to evaluate the average value of the output voltage for the two rectifiers shown in Fig. 4.14, the output voltages in Fig. 4.15a and b have to be integrated over a cycle. For the circuit shown in Fig. 4.14a, the integration yields the following:

$$V_o = \frac{3}{2\pi} \int_{\pi/6}^{5\pi/6} \sqrt{2} V_{L-N} \sin(wt) d(wt)$$

$$V_o = \frac{3 \times \sqrt{3} \times \sqrt{2} \times V_{L-N}}{2 \times \pi}$$

Similar operations can be performed to obtain the average output voltage for the circuit shown in Fig. 4.14b. This yields:

$$V_o = \frac{3}{\pi} \int_{\pi/3}^{2\pi/3} \sqrt{2} V_{L-L} \sin(wt) d(wt)$$

$$V_o = \frac{3 \times \sqrt{2} \times V_{L-L}}{\pi} = \frac{3 \times \sqrt{2} \times \sqrt{3} \times V_{L-N}}{\pi}$$

In other words, the average output voltage for the circuit in Fig. 4.14b is twice that for the circuit in Fig. 4.14a.

Influence of Three-Phase Rectification on the Power System

Events over the last several years have focused attention on certain types of loads on the electrical system that result in power quality problems for the user and utility alike. When the input current into the electrical equipment does not follow the impressed voltage across the equipment, then the equipment is said to have a nonlinear relationship between the input voltage and input current. All equipment that employs some sort of rectification (either single phase or three phase) are examples of nonlinear loads. Nonlinear loads generate voltage and current harmonics that can have adverse effects on equipment designed for operation as linear loads. Transformers that bring power into an industrial environment are subject to higher heating losses due to harmonic generating sources (nonlinear loads) to which they are connected. Harmonics can have a detrimental effect on emergency generators, telephones, and other electrical equipment. When reactive power compensation (in the form of passive power factor improving capacitors) is used with nonlinear loads, resonance conditions can occur that may result in even higher

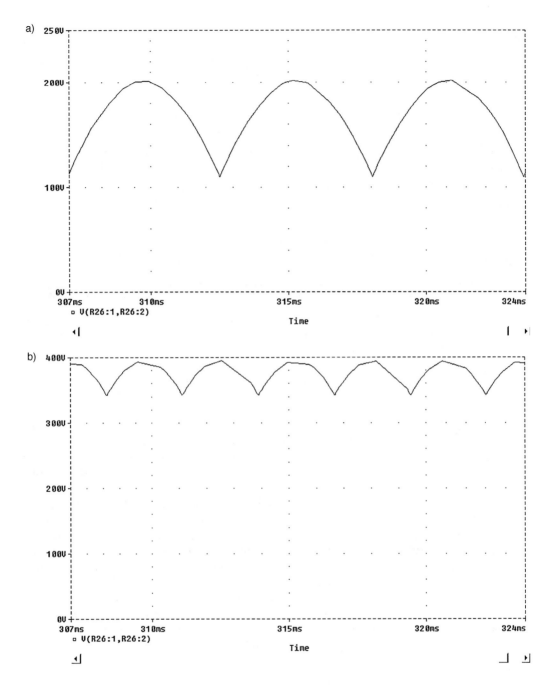

FIGURE 4.15 (a) Typical output voltage across a purely resistive network for the half-wave rectifier shown in Fig. 4.6a. (b) Typical output voltage across a purely resistive network for the full-wave rectifier shown in Fig. 4.6b.

levels of harmonic voltage and current distortion, thereby causing equipment failure, disruption of power service, and fire hazards in extreme conditions.

The electrical environment has absorbed most of these problems in the past. However, the problem has now reached a magnitude where Europe, the United States, and other countries have proposed standards to responsibly engineer systems considering the electrical environment. IEEE 519-1992 and IEC 1000 have evolved to become a common requirement cited when specifying equipment on newly engineered projects.

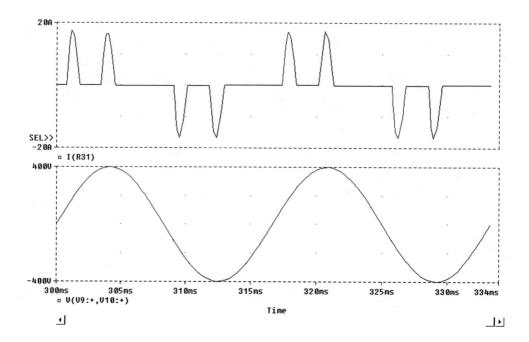

FIGURE 4.16 Typical pulsed-current waveform as seen at input of a three-phase diode rectifier with DC capacitor filter. The lower trace is input line-line voltage.

Why Diode Rectifiers Generate Harmonics

The current waveform at the inputs of a three-phase full-wave rectifier is not continuous. It has multiple zero crossings in one electrical cycle. The current harmonics generated by rectifiers having DC bus capacitors are caused by the pulsed current pattern at the input. The DC bus capacitor draws charging current only when it gets discharged due to the load. The charging current flows into the capacitor when the input rectifier is forward-biased, which occurs when the instantaneous input voltage is higher than the steady-state DC voltage across the DC bus capacitor. The pulsed current drawn by the DC bus capacitor is rich in harmonics due to the fact that it is discontinuous as shown in Fig. 4.16. Sometimes there are also voltage harmonics that are associated with three-phase rectifier systems. The voltage harmonics generated by three-phase rectifiers are due to the flat-topping effect caused by a weak AC source charging the DC bus capacitor without any intervening impedance. The distorted voltage waveform gives rise to voltage harmonics that could lead to possible network resonance.

The order of current harmonics produced by a semiconductor converter during normal operation is termed characteristic harmonics. In a three-phase, six-pulse rectifier with *no DC bus capacitor*, the characteristic harmonics are nontriplen odd harmonics (e.g., 5th, 7th, 11th, etc.). In general, the characteristic harmonics generated by a semiconductor recitifier are given by:

$$h = kq \pm 1$$

where h is the order of harmonics; k is any integer, and q is the pulse number of the semiconductor rectifier (six for a six-pulse rectifier). When operating a six-pulse rectifier system with a DC bus capacitor (as in voltage source inverters, or VSI), one may start observing harmonics of orders other than those given by the above equation. Such harmonics are called *noncharacteristic* harmonics. Though of lower magnitude, these also contribute to the overall harmonic distortion of the input current. The per-unit value of the characteristic harmonics present in the theoretical current waveform at the input of the semiconductor converter is given by $1/h$, where h is the order of the harmonics. In practice, the observed per-unit value of the harmonics is much greater than $1/h$. This is because the theoretical current waveform

Rectifiers	4-17

is a rectangular pattern made up of equal positive and negative halves, each occupying 120 electrical degrees. The pulsed discontinuous waveform observed commonly at the input of a three-phase full-wave rectifier system depends greatly on the impedance of the power system, the size of the DC bus capacitors, and the level of loading of the DC bus capacitors. Total harmonic current distortion is defined as:

$$THD_I = \frac{\sqrt{\sum_{n=2}^{n=\infty} I_n^2}}{I_1}$$

where I_1 is the rms value of the fundamental component of current; and I_n is the rms value of the nth harmonic component of current.

Harmonic Limits Based on IEEE Std. 519-1992

The IEEE Std. 519-1992 relies strongly on the definition of the point of common coupling or PCC. The PCC from the utility viewpoint will usually be the point where power comes into the establishment (i.e., point of metering). However, IEEE Std. 519-1992 also suggests that **"within an industrial plant, the point of common coupling (PCC) is the point between the nonlinear load and other loads"** (IEEE Std. 519-1992). This suggestion is crucial since many plant managers and building supervisors feel that it is equally, if not more important to keep the harmonic levels at or below acceptable guidelines within their facility. In view of the many recently reported problems associated with harmonics within industrial plants, it is important to recognize the need for mitigating harmonics at the point where the offending equipment is connected to the power system. This approach would minimize harmonic problems, thereby reducing costly downtime and improving the life of electrical equipment. If one is successful in mitigating individual load current harmonics, then the total harmonics at the point of the utility connection will in most cases meet or exceed the IEEE recommended guidelines. In view of this, it is becoming increasingly common for specifiers to require nonlinear equipment suppliers to adopt the procedure outlined in IEEE Std. 519-1992 to mitigate the harmonics to acceptable levels at the point of the offending equipment. For this to be interpreted equally by different suppliers, the intended PCC must be identified. If the PCC is not defined clearly, many suppliers of offending equipment would likely adopt the PCC at the utility metering point, which would not benefit the plant or the building, but rather the utility.

Having established that it is beneficial to adopt the PCC to be the point where the nonlinear equipment connects to the power system, the next step is to establish the short circuit ratio. Short circuit ratio calculations are key in establishing the allowable current harmonic distortion levels. For calculating the short circuit ratio, one has to determine the available short circuit current at the input terminals of the nonlinear equipment. The short-circuit current available at the input of nonlinear equipment can be calculated by knowing the value of the short-circuit current available at the secondary of the utility transformer supplying power to the establishment (building) and the series impedance in the electrical circuit between the secondary of the transformer and the nonlinear equipment. **In practice, it is common to assume the same short circuit current level as at the secondary of the utility transformer feeding the nonlinear equipment.** The next step is to compute the fundamental value of the rated input current into the nonlinear equipment (three-phase full-wave rectifier in this case). An example is presented here to recap the above procedure. A widely used industrial equipment item that employs a three-phase full-wave rectifier is the voltage source inverter (VSI). These are used for controlling speed and torque of induction motors. Such equipment is also known as an Adjustable Speed Drive (ASD) or Variable Frequency Drive (VFD).

A 100-hp ASD/motor combination connected to a 480-V system being fed from a 1500-kVA, three-phase transformer with impedance of 4% is required to meet IEEE Std. 519-1992 at its input terminals. The rated current of the transformer is $1500 \times 1000/(\sqrt{3} \times 480)$, and is calculated to be 1804.2 A. The short-circuit current available at the secondary of the transformer is equal to the rated current divided by the per unit impedance of the transformer. This is calculated to be 45,105.5 A. The short-circuit ratio,

TABLE 4.1 Current Distortion Limits for General Distribution Systems

	(120 V through 69,000 V) Maximum Harmonic Current Distortion in percent of I_L Individual Harmonic Order (Odd Harmonics)[a]					
I_{sc}/I_L	<11	$11 \leq h \leq 17$	$17 \leq h \leq 23$	$23 \leq h \leq 35$	$35 \leq h$	TDD[b]
<20[c]	4.0	2.0	1.5	0.6	0.3	5.0
20 < 50	7.0	3.5	2.5	1.0	0.5	8.0
50 < 100	10.0	4.5	4.0	1.5	0.7	12.0
100 < 1000	12.0	5.5	5.0	2.0	1.0	15.0
>1000	15.0	7.0	6.0	2.5	1.4	20.0

[a] Even harmonics are limited to 25% of the odd harmonic limits above.

[b] TDD is Total Demand Distortion and is defined as the harmonic current distortion in % of maximum demand load current. The maximum demand current could either be a 15-min or a 30-min demand interval.

[c] All power generation equipment is limited to these values of current distortion, regardless of actual I_{sc}/I_L; where I_{sc} is the maximum short circuit current at PCC and I_L is the maximum demand load current (fundamental frequency) at PCC.

Source: IEEE Std. 519-1992.

which is defined as the ratio of the short-circuit current at the PCC to the fundamental value of the nonlinear current is computed next. NEC amps for 100-hp, 460-V is 124 A. Assuming that the short-circuit current at the ASD input is practically the same as that at the secondary of the utility transformer, the short-circuit ratio is calculated to be: 45,105.5/124, which equals 363.75. On referring to IEEE Std. 519-1992, Table 10.3 (IEEE Std. 519-1992), the short-circuit ratio falls in the 100 to 1000 category. For this ratio, the total demand distortion (TDD) at the point of ASD connection to the power system network is recommended to be 15% or less. For reference, see Table 4.1.

Harmonic Mitigating Techniques

Various techniques of improving the input current waveform are discussed below. The intent of all techniques is to make the input current more continuous so as to reduce the overall current harmonic distortion. The different techniques can be classified into four broad categories:

1. Introduction of line reactors and/or DC link chokes
2. Passive filters (series, shunt, and low pass broadband filters)
3. Phase multiplication (12-pulse, 18-pulse rectifier systems)
4. Active harmonic compensation

The following paragraphs will briefly discuss the available technologies and their relative advantages and disadvantages. The term three-phase line reactor or just reactor is used in the following paragraphs to denote three-phase line inductors.

Three-Phase Line Reactors

Line reactors offer a significant magnitude of inductance that can alter the way the current is drawn by a nonlinear load such as a rectifier bridge. The reactor makes the current waveform less discontinuous, resulting in lower current harmonics. Since the reactor impedance increases with frequency, it offers larger impedance to the flow of higher order harmonic currents. Therefore, it is instrumental in impeding higher frequency current components while allowing the fundamental frequency component to pass through with relative ease.

On knowing the input reactance value, one can estimate the expected current harmonic distortion. A table illustrating the typically expected input current harmonics for various amounts of input reactance is shown in Table 4.2.

Input reactance is determined by the accumulated impedance of the AC reactor, DC link choke (if used), input transformer, and cable impedance. To maximize the input reactance while minimizing AC

Rectifiers

TABLE 4.2 Percent Harmonics vs. Total Line Impedance

Harmonic	Total Input Impedance						
	3%	4%	5%	6%	7%	8%	9%	10%
5th	40	34	32	30	28	26	24	23
7th	16	13	12	11	10	9	8.3	7.5
11th	7.3	6.3	5.8	5.2	5	4.3	4.2	4
13th	4.9	4.2	3.9	3.6	3.3	3.15	3	2.8
17th	3	2.4	2.2	2.1	0.9	0.7	0.5	0.4
19th	2.2	2	0.8	0.7	0.4	0.3	0.25	0.2
%THID	44	37	35	33	30	28	26	25
True rms	1.09	1.07	1.06	1.05	1.05	1.04	1.03	1.03

voltage drop, one can combine the use of both AC-input reactors and DC link chokes. One can approximate the total effective reactance and view the expected harmonic current distortion from Table 4.2. The effective impedance value in percent is based on the actual loading and is:

$$Z_{eff} = \frac{\sqrt{3} \times 2 \times \pi \times f \times L \times I_{act\,(fnd.)}}{V_{L-L}} \times 100$$

where $I_{act(fnd.)}$ is the fundamental value of the actual load current and V_{L-L} is the line-line voltage. The effective impedance of the transformer as seen from the nonlinear load is:

$$Z_{eff,\,x\text{-mer}} = \frac{Z_{eff,\,x\text{-mer}} \times I_{act\,(fnd.)}}{I_r}$$

where $Z_{eff,x\text{-mer}}$ is the effective impedance of the transformer as viewed from the nonlinear load end; $Z_{x\text{-mer}}$ is the nameplate impedance of the transformer; and I_r is the nameplate rated current of the transformer.

On observing one conducting period of a diode pair, it is interesting to see that the diodes conduct only when the instantaneous value of the input AC waveform is higher than the DC bus voltage by at least 3 V. Introducing a three-phase AC reactor in between the AC source and the DC bus makes the current waveform less pulsating because the reactor impedes sudden change in current. The reactor also electrically differentiates the DC bus voltage from the AC source so that the AC source is not clamped to the DC bus voltage during diode conduction. This feature practically eliminates flat topping of the AC voltage waveform caused by many ASDs when operated with weak AC systems.

DC-Link Choke

Based on the above discussion, it can be noted that any inductor of adequate value placed between the AC source and the DC bus capacitor of the ASD will help in improving the current waveform. These observations lead to the introduction of a DC-link choke, which is electrically present after the diode rectifier and before the DC bus capacitor. The DC-link choke performs very similar to the three-phase line inductance. The ripple frequency that the DC-link choke has to handle is six times the input AC frequency for a six-pulse ASD. However, the magnitude of the ripple current is small. One can show that the effective impedance offered by a DC-link choke is approximately half of that offered by a three-phase AC inductor. In other words, a 6% DC-link choke is equivalent to a 3% AC inductor from an impedance viewpoint. This can be mathematically derived equating AC side power flow to DC side power flow as follows:

$$P_{ac} = \frac{3 \times V_{L-N}^2}{R_{ac}}; \quad P_{ac} = P_{dc}$$

V_{L-N} is the line-neutral voltage at the input to the three-phase rectifier.

$$P_{dc} = \frac{V_{dc}^2}{R_{dc}}; \quad V_{dc} = \frac{3 \times \sqrt{3} \times \sqrt{2} \times V_{L-N}}{\pi}; \quad \text{Hence, } R_{dc} = 2\left(\frac{9}{\pi^2}\right) R_{ac}$$

Since $9/\pi^2$ is approximately equal to 1, the ratio of DC impedance to AC impedance can be said to be approximately 1:2. The DC link choke is less expensive and smaller than a three-phase line reactor and is often included inside an ASD. However, as the derivation shows, one has to keep in mind that the effective impedance offered by a DC link choke is only half its numerical impedance value when referred to the AC side. DC link chokes are electrically after the diode bridge and so they do not offer any significant spike or overvoltage surge protection to the diode bridge rectifiers. It is a good engineering practice to incorporate both a DC link choke and a three-phase line reactor in an ASD for better overall performance.

Passive Filters

Passive filters consist of passive components like inductors, capacitors, and resistors arranged in a predetermined fashion either to attenuate the flow of harmonic components through them or to shunt the harmonic component into them. Passive filters can be of many types. Some popular ones are series passive filters, shunt passive filters, and low-pass broadband passive filters. Series and shunt passive filters are effective only in the narrow proximity of the frequency at which they are tuned. Low-pass broadband passive filters have a broader bandwidth and attenuate almost all harmonics above their cutoff frequency. However, applying passive filters requires good knowledge of the power system because passive filter components can interact with existing transformers and power factor correcting capacitors and could create electrical instability by introducing resonance into the system. Some forms of low-pass broadband passive filters do not contribute to resonance but they are bulky, expensive, and occupy space. A typical low-pass broadband filter structure popularly employed by users of ASDs is shown in Fig. 4.17.

Phase Multiplication

As discussed previously, the characteristic harmonics generated by a full-wave rectifier bridge converter is a function of the pulse number for that converter. A 12-pulse converter will have the lowest harmonic order of 11. In other words, the 5th, and the 7th harmonic orders are theoretically nonexistent in a 12-pulse converter. Similarly, an 18-pulse converter will have harmonic spectrum starting from the 17th harmonic and upwards. The lowest harmonic order in a 24-pulse converter will be the 23rd. The size of the passive harmonic filter needed to filter out the harmonics reduces as the order of the lowest harmonic in the current spectrum increases. Hence, the size of the filter needed to filter the harmonics out of a 12-pulse converter is much smaller than that needed to filter out the harmonics of a 6-pulse converter. However, a 12-pulse

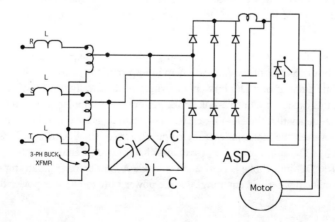

FIGURE 4.17 Schematic representation of a low-pass broadband harmonic filter connected to an ASD with diode rectifier front end. (U.S. Patent 5,444,609.)

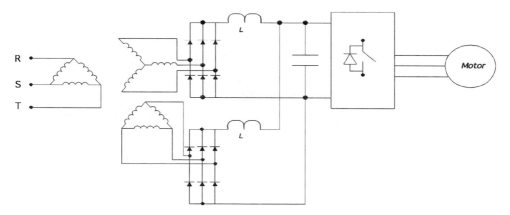

FIGURE 4.18 Schematic of a 12-pulse converter employing a three-winding transformer. Note that the input transformer has to be sized for rated power operation.

converter needs two 6-pulse bridges and two sets of 30° phase shifted AC inputs. The phase shift is achieved either by using an isolation transformer with one primary and two phase-shifted secondary windings or by using an autotransformer that provides phase-shifted outputs. Many different autotransformer topologies exist and the choice of a topology over the other involves a compromise between ease of construction, performance, and cost. An 18-pulse converter would need three 6-pulse diode bridges and three sets of 20° phase-shifted inputs; similarly, a 24-pulse converter would need four 6-pulse diode bridges and four sets of 15° phase-shifted inputs. The transformers providing the phase-shifted outputs for multipulse converters have to be properly designed to handle circulating harmonic flux.

A typical 12-pulse structure is shown in Fig. 4.18. In one electrical cycle, the DC voltage will have 12 humps and hence the name 12-pulse rectifier.

Active Harmonic Compensation

Most passive techniques discussed above aim to cure the harmonic problems once nonlinear loads have created them. However, motor-drive manufacturers are developing rectification techniques that do not generate low-order harmonics. These drives use active front ends. Instead of using diodes as rectifiers, the active front-end ASDs make use of active switches like IGBTs along with parallel diodes. Power flow through a switch becomes bidirectional and can be manipulated to recreate a current waveform that linearly follows the applied voltage waveform.

Apart from the active front ends, there also exist shunt active filters used for actively introducing a current waveform into the AC network, which, when combined with the harmonic current, results in an almost perfect sinusoidal waveform.

One of the most interesting active filter topologies for use in retrofit applications is the combination of a series active filter along with shunt tuned passive filters. This combination is also known as the hybrid structure.

Most active filter topologies are complicated and require active switches and control algorithms that are implemented using digital signal processing (DSP) chips. The active filter topology also needs current and voltage sensors and corresponding analog-to-digital (A/D) converters. This extra hardware increases the cost and component count, reducing the overall reliability and robustness of the design. Manufacturers of smaller power equipment like computer power supplies, lighting ballast, etc. have successfully employed active circuits, employing boost converter topologies.

Controlled Rectifiers

Controlled rectifier circuits make use of devices known as "thyristors." A thyristor is a four-layer (*pnpn*), three-junction device that conducts current only in one direction similar to a diode. The last (third) junction is utilized as the control junction and consequently the rectification process can be initiated at will provided the device is favorably biased and the load is of favorable magnitude. The operation of a

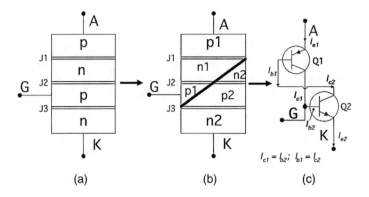

FIGURE 4.19 Virtual representation of a thyristor to explain its operation.

thyristor can be explained by assuming it to be made up of two transistors connected back-to-back as shown in Fig. 4.19.

Let α_1 and α_2 be the ratio of collector to emitter currents of transistors Q_1 and Q_2, respectively. In other words:

$$\alpha_1 = \frac{I_{c1}}{I_{e1}}; \quad \alpha_2 = \frac{I_{c2}}{I_{e2}};$$

Also, from Fig. 4.19: $I_{e1} = I_{e2} = I_A$ where I_A is the anode current flowing through the thyristor. From transistor theory, the value of I_{e2} is equal to $I_{c2} + I_{b2} + I_{lkg}$; where I_{lkg} is the leakage current crossing the $n_1 = p_2$-junction. From Fig. 4.19, $I_{b2} = I_{c1}$. Hence, the anode current can be rewritten as:

$$I_A = I_{c1} + I_{c2} + I_{lkg}$$

Substituting the collector currents by the product of ratio α and emitter current, the anode current becomes:

$$I_A = (\alpha_1 \times I_{e1}) + (\alpha_2 \times I_{e2}) + I_{lkg}$$
$$I_A = (\alpha_1 + \alpha_2)I_A + I_{lkg}$$
$$I_A = \frac{I_{lkg}}{1 - (\alpha_1 + \alpha_2)}$$

If the ratios of the collector current to base current (gain) of the transistors are assumed to be β_1 and β_2, respectively, then the relationship between to β_1, β_2 and α_1, α_2 can be written as:

$$\alpha_1 = \frac{\beta_1}{1+\beta_1}; \quad \alpha_2 = \frac{\beta_2}{1+\beta_2}$$

Substituting for α_1 and α_2 in the expression for I_A yields the following expression:

$$I_A = \frac{(1+\beta_1)(1+\beta_2)I_{lkg}}{1-\beta_1\beta_2}.$$

If the values of α_1 and α_2 are low (low gains), then the anode current is low and comparable to the leakage current. Under this condition, the thyristor is said to be in its OFF state. However, if the effective gain of the transistor is such that the product of the gains are close to 1 (i.e., sum of the ratios of α_1 and α_2 are close to 1), then there is a large increase in anode current and the thyristor is said to be in conduction. External circuit conditions can be changed to influence the product of the gains ($\beta_1\beta_2$). Some techniques of achieving this are briefly discussed next.

Increasing Applied Voltage
On applying a voltage across the anode to cathode terminals of the thyristor (anode being more positive than the cathode), one can see that junctions J_1 and J_3 in Fig. 4.19 are forward–biased while junction J_2 is reverse-biased. The thyristor does not conduct any current and is said to be in a blocking state. On increasing the applied voltage, minority carriers in junction J_2 (i.e., holes in n_1, n_2 and electrons in p_1, p_2) start acquiring more energy and hence start to migrate. In the process, these holes could dislodge more holes. Recombination of the electrons and holes also occur, which creates more motion. If the voltage is increased beyond a particular level, the movement of holes and electrons becomes great and junction J_2 ceases to exist. The product of the gains of the two transistors in the two-transistor model is said to achieve values close to unity. This method of forcing current to flow through the thyristor is not recommended since junction J_2 gets permanently damaged and the thyristor ceases to block forward voltage. Hence, this method is a destructive method.

High dv/dt
As explained earlier, junction J_2 is the forward blocking junction when a forward voltage is applied across anode to cathode of a thyristor. Any pn-junction behaves like a depletion region when it is reverse biased. Since J_2 is reverse-biased, this junction behaves like a depletion region. Another way of looking at a depletion region is that the boundary of the depletion region has abundant holes and electrons while the region itself is depleted of charged carriers. This characteristic is similar to that of a capacitor. If the voltage across the junction (J_2) changes very abruptly, then there will be rapid movement of charged carriers through the depleted region. If the rate of change of voltage across this junction (J_2) exceeds a predetermined value, then the movement of charged carriers through the depleted region is so high that junction J_2 is again annihilated. After this event, the thyristor is said to have lost its capability to block forward voltage and even a small amount of forward voltage will result in significant current flow, limited only by the load impedance. This method is destructive too, and is hence not recommended.

Temperature
Temperature affects the movement of holes and electrons in any semiconductor device. Increasing the temperature of junction J_2 will have a very similar effect. More holes and electrons will begin to move, causing more dislodging of electrons and holes from neighboring lattice. If a high temperature is maintained, this could lead to an avalanche breakdown of junction J_2 and again render the thyristor useless since it would no longer be able to block forward voltage. Increasing temperature is yet another destructive method of forcing the thyristor to conduct.

Gate Current Injection
If a positive voltage is applied across the gate to cathode of a thyristor, then one would be forward-biasing junction J_3. Charged carriers will start moving. The movement of charged carriers in junction J_3 will attract electrons from n_2 region of the thyristor (Fig. 4.19). Some of these electrons will flow out of the gate terminal but there would be ample of electrons that could start crossing junction J_2. Since electrons in p_2 region of junction J_2 are minority carriers, these can cause rapid recombination and help increase movement of minority carriers in junction J_2. By steadily increasing the forward-biasing potential of junction J_3, one could potentially control the depletion width of junction J_2. If a forward-biasing voltage is applied across anode to cathode of the thyristor with its gate to cathode favorably biased at the same time, then the thyristor can be made to conduct current. This method achieves conduction by increasing the leakage current in a controlled manner. The gain product in the two-transistor equivalent is made to achieve a value of unity in a controlled manner and the thyristor is said to turn ON. This is the only recommended way of turning ON a thyristor. When the gate–cathode junction is sufficiently forward-biased, the current through the thyristor depends on the applied voltage across the anode–cathode and

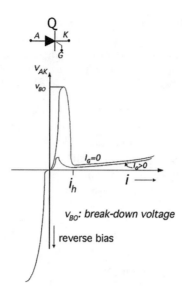

FIGURE 4.20 v–i characteristic of a thyristor along with its symbol.

the load impedance. The load impedance and the externally applied anode–cathode voltage should be such that the current through the thyristor is greater than a minimum current known as *latching current*, I_l. Under such a condition, the thyristor is said to have *latched ON*. Once it has latched ON, the thyristor remains ON. In other words, even if the forward-biasing voltage across the gate–cathode terminals is removed, the thyristor continues to conduct. Junction J_2 does not exist during the ON condition. The thyristor reverts to its blocking state only when the current through it falls below a minimum threshold value known as *holding current*, I_h. Typically, holding current is lower than latching current ($I_h < I_l$). There are two ways of achieving this. They are either (1) increase the load impedance to such a value that the thyristor current falls below I_h or (2) apply reverse-biasing voltage across the anode-cathode of the thyristor.

An approximate v–i characteristic of a typical thyristor and its symbol are shown in Fig. 4.20.

Since the thyristor allows flow of current only in one direction like a diode and the instant at which it is turned ON can be controlled, the device is a key component in building a controlled rectifier unit. One can replace the diode in all the circuits discussed so far with the thyristor. Because of its controllability, the instant at which the thyristor conducts can be delayed to alter the average and rms output voltages. By doing so, one can choose to control the output voltage and power of a rectifier circuit. Hence, rectifiers that employ thyristors are also known as silicon controlled rectifiers or SCR.

A typical single-phase, R-L rectifier circuit with one thyristor as the rectifier is shown in Fig. 4.21. The figure also shows the relevant circuit waveforms. The greatest difference between this circuit and its diode counterpart is also shown for comparison. Both circuits conduct beyond π radians due to the presence of the inductor L since the average voltage across an inductor is zero. If the value of the circuit components and the input supply voltage are the same in both cases, the duration for which the current flows into the output R-L load depends on the values of R and L. In the case of the diode circuit, it does not depend on anything else; while in the case of the thyristor circuit, it also depends on the instant the thyristor is given a gate trigger.

From Fig. 4.21, it is interesting to note that the energy stored in the inductor during the conduction interval can be controlled in the case of a thyristor is such a manner so as to reduce the conduction interval and thereby alter (reduce) the output power. Both the diode and the thyristor show reverse recovery phenomenon. The thyristor, like the diode, can block reverse voltage applied across it repeatedly, provided the voltage is less than its breakdown voltage.

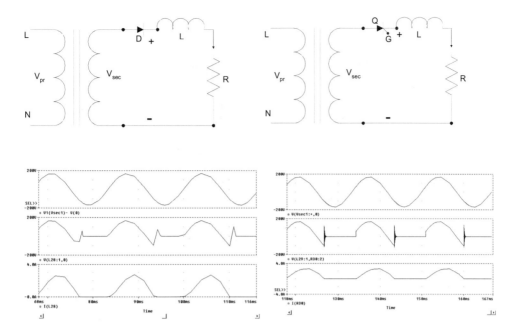

FIGURE 4.21 Comparing a single diode rectifier circuit with a single thyristor rectifier circuit. Note that the thyristor conduction is delayed deliberately to bring out the differences.

Gate Circuit Requirements

The trigger signal should have voltage amplitude greater than the minimum gate trigger voltage of the thyristor being turned ON. It should not be greater than the maximum gate trigger voltage, either. The gate current should likewise be in between the minimum and maximum values specified by the thyristor manufacturer. Low gate current driver circuits can fail to turn ON the thyristor. The thyristor is a current controlled switch and so the gate circuit should be able to provide the needed turn ON gate current into the thyristor. Unlike the bipolar transistor, the thyristor is not an amplifier and so the gate current requirement does not absolutely depend on the voltage and current rating of the thyristor. Sufficient gate trigger current will turn ON the thyristor and current will flow from the anode to the cathode provided that the thyristor is favorably biased and the load is such that the current flowing is higher than the latching current of the thyristor. In other words, in single phase AC to DC rectifier circuits, the gate trigger will turn ON the thyristor only if it occurs during the positive part of the AC cycle (Fig. 4.21). Any trigger signal during the negative part of the AC cycle will not turn ON the thyristor and the thyristor will remain in blocking state. Keeping the gate signal ON during the negative part of the AC cycle does not typically damage a thyristor.

Single-Phase H-Bridge Rectifier Circuits with Thyristors

Similar to the diode H-bridge rectifier topology, there exist SCR-based rectifier topologies. Because of their unique ability to be controlled, the output voltage and hence the power can be controlled to desired levels. Since the triggering of the thyristor has to be synchronized with the input sinusoidal voltage in an AC to DC rectifier circuit, one can achieve a soft-charge characteristic of the filter capacitor. In other words, there is no need for employing soft-charge resistor and contactor combination as is required in single-phase and three-phase AC to DC rectifier circuits with DC bus capacitors.

In controlled AC-to-DC rectifier circuits, it is important to discuss control of resistive, inductive, and resistive-inductive load circuits. DC motor control falls into the resistive-inductive load circuit. DC motors are still an important part of the industry. However, the use of DC motors in industrial applications is

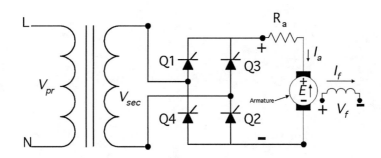

FIGURE 4.22 Single-phase DC motor control circuit for controlling a separately excited DC motor. R_a indicates equivalent armature resistance and E is the back emf.

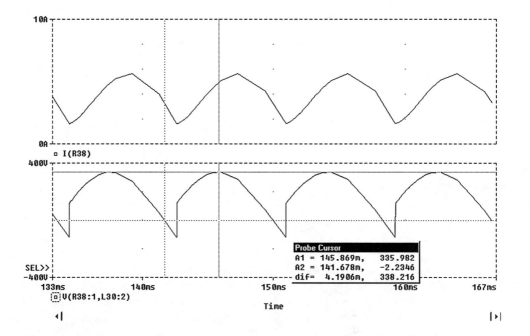

FIGURE 4.23 Armature current and output voltage of AC-to-DC rectifier employed to control a DC motor.

declining rapidly. Control of DC motors are typically achieved by controlled rectifier circuits employing thyristors. Small motors of less than 3 kW (approximately 5 hp) rating can be controlled by single-phase SCR circuits while larger ratings require three-phase versions. A typical single-phase H-bridge SCR-based circuit for the control of a DC motor is shown in Fig. 4.22. Typical output waveforms are shown in Fig. 4.23. The current in the load side can be assumed continuous due to the large inductance of the armature of the DC motor.

In Fig. 4.22, V_f is the field voltage, which is applied externally and generally is independent of the applied armature voltage. Such a DC motor is known as a separately excited motor. I_a is the armature current while I_f is the field current. The output of the controlled rectifier is applied across the armature. Since the output voltage can be controlled, one can effectively control the armature current. Since the torque produced by a DC motor is directly proportional to the armature current, the torque developed can thus be controlled.

$$T = K\phi I_a$$

where K is the motor constant and depends on the number of armature conductors, number of poles, and type of winding employed in the DC machine. ϕ is flux produced by the field and is proportional to the field current, I_f. Hence, the torque produced by a DC machine can be rewritten as $T = K(K_1 I_f) I_a$. By keeping the field current constant, the torque then becomes directly proportional to the armature current, which is controlled by controlling the output voltage of the AC-to-DC controlled rectifier. In the circuit shown in Fig. 4.22, it is interesting to note that the current I_a, cannot flow in the opposite direction. Hence, the motor cannot generate negative torque. In order to make the motor run in the opposite direction, the direction of the field has to be changed. Speed control within the base speed can also be accomplished by controlling the armature voltage as is shown below.

$$E = K\phi\omega = K(K_1 I_f)\omega$$

ω is the speed of the armature in rad/s. The back emf, E, is the difference between the output DC voltage of the AC-to-DC controlled rectifier and the drop across the equivalent armature resistance. Hence, E can be rewritten as:

$$E = V_a - (I_a R_a); \qquad \omega = \frac{V_a - (I_a R_a)}{K K_1 I_f}$$

For control of speed beyond base speed, the field current has to be altered. Hence, it can be shown that controlling the armature current can control the speed and torque produced by a DC machine. Controlling the output DC voltage can control the armature current. Because of the large inductance of the armature circuit, the current through it can be assumed to be continuous for a practical operating region. The average output voltage of a single-phase AC-to-DC rectifier circuit for continuous current operation is given by (referring to Fig. 4.23):

$$V_o = \frac{1}{\pi}\int_{\alpha}^{\pi+\alpha}(\sqrt{2}\times V_{rms})\,d(wt) = \frac{2\times\sqrt{2}\times V_{rms}\times\cos(\alpha)}{\pi}.$$

for continuous current condition. By controlling the triggering angle, α, one can control the average value of the output voltage, V_o. If armature current control is the main objective (to control output torque), then one can configure the controller of Fig. 4.22 with a feedback loop. The measured current can be compared with a set reference and the error can be used to control the triggering angle, α. Since the output voltage and hence the armature current are not directly proportional to α but to $\cos(\alpha)$, the above method will yield a nonlinear (co-sinusoidal) relationship between the output voltage and control angle, α. However, one could choose to use the error signal to control $\cos(\alpha)$ instead of α. This would then yield a linear relationship between the output voltage and cos of control angle, α.

It is interesting to note from the equation for the output average voltage that the output average voltage can become negative if the triggering angle is greater than 90 electrical degrees. This leads us to the topic of regeneration. AC-to-DC controlled rectifiers employing thyristors and having large inductance on the DC side can be made to operate in the regeneration mode by simply delaying the trigger angle. This is quite beneficial in overhauling loads like cranes. When the load on a hook of the crane has to be lifted up, electrical energy is supplied to the motor. The voltage across the motor is positive and the current through the armature is positive. Positive torque is generated and hence the load moves up. When the load is to be brought down, the load starts to rotate the motor in the opposite direction due to gravity. The voltage at the terminals becomes negative since speed is negative. The thyristors are gated at an angle greater than 90 electrical degrees to match the generated (negative) voltage of the DC motor. Since current through the thyristors cannot go negative, current is forced to flow into the DC motor in the positive direction. The large inductance of the motor helps to maintain the positive direction of current through the armature. Positive torque is still produced since the direction of current is still positive and the field

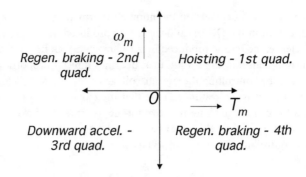

FIGURE 4.24 Four-quadrant operation of a crane or hoist.

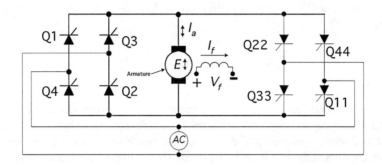

FIGURE 4.25 Two rectifier-bridge arrangements for four-quadrant operation of DC motor.

remains unchanged. In other words, the motor develops positive torque and tries to move the load up against gravity but the gravity is pushing the motor down. The product of current through the motor and the voltage across it is negative, meaning that the motor is not consuming energy, and on the contrary, is producing electrical energy—the kinetic energy due to the motor's downward motion is partly converted to electrical energy by the field and armature. This energy produced by the motor is routed out to the supply via the appropriately gated thyristors. Conversion of kinetic energy to electrical energy acts like a *dynamic-brake* and slows the rapid downward descent of the load.

A typical crane is required to operate in all four quadrants (Fig. 4.24). In the first quadrant, the motor develops positive torque and the motor runs in the positive direction, meaning its speed is positive—the product of torque and speed is power, and so positive electric power is supplied to the motor from the AC-to-DC rectifier. When the crane with a load is racing upward, close to the end of its travel, the AC-to-DC controlled rectifier is made to stop powering the motor. The rectifier generates practically no voltage. The inertia of the load moving upward generates a voltage in the form of a back emf. This voltage is fed into a second rectifier bridge arranged in the opposite direction as shown in Fig. 4.25. The second bridge is turned ON to let the generated voltage across the still upwardly mobile motor flow into the utility, thereby converting the inertial motion to electric power. In the second quadrant, speed remains positive but torque becomes negative, since the current through the motor flows in the opposite direction into the second rectifier bridge arrangement (Fig. 4.25). The product of speed and torque is negative, meaning that the motor behaves like a generator during this part of the travel.

The third quadrant, as explained earlier, occurs at the beginning of the lowering action. Both torque and speed are negative and so the product of torque and speed is positive. Power is applied to the motor to overcome static friction and accelerate the rotating parts of the mechanism to move the load downward. In this case, the direction of armature current through the motor is opposite to that in quadrant 1, and the electrical power needed by the motor is supplied by the second rectifier bridge arrangement (Fig. 4.25).

The mechanical load and motor arrangement goes into the fourth quadrant of operation for the larger part of the downward motion. This is the period during which the motor resists the tendency of the load to accelerate downward by developing positive torque. Since motion is downward, speed is negative and the product of torque and speed is negative. This means the motor behaves like a generator as explained earlier.

Since the thyristors cannot conduct in the opposite direction, a new inverter section had to be provided to enable the four-quadrant operation needed in cranes and hoists. The method by which unidirectional electrical power was routed to the bidirectional AC utility lines is known as inversion (opposite of rectification). Since no external means of switching OFF the thyristors was employed, the process of inversion was achieved by natural commutation. Such an inverter is also known as a line commutated inverter.

Three-Phase Controlled AC-to-DC Rectifier Systems

The observations made so far for the single-phase controlled AC-to-DC rectifiers can be easily extended to three-phase versions. An important controlled rectification scheme that was not mentioned in the single-phase case is the semiconverter circuit. In Fig. 4.22, if the thyristors Q_2 and Q_4 are replaced by diodes (D_2 and D_4), then the circuit of Fig. 4.22 is converted into a semiconverter circuit. Such a circuit does not have the potential to provide regeneration capability and hence is of limited use. However, in dual converter applications, especially in three-phase versions, there are a few instances where a semiconverter can be employed to reduce cost. A typical three-phase semiconverter circuit will consist of three thyristors and three diodes arranged in an H-bridge configuration as shown in Fig. 4.26.

Three-phase dual converter schemes similar to the one shown in Fig. 4.25 are still employed to operate large steel mills, hoists, and cranes. However, the advent of vector-controlled AC drives has drastically changed the electrical landscape of the modern industry. Most DC motor applications are being rapidly replaced by AC motors with field-oriented control schemes. DC motor application in railway traction has also seen significant reduction due to the less expensive and more robust AC motors.

However, there are still a few important applications where three-phase controlled rectification (inversion) is the most cost-effective solution. One such application is the regenerative converter module that many inverter-drive manufacturers provide as optional equipment to customers with overhauling loads. Under normal circumstances, during the motoring mode of operation of an AC drive, the regenerative unit does not come into the circuit. However, when the DC bus voltage tends to go higher than a predetermined level due to overhauling of the load, the kinetic energy of the load is converted into electrical energy and is fed back into the AC system via a six-pulse thyristor-based inverter-bridge. One such scheme is shown in Fig. 4.27.

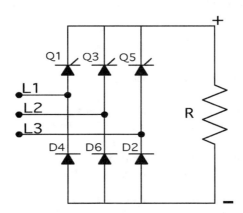

FIGURE 4.26 A typical three-phase semiconverter. Rarely employed in modern industry.

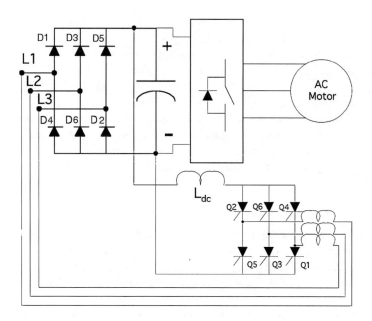

FIGURE 4.27 Use of six-pulse thyristor bridge in the inverter mode to provide regeneration capability to an existing AC drive system.

Average Output Voltage

In order to evaluate the average value of the output voltage for a three-phase full-bridge converter, the process of integrating the output voltage similar to the one in Fig. 4.15b has to be undertaken. For the circuit shown in Fig. 4.14b, where the diodes are replaced by thyristors, the integration yields the following:

$$V_o = \frac{3}{\pi} \int_{\alpha+(\pi/3)}^{\alpha+(2\pi/3)} \sqrt{2} V_{L-L} \sin(wt) d(wt)$$

$$V_o = \frac{3 \times \sqrt{2} \times V_{L-L} \times \cos(\alpha)}{\pi} = \frac{3 \times \sqrt{2} \times \sqrt{3} \times V_{L-N} \times \cos(\alpha)}{\pi}$$

The average output voltage for the circuit in Fig. 4.14b with the diodes being replaced by thyristors is only different in the cosine of the triggering angle, α. If the triggering angle is zero, the circuit performs similar to a three-phase diode rectifier and the average output voltages become the same.

HVDC Transmission Systems

One area where it is difficult to replace the use of high voltage, high current carrying thyristors is high voltage DC (HVDC) transmission systems. When a large amount of power is to be transported over long distances, or under water, it has been found that high voltage DC transmission is more economical. HVDC systems are in reality back-to-back rectifier systems. The sending end rectifier system consists typically of 12- or 24-pulse thyristor bridges while the receiving end consists of a similar configuration but in the opposite direction. The receiving end 12- or 24-pulse bridge operates in the inverter mode while the sending end operates in the rectifier mode. 12-pulse configuration is achieved by cascading two six-pulse bridges in series while 24-pulse configuration needs four six-pulse bridges cascaded in series. Typical advantages of high voltage DC transmission over high voltage AC transmission is briefly listed below:

1. No stability problems due to transmission line length since no reactive power needs to be transmitted.
2. No limitation of cable lengths for underground cable or submarine cable transmission due to the fact that no charging power compensation need be done.

3. AC power systems can be interconnected employing a DC tie without reference to system frequencies, short circuit power, etc.
4. High-speed control of DC power transmission is possible due to the fact that the control angle, α, has a relatively short time constant.
5. Fault isolation between receiving end and sending end can be dynamically achieved due to fast efficient control of the high voltage DC link.
6. Employing simple control logic can change energy flow direction very fast. This can help in meeting peak demands at either the sending or the receiving station.
7. High reliability of thyristor converter and inverter stations makes this mode of transmission a viable solution for transmission lengths typically over 500 km.
8. The right-of-way needed for high voltage DC transmission is much lower than that of AC transmission of the same power capacity.

The advantages of DC transmission over AC should not be misunderstood and DC should not be considered as a general substitute for AC power transmission. In a power system, it is generally believed that both AC and DC should be considered as complementary to each other, so as to bring about the integration of their salient features to the best advantage in realizing a power network that ensures high quality and reliability of power supply. A typical rectifier-inverter system employing a 12-pulse scheme is shown in Fig. 4.28.

Typical DC link voltage can be as high as 400 to 600 kV. Higher voltage systems are also in use. Typical operating power levels are over 1000 MW. There are a few systems transmitting close to 3500 MW of power through two bipolar systems. Most thyristors employed in large HVDC transmission systems are liquid cooled to improve their performance.

Power System Interaction with Three-Phase Thyristor AC-to-DC Rectifier Systems

Similar to the diode rectifiers, the thyristor based AC-to-DC rectifiers also suffer from low order current harmonics. In addition to current harmonics, there is a voltage notching phenomenon occurring at the input terminals of an AC-to-DC thyristor based rectifier system. The voltage notching is a very serious problem. Since thyristors are generally slower to turn ON and turn OFF compared to power semiconductor diodes, there are nontrivial durations during which an outgoing thyristor and an incoming

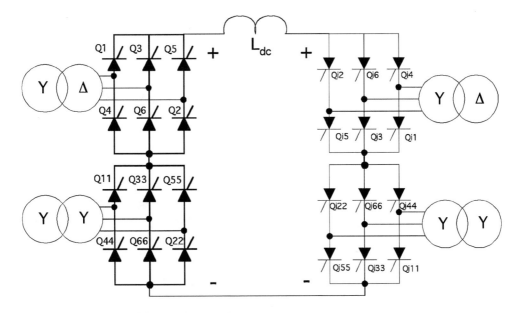

FIGURE 4.28 Schematic representation of a bipolar HVDC system employing 12-pulse rectification/inversion scheme.

thyristor remain in conduction, thereby creating a short-circuit across the power supply phases feeding the corresponding thyristors. Thyristors used in rectifiers are generally known as phase control type thyristors and have typical turn OFF times of 50 to 100 μs. Thyristors employed in inverter circuits typically are faster and have turn OFF times in the 10 to 50 μs range.

Notching can create major disturbances in sensitive electronic equipment that rely on the zero-crossing of the voltage for satisfactory operation. Multiple pseudo-zero-crossings of the voltage waveform can occur due to the notching effect of thyristor-based rectifier systems. Notching can create large magnitudes of currents to flow into power-factor correcting capacitors, thereby potentially causing permanent damage to them. IEEE Std. 519-1992 in the United States has strict regulations regarding the depth of the notch as well as the duration of the notch. AC line inductors in series with the supply feeding power to the three-phase bridge help to minimize the notching effect on the power system. The theory behind this phenomenon is discussed next.

When an external inductance is added in front of a three-phase AC-to-DC rectifier employing thyristors, the duration of commutation increases. In other words, the time period for which the outgoing thyristor remains in conduction along with the incoming thyristor increases. This overlap period causes the average output voltage to reduce because during this period, the output voltage is composed of two shorted phases and a healthy phase. The extent of reduction in the output voltage depends on the duration of overlap in electrical degrees. The duration of overlap in electrical degrees is commonly represented by μ. The overlap duration is directly proportional to the value of the external inductance used. If no external line inductor is used, then this duration will depend on the existing inductance of the system including the wiring inductance. In order to compute the factors influencing the overlap duration, a simple model can be assumed. Assume that the line is comprised of inductance L in each phase. Let the DC load current be I_{dc} and let it be assumed that this current does not change during the overlap interval. The current in the incoming thyristor is zero at start and by the end of the overlap interval, it rises to I_{dc}. Based on this assumption, the relationship between current and voltage can be expressed as:

$$v_{ab} = \sqrt{2} \times V_{L-L} \times \sin(wt) = 2 \times L \times (di/dt)$$

$$\sqrt{2} \times V_{L-L} \times \int_{\alpha+(\pi/3)}^{\alpha+(\pi/3)+\mu} \sin(wt)\, d(t) = 2 \times L \times \int_0^{I_{dc}} di$$

$$I_{dc} = \frac{\sqrt{2} \times V_{L-L} \times (\cos(\alpha+\pi/3) - \cos(\alpha+\pi/3+\mu))}{2wL}$$

$$= \frac{\sqrt{2} \times V_{L-L} \times \sin(\alpha+\pi 3+\mu/2) \times \sin(\mu/2)}{wL}$$

For small values of overlap angle μ, $\sin(\mu/2) = \mu/2$ and $\sin(\alpha+\pi/3+(\mu/2)) = \sin(\alpha+\pi/3)$. Rearranging the above equation yields:

$$\mu = \frac{2wL \times I_{dc}}{\sqrt{2} \times V_{L-L} \times \sin(\alpha+\pi/3)}$$

From the above expression, it is interesting to note the following:

1. If the inductance L in the form of either external inductance or leakage inductance of transformer or lead length is large, the overlap duration will be large.
2. If the load current I_{dc} is large, the overlap duration will be large.
3. If the delay angle is small, then the inductance will store more energy and so the duration of overlap will be large. The minimum value of delay angle α is 0° and the maximum value typically is 60°.

The average output voltage will reduce due to the overlap angle as mentioned before. In order to compute the average output voltage with a certain overlap angle, the limits of integration have to be changed. This exercise yields the following:

$$V_o = \frac{3}{\pi} \int_{\alpha+\mu+(\pi/3)}^{\alpha+\mu+(2\pi/3)} \sqrt{2} V_{L-L} \sin(wt) \, d(wt)$$

$$V_o = \frac{3 \times \sqrt{2} \times V_{L-L} \times \cos(\alpha+\mu)}{\pi} = \frac{3 \times \sqrt{2} \times \sqrt{3} \times V_{L-N} \times \cos(\alpha+\mu)}{\pi}$$

Thus, it can be seen that the overlap angle has an equivalent effect of advancing the delay angle, thereby reducing the average output voltage. From the discussions in the previous paragraphs on notching, it is interesting to note that adding external inductance increases the duration of the overlap and reduces the average value of the output DC voltage. However, when viewed from the AC source side, the notching effect is conspicuously reduced and in some cases not observable. Since all other electrical equipment in the system will be connected to the line side of the AC inductor (in front of a thyristor-based AC-to-DC rectifier), these equipment will not be affected by the notching phenomenon of thyristors. The external inductance also helps limit the circulating current between the two thyristors during the overlap duration.

Conclusion

Uncontrolled and controlled rectifier circuits have been discussed in this section. An introduction to the theory of diode and thyristor conduction has been presented to explain the important operating characteristics of these devices. Rectifier topologies employing both diodes and thyristors and their relative advantages and disadvantages have been discussed. Use of a dual thyristor bridge converter to achieve four-quadrant operation of a DC motor has been discussed. The topic of high-voltage DC (HVDC) transmission has been briefly introduced. Power quality issues relating to diode and thyristor-based rectifier topologies has also been addressed. To probe further into the various topics briefly discussed in this section, the reader is encouraged to refer to the references listed below.

References

Dewan, S. B. and Straughen, A., *Power Semiconductor Circuits*, John Wiley & Sons, New York, 1975.
Hoft, R. G., *Semiconductor Power Electronics*, Van Nostrand Reinhold, New York, 1986.
IEEE Recommended Practices and Requirements for Harmonic Control in Electrical Power Systems, IEEE Std. 519–1992.
Laughton M. A. and Say, M. G., Eds., *Electrical Engineer's Reference Book*, 14th ed., Butterworths, Boston, 1985.
Passive Harmonic Filter Systems for Variable Frequency Drives, U.S. Patent 5,444,609, 1995.
Sen, P. C., *Principles of Electric Machines and Power Electronics*, John Wiley & Sons, New York, 1997.

4.3 Three-Phase Pulse-Width-Modulated Boost-Type Rectifiers

Ana Stankovic

Introduction

The boost-type rectifier has been extensively developed and analyzed in recent years [1, 3, 6]. It offers advantages over traditionally used phase-controlled thyristor rectifiers in AC-DC-AC converters for variable-control drives because of its capability for nearly instantaneous reversal of power flow, power factor management, and reduction of input harmonic distortion. Figure 4.29 shows the structure of the pulse-width-modulated (PWM) boost-type rectifier.

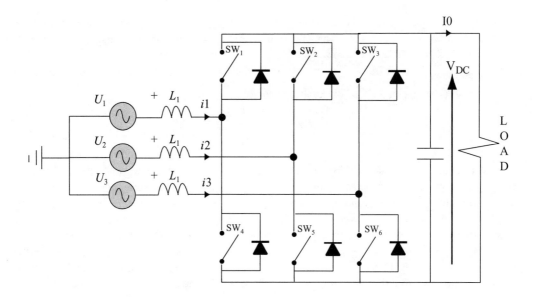

FIGURE 4.29 PWM boost-type rectifier.

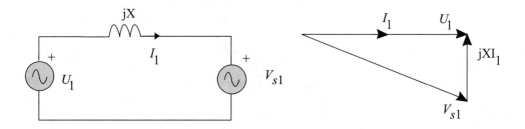

FIGURE 4.30 The per-phase equivalent circuit and phasor diagram.

Power flow in the PWM converter is controlled by adjusting the phase shift angle δ between the source voltage U_1 and the respective converter reflected input voltage V_{s1} [2].

When U_1 leads V_{s1} the real power flows from the AC source into the converter. Conversely, if U_1 lags V_{s1}, power flows from the DC side of the converter into the AC source. The real power transferred is given by the Eq. (4.10).

$$P = \frac{U_1 V_{s1}}{X_1} \sin(\delta) \quad (4.10)$$

The AC power factor is adjusted by controlling the amplitude of V_{s1}. The phasor diagram in Fig. 4.30 shows that, to achieve a unity power factor, V_{s1} must be

$$V_{s1} = \sqrt{U_1^2 + (X_1 I_1)^2} \quad (4.11)$$

Indirect Current Control of a Unity Power Factor Sinusoidal Current Boost-Type Rectifier

To control the DC output voltage of the PWM boost-type rectifier, the input line currents must be regulated [4, 5]. In typical rectifier controllers presented to date, the DC bus voltage error is used to synthesize a line current reference. Specifically, the line current reference is derived through the multiplication of a term proportional to the bus voltage error by a template sinusoidal waveform. The sinusoidal

Rectifiers

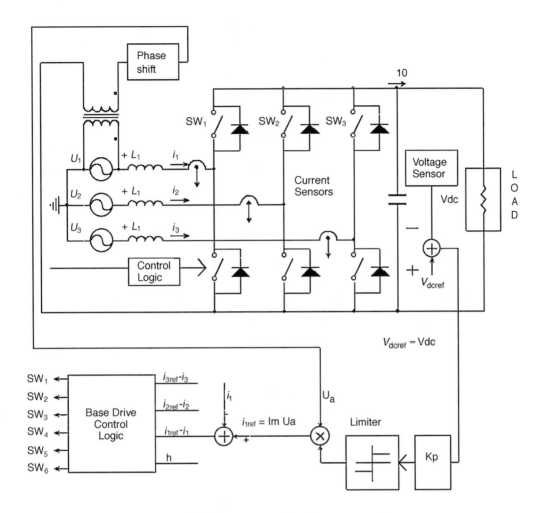

FIGURE 4.31 Indirect current control of the unity power factor boost-type rectifier.

template is directly proportional to the input voltage, resulting in a unity power factor. The line current is then controlled to track this reference. Current regulation is accomplished through the use of hysteresis controllers [5]. A proposed control method [4] is shown in Fig. 4.31.

To explain the closed-loop operation of the PWM boost-type rectifier, the switch matrix theory is used. The output current I_0 of the matrix converters is a function of the converter transfer function vector T and the input current vector i and is given by,

$$I_0 = Ti \tag{4.12}$$

The converter transfer function vector T is composed of three independent line-to-neutral switching functions: SW_1, SW_2, SW_3.

$$T = [SW_1 \quad SW_2 \quad SW_3] \tag{4.13}$$

The input current vector is given by

$$i = \begin{bmatrix} i_1 \\ i_2 \\ i_3 \end{bmatrix} \tag{4.14}$$

The line-to-neutral switching functions are balanced and are represented by their fundamental components only.

$$SW_1(t) = S_1 \sin(wt - \Theta)$$
$$SW_2(t) = S_1 \sin(wt - \Theta - 120°) \quad (4.15)$$
$$SW_3(t) = S_1 \sin(wt + 120° - \Theta)$$

Therefore, converter synthesized line-to-neutral voltages can be expressed as

$$V_{s1} = \frac{1}{2} V_{dc} S_1 \sin(wt - \Theta)$$
$$V_{s2} = \frac{1}{2} V_{dc} S_1 \sin(wt - 120° - \Theta) \quad (4.16)$$
$$V_{s3} = \frac{1}{2} V_{dc} S_1 \sin(wt + 120° - \Theta)$$

Eq. (4.15) shows the rectifier synthesized voltages. V_{dc} represents the output DC voltage.

In the time domain, the fundamental components of the three-phase input currents are given by

$$i_1(t) = I_1 \sin(wt - \varphi_1)$$
$$i_2(t) = I_1 \sin(wt - 120° - \varphi_1) \quad (4.17)$$
$$i_3(t) = I_1 \sin(wt + 120° - \varphi_1)$$

By combining Eqs. (4.12), (4.15), and (4.17), the output current $I_0(t)$ is obtained, given by

$$I_0(t) = I_1 \sin(wt - \varphi_1) S_1 \sin(wt - \Theta) + I_1 \sin(wt - 120° - \varphi_1) S_1 \sin(wt - 120° - \Theta)$$
$$+ I_1 \sin(wt + 120° - \varphi_1) S_1 \sin(wt + 120° - \Theta) \quad (4.18)$$

By using a trigonometric identity, $I_0(t)$ becomes

$$I_0(t) = \frac{3}{2} I_1 S_1 \cos(\Theta - \varphi_1) \quad (4.19)$$

Because the angle $(\Theta - \varphi_1)$ is constant for any set value of the input power factor, the output DC current, $I_0(t)$, is proportional to the magnitude of the input current, $I_1(t)$, and so is the output voltage, V_{dc}. For unity power factor control, angle φ_1 is equal to zero.

The output voltage, V_{dc} is

$$V_{dc} = RI_0 \quad (4.20)$$

$$(V_{dcref} - V_{dc}) = KI_1 \quad (4.21)$$

Figure 4.30 shows that the DC bus error, $(V_{dcref} - V_{dc})$, is used to set the reference for the input current magnitude. The input sinusoidal voltage, U_a, is multiplied by the DC bus error and it becomes a reference for the input current in phase 1. The reference value for current in phase 2 is phase-shifted by 120° with respect to the current in phase 1. Since the sum of three input currents is always zero, the reference for current in phase 3 is obtained from the following equation:

$$i_{3ref}(t) = -i_{1ref}(t) - i_{2ref}(t) \quad (4.22)$$

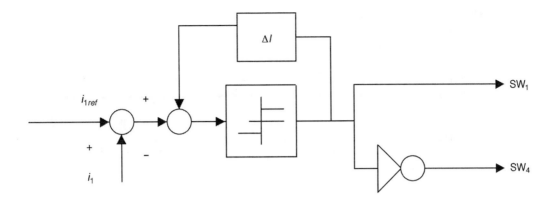

FIGURE 4.32 Hysteresis controller for one phase.

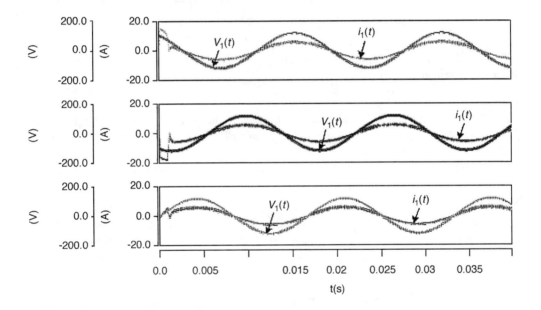

FIGURE 4.33 Input voltage and input current.

The input currents, $i_1(t)$, $i_2(t)$, $i_3(t)$ are measured and compared with the reference currents, $i_{1\text{ref}}(t)$, $i_{2\text{ref}}(t)$, $i_{3\text{ref}}(t)$. The error is fed to a comparator with a prescribed hysteresis band $2\Delta I$. Switching of the leg of the rectifier (SW$_1$ off and SW$_4$ on) occurs when the current attempts to exceed a set value corresponding to the desired current $i_{\text{ref}} + \Delta I$. The reverse switching (SW$_1$ on and SW$_4$ off) occurs when the current attempts to become less than $i_{\text{ref}} - \Delta I$. The hysteresis controller produces a very good quality waveform and is simple to implement. Unfortunately, with this type of control (hysteresis controller) the switching frequency does not remain constant but varies along different portions of the desired current. A hysteresis current controller for one phase is shown in Fig. 4.32.

Example

Simulate the three-phase PWM rectifier of Fig. 4.31 using SABER with the following parameters: $U_1 = U_2 = U_3 = 120$ V at 60 Hz, $L_1 = 1$ mH, $C = 100$ μF, and $R_{\text{load}} = 100$ Ω. Set the reference value for the output DC voltage, V_{dcref}, to 300 V and input a power factor to unity. Assume the switches to be ideal and choose the time step $\Delta t = 20$ μs.

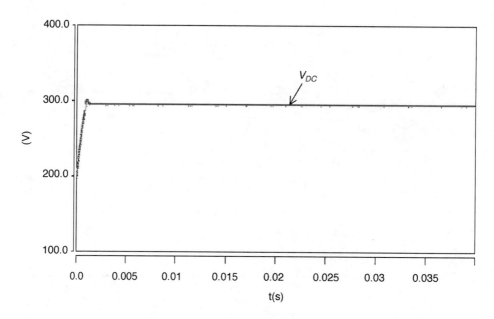

FIGURE 4.34 Output voltage V_{dc}.

Solution

The SABER program listing is included in the Appendix to this section, and the results are shown in Figs. 4.33 and 4.34. Figure 4.33 shows input currents as well as input voltages of the PWM boost-type rectifier shown in the Fig. 4.31. Figure 4.33 shows controlled output voltage, V_{dc}.

High-quality input currents and output DC voltage are obtained at a unity power factor.

References

1. L. Moran, P. D. Ziogas, and G. Joos, Design aspects of synchronous PWM rectifier-inverter system under unbalanced input voltages conditions, *IEEE Trans. Ind. Appl.*, 28(6), 1286–1293, Nov./Dec. 1992.
2. J. W. Wilson, The forced-commutated inverter as a regenerative rectifier, *IEEE Trans. Ind. Appl.*, IA-14(4), 335–340, July/Aug. 1978.
3. A. V. Stankovic and T. A. Lipo, A novel control method for input-output harmonic elimination of the PWM boost type rectifier under unbalanced operating conditions, *IEEE Trans. Power Electron.*, 16(5), Sept. 2001.
4. J. W. Dixon, and B. T. Ooi, Indirect current control of a unity power factor sinusoidal current boost type three-phase rectifier, *IEEE Trans. Ind. Electron.*, 35(4), 508–515, Nov./Dec. 1988.
5. D. M. Brod and D. W. Novotny, Current control of VSI-PWM inverters, *IEEE Trans. Ind. Appl.*, IA-21(4), 769–775, Nov./Dec. 1984.
6. T. A. Lipo, Recent progress and development of solid state AC motor drives, *IEEE Trans. Power Electron.*, 3(2), 105–117, April 1988.

Appendix

```
# Hysteresis controller-Balanced three-phase system
# Reference voltage = 300 V
# Small resistor is added in series with the inductor for current measurements
# igbt pwm rectifier
# main voltage sources
```

Rectifiers

```
# Program for balanced voltages and balanced impedances-Closed-loop solution
# Hysteresis Controller + P-I controller

vac.va n11 n17 = va = 120, theta = 0,     f = 60
vac.vb n12 n17 = va = 120, theta = -120,  f = 60
vac.vc n13 n17 = va = 120, theta = 120,   f = 60

r.111 n11 n111 = 0.001
r.222 n12 n122 = 0.001
r.333 n13 n133 = 0.001

v.vpl vp 0 = dc = 10
v.vmi vn 0 = dc = -10
v.vco vnn 0 = dc = 5
v.vcont vc1 0 = 7
# reference voltage source
vac.va1 na 0 = va = 6.6655, theta = -1.0803, f = 50
vac.vb1 nb 0 = va = 6.6655, theta = -121.0803, f = 50
vac.vc1 nc 0 = va = 6.6655, theta = -118.9197, f = 50
# sow generator
v.vsow vs 0 = tran = (pulse = (v1 = -10,v2 = 10,td = 0,tr = 250u,tf =
250u,pw = 0,per = 500u))
#r.r30 n30 n18 = 1
r.rou1 vp ou1 = 2.2k
r.rou2 vp ou2 = 2.2k
r.rou3 vp ou3 = 2.2k
r.routl1 vp out1 = 2.2k
r.rout22 vp out2 = 2.2k
r.rout33 vp out3 = 2.2k

#cmp1_oc.u1 na vs ou1 vp vn 0 = model = (vos=0, tdr=2n, tdf=2n, av=10K)
#cmp1_14.u2 ts1 vnn t4 vp 0 0 = model = (vos = 0, tdr = 2n, tdf = 2n)
#cmp1_oc.u3 nb vs ou2 vp vn 0 = model = (vos = 0, tdr = 2n, tdf = 2n, av = 10K)
#cmp1_14.u4 ts3 vnn t6 vp 0 0 = model = (vos = 0, tdr = 2n, tdf = 2n)
#cmp1_oc.u5 nc vs ou3 vp vn 0 = model = (vos = 0, tdr = 2n, tdf = 2n,av = 10K)
#cmp1_14.u6 ts5 vnn t2 vp 0 0 = model = (vos = 0, tdr = 2n, tdf = 2n)

#Comparison between two current signals
cmp1_oc.u20 n2000 n2001 out1 vp vn 0 = model = (vos = 0, tdr = 2n, tdf
= 2n, av = 100K)
cmp1_14.u21 out1 vnn t4 vp 0 0 = model = (vos = 0, tdr = 2n, tdf = 2n)
cmp1_oc.u22 n2002 n2003 out2 vp vn 0 = model = (vos = 0, tdr = 2n, tdf
= 2n, av = 100K)
cmp1_14.u23 out2 vnn t6 vp 0 0 = model = (vos = 0, tdr = 2n, tdf = 2n)
cmp1_oc.u24 n2004 n2005 out3 vp vn 0 = model = (vos = 0, tdr = 2n, tdf
= 2n, av = 100K)
cmp1_14.u25 out3 vnn t2 vp 0 0 = model = (vos = 0, tdr = 2n, tdf = 2n)
inv_14.in1 t4 t1
inv_14.in2 t6 t3
inv_14.in3 t2 t5
```

```
ide_d2an.vd1 t1 tt1 0 = model = (voh = 5,vol = 0,tr = 500n,tf = 250n)
ide_d2an.vd2 t2 tt2 0 = model = (voh = 5,vol = 0,tr = 500n,tf = 250n)
ide_d2an.vd3 t3 tt3 0 = model = (voh = 5,vol = 0,tr = 500n,tf = 250n)
ide_d2an.vd4 t4 tt4 0 = model = (voh = 5,vol = 0,tr = 500n,tf = 250n)
ide_d2an.vd5 t5 tt5 0 = model = (voh = 5,vol = 0,tr = 500n,tf = 250n)
ide_d2an.vd6 t6 tt6 0 = model = (voh = 5,vol = 0,tr = 500n,tf = 250n)

# optical isolation : voltage-controlled voltage-source
#clipv.vgg1 tt1 0 vg1 0 = a = 100, vomax = 10, vomin = 0
#clipv.vgg2 tt2 0 vg2 0 = a = 100, vomax = 10, vomin = 0
#clipv.vgg3 tt3 0 vg3 0 = a = 100, vomax = 10, vomin = 0
#clipv.vgg4 tt4 0 vg4 0 = a = 100, vomax = 10, vomin = 0
#clipv.vgg5 tt5 0 vg5 0 = a = 100, vomax = 10, vomin = 0
#clipv.vgg6 tt6 0 vg6 0 = a = 100, vomax = 10, vomin = 0
# three-phase bridge
l.la n111 n14 = 1m
l.lb n122 n15 = 1m
l.lc n133 n16 = 1m
sdr_thr2.s1 tt1 0 ps1 = vpull = 4.9,vdrop = 0.1,tdelay = 0m
sdr_thr2.s2 tt2 0 ps2 = vpull = 4.9,vdrop = 0.1,tdelay = 0m
sdr_thr2.s3 tt3 0 ps3 = vpull = 4.9,vdrop = 0.1,tdelay = 0m
sdr_thr2.s4 tt4 0 ps4 = vpull = 4.9,vdrop = 0.1,tdelay = 0m
sdr_thr2.s5 tt5 0 ps5 = vpull = 4.9,vdrop = 0.1,tdelay = 0m
sdr_thr2.s6 tt6 0 ps6 = vpull = 4.9,vdrop = 0.1,tdelay = 0m
sw_1pno.1 ps1 n10 n14 = ron = 0,roff = inf,tdbrk = 0,tdmk = 0,rfunc = cont
sw_1pno.2 ps2 n16 0 = ron = 0,roff = inf,tdbrk = 0,tdmk = 0,rfunc = cont
sw_1pno.3 ps3 n10 n15 = ron = 0,roff = inf,tdbrk = 0,tdmk = 0,rfunc = cont
sw_1pno.4 ps4 n14 0 = ron = 0,roff = inf,tdbrk = 0,tdmk = 0,rfunc = cont
sw_1pno.5 ps5 n10 n16 = ron = 0,roff = inf,tdbrk = 0,tdmk = 0,rfunc = cont
sw_1pno.6 ps6 n15 0 = ron = 0,roff = inf,tdbrk = 0,tdmk = 0,rfunc = cont

d.d1 n14 n10
d.d2 0 n16
d.d3 n15 n10
d.d4 0 n14
d.d5 n16 n10
d.d6 0 n15
r.r10 n10 0 = 100
c.cr n10 0 = 100u
# Actual output voltage converted to nonelectrical value output n50
elec2var.1 n10 0 n50 = 1
constant.1 n51 = 300
#Difference between the set value and the actual value
sum.1 n51 n50 n52 = 1,-1
#Proportional-integral controller
gain.96 n52 n98 = 1
#prop_int.1 n52 n98 = 1,0,0
#Limit the input for comparators
limit.1 n98 n99 = 15,1
# Multiply the gain with three voltages Ea,Eb, and Ec
# Set values for currents
```

Rectifiers 4-41

```
elec2var.2  n11 n17 n100 = 1/100
elec2var.3  n12 n17 n101 = 1/100
elec2var.4  n13 n17 n102 = 1/100
mult.2  n99 n100 n200 = 1
mult.3  n99 n101 n201 = 1
mult.4  n99 n102 n202 = 1
#Nonelectrical current values
#Current sensor
elec2var.12  n11 n111 n82 = 1000
elec2var.13  n12 n122 n83 = 1000
elec2var.14  n13 n133 n84 = 1000
#Current in phase 1. and current in phase 1 .Transformation
#to electrical quantities
var2elec.1  n200 n2000  0 = 1
var2elec.2  n82  n2001   0 = 1
# Conversion of peak values to nonelectrical quantities
var2elec.3  n201 n2002  0 = 1
var2elec.4  n83  n2003   0 = 1
#Addition current error + actual current
#Current and voltage in phase 3 Transformation to electrical
#quantities
var2elec.5  n202 n2004  0 = 1
var2elec.6  n84  n2005   0 = 1
```

5
Inverters

Michael Giesselmann
Texas Tech University

Attila Karpati
Budapest University of Technology and Economics

István Nagy
Budapest University of Technology and Economics

Dariusz Czarkowski
Polytechnic University, Brooklyn

Michael E. Ropp
South Dakota State University

Eric Walters
P. C. Krause and Associates

Oleg Wasynczuk
Purdue University

5.1 Overview 5-1
   Fundamental Issues • Single-Phase Inverters • Three-Phase Inverters • Multilevel Inverters • Line Commutated Inverters

5.2 DC-AC Conversion 5-8
   Basic DC-AC Converter Connections (Square-Wave Operation) • Control of the Output Voltage • Harmonics in the Output Voltage • Filtering of Output Voltage • Practical Realization of Basic Connections • Special Realizations (Application of Resonant Converter Techniques)

5.3 Resonant Converters 5-25
   Survey of Second-Order Resonant Circuits • Load Resonant Converters • Resonant Switch Converters • Resonant DC-Link Converters with ZVS

5.4 Series-Resonant Inverters 5-42
   Voltage-Source Series-Resonant Inverters • Voltage-Source Parallel-Resonant Inverters • Voltage-Source Series–Parallel-Resonant Inverters • Summary

5.5 Resonant DC-Link Inverters 5-56
   The Resonant DC-Link Inverter • The Parallel-Resonant DC-Link Inverter • Current Research Trends

5.6 Auxiliary Resonant Commutated Pole Inverters 5-67
   Losses in Hard-Switched Inverters • Analysis of ARCP Phase Leg • Analysis of ARCP H-Bridge • Analysis of ARCP Three-Phase Inverter • Summary

5.1 Overview

Michael Giesselmann

Inverters are used to create single or polyphase AC voltages from a DC supply. In the class of polyphase inverters, three-phase inverters are by far the largest group. A very large number of inverters are used for adjustable speed motor drives. The typical inverter for this application is a "hard-switched" voltage source inverter producing pulse-width modulated (PWM) signals with a sinusoidal fundamental [Holtz, 1992]. Recently research has shown detrimental effects on the windings and the bearings resulting from unfiltered PWM waveforms and recommend the use of filters [Cash and Habetler, 1998; Von Jouanne et al., 1996]. A very common application for single-phase inverters are so-called "uninterruptable power supplies" (UPS) for computers and other critical loads. Here, the output waveforms range from square wave to almost ideal sinusoids. UPS designs are classified as either "off-line" or "online". An off-line UPS will connect the load to the utility for most of the time and quickly switch over to the inverter if the utility fails. An online UPS will always feed the load from the inverter and switch the supply of the DC bus instead. Since the DC bus is heavily buffered with capacitors, the load sees virtually no disturbance if the power fails.

In addition to the very common hard-switched inverters, active research is being conducted on soft-switching techniques. Hard-switched inverters use controllable power semiconductors to connect an output terminal to a stable DC bus. On the other hand, soft switching inverters have an oscillating intermediate circuit and attempt to open and close the power switches under zero-voltage and or zero-current conditions.

A separate class of inverters are the line commutated inverters for multimegawatt power ratings, that use thyristors (also called silicon controlled rectifiers, SCRs). SCRs can only be turned "on" on command. After being turned on, the current in the device must approach zero in order to turn the device off. All other inverters are self-commutated, meaning that the power control devices can be turned on and off. Line commutated inverters need the presence of a stable utility voltage to function. They are used for DC-links between utilities, ultralong distance energy transport, and very large motor drives [Ahmed, 1999; Barton, 1994; Mohan et al., 1995; Rashid, 1993; Tarter, 1993]. However, the latter application is more and more taken over by modern hard-switched inverters including multilevel inverters [Brumsickle et al., 1998; Tolbert et al., 1999].

Modern inverters use insulated gate bipolar transistors (IGBTs) as the main power control devices [Mohan et al., 1995]. Besides IGBTs, power MOSFETs are also used especially for lower voltages, power ratings, and applications that require high efficiency and high switching frequency. In recent years, IGBTs, MOSFETs, and their control and protection circuitry have made remarkable progress. IGBTs are now available with voltage ratings of up to 3300 V and current ratings up to 1200 A. MOSFETs have achieved on-state resistances approaching a few milliohms. In addition to the devices, manufacturers today offer customized control circuitry that provides for electrical isolation, proper operation of the devices under normal operating conditions and protection from a variety of fault conditions [Mohan et al., 1995]. In addition, the industry provides good support for specialized passive devices such as capacitors and mechanical components such as low inductance bus-bar assemblies to facilitate the design of reliable inverters. In addition to the aforementioned inverters, a large number of special topologies are used. A good overview is given by Gottlieb [1984].

Fundamental Issues

Inverters fall in the class of power electronics circuits. The most widely accepted definition of a power electronics circuit is that the circuit is actually processing electric energy rather than information. The actual power level is not very important for the classification of a circuit as a power electronics circuit. One of the most important performance considerations of power electronics circuits, like inverters, is their energy conversion efficiency. The most important reason for demanding high efficiency is the problem of removing large amounts of heat from the power devices. Of course, the judicious use of energy is also paramount, especially if the inverter is fed from batteries such as in electric cars. For these reasons, inverters operate the power devices, which control the flow of energy, as switches. In the ideal case of a switching event, there would be no power loss in the switch since either the current in the switch is zero (switch open) or the voltage across the switch is zero (switch closed) and the power loss is computed as the product of both. In reality, there are two mechanisms that do create some losses, however; these are on-state losses and switching losses [Bird et al., 1993; Kassakian et al., 1991; Mohan et al., 1995; Rashid, 1993]. On-state losses are due to the fact that the voltage across the switch in the on state is not zero, but typically in the range of 1 to 2 V for IGBTs. For power MOSFETs, the on-state voltage is often in the same range, but it can be substantially below 0.5 V due to the fact that these devices have a purely resistive conduction channel and no fixed minimum saturation voltage like bipolar junction devices (IGBTs). The switching losses are the second major loss mechanism and are due to the fact that, during the turn-on and turn-off transition, current is flowing while voltage is present across the device. In order to minimize the switching losses, the individual transitions have to be rapid (tens to hundreds of nanoseconds) and the maximum switching frequency needs to be carefully considered.

In order to avoid audible noise being radiated from motor windings or transformers, most modern inverters operate at switching frequencies substantially above 10 kHz [Bose, 1992; 1996].

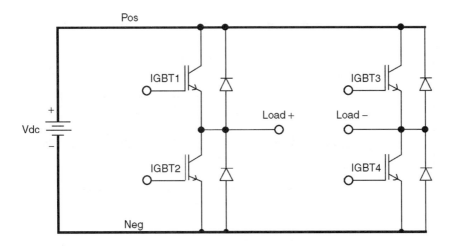

FIGURE 5.1 Topology of a single-phase, full-bridge inverter.

Single-Phase Inverters

Figure 5.1 shows the basic topology of a full-bridge inverter with single-phase output. This configuration is often called an H-bridge, due to the arrangement of the power switches and the load. The inverter can deliver and accept both real and reactive power. The inverter has two legs, left and right. Each leg consists of two power control devices (here IGBTs) connected in series. The load is connected between the midpoints of the two phase legs. Each power control device has a diode connected in antiparallel to it. The diodes provide an alternate path for the load current if the power switches are turned off. For example, if the lower IGBT in the left leg is conducting and carrying current towards the negative DC bus, this current would "commutate" into the diode across the upper IGBT of the left leg, if the lower IGBT is turned off. Control of the circuit is accomplished by varying the turn on time of the upper and lower IGBT of each inverter leg, with the provision of never turning on both at the same time, to avoid a short circuit of the DC bus. In fact, modern drivers will not allow this to happen, even if the controller would erroneously command both devices to be turned on. The controller will therefore alternate the turn on commands for the upper and lower switch, i.e., turn the upper switch on and the lower switch off, and vice versa. The driver circuit will typically add some additional blanking time (typically 500 to 1000 ns) during the switch transitions to avoid any overlap in the conduction intervals.

The controller will hereby control the duty cycle of the conduction phase of the switches. The average potential of the center-point of each leg will be given by the DC bus voltage multiplied by the duty cycle of the upper switch, if the negative side of the DC bus is used as a reference. If this duty cycle is modulated with a sinusoidal signal with a frequency that is much smaller than the switching frequency, the short-term average of the center-point potential will follow the modulation signal. "Short-term" in this context means a small fraction of the period of the fundamental output frequency to be produced by the inverter. For the single phase inverter, the modulation of the two legs are inverse of each other such that if the left leg has a large duty cycle for the upper switch, the right leg has a small one, etc. The output voltage is then given by Eq. (5.1) in which m_a is the modulation factor. The boundaries for m_a are for linear modulation. Values greater than 1 cause overmodulation and a noticeable increase in output voltage distortion.

$$V_{ac1}(t) = m_a \cdot V_{dc} \cdot \sin(\omega_1 \cdot t) \quad 0 \leq m_a \leq 1 \quad (5.1)$$

This voltage can be filtered using a LC low-pass filter. The voltage on the output of the filter will closely resemble the shape and frequency of the modulation signal. This means that the frequency, wave-shape, and amplitude of the inverter output voltage can all be controlled as long as the switching frequency is

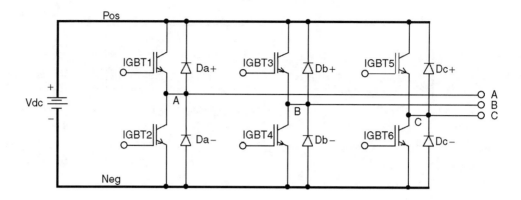

FIGURE 5.2 Topology of a three-phase inverter.

at least 25 to 100 times higher than the fundamental output frequency of the inverter [Holtz, 1992]. The actual generation of the PWM signals is mostly done using microcontrollers and digital signal processors (DSPs) [Bose, 1987].

Three-Phase Inverters

Figure 5.2 shows a three-phase inverter, which is the most commonly used topology in today's motor drives. The circuit is basically an extension of the H-bridge-style single-phase inverter, by an additional leg. The control strategy is similar to the control of the single-phase inverter, except that the reference signals for the different legs have a phase shift of 120° instead of 180° for the single-phase inverter. Due to this phase shift, the odd triplen harmonics (3rd, 9th, 15th, etc.) of the reference waveform for each leg are eliminated from the line-to-line output voltage [Shepherd and Zand, 1979; Rashid, 1993; Mohan et al., 1995; Novotny and Lipo, 1996]. The even-numbered harmonics are canceled as well if the waveforms are pure AC, which is usually the case. For linear modulation, the amplitude of the output voltage is reduced with respect to the input voltage of a three-phase rectifier feeding the DC bus by a factor given by Eq. (5.2).

$$\frac{3}{(2 \cdot \pi)} \cdot \sqrt{3} = 82.7\% \tag{5.2}$$

To compensate for this voltage reduction, the fact of the harmonics cancellation is sometimes used to boost the amplitudes of the output voltages by intentionally injecting a third harmonic component into the reference waveform of each phase leg [Mohan et al., 1995].

Figure 5.3 shows the typical output of a three-phase inverter during a startup transient into a typical motor load. This figure was created using circuit simulation. The upper graph shows the pulse-width modulated waveform between phases A and B, whereas the lower graph shows the currents in all three phases. It is obvious that the motor acts a low-pass filter for the applied PWM voltage and the current assumes the waveshape of the fundamental modulation signal with very small amounts of switching ripple.

Like the single-phase inverter based on the H-bridge topology, the inverter can deliver and accept both real and reactive power. In many cases, the DC bus is fed by a diode rectifier from the utility, which cannot pass power back to the AC input. The topology of a three-phase rectifier would be the same as shown in Fig. 5.2 with all IGBTs deleted.

A reversal of power flow in an inverter with a rectifier front end would lead to a steady rise of the DC bus voltage beyond permissible levels. If the power flow to the load is only reversing for brief periods of time, such as to brake a motor occasionally, the DC bus voltage could be limited by dissipating the power in a so-called brake resistor. To accommodate a brake resistor, inverter modules with an additional seventh

Inverters

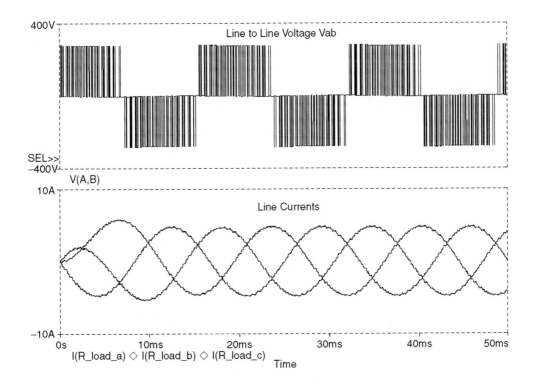

FIGURE 5.3 Typical waveforms of inverter voltages and currents.

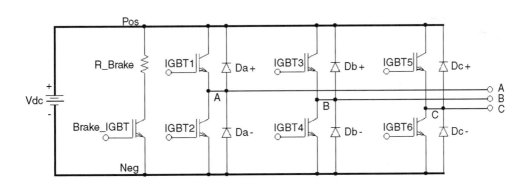

FIGURE 5.4 Topology of a three-phase inverter with brake-chopper IGBT.

IGBT (called "brake-chopper") are offered. This is shown in Fig. 5.4. For long-term regeneration, the rectifier can be replaced by an additional three-phase converter [Mohan et al., 1995]. This additional converter is often called a controlled synchronous rectifier. The additional converter including its controller is of course much more expensive than a simple rectifier, but with this arrangement bidirectional power flow can be achieved. In addition, the interface toward the utility system can be managed such that the real and reactive power that is drawn from or delivered to the utility can be independently controlled. Also, the harmonics content of the current in the utility link can be reduced to almost zero. The topology for an arrangement like this is shown in Fig. 5.5.

The inverter shown in Fig. 5.2 provides a three-phase voltage without a neutral point. A fourth leg can be added to provide a four-wire system with a neutral point. Likewise four-, five-, or *n*-phase inverters can be realized by simply adding the appropriate number of phase legs.

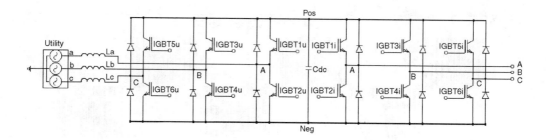

FIGURE 5.5 Topology of a three-phase inverter system for bidirectional power flow.

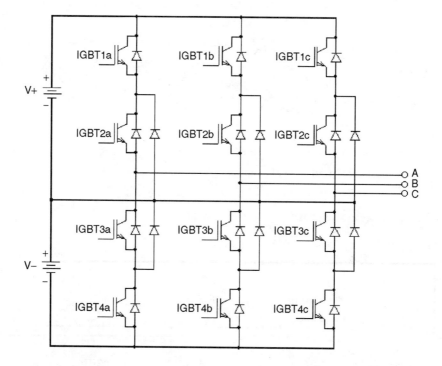

FIGURE 5.6 Topology of a three-level inverter.

As in single-phase inverters, the generation of the PWM control signals is done using modern microcontrollers and DSPs. These digital controllers are typically not only controlling just the inverter, but through the controlled synthesis of the appropriate voltages, motors and attached loads are controlled for high-performance dynamic response. The most commonly used control principle for superior dynamic response is called field-oriented or vector control [Bose, 1987; 1996; DeDonker and Novotny, 1988; Lorenz and Divan, 1990; Trzynadlowski, 1994].

Multilevel Inverters

Multilevel inverters are a class of inverters where a DC source with several tabs between the positive and negative terminal is present. The two main advantages of multilevel inverters are the higher voltage capability and the reduced harmonics content of the output waveform due to the multiple DC levels. The higher voltage capability is due to the fact that clamping diodes are used to limit the voltage stress on the IGBTs to the voltage differential between two tabs on the DC bus. Figure 5.6 shows the topology of a three-level inverter. Here, each phase leg consists of four IGBTs in series with additional antiparallel

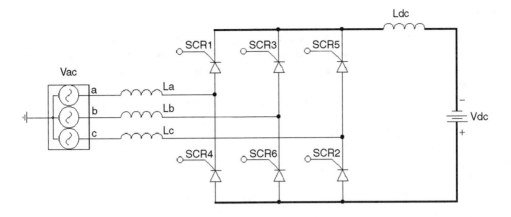

FIGURE 5.7 Line commutated converter in inverter mode.

and clamping diodes. The output is again at the center-point of the phase leg. The output of each phase can be connected to the top DC bus, the center connection of the DC supply, or the negative DC bus. This amounts to three distinct voltage levels for the voltage of each phase, which explains the name of the circuit. It turns out that the resulting line-to-line voltage has five distinct levels in a three-phase inverter.

Line-Commutated Inverters

Figure 5.7 shows the topology of a line commutated inverter. In Fig. 5.7 the SCRs are numbered according to their firing sequence. The circuit can operate both as a rectifier and an inverter. The mode of operation is controlled by the firing angle of the SCRs in the circuit [Ahmed, 1999; Barton, 1994; Mohan et al., 1995]. The reference value for the firing angle α is the instant when the voltage across each SCR becomes positive; i.e., when an uncontrolled diode would turn on. This time corresponds to 30° past the positive going zero crossing of each phase. By delaying the turn-on angle α more than 90° past this instant, the polarity of the average DC bus voltage reverses and the circuit enters the inverter mode. The DC source in Fig. 5.7 shows the polarity of the DC voltage for inverter operation. The firing delay angle corresponds to the phase of the utility voltage. The maximum delay angle must be limited to less than 180°, to provide enough time for the next SCR in the sequence to acquire the load current. Equation (5.3) gives the value of the DC output voltage of the converter as a function of the delay angle α and the DC current I_{dc}, which is considered constant.

$$V_{dc} = \frac{3}{\pi}(\sqrt{2} \cdot V_{LL} \cdot \cos(\alpha) - \omega \cdot L_S \cdot I_{dc}) \tag{5.3}$$

V_{LL} is the rms value of the AC line-to-line voltage, ω is the radian frequency of the AC voltage, and L_s is the value of the inductors L_a, L_b, and L_c in Fig. 5.7. Line commutated inverters have a negative impact on the utility voltage and a relatively low total power factor. Equation (5.4) gives an estimate of the total power factor of the circuit shown in Fig. 5.7 for constant DC current and negligible AC line reactors.

$$PF = \frac{3}{\pi} \cdot \cos(\alpha) \tag{5.4}$$

References

Ahmed, A., *Power Electronics for Technology,* Prentice-Hall, Upper Saddle River, NJ, 1999.
Barton, T. H., *Rectifiers, Cycloconverters, and AC Controllers,* Oxford University Press, New York, 1994.
Bird, B. M., King, K. G., and Pedder, D. A. G., *An Introduction to Power Electronics,* 2nd ed., John Wiley & Sons, New York, 1993.
Bose, B. K., *Modern Power Electronics, Evolution, Technology, and Applications,* IEEE Press, Piscataway, NJ, 1992.
Bose, B. K., *Microcomputer Control of Power Electronics and Drives,* IEEE Press, Piscataway, NJ, 1987.
Bose, B. K., *Power Electronics and Variable Frequency Drives,* IEEE Press, Piscataway, NJ, 1996.
Brumsickle, W. E., Divan, D. M., and Lipo, T. A., Reduced switching stress in high-voltage IGBT inverters via a three-level structure, *IEEE-APEC 2.* 544–550, Feb. 1998.
Cash, M. A. and Habetler, T. G., Insulation failure prediction in induction machines using line-neutral voltages, *IEEE Trans. Ind. Appl.,* 34(6), 1234–1239, Nov./Dec. 1998.
De Donker, R. and Novotny, D. W., The universal field-oriented controller, *Conf. Rec. IEEE-IAS* 1988, 450–456.
Gottlieb, I. M., *Power Supplies, Switching Regulators, Inverters and Converters,* TAB Books, Blue Ridge Summit, PA, 1984.
Holtz, J., Pulsewidth modulation—a survey, *IEEE Trans. Ind. Electr.,* 39(5), 410–420, 1992.
Kassakian, J. G., Schlecht, M. F., and Verghese, G. C., *Principles of Power Electronics,* Addison-Wesley, Reading, MA, 1991.
Lorenz, R. D. and Divan, D. M., Dynamic analysis and experimental evaluation of delta modulators for field oriented induction machines, *IEEE Trans. Ind. Appl.,* 26(2), 296–301, 1990.
Mohan, N., Undeland, T., and Robbins, W., *Power Electronics: Converters, Applications, and Design,* 2nd ed., John Wiley & Sons, New York, 1995.
Novotny, D. W. and Lipo, T. A., *Vector Control and Dynamics of AC Drives,* Oxford Science Publications, New York, 1996.
Rashid, M. H., *Power Electronics, Circuits, Devices, and Applications,* Prentice-Hall, Englewood Cliffs, NJ, 1993.
Shepherd, W. and Zand, P., *Energy Flow and Power Factor in Nonsinusoidal Circuits,* Cambridge University Press, London, 1979.
Tarter, R. E., *Solid State Power Conversion Handbook,* John Wiley & Sons, New York, 1993.
Tolbert, L. M., Peng, F. Z., and Habetler, T. G., Multilevel converters for large electric drives, *IEEE Trans. Ind. Appl.,* 35(1), 36–44, Jan./Feb. 1999.
Trzynadlowski, A. M., *The Field Orientation Principle in Control of Induction Motors,* Kluwer Academic Publishers, Dordrecht, the Netherlands, 1994.
Von Jouanne, A., Rendusara, D., Enjeti, P., and Gray, W., Filtering techniques to minimize the effect of long motor leads on PWM inverter fed AC motor drive systems, *IEEE Trans. Ind. Appl.,* 32(4), 919–926, July/Aug. 1996.

5.2 DC-AC Conversion

Attila Karpati

The DC-AC converters, also known as inverters and shown in Fig. 5.8, produce an AC voltage from a DC input voltage. The frequency and amplitude produced are generally variable. In practice, inverters with both single-phase and three-phase outputs are used, but other phase numbers are also possible. Electric power usually flows from the DC to the AC terminal, but in some cases reverse power flow is possible. These types of inverters, where the input is a DC voltage source, are also known as voltage-source inverters (VSI). The other type of inverter is the current-source inverters (CSI), where the DC input is a DC current source. These converters are used primarily in high-power AC motor drives.

Inverters

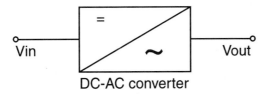

FIGURE 5.8 DC-AC converter.

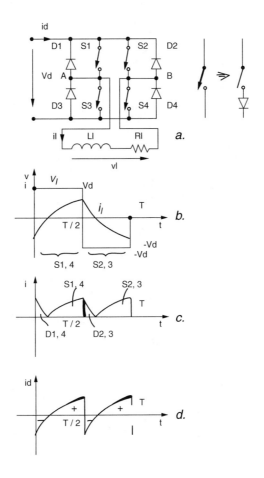

FIGURE 5.9 Voltage-source, single-phase, full-bridge inverter connection.

Basic DC-AC Converter Connections (Square-Wave Operation)

This section presents a short summary of the main types of voltage-source DC-AC converter connections and a brief description of their functions. At the end of this subsection is also given a current-source converter configuration with its short description. It is assumed that the circuits incorporate ideal semiconductor switches.

The most frequently used types of single-phase inverters are full-bridge inverters, as shown in Fig. 5.9a, the half-bridge inverters, as shown in Fig. 5.10a, and push-pull inverters, as shown in Fig. 5.11a.

The switching sequences for the switches and the most important time functions for the full-bridge, half-bridge, and push-pull inverters during square-wave operation can be seen in Figs. 5.9 through 5.11.

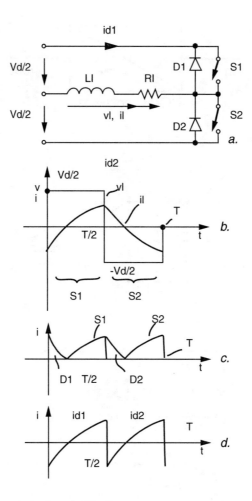

FIGURE 5.10 Voltage-source, single-phase, half-bridge inverter connection.

It is assumed that the load on the output consists of a series resistance and inductance. The three-phase basic inverter configuration is the full-bridge connection shown in Fig. 5.12a. The loads are assumed to be symmetrical inductances in the three phases. The switching sequences of the switches and the most important time functions at square-wave operation are demonstrated in Fig. 5.12b through g.

One can draw the following conclusions from these figures:

- The output voltage is nonsinusoidal.
- Due to the presence of the freewheeling diodes, the output voltage is independent of the direction of the load current, and is only dependent on the on and off state of the switches.
- The semiconductor switches and freewheeling diodes form two rectifiers. They are connected in inverse parallel. The semiconductor switches make the energy flow from the DC side to the AC side possible. The freewheeling diodes allow the reverse situation.
- Accordingly, the freewheeling diodes are necessary if the converter outputs are connected to loads, which require either reactive power or effective power feedback. In the case of reactive power, the direction of the power flow in the converter changes periodically (see the i_B currents in Figs. 5.9 through 5.12).

A three-phase current-source inverter configuration is shown in Fig. 5.13a. The switching sequences of the switches and the most important time functions are demonstrated in Fig. 5.13b.

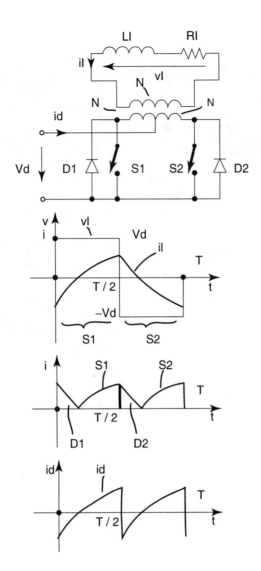

FIGURE 5.11 Voltage-source, single-phase, push-pull inverter connection.

Because of jumps in the output current, capacitors must be used, which are connected in parallel to the load. In most cases, the current-source inverters use thyristors as switching devices, and the aforementioned capacitances are the energy storage elements of the quenching circuits.

Control of the Output Voltage

In voltage-source inverters, the output voltage is controlled by following methods:

- In inverters with square-wave operation, voltage changes on the DC side
- Voltage cancellation, which is feasible in single-phase full-bridge inverters
- Sinusoidal pulse-width modulation (sinusoidal PWM), with bipolar and unipolar voltage switching
- Programmed harmonic elimination switching
- Tolerance band control
- Fixed-frequency control

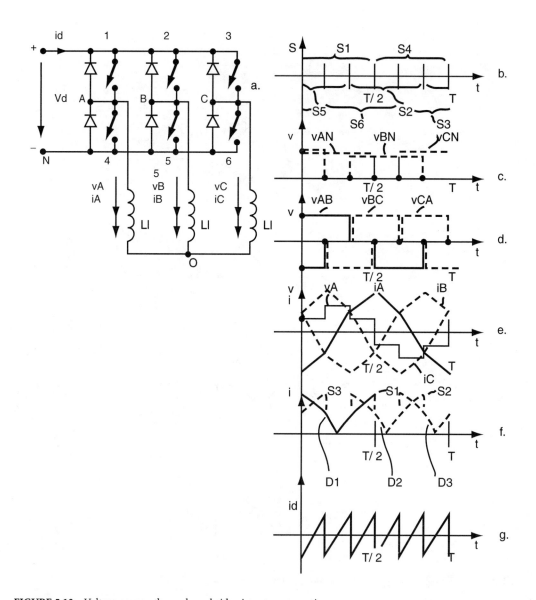

FIGURE 5.12 Voltage-source, three-phase, bridge inverter connection.

In current-source inverters, the output is controlled by changing the input DC voltage. In most cases the DC voltage is changed by a controlled rectifier or DC chopper. In voltage cancellation the T_{on} and T_{off} times of the switches in the two legs of the full-bridge connection are shifted to one another as shown in Fig. 5.14. The rms value of the AC voltage can be changed between 0 and a maximum value, as defined by the square-wave operation. This is a very simple method, in which the switching frequency of the semiconductor elements is equal to the output frequency, but the harmonic content of the AC side voltage is rather high. Therefore, it is the preferred method used in converters with high-frequency output. At lower output frequency, i.e., at 60 Hz, other methods are used, and the switching frequency of the semiconductors is much higher than the output frequency. This method allows for extensive reduction of the harmonic content in the output voltage or current. In inverter circuits the sinusoidal PWM is used to minimize the output harmonic content. The basic principle employed in a one-phase half-bridge converter with bipolar voltage switching is demonstrated in Fig. 5.15. The switches S_+ and S_- work with an internal frequency, which is much higher than the output frequency. The on and off state of the

Inverters

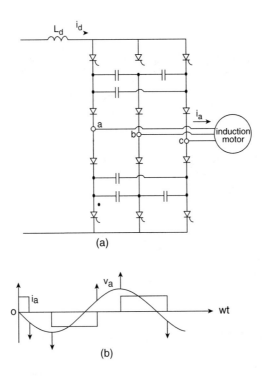

FIGURE 5.13 Three-phase current-source inverter circuit.

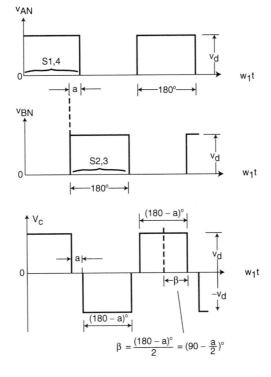

FIGURE 5.14 Voltage cancellation by full-bridge connection.

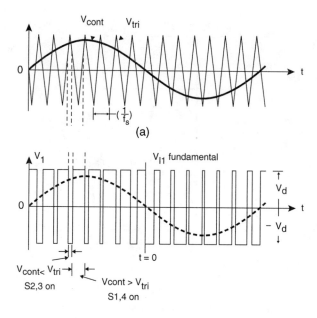

FIGURE 5.15 Pulse-width modulation with bipolar voltage switching.

switches is determined by the crossover points of the triangular comparison signal V_{tri} and the sinusoidal control signal V_{cont}. The sinusoidal control signal causes constant changes in the duty ratio of the switches S_+ and S_- during the half-period of the output so that the harmonic content of the output is minimized. The output voltage or current can be changed by varying V_{cont}.

The most important definitions are as follows:

The amplitude-modulation ratio: $m_a = V_{contM}/V_{tri}$
The frequency-modulation ratio: $m_f = f_s/f_1$

where f_s is the internal switching frequency and f_1 is the frequency of the fundamental of the output.

At small m_f ($m_f \leq 21$), synchronous PWM should be used, namely, m_f should be an integer and V_{cont} and V_{tri} are synchronized to one another. (Asynchronous PWM in the $m_f \neq$ integer output produces subharmonics of the fundamental frequency, which are generally undesirable.)

At large values of m_f ($m_f > 21$), the amplitudes of subharmonics caused by asynchronous PWM are small. Therefore asynchronous PWM may be used, except in AC motor drives, if the frequency approaches zero. In this case, small subharmonic voltages can also occur as well as high and undesirable currents.

In the case of $m_a < 1.0$, the sinusoidal PWM operates in the linear range. The amplitude of the fundamental frequency component varies linearly with m_a. In this range, the maximum value of the fundamental is less than the allowable maximum, which is achieved by overmodulation, with $m_a > 1$. In this range, the relation is not linear between m_a and the fundamental. The allowable maximum value is given by square-wave operation. The relation between the fundamental and m_a is illustrated in Fig. 5.16.

The operating principles for sinusoidal PWM with unipolar voltage switching for a full-bridge inverter can be seen in Fig. 5.17. The two legs of the inverter are not switched simultaneously, and are controlled separately. For this reason, two control signals, V_{cont} and $-V_{cont}$ are used. The advantage of this method is that of "effectively" doubling the switching frequency, which results from the cancellation of certain harmonic components.

The operating principles for sinusoidal PWM with three-phase inverters are shown in Fig. 5.18. To control the three legs of the bridge connection, three control signals are used, $V_{cont,A}$ $V_{cont,B}$ and $V_{cont,C}$. The fundamental of the output as a function of m_a is given in Fig. 5.19.

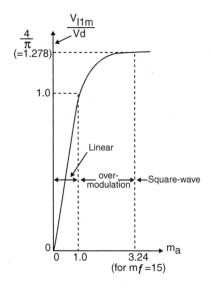

FIGURE 5.16 Voltage control by varying m_a.

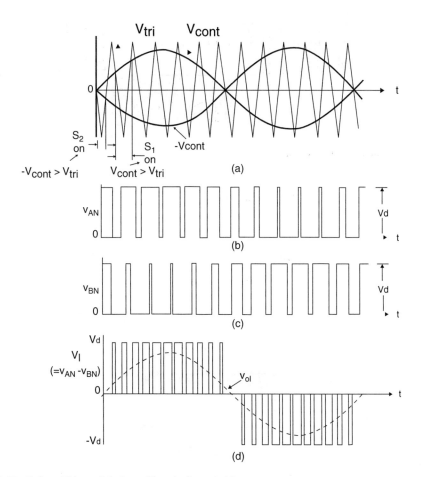

FIGURE 5.17 Pulse-width modulation with unipolar switching.

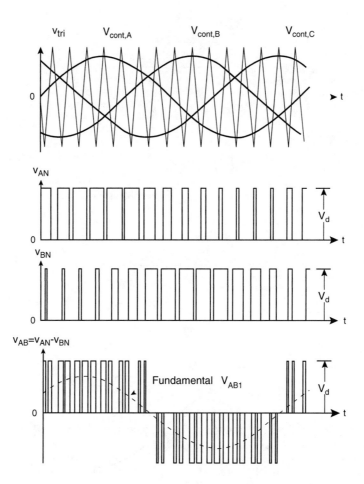

FIGURE 5.18 Three-phase PWM waveforms.

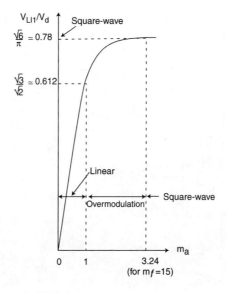

FIGURE 5.19 Three-phase inverter $V_{Li1} = V_{Li1}\,(m_a)$.

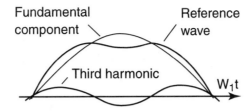

FIGURE 5.20 Reference wave with third harmonic.

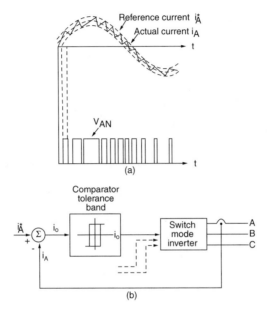

FIGURE 5.21 Tolerance-band current control.

In the case of overmodulation, low-order harmonics appear, and therefore the above-mentioned natural sampling method is used only until the output fundamental voltage becomes equal to 78.5% of its maximum possible value. For a three-phase system, this situation can be improved by using a reference wave with the addition of the third harmonic, as shown in Fig. 5.20. In the output phase voltages, the third harmonics have in all of the phases the same time functions (zero sequence components), and, therefore, cannot produce current.

For programmed harmonic elimination switching, the moments of the semiconductor switching are calculated so that the lower harmonics will be eliminated. This method permits the elimination of undesirable lower harmonics, without a very high resulting switching frequency. Therefore, the power losses in the converter can be reduced.

The principles of tolerance band control (following control) can be seen in Fig. 5.21. The difference between the reference value and the actual value will be directed to one comparator with a tolerance band. The output of the comparator controls the switches in the inverter so that the above-mentioned difference will not be greater than that required. At the sinusoidal output, the reference value has the required sinusoidal form, and the actual value fluctuates along the curve. The switching frequency varies in a large interval and depends on the AC side load and the input DC voltage. The controlled variable can be the output voltage or current.

The principles of fixed-frequency control are shown in Fig. 5.22. The difference between the reference value and actual value will be directed to a regulator. The regulator output is the control signal, V_{contr}, which is compared to a triangular waveform, V_{tri}, with the switching frequency f_s. The switching moments

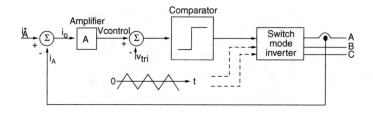

FIGURE 5.22 Fixed-frequency current control.

are specified by the crossover points of the two signals. This type of control circuit is also used in following control. At the sinusoidal output, the reference value has the required sinusoidal form.

Harmonics in the Output Voltage

The harmonics in the output voltage depend primarily on the control method for the output voltage. For inverters with square-wave operation, the harmonic content is constant. For single-phase inverters the harmonic numbers are

$$n = 1, 3, 5, 7, \ldots$$

The amplitude of the nth harmonic can be calculated for the full-bridge inverter by the following formula:

$$V_{onrmso} = 0.9 V_d/n$$

For three-phase inverters the harmonic numbers are

$$n = 6c \pm 1, \quad \text{where } c = 1, 2, 3, \ldots$$

The rms value of the line voltage can be calculated as follows:

$$V_{onrms} = 0.78 V_d/n$$

For voltage cancellation in a single-phase full-bridge inverter, the harmonic numbers are the same as those for square wave operation, but the amplitude of the output voltage harmonics varies with the control angle in the following form:

$$V_{onrms} = V_{onrmso} \sin(n \cdot \beta)$$

For sinusoidal PWM with bipolar voltage switching and $m_a \leq 1.0$, the harmonic numbers are

$$n = jm_f \pm k$$

where the fundamental frequency is denoted by $n = 1$. For odd values of j, only even values of k are possible, and vice versa.

The harmonic spectrum is presented in Fig. 5.23. In case of overmodulation, the harmonic content will be higher, as shown in Fig. 5.24.

For sinusoidal PWM with unipolar voltage switching, the harmonic content is less than that for bipolar voltage switching, due to the cancellation of some harmonics, as shown in Fig. 5.25.

For sinusoidal PWM with three-phase inverters, the harmonic spectrum of the output voltage is given in Fig. 5.26.

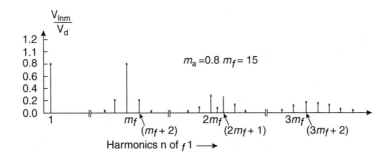

FIGURE 5.23 Single-phase full-bridge.

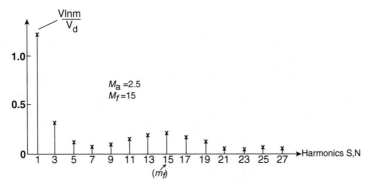

Harmonics due to overmodulation; figure is drawn for $m_a = 2.5$ and $m_f = 15$ (single-phase full-bridge)

FIGURE 5.24 Single-phase full-bridge with bipolar switching, harmonic spectrum.

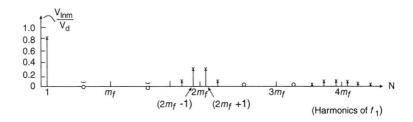

FIGURE 5.25 Single-phase full-bridge with unipolar switching, harmonic spectrum.

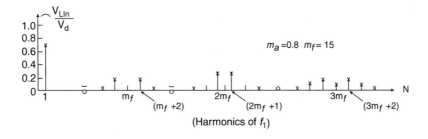

FIGURE 5.26 Three-phase PWM, harmonic spectrum.

For programmed harmonic elimination switching the harmonics of lower order are eliminated. In three-phase bridge inverters, the 5th, 7th, 11th, and 13th harmonics are usually eliminated.

For tolerance band control, the switching frequency varies in a large interval. Therefore, the frequencies of the harmonic spectrum and the harmonic amplitudes are not constant.

Filtering of Output Voltage

As was demonstrated in the previous section, the output voltage is not sinusoidal. If AC voltage with low distortion is necessary, and the output frequency is constant (for example, in uninterruptible power supplies), output voltage filter circuits are used to decrease distortion. Reducing the internal frequency of the inverters results in greater filtering problems. The solution of the filtering problems is most difficult in line frequency inverters with voltage cancellation. In this case, the use of large and complicated output filters is necessary, as shown in Fig. 5.27. The basic principle is simple. The filter circuit is a frequency-dependent voltage divider. Under ideal conditions, the transfer ratio (V_{out}/V_{in}) for the fundamental is equal to one, and for the other harmonics it is equal to zero. In the basic version of the filter circuit (Fig. 5.27) the ideal behavior is approximated using a series resonant circuit in the input of the filter, and a parallel resonant circuit in the output. Both circuits are tuned to the fundamental frequency. Therefore, the transfer ratio for the fundamental is equal to one, and the inverter is not loaded with the reactive power of the parallel output capacitance. For the harmonics, the series impedance increases with frequency, and the parallel impedance decreases. This effect ensures a certain reduction in the harmonic voltages. If this reduction is not adequate, series resonant circuits, which are tuned to various harmonic frequencies, will be connected in parallel with the output. The resulting output will be short-circuited at the chosen frequencies. The dynamic behavior of this filter circuit is not good at load jumps because of the large number of energy-storage elements. Since modern converter circuits are used with a high internal frequency (e.g., 20 kHz at PWM), the necessary filter circuit is simpler. The simplified filter circuit in Fig. 5.28 is currently utilized. If an output transformer is also used, the transformer values are calculated such that the series inductance of the filter circuit is given by the transformer's leakage inductance and the parallel inductance is equal to the transformer's magnetizing inductance. To ensure the required magnetizing inductance, the application of an air gap in the iron core is necessary. Using modern converter techniques, low distortion levels (a few percent) and very good dynamic behavior (5 to 10% overshoot at load jumps) can be achieved.

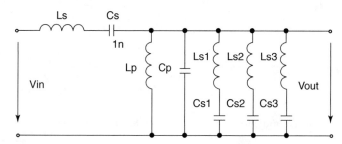

FIGURE 5.27 Basic filter circuit.

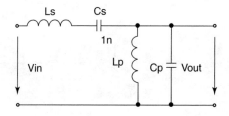

FIGURE 5.28 Simplified filter circuit.

Inverters

Practical Realization of Basic Connections

Bipolar transistors, IGBTs, and FETs are generally used in modern converters with ≤100 kW output power. At higher power, the application of GTOs and thyristors are common. If thyristors are used, the connection must be completed by quenching circuits to turn off the current conducting thyristor. The energy necessary to turn off the thyristor is stored in capacitors. The basic connections with thyristors are used for frequencies up to 1 to 2 kHz. With IGBTs a frequency of ~20 kHz is attainable. If FETs are used, 100 kHz frequency is normal, but equipment with 500 kHz frequency is also possible.

Special Realizations (Application of Resonant Converter Techniques)

Certain types of DC-AC converters use series or parallel resonant circuits. They are known as resonant converters, which can be subdivided into the following groups:

- Load resonant converters, i.e., current-source parallel-resonant and voltage-source series resonant DC-to-AC inverters
- Resonant switch converters; ZVS-CV DC-to-AC inverters
- Resonant converter connections, used in electrical drives; auxiliary resonant-commutated pole inverters; parallel- and series-resonant DC-link converters; active clamped parallel-resonant DC-link inverters; parallel- and series-resonant AC-link converters

In load resonant converters the load is completed by capacitance to a resonant circuit. In the current-source inverters a capacitance is connected in parallel with the load. The circuit and time functions are in steady state as shown in Fig. 5.29. The connection operates as a line-commutated circuit in the inverter working mode; however, the voltage on the parallel resonant circuit ensures commutation. The power can be controlled by changing the value of V_d. A controlled rectifier is generally used for this purpose. This connection is typically applied in induction heating.

For voltage-source inverters the capacitance is connected in series with the load. If converter thyristors are used, the circuit and the time functions in steady-state are shown in Fig. 5.30. The quenching of the thyristor is ensured by the voltage drop across the freewheeling diode, which is connected to the thyristor in inverse parallel. The output frequency is less than the series resonant frequency. The output power is usually controlled by changing the output frequency. If semiconductor elements, e.g., IGBT, FET etc., which can be turned off by a gate signal are used, the output frequency can be equal to or greater than the resonant frequency. In the latter case, the switching losses are smaller. The output power can be controlled by changing the output frequency or voltage, V_1. In the latter case, voltage cancellation can be utilized.

In resonant switch converters, resonant circuits are connected to the semiconductors to ensure soft switching and to reduce the switching losses. In practice, zero current switching (ZCS) and zero voltage switching (ZVS) are possible. Because the voltage on the semiconductors increases with simple ZVS,

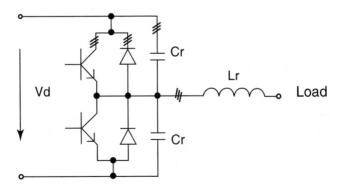

FIGURE 5.29 ZVS-CV DC-to-AC inverter.

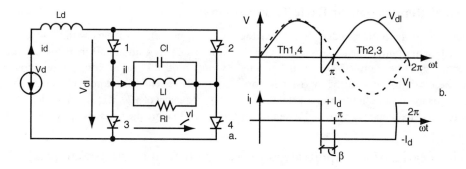

FIGURE 5.30 Current source parallel resonant inverter.

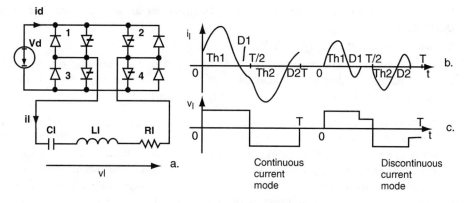

FIGURE 5.31 Voltage-source series resonant converter.

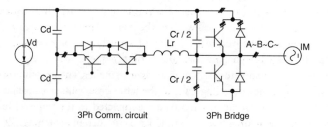

FIGURE 5.32 Auxiliary resonant commutated pole inverter.

clamped voltage (CV) versions are used. A simplified version of a three-phase ZVS-CV DC-to-AC inverter is shown in Fig. 5.31. The transistor's switching is done at zero voltage on the capacitances, which are connected in parallel to the transistors.

The most important types of inverters used with electrical drives are the three-phase bridge connections. Solutions for the realization of soft switching are briefly described below. The auxiliary resonant-commutated pole inverter is shown in Fig. 5.32. It is a traditional voltage-source inverter, which contains switched resonant circuits, with components L_r, C_r, $T_{r,1,2}$, for each leg. The resonant circuits and the switch control ensure that the additional circuits operate only during switching in the main bridge, which guarantee soft switching for the semiconductor elements.

The parallel resonant DC-link converter is shown in Fig. 5.33. An AC voltage on the input DC voltage is superimposed using the resonant circuit L_r, C_r, so that V_r will be periodically zero. When V_r equals zero, the semiconductor elements in the output bridge are switched (ZVS) which results in soft switching. The resonant circuit is excited by the periodic common turn-on of all elements in the output bridge.

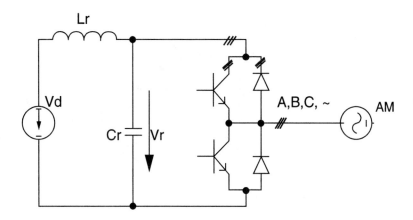

FIGURE 5.33 Parallel resonant DC-link converter.

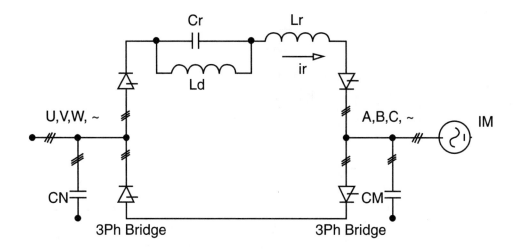

FIGURE 5.34 Series resonant DC-link converter.

The series resonant DC-link converter is shown in Fig. 5.34. It is a traditional current-source inverter that contains a series resonant circuit. Therefore, an AC component is superimposed on the DC current which ensures that the current in the bridges will be periodically zero. The semiconductor elements are switched when the current is equal to zero (ZCS). A suitable control strategy ensures that the network and output current are approximately sinusoidal. Thyristors or GTOs are used as semiconductor elements that can also operate in the reverse voltage direction.

The active clamped parallel resonant DC-link inverter is shown in Fig. 5.35. (It is a parallel resonant DC-link inverter containing a clamping circuit, C_{cl}, T_{cl}, to limit the maximum voltage on the semiconductor elements.)

The AC-link resonant converter is a special type of converter. The parallel resonant AC-link converter is shown in Fig. 5.36. Suitable operation of the switches and parallel resonant circuit ensure that there is a high-frequency AC voltage on the input of the output bridge. The output voltage with the required frequency and small harmonic content is defined by suitably linking the half-periods of the input pulses.

The series resonant AC-link converter is shown in Fig. 5.37. The suitable operation of the switches and series resonant circuit ensures that a high-frequency AC current is present in the input of the output bridge. The output current with the required frequency and small harmonic content is defined by suitably linking the half-periods of the input pulses.

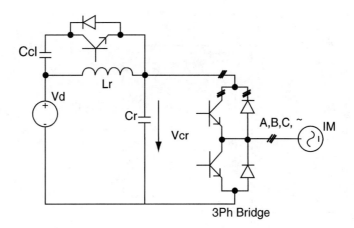

FIGURE 5.35 Active clamped parallel resonant DC-link converter.

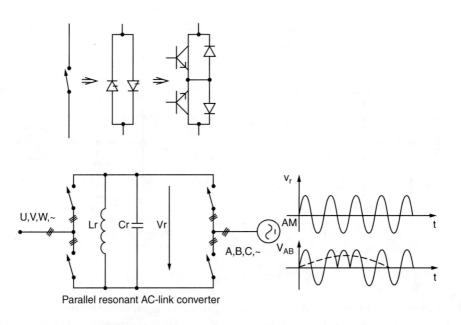

FIGURE 5.36 Parallel resonant AC-link converter.

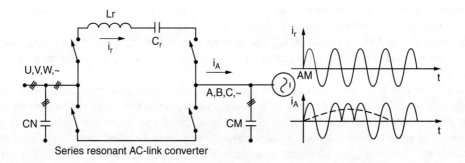

FIGURE 5.37 Series resonant AC-link converter.

5.3 Resonant Converters

István Nagy

Resonant converters connect a DC system to an AC system or another DC system and control both the power transfer between them and the output voltage or current. They are used in such applications as: induction heating, very high frequency DC-DC power supplies, sonar transmitters, ballasts for fluorescent lamps, power supplies for laser cutting machines, ultrasonic generators, etc.

There are some common features characterizing the behavior of most, or at least some, of these elements. DC-DC and DC-AC converters have two basic shortcomings when their switches are operating in the switch mode. During the turn-on and turn-off time, high current and voltage appear simultaneously in and across the switches producing high power losses in them, that is, high switching stresses. The power loss increases linearly with the switching frequency. To ensure reasonable efficiency of the power conversion, the switching frequency has to be kept under a certain maximum value. The second shortcoming in a switching mode operation is the electromagnetic interference (EMI) generated by the large dv/dt and di/dt values of the switching variables. The drawbacks have been accentuated by the trend which is pushing the switching frequency to higher and higher range in order to reduce the converter size and weight.

The resonant converters can minimize these shortcomings. The switches in resonant converters create a square-wave-like voltage or current pulse train with or without a DC component. A resonant L-C circuit is always incorporated. Its resonant frequency could be close to the switching frequency or could deviate substantially. If the resonant L-C circuit is tuned to approximately the switching frequency, the unwanted harmonics are removed by the circuit. In both cases the variation of the switching frequency is one of the means for controlling the output power and voltage.

The advantages of resonant converters are derived from their L-C circuit and they are as follows: sinusoidal-like wave shapes, inherent filter action, reduced dv/dt and di/dt and EMI, facilitation of the turn-off process by providing zero current crossing for the switches and output power and voltage control by changing the switching frequency. In addition, some resonant converters e.g., quasi-resonant converters, can accomplish zero current and/or zero voltage across the switches at the switching instant and reduce substantially the switching losses. The literature categorizes these converters as hard switched and soft switched converters. Unlike hard switched converters the switches in soft switched converters, quasi-resonant and some resonant converters are subjected to much lower switching stresses. Note that not all resonant converters offer zero current and/or zero voltage switchings, that is, reduced switching power losses. In return for these advantageous features, the switches are subjected to higher forward currents and reverse voltages than they would encounter in a nonresonant configuration of the same power. The variation in the operation frequency can be another drawback.

First, a short review of the two basic resonant circuits, series and parallel, are given. Then the following three types of resonant converters are discussed:

- Load resonant converters
- Resonant switch converters
- Resonant DC-link converters

Survey of Second-Order Resonant Circuits

The parallel resonant circuit is the dual of the series-resonant circuit (Fig. 5.38). The series (parallel) circuit is driven by a voltage (current) source. The analog variables for the voltages and currents are the corresponding currents and voltages (Fig. 5.38). Kirchhoff's voltage law for the series circuit

$$v_i = v_L + v_R + v_C = i_i\left(sL + R + \frac{1}{sC}\right) \tag{5.5}$$

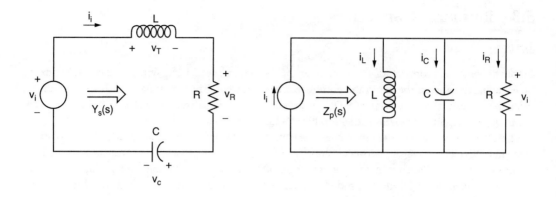

FIGURE 5.38 Dual circuits.

and Kirchhoff's current law for the parallel circuit

$$i_i = i_L + i_R + i_C = v_i\left(\frac{1}{sL} + \frac{1}{R} + sC\right) \quad (5.6)$$

have to be used. The analog parameters for the impedances are the corresponding admittances (Fig. 5.38). The input current for the series circuit is

$$i_i = Y_s(s)v_i = \frac{1}{Z_s(s)}v_i \quad (5.7)$$

and the input voltage for the parallel circuit is

$$v_i = Z_p(s)i_i \quad (5.8)$$

where the input admittance is

$$Y_s(s) = \frac{1}{R}\frac{2\xi_s Ts}{T^2s^2 + 2\xi_s Ts + 1} \quad (5.9)$$

Inverters

TABLE 5.1 Parameters

	Series	Parallel
Time constant	$T = \sqrt{LC}$	
Resonant angular frequency	$\omega_0 = 2\pi f_0 = \dfrac{1}{T}$	
Damping factor	$\xi_s = \dfrac{1}{2}\dfrac{R}{\omega_0 L} = \dfrac{1}{2}\omega_0 CR$	$\xi_p = \dfrac{1}{2}\dfrac{\omega_0 L}{R} = \dfrac{1}{2}\dfrac{1}{\omega_0 CR}$
Characteristic impedance	$Z_0 = \sqrt{L/C}$	
Damped resonant angular frequency	$\omega_d = \omega_0\sqrt{1-\xi_s^2}$	$\omega_d = \omega_0\sqrt{1-\xi_p^2}$
Quality factor	$Q_s = \dfrac{1}{2\xi_s}$	$Q_p = \dfrac{1}{2\xi_p}$

and the input impedance is

$$Z_p(s) = R\,\frac{2\xi_p Ts}{T^2 s^2 + 2\xi_p Ts + 1} \tag{5.10}$$

The time constant and the damping factor ξ together with some other parameters are given in Table 5.1. ξ must be smaller than unity in Eqs. (5.9) and (5.10) to have complex roots in the denominators, that is, to obtain an oscillatory response.

When v_i is a unit step function, $v_i(s) = 1/s$, the time response of the voltage across R in the series resonance circuit from Eqs. (5.7) and (5.9) is

$$Ri_i(t) = 2\xi_s T\left[\frac{1}{T\sqrt{1-\xi_s^2}}\,e^{-\xi_s t/T}\sin(\sqrt{1-\xi_s^2})t/T\right] \tag{5.11}$$

$$= 2\xi_s Tf(t/T)$$

or for $\xi_s = 0$

$$i_i(t) = \frac{1}{\omega_0 L}\sin\omega_0 t \tag{5.11a}$$

that is, the response is a damped, or for $\xi_s = 0$ undamped, sinusoidal function.

When the current changes as a step function in the parallel circuit, $Ri_i(s) = 1/s$, the expression for the voltage response v_i is given by the right side of Eq. 5.11, as well, since $RY_s = Z_p/R$. Of course, now ξ_s has to be replaced by ξ_p. The time function $f(t/T)$ for various damping factors ξ is shown in Fig. 5.39.

Assuming sinusoidal input variables, the frequency response for series circuit is

$$\frac{R\bar{i}_i}{\bar{v}_i} = R\bar{Y}_s(jv) = \frac{1}{1+jQ_s(v-1/v)} = \frac{1}{\bar{D}_s(v)} \tag{5.12}$$

and for parallel circuit is

$$\frac{\bar{v}_i}{R\bar{i}_i} = \frac{1}{R}\bar{Z}_p(jv) = \frac{1}{1+jQ_p(v-1/v)} = \frac{1}{\bar{D}_p(v)} \tag{5.13}$$

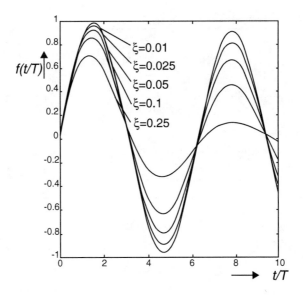

FIGURE 5.39 Time response of $f(t/T)$. ($T = 1$).

where

$$v = \omega/\omega_0.$$

Both circuits are purely resistive at resonance: $\bar{v}_i = R\bar{i}_i$ when $v = 1$.

The plot of the amplitude and phase of the right side of Eqs. (5.12) and (5.13) as a function of v are shown in Fig. 5.40. The voltage across R and its power can be changed by varying v. When Q is high, a small change in v can produce a large variation in the output.

The voltage across the energy storage components, for instance, across L in the series circuit, is

$$\frac{\bar{v}_L}{\bar{v}_i} = \frac{jvQ_s}{N_s(v)} \tag{5.14}$$

and the currents in the energy storage components, for instance, in L in the parallel circuit, is

$$\frac{\bar{i}_L}{\bar{i}_i} = \frac{Q_p}{jv\bar{N}_p(v)} \tag{5.15}$$

The voltages (currents) of the energy storage components in series (parallel) resonant circuits at $v = 1$ is Q times as high as the input voltage (current) (Table 5.2). If $Q = 10$ the capacitor or inductor voltage (current) is ten times the source voltage (current).

The value of L and C and their power rating is tied to the quality factor. The higher the value of Q, the better the filter action, that is, the attenuation of the harmonics is better and it is easier to control the output voltage and power by a small change in the switching frequency. The definition of Q is

$$Q = \frac{2\pi \times \text{Peak stored energy}}{\text{Energy dissipated per cycle}} \tag{5.16}$$

Using this definition, the expressions for Q are given in Table 5.2 where I_p and V_p are the peak current in the inductor and peak voltage across the capacitor, respectively. For a given output power, the energy

TABLE 5.2 Resonance, $\omega = \omega_o$

Series	Parallel
$\dfrac{\bar{v}_c}{\bar{v}_i} = -jQ_s$	$\dfrac{\bar{i}_C}{\bar{i}_i} = jQ_p$
$\dfrac{\bar{v}_L}{\bar{v}_i} = jQ_s$	$\dfrac{\bar{i}_L}{\bar{i}_i} = -jQ_p$
$Q_s = \dfrac{2\pi(\frac{1}{2}LI_P^2)}{(\frac{1}{2}RI_P^2)\frac{1}{f_0}}$	$Q_p = \dfrac{2\pi(\frac{1}{2}CV_P^2)}{\left(\frac{1}{2}\frac{V_P^2}{R}\right)\frac{1}{f_0}}$

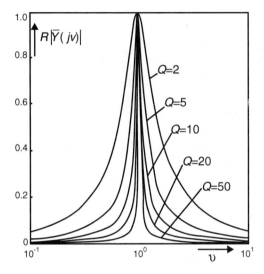

a.

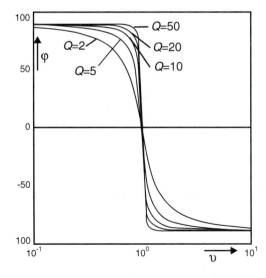

b.

FIGURE 5.40 Frequency response of $[R\bar{Y}(jv)]$. Amplitude (a), phase (b).

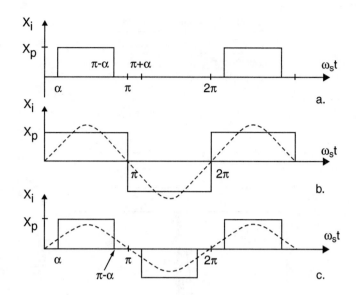

FIGURE 5.41 Frequently used input time functions.

dissipated per cycle is specified. The only way to obtain a higher Q is to increase the peak stored energy. The price paid for a high Q is the high peak energy storage requirements in both the inductor and capacitor.

Load Resonant Converters

In these converters the resonant L-C circuit is connected in the load. The currents in the switching semiconductors decay to zero due to the oscillation in the load circuit. Four typical converters are discussed:

1. Voltage source series-resonant converters
2. Current source parallel-resonant converters
3. Class E resonant converters
4. Series- and parallel-loaded resonant DC-DC converters

Input Time Functions

As a result of the on–off action of the switching devices, the frequently produced time functions of the input variable at the terminals of the ringing load circuit are shown in Fig. 5.41. The input variable x_i can be either voltage in series-resonant converters (SRC) or current in parallel-resonant converters (PRC) and it can be unidirectional (Fig. 5.41a) or bidirectional (Figure 5.41b and c). The ringing load is excited by a variable (Fig. 5.41a) which is constant in the interval $\alpha \leq \omega_s t \leq \pi - \alpha$ and short-circuited in the interval $\pi + \alpha \leq \omega_s t < 2\pi - \alpha$, where ω_s is the switching angular frequency. The circuit is interrupted during the rest of the period. The interruption interval shrinks to zero when $\omega_s \geq \omega_d$. The input variable is square-wave and a quasi-square-wave in Fig. 5.41b and c, respectively. The rms value of the fundamental component is

$$X_{irms} = \frac{4}{\pi\sqrt{2}} X_p \cos\alpha \tag{5.17}$$

The output variable changes in proportion to the input. Varying the angle α provides another means of controlling the output besides the switching frequency f_s.

Inverters

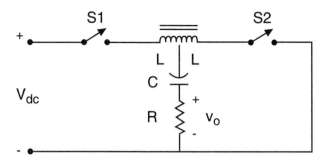

FIGURE 5.42 SRC with unidirectional switches.

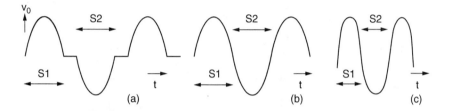

FIGURE 5.43 Output voltage waveforms for Fig. 5.42, $f_s < f_d$ (a), $f_s = f_d$ (b), $f_s > f_d$ (c).

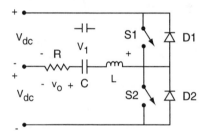

FIGURE 5.44 SRC with bidirectional switches.

Series-Resonant Converters

Series-resonant converters (SRC) can be implemented by employing either unidirectional (Fig. 5.42) or bidirectional (Fig. 5.43) switches. The unidirectional switch can be a thyristor, GTO, bipolar transistor, IGBT, etc., while these devices with an antiparallel diode or RCT (reverse conducting thyristor) can be used as a bidirectional switch.

Depending on the switching frequency f_s, the wave shape of the output voltage v_o can take any one of the forms shown in Fig. 5.44 using the circuit in Fig. 5.42. The damped resonant frequency f_d is greater than f_s in Fig. 5.44a, $f_s < f_d$; equal to f_s in Fig. 5.44b; $f_s = f_d$; and smaller than f_s in Fig. 5.44c, $f_s > f_d$. S_1 and S_2 are alternately turned on. The terminals of the series resonant circuit are connected to the source voltage V_{dc} by S_1 or short-circuited by S_2. When both switches are off, the circuit is interrupted. The voltage across the terminals of the series-resonant circuit follows the time function shown in Fig. 5.41a for $f_d > f_s$, and in Fig. 5.41b for $f_d \leq f_s$, respectively. By turning on one of the switches, the other one will be force commutated by the close coupling of the two inductances.

The configuration shown in Fig. 5.43 can be operated below resonance, $f_s < f_d$ (Fig. 5.45a); at resonance, $f_s = f_d$ (Fig. 5.45b); and above resonance, $f_s > f_d$ (Fig. 5.45). The voltage, v_i, across the terminals of the series-resonant circuit is square wave. The harmonics of the load current can be neglected for high Q value. The output voltage v_o equals its fundamental component v_{o1}. The L-C network can be replaced by an equivalent capacitor (inductor) below (above) resonance and by a short-circuit at resonance. The circuit is capacitive

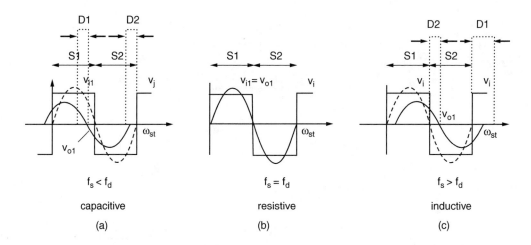

FIGURE 5.45 Output voltage waveforms for Fig. 5.43.

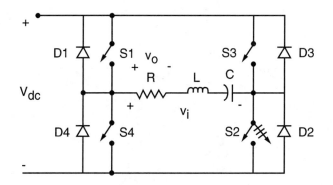

FIGURE 5.46 SRC in bridge topology.

(inductive) below (above) resonance and purely resistive at resonance (Fig. 5.45). The output voltage $v_o \cong v_{o1}$ is leading (lagging) the input voltage v_i below (above) resonance and in phase at resonance. Negative voltage develops across switches S_1 and S_2 during diode conduction and can be utilized to assist the turn-off process of switches S_1 and S_2.

No switching loss develops in the switches at $f_s = f_d$ (Fig. 5.45b) since the load current will be passing through zero exactly at the time when the switches change state (zero current switching). However, when $f_s < f_d$ or $f_s > f_d$ the switches are subjected to lossy transitions. For instance, if $f_s < f_d$ the load current will flow through the switch at the beginning of each half-cycle and then commutate to the diode when the current changes polarity (Fig. 5.45a). These transitions are lossless. However, when the switch turns on or when the diode turns off, they are subjected to simultaneous step changes in voltage and current. These transitions therefore are lossy ones. As a result, each of the four devices is subjected to only one lossy transition per cycle.

The bridge topology (Fig. 5.46) extends the output power to a higher range and provides another control mode for changing the output power and voltage (Fig. 5.47).

Discontinuous Mode

Converters with either unidirectional or bidirectional switches can be controlled in a discontinuous mode as well. In this mode, the resonant current is interrupted in every half-cycle when using unidirectional switches (Fig. 5.44a) and in every cycle when using bidirectional switches (Fig. 5.48). The power is controlled by varying the duration of the current break as it is done in duty ratio control of DC-DC converters. Note that this control mode theoretically avoids switching losses because whenever a switch

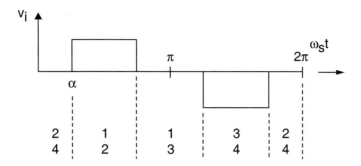

FIGURE 5.47 Quasi-square-wave voltage for output control.

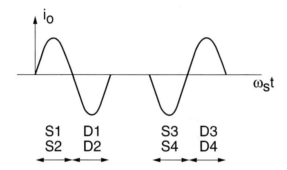

FIGURE 5.48 Discontinuous mode for bridge topology.

turns on or off its current is zero and no step change can occur in its current as a result of the inductance L. The shortcoming of this control mode is the distorted current waveform. In some applications, such as induction heating and ballasts for fluorescent lamps, the sinusoidal waveform is not necessary.

Parallel-Resonant Converters

The parallel-resonant converters (PRC) are the dual of the SRCs (Fig. 5.49). The bidirectional switches must block both positive and negative voltages rather than conduct bidirectional current. They are supplied by a current source and the converters generate a square wave input current i_i that flows through the parallel resonant circuit (Fig. 5.50). They offer better short-circuit protection under fault conditions than the SRCs with a voltage source.

When the quality factor Q is high and f_s is near resonance, the harmonics in the R-C-L circuit can be neglected. For $f_s < f_d$, the parallel L-C network is, in effect, inductive. The effective inductance shunts some of the fundamental components of the input current i_{i1} and a reduced leading current i_{i1} flows in the load resistance (Fig. 5.50a). For $f_s = f_d$, the parallel L-C filter looks like an infinitely large impedance. The total current i_{i1} passes through R and the output voltage v_{o1} is in phase with i_{i1} (Fig. 5.50b). Since $v_{o1} = 0$ at switching instants, no switching loss develops in the switching devices. For $f_s > f_d$, the L-C network is an equivalent capacitor at the fundamental component of i_{i1}. A part of the input current flows through the equivalent capacitor and only the remaining portion passes through the resistor R developing the lagging voltage v_{o1} (Fig. 5.50c). As a result of the current shunting through the equivalent L_e and C_e, the voltage v_{o1} is smaller in Fig. 5.50a and c than in Fig. 5.50b, although i_{i1} is the same in all three cases. The current source is usually implemented by the series connection of a DC voltage source and a large inductor (Fig. 5.51a). The bidirectional switch is implemented in practice for SRCs with the anti-parallel connection of a transistor-diode or thyristor–diode pair (Fig. 5.51b) and for PRCs with the series connection of a transistor-diode pair or thyristor. The condition $f_s > f_d$ must be met for PRCs in order for the thyristor to be commutated. By turning on one of the thyristors, a negative voltage is imposed across the previously conducting one, forcing it to turn off (Figs. 5.49b and 5.50c). If $f_s > f_d$ and a series transistor–diode pair

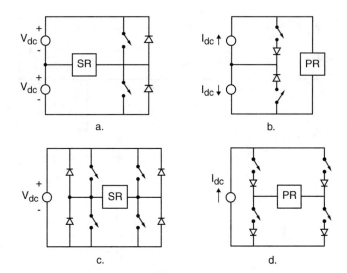

FIGURE 5.49 SRC and PRC are duals.

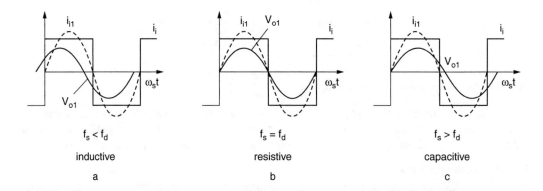

FIGURE 5.50 Waveforms for PRC.

is used, the diode will experience switching losses at turn-off and the transistor will experience losses at turn-on (Fig. 5.50c).

Class E Converter

The class E converter is supplied by a DC current source (Fig. 5.51a) and its load R is fed through a sharply tuned series resonant circuit ($Q \geq 7$) (Fig. 5.52a). The output current i_o is practically sinusoidal. It uses a single switch (transistor) that is turned on and off at zero voltage. The converter has low—theoretically zero—switching losses and a high efficiency of more than 95% at an operating frequency of several ten kHz. Its output power is usually low, less than 100 W, and it is used mostly in high-frequency electronic lamp ballasts.

The converter can be operated in optimum and in suboptimum modes. The first mode is explained in Fig. 5.52. When the switch is on (off) the equivalent circuit is shown in Fig. 5.52b (5.52c). In the optimum mode of operation the switch (capacitor) voltage, $v_T = v_{C1}$, decays to zero with a zero slope; $I_{dc} + i_o = i_{C1} = 0$. Turning on the switch at t_0, a current pulse $i_T = I_{dc} + i_o$ will flow through the switch with a high peak value; $\hat{I}_T \cong 3I_{dc}$ (Fig. 5.52d). Turning off the switch at $t = t_1$, the capacitor voltage builds up reaching a rather high value: $\hat{V}_C = 3.5V_{dc}$ and eventually falls back to zero at $t = t_0 + T$ (Fig. 5.52e and d). The average value of v_T, and that of the capacitor voltage v_C, is V_{dc}. The average value of i_T is I_{DC} while

Inverters

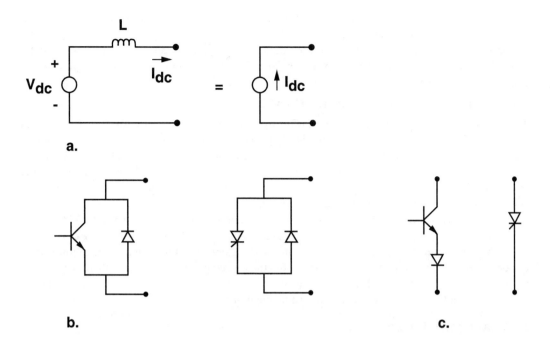

FIGURE 5.51 Implementation of current source (a). Implementation of bidirectional switch for SRC (b) and for PRC (c).

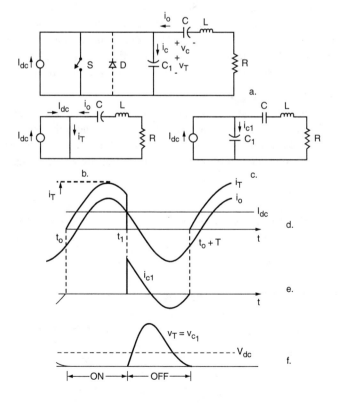

FIGURE 5.52 Class E resonant converter (optimum mode).

there is no DC current component in i_o. In the non-optimum mode of operation, $i_{C1} < 0$ when v_T reaches zero value and the diode D is needed.

The advantage of the class E converter is the simple configuration, the sinusoidal output current, the high efficiency, the high output frequency and the low EMI. Its shortcomings are the high peak voltage and current of the switch and the large voltages across the resonant L-C components.

Series- and Parallel-Loaded Resonant DC-DC Converters

The load R can be connected in series with L-C or in parallel with C in series resonant converters. The first case is called a series-loaded resonant (SLR) converter while the second one is called a parallel-loaded resonant (PLR) converter. When the converter is used as a DC-DC converter, the load circuit is built up by a transformer followed by a diode rectifier, a low-pass filter and finally the actual load resistance. The resonant circuit makes possible the use of a high-frequency transformer reducing its size and the size of the filter components in the low-pass filter.

The properties of the SLR and PLR converters are quite different in some respects. Without the transformers action, the SLR converter can only step-down the voltage Eq. (5.12) while the PLR converter can both step-up and step-down (in discontinuous mode of operation) the voltage. The step-up action can be understood by noting that the voltage across the capacitor is Q times higher than that across R in the SRC. The PLR converter has an inherent short-circuit protection when the capacitor is shorted due to a fault in the load. The current is limited by the inductor L.

Resonant Switch Converters

The trend to push the switching frequency to higher values, to reduce size and weight and to suppress EMI led to the development of switch configurations providing zero-current-switching (ZCS) or zero-voltage-switching (ZVS). As a result of having zero current (voltage) during turn-on and turn-off in ZCS (ZVS), the switching power loss is greatly reduced. The L-C resonant circuit is built around the semiconductor switch to ensure ZCS or ZVS. Sometimes the undesirable parasitic components, such as the leakage inductance of the transformer and the capacitance of the seminconductor switch, are utilized as components of the resonant circuit. Two ZCS and one ZVS configurations are shown in Fig. 5.53. The switch S can be implemented for unidirectional and bidirectional current (Fig. 5.54). Converters using ZCS or ZVS topology are termed resonant switch converters or quasi-resonant converters.

ZCS Resonant Converters

A step-down DC-DC converter using the ZCS configuration shown in Fig. 5.53a is presented in Fig. 5.55a. Switch S is implemented as shown in Fig. 5.54a. The L_f – C_f are sufficiently large to filter the harmonic current components. Current I_o can be assumed to be constant in one switching cycle. Four equivalent circuits associated with the four intervals of each cycle of operation are shown in Fig. 5.55b and c together with the waveforms.

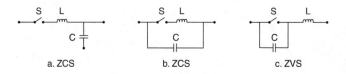

FIGURE 5.53 ZCS (a and b) and ZVS (c) configurations.

FIGURE 5.54 Switch for unidirectional (a) and for bidirectional (b) current.

Inverters

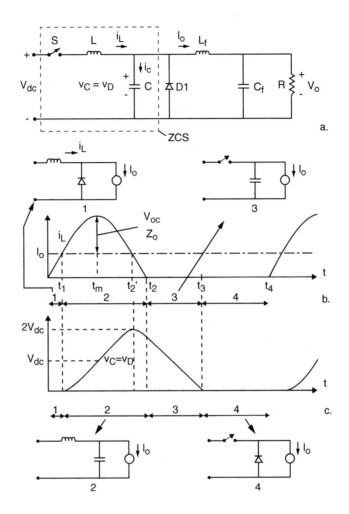

FIGURE 5.55 ZCS resonant converter.

Interval 1 $(0 \leq t \leq t_1)$: Both the current i_L in L and the voltage v_C across C are zero prior to turning the switch on at $t = 0$. The output current flows through the freewheeling diode D_1. After turning the switch on, the total input voltage develops across L and i_L rises linearly *ensuring ZCS* and soft current change. The interval 1 ends when i_s reaches I_o and the current conduction stops in D_1 at t_1.

Interval 2 $(t_1 \leq t \leq t_2)$: The L-C resonant circuit starts resonating and the change in i_L and v_C will be sinusoidal (Fig. 5.55b and c). Interval 2 has two subintervals. The capacitor current $i_C = i_L - I_o$ is positive in $t_1 \leq t \leq t'_2$ and v_C rises; while it is negative in $t'_2 \leq t \leq t_2$, v_C falls. The peak current is $\hat{I}_L = I_o + V_{dc}/Z_o$ at $t = t_m$ and peak voltage is $\hat{V}_c = 2V_{dc}$ at $t = t'_2$. V_{dc}/Z_o must be larger then I_o otherwise i_L will not swing back to zero.

Interval 3 $(t_2 \leq t \leq t_3)$: Current I_L reaches zero at t_2 and the switch is turned off *by ZCS*. The capacitor supplies the load current and its voltage falls linearly.

Interval 4 $(t_3 \leq t \leq t_4)$: The output current freewheels through D_1. The switch is turned on at t_4 again and the cycle is repeated.

The output voltage V_o will equal the average value of voltage v_C. V_o can be varied by changing the interval $t_4 - t_3$, that is, the switching frequency.

Applying the ZCS configuration shown in Fig. 5.53b, rather than that shown in Fig. 5.53a, the operation of the converter remains basically the same. The time function of the switch current and the D_1 diode voltage will be unchanged. The C capacitor voltage will be $v_C = V_{dc} - v_D$.

ZVS Resonant Converter

A ZVS resonant and step-down DC-DC converter is shown in Fig. 5.56a and is obtained from Fig. 5.55a by replacing the ZCS configuration with the ZVS configuration shown in Fig. 5.53c. Note, that the bidirectional current switch is used. This converter's operation is very similar to that of the ZCS converter. The waveform of v_C is the same as the one for i_L in Fig. 5.55b and the waveform of i_L is the same as the one for v_C when the ZCS configuration shown in Fig. 5.53b is used. I_o = const in one cycle can be assumed again.

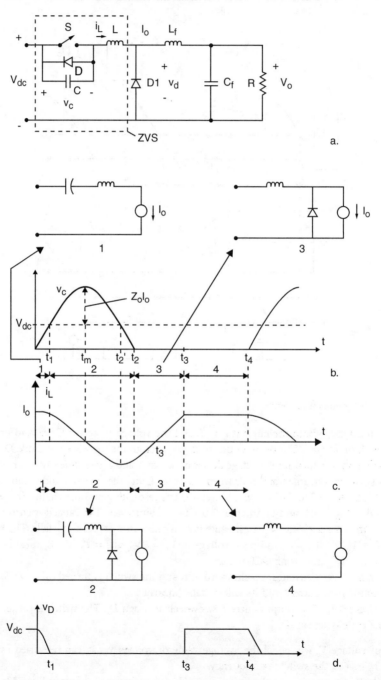

FIGURE 5.56 ZVS resonant converter.

Inverters

Interval 1 $(0 \le t \le t_1)$: S is turned off at $t = 0$. The constant $i_L = I_o$ current starts passing through the capacitor C. Its voltage v_C rises linearly from zero to V_{dc}. ZVS occurs.

Interval 2 $(t_1 \le t \le t_2)$: Diode D_1 turns on at t_1. The L-C circuit starts resonating through D_1 and the source. Both v_C and i_L are changing sinusoidally. When i_L drops at zero v_C reaches its peak value: $\hat{V}_C = V_{dc} + Z_o I_o$. The voltage v_C reaches zero at t_2. The load current must be high enough so that $Z_o I_o > V_{dc}$; otherwise v_C will not reach zero and the switch will have to be turned on at nonzero voltage.

Interval 3 $(t_2 \le t \le t_3)$: Diode D turns on. It clamps v_C to zero and conducts i_L. The gate signal is reapplied to the switch. V_{dc} develops across L and i_L increases linearly up to I_o, which is reached at t_3. Prior to that, the current i_L changes its polarity at t'_3 and S begins to conduct it.

Interval 4 $(t_3 \le t \le t_4)$: Freewheeling diode D_1 turns off at t_3. It is a soft transition because of the small negative slope of the current i_D. Current I_o flows through S at t_4 when S is turned off and the next cycle begins.

Diode voltage v_D develops across D_1 only in intervals 1 and 4 (Fig. 5.56d). Its average value is equal to V_o which can be varied by interval 4, or in other words, by the switching frequency.

Summary and Comparison of ZCS and ZVS Converters

The main properties of ZCS and ZVS are highlighted as follows:

- The switch turn-on and turn-off occurs at zero current or at zero voltage which significantly reduces the switching losses.
- Sudden current and voltage changes in the switch are avoided in ZCS and in ZVS, respectively. The di/dt and dv/dt values are rather small. EMI is reduced.
- In the ZCS, the peak current $I_o + V_{dc}/Z_o$ conducted by S must be more than twice as high as the maximum of the load current I_o.
- In the ZVS, the switch must withstand the forward voltage $V_{dc} + Z_o I_o$ and $Z_o I_o$ must exceed V_{dc}.
- The output voltage can be varied by the switching frequency.
- The internal capacitances of the switch are discharged during turn-on in ZCS which can produce significant switching loss at high switching frequency. No such loss occurs in ZVS.

Two-Quadrant ZVS Resonant Converters

One drawback in the ZVS converter, shown in Fig. 5.56, is that the switch peak forward voltage is significantly higher than the supply voltage. This drawback does not appear in the two-quadrant ZVS resonant converter where the switch voltage is clamped at the input voltage. In addition, this technique can be extended to the single phase and the three-phase DC-to-AC converter to supply an inductive load.

The basic principle will be presented by means of the DC-DC stepdown converter shown in Fig. 5.57a. Two switches, two diodes and two resonant capacitors $C_1 = C_2 = C$ are used. The voltage V_o can be assumed to be constant in one switching period because C_f is large. The current i_L must fluctuate in large scale and must take both positive and negative values in one switching cycle. To achieve this operation L must be rather small. One cycle consists of six intervals.

Interval 1. S_1 is on. The inductor voltage is $v_L = V_{dc} - V_o$. i_L rises linearly from zero.

Interval 2. S_1 is turned off at t_1. None of the four semiconductors conducts. The resonant circuit consisting of L and the two capacitors connected in parallel is ringing through the source and the load. Now the impedance $Z_o = \sqrt{2L/C}$ is high (C is small) and the peak current will be small. The voltage across C_2 approximately changes linearly and reaches zero at t_2. As a result of C_1 the voltage across S_1 changes slowly from zero.

Interval 3. D_2 conducts i_L. The inductor voltage v_L is $-V_o$. i_L is reduced linearly to zero at t_3. S_2 is turned on in this interval when its voltage is zero.

Interval 4. S_2 begins to conduct, v_L is still $-V_o$ and i_L increases linearly in a negative direction.

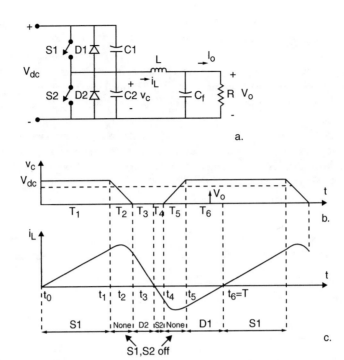

FIGURE 5.57 Two-quadrant ZVS resonant converter.

Interval 5. S_2 is turned off at t_4. None of the four semiconductors conducts. A similar resonant process occurs as in interval 2. As a result of C_2, the voltage across S_2 rises slowly from zero to V_{dc}.

Interval 6. v_C reaches V_{dc} at t_5. D_1 begins to conduct i_L. The inductor voltage $v_L = V_{dc} - V_o$ and i_L rises linearly with the same positive slope as in interval 1 and reaches zero at t_6. The cycle is completed.

The output voltage can be controlled by PWM at a constant switching frequency. Assuming that the intervals of the two resonant processes, that is, interval T_2 and T_5, are small compared to the period T, the wave shape of v_C is of a rectangular form. V_o is the average value of v_C and, therefore, $V_o = DV_{dc}$, where D the duty ratio: $D = (T_1 + T_6)/T$. Here T is the period: $T \cong T_1 + T_3 + T_4 + T_6$. During the time DT either S_1 or D_1 is on. Similarly, the output current is equal to the average value of i_L.

Resonant DC-Link Converters with ZVS

To avoid the switching losses in the converters, a resonant circuit is connected between the DC source and the PWM inverter. The basic principle is illustrated by the simple circuit shown in Fig. 5.58a. The resonant circuit consist of the L-C-R components. The load of the inverter is modelled by the I_o current source. I_o is assumed to be constant in one cycle of the resonant circuit.

Switch S is turned off at $t = 0$ when $i_L = I_{Lo} > I_o$. First, assuming a lossless circuit ($R = 0$), the equations for the resonant circuit are as follows:

$$i_L = I_o + \frac{V_{dc}}{Z_o} \sin \omega_0 t + (I_{L0} - I_0) \cos \omega_0 t \tag{5.18}$$

$$v_C = V_{dc}(1 - \cos \omega_0 t) + Z_0(I_{L0} - I_0) \sin \omega_0 t \tag{5.19}$$

where

$$\omega_o = 1/\sqrt{LC} \quad \text{and} \quad Z_o = \sqrt{L/C}$$

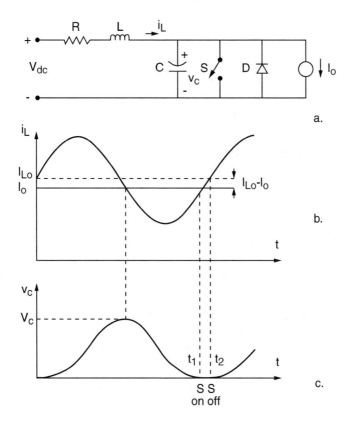

FIGURE 5.58 Circuit of a Class D voltage source half-bridge SRI.

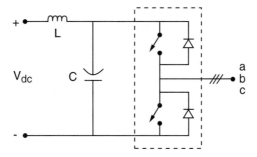

FIGURE 5.59 Idealized waveforms in the Class D voltage-source SRI (a) for $f < f_o$. (b) $f > f_o$.

To turn on and off the switch at zero voltage the capacitor voltage v_c must start from zero at the beginning and must return to zero at the end of each cycle (Fig. 5.58c). Without losses and when $I_{Lo} = I_o$, the voltage swing must start off and return to zero peaking at $2V_{dc}$. However, when $R \neq 0$ which represents the losses, the voltage swing is damped and v_C would never return to zero under the condition $I_{Lo} = I_o$. To force v_C back to zero a value of $I_{Lo} > I_o$ must be chosen (Fig. 5.58b). This condition adds the term $Z_o(I_{Lo} - I_o)\sin \omega_o t$ into the right side of Eq. (5.18) and thus v_C can reach zero again. By controlling the time $t_2 - t_1$, in other words, the on-time of switch S, both $I_{Lo} - I_o$ and the peak voltage $\hat{V}_c$ are regulated (Fig. 5.58c).

This principle can be extended to the three-phase PWM voltage source inverter (VSI) shown in Fig. 5.59. The three cross lines indicate that the configuration has three legs. Any of the two switches and two diodes in one leg can perform the same function which is done by the antiparallel connected S-D circuit in Fig. 5.58a. All of the six switches can be turned on and off at zero voltage in Fig. 5.59.

References

Kassakaian, J. G., Schlecht, M. F., and Verghese, G. C. 1992. *Principles of Power Electronics*, Addison-Wesley, Reading, MA.

Mohan, N., Undeland, T. M., and Robbins, W. P. 1989. *Power Electronics*, John Wiley & Sons, New York.

Ohno, E. 1988. *Introduction to Power Electronics*, Clarendon Press, Oxford, U.K.

Rashid, M. H. 1993. *Power Electronics*, Prentice-Hall International, London, U.K.

5.4 Series-Resonant Inverters

Dariusz Czarkowski

Class D DC-AC resonant inverters were proposed by Baxandall [1] in 1959. They have been widely applied to convert DC energy into AC energy in such areas as DC-DC resonant conversion, radio transmitters, solid-state electronic ballasts for fluorescent lamps, high-frequency electric process heating for induction welding, dielectric heating for plastic welding, surface hardening, soldering and annealing, induction sealing for tamper-proof packaging, and fiber-optics manufacturing. Class D inverters can be divided into two groups:

1. Class D series (or voltage-source) inverters
2. Class D parallel (or current-source) inverters.

Class D voltage-source inverters are fed by a DC voltage source. They employ a series-resonant circuit or a resonant circuit that is derived from the series-resonant circuit. For a sufficiently high loaded quality factor, the current through the resonant circuit is sinusoidal and the currents through the switches are half-wave sinusoids. The voltages across the switches are square waves.

On the other hand, the Class D current-source inverters are fed by a DC current source. They include a parallel resonant circuit or a resonant circuit that is derived from the parallel resonant circuit. The voltage across the resonant circuit is sinusoidal for high values of the loaded quality factor. The voltages across the switches are half-wave sinusoids and the currents through the switches are square waves.

One of the main advantages of Class D voltage-source inverters is the low voltage across the semiconductor switches. It is equal to the supply voltage. Hence, low-voltage-rated devices can be used, which increases an inverter efficiency and decreases the cost. The output voltage or the output power of resonant inverters is often controlled by varying the operating frequency f (FM control). In full-bridge configurations, the output can be adjusted by changing the duty ratio of pulses that fed the resonant circuit (PM control).

Voltage-Source Series-Resonant Inverters

Circuit and Waveforms

A circuit of the Class D voltage-source half-bridge series-resonant inverter (SRI) is presented in Fig. 5.60. It is composed of two bidirectional switches S_1 and S_2 and a series-resonant circuit L-C-R. Each switch consists of a transistor (power MOSFET, IGBT, or BJT) and an antiparallel diode. The switch can conduct either positive or negative current. It can only support, however, voltages higher than about −1 V. A positive or negative switch current can flow through the transistor if the transistor is ON. If the transistor is OFF, the switch can conduct only a negative current, which flows through the diode. The transistors are driven by nonoverlapping gating signals with a small dead time at the operating frequency $f = 1/T$. Switches S_1 and S_2 are alternately ON and OFF with a duty ratio of 50% or slightly less. The dead time is the time interval when both controllable devices are off. Resistance R is an AC load. If the inverter is a part of a DC-DC resonant converter, R represents an input resistance of a rectifier.

Selected waveforms in the Class D voltage-source resonant inverter are shown in Fig. 5.61. The voltage that feeds the series-resonant circuit is a square wave. If the loaded quality factor $Q = (\sqrt{L/C})/R$ of the resonant circuit is high enough, the current through the circuit is nearly a sine wave. When $f = f_o$

Inverters 5-43

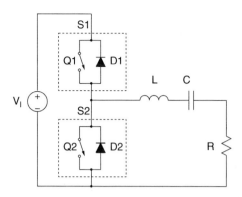

FIGURE 5.60 Circuit of a Class D voltage-source half-bridge series-resonant inverter.

the controllable devices turn on and off at zero current. The antiparallel diodes do not conduct. This yields zero switching losses and high efficiency. The operating frequency f is, however, rarely equal to the resonant frequency $f_o = 1/(2\pi\sqrt{LC})$ because the output power or the output voltage is controlled by changing f. Fig. 5.61a and b shows the waveforms for $f < f_o$ and $f > f_o$, respectively. Transistors should be turned off for $f < f_o$ or turned on for $f > f_o$ during the time interval when the switch current is negative. During this time interval, the switch current can flow through the antiparallel diode. To prevent shorting of the input voltage source (or, in other words, cross conduction or a shoot-through current), the gating signals of transistors cannot overlap and, additionally, must have a sufficient dead time. MOSFETs exhibit a delay time and bipolar devices (IGBTs and BJTs) have a storage time at turn-off. If the dead time is too short, one transistor still remains on while the other turns on. Hence, both transistors may be ON at the same time, which results in short-circuiting the input voltage source by small transistor on-resistances. To allow for a positive current flow through the switches, the dead time should not be too long.

Operation below Resonance

For $f < f_o$, the inductor current leads the fundamental component of the S_2 voltage by a phase angle $-\psi$. It is said that the series-resonant circuit represents a capacitive load to the switches. Therefore, the switch current is positive after switch turn-on and is negative before switch turn-off. The semiconductor devices conduct in a sequence $Q_1 - D_1 - Q_2 - D_2$. The inductor current is diverted from the diode of one switch to the transistor of the other switch (Fig. 5.58). Consider the turn-on of switch S_2. Prior to this transition, the inductor current flows through antiparallel diode D_1 of switch S_1. When transistor Q_2 is turned on by its gating signal, the voltage across S_2 decreases, causing the voltage across S_1 to increase. Therefore, diode D_1 turns off and the inductor current is diverted from D_1 to Q_2. There are three nondesirable effects at a turn-on of the transistor: reverse recovery of the antiparallel diode of the opposite switch, discharge of the transistor output capacitance, and Miller's effect.

The diode reverse-recovery stress at the turn-off seems to be the most detrimental effect of operation below resonance. Each diode turns off at a very large dv/dt and therefore at a very large di/dt, generating a high reverse-recovery current spike (turned upside down). Since the resonant inductor L does not allow for abrupt current changes, this spike flows through the other transistor. Consequently, the spikes occur in the switch current waveform at both the turn-on and turn-off transitions of the switch. The magnitude of these spikes can be several times higher than the magnitude of the steady-state switch current. High current spikes may destroy the transistors and always cause a considerable increase in switching losses and noise. During a part of the reverse-recovery interval, the diode voltage increases from about −1 V to the full power supply voltage V_I. Both the diode current and voltage are simultaneously high, causing a high reverse-recovery power loss. The current spikes can be reduced by using fast reverse-recovery diodes or, in low voltage applications, Schottky diodes as antiparallel diodes. Snubbers can be used to slow the switching process, and reverse-recovery spikes can be reduced by connecting small inductances in series with the switches.

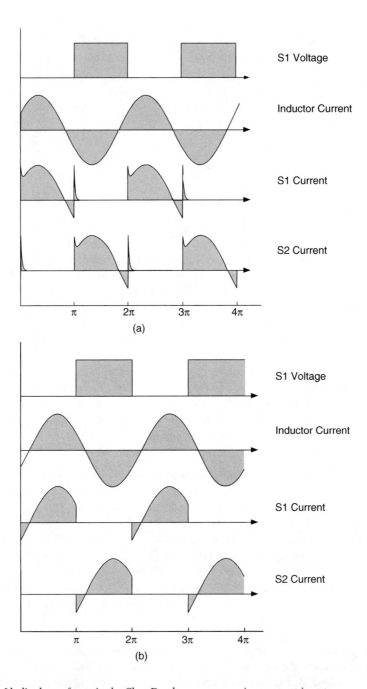

FIGURE 5.61 Idealized waveforms in the Class D voltage-source, series-resonant inverter.

For $f < f_o$, the turn-off switching loss is zero, but the turn-on switching loss is not zero. The transistors are turned on at a high voltage, equal to the power supply voltage. When the transistor is turned on, its output capacitance C_{out} is discharged, causing a switching loss. The total power loss associated with charging and discharging the output capacitance is

$$P = fC_{out}V_I^2 \tag{5.20}$$

per transistor.

Another effect that should be considered at turn-on of a transistor is Miller's effect. It increases the transistor input capacitance, and the gate drive charge and power requirements, and reduces the turn-on switching speed.

An advantage of operation below resonance is that the transistors are turned off at nearly zero voltage, resulting in zero turn-off switching loss. Since the switch voltage is constant, Miller's effect is absent during turn-off, the transistor input capacitance is not increased by Miller's effect, the gate drive requirement is reduced, and the turn-off switching speed is enhanced.

In summary, for $f < f_o$, there is a turn-on switching loss in the transistor and a turn-off (reverse-recovery) switching loss in the diode. The transistor turn-off and the diode turn-on are lossless.

Operation above Resonance

For $f > f_o$, the series-resonant circuit represents an inductive load to the switches. The inductor current lags behind the fundamental component of the S_1 voltage by a phase angle ψ. The switch current is negative after turn-on (for part of the switch "on" interval) and positive before turn-off. The semiconductor devices conduct in a sequence D_1–Q_1–D_2–Q_2. Consider the turn-off of switch S_1. When transistor Q_1 is turned off by its gating signal, the voltage across it increases, causing the voltage across the switch S_2 to decrease. Eventually, D_2 turns on and the inductor current is diverted from transistor Q_1 to diode D_2.

The transistors are turned on at almost zero voltage. There is a small negative voltage of the antiparallel diode, but this voltage is negligible comparison with the input voltage. Hence, the turn-on switching loss is eliminated, Miller's effect is absent, transistor input capacitance not increased by Miller's effect, the gate drive power low, and the turn-on switching speed high. The diodes turn on at a very low di/dt. The diode reverse-recovery current is a fraction of a sine wave and becomes a part of the switch current when the switch current is positive. Therefore, the antiparallel diodes can be slow. The diode voltage is kept at a low voltage of the order of 1 V by the transistor in the on-state during the reverse-recovery interval, reducing the diode reverse-recovery power loss. The transistor can be turned on not only when the switch current is negative but also when the switch current is positive and diode is still conducting because of the reverse recovery. Therefore, the range of the on-duty cycle of the gate-source voltages and the dead time can be larger. If, however, the dead time is too long, the current will be diverted from the recovered diode D_2 to diode D_1 of the opposite transistor until transistor Q_2 is turned on, causing extra transitions of switch voltages, current spikes, and switching losses. For $f > f_o$, the turn-on switching loss is zero, but there is a turn-off loss in the transistor. Both the switch voltage and current waveforms overlap during turn-off, causing a turn-off switching loss. Also, Miller's effect is considerable, increasing the transistor input capacitance, the gate drive requirements, and reducing the turn-off speed.

In summary, for $f > f_o$, there is a turn-off switching loss in the transistor, while turn-on of the transistor and the diode are lossless. The turn-off switching loss can be eliminated by adding a shunt capacitor to one of the transistors and using a dead time in the drive voltages.

Voltage Transfer Function

Class D inverters can be functionally divided into two parts: the switching part and the resonant part. A block diagram of an inverter is shown in Fig. 5.62. The switching part comprises a DC input voltage

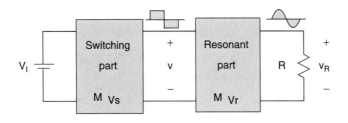

FIGURE 5.62 Block diagram of a Class D voltage-source inverter.

source V_I and a set of switches. The switches are controlled to produce a square-wave voltage v. Since a resonant circuit forces a sinusoidal current, only the power of the fundamental component is transferred from the switching part to the resonant part. Hence, consideration of only the fundamental component of the voltage at the input to the resonant circuit yields proper power relationships. A voltage transfer function of the switching part can be defined as

$$M_{V_s} \equiv \frac{V_{rms}}{V_I} \tag{5.21}$$

where V_{rms} is the rms value of the fundamental component of the voltage v. For a half-bridge inverter, $M_{V_s} = \sqrt{2}/\pi = 0.45$. In a full-bridge configuration, $M_{V_s} = 0.9$. The resonant part of an inverter converts a square-wave voltage v into sinusoidal current or voltage signal.

The analysis of the Class D inverter of Fig. 5.60 is performed under an assumption that the semiconductor devices and reactive components are ideal. Effects of nonidealities on inverter operation can be found in the literature, e.g., Ref. 3.

The parameters of the series-resonant circuit are defined as:

- The resonant frequency

$$\omega_o = \frac{1}{\sqrt{LC}} \tag{5.22}$$

- The characteristic impedance

$$Z_o = \sqrt{\frac{L}{C}} = \omega_o L = \frac{1}{\omega_o C} \tag{5.23}$$

and
- The loaded quality factor

$$Q = \frac{\omega_o L}{R} = \frac{1}{\omega_o CR} = \frac{Z_o}{R} = \frac{\sqrt{\frac{L}{C}}}{R} \tag{5.24}$$

The input impedance of the series-resonant circuit is

$$\mathbf{Z} = Ze^{j\psi}R + j\left(\omega L - \frac{1}{\omega C}\right) = R\left[1 + jQ\left(\frac{\omega}{\omega_o} - \frac{\omega_o}{\omega}\right)\right] \tag{5.25}$$

where

$$\psi = \tan^{-1}\left[Q\left(\frac{\omega}{\omega_o} - \frac{\omega_o}{\omega}\right)\right] \tag{5.26}$$

For $f < f_o$, ψ is less than zero, which means that the resonant circuit represents a capacitive load to the switching part of the inverter. For $f > f_o$, ψ is greater than zero, which indicates that the resonant circuit represents an inductive load.

Inverters

If the resonant circuit is loaded by a resistance R, a voltage transfer function of the resonant part is

$$\mathbf{M}_{V_r} = M_{V_r} e^{-j\psi} = \frac{V_R}{V_{rms}} e^{-j\psi} = \frac{R}{R + j\left(\omega L - \frac{1}{\omega C}\right)} = \frac{1}{1 + jQ\left(\frac{\omega}{\omega_o} - \frac{\omega_o}{\omega}\right)} \qquad (5.27)$$

where V_R is the rms value of the voltage across the load resistance R. Eq. (5.27) yields

$$M_{V_r} = \frac{1}{\sqrt{1 + jQ^2\left(\frac{\omega}{\omega_o} - \frac{\omega_o}{\omega}\right)^2}} \qquad (5.28)$$

The voltage transfer function M_{V_r} is illustrated in Fig. 5.63.
A voltage transfer function of the entire inverter is defined as

$$M_{V_I} = \frac{V_R}{V_I} \qquad (5.29)$$

The DC-to-AC voltage transfer function for a Class D voltage-source SRI is

$$M_{V_I} = \frac{V_R}{V_I} = \frac{V_R}{V_{rms}} \frac{V_{rms}}{V_I} = M_{V_s} M_{V_r} = \frac{M_{V_s}}{\sqrt{1 + Q^2\left(\frac{\omega}{\omega_o} - \frac{\omega_o}{\omega}\right)^2}} \qquad (5.30)$$

where the value of M_{V_s} depends on the switching part topology (full-bridge or half-bridge).

The Class D voltage-source SRI can operate safely with an open circuit at the output. It is, however, exposed to large currents and voltages if the output is short-circuited at an operating frequency f close to the resonant frequency f_o.

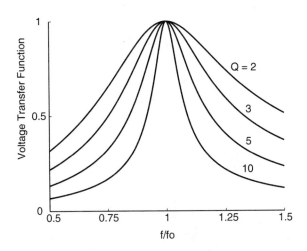

FIGURE 5.63 Voltage transfer function of the resonant circuit of a voltage-source SRI.

Voltage-Source Parallel-Resonant Inverters

Circuit and Waveforms

In the Class D voltage-source SRI presented in the preceding section, the load resistance is connected in series with the *LC* components. When the load resistance is increased, the current through the resonant circuit and the switches decreases. Consequently, the output power also decreases. In this section, a Class D voltage-source parallel-resonant inverter (PRI) is discussed. The load resistance in this inverter is connected in parallel with the resonant capacitor. As a result, if the load resistance is much higher than the reactance of the resonant capacitor, the current through the resonant inductor and the switches is almost independent of the load. As the load resistance is increased, the voltage across the resonant capacitor and the load increases, causing the output power to increase.

A circuit of a Class D voltage-source half-bridge PRI is shown in Fig. 5.64. It consists of two switches S_1 and S_2, a resonant inductor *L*, and a resonant capacitor *C*. Resistance *R* represents a load to which the AC power is to be delivered and is connected in parallel with the resonant capacitor *C*. In practical realizations, a large DC-blocking capacitor should be connected in series with the load to cut off a DC current flow through the resistance *R*. The two bidirectional two-quadrant switches S_1 and S_2 and the DC input voltage source V_I form a square-wave voltage source that drives the resonant circuit *L-C-R*. Each switch consists of a transistor and an antiparallel diode. Hence, they can be controlled only at a positive current. The switches S_1 and S_2 are turned ON and OFF alternately at the switching frequency $f = \omega/2\pi$. The duty cycle of the switches should be slightly less than 50% to avoid a shoot-through current.

The resonant circuit in the inverter of Fig. 5.64 is a second-order low-pass filter and can be described by the following normalized parameters:

- The corner frequency (or the undamped natural frequency)

$$\omega_o = \frac{1}{\sqrt{LC}} \tag{5.31}$$

- The characteristic impedance

$$Z_o = \omega_o L = \frac{1}{\omega_o C} = \sqrt{\frac{L}{C}} \tag{5.32}$$

and
- The loaded quality factor at the corner frequency f_o

$$Q = \omega_o CR = \frac{R}{\omega_o L} = \frac{R}{Z_o} \tag{5.33}$$

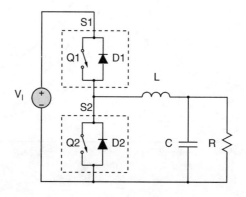

FIGURE 5.64 Circuit of a Class D voltage-source half-bridge PRI.

Inverters

Using Eqs. (5.31) and (5.32), the input impedance of the resonant circuit can be expressed as

$$\mathbf{Z} = Ze^{j\psi} = j\omega L + \frac{R\dfrac{1}{j\omega C}}{R + \dfrac{1}{j\omega C}} = \frac{R\left[1 - \left(\dfrac{\omega}{\omega_o}\right)^2 + j\dfrac{1}{Q}\left(\dfrac{\omega}{\omega_o}\right)\right]}{1 + jQ\left(\dfrac{\omega}{\omega_o}\right)} \tag{5.34}$$

where

$$\psi = \tan^{-1}\left\{Q\left(\frac{\omega}{\omega_o}\right)\left[\left(\frac{\omega}{\omega_o}\right)^2 + \frac{1}{Q^2} - 1\right]\right\} \tag{5.35}$$

At $f = f_o$,

$$Z(f_o) = \frac{Z_o}{\sqrt{Q^2 + 1}} \approx \frac{Z_o^2}{R} \quad \text{for } Q^2 \gg 1 \tag{5.36}$$

As R is increased, $Z(f_o)$ decreases.

The resonant frequency f_r is defined as a frequency at which the phase shift ψ is zero. Hence, from Eq. (5.35), the ratio of the resonant frequency f_r to the corner frequency f_o is

$$\frac{f_r}{f_o} = \sqrt{1 - \frac{1}{Q^2}} \quad \text{for } Q \geq 1 \tag{5.37}$$

Frequency f_r forms the boundary between inductive and capacitive loads. The following conclusions can be drawn from Eq. (5.37):

1. For $Q \leq 1$, the resonant frequency f_r does not exist and the resonant circuit represents an inductive load at any operating frequency.
2. For $Q > 1$, f_r/f_o increases with Q.

Current and voltage waveforms for $f > f_r$ are similar to those of Fig. 5.61b. The input voltage of the resonant circuit is a square wave with a low-level value equal to zero (or to $-V_I$ in the full-bridge configuration) and a high-level value equal to V_I. The analysis is simplified by assuming sinusoidal currents in L, C, and R. This approximation is valid if the loaded quality factor Q of the resonant circuit is high (e.g., $Q \geq 2.5$). If $Q < 2.5$, the inductor current waveform differs from a sine wave and an accurate analytical solution is more difficult to obtain. However, the predicted results are still qualitatively correct. A sinusoidal inductor current is assumed in the subsequent analysis. The inductor current i is conducted alternately by switches S_1 and S_2. Each transistor should be turned on when the switch current is negative and flows through the diode. To achieve high efficiency and reliability, the switching frequency f should be higher than f_r under all operating conditions.

For $f > f_r$, the phase shift $\psi > 0$, the resonant circuit represents an inductive load and the current i lags behind the fundamental component of voltage v. Hence, the switch current is negative after turn-on and positive before turn-off. Consider the turn-off of switch S_1. When transistor Q_1 is turned off, its voltage increases, causing the decrease of the voltage of the other switch. As S_2 voltage reaches -0.7 V, D_2 turns on and therefore the current i is diverted from transistor Q_1 to diode D_2. The turn-off switch transition is forced by the driver, while the turn-on transition of the switch is caused by the turn-off transition of the opposite transistor, not by the driver. Only the turn-off transition is directly controllable by the driver. The transistor should be turned on by the driver when the switch current is negative and flows through the antiparallel diode. Therefore, the transistor is turned on at nearly zero voltage, reducing the turn-on switching loss to a negligible level.

For $f < f_r$, the phase shift $\psi < 0$, the resonant circuit represents a capacitive load for the switching part, and the inductor current i leads the fundamental component of the voltage v. Therefore, the antiparallel diodes turn off at high di/dt, causing high current spikes in the switches and reducing efficiency and reliability. This problem can be alleviated by adding external diodes; however, the efficiency will be reduced.

Voltage Transfer Function

The voltage transfer function of the resonant circuit of Fig. 5.64 is

$$M_{V_r} \equiv \frac{V_R}{\sqrt{2}V_{rms}} = \frac{R + \frac{1}{j\omega C}}{j\omega L + \frac{\frac{R}{j\omega C}}{R + \frac{1}{j\omega C}}} = \frac{1}{1 - \left(\frac{\omega}{\omega_o}\right)^2 + j\frac{1}{Q}\left(\frac{\omega}{\omega_o}\right)} = M_{V_r} e^{j\varphi} \quad (5.38)$$

where

$$M_{V_r} = \frac{V_R}{V_{rms}} = \frac{1}{\sqrt{\left[1 - \left(\frac{\omega}{\omega_o}\right)^2\right]^2 + \frac{1}{Q^2}\left(\frac{\omega}{\omega_o}\right)^2}} \quad (5.39)$$

$$\varphi = -\tan^{-1}\left[\frac{\frac{1}{Q}\left(\frac{\omega}{\omega_o}\right)}{1 - \left(\frac{\omega}{\omega_o}\right)^2}\right] \quad (5.40)$$

$\mathbf{V}_R$ is the phasor of the voltage across R, and V_R is the rms value of $\mathbf{V}_R$. Figure 5.65 shows the voltage transfer function of the resonant circuit given by Eq. (5.39). From Eq. (5.39), $M_{V_r} = Q$ at $f/f_o = 1$ and

$$M_{V_r} \to \frac{1}{1 - \left(\frac{\omega}{\omega_o}\right)^2} \quad \text{as} \quad Q \to \infty \quad (5.41)$$

With an open circuit at the output, M_{V_r} increases from 1 to ∞ as ω/ω_o is increased from zero to 1, and M_{V_r} decreases from ∞ to zero as ω/ω_o increases from 1 to ∞.

The maximum value of M_{V_r} is obtained by differentiating the quantity under the square-root sign with respect to f/f_o and setting the result equal to zero. Notice, however, that the maximum value of M_{V_r} occurs at the frequency equal to zero for $Q < 1/\sqrt{2}$. Hence, the normalized peak frequency is

$$\frac{f_{pk}}{f_o} = \begin{cases} 0, & \text{for } 0 \leq Q \leq \frac{1}{\sqrt{2}} \\ \sqrt{1 - \frac{1}{2Q^2}} & \text{for } Q > \frac{1}{\sqrt{2}} \end{cases} \quad (5.42)$$

Inverters

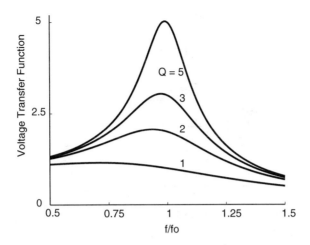

FIGURE 5.65 Voltage transfer function of the resonant circuit of a voltage-source PRI for $A = 1$.

resulting in the maximum magnitude of the voltage transfer function of the resonant circuit

$$M_{V_{r(max)}} = \begin{cases} 1, & \text{for } 0 \leq Q \leq \dfrac{1}{\sqrt{2}} \\ \dfrac{Q}{\sqrt{1 - \dfrac{1}{4Q^2}}} & \text{for } Q \geq \dfrac{1}{\sqrt{2}} \end{cases} \qquad (5.43)$$

For $Q^2 \gg 1$, $f_{pk} \approx f_o$ and $M_{V_{r(max)}} \approx Q$.

The magnitude of the DC-to-AC voltage transfer function of the Class D voltage-source PRI is obtained from Eq. (5.39) as

$$M_{V_I} \equiv \dfrac{V_R}{V_I} = M_{V_s} M_{V_r} = \dfrac{M_{V_s}}{\sqrt{\left[1 - \left(\dfrac{\omega}{\omega_o}\right)^2\right]^2 + \dfrac{1}{Q^2}\left(\dfrac{\omega}{\omega_o}\right)^2}} \qquad (5.44)$$

The range of M_{V_I} can be theoretically, from zero to ∞.

Voltage-Source Series–Parallel-Resonant Inverters

Circuit and Waveforms

This section presents the circuit and major characteristics of a series–parallel-resonant inverter (SPRI). The topology of this inverter is the same as that of the PRI except for an additional capacitor in series with the resonant inductor, or the same as that of the SRI except for an additional capacitor in parallel with the load. As a result, the inverter exhibits the characteristics that are intermediate between those of the SRI and the PRI. In particular, it has high part-load efficiency.

A circuit of the Class D series–parallel-resonant inverter is shown in Fig. 5.66. The inverter is composed of two bidirectional two-quadrant switches S_1 and S_2 and a resonant circuit L-C_1-C_2-R, where R is the AC load resistance. Capacitor C_1 is connected in series with resonant inductor L as in the SRI and capacitor C_2 is connected in parallel with the load as in the PRI. The switches consist of transistors and antiparallel diodes. Each switch can conduct a positive or a negative current. Switches S_1 and S_2 are alternately turned ON and OFF at the switching frequency $f = \omega/2\pi$ with a duty cycle of 50%. If capacitance C_1 becomes

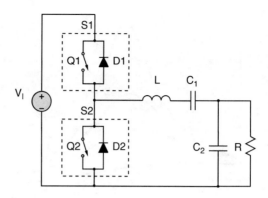

FIGURE 5.66 Circuit of a Class D voltage-source SPRI.

very large (that is, capacitor C_1 is replaced by a DC-blocking capacitor), the SPRI becomes the PRI. If capacitance C_2 becomes zero (that is, capacitor C_2 is removed from the circuit), the SPRI becomes the SRI. Basic waveforms in the Class D voltage-source SPRI are similar to those in the SRI, shown in Fig. 5.61.

Voltage Transfer Function

The resonant circuit in the inverter of Fig. 5.66 is a third-order low-pass filter. Under assumption of a nearly sinusoidal inductor current, the circuit can be described by the following normalized parameters:

- The ratio of the capacitances

$$A = \frac{C_2}{C_1} \tag{5.45}$$

- The equivalent capacitance of C_1 and C_2 in series

$$C = \frac{C_1 C_2}{C_1 + C_2} = \frac{C_2}{1 + A} = \frac{C_1}{1 + 1/A} \tag{5.46}$$

- The corner frequency (or the undamped natural frequency)

$$\omega_o = \frac{1}{\sqrt{LC}} = \sqrt{\frac{C_1 + C_2}{LC_1 C_2}} \tag{5.47}$$

- The characteristic impedance

$$Z_o = \omega_o L = \frac{1}{\omega_o C} = \sqrt{\frac{L}{C}} \tag{5.48}$$

and
- The loaded quality factor at the corner frequency f_o

$$Q = \omega_o CR = \frac{R}{\omega_o L} = \frac{R}{Z_o} \tag{5.49}$$

Inverters

The input impedance of the resonant circuit shown in Fig. 5.64 is

$$\mathbf{Z} = Ze^{j\psi} = \frac{R\left\{(1+A)\left[1-\left(\frac{\omega}{\omega_o}\right)^2\right] + j\frac{1}{Q}\left(\frac{\omega}{\omega_o} - \frac{\omega_o}{\omega}\frac{A}{A+1}\right)\right\}}{1 + jQ\left(\frac{\omega}{\omega_o}\right)(1+A)} \tag{5.50}$$

where

$$Z = Z_o Q \sqrt{\frac{(1+A)^2\left[1-\left(\frac{\omega}{\omega_o}\right)^2\right]^2 + \frac{1}{Q^2}\left(\frac{\omega}{\omega_o} - \frac{\omega_o}{\omega}\frac{A}{A+1}\right)^2}{1 + \left[Q\left(\frac{\omega}{\omega_o}\right)(1+A)\right]^2}} \tag{5.51}$$

and

$$\psi = \tan^{-1}\left\{\frac{1}{Q}\left(\frac{\omega}{\omega_o} - \frac{\omega_o}{\omega}\frac{A}{A+1}\right) - Q(1+A)^2\left(\frac{\omega}{\omega_o}\right)\left[1-\left(\frac{\omega}{\omega_o}\right)^2\right]\right\} \tag{5.52}$$

At $f/f_o = 1$,

$$Z = \frac{Z_o}{(1+A)\sqrt{1+Q^2(1+A)^2}} \approx \frac{Z_o^2}{R(1+A)^2} \quad \text{for} \quad Q^2(1+A)^2 \gg 1 \tag{5.53}$$

and

$$\psi = \tan^{-1}\left[\frac{1}{Q(1+A)}\right] \tag{5.54}$$

Thus, Z decreases with increasing A and R at $f = f_o$. Since $\psi > 0$, the resonant circuit always represents an inductive load for the switches at $f = f_o$.

The resonant frequency f_r is defined as the frequency at which the phase shift ψ is equal to zero. This frequency forms the boundary between capacitive and inductive loads. For $f < f_r$, ψ is less than zero and the resonant circuit represents a capacitive load. Operation in this frequency range is not recommended because the antiparallel diodes of the switches turn off at high di/dt, generating high reverse-recovery current spikes. For $f > f_r$, $\psi > 0$ and the resonant circuit represents an inductive load. Consequently, the inductor current i lags behind the fundamental component of the voltage v_{DS2}. The antiparallel diodes turn off at low di/dt and do not generate reverse-recovery current spikes. Operation in this frequency range is recommended for practical applications. Setting ψ given in Eq. (5.52) to zero yields

$$\frac{f_r}{f_o} = \sqrt{\frac{Q^2(1+A)^2 - 1 + \sqrt{[Q^2(1+A)^2 - 1]^2 + 4Q^2 A(1+A)}}{2Q^2(1+A)^2}} \tag{5.55}$$

The resonant frequency f_r depends on Q and A. As $Q \to 0$, $f_r/f_o \to 1/\sqrt{1+1/A}$.

Referring to Fig. 5.66, the voltage transfer function of the resonant circuit is

$$\mathbf{M}_{V_r} = \frac{1}{(1+A)\left[1-\left(\frac{\omega}{\omega_o}\right)^2\right]+j\frac{1}{Q}\left(\frac{\omega}{\omega_o}-\frac{\omega_o}{\omega}\frac{A}{A+1}\right)} = M_{V_r}e^{j\varphi} \tag{5.56}$$

Where

$$M_{V_r} \equiv \frac{V_R}{V_{\text{rms}}} = \frac{1}{\sqrt{(1+A)^2\left[1-\left(\frac{\omega}{\omega_o}\right)^2\right]^2+\frac{1}{Q^2}\left(\frac{\omega}{\omega_o}-\frac{\omega_o}{\omega}\frac{A}{A+1}\right)^2}} \tag{5.57}$$

$$\varphi = -\tan^{-1}\left\{\frac{\frac{1}{Q}\left(\frac{\omega}{\omega_o}-\frac{\omega_o}{\omega}\frac{A}{A+1}\right)}{(1+A)\left[1-\left(\frac{\omega}{\omega_o}\right)^2\right]}\right\} \tag{5.58}$$

V_R is the rms value of the voltage across R, and V_{rms} is the rms value of the fundamental component of the voltage at the input of the resonant circuit. In Fig. 5.67, M_{V_r} is plotted as a function of f/f_o at selected values of Q for $A = 1$.

By using Eq. (5.57), the magnitude of the DC-to-AC voltage transfer function of the Class D voltage-source SPRI can be expressed as

$$M_{V_I} \equiv \frac{V_R}{V_I} = M_V M_{V_r} = \frac{M_{V_s}}{\sqrt{(1+A)^2\left[1-\left(\frac{\omega}{\omega_o}\right)^2\right]^2+\frac{1}{Q^2}\left(\frac{\omega}{\omega_o}-\frac{\omega_o}{\omega}\frac{A}{A+1}\right)^2}} \tag{5.59}$$

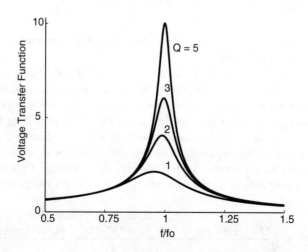

FIGURE 5.67 Voltage transfer function of the resonant circuit of a voltage-source SPRI for $A = 1$.

where M_{V_s} is the voltage transfer of the switching part of the inverter and is constant for a particular topology.

The inverter is not safe under short-circuit and the open-circuit conditions. At $R = 0$, the capacitor C_2 is shorted-circuited and the resonant circuit consists of L and C_1. If the switching frequency f is equal to the resonant frequency of this circuit $f_{r_s} = 1/(2\pi\sqrt{LC_1})$, the magnitude of the current through the switches and the L-C_1 resonant circuit is limited only by a small parasitic resistance of the switches and the reactive components. This current may become excessive and may destroy the circuit. If f is far from f_{r_s}, the amplitude of the current is limited by the reactance of the resonant circuit. Because $f_{r_s} < f_o$, the inverter is safe for switching frequencies above f_o. At $R = \infty$, the resonant circuit consists of L and the series combination of C_1 and C_2. Consequently, its resonant frequency is equal to f_o and the inverter is not safe at or close to this frequency as discussed in the section on the SRI.

Summary

Three main types of Class D voltage-source resonant inverters (known also as series-loaded or simply series-resonant inverters) have been presented, namely, the series-resonant inverter (SRI), the parallel-resonant inverter (PRI), and the series–parallel-resonant inverter (SPRI). The maximum voltage across the switches in Class D voltage-source inverters (both half-bridge and full-bridge) is low and equal to the DC input voltage V_I. Operation above the resonant frequency f_r is preferred for Class D inverters. Such an operation results in an inductive load for semiconductor switches. The transistors turn on at zero voltage, the turn-on switching loss is reduced, Miller's effect is absent, the transistor input capacitance is low, the transistor drive-power requirement is low, and turn-on speed is high. However, the transistor turn-off is lossy. The antiparallel diodes turn off at a low di/dt. During operation below resonance the antiparallel diodes turn off at a high di/dt and, if they are not sufficiently fast, can generate high reverse-recovery current spikes. These spikes are present in the switch current waveforms at both the switch turn-on and turn-off and may destroy the transistor. For operation below resonance, the transistors are turned on at a high voltage equal to V_I and the transistor output capacitance is discharged into a low transistor on-resistance. Hence, the turn-on switching loss is high. The resonant frequency f_r is constant in the SRI and depends on the load in the PRI and the SPRI.

The SRI can operate safely with an open circuit at the output. It is, however, exposed to excessive currents if the output is short-circuited at the operating frequency f close to the resonant frequency $f_r = f_o$. The PRI is protected by the impedance of the inductor for short circuit at the output at any switching frequency. It can be damaged by large currents when the output is open-circuited at an switching frequency close to the corner frequency f_o. The SPRI operation is not safe with an open-circuited output at frequencies close to the corner frequency f_o and with a short-circuited output at frequencies close to the resonant frequency f_r.

The output voltage of the resonant inverters can be regulated by changes in the switching frequency. However, the required frequency changes in the SRI are very large for no-load or light-load conditions. The PRI exhibits a good light-load regulation. It has, however, a low, light-load efficiency due to a relatively constant current through the resonant circuit. The SPRI combines the advantages of the SRI and PRI topologies at the expense of an additional resonant capacitor.

The input voltage of the resonant circuit in the Class D full-bridge inverters is a square wave with the low level of $-V_I$ and the high level is V_I. The peak-to-peak voltage across the resonant circuit in the full-bridge inverter is two times higher than in the half-bridge inverter. Therefore, the output voltage of the full-bridge inverter is also two times higher and the output power is four times higher than in the half-bridge inverter at the operating conditions (load, input voltage, and switching frequency).

Several issues were not addressed in this brief overview of SRIs. To explore such topics as operation with a nonsinusoidal resonant current, phase control, and Class E inverters with series-resonant circuits, the reader is referred to Refs. 2 through 8 and to periodicals on power electronics.

References

1. P. J. Baxandall, Transistor sine-wave *LC* oscillators, some general considerations and new developments, *Proc. IEE 13,* 106(suppl. 16), 748–758, 1959.
2. M. H. Rashid, *Power Electronics,* 2nd ed., Prentice-Hall, Englewood Cliffs, NJ, 1993.
3. M. K. Kazimierczuk and D. Czarkowski, *Resonant Power Converters,* Wiley Interscience, New York, 1995.
4. N. Mohan, T. M. Undeland, and W. P. Robbins, *Power Electronics: Converters, Applications and Design,* 2nd ed., John Wiley & Sons, New York, 1995.
5. R. Erickson, *Fundamentals of Power Electronics,* Chapman & Hall, New York, 1997.
6. D. W. Hart, *Introduction to Power Electronics,* Prentice-Hall, Englewood Cliffs, NJ, 1997.
7. P. T. Krein, *Elements of Power Electronics,* Oxford University Press, New York, 1998.
8. A. M. Trzynadlowski, *Introduction to Modern Power Electronics,* Wiley Interscience, New York, 1998.

5.5 Resonant DC-Link Inverters

Michael E. Ropp

Inverters are DC-to-AC converters; they create an AC output waveform from a DC input. They are required in any application that involves a DC source, such as batteries, photovoltaic arrays, or fuel cells, supplying power to an AC load. Inverters are also frequently used in applications in which AC with a controllable frequency is required. Probably the most common application of inverters is in variable-speed drives for motors, in which the frequency of the AC output of the inverters is controlled to obtain a desired rotational speed from an induction machine. In this application, the DC source is usually a rectifier that converts the fixed-frequency utility AC power to DC, which can then be converted to AC at the desired frequency by the inverter. This configuration is illustrated in Fig. 5.68. Another interesting application of inverters is in interfacing wind turbines to the utility system. This application is in a sense the dual of the previous one; the wind turbine can produce maximum output power if it operates at a variable frequency, but then its output AC power is at a variable frequency as well. This variable-frequency AC is usually first rectified to DC and then converted to fixed-frequency utility-compatible AC power by an inverter, and the system configuration is the same as that shown in Fig. 5.68.

One feature of this configuration is the large capacitor between the rectifier and the inverter. This capacitor is commonly called the DC-link capacitor, and it is usually quite large with values in the high hundreds of microfarads. The purpose of this capacitor is to absorb, or filter, the large ripple in the input current drawn by the inverter, so that it is not presented to the DC source. Because the DC-link capacitor is so large, it maintains a nearly constant voltage at the input terminals of the inverter, and thus the combination of this large capacitor with the inverter circuit is referred to as a voltage-source inverter (VSI) topology [1,2]. This topology is shown in Fig. 5.67, in both the single-phase (H-bridge, Fig. 5.69a) and three-phase (three-phase bridge, Fig. 5.69a) configurations.

The switches in Fig. 5.69a and b are implemented using semiconductor switching devices such as IGBTs, MOSFETs, and, at higher power levels, GTOs. In the VSI topologies shown, the semiconductor switches must interrupt the load current at high voltage, a technique known as hard switching. Hard switching results in energy dissipation in the semiconductor switches at each on–off or off–on transition.

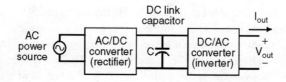

FIGURE 5.68 Back-to-back AC-DC-AC converter used in motor drives and wind energy conversion systems.

Inverters

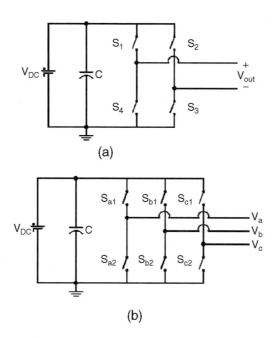

FIGURE 5.69 (a) Single-phase DC-AC inverter (H-bridge); (b) three-phase DC-AC inverter (three-phase bridge).

While a semiconductor switch is off, there is essentially zero current flowing through the switch, so the total power dissipation in the switch is nearly zero. When the switch is on, the current through the switch is large, but the voltage is small, so the amount of power dissipated in the switch is relatively low. However, during a transition from on to off, the current is decreasing from its "on"-state value while the voltage is increasing to its "off"-state value, and both are simultaneously high for a short period. During this period, the power dissipation in the switch is high. A similar situation occurs during the off-to-on transition. Such losses occurring during switch state transitions are commonly called switching losses, and they can make up as much as 50% of the total losses in the VSI [3]. Switching losses obviously reduce the efficiency of the converter. However, such losses also significantly impact the thermal design and packaging of VSIs. The power lost in the switches is dissipated as heat, which must be removed from the switch to avoid damage. Removing the heat requires large heat sinks, which increase converter size and cost, and possibly fans, which decrease reliability and efficiency.

As the switching frequency (the inverse of the period of the switching cycle) increases, the number of switching transitions per unit time also increases. More energy is dissipated in the switch during a shorter time, and therefore the power dissipation in the switches increases as the switching frequency increases. Thus, it might be thought that decreasing the switching frequency would ease the switching loss problem. Unfortunately, it is desirable to significantly increase the switching frequency for many reasons. Primary among these is that the energy storage elements (inductors and capacitors) can be made smaller if the switching frequency is higher [4], allowing for increased power density and lower cost. Higher switching frequencies also improve power quality in pulse width–modulated (PWM) inverters because the harmonics produced are concentrated at multiples of the switching frequency and closely spaced sidebands [5]. High switching frequencies also enable the inverter to respond more rapidly to control signals, increasing the "control bandwidth" of the converter. The factor in most VSIs that prevents switching frequencies from being increased to realize these benefits is the switching loss and subsequent thermal management of the semiconductor switches (hard-switched VSIs are often said to be "thermally limited" for this reason).

It is evident that reducing or eliminating switching loss would be highly beneficial to VSI design, not only for efficiency and thermal management reasons, but also because it would improve reliability, increase power density, and improve performance. There are two basic methods for eliminating switching loss. The first is through the use of snubbers. A snubber is a circuit designed to modify the waveform of the voltage across or the current through the switch [1]. Turn-on snubbers limit the rise of the current during switch turn-on, and turn-off snubbers reduce the voltages experienced by the switch during turn-off. Snubbers can be effective in reducing switch power dissipation and allowing higher switch frequencies, but most snubber configurations dissipate power themselves, and thus the efficiency of the inverter can still suffer as the switching frequency increases [1]. Snubbers can also be quite complex and add many additional parts to the inverter.

The other method for reducing or eliminating power dissipation in switches is to use a resonant converter topology. Resonant topologies involve the use of resonant L-C tanks that are configured in such a way that the current through or the voltage across the semiconductor switches oscillates and crosses zero periodically. At those moments at which the current and/or voltage is zero, the switch may be turned on or off without any power dissipation. Zero-current switching (ZCS) and zero-voltage switching (ZVS) have both been heavily investigated.

In the VSI, the DC-link capacitor can be used to form a resonant tank, causing the DC-link voltage periodically to go to zero. During these moments of zero DC-link voltage, ZVS can be accomplished.[*] Two basic resonant variants of the VSI have been proposed: the resonant DC-link inverter (RDCLI), sometimes also called the series-resonant DC-link inverter, and the parallel-resonant DC-link inverter (PRDCLI). In the following, the operation of both of these will be described. For detailed design procedures, the reader is referred to the literature [6].

The Resonant DC-Link Inverter

The RDCLI topology is shown in Fig. 5.70 [1, 2, 7]. It is formed by adding the inductor L and a single switch S_r to the VSI. An antiparallel diode D is connected across the switch. The resistance R can represent the series resistance in the inductor, but to simplify the first explanation of the circuit this resistor will be neglected for now ($R = 0$). The current source I_{out} represents the current drawn by a load. This load in practice will be a motor or will include a sizable LC filter, and thus will contain an inductive component much larger than the resonant inductor L. Because the current in this inductance will change slowly, the output current is nearly constant over one period of the switch S_r, and the representation of the load as a constant-current source is reasonable. At the beginning of a resonant cycle, the capacitor is fully discharged and the voltage across the switch, the DC bus voltage v_{bus}, is zero. The switch S_r is turned on (closed). If an ideal switch is assumed, the voltage across the inductor is V_{DC} (assumed to be positive), so the inductor current begins to ramp up linearly according to

$$\frac{di_L}{dt} = \frac{v_L}{L} \quad (5.60)$$

The difference between i_L and I_{out} flows through the switch, and v_{bus} remains at zero. When i_L becomes equal to I_{out}, the switch is opened at zero voltage. The inductor current continues to increase, because the voltage across it is still positive, but the difference between i_L and I_{out} now flows into C, and the LC

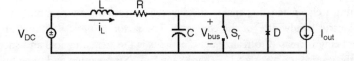

FIGURE 5.70 Simplified circuit to explain the operation of the RDCLI.

---

[*]The terms "DC link" and "DC bus" are still used, even though the voltage on the DC bus is now oscillatory.

circuit begins to oscillate. An analysis of the circuit [2] gives the voltage and the current in the lossless resonant tank as

$$v_{bus}(t) = V_{DC}[1 - \cos \omega_0 t] \qquad (5.61)$$

$$i_L(t) = I_{out} + \frac{V_{DC}}{\omega_0 L} \sin(\omega_0 [t - t_0]) \qquad (5.62)$$

where $\omega_0 = (LC)^{-0.5}$ is the resonant frequency of the tank and t_0 is the time at which the switch was turned off (the time at which $i_L = I_{out}$). Eqs. (5.61) and (5.62) are valid until the voltage across the switch returns to zero. At that moment, the antiparallel diode D becomes forward-biased and clamps the DC bus voltage v_{bus} to (approximately) zero, allowing a ZVS turn-on of the switch. These equations show that in this idealized lossless case the switch actually need not be turned on once oscillations have started, as they would continue indefinitely.

In practice, of course, $R \neq 0$, although it is typically quite small in relation to other impedances in the circuit. Because of the dissipation in R, it will be necessary to turn the switch on in each cycle to store extra energy in L to keep oscillations going. Also, to store extra energy in L, the switch must be turned off not at $i_L = I_{out}$ but at some higher current $i_L = I_{L0}$, since the energy stored in the inductor is $W_L = 0.5 L i_L^2$. Analyzing the circuit with a small nonzero R results in the appearance of a damped exponential term multiplying the sinusoidal terms in Eqs. (5.61) and (5.62):

$$v_{bus}(t) \cong V_{DC} + e^{-\alpha t}[-V_{DC} \cdot \cos(\omega t) + \omega L (I_{L0} - I_{out}) \cdot \sin(\omega t)] \qquad (5.63)$$

$$i_L(t) \cong I_{out} + e^{-\alpha t}\left[(I_{L0} - I_{out})\cos(\omega t) + \frac{V_{DC}}{\omega L}\sin(\omega t)\right] \qquad (5.64)$$

where

$$\alpha = \frac{R}{2L}$$

$$\omega = \sqrt{\omega_0^2 - \alpha^2}$$

In this analysis, it has been assumed that $R \ll \omega L$ [2].

To achieve ZVS, it is necessary that v_{bus}, described by Eq. (5.63), return to zero at the end of the cycle. Note that the term that determines whether this happens is the last term, which is dependent on the difference between the output current I_{out} and the inductor current at which the switch is turned off, I_{L0}. This shows that oscillation can be maintained simply by turning the switch off at the desired value of I_{L0}.

This single-switch tank may be added to an H-bridge inverter as shown in Fig. 5.71. However, notice that in this configuration the switch in the resonant tank is in parallel with the bridge switches. Thus, the extra

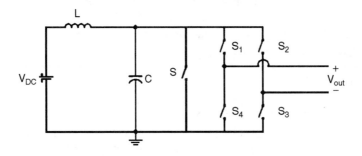

FIGURE 5.71 Single-phase H-bridge DC-AC inverter with the RDCLI resonant tank.

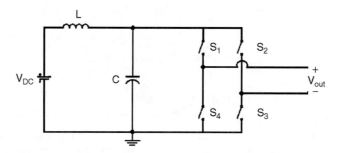

FIGURE 5.72 Single-phase H-bridge DC-AC inverter with the RDCLI resonant tank but without the redundant switch.

switch in the tank is not needed; its function may be accomplished by appropriate control of the switches in the H-bridge. The final result is an RDCLI that has only one more part than its hard-switched counterpart, as shown in Fig. 5.72.

Disadvantages of the RDCLI

Figure 5.72 shows an H-bridge ZVS inverter that can take advantage of very high switching frequencies without suffering the increased switching losses. Unfortunately, as attractive as this topology is, it has some serious drawbacks [3, 8]. One of the disadvantages is readily apparent from Eq. (5.61). The maximum DC-link voltage now reaches a value twice the value of the source V_{DC}. In the case with nonzero R, the peak of v_{bus} is well over $2 \cdot V_{DC}$. This means that the switches, when off, must now block a voltage more than twice as large as in the hard-switched case, and the DC-link capacitor must operate at the higher voltage as well. This increased voltage stress necessitates more expensive switches and capacitors, and can decrease the lifetimes of components, thus degrading reliability.

Another problem with the RDCLI circuit is that the instants at which ZVS may be accomplished happen only at the zero crossings of the DC-link voltage, which occur only once per resonant cycle. Thus, the moments at which ZVS can be done are, in a sense, discretized. In other words, continuous PWM control of the inverter switches and ZVS switching may not be simultaneously accomplished. In addition, when the inverter switches are on, they apply the resonant pulses of the oscillating DC bus voltage to the output inductance (filter), as opposed to the alternating-polarity DC that is applied to the inductor in the nonresonant VSI. (This is similar to the operation of a high-frequency transformer-isolated DC-DC converter, in which pulses of energy are applied to the primary side of the transformer and filtered on the secondary side to recover the DC.)

It is possible to operate the RDCLI in this mode, synthesizing desired output waveforms by controlling the number and polarity of DC-link oscillatory pulses applied to an output filter. The technique of synthesizing a desired output waveform from a filtered series of resonant pulses is called discrete pulse width modulation (DPWM). DPWM techniques that give quite good converter performance have been demonstrated. Unfortunately, these techniques generally result in output waveforms with significant amounts of harmonic content at frequencies lower than the link frequency [9]. This is often called "subharmonic" spectral content [10, 11], even though these frequencies are usually still higher than the fundamental of the AC output waveform.

Many solutions to each of these problems have been proposed, with varying degrees of success. Each problem will be considered in turn.

Reducing the Peak Voltage on the DC-Link

First consider how the DC-link voltage may be clamped to a value nearer to V_{DC}. Most of the solutions to this problem are based on an active clamp like that shown in Fig. 5.73[3, 8]. The active clamp allows limiting the DC bus voltage to $K \cdot V_{DC}$, where K is selected by the designer and is usually in the range

Inverters

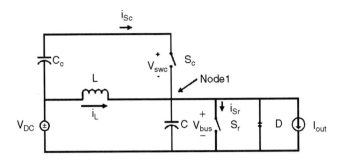

FIGURE 5.73 The ACRDCLI circuit.

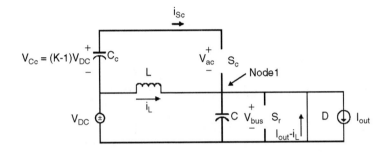

FIGURE 5.74 The ACRDCLI circuit, Mode 1 with diode conducting.

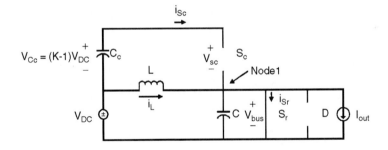

FIGURE 5.75 The ACRDCLI circuit, Mode 1 with switch S_r conducting.

of 1.2 to 1.4 for reasons to be explained shortly [3]. An RDCLI equipped with an active clamp like this one is referred to as an actively clamped resonant DC-link inverter (ACRDCLI).

The operation of the clamp will be described by separating a resonant DC-link cycle into three modes. Before the cycle begins, the clamp capacitor C_c must be precharged to a voltage of $(K-1) \cdot V_{DC}$ [3]. During Mode 1, the first part of the resonant cycle, the diode D conducts because $I_{out} > i_L$. The DC bus is shorted, and the current in the inductor L ramps up linearly according to Eq. (5.60). This configuration is shown in Fig. 5.74. The operation of the circuit in this part of the cycle is the same as that of the unclamped RDCLI. The DC-link capacitor C does not charge, and thus it clamps the voltage across S_r so that ZVS of S_r can be accomplished. (The conducting voltage drop of the switch in each direction is neglected.) The inductor current i_L continues to ramp up to a value larger than I_{out}, and i_{S_r} becomes positive. At the zero crossing of i_{S_r}, S_r turns on, as shown in Fig. 5.75. When i_L reaches the level required for maintenance of DC-link oscillation, I_{L0}, Mode 2 operation begins. S_r is turned off, as shown in Fig. 5.76, and the DC-link capacitor C begins to charge and resonate with L. The DC bus voltage begins to rise as C charges. The moment

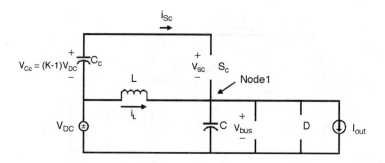

FIGURE 5.76 The ACRDCLI circuit, Mode 2.

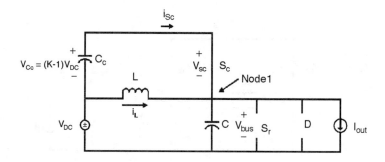

FIGURE 5.77 The ACRDCLI circuit, Mode 3.

at which the DC bus voltage v_{bus} is equal to $K \cdot V_{DC}$ marks the beginning of Mode 3. At that time, S_C, the switch in the active clamp circuit, turns on. The new circuit configuration is shown in Fig. 5.77. Note that the voltage across S_C, v_{SC}, is zero when $v_{bus} = K \cdot V_{DC}$. Therefore, the turn-on of S_C is also a ZVS transition. The current i_{SC} is negative, and the clamp capacitor C_c begins to charge. However, if C_c is very large, the voltage across it remains relatively constant at $(K-1) \cdot V_{DC}$. Thus, while S_C is on (again, neglecting the on-state voltage drop across the switch), summing voltages around the loop formed by S_C, C_c, and V_{DC} shows that the voltage at the node labeled Node 1 in Fig. 5.73 is clamped to $(K-1) \cdot V_{DC} + V_{DC} = K \cdot V_{DC}$. The DC bus voltage is thus clamped to that value. It should be borne in mind that, to achieve this clamping, the capacitor C_c must be large enough to minimize the ripple in its voltage.

Divan and Skibinski [3] showed that the practically achievable values of K are somewhat limited. Clearly, it would be desirable to make K as close to 1 as possible, thereby clamping the DC bus voltage to V_{DC}. However, there is a relationship between K and the period of the DC-link oscillation that indicates a longer oscillation period for decreasing K values. A long period of oscillation of the DC-link would be undesirable because it would require long time intervals between transitions of the inverter switches to maintain ZVS, and the primary benefit of the resonant converter, increased switching frequency, would be mitigated. Because of this, the minimum practically useful K value with this circuit is in the range of 1.2 to 1.4 [3]. Although less than might be desired, this still represents a reduction in the peak DC-link voltage of a factor of two or more, while requiring only two additional devices in the power circuit.

However, the addition of the active clamp also has an impact on the design of the circuit. The addition of C_c to the circuit changes the characteristics of the resonant DC-link, including the frequency of oscillation and the level of current required through L to maintain the oscillation. The authors of Ref. 3 state that "detailed design of the actively clamped circuit is extremely complex as it is dependent on a large number of variables." It is clear that the conversion of an existing inverter design to an ACRDCLI design would be a nontrivial task. In addition to the design challenge, there are also significant ramifications for converter control. To maintain the steady-state voltage on C_c, S_C must be controlled in such a way that the charge balance on C_c is maintained. This control requirement is in addition to the previously described requirements on S_r to maintain the oscillation of the DC-link. Suitable control methods are described in the literature [3], but clearly the complexity of the controller will be increased somewhat by the addition of the active clamp. In summary, the ACRDCLI does reduce the problem of the high DC bus voltage found in the RDCLI, but its design and control are much more difficult.

A significant number of variants of this circuit have been described. Many ACRDCLI topologies have been recently devised with the objective of clamping the maximum DC-link voltage to V_{DC}, thus realizing a K value of 1. Such a converter is typically termed a source-clamped RDCLI [12]. Other ACRDCLI variants have been designed to eliminate the need for precharging the clamp capacitor [13,14]. As might be expected, most of these variants improve the converter performance at the expense of a more-complicated circuit, and more difficult design and control.

Achieving Continuous PWM Control with the RDCLI: The Quasi-RDCLI

As was previously described, the necessity of synchronizing the transitions of the inverter bridge switches with the zero-voltage periods of the oscillating DC-link voltage creates a conflict that leads to a loss of output waveform quality and a narrowly limited range of controllability. Early attempts to solve this problem within the constraints of DPWM focused on the use of advanced control techniques [15, 16] to improve the converter performance without modification of the circuit topology. Such control techniques were used with considerable success and were shown to be capable of partially eliminating the problems of DPWM and giving RDCLI performance sufficient for many applications. Another technique intended to give PWM control while only slightly modifying the circuit and not adding semiconductor switches has been described [10]. This method requires placing a snubber capacitor in parallel with each semiconductor switch, and then using these snubber capacitors collectively to replace the DC-link capacitor. Unfortunately, this circuit had significantly higher losses than the original RDCLI [15], and the range of PWM control possible was found to be rather limited.

To obtain true continuous PWM control, the ACRDCLI circuit must be modified in such a way that the zero crossings of the DC bus voltage are controllable. Clearly, such a circuit must interrupt or modify the oscillation of the resonant tank, and for this reason circuits that implement this concept are called quasi-resonant DC-link inverters (QRDCLIs). QRDCLIs use a network of semiconductor switches, inductors, capacitors, and sometimes transformers to shape and control the resonant cycle of the DC bus. Several QRDCLI topologies have been proposed [17–23], all of which also contain an active (or passive) clamp. However, no single superior QRDCLI topology has yet been demonstrated. Because of the myriad requirements of achieving charge balance on all capacitors, volt-second balance on all inductors, and ZVS or ZCS of all switches in the added circuit, all the QRDCLI devices proposed thus far involve significant numbers of components, complicated control, and difficult design. To summarize, then, the QRDCLI enables ZVS or ZCS operation of all the switches in the circuit, with lower voltage stresses than the RDCLI, and with true PWM capability. It does so at the expense of an increased parts count, a higher level of difficulty in design, and greater complexity of control.

The Parallel-Resonant DC-Link Inverter

Shortly after the initial proposal of the RDCLI, a different topology was proposed to allow ZVS operation of VSIs. This topology, shown in Fig. 5.78, is the parallel-resonant DC-link inverter (PRDCLI) [24]. As before, the inverter current is represented as a constant current source.

Consider the operation of the PRDCLI [24]. The circuit has seven modes of operation. Initially, the circuit is in Mode 1, switches S_{r1} and S_{r3} are both on, and S_{r2} and S_{r4} are both off, as shown in Fig. 5.79. The source supplies the inverter current through S_{r1}, and the two capacitors C_1 and C_2 appear in parallel and are both charged to the DC bus voltage v_{bus}. Before an inverter switch transition is commanded, the parallel-resonant circuit drives the DC bus voltage to zero. This is accomplished as follows. The circuit enters Mode 2 when switch S_{r2} is turned on, as shown in Fig. 5.80. This is a ZCS transition, because during the switch transition the current i_L cannot change instantaneously and is clamped to zero. The voltage across L is v_{bus} because both S_{r2} and S_{r3} are on, so i_L begins to ramp up according to Eq. (5.60).

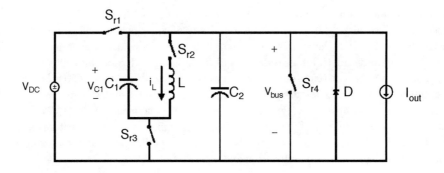

FIGURE 5.78 The PRDCLI resonant circuit.

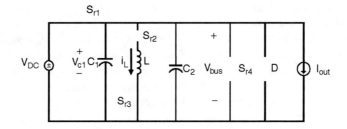

FIGURE 5.79 The PRDCLI circuit, Mode 1.

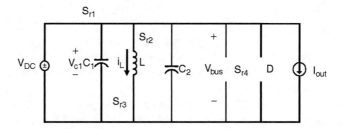

FIGURE 5.80 The PRDCLI circuit, Modes 2 and 7.

The inductor current ramps up until it reaches a critical value I_{Lo}. As was the case in the RDCLI, I_{Lo} must be chosen such that there is sufficient energy stored in the inductor to allow v_{bus} to go all the way to zero during the resonant cycle to follow.

As soon as $i_L = I_{Lo}$, S_{r1} is turned off, and the circuit configuration becomes that shown in Fig. 5.81. This initiates Mode 3. C_1 and C_2 begin to discharge and resonate with L. Because the voltage across L, v_{bus}, is still positive, i_L continues to ramp up to a resonant peak current higher than I_{Lo}. When i_L reaches its peak current, the DC bus voltage v_{bus} will be zero. At that moment, Mode 4 begins. Switch S_{r3} turns off and S_{r4} turns on, as shown in Fig. 5.82. This effectively separates the PRDCLI into two subcircuits, the resonant circuit with C_1 and L and the output circuit consisting of C_2, S_{r4}, and the inverter. In this configuration, since S_{r4} shorts the DC bus, v_{bus} is zero and is clamped to zero by C_2. The inverter devices may now be switched under ZVS conditions. Capacitor C_1 and inductor L continue to resonate, with v_{C1} and i_L passing through negative peaks. When i_L reaches its negative peak value, Mode 5 operation begins. In this mode, the energy stored in the inductor is used to recharge the two capacitors and reset the circuit to its initial condition so that the resonant cycle may begin again when needed. To do this, S_{r3} turns on under ZCS and S_{r4} turns off under ZVS, returning the circuit to the configuration shown in Fig. 5.81 and initiating Mode 6. The inductor current i_L is negative, so current flows back into the DC bus and

Inverters

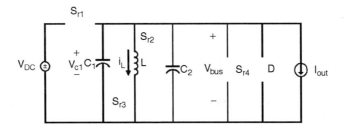

FIGURE 5.81 The PRDCLI circuit, Modes 3 and 6.

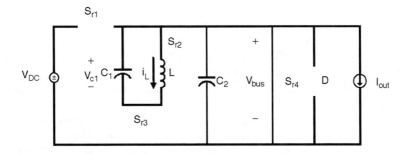

FIGURE 5.82 The PRDCLI circuit, Mode 4.

recharges C_1 and C_2. When the capacitors are charged such that v_{bus} is once again equal to V_{DC}, S_{r1} turns back on under ZVS conditions, initiating Mode 7 and returning the circuit to the configuration shown in Fig. 5.80. It remains in this state until the inductor current returns to zero, at which time S_{r2} turns off under ZCS, and the cycle is complete (the circuit is again in Mode 1).

The PRDCLI has several advantages over the RDCLI. First, it is clear from the above explanation that the ZVS instants for the inverter switches are controllable, meaning that true PWM control is possible with the PRDCLI, without any circuit modification. Also, the maximum voltage ever reached by the DC bus is V_{DC}, so no additional clamping is necessary.

However, there are several disadvantages of the PRDCLI. The most obvious is that it has many more parts than the RDCLI. Comparison of Fig. 5.78 with Fig. 5.70 shows that the PRDCLI requires not just the resonant inductor L (which is now in parallel with the DC-link capacitor C) and a control modification, as in the case of the RDCLI, but also four additional semiconductor switches and an additional capacitor. The control is also more complicated than for the RDCLI. This might partially explain why researchers in the early 1990s seem to have paid more attention to the RDCLI, as indicated by the literature. (It should be noted that these "disadvantages" may not be important, as the modifications to the RDCLI required to clamp the DC bus voltage and enable PWM control, i.e., the QRDCLI, also add many parts and complexity.) Another problem with the PRDCLI is that, although switching transition losses are minimized, overall converter losses, including both switching and conduction losses, may not be significantly reduced over the hard-switched PWM VSI [24]. This is mainly because of the additional conduction losses in S_{r1}. These arise because of the on-state resistance of S_{r1}, which appears in the current path from the source to the inverter at all times.

Current Research Trends

The current trend in research in resonant DC-link inverters seems to be toward a blending of the QRDCLI and the PRDCLI to form a quasi-resonant parallel RDCLI (QPRDCLI) [11, 25]. QPRDCLI designs generally attempt to maintain the PWM controllability and DC bus voltage limitation of the RDCLI while escaping the power loss across switch S_{r1} in the PRDCLI circuit. It is also noteworthy that ACRDCLI technology has moved beyond the laboratory; ACRDCLIs with power ratings as high as 200 kVA were commercially available as early as 1997 [26] and are already being used in uninterruptible power supplies, variable-speed drives, battery chargers, and several other products [27].

References

1. Mohan, N., Undeland, T., and Robbins, W., *Power Electronics: Converters, Applications, and Design*, 2nd ed., John Wiley & Sons, New York, 1995.
2. Divan, D., The resonant DC link converter—a new concept in static power conversion, *IEEE Trans. Ind. Appl.*, 25, 317, 1989.
3. Divan, D. and Skibinski, G., Zero-switching-loss inverters for high-power applications, *IEEE Trans. Ind. Appl.*, 25, 634, 1989.
4. Hart, D.W., *Introduction to Power Electronics*, Prentice-Hall, Upper Saddle River, NJ, 1997, chap. 6.
5. Hart, D.W., *Introduction to Power Electronics*, Prentice-Hall, Upper Saddle River, NJ, 1997, chap. 8.
6. Divan, D.M., Venkataramanan, G., and DeDoncker, R.W.A.A., Design methodologies for soft switched inverters, *IEEE Trans. Ind. Appl.*, 29, 126, 1993.
7. Rashid, M., *Power Electronics: Circuits, Devices, and Applications*, 2nd ed., Prentice-Hall, Upper Saddle River, NJ, 1993.
8. Garcia, P.D. and Barbi, I., A family of resonant DC-link voltage source inverters, in *Transactions of the IEEE Annual Conference on Industrial Electronics (IECON)*, 1990, 844.
9. Divan, D. M., Malesani, L., Tenti, P., and Toigo, V., A synchronized resonant DC link converter for soft-switched PWM, *IEEE Trans. Ind. Appl.*, 29, 940, 1993.
10. Venkataramanan, G. and Divan, D. M., Pulse width modulation with resonant DC link converters, *IEEE Trans. Ind. Appl.*, 29, 112, 1993.
11. Kwon, K.-A., Kim, K.-H., Jung, Y.-C., and Park, M., New low-loss quasi-resonant DC link inverter with lossless variable zero voltage duration, in *Proceedings of the IEEE Annual Conference on Industrial Electronics (IECON)*, 1997, 459.
12. Oh, I.-H., Jung, Y.-S., and Youn, M.-J., Source voltage-clamped resonant link inverter for a PMSM using a predictive current control technique, *IEEE Trans. Power Electron.*, 14, 1122, 1999.
13. Simonelli, J.M. and Torrey, D.A., Alternative bus clamp for resonant DC-link converters, *IEEE Trans. Ind. Appl.*, 9, 56, 1994.
14. Dahono, P.A., Kataoka, T., and Sato, Y., New clamping circuits for resonant DC link inverters, in *Proceedings of the IEEE International Conference on Power Electronics, Drives, and Energy Systems for Industrial Growth (PEDES)*, 1998, 833.
15. Venkataramanan, G., Divan, D., and Jahns, T., Discrete pulse modulation strategies for high-frequency inverter systems, *IEEE Trans. Power Electron.*, 8, 279, 1993.
16. Kheraluwala, M.H. and Divan, D.M., Delta modulation strategies for resonant link inverters, *IEEE Trans. Power Electron.*, 5, 220, 1989.
17. Hui, S.Y.R., Gogani, E.S., and Zhang, J., Analysis of a quasi-resonant circuit for soft-switched inverters, *IEEE Trans. Power Electron.*, 11, 106, 1996.
18. Jafar, J.J. and Fernandes, B.G., Novel quasi-resonant DC-link PWM inverter for induction motor drive, in *Record of the IEEE Annual Power Electronics Specialists Conference*, 1999, 482.
19. Oh, I.-H. and Youn, M.-J., A simple soft-switched PWM inverter using source voltage clamped resonant circuit, *IEEE Trans. Ind. Electron.*, 46, 468, 1999.
20. Ba-Thunya, A.S. and Toliyat, H.A., High frequency transformer assisted a new passive clamp ZVS quasi-resonant DC link PWM inverter with low voltage stress across the switches, in *Record of the IEEE Annual Power Electronics Specialists Conference (PESC)*, 1999, 981.
21. Lee, J.-W., Sul, S.-K., and Park, M.-H., A novel three-phase quasi-resonant DC link inverter, in *IEEE Industry Applications Society Annual Meeting*, 1992, 803.
22. Aoki, T., Nozaki, Y., and Kuwata, Y., Characteristics of a quasi-resonant DC link PWM inverter, in *Proceedings of the IEEE International Telecommunications Energy Conference (INTELEC)*, 1995, 602.

23. Lai, J.-S. and Bose, B.K., High frequency quasi-resonant DC voltage notching inverter for AC motor drives, in *Conference Record of the IEEE Industry Applications Society Annual Meeting*, 1990, 1202.
24. Jin, H. and Mohan, N., Parallel resonant DC link circuit—a novel zero switching loss topology with minimum voltage stresses, *IEEE Trans. Power Electron.*, 6, 687, 1991.
25. Chen, Y.-T., New quasi-parallel resonant DC link for soft-switching PWM inverters, *IEEE Trans. Power Electron.*, 13, 427, 1998.
26. Divan, D. and Wallace, I., New developments in resonant DC link inverters, in *Proceedings of the IEEE Power Conversion Conference*, 1997, 311.
27. WWWeb site of Soft Switching Technologies Corporation, available at http://www.softswitch.com.

5.6 Auxiliary Resonant Commutated Pole Inverters

Eric Walters and Oleg Wasynczuk

The auxiliary resonant commutated pole (ARCP) circuit was developed by General Electric Corporation R&D to be used in high-efficiency inverters. By increasing the efficiency of the inverter, not only is the power loss in the inverter reduced, but the size and weight of the inverter can also be greatly reduced. This fact makes the ARCP technology extremely valuable in applications where size constraints are a primary concern. The ARCP achieves high efficiency by soft switching, that is, by turning on or off the primary switches when the switch voltage or current are zero. Therefore, the switching loss, the product of voltage and current, is zero. This is similar to LC snubber circuits; however, in snubber circuits, load current constraints determine whether zero voltage or zero current switching can be obtained. With the ARCP circuit topology, zero switching losses are independent of the load current. In this section, conventional hard-switched and ARCP phase legs are both discussed, a comparison between the losses in a hard-switched H-bridge and an ARCP H-bridge is presented, and an application using the ARCP in a current-controlled induction motor is presented.

Losses in Hard-Switched Inverters

Introduction

The ARCP is a soft switching device that eliminates turn-on and turn-off losses, as is discussed in the next section. However, prior to examining the elimination of these losses, an explanation of the switching losses of a conventional hard-switched phase leg is presented. The first losses to be discussed in this section involve the conduction losses associated with the four states of a hard-switched phase leg. Then, the analysis of two switching scenarios will be explored in which the turn-on and turn-off losses associated with the transistors will be discussed.

Hard-Switched Phase Leg

The circuit shown in Fig. 5.83 is an example of a hard-switched phase leg. In this circuit, bipolar junction transistors (BJTs) are used as switches. Although BJTs are shown in this figure, other solid-state devices can also be used. Examples of other solid-state devices that are typically used include field effect transistors (FETs), metal oxide semiconductor FETs (MOSFETs), gate turnoff thyristors (GTOs), and insulated gate transistors (IGTs). Analysis of the losses associated with each of these devices can vary greatly; however, the concept of switching loss that is explored is universal.

In the circuit shown in Fig. 5.83, the positive and negative DC rails are labeled as $+V_{dc}$ and $-V_{dc}$, respectively. The BJTs are labeled Q_1 and Q_2 with their collector currents denoted as I_{C1} and I_{C2}, respectively. The base currents associated with the BJTs are labeled I_{B1} and I_{B2}. It will be assumed that the base currents, which are controlled by independent current sources, will determine if the transistors are on or off. If the base current associated with one of the transistors is zero, the associated transistor will not conduct and can be omitted from the circuit; therefore, the transistor is considered to be off. The DC

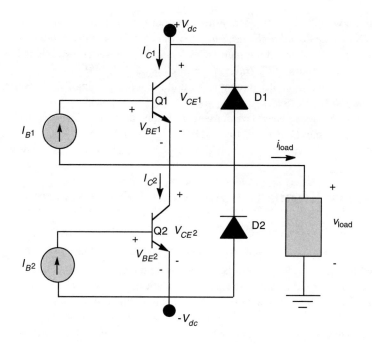

FIGURE 5.83 A hard-switched phase leg using BJTs.

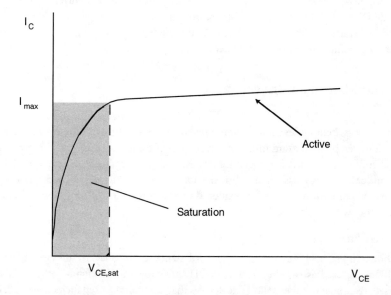

FIGURE 5.84 I_C vs. V_{CE} with a constant I_B.

current gain of each transistor is h_{FE}, which is defined as the ratio of the collector current to the base current, I_{C1}/I_{B1}, in the active region. If the collector current is assumed to be less in magnitude than I_{max} and the base current is set equal to I_{max}/h_{FE}, then the transistor is in the saturation region of operation as can be observed from the plot of the collect current vs. the collector to emitter voltage, V_{CE}, with a constant base current shown in Fig. 5.84. From the plot shown in Fig. 5.84, it can be seen that if the transistor is supplying a load current that is less than I_{max}, then the voltage drop across the transistor will be less than

Inverters

$V_{CE,sat}$ which is approximately 2 V for high-voltage BJTs. Therefore, the load voltage will be within 2 V of the potential of the DC rail to which the transistor is connected. Specifically,

$$v_{load} = V_{dc} - V_{CE,sat} \tag{5.65}$$

If the source voltage is large compared to $V_{CE,sat}$, i.e.,

$$V_{dc} \gg V_{CE,sat} \tag{5.66}$$

then

$$v_{load} \approx V_{dc} \tag{5.67}$$

The previous equations apply only when Q_1 is on and the load current is positive. However, if the load current is negative and transistor Q_2 is off, the only path for the current is through diode D_1. In this case,

$$v_{load} = V_{dc} + V_{D1} \tag{5.68}$$

Typically, V_{D1} is on the order of 1 V, where

$$V_{dc} \gg V_{D1} \tag{5.69}$$

Thus,

$$v_{load} \approx V_{dc} \tag{5.70}$$

It can be seen that the output voltage is approximately equal to V_{dc} when Q_1 is on or when Q_2 is off and the load current is negative.

If transistor Q_2 is on and the load current is negative

$$v_{load} = V_{CE,sat} - V_{dc} \tag{5.71}$$

Hence,

$$v_{load} \approx -V_{dc} \tag{5.72}$$

With Q_1 off and the load current positive, the load current must flow through diode D_2. In this case,

$$v_{load} = -V_{dc} - V_{D2} \tag{5.73}$$

Therefore,

$$v_{load} \approx -V_{dc} \tag{5.74}$$

Hence, when Q_2 is on or when Q_1 is off and the load current is positive, the load voltage is approximately equal to $-V_{dc}$.

The hard-switched phase leg has four states: (1) Q_1 on and the load current is positive, (2) Q_2 off and the load current negative, (3) Q_2 on and the load current is negative, and (4) Q_1 off and the load current positive. In each of these states, the load current is flowing through one of the four solid-state devices composing the phase leg. With the conduction of current through these physical devices, there is an associated power loss. This conduction power loss can be calculated for each state as the product of the load current and the voltage across the device supplying the current. When Q_1 is on and the load current is positive, the conduction power loss is

$$P_{Q1,\text{con}} = i_{\text{load}} \cdot V_{CE,\text{sat}} \tag{5.75}$$

With Q_2 off and the load current negative, the load current flows through diode D_1 and introduces a conduction power loss of

$$P_{D1,\text{con}} = -i_{\text{load}} \cdot V_{D1} \tag{5.76}$$

In the third state with Q_2 on and a negative load current, the conduction loss is

$$P_{Q2,\text{con}} = -i_{\text{load}} \cdot V_{CE,\text{sat}} \tag{5.77}$$

When Q_1 is off and the load current is positive, the conduction power loss associated with D_2 is

$$P_{D2,\text{con}} = i_{\text{load}} \cdot V_{D2} \tag{5.78}$$

Transistor Turn-On Losses

Two examples of switching losses are explored in this section. Both examples involve the load voltage being switched from the lower DC rail, $-V_{dc}$, to the upper DC rail, $+V_{dc}$. In the first example, the load current, i_{load}, is assumed to be positive. In the second case, the load current is assumed to be negative. In both examples, the load is assumed to be inductive whereby the load current is essentially constant during the switching interval.

It is assumed, initially, that the load is latched to the lower DC rail (transistor Q_2 is on and Q_1 is off) and the load current is positive. Although Q_2 is gated on, the load current flows through the diode D_2, whereupon a small conduction loss, P_{D2}, is associated with D_2. The value of diode conduction power loss is the product of the load current and the forward voltage drop across D_1 (approximately 1 V). The conduction energy loss for D_2 can be calculated as the integral of the conduction power loss with respect to time, i.e.,

$$E_{D2} = \int P_{D2}\, dt = V_{D2} \int i_{\text{load}}\, dt \tag{5.79}$$

With all the load current flowing through the diode, the transistor Q_2 will not have any conduction losses. Therefore, the only loss initially associated with this state of the phase leg is the conduction loss in the diode.

When the commutation process begins, transistor Q_2 is turned off, and after a brief delay transistor Q_1 is turned on. Since transistor Q_2 is switched off under zero current conditions, there is no power loss associated with Q_2 in the turn-off process. However, when Q_1 is turned on, the load current does not immediately commute from diode D_2 to Q_1 because the minority carriers in the base region of Q_1 must be supplied before conduction through Q_1 can begin. As the base current, I_{B1}, adds minority carriers to the base of Q_1, the collector current starts increasing and displacing the current through D_2 as the source of the load current. This increase in the collector current can be observed in the simplified plots shown in Fig. 5.85. During this interval, both diode D_2 and transistor Q_1 are conducting; however, with D_2 conducting, the load voltage is still clamped to the lower DC rail. This results in the rail-to-rail voltage being placed across Q_1 while current is flowing through Q_1; thus, a large amount of power is lost in Q_1 during this phase of the commutation. This increase in the collector current, I_{C1}, and the constant V_{CE1} can be observed in Fig. 5.85 [1]. The resulting power loss in Q_1 during this interval is approximated as a straight line increasing to P_{max} (Fig. 5.85). However, since current is also flowing through diode D_2 during this interval, there is an additional loss term associated with the conduction of some of the load current through D_2.

Inverters

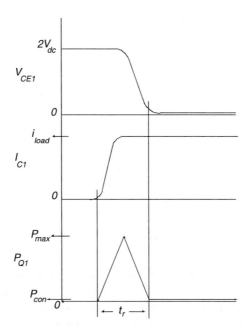

FIGURE 5.85 Simplified turn-on switching waveforms for a typical BJT.

The second interval of high loss begins when the collector current of Q_1 displaces all the diode current, at which point the load voltage switches from the lower rail to the upper rail. During this swing, transistor Q_1 conducts all of the load current and the load voltage goes from the rail-to-rail voltage to $V_{CE,sat}$ (Fig. 5.85). Hence, the product of the transistor current and the collector-to-emitter voltage is large during this interval and may be approximated as a straight line going from P_{max} to P_{con}.

The energy loss associated with the turning on of transistor Q_1 is the integral of the power loss during turn-on. This loss is divided into two parts: the loss in Q_1, and the conduction loss in D_2 during commutation. The energy loss in Q_1 is approximated as the area of the triangle in the power loss diagram (Fig. 5.85).

$$E_{Q1,turnon} = \frac{1}{2}P_{max}t_r = \frac{1}{2}(2V_{dc}i_{load})t_r \qquad (5.80)$$

The time required to complete the turn-on process is called the rise time, t_r. The current flowing through D_2 during the commutation of the load current is

$$I_{D2} = i_{load} - I_{C1} \qquad (5.81)$$

The diode current can be approximated as a straight line going from a value of i_{load} to zero in a time of $\frac{1}{2}t_r$. Therefore, the energy loss associated with the conduction of current through D_2 during the commutation of the load current can be expressed as

$$E_{D2,turnon} = \left(\frac{1}{2}i_{load}\right)V_{D2}\left(\frac{1}{2}t_r\right) \qquad (5.82)$$

Thus, the total turn-on energy loss is

$$E_{turnon} = \frac{1}{2}(2V_{dc}i_{load})t_r + \left(\frac{1}{2}i_{load}\right)V_{D2}\left(\frac{1}{2}t_r\right) \qquad (5.83)$$

When the load voltage reaches the upper rail, the switching sequence is complete. The only loss during this phase is the conduction loss in transistor Q_1. This loss is small because the transistor is in the saturation region where the collector-to-emitter voltage is $V_{CE,sat}$. The conduction energy loss for Q_1 is

$$E_{Q1,con} = \int P_{con} \, dt \approx V_{CE,sat} \int i_{load} \, dt \tag{5.84}$$

Transistor Turn-Off Losses

In this analysis, it is assumed that the load current is initially negative and the load voltage is to be switched from the lower rail to the upper rail. With a negative load current and the load voltage initially latched to the lower rail, transistor Q_2 supplies the load current. Initially, the only loss associated with this phase leg is the conduction loss in transistor Q_2. The conduction energy loss for Q_2 is

$$E_{Q2,con} = \int P_{Q2} \, dt = -V_{CE,sat} \int i_{load} \, dt \tag{5.85}$$

The energy loss has a negative sign in the third term because the load current is negative in this example and the energy loss is always considered to be positive. The collector current, I_{C2}, the collector-to-emitter voltage, V_{CE2}, and the conduction power loss for Q_2 can be observed in Fig. 5.86 [1].

To switch the output voltage, transistor Q_2 is switched off and after a brief delay transistor Q_1 is gated on. However, because minority carriers are still in the base of transistor Q_2, the collector current remains constant initially. As the minority carriers are collected, the minority carrier concentration in the base will begin to diminish. Consequently, the collector-to-emitter voltage for Q_2, V_{CE2}, will start to increase. As a result of the inductive nature of the load, the collector current will remain constant until V_{CE2} reaches a value of $2V_{dc}$. Therefore, during this period of the switching, there is significant power dissipated in transistor Q_2 because the load current flows through Q_2 and the voltage across Q_2 reaches $2V_{dc}$. This increase in V_{CE2} during which I_{C2} is constant can be observed in Fig. 5.85 along with the increasing power loss associated with this interval.

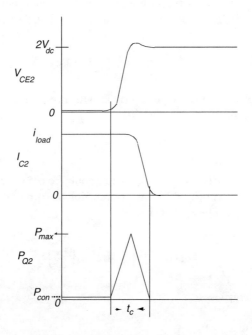

FIGURE 5.86 Simplified turn-off switching waveform for a typical BJT.

Inverters

Once V_{CE2} reaches a value of $2V_{dc}$, diode D_1 becomes forward-biased and begins to assume part of the load current. This commutation of the load current from Q_2 to D_1 causes I_{C2} to decrease. During this decrease, large amounts of power are still being dissipated in transistor Q_2. In addition, there is also a power loss associated with the conduction of current through diode D_1.

The energy loss associated with the turn-off process of transistor Q_2 is expressed as the sum of the energy loss in Q_2 and the conduction energy loss in D_1 during the commutation of the load current from Q_2 to D_1. The energy loss in Q_2 can be approximated as the area of the triangle in the power loss diagram for Q_2 (Eq. 5.86), where t_c is called the commutation time.

$$E_{Q2,\text{turnoff}} = \frac{1}{2}(P_{\max} \cdot t_c) = -\frac{1}{2}(2V_{dc}i_{\text{load}})t_c \qquad (5.86)$$

During the commutation of the load current, the diode current, I_{D1}, can be expressed as

$$I_{D1} = I_{C1} - i_{\text{load}} \qquad (5.87)$$

With a negative load current and I_{C1} going from a value of $-i_{\text{load}}$ to zero, the diode current can be approximated as a straight line going from zero to i_{load}. Therefore, the conduction energy loss associated with the commutation of the load current can be approximated as the product of the average diode current, the forward diode voltage drop, and the time of the commutation, $\frac{1}{2}t_c$.

$$E_{D1,\text{turnoff}} = \left(-\frac{1}{2}i_{\text{load}}\right)V_{D1}\left(\frac{1}{2}t_c\right) \qquad (5.88)$$

The total turn-off energy loss is

$$E_{\text{turnoff}} = -\frac{1}{2}(2V_{dc}i_{\text{load}})t_c + \left(-\frac{1}{2}i_{\text{load}}\right)V_{D1}\left(\frac{1}{2}t_c\right) \qquad (5.89)$$

When all the load current is conducting through D_1, the commutation of the load current from Q_2 to D_1 is complete. The only power loss in this final state of the switching sequence is the commutation loss associated with diode D_1. The conduction energy loss for D_1 is

$$E_{D1,\text{con}} = \int P_{D1}\,dt = -V_{D1}\int i_{\text{load}}\,dt$$

Analysis of ARCP Phase Leg

An analysis of an ARCP phase leg is presented in this section. This analysis begins with a description of the ARCP circuit. The switching of the load voltage from the lower rail to the upper rail is then explored for three separate cases: (1) commutation from a diode, (2) commutation from a transistor with low load current, and (3) commutation from a transistor with high load current. The elimation of the switching losses is discussed in each case.

ARCP Circuit Description

A circuit diagram of the ARCP phase leg is illustrated in Fig. 5.87 [2]. The ARCP contains snubber capacitors C_r between the load and the DC rails. The snubber capacitors serve the purpose of holding the voltage across the switches constant during turn-off. This makes it possible for the switch being turned-off to have zero voltage across it during turn-off, thus eliminating switching losses. The ARCP also includes an auxiliary circuit connected between the DC neutral and phase connection. The auxiliary circuit helps enable the load to be swung to the opposite rail to ensure zero turn-on voltage. If the auxiliary circuit is not included, then there is a load current constraint to ensure zero turn-on voltage.

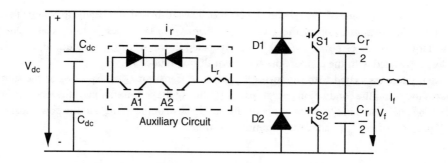

FIGURE 5.87 The auxiliary resonant commutated pole (ARCP).

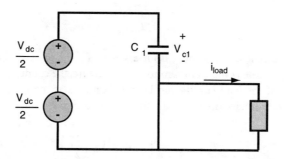

FIGURE 5.88 Circuit diagram of ARCP in state 1.

Low-to-High Commutation from Diode

An example of an ARCP single-phase leg commutation will now be examined in the case where the load will be switched from the lower rail to the upper rail with the diode D_2 initially conducting. Initially, the load is connected to the lower DC rail and switch S_2 is on, as illustrated in Fig. 5.88 (state 1). With the lower switch gated on, the capacitor C_1 has a DC voltage across it. Therefore, the capacitor voltage, V_{C1}, will be constant and the governing differential equation for this state is shown in (Eq. 5.90). Since the auxiliary circuit is gated off, the auxiliary current, i_r, is zero. The differential equation describing the auxiliary current in state 1 is shown in (Eq. 5.91).

$$pV_{C1} = 0 \tag{5.90}$$

$$pi_r = 0 \tag{5.91}$$

Upon request to switch to the upper rail, the load current is checked. Since the diode is assumed to be conducting in this example, the auxiliary switch A_2 is gated on. When A_2 is gated on, the inductance of the auxiliary circuit does not allow the auxiliary current, i_r, to change instantaneously; hence, the current through A_2 remains zero while the gate is being turned on, which eliminates losses associated with the turning on of A_2. With A_2 gated on, the auxiliary circuit is introduced to the circuit as illustrated in Fig. 5.89 (state 2). In state 2, L_r has a DC voltage across it, resulting in the auxiliary current ramping up linearly as described by Eq. (5.93). The voltage V_{dc} remains across the capacitor C_1 in this state; thus, the governing differential equation for the capacitor voltage is given by Eq. (5.92).

$$pV_{C1} = 0 \tag{5.92}$$

$$pi_r = \frac{V_{dc}}{2L_r} \tag{5.93}$$

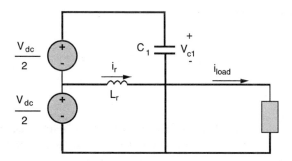

FIGURE 5.89 Circuit diagram of ARCP in state 2.

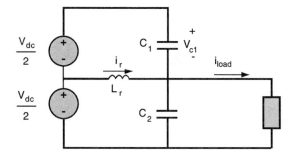

FIGURE 5.90 Circuit diagram of ARCP in state 3.

This increase in i_r displaces the diode current causing the load current to flow through the auxiliary circuit. When i_r exceeds the load current, the excess current flows through switch S_2, $i_r - i_{load}$. When the auxiliary current exceeds the load current plus a boost current, $i_{load} + i_{boost}$, the switch S_2 is gated off. The boost current acts to give the auxiliary circuit additional energy to help ensure that the load will swing completely to the upper rail, thus achieving zero voltage turn-on.

Although S_2 is gated off, the current through S_2 will not immediately go to zero; therefore, if the voltage across S_2 is allowed to swing immediately to the upper rail, there will be power loss in the switch because the power loss is the product of the current through the switch and the voltage across the switch. The capacitors are used to eliminate this loss. Since the voltage across a capacitor cannot change instantaneously, the capacitors hold the voltage across S_2 at zero until the current in S_2 goes to zero; therefore, the product of voltage times current for the switch is zero.

With both switches off and the auxiliary circuit on, the ARCP circuit enters state 3 (Fig. 5.90). Since the load is inductive, the load current is assumed to be constant during a switch; therefore, the excess auxiliary current flows into the snubber capacitors. This places positive charge on the upper plate of the lower capacitor and on the lower plate of the upper capacitor. This charge causes the load voltage to begin to increase. The auxiliary current will continue to increase until the load voltage exceeds the DC neutral voltage, then causing the auxiliary current to decrease. When the load voltage reaches the upper rail DC voltage, the diode D_1 will become forward-biased and switch S_1 is gated on; hence, the load is latched to the upper rail.

The differential equation for V_{C1} in state 3 can be derived by writing Kirchhoff's current law (KCL) at the load node (Eq. 5.94) and Kirchhoff's voltage law (KVL) around the capacitors and the DC voltage sources (Eq. 5.95).

$$i_r + C_1(pV_{C1}) - C_2(pV_{C2}) - i_{load} = 0 \quad (5.94)$$

$$V_{C1} + V_{C2} = V_{dc} \quad (5.95)$$

A relationship between pV_{C1} and pV_{C2} can be established (Eqs. 5.96 and 5.97) by taking the derivative of both sides of Eq. (3–6).

$$p(V_{C1} + V_{C2}) = pV_{dc} = 0 \tag{5.96}$$

$$pV_{C1} = -pV_{C2} \tag{5.97}$$

Substituting Eq. (5.97) into Eq. (5.94) and simplifying yields the differential equation for V_{C1} shown in Eq. (5.99). The differential equation describing i_r in state 3 can be derived by simplifying the KVL equation around the loop including the auxiliary circuit, C_1, and the upper DC voltage source (Eqs. 5.98 and 5.100).

$$\frac{V_{dc}}{2} + L_r(pi_r) - V_{C1} = 0 \tag{5.98}$$

$$pV_{C1} = \frac{i_{load} - i_r}{C_1 + C_2} \tag{5.99}$$

$$pi_r = \frac{V_{C1} - \frac{V_{dc}}{2}}{L_r} \tag{5.100}$$

When diode D_1 is forward-biased and switch S_1 is gated on, the snubber capacitors ensure that the voltage across S_1 remains zero, thereby eliminating any turn-on loss in switch S_1. With the load latched to the upper rail, the auxiliary circuit has a negative voltage across it. Therefore, the auxiliary current will linearly ramp down as described by Eq. (5.102). With switch S_1 gated on, the capacitor voltage V_{C1} is held at zero; thus, the derivative of the capacitor voltage will also be zero (Eq. 5.101).

$$pV_{C1} = 0 \tag{5.101}$$

$$pi_r = -\frac{V_{dc}}{2L_r} \tag{5.102}$$

When the auxiliary current reaches zero, switch A_2 is gated off. Since the current through A_2 is zero when switch A_2 is gated off, there is no switching losses associated with A_2.

When A_2 is gated off, the auxiliary circuit is taken off the circuit and the final state is reached with the load latched to the upper rail. With the auxiliary circuit removed, the auxiliary current is zero; hence, the derivative of the auxiliary current will also be zero Eq. (5.104). Since switch S_1 is still latched on, the derivative of the capacitor voltage will remain zero Eq. (5.103).

$$pV_{C1} = 0 \tag{5.103}$$

$$pi_r = 0 \tag{5.104}$$

A computer-simulated plot of the upper capacitor voltage V_{C1} and the auxiliary current i_r is shown in Fig. 5.93 for a switch from low to high with the diode initially conducting. The ARCP parameter values used in the simulation were derived from Ref. 2 and are listed in Table 5.3.

The computer code used in the simulation was written in Advanced Continuous Simulation Language (ACSL). From the graph, the individual switching states can be observed along with the transition points

Inverters

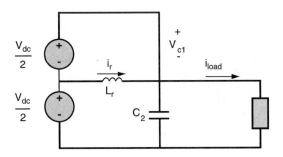

FIGURE 5.91 Circuit diagram of ARCP in state 4.

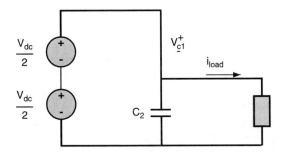

FIGURE 5.92 Circuit diagram of ARCP in state 5.

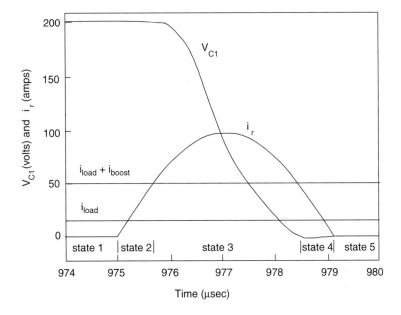

FIGURE 5.93 ARCP commutation low-to-high from diode.

between states. Initially, the circuit is latched to the lower rail and the auxiliary circuit is off (state 1). When the auxiliary circuit is gated on, the auxiliary current begins to increase, thus representing the transition into state 2. When the auxiliary current exceeds the sum of the load current and the boost current, state 3 is entered. In state 3, the load voltage, $V_{dc} - V_{C1}$, is swung from the lower rail to the upper rail. Upon the load voltage reaching the upper rail ($V_{C1} = 0$), the circuit passes into state 4.

TABLE 5.3 ARCP Parameters

ARCP Parameter	Value Used in Simulation
Upper capacitor, C_1	0.159 µF
Lower capacitor, C_2	0.159 µF
Resonant inductor, L_r	0.159 µH
Threshold current, i_{th}	60 A
Boost current, i_{boost}	30 A
Rail-to-rail voltage, V_{dc}	200 V

In state 4, the auxiliary current decreases linearly. When the auxiliary current equals zero, state 5 is reached. The capacitor voltage, V_{C1}, is not constant during states 2 and 4 as discussed earlier because in the simulation the diodes and switches were not modeled as ideal; thus, a small deviation was introduced.

Low-to-High Commutation from Switch, Low Current

In the second example of an ARCP single-leg commutation, it is assumed that the load voltage is to be switched from the lower to upper rail with switch S_2 initially conducting a small current. The governing differential equations describing each state are the same as in the previous example.

Initially, the load is latched to the lower rail (state 1). When the ARCP phase leg is commanded to switch to the upper rail, the load current is checked. Since the load current is assumed to be flowing through the switch, the magnitude of the current is compared with a threshold value. In this case, it is assumed that the current is less than the threshold value; therefore, the load does not contain enough energy to overcome losses to drive the load to the opposite rail without the introduction of the auxiliary circuit. Thus, the auxiliary circuit is turned on by gating switch A_2 on.

With the auxiliary circuit gated on, the circuit is in state 2. The auxiliary current will ramp up because of the DC voltage placed across it. When the auxiliary current reaches the value of the boost current plus the load current, the auxiliary circuit has substantial energy to drive the load to the opposite rail. Therefore, the lower switch, S_2, is gated off. The snubber capacitors prevent switching losses in S_2 because the capacitors hold the voltage across the switch to zero while the switch current diminishes.

State 3 is entered after turning off switch S_2. The load current and the auxiliary current will charge the snubber capacitors, hence driving the load voltage to the upper rail. When the load voltage reaches the upper rail, the diode D_1 will become forward-biased and will stop further charging of the capacitors by conducting the excess current. When the diode becomes forward-biased, the switch S_1 is gated on with zero volts across the switch, thereby preventing any turn-on losses in S_1.

When the load voltage is equal to the upper rail, state 4 (Fig. 5.91) is obtained. The auxiliary circuit has a negative voltage across it. Therefore, the auxiliary current will ramp down. When the auxiliary current reaches zero, switch A_2 is gated off to disconnect the auxiliary circuit. With zero current through A_2 during turn-off, switching losses associated with A_2 are avoided. The load is now latched to the upper rail and the auxiliary circuit is removed; therefore, the commutation is completed and the final state, state 5, is reached.

The upper capacitor voltage, V_{C1}, and the auxiliary circuit, i_r, are plotted in Fig. 5.94 for commutation from the switch in the low-current case. Commutation from the switch at low current levels is similar to commutation from the diode except that the auxiliary current does not have to become as large in the switch example. This is because the load current aids the commutation process when the switch is initially conducting and hinders commutation when the diode is initially conducting.

Low-to-High Commutation from Switch, High Current

The final case to be explored involves switching of the load from the lower to upper rail with the switch initially conducting a current larger than the threshold current. Initially, the load is latched to the lower rail and the circuit is in state 1 (Fig. 5.88). For a switch from the lower rail to the upper rail when switch S_2 is conducting a current larger than the threshold value, the auxiliary circuit does not need to be included in the switching sequence. This is a result of the load inductor having sufficient energy to

Inverters

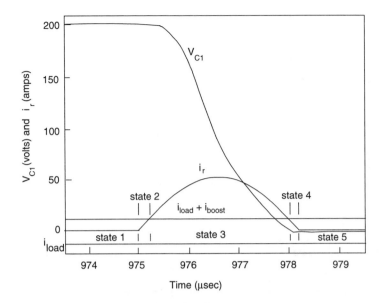

FIGURE 5.94 ARCP commutation low-to-high from switch (low current).

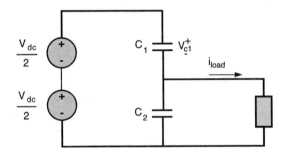

FIGURE 5.95 Circuit diagram of ARCP in state 6.

overcome any switching losses and driving the load voltage to the opposite rail. In this case, the ARCP acts exactly like a snubber circuit, because switch S_2 is gated off without introducing the auxiliary circuit.

Once S_2 is gated off, state 6 is entered. In state 6, the load current charges the snubber capacitors and drives the load to the upper rail. When the load voltage reaches the upper rail, diode D_1 is forward-biased and switch S_1 is gated on under zero voltage conditions.

The differential equation characterizing V_{C1} in state 6 can be derived exactly as in state 3 with the auxiliary current neglected (Eq. 5.105). Since the auxiliary circuit is not gated on in state 6, the derivative of the auxiliary current is zero (Eq. 5.106).

$$pV_{C1} = \frac{i_{load}}{C_1 + C_2} \tag{5.105}$$

$$pi_r = 0 \tag{5.106}$$

With the load gated to the upper rail, the switching transition is completed. and the circuit is in state 5 (Fig. 5.92).

A plot of the capacitor voltage, V_{C1}, is displayed in Fig. 5.96. Since the load is assumed to be inductive, the load current is assumed to be constant over a switching cycle. Therefore, during state 6 (Fig. 5.95) the capacitor voltage decreases linearly until diode D_1 is forward-biased and conducts the load current.

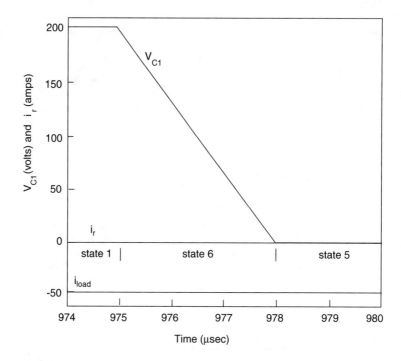

FIGURE 5.96 ARCP commutation low-to-high from switch (high current).

Simulation of ARCP Phase Leg

The equations describing the ARCP phase leg have been implemented digitally using ACSL. The ARCP phase leg switch command is given by the ACSL variable SW1. When SW1 = 5, the output voltage is connected to the upper rail (state 5) or is switching to the upper rail. Similarly, when SW1 = 1, the output voltage is connected to the lower rail (state 1) or is switching to the lower rail. The derivatives of the state variables are dependent upon the present state of the phase leg. The present state is given by the variable state1. Logic is used to detect a change from one state into the next state, at which a flag is set. This flag calls a discrete block that changes state1 to its new value, calls a data file to log the present value of all the prepare variable, and resets the value of the flag. Discrete blocks were used for changes in state because this allowed the state equations to be changed at discrete instances in time. With a switching frequency of 20 kHz, 1.36 s of central processor time on a Sun Sparcstation 5 was required to run the computer simulation for 5 ms. Computer studies involving the ARCP phase leg are described in subsequent sections.

Analysis of ARCP H-Bridge

A comparison of losses is made between an H-bridge circuit using conventional hard-switched phase legs and an H-bridge circuit using ARCP phase legs. Prior to discussing the losses, an H-bridge circuit is discribed along with the pulse width modulation control used in this study. Analysis of the ARCP H-bridge is also presented.

Circuit Description and Pulse Width Modulation

An H-bridge circuit is a load connected between two phase legs. An example of an H-bridge using ARCP phase legs is shown in Fig. 5.97. The load voltage, v_{load}, in an H-bridge can be swung from $+V_{dc}$ to $-V_{dc}$. A positive load voltage can be achieved by having the first phase leg latched to the upper rail and the second phase leg latched to the lower phase leg. A negative load voltage can be obtained by having the first phase leg latched to the lower rail and the second phase leg latched to the upper rail.

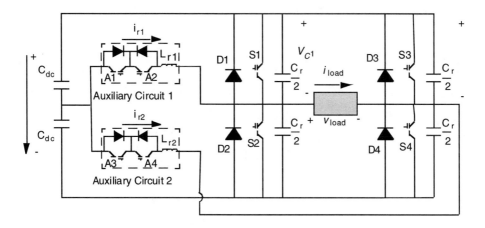

FIGURE 5.97 H-bridge using ARCP phase legs.

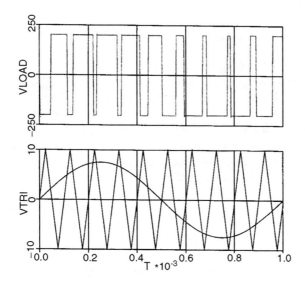

FIGURE 5.98 v_{load} vs. time and PVM waveforms.

Many control strategies may be used to control the load voltage or load current. In this section, pulse width modulation (PWM) will be uses to control the load voltage. In PWM, a controlled sinusoidal waveform that oscillates at the fundamental frequency of the load, f_1 (1 kHz), is compared with a triangle wave whose frequency is on the order of ten times larger than f_1 to determine whether the load voltage is high or low. If the controlled waveform is larger than the triangle wave, the load voltage is switched positive. If the controlled waveform becomes smaller than the triangle wave, the load voltage is switched negative. Plots showing the controlled waveform, the triangle waveform, and the load voltage are shown in Fig. 5.98 [3].

H-Bridge Loss Analysis

In this section, a comparison is made between the losses of a hard-switched H-bridge and an H-bridge using ARCP phase legs. In the standard hard-switched converter, the losses will be conduction losses of the transistors and diodes and the switching losses in the transistors. However, in the ARCP converter, the switching losses in the transistors are eliminated at the cost of introducing conduction losses from the auxiliary circuits. The energy loss in the hard-switched converter is explained in detail in the section on Losses in Hard-Switched Inverters with the equations describing the losses shown in Eqs. (5.107) through (5.110).

$$E_D = \int P_D dt = V_D \int i_{\text{load}} dt \tag{5.107}$$

$$E_{Q,\text{con}} = \int P_{\text{con}} dt = V_{CE,\text{sat}} \int i_{\text{load}} dt \tag{5.108}$$

$$E_{\text{turnon}} = \frac{1}{2}(2V_{\text{dc}} i_{\text{load}}) t_r + \left(\frac{1}{2} i_{\text{load}}\right) V_{D2}\left(\frac{1}{2} t_r\right) \tag{5.109}$$

$$E_{\text{turnoff}} = -\frac{1}{2}(2V_{\text{dc}} i_{\text{load}}) t_c + \left(-\frac{1}{2} i_{\text{load}}\right) V_{D1}\left(\frac{1}{2} t_c\right) \tag{5.110}$$

In the ARCP converter, the conduction losses will be the same; however, the switching losses shown in Eqs. (5.109) and (5.110) are replaced by the auxiliary circuit conduction loss (Eq. 5.11).

$$E_{\text{aux,con}} = \int P_{\text{aux,con}} dt = V_{CE,\text{sat}} \int i_r dt \tag{5.111}$$

Plots of the energy loss during one cycle using PWM and an H-bridge with a resistive and inductive load for both the hard-switched example, losshs, and the ARCP example, lossarcp, are shown in Fig. 5.99. In this example, the turn-on commutation interval, t_c, and the turn-off rise time interval, t_r, were both set equal to 5 μs, which is a typical value for high-power transistors. From the plot, it is shown that through one PWM cycle the ARCP H-bridge dissipated about 1/9 the amount of energy of the hard-switched example used. This drastic savings in energy not only allows the application to be more efficient, but also allows the switching circuitry to be reduced is size because of the elimination of heat sinks. This reduction in size makes the ARCP technology extremely useful in applications where size constraints are the major driving factors.

Plots of the capacitor voltage, V_{C1}, and the auxiliary circuit current, i_{r1}, are shown in Fig. 5.100. The capacitor voltage and auxiliary current for the second phase leg is shown in Fig. 5.101. The plots show that the auxiliary circuits are only on during a capacitor voltage swing. Therefore, the only time the additional

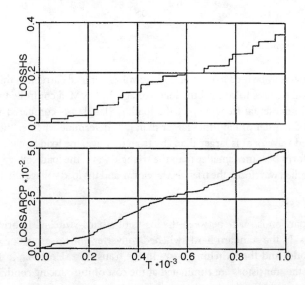

FIGURE 5.99 Energy losses for hard-switched and ARCP converters.

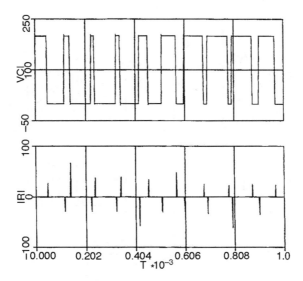

FIGURE 5.100 ARCP H-bridge example, V_{C1} and i_{r1} vs. time.

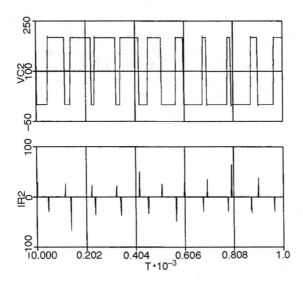

FIGURE 5.101 ARCP H-bridge converter, V_{C2} and i_{r2} vs. time.

circuitry introduces conduction losses is during the switching interval. Since the auxiliary circuits are turned on during every switching interval, the high current switching case of the ARCP, which allows the voltage to swing from one rail to the opposite rail without the introduction of the auxiliary circuit, is not entered.

Plots of the load current, i_{load}, and the load voltage, v_{load}, for the RL-load are shown in Fig. 5.102. From the plot of the load current, the fundamental frequency of the load is shown to be 1 kHz. This is the same frequency of the control sinusoidal wave used in the PWM control.

Analysis of ARCP Three-Phase Inverter

In this section, a variable-speed drive system that includes a three-phase ARCP inverter, a current controller, and an induction motor, is described. A computer simulation of this system has been implemented using ACSL. Results of a computer study using the variable-speed drive system are presented.

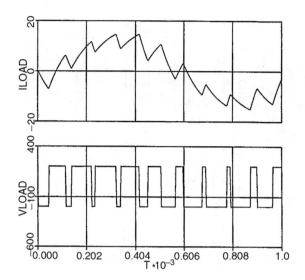

FIGURE 5.102 ARCP H-bridge converter, i_{load} and v_{load} vs. time.

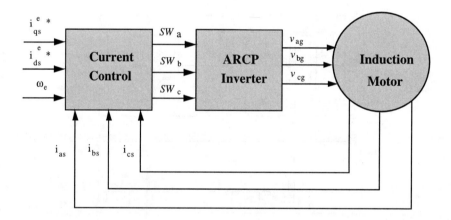

FIGURE 5.103 Three-phase example using ARCP inverter.

Description of ARCP Three-Phase Circuit

A block diagram of the system configuration studied is shown in Fig. 5.103. The current controller is depicted in Fig. 5.104. Therein, i_{qs}^{e*} and i_{ds}^{e*} are the commanded currents in the synchronously rotating reference frame. The speed of the synchronous reference frame is given by ω_e. The synchronous reference frame variables are transformed into the stationary reference frame by

$$\begin{bmatrix} i_{as}^* \\ i_{bs}^* \\ i_{cs}^* \end{bmatrix} = {}^eK_s^s \begin{bmatrix} i_{qs}^{e*} \\ i_{ds}^{e*} \end{bmatrix} = \begin{bmatrix} \cos\Theta_e & \sin\Theta_e \\ \cos\left(\Theta_e - \frac{2\pi}{3}\right) & \sin\left(\Theta_e - \frac{2\pi}{3}\right) \\ \cos\left(\Theta_e + \frac{2\pi}{3}\right) & \sin\left(\Theta_e + \frac{2\pi}{3}\right) \end{bmatrix} \begin{bmatrix} i_{qs}^{e*} \\ i_{ds}^{e*} \end{bmatrix} \quad (5.112)$$

The actual *as*, *bs*, and *cs* currents (i_{as}, i_{bs}, and i_{cs}) are then subtracted from the commanded *as*, *bs*, and *cs* currents (i_{as}^*, i_{bs}^*, and i_{cs}^*) to produce an error value i_ε for each phase [3]. If the error value is larger

Inverters

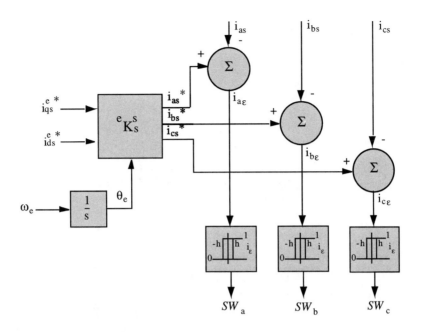

FIGURE 5.104 Current control block diagram.

than a hysteresis value, the switch signal for that phase (SW_a, SW_b, or SW_c) will command that phase leg to switch to the upper rail to increase the current of that phase and to reduce the error value. If the error value becomes more negative than a negative hysteresis value, the switch signal for that phase will command the phase leg to switch to the lower rail; thus, the phase current will decrease and the error value will become smaller in magnitude.

The switch signals SW_a, SW_b, and SW_c are used to control the a, b, and c phase legs of the inverter, respectively. A circuit diagram of the inverter using ARCP phase legs is shown in Fig. 5.105. The inverter is identical to a conventional inverter except for the auxiliary circuit and the snubber capacitors associated with each phase. The output voltages of the phase legs is used as the three-phase input voltage for the induction motor. These voltages are the voltages between the output phase leg and the lower rail of the inverter. Since the neutral of the induction motor is internal to the motor, the neutral voltage v_n is not equal to the lower rail voltage v_g of the inverter. Therefore, to obtain v_{as}, v_{bs}, and v_{cs} for the induction motor the following algebra must be applied [4]:

$$v_{ng} = v_n - v_g = \frac{1}{3}(v_{ag} + v_{bg} + v_{cg}) \tag{5.113}$$

$$v_{as} = v_{ag} - v_{ng} \tag{5.114}$$

$$v_{bs} = v_{bg} - v_{ng} \tag{5.115}$$

$$v_{cs} = v_{cg} - v_{ng} \tag{5.116}$$

With v_{as}, v_{bs}, and v_{cs} as inputs to the induction motor the phase currents (i_{as}, i_{bs}, and i_{cs}) can be calculated [4]. These phase currents serve as inputs to the current controller.

Computer Study

In this study, the steady-state characteristics of the three-phase system shown in Fig. 5.103 are established by computer simulation. The parameters of an ARCP phase leg were shown in Table 5.3, and the parameters of the induction motor are listed in Table 5.4. It is assumed that the induction machine is operating at

TABLE 5.4 Induction Motor Parameters

Induction Motor Parameter	Parameter Value
r_s	0.087 Ω
X_{ls}	0.302 Ω
X_M	13.08 Ω
X'_{lr}	0.302 Ω
r'_r	0.228 Ω
J	1.662 kg/m^2
Rated line-to-line voltage	460 V
Rated slip	0.05278

Source: Krause, P.C. et al., *Analysis of Electric Machinery*, McGraw-Hill, New York, 1986. With permission.

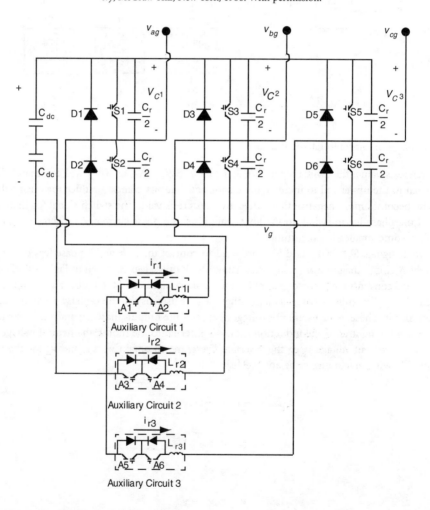

FIGURE 5.105 Inverter using ARCP phase legs.

half of rated or base frequency (188.5 rad/s); therefore, the required line-to-neutral voltage is approximately 133 V rms to produce rated torque. The rated slip is 0.05278. Hence, the rotor speed for this study is set to a constant value of 178.6 rad/s.

Plots of the simulated current i_{as} and the commanded current i^*_{as} are shown in Fig. 5.106. These plots were made with a hysteresis value of 7.5 A or a tolerance band of 15 A [3]. The smaller the tolerance

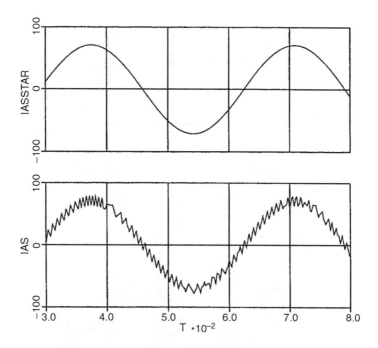

FIGURE 5.106 Plots of i_{as}^* and i_{as} vs. time.

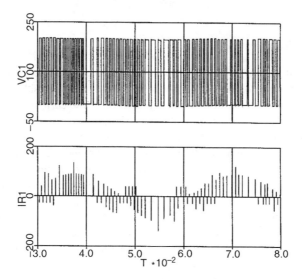

FIGURE 5.107 Auxiliary circuit current, i_{r1}, and capacitor voltage, V_{C1}, for phase a.

band, the more closely the actual currents will track the commanded currents. However, with this smaller tolerance band, the switching frequency of the phase legs will increase because of the increased restrictions on the controls.

The upper capacitor voltage and the auxiliary circuit current for the a-phase leg is shown in Fig. 5.107. Each spike in the auxiliary current corresponds to commutation from a diode or commutation from a switch in the low-current switch in the a-phase leg. Although the capacitor voltage waveform appears to be a square wave with instantaneous switching, this is because timescale is too large to observe the soft-switching transitions of the capacitor voltages as discussed in earlier. From the plot of the capacitor voltage, it can

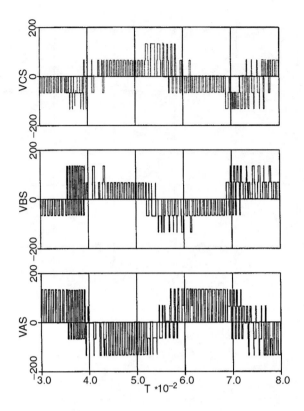

FIGURE 5.108 Stator voltages for the induction motor.

be observed that the switching frequency of the phase leg does not remain constant. This is result of the switching frequency depending on how fast the current changes from one side of the tolerance band to the other, which is not constant because of the dependence of the current changes on V_{dc}, the back-electromotive force, and the load of the induction motor [3].

Comparing i_{as} from Fig. 5.106 with the auxiliary circuit current for the *a*-phase in Fig. 5.107, a dependence of phase current on the auxiliary current can be examined. When the phase current is smaller than the threshold value for the ARCP (60 A), every switch is in the low-current case. In this case, the auxiliary current spikes oscillate from positive to negative since one of the commutations is from a diode and the other is from a switch. In the other case, were the phase current is positive and larger than the threshold value, every auxiliary current spike is positive. This is a result of the commutation from the switch (upper-to-lower transition) not requiring the auxiliary circuit to be turned on. Therefore, the spikes are all results of lower-to-upper transitions when the current is commutating from the lower diode. When the phase current is negative and larger in magnitude than the threshold voltage, all the auxiliary current spikes are negative. This is for the same reason as in the previous case except that commutation here is from the upper diode.

Plots of the stator voltages (v_{as}, v_{bs}, and v_{cs}) for the induction motor are shown in Fig. 5.108. The peak value of the stator voltages is 133 V, two thirds of the rail-to-rail voltage of the ARCP. This value is comparable to the rms voltage for the induction motor at half speed.

Summary

In this section, the ARCP phase leg was described and analyzed. For purposes of comparison, the switching and conduction losses of a conventional hard-switched phase leg were also described. Losses associated with hard-switching include conduction loss and switching loss associated with transistor turn-on and turn-off.

In an ARCP phase leg, all switching losses are eliminated by turning transistors on and off under either zero current or zero voltage conditions. This is accomplished with the introduction of anauxiliary circuit that aids the commutation process. With the switching losses eliminated, the only losses in an ARCP phase leg involve the conduction losses in the phase leg and in the auxiliary circuit.

The switching sequence of the ARCP and the corresponding state equations are described in detail in this section. Based on these state equations, computer models of the ARCP and hard-switched phase legs were developed and implemented using ACSL.

A computer study was performed to compare the losses of ARCP and hard-switched switching strategies. In the study, an H-bridge using conventional PWM was analyzed using both types of phase legs and a comparison was made between the energy loss in both cases. It was shown that the ARCP, by eliminating switching losses, used approximately 1/9 the energy per PWM cycle of the hard-switched converter.

Finally, a variable-speed induction motor drive system using a three-phase ARCP inverter was simulated. The dependence of the auxiliary current of an ARCP phase leg on the phase current was presented along with a discussion on the relationship between current control parameters and switching frequency.

References

1. R. E. Tarter, *Principles of Solid-State Power Conversion*, Howard W. Sams & Co., Indianapolis, IN, 1985, 68–76.
2. R. W. DeDoncker, Resonant pole converters, presented at *EPE*-1993, 4.1–4.45.
3. N. Mohan, T. M. Undeland, and W. P. Robbins, *Power Electronics: Converters, Applications, and Design*, John Wiley & Sons, New York, 1989, 104–121, 146–148.
4. P. C. Krause, O. Wasynczuk, and S. D. Sudhoff, *Analysis of Electric Machinery*, McGraw-Hill, New York, 1986.

6
Multilevel Converters

Keith Corzine
University of Wisconsin–Milwaukee

6.1 Introduction 6-1
6.2 Multilevel Voltage Source Modulation 6-2
6.3 Fundamental Multilevel Converter Topologies................ 6-7
   Diode-Clamped Multilevel Converters • Flying-Capacitor Multilevel Converters • Cascaded H-Bridge Multilevel Converters • Multilevel H-Bridge Converters
6.4 Cascaded Multilevel Converter Topologies 6-15
   Cascaded Multilevel Converters • Cascaded Multilevel H-Bridge Converters
6.5 Multilevel Converter Laboratory Examples 6-17
   Three-Level Diode-Clamped Inverter • The Cascade-3/2 Inverter • The Cascade-5/3H Inverter
6.6 Conclusion 6-21

6.1 Introduction

Multilevel power conversion was first introduced 20 years ago [1]. The general concept involves utilizing a higher number of active semiconductor switches to perform the power conversion in small voltage steps. There are several advantages to this approach when compared with traditional (two-level) power conversion. The smaller voltage steps lead to the production of higher power quality waveforms and also reduce the dv/dt stresses on the load and reduce the electromagnetic compatibility (EMC) concerns. Another important feature of multilevel converters is that the semiconductors are wired in a series-type connection, which allows operation at higher voltages. However, the series connection is typically made with clamping diodes, which eliminates overvoltage concerns. Furthermore, since the switches are not truly series connected, their switching can be staggered, which reduces the switching frequency and thus the switching losses.

One clear disadvantage of multilevel power conversion is the larger number of semiconductor switches required. It should be pointed out that lower voltage rated switches can be used in the multilevel converter and therefore the active semiconductor cost is not appreciably increased when compared with the two-level case. However, each active semiconductor added requires associated gate drive circuitry and adds further complexity to the converter mechanical layout. Another disadvantage of multilevel power converters is that the small voltage steps are typically produced by isolated voltage sources or a bank of series capacitors. Isolated voltage sources may not always be readily available and series capacitors require voltage balance. To some extent, the voltage balancing can be addressed by using redundant switching states, which exist due to the high number of semiconductor devices. However, for a complete solution to the voltage-balancing problem, another multilevel converter may be required [2–4].

In recent years, there has been a substantial increase in interest in multilevel power conversion. This is evident by the fact that some Institute of Electrical and Electronic Engineers (IEEE) conferences are now holding entire sessions on multilevel converters. Recent research has involved the introduction of novel converter topologies and unique modulation strategies. Some applications for these new converters include industrial drives [5–7], flexible AC transmission systems (FACTS) [8–10], and vehicle propulsion [11, 12]. One area where multilevel converters are particularly suitable is that of medium-voltage drives [13].

This chapter presents an overview of multilevel power conversion methods. The first section describes a general multilevel power conversion system. Converter performance is discussed in terms of voltage levels without regard to the specific topology of the semiconductor switches. A general method of multilevel modulation is described that may be extended to any number of voltage levels. The next section discusses the switching state details of fundamental multilevel converter topologies. The concept of redundant switching states is introduced in this section as well. The next section describes cascaded multilevel topologies, which involve alternative connections of the fundamental topologies. The final section shows example multilevel power conversion systems including laboratory measurements.

6.2 Multilevel Voltage Source Modulation

Before proceeding with the discussion of multilevel modulation, a general multilevel power converter structure will be introduced and notation will be defined for later use. Although the primary focus of this chapter is on power conversion from DC to an AC voltages (inverter operation), the material presented herein is also applicable to rectifier operation. The term *multilevel converter* is used to refer to a power electronic converter that may operate in an inverter or rectifier mode.

Figure 6.1 shows the general structure of the multilevel converter system. In this case, a three-phase motor load is shown on the AC side of the converter. However, the converter may interface to an electric utility or drive another type of load. The goal of the multilevel pulse-width modulation (PWM) block is to switch the converter transistors in such a way that the phase voltages v_{as}, v_{bs}, and v_{cs} are equal to commanded voltages v_{as}^*, v_{bs}^*, and v_{cs}^*. The commanded voltages are generated from an overall supervisory

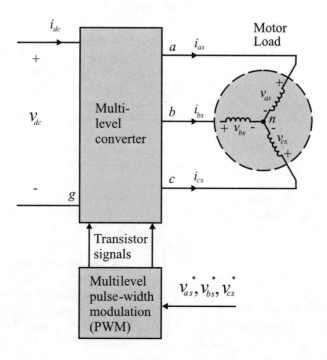

FIGURE 6.1 Multilevel converter structure.

control [14] and may be expressed in a general form as

$$v_{as}^* = \sqrt{2}\, v_s^* \cos(\theta_c) \tag{6.1}$$

$$v_{bs}^* = \sqrt{2}\, v_s^* \cos\left(\theta_c - \frac{2\pi}{3}\right) \tag{6.2}$$

$$v_{cs}^* = \sqrt{2}\, v_s^* \cos\left(\theta_c + \frac{2\pi}{3}\right) \tag{6.3}$$

where v_s^* is a voltage amplitude and θ_c is an electrical angle. To describe how the modulation is accomplished, the converter AC voltages must be defined. For convenience, a line-to-ground voltage is defined as the voltage from one of the AC points in Fig. 6.1 (a, or b, or c) to the negative pole of the DC voltage (labeled g in Fig. 6.1). For example, the voltage from a to g is denoted v_{ag}. It is important to note that the converter has direct control of the voltages v_{ag}, v_{bg}, and v_{cg}. The next step in defining the control of the line-to-ground voltages is expressing a relationship between these voltages and the motor phase voltages. Assuming a balanced wye-connected load, it can be shown that [15]

$$\begin{bmatrix} v_{as} \\ v_{bs} \\ v_{cs} \end{bmatrix} = \frac{1}{3} \begin{bmatrix} 2 & -1 & -1 \\ -1 & 2 & -1 \\ -1 & -1 & 2 \end{bmatrix} \begin{bmatrix} v_{ag} \\ v_{bg} \\ v_{cg} \end{bmatrix} \tag{6.4}$$

Because an inverse of the matrix in Eq. (6.4) does not exist, there is no direct relationship between commanded phase voltages and line-to-ground voltages. In fact, there are an infinite number of voltage sets $\{v_{ag}\ v_{bg}\ v_{cg}\}$ that will yield a particular set of commanded phase voltages because any zero sequence components of the line-to-ground voltages will not affect the phase voltages according to Eq. (6.4). In a three-phase system, zero sequence components of $\{v_{ag}\ v_{bg}\ v_{cg}\}$ include DC offsets and triplen harmonics of θ_c. To maximize the utilization of the DC bus voltage, the following set of line-to-ground voltages may be commanded [16]

$$v_{ag}^* = \frac{v_{dc}}{2}\left[1 + m\cos(\theta_c) - \frac{m}{6}\cos(3\theta_c)\right] \tag{6.5}$$

$$v_{bg}^* = \frac{v_{dc}}{2}\left[1 + m\cos\left(\theta_c - \frac{2\pi}{3}\right) - \frac{m}{6}\cos(3\theta_c)\right] \tag{6.6}$$

$$v_{cg}^* = \frac{v_{dc}}{2}\left[1 + m\cos\left(\theta_c + \frac{2\pi}{3}\right) - \frac{m}{6}\cos(3\theta_c)\right] \tag{6.7}$$

where m is a modulation index. It should be noted that the power converter switching will yield line-to-ground voltages with a high-frequency component and, for this reason, the commanded voltages in Eqs. (6.5) to (6.7) cannot be obtained instantaneously. However, if the high-frequency component is neglected, then the commanded line-to-ground voltages may be obtained on a fast-average basis. By substitution of Eqs. (6.5) to (6.7) into Eq. (6.4), it can be seen that commanding this particular set of line-to-ground voltages will result in phase voltages of

$$\hat{v}_{as} = \frac{m v_{dc}}{2}\cos(\theta_c) \tag{6.8}$$

$$\hat{v}_{bs} = \frac{m v_{dc}}{2}\cos\left(\theta_c - \frac{2\pi}{3}\right) \tag{6.9}$$

$$\hat{v}_{cs} = \frac{m v_{dc}}{2}\cos\left(\theta_c + \frac{2\pi}{3}\right) \tag{6.10}$$

where the ^ symbol denotes fast-average values. By comparing Eqs. (6.8) to (6.10) with Eqs. (6.1) to (6.3), it can be seen that the desired phase voltages are achieved if

$$m = \frac{2\sqrt{2}\, v_s^*}{v_{dc}} \qquad (6.11)$$

It should be noted that in H-bridge-based converters, the range of line-to-ground voltage is twice that of converters where one DC voltage supplies all three phases (as in Fig. 6.1). The modulation method here can accommodate these converters if the modulation index is related to the commanded voltage magnitude by

$$m_H = \frac{\sqrt{2}\, v_s^*}{v_{dc}} \qquad (6.12)$$

The modulation process described here may be applied to H-bridge converters by substituting m_H for m in the equations that follow. The benefit of including the third harmonic terms in Eqs. (6.5) to (6.7) is an extended range of modulation index [16]. In particular, the range of the modulation index is

$$0 \leq m \leq \frac{2}{\sqrt{3}} \qquad (6.13)$$

It is sometimes convenient to define a modulation index that has an upper limit of 100% or

$$\overline{m} = \frac{\sqrt{3}}{2} m \qquad (6.14)$$

The next step in the modulation process is to define normalized commanded line-to-ground voltages, which will be referred to as duty cycles. In terms of the modulation index and electrical angle, the duty cycles may be written:

$$d_a = \frac{1}{2}\left[1 + m\cos(\theta_c) - \frac{m}{6}\cos(3\theta_c)\right] \qquad (6.15)$$

$$d_b = \frac{1}{2}\left[1 + m\cos\left(\theta_c - \frac{2\pi}{3}\right) - \frac{m}{6}\cos(3\theta_c)\right] \qquad (6.16)$$

$$d_c = \frac{1}{2}\left[1 + m\cos\left(\theta_c + \frac{2\pi}{3}\right) - \frac{m}{6}\cos(3\theta_c)\right] \qquad (6.17)$$

To relate the duty cycles to the inverter switching operation, switching states must be defined that are valid for any number of voltage levels. Here, the switching states for the a-, b-, and c-phase will be denoted s_a, s_b, and s_c, respectively. Although the specific topology of the multilevel converter is covered in the next section, it may be stated in general for an n-level converter that the AC output consists of a number of voltage levels related to the switching state by

$$v_{ag} = \frac{s_a v_{dc}}{(n-1)} \qquad s_a = 0, 1, \ldots(n-1) \qquad (6.18)$$

$$v_{bg} = \frac{s_b v_{dc}}{(n-1)} \qquad s_b = 0, 1, \ldots(n-1) \qquad (6.19)$$

$$v_{cg} = \frac{s_c v_{dc}}{(n-1)} \qquad s_c = 0, 1, \ldots(n-1) \qquad (6.20)$$

As can be seen, a higher number of levels n leads to a larger number of switching state possibilities and smaller voltage steps. An overall switching state can be defined by using the base n mathematical expression

$$sw = n^2 s_a + n s_b + s_c \qquad (6.21)$$

Figure 6.2 shows the a-phase commanded line-to-ground voltage according to Eq. (6.5) as well as line-to-ground voltages for two-level, three-level, and four-level converters. In each case, the fast-average of v_{ag} will equal the commanded value v_{ag}^*. However, it can be seen that as the number of voltage levels increases, the converter voltage yields a closer approximation to the commanded value, resulting in lower harmonic distortion.

The next step in multilevel modulation is to relate the switching states s_a, s_b, and s_c to the duty cycles defined in Eqs. (6.14) through (6.16). Here, the multilevel sine-triangle technique will be used for this purpose [17, 18, 19]. The first step involves scaling the duty cycles for the n-level case as

$$d_{am} = (n-1)d_a \qquad (6.22)$$

$$d_{bm} = (n-1)d_b \qquad (6.23)$$

$$d_{cm} = (n-1)d_c \qquad (6.24)$$

The switching state may then be directly determined from the scaled duty cycles by comparing them to a set of high-frequency triangle waveforms with a frequency of f_{sw}. For an n-level converter, $n-1$ triangle waveforms of unity amplitude are defined. As an example, consider the four-level case. Figure 6.3a shows the a-phase duty cycle and the three triangle waveforms offset so that their peaks correspond to the nearest switching states. In general, the highest triangle waveform has a minimum value of $(n-2)$ and a peak value of $(n-1)$. The switching rules for the four-level case are fairly straightforward and may be specifically stated as

$$s_a = \begin{cases} 0 & d_{am} < v_{tr1} \\ 1 & v_{tr1} \leq d_{am} < v_{tr2} \\ 2 & v_{tr2} \leq d_{am} < v_{tr3} \\ 3 & v_{tr3} \leq d_{am} \end{cases} \qquad (6.25)$$

Figure 6.3b shows the resulting switching state based on the switching rules. As can be seen, the form is similar to that of Fig. 6.2d and, therefore, the resulting line-to-ground voltage according to Eq. (6.17) will have a fast-average value equal to its commanded value. These switching rules may be extended to

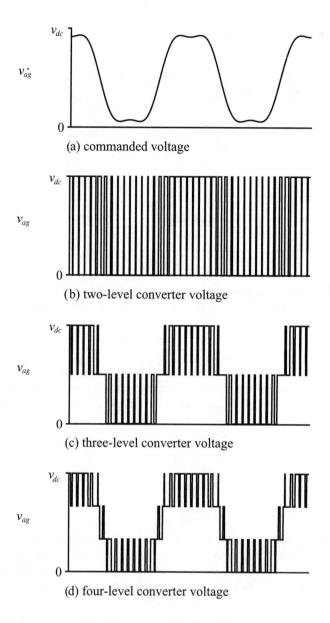

FIGURE 6.2 Power converter line-to-ground output voltages.

any number of levels by incorporating the appropriate number of triangle waveforms and defining switching rules similar to Eq. (6.25). It should be pointed out that the sine-triangle method is shown here since it is depicts a fairly straightforward method of accomplishing multilevel switching. In practice, the modulation is typically implemented on a digital signal processor (DSP) or erasable programmable logic device (EPLD) without using triangle waveforms. One common method for implementation is space-vector modulation [20–22], which is a method where the switching states are viewed in the voltage reference frame. Another method that may be used is duty-cycle modulation [23], which is a direct calculation method that uses duty cycles instead of triangle waveforms and is more readily implementable on a DSP. It is also possible to perform modulation based on a current-regulated approach [22, 24], which is fundamentally different than voltage-source modulation and results in a higher bandwidth control of load currents.

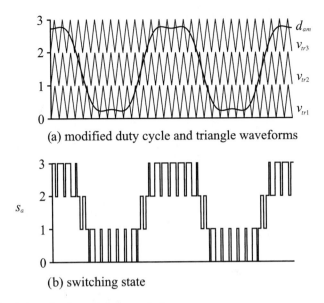

FIGURE 6.3 Four-level sine-triangle modulation technique.

6.3 Fundamental Multilevel Converter Topologies

This section describes the most common multilevel converter topologies. In particular, the diode-clamped [25–28], flying capacitor [29, 30], cascaded H-bridge [31–33], and multilevel H-bridge [34] structures are described. In each case, the process of creating voltage steps is illustrated and the relation to the generalized modulation scheme in the previous section is defined. For further study, the reader may be interested in other topologies not discussed here, such as the parallel connected phase poles [35], AC magnetically combined converters [36, 37], or soft-switching multilevel converters [38].

Diode-Clamped Multilevel Converters

One of the most common types of multilevel topologies is the diode-clamped multilevel converter [25–28]. Figure 6.4 shows the structure for the three-level case. Comparing this topology with that of a standard two-level converter, it can be seen that there are twice as many transistors as well as added diodes. However, it should be pointed out that the voltage rating of the transistors is half that of the transistors in a two-level converter. Although the structure appears complex, the switching is fairly straightforward. Figure 6.5 shows the a-phase leg of the three-level diode clamped converter along with the corresponding switching states. Here, it is assumed that the transistors act as ideal switches and that the capacitor voltages are charged to half of the DC-link voltage. As can be seen in Fig. 6.5b, in switching state $s_a = 0$, transistors T_{a3} and T_{a4} are gated on and the output voltage is $v_{ag} = 0$. Similarly, switching state $s_a = 2$ involves gating on transistors T_{a1} and T_{a2} and the output voltage is $v_{ag} = v_{dc}$. These switching states produce the same voltages as a two-level converter. Switching state $s_a = 1$ involves gating on transistors T_{a2} and T_{a3} as shown in Fig. 6.5c. In this case, the point a is connected to the capacitor junction through the added diodes and the output voltage is $v_{ag} = v_{dc}/2$. Note that for each of the switching states, the transistor blocking voltage is one half the DC-link voltage. When compared with the two-level converter, the additional voltage level allows the production of line-to-ground voltages with lower harmonic distortion, as illustrated in Fig. 6.2. Furthermore, the switching losses for this converter will be lower than that of a two-level converter. Switching losses are reduced by the lower transistor blocking voltage and increased by the higher number of transistors. However, it can be seen by inspection of Figs. 6.2 and 6.5 that each transistor is switching only during a portion of the period of d_a, which again reduces the switching losses. Maintaining voltage balance on the capacitors can be accomplished through selection of the

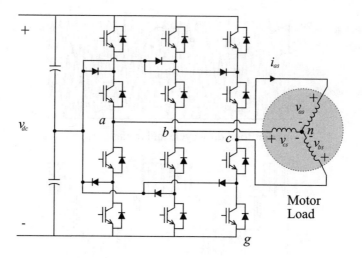

FIGURE 6.4 Four-level converter topology.

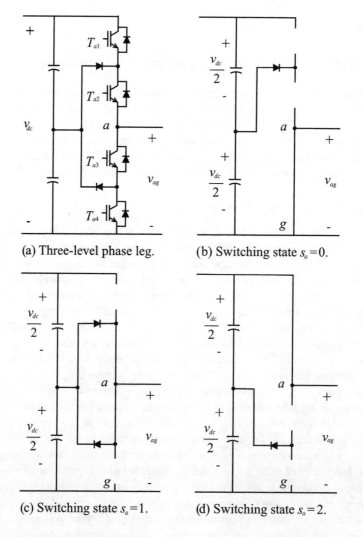

(a) Three-level phase leg.

(b) Switching state $s_a = 0$.

(c) Switching state $s_a = 1$.

(d) Switching state $s_a = 2$.

FIGURE 6.5 Three-level converter switching states.

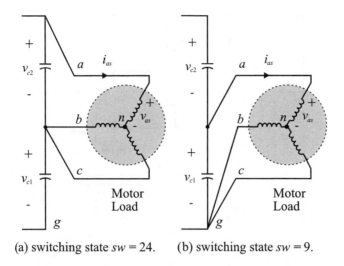

(a) switching state sw = 24.   (b) switching state sw = 9.

FIGURE 6.6 Redundant switching state example.

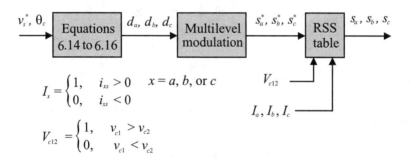

FIGURE 6.7 Redundant switching state example.

redundant states [27]. Redundant switching states are states that lead to the same motor voltages, but yield different capacitor currents. As an example, consider the three-level converter redundant switching states $sw = 24$ and $sw = 9$ shown in Fig. 6.6. It can be shown through Eq. (6.4) that either switching state will produce the same voltages on the load (assuming the capacitor voltages are nearly balanced). However, from Fig. 6.6 it can be seen that the current drawn from the capacitor bank will be different in each case. In particular, if the a-phase current is positive, the load will discharge the capacitor that it is connected to. In this case, the load should be connected across the capacitor with the highest voltage. On the other hand, if the a-phase current is negative, it will have a charging effect and the load should be connected across the capacitor with the lowest voltage. Therefore, capacitor voltage balancing through redundant state selection (RSS) is a straightforward matter of selecting between the redundant states based on which capacitor is overcharged with respect to the other and the direction of the phase currents. This information may be stored in a lookup table for inclusion in the modulation scheme [25, 27]. Figure 6.7 shows a block diagram of how an RSS table may be included in the modulation control. There, the modulator determines the desired switching states as described in the previous section. The desired switching state as well as the capacitor imbalance and phase current direction information are used as inputs to the lookup table, which determines the final switching state. As a practical matter, this table may be implemented in a DSP along with the duty-cycle calculations or may be programmed into an EPLD as a logic function.

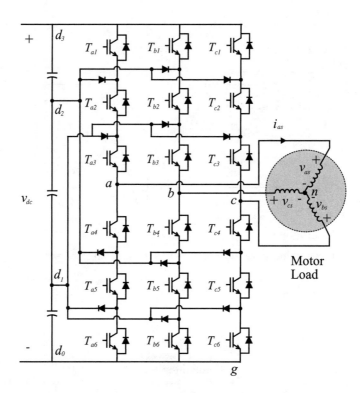

FIGURE 6.8 Four-level converter topology.

Figure 6.8 shows the topology for the four-level diode-clamped converter. Proper operation requires that each capacitor be charged to one third of the DC-link voltage. The transistor switching is similar to that of a three-level converter in that there are $(n-1)$ adjacent transistors gated on for each switching state. The switching results in four possibilities for the output voltage $v_{ag} = \{0 \ \frac{1}{3}v_{dc} \ \frac{2}{3}v_{dc} \ v_{dc}\}$. In this topology, each transistor need only block one third of the DC-link voltage. The diode-clamped concept may be extended to a higher number of levels by the expansion of the capacitor bank, switching transistors, and clamping diodes. However, there are some practical problems with diode-clamped converters of four voltage levels or more. The first difficulty is that some of the added diodes will need to block $(n-2)/(n-1)$ of the DC-link voltage [28]. As the number of levels is increased, it may be necessary to connect clamping diodes in series to block this voltage. It should also be pointed out that capacitor voltage balance through RSS works well for the three-level topology, but for converters with a higher number of voltage levels there are not enough redundant states to balance the capacitor voltages when the modulation index $\overline{m}$ becomes greater than 60% [27]. In these cases, another multilevel converter, such as a multilevel rectifier [25] of multilevel DC-DC converter [25] must be placed on the input side for voltage balance.

Flying-Capacitor Multilevel Converters

Figure 6.9 shows one phase of a three-level flying-capacitor multilevel converter. The general concept behind this converter is that the added capacitor is charged to one half of the DC-link voltage and may be inserted in series with the DC-link voltage to form an additional voltage level [29, 30]. Figure 6.10 shows how this is accomplished through the transistor switching. As can be seen, switching states $s_a = 0$ and $s_a = 2$ involve gating on the two lower and upper transistors as was done with the diode-clamped structure. In this topology, there are two options for switching to the state $s_a = 1$, as can be seen in Fig. 6.10c. The capacitor voltage may be either added to the converter ground or subtracted from the DC-link voltage. In essence, there is switching redundancy within the phase leg. Since the direction of the current through the capacitor changes depending on which redundant state is selected, the capacitor

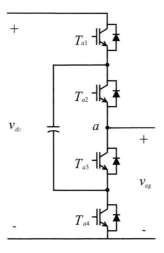

FIGURE 6.9 Three-level flyback converter topology.

voltage may be maintained at one half the DC-link voltage through the redundant state selection within the phase.

Cascaded H-Bridge Multilevel Converters

Cascaded H-bridge converters consist of a number of H-bridge power conversion cells, each supplied by an isolated source on the DC side and series-connected on the AC side [31–33]. Figure 6.11 shows the *a*-phase of a cascaded H-bridge converter, where two H-bridge cells are utilized. It should be pointed out that, unlike the diode-clamped and flying-capacitor topologies, isolated sources are required for each cell in each phase. In some systems these sources may be available through batteries or photovoltaic cells [32], but in most drive systems transformer/rectifier sources are used. Figure 6.12 illustrates the switching state detail for one H-bridge cell. As can be seen, three unique output voltages are possible. In accordance with the convention used here, the lowest switching state will be labeled state 0. When these cells are combined in series, an effective switching state can be related to the switching states of the individual cells. By defining switching states in this way, the modulation scheme of the previous section may be applied to this converter as well. The output voltage of the inverter may be determined from the switching states of the individual cells by

$$v_{ag} = \sum_{i=0}^{p}(S_{ai} - 1)v_{dci} \qquad (6.26)$$

where p is the number of series H-bridge cells.

If the DC voltage applied to each cell is set to the same value, then the effective number of voltage levels may be related to the number of cells by

$$n = 3 + 2(p-1) \qquad (6.27)$$

Therefore, the converter shown in Fig. 6.10 would operate with five voltage levels. To obtain a clearer comprehension of how the voltage levels are produced, Table 6.1 shows the overall switching state as well as the switching states of the individual cells and the resulting output voltage. As can be seen, there is quite a bit of switching state redundancy within one phase leg of the cascaded H-bridge converter for states $s_a = 1$, $s_a = 2$, and $s_a = 3$. This redundancy may be exploited to increase the number of voltage levels

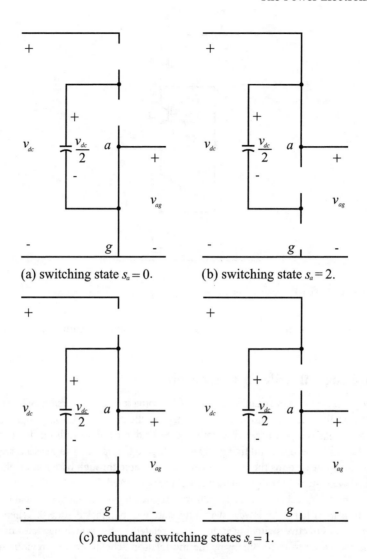

FIGURE 6.10 Three-level flyback converter switching states.

by selecting different DC voltage values for each cell [33]. As an example, consider the case where $v_{dc1a} = 3v_{dc2a}$. Table 6.2 defines the overall switching state and demonstrates a method for obtaining nine-level performance in this case. As can be seen, the number of voltage levels is greatly improved by setting the DC voltages to different values. In particular, if a ratio of three is used for each cell added, the number of voltage levels for a given number of cells may be computed as

$$n = (3)^p \tag{6.28}$$

Besides an improvement in power quality, the DC voltage ratio used in Table 6.2 will also split the power conversion process into a high-voltage, low-switching frequency converter and a low-voltage, high-switching frequency converter [33]. For this type of converter, one cell may utilize GTOs and the other cell may utilize IGBTs to make the best use of the switching devices. One disadvantage of the DC voltage ratio used in Table 6.2 is that the system is not as modular as before, requiring two types of converter cells. Another disadvantage is that the DC voltage source on the lower voltage cell may be required to absorb a negative current because it is supplying a negative output voltage when the AC current is positive [33].

Multilevel Converters

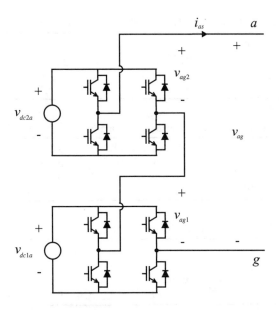

FIGURE 6.11 Cascaded H-bridge topology with two cells.

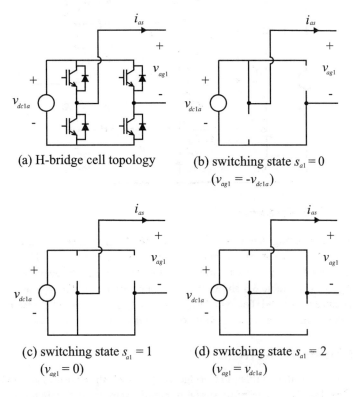

FIGURE 6.12 Switching states of the H-bridge cell.

TABLE 6.1 Cascaded H-Bridge Switching States ($v_{dc1a} = v_{dc2a} = E$)

s_a	0	1		2			3		4
s_{a1}	0	0	1	0	1	2	2	1	2
s_{a2}	0	1	0	2	1	0	1	2	2
v_{ag}	$-2E$	$-E$	$-E$	0	0	0	E	E	$2E$

TABLE 6.2 Cascaded H-Bridge Switching States ($v_{dc1} = 3v_{dc2a} = 3E$)

s_a	0	1	2	3	4	5	6	7	8
s_{a1}	0	0	0	1	1	1	2	2	2
s_{a2}	0	1	2	0	1	2	0	1	2
v_{ag}	$-4E$	$-3E$	$-2E$	$-E$	0	E	$2E$	$3E$	$4E$

TABLE 6.3 Cascaded H-Bridge Switching States ($v_{dc1a} = v_{dc2a} = E$)

s_a	0	1	2		3	4		5	6
s_{a1}	0	0	0	1	1	1	2	2	2
s_{a2}	0	1	2	0	1	2	0	1	2
v_{ag}	$-3E$	$-2E$	$-E$	$-E$	0	E	E	$2E$	$3E$

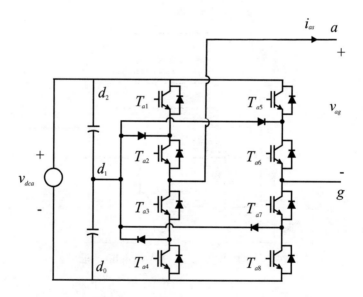

FIGURE 6.13 Five-level H-bridge inverter.

For example, obtaining switching state $s_a = 6$ requires that the lower voltage cell output a negative voltage (i.e., $s_{a2} = 0$). If a high power factor load is assumed, then the phase current i_{as} is likely positive when this switching state occurs resulting in a current into v_{dc1a}. This situation may be avoided by lowering the DC voltage ratio to $v_{dc1a} = 2v_{dc2a}$. Table 6.3 shows the switching pattern for this operation. As can be seen, seven-level performance is ensured. For switching state $s_a = 4$, the redundant choice may now be made dependent on the direction of the phase current to ensure that the current from v_{dc2a} is positive [33].

Multilevel H-Bridge Converters

Another possible topology is the multilevel H-bridge converter, which consists of an H-bridge made from diode-clamped phase legs [34]. The most straightforward example of this is the five-level H-bridge converter shown in Figure 6.13. As can be seen, the structure is made from two three-level diode-clamped phase legs. Based on the previous discussion on the diode-clamped structure, it may be stated that the points a and g can be connected to any of the junction points d_0, d_1, or d_2. If the capacitors are charged to one half of the DC-link voltage, then the possible output voltages are $v_{ag} = \{-v_{dca}\ -\frac{1}{2}v_{dca}\ 0\ \frac{1}{2}v_{dca}\ v_{dca}\}$. In general, the effective number of levels for a multilevel H-bridge made from phase legs with n_{leg} levels is

$$n = 2(n_{leg} - 1) + 1 \tag{6.29}$$

Redundancy exists within the structure which may be used to ensure capacitor voltage balance. An isolated source is required for each phase.

6.4 Cascaded Multilevel Converter Topologies

Recent research in multilevel power conversion includes the introduction of the concept of cascading multilevel converters [39,40]. In this section, two types of cascaded multilevel converters are considered. The first type consists of three-phase multilevel converters cascaded through splitting of the neutral point of a wye-connected load. The second type focuses on the cascaded H-bridge converter utilizing multilevel cells.

Cascaded Multilevel Converters

Figure 6.14 shows the topology of the cascaded multilevel converter [39]. In the figure, a three-phase multilevel converter with n_1 voltage levels is connected to a motor load. The neutral point of the motor has been split and connected to another three-phase multilevel converter. The operation of the cascaded multilevel converter requires that each converter be supplied by an isolated DC voltage source (labeled v_{dc1} and v_{dc2} in Fig. 6.14). The analysis of this converter begins by deriving equations that relate the motor phase voltages to the line-to-ground voltages of the individual converters. If the point n_{12} in Fig. 6.14

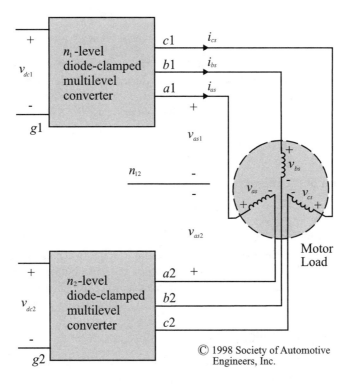

FIGURE 6.14 The cascaded multilevel converter.

indicates a fictitious neutral point, then the phase voltages from converter 1 may be written similar to Eq. (6.4). In particular [39]

$$\begin{bmatrix} v_{as1} \\ v_{bs1} \\ v_{cs1} \end{bmatrix} = \frac{1}{3} \begin{bmatrix} 2 & -1 & -1 \\ -1 & 2 & -1 \\ -1 & -1 & 2 \end{bmatrix} \begin{bmatrix} v_{a1g1} \\ v_{b1g1} \\ v_{c1g1} \end{bmatrix} \quad (6.30)$$

where the line-to-ground voltages are defined with respect to the negative DC rail of v_{dc1}. For example v_{a1g1} is the voltage from the point $a1$ to $g1$. Similarly for converter 2,

$$\begin{bmatrix} v_{as2} \\ v_{bs2} \\ v_{cs2} \end{bmatrix} = \frac{1}{3} \begin{bmatrix} 2 & -1 & -1 \\ -1 & 2 & -1 \\ -1 & -1 & 2 \end{bmatrix} \begin{bmatrix} v_{a2g2} \\ v_{b2g2} \\ v_{c2g2} \end{bmatrix} \quad (6.31)$$

By applying KVL equations to the circuit topology, it can be seen that

$$v_{as} = v_{as1} - v_{as2} \quad (6.32)$$

$$v_{bs} = v_{bs1} - v_{bs2} \quad (6.33)$$

$$v_{cs} = v_{cs1} - v_{cs2} \quad (6.34)$$

Substitution of Eqs. (6.30) and (6.31) into Eqs. (6.32) to (6.34) yields

$$\begin{bmatrix} v_{as} \\ v_{bs} \\ v_{cs} \end{bmatrix} = \frac{1}{3} \begin{bmatrix} 2 & -1 & -1 \\ -1 & 2 & -1 \\ -1 & -1 & 2 \end{bmatrix} \begin{bmatrix} v_{a1g1} - v_{a2g2} \\ v_{b1g1} - v_{b2g2} \\ v_{c1g1} - v_{c2g2} \end{bmatrix} \quad (6.35)$$

By comparison of Eq. (6.35) to Eq. (6.4), it can be seen that the effective line-to-ground voltage is the difference of the line-to-ground voltages from converter 1 to converter 2. The effective number of voltage levels in this cascaded topology depends on the ratio of the DC voltages. It has been shown that the maximum number of voltage levels achievable is the product of the voltage levels of the individual converters or [39]

$$n = n_1 n_2 \quad (6.36)$$

To obtain this number of effective voltage levels, the DC voltage ratio must be set to [39]

$$\frac{v_{dc2}}{v_{dc1}} = \frac{n_2 - 1}{n_2(n_1 - 1)} \quad (6.37)$$

As with the cascaded H-bridge inverter described in Section 6.3, the switching states of each converter may be related to an overall switching state in order to utilize the same multilevel modulation technique of Section 6.2. A specific example of this is given in the following section where a three-level diode-clamped converter is cascaded with a two-level converter to form an effective six-level converter.

Cascaded Multilevel H-Bridge Converters

Combining the concept of cascading converters with the cascaded H-bridge converter described in Section 6.3, it stands to reason that a cascaded H-bridge converter can be made from multilevel H-bridge cells. The general concept is shown in Fig. 6.15, which depicts the a-phase of a cascaded H-bridge

Multilevel Converters

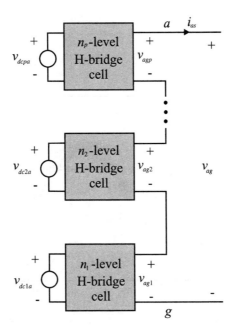

FIGURE 6.15 Cascaded multilevel H-bridge converter.

converter having p multilevel cells. Each cell consists of a multilevel H-bridge converter (for example, a five-level H-bridge cell would be as shown in Fig. 6.13). For this converter, the effective number of voltage levels is the product of the voltage levels of the individual cells or

$$n = \prod_{i=1}^{p} n_i \quad (6.38)$$

assuming that the DC voltage of cell i is set based on the adjacent cell as [40]

$$v_{dc(i-1)x} = \frac{n_i - 1}{n_i(n_{i-1} - 1)} v_{dcix} \quad i = 2, 3, \ldots m \quad (6.39)$$

The switching states of the individual converters may be related to an overall switching state so that the standard modulation scheme may be used. As with the cascaded H-bridge converter discussed in Section 6.3, it may be necessary to utilize a DC voltage ratio different from that of Eq. (6.39) to increase the redundancy within the phase leg and avoid negative DC currents when the DC sources are supplied from transformer/rectifier circuits. A specific example is given in the following section, where a five-level H-bridge cell is cascaded with a three-level H-bridge cell.

6.5 Multilevel Converter Laboratory Examples

Three-Level Diode-Clamped Inverter

The first laboratory example in this section involves the three-level diode-clamped inverter as shown in Fig. 6.4. For this study, both of the capacitors on the DC link were supplied by isolated voltage sources so that the total voltage was 187.5 V. The induction motor load was a 3.7 kW machine with a rated line voltage of 230 V, a rated line frequency of 60 Hz, and a rated mechanical speed of 183 rad/s. The modulation strategy for this study was space-vector modulation. However, similar performance can be achieved

TABLE 6.4 Cascaded-3/2 Inverter Switching States ($v_{dc1} = 4v_{dc2} = 4E$)

s_a	0	1	2	3	4	5
s_{a1}	0	0	1	1	2	2
s_{a2}	1	0	1	0	1	0
$v_{a1g1} - v_{a2g2}$	$-E$	0	E	$2E$	$3E$	$4E$

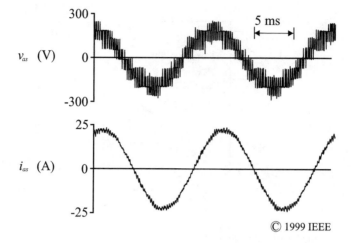

FIGURE 6.16 Three-level inverter laboratory measurements. (From IEEE, 1999. With permission.)

with the method described in Section 6.2 if the modulation index is set to $\overline{m} = 0.87$ and the switching frequency is set to $f_{sw} = 1.8$ kHz.

Figure 6.16 shows the a-phase voltage and current for steady-state operation at rated conditions [41]. As can be seen, the phase voltage contains more steps than would be present in a standard two-level converter [42]. The voltage and current total harmonic distortion (THD) in this example were 34 and 5.1%. These waveforms are used as a reference point for the cascaded converter performance shown below.

The Cascade-3/2 Inverter

Figure 6.17 shows a special case of the cascaded multilevel converter where a three-level diode-clamped converter is cascaded with a two-level converter. This topology is referred to here as a cascade-3/2 converter. In this study, the DC voltages were set to $v_{dc1} = 250$ V and $v_{dc2} = 125$ V in accordance with Eq. (6.37). As with the previous example, the DC voltages on the three-level converter were supplied by isolated sources. Under these operating conditions, six voltage levels are achievable according to Eq. (6.36). Table 6.4 relates the switching states of the individual converters to an overall switching state in a way that six-level performance can be demonstrated.

The modulation strategy chosen for this study was space-vector modulation. However, the generalized switching state s_a from Table 6.4 may be used in the modulation strategy discussed in Section 6.2 with $\overline{m} = 0.87$ and $f_{sw} = 540$ Hz for similar results. It should be pointed out that RSS switching involving all three phases was performed in this study to ensure that the average current drawn from v_{dc2} was positive. The details of this RSS [41] will not be covered here. The induction motor used in this study is identical to that for the three-level inverter study discussed above. Figure 6.18 shows the motor a-phase voltage and current in the steady state at rated operation as measured in the laboratory [39]. As can be seen, the voltage has more steps when compared with the three-level example and therefore more closely resembles an ideal sine wave. The voltage and current THD in this example were 16.8 and 5.4%, respectively. The switching frequency was intentionally set to a low value to compare the voltage THD with that of the

Multilevel Converters

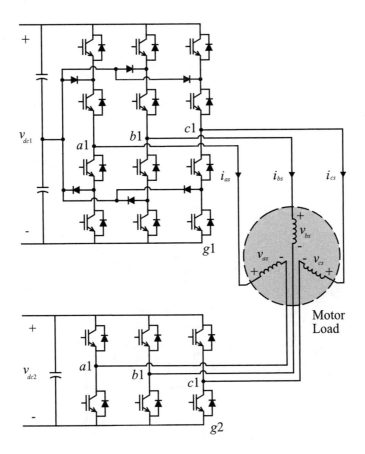

FIGURE 6.17 The cascade-3/2 multilevel inverter.

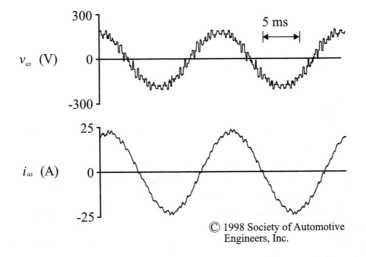

© 1998 Society of Automotive Engineers, Inc.

FIGURE 6.18 Cascade-3/2 inverter laboratory measurements.

TABLE 6.5 Cascade-5/3H Inverter Switching States ($v_{dc1a} = 4v_{dc2a} = 4E$)

S_a	0	1	2	3	4	5	6	7	8	9	10
S_{a1}	0	0	0	1	1	1	2	2	3	3	4	4
S_{a2}	0	1	2	0	1	2	0	1	2	0	1	2
v_{ag}	$-5E$	$-4E$	$-3E$	$-3E$	$-2E$	$-E$	$-E$	0	E	E	$2E$	$3E$	$3E$	$4E$	$5E$

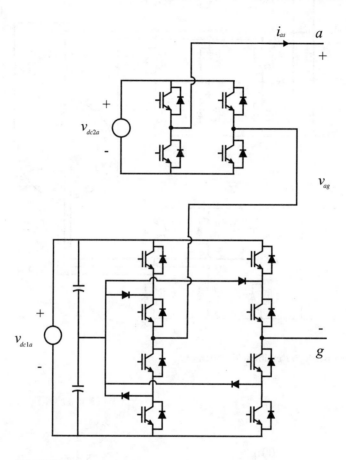

FIGURE 6.19 Cascaded-5/3H inverter topology.

three-level example [41], yielding a measure of performance improvement. However, increasing the switching frequency would lead to a significant reduction in the current ripple.

The Cascade-5/3H Inverter

Figure 6.19 shows the a-phase topology for a specific case of the cascaded multilevel H-bridge inverter. In this example, a five-level H-bridge cell is cascaded with a three-level H-bridge cell. The DC voltages were set to $v_{dc1a} = 260$ V and $v_{dc2a} = 65$ V. Although this ratio does not yield the maximum number of voltage levels as depicted by Eq. (6.39), it allows redundancy within each phase that may be used for ensuring positive DC current from v_{dc2a} as discussed in Section 6.3. For this example, redundant state selection within the five-level converter was used to balance the capacitor voltages so that only two isolated voltage sources are needed per phase. Table 6.5 shows the overall switching state and its relation to the switching of the individual converters. For this study, duty-cycle modulation was used. However, similar results may be obtained by using the modulation scheme outlined in Section 6.2 with a modulation index of $\overline{m} = 0.91$ and a switching frequency of $f_{sw} = 10$ kHz. The induction motor is a 5.2-kW machine with

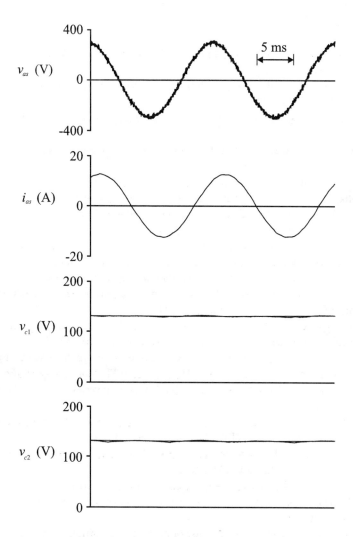

FIGURE 6.20 Cascade-5/3H inverter laboratory measurements.

a rated line voltage of 460 V. Figure 6.20 shows the measured performance of the inverter operating at rated voltage and a power level of 3.7 kW. Therein, the a-phase voltage, current, and capacitor voltages are shown. As can be seen, from the motor voltage, the number of steps is greatly increased from the three- and six-level inverter examples shown above. The voltage and current THD in this example are 6.5 and 3.7%, respectively. It can also be seen from the capacitor voltages that the RSS algorithm is effective.

6.6 Conclusion

This chapter has presented an overview of multilevel power conversion. Multilevel voltage-source modulation was introduced and formulated in terms of a per-phase switching state which related to the converter line-to-ground voltage. Several topologies for performing multilevel power conversion were presented and the switching states were defined to match that of the modulation. In this way, the generalized multilevel modulation technique can be readily applied to any of the specific topologies. Two types of cascaded multilevel topologies were discussed. The advantage of cascading multilevel converters is the large number of voltage levels available as the result of a compounding effect of cascading process. A few laboratory examples were given that demonstrated the performance achievable through multilevel power converters.

References

1. Nabe, A., Takahashi, I., and Akagi, H., A new neutral-point clamped PWM inverter, *IEEE Trans. Ind. Appl.*, 17, 518, 1981.
2. Corzine, K. A. and Majeethia, S. K., Analysis of a novel four-level DC/DC boost converter, *IEEE Trans. Ind. Appl.*, 36, 1342, 2000.
3. Tolbert, L. M. and Peng, F. Z., Multilevel converters for large electric drives, *IEEE Trans. Ind. Appl.*, 35, 36, 1999.
4. Rojas, R., Ohnishi, T., and Suzuki, T., PWM control method for a four-level inverter, in *Proc. Electric Power Applications*, IEE, 1995, 390.
5. Tenconi, S. M., Carpita, M., Bacigalupo, C., and Cali, R., Multilevel voltage source converter for medium voltage adjustable speed drives, in *Proc. ISIE*, IEEE, Athens, 1995, 91.
6. Osman, R. H., A medium-voltage drive utilizing series-cell multilevel topology for outstanding power quality, in *Proc. IAS*, IEEE, Phoenix, 1999, 2662.
7. Mahfouz, A., Holtz, J., and El-Tobshy, A., Development of an integrated high-voltage 3-level converter-inverter system with sinusoidal input-output for feeding 3-phase induction motors, in *Proc. 5th European Conference on Power Electronics and Applications*, Brighton, U.K., 1993, 134.
8. Seki, N. and Uchino, H., Converter configurations and switching frequency for a GTO reactive power compensator, *IEEE Trans. Ind. Appl.*, 33, 1011, 1997.
9. Chen, Y., Mwinyiwiwa, B., Wolanski, Z., and Ooi, B. T., Regulating and equalizing DC capacitive voltages in multilevel statcom, *IEEE Trans. Power Delivery*, 12, 901, 1997.
10. Hatziadoniu, C. J. and Chalkiadakis, F. E., A 12-pulse static synchronous compensator for the distribution system employing the 3-level GTO-inverter, *IEEE Trans. Power Delivery*, 12, 1830, 1997.
11. Nakata, K., Nakamura, K., and Ito, S., A three-level traction inverter with IGBTS for EMU, in *Proc. IAS*, IEEE, Denver, 1994, 667.
12. Willkinson, R. H., Horn, A., and Enslin, J. H. R., Control options for a bi-directional multilevel traction chopper, in *Proc. PESC*, IEEE, Maggiore, 1996, 1395.
13. Lyons, J. P., Vlatkovic, V., Espelage, P. M., Boettner, F. H., and Larsen, E., Innovation IGCT main drives, in *Proc. IAS*, IEEE, Phoenix, 1999, 2655.
14. Vas, P., *Sensorless Vector and Direct Torque Control*, Oxford University Press, Oxford, 1998, chap. 2.
15. Krause, P. C., Wasynczuk, O., and Sudhoff, S. D., *Analysis of Electric Machinery*, IEEE Press, Piscataway, NJ, 1995, chap. 13.
16. Houldsworth, J. A. and Grant, D. A., The use of harmonic distortion to increase the output voltage of a three-phase pwm inverter, *IEEE Trans. Ind. Appl.*, 20, 1224, 1984.
17. Corzine, K. A., Sudhoff, S. D., Lewis, E. A., Schmucker, D. H., Youngs, R. A., and Hegner, H. J., Use of multi-level converters in ship propulsion drives, in *Proc. All Electric Ship Conference*, IMarE, London, 1998, 155.
18. Tolbert, L. M. and Peng, F. Z., Multilevel PWM methods at low modulation indices, *IEEE Trans. Power Electron.*, 15, 719, 2000.
19. Menzies, R. W., Steimer, P., and Steknke, J. K., Five-level GTO inverter for large induction motor drives, *IEEE Trans. Ind. Appl.*, 30, 1994.
20. Liu, H. L. and Cho, G. H., Three-level space vector PWM in low index modulation region avoiding narrow pulse problem, *IEEE Trans. Power Electron.*, 9, 481, 1994.
21. Lee, Y. H., Suh, B. S., and Hyun, D. S., A novel PWM scheme for a three-level voltage source inverter with GTO thyristors, *IEEE Trans. Ind. Appl.*, 32, 260, 1996.
22. Corzine, K. A., A hysteresis current-regulated control for multi-level converters, *IEEE Trans. Energy Conversion*, 15, 169, 2000.
23. Corzine, K. A. and Majeethia, S. K., Analysis of a novel four-level DC/DC boost converter, in *Proc. IAS*, IEEE, Phoenix, 1999, 1964.
24. Marchesoni, M., High-performance current control techniques for applications to multilevel high-power voltage source inverters, *IEEE Trans. Power Electron.*, 7, 189, 1992.

25. Corzine, K. A., Delisle, D. E., Borraccini, J. P., and Baker, J. R., Multi-level power conversion: present research and future investigations, in *Proc. All Electric Ship Conference,* IMarE, Paris, France, 2000.
26. Corzine, K. A., Topology and Control of Cascaded Multi-Level Converters, Ph.D. dissertation, University of Missouri, Rolla, 1997.
27. Fracchia, M., Ghiara, T., Marchesoni, M., and Mazzucchelli, M., Optimized modulation techniques for the generalized n-level converter, in *Proc. PESC,* IEEE, Madrid, 1992, 1205.
28. Horn, A., Wilkinson, R. H., and Enslin, J. H. R., Evaluation of converter topologies for improved power quality in DC traction, in *Proc. ISIE,* IEEE, Warsaw, 1996, 802.
29. Tourkhani, F., Viarouge, P., and Meynard, T. A., A simulation-optimization system for the optimal design of a multilevel inverter, *IEEE Trans. Power Electron.,* 14, 1037, 1999.
30. Yuan, X. and Barbi, I., Zero-voltage switching for three-level capacitor clamping inverter, *IEEE Trans. Power Electron.,* 14, 771, 1999.
31. Hammond, P. W., Medium Voltage PWM Drive and Method, U.S. Patent 5,625,545, 1997.
32. Calais, M., Agelidis, V. G., Borle, L. J., and Dymond, M. S., A transformerless five level cascaded inverter based single phase photovoltaic system, in *Proc. PESC,* IEEE, Galway, 2000, 1173.
33. Lipo, T. A. and Menjrekar, M. D., Hybrid Topology for Multi-Level Power Conversion, U.S. Patent 6,005,788, 1999.
34. Hill, W. A. and Harbourt, C. D., Performance of medium voltage multi-level inverters, in *Proc. IAS,* IEEE, Phoenix, 1999, 1186.
35. Matsui, K., Asao, M., Ueda, F., Tsuboi, K., and Iwata, K., A technique of parallel-connections of pulsewidth modulated NPC inverters by using current sharing reactors, in *Proc. IECON,* IEEE, Maui, 1993, 1246.
36. Iturriz, F. and Ladoux, P., Phase-controlled multilevel converters based on dual structure associations, *IEEE Trans. Power Electron.,* 15, 92, 2000.
37. Masukawa, S. and Iida, S., A method of reducing harmonics in output voltages of a double-connected inverter, *IEEE Trans. Power Electron.,* 9, 543.
38. Yamamoto, M., Hiraki, E., Iwamoto, H. Sugimoto, S., Kouda, I., and Nakaoka, M., Voltage-fed NPC soft-switched inverter with new space voltage vector modulation scheme, in *Proc. IAS,* IEEE, Phoenix, 1999, 1178.
39. Corzine, K. A. and Sudhoff, S. D., High state count power converters: an alternate direction in power electronics technology, *SAE Trans. J. Aerospace,* SAE paper number 981266 © 1998 Society of Automotive Engineers, Inc., Williamsburg, 1, 124, 1998.
40. Corzine, K. A., Use of multi-level inverters to reduce high-frequency effects in induction motor drives, in *Proc. Naval Symposium on Electric Machines,* ONR/NSWC, Philadelphia, 2000.
41. Corzine, K. A., Sudhoff, S. D., and Whitcomb, C. A., Performance characteristics of a cascaded two-level converter, *IEEE Trans. Energy Conversion,* 14, 433, 1999.
42. Hart, D. W., *Introduction to Power Electronics,* Prentice-Hall, Upper Saddle River, NJ, 1997, chap. 8.

7
Modulation Strategies

Michael Giesselmann
Texas Tech University

Hossein Salehfar
University of North Dakota

Hamid A. Toliyat
Texas A&M University

Tahmid Ur Rahman
Texas A&M University

7.1 Introduction 7-1
7.2 Six-Step Modulation 7-2
7.3 Pulse Width Modulation 7-2
   PWM Signals with DC Average • PWM Signals for AC Output
7.4 Third Harmonic Injection for Voltage Boost of SPWM Signals 7-9
7.5 Generation of PWM Signals Using Microcontrollers and DSPs 7-11
7.6 Voltage Source–Based Current Regulation 7-12
7.7 Hysteresis Feedback Control 7-14
   Introduction • Principles of the Hysteresis Feedback Control Circuits • Design Procedure • Experimental Results • Conclusions
7.8 Space-Vector Pulse Width Modulation 7-28
   How the SVPWM Works • Implementation • Switching Signals

7.1 Introduction

Michael Giesselmann

In this chapter, modulation techniques for power electronics circuits are discussed. Modulation techniques are strategies to control the state of switches in these circuits. Switch mode is preferred to linear operation since switches ideally do not dissipate any power in either the ON or OFF state. Depending on the switches that are being used, it may only be possible to control the turn on instants. However, most modern power semiconductors such as IGBTs can be turned on and off tens of thousands of times per second on command. In parallel with the development of these modern power semiconductors, new modulation techniques have emerged. In the following sections, a number of modulation techniques along with their advantages and disadvantages will be discussed. Most figures have been generated using MathCAD® 2000 [1]. The examples for the digital modulation techniques and the voltage source–based current control techniques have been generated using PSpice®[2].

References

1. MathCAD® 2000 Professional, MathSoft Engineering & Education, Inc., 101 Main Street, Cambridge, MA 02142-1521, http://www.mathsoft.com.
2. PSpice® Documentation, 555 River Oaks Parkway, San Jose, CA 95134, U.S.A.; (408)-943-1234; http://pcb.cadence.com/.

7.2 Six-Step Modulation

Michael Giesselmann

Six-step modulation represents an early technique to control a three-phase inverter. Six-step modulation uses a sequence of six switching patterns for the three phase legs of a full-bridge inverter to generate a full cycle of three-phase voltages. A switch pair connected between the positive DC bus and the negative DC bus represents a phase leg. The output terminal is the midpoint of the two switches. Only one switch of a phase leg may be turned on at any given time to prevent a short circuit between the DC buses. One state of the inverter leg represents the case when the upper switch is turned on whereas the opposite state is represented by the lower switch being turned on. If each phase leg has these two states, the inverter has $2^3 = 8$ possible switching states. Six of these states are active states, whereas the two states in which either all of the upper or all of the lower switches are turned on are called zero states, because the line-to-line output voltage is zero in these cases. The six discrete switching patterns for six-step modulation are shown in Fig. 7.1a to f. For clarity, free-wheeling diodes have been omitted. After the switching pattern shown in Fig. 7.1f, the cycle begins anew with the switching pattern shown in Fig. 7.1a. Note that in subsequent patterns, only a single inverter leg changes states. The switching patterns shown in Fig. 7.1a to f represent the following inverter states in the following order:

- Positive peak of Phase A
- Negative peak of Phase C
- Positive peak of Phase B
- Negative peak of Phase A
- Positive peak of Phase C
- Negative peak of Phase B

The aforementioned inverter states are equally spaced in a circle with 60° of phase shift between them. This is illustrated in Fig. 7.2. The hexagon in Fig. 7.2 represents the trace of a voltage vector around a circle for six-step modulation. This scheme could be extended to space vector modulation, if the voltage vector would not make discrete 60° steps, but would alternate at high speed between two adjacent states. The switching control would be such that the average time spend in the previous state is gradually decreasing, whereas the average time spent in the next state is gradually increasing. Also by inserting zero states, the magnitude of the output voltage could be controlled.

Figure 7.3 shows the phase to neutral waveform of one inverter leg for six-step operation if the neutral point is considered the midpoint between the positive and negative bus. The resulting line-to-line output voltage is shown in Fig. 7.4. This waveform is closer to a sinusoid than the phase to neutral voltage but it still has a considerable amount of harmonics. Figure 7.5 shows the spectrum of the line-to-line voltage for six-step operation normalized to the fundamental frequency. The lowest harmonic component is the 5th harmonic.

The advantages of six-step modulation are the simplicity of the procedure and the ability to use slow-switching, high-power devices like GTOs. However, the harmonic content of the output voltage and the inability to control the magnitude of the output voltage are serious drawbacks. Because of these drawbacks and due to the recent advances in high-power IGBT technology, this modulation scheme is today seldom considered for new designs.

7.3 Pulse Width Modulation

Michael Giesselmann

Pulse width modulation (PWM) is the method of choice to control modern power electronics circuits. The basic idea is to control the duty cycle of a switch such that a load sees a controllable average voltage. To achieve this, the switching frequency (repetition frequency for the PWM signal) is chosen high enough that the load cannot follow the individual switching events. Switching, rather than linear operation of the

Modulation Strategies

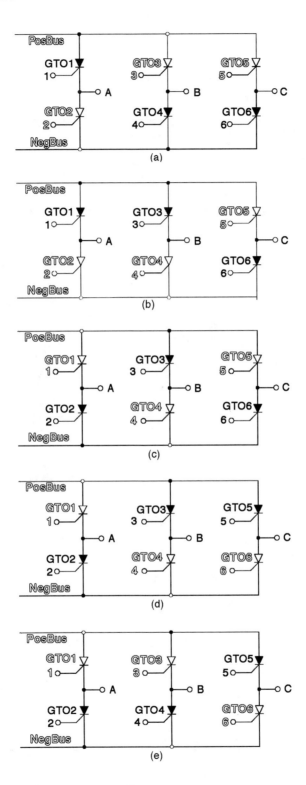

FIGURE 7.1 (a) GTO inverter indicating conducting switches for step 1 in six step sequence. (b) GTO inverter indicating conducting switches for step 2 in six-step sequence. (c) GTO inverter indicating conducting switches for step 3 in six-step sequence. (d) GTO inverter indicating conducting switches for step 4 in six-step sequence. (e) GTO inverter indicating conducting switches for step 5 in six-step sequence. (Continued)

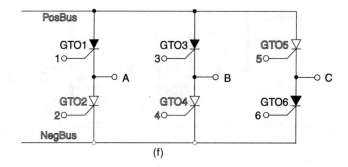

FIGURE 7.1 (Continued.) (f) GTO inverter indicating conducting switches for step 6 in six-step sequence.

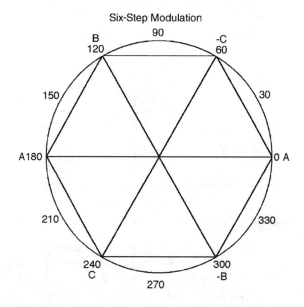

FIGURE 7.2 Graphical representation of the vector positions of the inverter states in a circle for six-step modulation.

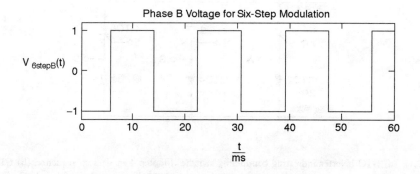

FIGURE 7.3 Phase to neutral waveform of the inverter for six-step operation.

Modulation Strategies

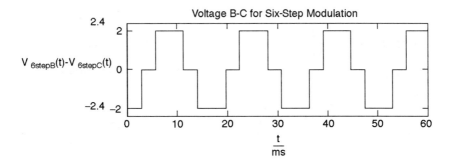

FIGURE 7.4 Line-to-line waveform of the inverter for six-step operation.

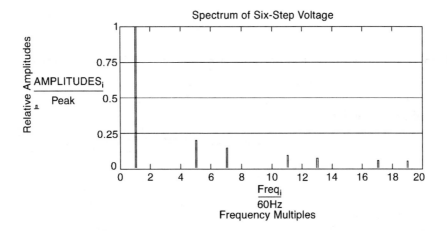

FIGURE 7.5 Spectrum of the line-to-line voltage for six-step operation normalized to the fundamental frequency.

power semiconductors, is of course done to maximize the efficiency because the power dissipation in a switch is ideally zero in both states. In a typical case, the switching events are just a "blur" to the load, which reacts only to the average state of the switch.

PWM Signals with DC Average

There are a number of different methods to generate periodic rectangular waveforms with varying duty cycle. A standard method is the so-called carrier-based PWM technique, which compares a control signal with a triangular (or sawtooth shaped) waveform. Figure 7.6 shows an example of a triangular waveform with 10-kHz repetition (switching) frequency. By comparing this signal with a reference level, which can vary between 0 and 1 V, a PWM signal with a duty cycle between 0 and 100% is generated. Because of the triangular carrier, the relation between the reference level and the resulting duty cycle is linear.

Figure 7.7 shows an example where a PWM signal with 80% duty cycle is created. This method works very well for duty cycles in the range from 5% up to 95% as shown in Figs. 7.8 and 7.9. However, if the reference signal exceeds 100% or falls below 0%, the resulting PWM signal would be always on or always off, respectively. This is called overmodulation. This regime must be avoided by proper conditioning of the control signal. In addition, for control signals resulting in PWM signals with duty cycle values as high as 99% or as low as 1%, the switch may never fully reach the opposite state and spend an undue amount of time in transitions. Therefore, it is typically recommended to limit the control signal to a range, which avoids overmodulation as well as extremely narrow pulses.

freq$_{sw}$:= 10.kHz   Definition of Switching Frequency

Triangle(t) := $\dfrac{a\cos(\cos(2\cdot\pi\cdot freq_{sw}\cdot t))}{\pi}$ Definition of Triangle Wave

t := 0·μs, 0.1·μs..300·μs   Definition of Time Window

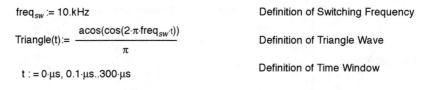

FIGURE 7.6 Triangular carrier wave for PWM modulation with a duty cycle between 0 and 100%.

DutyCycle := 80%

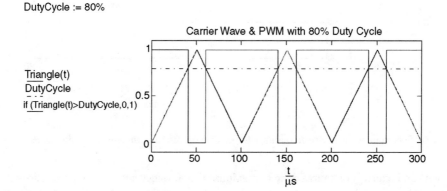

FIGURE 7.7 Triangular carrier wave and PWM signal for 80% duty cycle.

DutyCycle := 95%

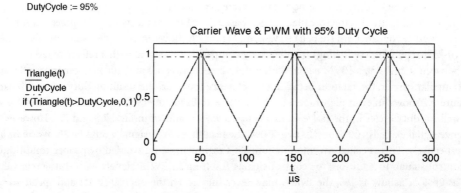

FIGURE 7.8 Triangular carrier wave and PWM signal for 95% duty cycle.

Modulation Strategies

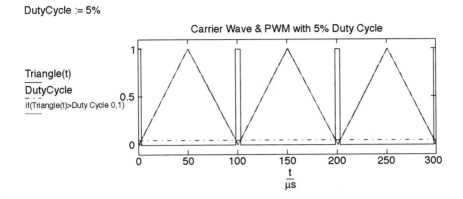

FIGURE 7.9 Triangular carrier wave and PWM signal for 5% duty cycle.

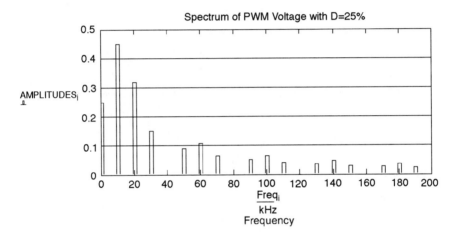

FIGURE 7.10 Spectrum of a PWM signal with 25% duty cycle.

The spectrum of a typical PWM signal with 25% duty cycle with a switching frequency of 10 kHz is shown in Fig. 7.10. The DC magnitude of 25% is clearly visible. The harmonics are multiples of the carrier frequency. The lowest harmonic is located at 10 kHz. This spectrum might look dramatic, especially in comparison with Fig. 7.5, but the reader should be reminded that, due to the switching speed of modern power semiconductors, the carrier frequency can be chosen sufficiently high that the harmonics can be easily filtered with capacitors and inductors of small size.

PWM Signals for AC Output

In addition to a DC reference signal, any other waveform could be used as the modulation signal as long as the highest frequency of its AC components are at least an order of magnitude less than the frequency of the carrier signal. Figure 7.11 shows an example of a carrier waveform, which is symmetrical with respect to the zero level. To generate a sinusoidal output voltage for an inverter, which is often desired, this carrier can be modulated with a sinusoidal reference signal. An example is shown in Fig. 7.12. Note that for clarity, the ratio between the carrier frequency and the frequency of the modulation signal is lower than recommended for actual implementation. The resulting sinusoidal PWM (SPWM) voltage drives one phase leg of an inverter. If the voltage level is +1, the upper switch is on, and vice versa. After filtering out the switching frequency components, the resulting output voltage has the shape and frequency of the modulation signal. For the remaining phase legs, the same technique, with reference signals

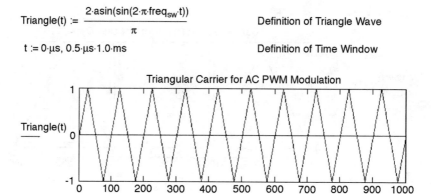

FIGURE 7.11 Triangular carrier wave AC modulation.

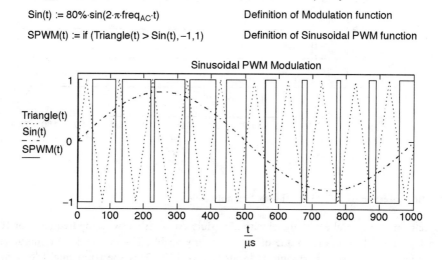

FIGURE 7.12 Illustration of the generation of sinusoidal PWM (SPWM) signals.

that are phase shifted by 120 and 240°, is used. The amplitude of the output voltage can be controlled by varying the ratio between the peak of the modulation signal and the peak of the carrier wave. If the amplitude of the modulation signal exceeds the amplitude of the carrier, overmodulation occurs and the shape of the fundamental of the output voltage deviates from the modulation signal.

To appreciate the spectral content of sinusoidal PWM signals, a 20-kHz triangular carrier has been modulated with a 500-Hz sinusoid with an amplitude of 80% of the carrier signal. The resulting SPWM signal is shown in Fig. 7.13. The spectrum of this PWM signal is shown in Fig. 7.14. The fundamental with an amplitude of 0.8 is located at 500 Hz. The harmonics are grouped around multiples of the carrier frequency [1].

It should be pointed out that this modulation scheme is far superior to the six-step technique described earlier, because the difference between the switching frequency and the fundamental is much larger. Therefore, the carrier frequency components can be easily removed with LC filters of small size [2]. In addition, the amplitude of the output voltage can be controlled simply by varying the amplitude ratio between the modulation signal and the carrier. If six-step modulation is used, the DC bus voltage would have to be controlled in order to control the amplitude of the output voltage.

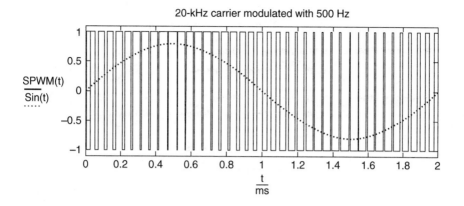

FIGURE 7.13 20-kHz carrier modulated with 500 Hz.

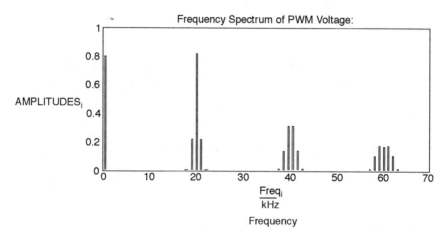

FIGURE 7.14 Spectrum of the SPWM signal shown in Fig. 7.13.

References

1. Mohan, N., Undeland, T., and Robbins, W., *Power Electronics: Converters, Applications, and Design*, 2nd ed., John Wiley & Sons, New York, 1995.
2. Von Jouanne, A., Rendusara, D., Enjeti, P., and Gray, W., Filtering techniques to minimize the effect of long motor leads on PWM inverter fed AC motor drive systems, *IEEE Trans. Ind. Appl.*, July/Aug., 919–926, 1996.

7.4 Third Harmonic Injection for Voltage Boost of SPWM Signals

Michael Giesselmann

It can be shown (Mohan et al.[1], p. 105) that if a three-phase input voltage is rectified using a standard three-phase rectifier, the resulting DC voltage is equal to 1.35 times the rms value of the AC line–line input voltage. If this DC voltage is used to feed a three-phase inverter using the SPWM modulation technique described above, the theoretical maximum AC line–line output voltage is only 82.7% of the AC line–line input voltage feeding the rectifier (Mohan et al.[1], p. 228). To boost the output voltage without resorting to overmodulation, the third harmonic of the fundamental frequency can be added to the modulation signal. Figure 7.15 shows an example, where a third harmonic with an amplitude of 21.1% has been added to the fundamental modulation signal.

Sinusoids for Phase a: $V_a(t) := V \cdot \sin(\omega_0 \cdot t)$
$V_{a3}(t) := 1.12 \cdot V \cdot [\sin(\omega_0 \cdot t) + m_{3h} \cdot \sin[3 \cdot (\omega_0 \cdot t)]]$

Sinusoids for Phase b: $V_b(t) := V \cdot \sin(\omega_0 \cdot t - 120 \cdot \deg)$
$V_{b3}(t) := 1.12 \cdot V \cdot [\sin(w_0 \cdot t - 120 \cdot \deg) + m_{3h} \cdot \sin[3 \cdot (w_0 \cdot t - 120 \cdot \deg)]]$

Sinusoids for Phase c: $V_c(t) := V \cdot \sin(\omega_0 \cdot t + 120 \cdot \deg)$
$V_{c3}(t) := 1.12 \cdot V \cdot [\sin(\omega_0 \cdot t + 120 \cdot \deg) + m_{3h} \cdot \sin[3 \cdot (\omega_0 \cdot t + 120 \cdot \deg)]]$

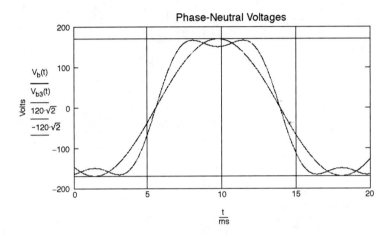

FIGURE 7.15 Sinusoidal modulation signal with and without added 3rd harmonic.

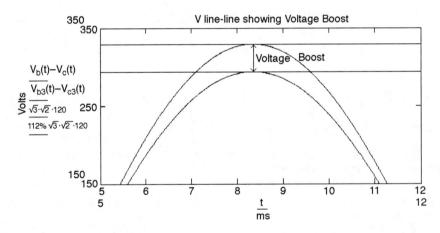

FIGURE 7.16 Line-to-line signal showing the voltage boost obtained by 3rd harmonic injection.

The amplitude of the fundamental has been increased to 112% in this example. It can be seen, that the peak amplitude of the resulting signal does not exceed the amplitude of the pure sinusoid with 100% amplitude. By inspection of Fig. 7.15 it is easy to see that the voltage–time integral will be higher if a 3rd harmonic is added to the reference signal for the phase to neutral voltage. This voltage boost beyond the previously mentioned value of 82.7% is very desirable, to retrofit induction motors with adjustable speed drives in existing installations. The 3rd harmonic components exactly cancel each other in the line-to-line voltages of the inverter. This is because the phase shift of the fundamental signals is 120° and therefore the phase shift of the 3rd harmonic is 3 × 120 = 360°. Therefore, the 3rd harmonic voltages precisely cancel and result in a pure sinusoidal output voltage being applied to the motor. This is shown in Fig. 7.16, which illustrates the voltage boost that is obtained.

7.5 Generation of PWM Signals Using Microcontrollers and DSPs

Michael Giesselmann

Modern power electronics controllers are rapidly moving toward digital implementation. Typical solutions consist of microcontrollers or DSPs. In addition, coprocessors, such as the ADMC200/201 from Analog Devices, are available that are specifically designed to support inverter control. Most of the processors, such as the 68HC12B32 from Motorola, that are commonly used to control power electronics have built-in hardware support for PWM generation. Figure 7.17 shows the basic principle of their digital PWM generation.

For clarity, the circuit shown in Fig. 7.17 has only 4-bit resolution for the duty cycle of the generated PWM signals, resulting in only 16 discrete duty cycles. In actual applications, 8 to 12 bits of resolution is typical. In Fig. 7.17, a digital counter (74163) counts from zero to its maximum value and repeats the cycle afterward. The count is continuously compared with a digital value representing the duty cycle using a hardware comparator (7485). The PWM signal is available on the output of the comparator. Figure 7.18 shows the simulation results from the example circuit shown in Fig. 7.17. The duty cycle in this example is 3/16.

If more than one channel is present, the PWM signals can be left, right, or center aligned. To be center aligned, up–down counters are used, which count up to their maximum count and then back to zero before starting the next cycle. The maximum count ($2^{bits} - 1$) is determined by the number of stages (bits) the digital counter has. In a digital PWM modulator each counter has an associated period register. The content of this register determines the maximum count at which the counter resets. If this number is less than the maximum count ($2^{bits} - 1$), the repetition (switching) frequency is increased and the

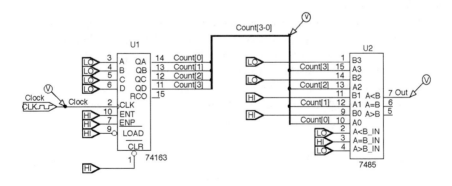

FIGURE 7.17 Principle of digital PWM signal generation.

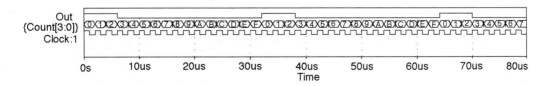

FIGURE 7.18 Simulation results from the circuit shown in Fig. 7.17.

resolution of the duty cycle is decreased for a given clock speed. It is often important to make the correct trade-off between the switching frequency and the resolution.

The advantage of hardware support for PWM generation is that the processor typically only needs to access any registers if the duty cycle is to be changed, since the period is typically only initialized once upon program start-up. It should also be mentioned that the duty cycle registers are typically "double-buffered," meaning that an update of a duty cycle does not need to be synchronized with the current state of the counter. In double-buffered systems, the new duty cycle will only be chosen once the previous period is completed to avoid truncated PWM signals. If necessary, a software override can disable this feature.

7.6 Voltage Source–Based Current Regulation

Michael Giesselmann

In motor drive applications, it is often desired to control directly the input current of the motor to control the torque. DC control also limits dynamics resulting from the electrical characteristics of the machine. Controlling the torque provides direct control over the angular acceleration, which is essential for precise motion control. Current control is typically performed in the innermost loop of a cascaded feedback control loop arrangement [1]. However, most power electronics converters are circuits with controllable voltage output. To achieve current control, the voltage of the power electronics converter can be controlled in such a way, that the desired current is obtained. Several methods can be used to achieve this:

- A feedback control loop, typically using a PI controller can be used control the current.
- The necessary voltage can be calculated in real time and applied to the motor.
- The necessary voltage for fast transients can be calculated in real time and applied to the motor and the residual error can be corrected by a PI controller.

Examples illustrating each of the schemes are described in the following. Figure 7.19 shows an example of a DC motor in which the current is controlled by adjusting the applied voltage using a PI controller such that the current follows the desired trajectory. The result is presented in Fig. 7.20, which shows that the current indeed follows the desired value at all times.

Sometimes even better results and higher loop bandwidth can be obtained if known information about the motor and the load is used to calculate the required voltage in real time. Figure 7.21 shows some fundamental equations of a permanent magnet (PM) DC motor. Here a capacitor is used to represent the kinetic energy stored in the machine. Therefore, the second voltage loop equation in Fig. 7.21 represents the voltage across the motor at all times. To test this theory, the "compensator" in the circuit shown in Fig. 7.22 calculates this voltage and applies the result to the DC motor. The subcircuit of the

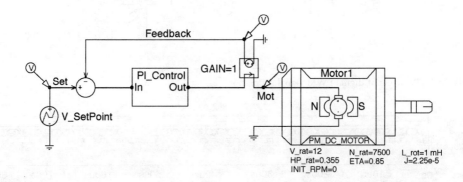

FIGURE 7.19 Voltage source–based current control using a PI feedback loop.

Modulation Strategies

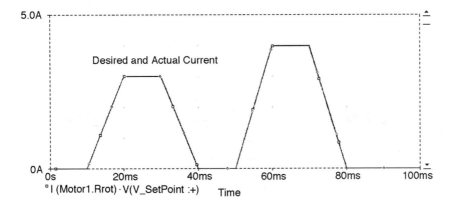

FIGURE 7.20 Simulation result for the circuit shown in Fig. 7.19.

$$V_s = R_{rot} \cdot I_{rot} + L_{rot} \cdot \frac{d}{dt} I_{rot} + K_m \cdot \omega \quad \text{Voltage loop equation}$$

$$\frac{1}{2} J \cdot \omega^2 = \frac{1}{2} C_{eq} \cdot E^2 \qquad C_{eq} = J \cdot \frac{\omega^2}{E^2} = \frac{J}{K_m^2} \quad \text{Equivalent Capacitance}$$

$$V_s = R_{rot} \cdot I_{rot} + L_{rot} \frac{d}{dt} I_{rot} + \frac{1}{C_{eq}} \cdot \int I_{rot} \, dt \quad \text{Voltage loop equation}$$

FIGURE 7.21 Fundamental equations for the PM DC machine.

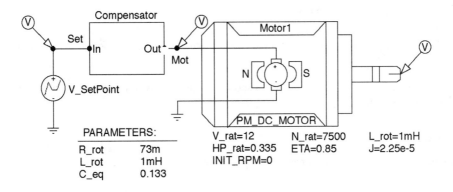

FIGURE 7.22 Voltage source–based current control using a feedforward approach.

compensator is shown in Fig. 7.23. The result is identical to that shown in Fig. 7.20. However, for the correct implementation of this scheme, the load and the inertia of the system needs to be known precisely, which is not realistic. Therefore, the best strategy is to implement the first two terms on the right side of the second voltage loop equation in Fig. 7.21 using a compensator and to use a PI controller to eliminate the residual error.

The advantage of the mixed (compensator and PI residual controller) approach is that the fast dynamics are covered by the feedforward path through the compensator, whereas the effect of the slower integral term is taken care of by the PI controller. The compensator will immediately apply the correct voltage to overcome the ohmic resistance of the winding and to establish the correct current slope in the rotor inductance. As the machine accelerates, the PI controller adds the appropriate voltage to offset the back-emf

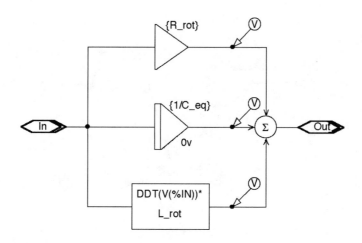

FIGURE 7.23 Subcircuit for the compensator shown in Fig. 7.22.

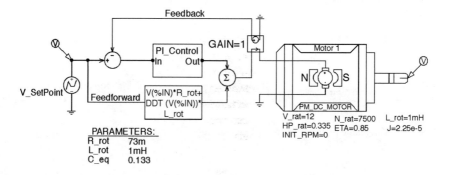

FIGURE 7.24 Voltage source–based current control using a feedforward approach for the fast transients and a PI controller for the residual error.

of the motor. An example for this approach is shown in Fig. 7.24. Again, the results are identical to the ones shown in Fig. 7.20.

Reference

1. Mohan, N., *Electric Drives, An Integrative Approach*, MNPERE, Minneapolis, MN, 2001.

7.7 Hysteresis Feedback Control

Hossein Salehfar

This section presents a hysteresis feedback control technique for DC-DC buck converters operating in both continuous and discontinuous conduction modes. A dead band with a high boundary above the voltage reference and a low boundary below the voltage reference are used to avoid chattering in the switch. The output voltage is regulated by comparing it to a reference voltage and the difference between the two voltages (error) is used to turn ON or OFF the switch. Regardless of where the voltage starts, switching takes place as soon as a boundary is encountered.

Initially, the switch turns ON because the output voltage of the converter is below the turn-on boundary. The output voltage then rises at a rate limited only by the inductor, the capacitor, and the load. The switch then turns OFF when the output voltage crosses the upper boundary and it remains OFF until the output voltage falls below and crosses the lower boundary where the switch is turned ON again. Once the voltage is between the upper and lower boundaries, switching actions will keep it in that vicinity under all conditions. This type of operation becomes independent of line, load, inductor, and capacitor values. Hysteresis control in principle eliminates the output variations other than the ripples. The system will stay close to the desired output voltage even if the input voltage, the load, or the component values change. Hysteresis control also provides an immediate response to dynamic disturbances. The control circuit is very simple and relatively straightforward to design and physically implement.

Introduction

Because of their high efficiency, compact size, and low cost, switching power supplies continue to gain popularity. Switching power supplies could be as high as three times more efficient than linear power supplies and in some cases eight times smaller in size. The heart of a switching power supply is its switch control circuit. Generally, the control circuit is a negative-feedback control loop connected to the switch through a comparator and a pulse width modulator (PWM). This control circuit regulates the output voltage against changes in the load and the input voltage. A feedforward loop may also be used to compensate for changes in the input voltage. Several topologies of switching power supplies have been developed and used. DC-DC buck converters are one of the widely used topologies.

The PWM-based voltage and current mode feedback controllers used in buck converters are widely used to improve line regulation [1]. Apparently, however, PWM voltage mode controllers have disadvantages. Since the input voltage is a significant parameter in the loop gain, any changes in the input voltage will alter the gain and will change the dynamics of the system. The central issue is that a voltage mode controller alone cannot correct any disturbances or changes until they are detected at the output. In the voltage-based controllers the compensation loop is difficult to implement. A limitation of the current mode controllers is the limit on the duty ratio. If the duty ratio exceeds 50%, then instability occurs [1]. In the current mode controllers a sensing resistor is used in the current loop, which increases the power losses in the converter. Most feedback controllers in buck converters use both the PWM voltage and current mode controllers to produce a better steady-state response and to reduce the voltage overshoots during start-ups.

Feedforward controllers may also be used to improve the line regulation in applications with a wide range of input voltages and loads [1, 2]. An apparent disadvantage of these types of controllers, however, is that the feedforward scheme with its direct sensing of the input quantities may have adverse effects when the converter is subjected to abrupt line transients. The sensing of input voltage through the feedforward loop may induce large-signal disturbances that could upset the normal duty cycle of the control mode. These concerns appear to be legitimate in light of the fact that such an effect is often observed with other forms of feedforward controls as well [2].

Most buck converters are designed for a continuous-current mode operation [3]. What if changes in the load or input voltage cause the system to operate in a discontinuous-current mode? In these situations, a voltage-based hysteresis feedback control circuit is a viable alternative [4–6]. Hysteresis feedback controllers enable buck converters to operate in both continuous- and discontinuous-current modes. With the ability of the system to operate in both modes, the size of the inductor and the capacitor are minimized compared with the other standard converters where a minimum size of the inductor is required to produce a continuous-current mode [3]. A minimum size of the capacitor is also required to limit the output voltage ripples as well. The output voltage of a hysteresis controller is stable and exhibits a robust behavior. The output is maintained even under extreme changes in the load, the line, or in the component values [4]. In some cases, the hysteresis controller determines the output ripple. The voltage-based hysteresis feedback controller presented in this section combines the advantages of both the PWM voltage and current-based controllers but it does not require a PWM circuit. It needs only a comparator circuit, which make it simpler and cheaper to build. Hysteresis controllers work well with DC-DC buck converters. But they do not work with other types of DC converters. If they are used, for example, with a boost converter, the results could be disastrous [4].

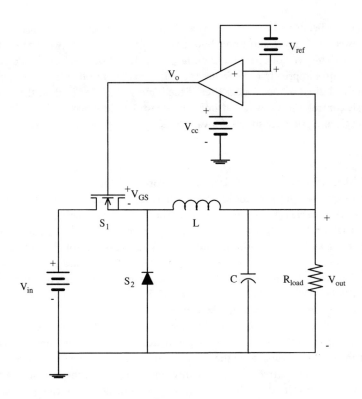

FIGURE 7.25 The basic circuit of a hysteresis controller for buck converters.

Principles of the Hysteresis Feedback Control Circuits

The main function of a DC-DC converter is to provide a good output voltage regulation. The output voltage must be maintained at the desired level against all changes in the input voltage or the load. The converter must act quickly to correct errors in the output voltage due to changes in the input voltage or in the load.

Figure 7.25 shows the basic circuit of the hysteresis controller for a buck converter. The circuit consists of a comparator and a switch (transistor). The comparator compares the output voltage V_{out} to a reference voltage V_{ref}. If $V_{out} > V_{ref}$, the switch is turned OFF. If $V_{out} < V_{ref}$, the switch is turned ON. This process repeats and V_{out} is maintained at a value close to V_{ref}. However, the circuit in Fig. 7.25 leads to chattering in the switch. In an attempt to keep V_{out} equal to V_{ref}, the switch chatters when it rapidly turns ON and OFF as V_{out} moves back and forth across V_{ref}. In power converters, an excessively fast switching action associated with chattering is destructive and it is therefore essential to avoid this condition.

Chattering is eliminated in hysteresis controllers by creating a dead band around V_{ref}. The dead band is created by using an upper boundary above V_{ref} and a lower boundary below V_{ref}. The space between the two boundaries is the dead band. Figure 7.26 shows the circuit of the hysteresis controller with a dead band. R_1 and R_2 resistors are added to the control circuit to provide the required dead band. The values of R_1 and R_2 determine the upper boundary and the lower boundary of the dead band.

The comparator in Fig. 7.26 is a Schmitt trigger. The input voltage V_+ of the positive terminal of the op-amp is no longer a fixed reference voltage V_{ref} as is the case in the basic comparator (without R_1 and R_2) in Fig. 7.25. V_+ now depends on V_{ref}, V_o, R_1, and R_2. It switches from one boundary of the dead band to another. During the initial start-up of the buck converter, the input to the negative terminal (−) of the op-amp (V_{out}) is a small positive value and is less than V_+. The amplifier will be saturated so that

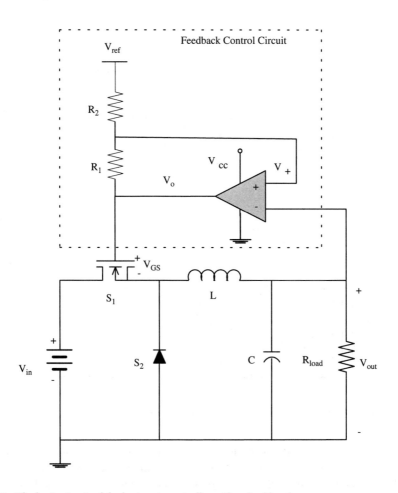

FIGURE 7.26 The basic circuit of the hysteresis controller with a dead band.

$V_o = V_{cc}$, thus the switch is turned ON. V_+ then becomes

$$V_+ = \frac{R_2 V_o}{R_1 + R_2} + \frac{R_1 V_{ref}}{R_1 + R_2} \quad (7.1)$$

As time progresses, V_{out} will increase gradually in the positive direction. The output voltage, V_o, remains unchanged at V_{cc} until V_{out} is equal to V_+. The op-amp will then enter its linear region and V_o will decrease and thus V_+ will decrease as well. This process will continue until V_o reaches zero and the op-amp will again saturate. The voltage V_+ will no longer be given by Eq. (7.1) but is given:

$$V_+ = \frac{R_1 V_{ref}}{R_1 + R_2} \quad (7.2)$$

Equations (7.1) and (7.2) represent the control boundaries of the control circuit. Equation (7.1) defines the upper boundary and Eq. (7.2) gives the lower boundary of the dead band. The dead band is computed by subtracting Eq. (7.2) from Eq. (7.1). The dead band is given by Eq. (7.3).

$$\Delta D_b = \frac{R_2 V_o}{R_1 + R_2} \quad (7.3)$$

R_1 and R_2 may be chosen to give the required value of $\Delta D_b / V_o$.

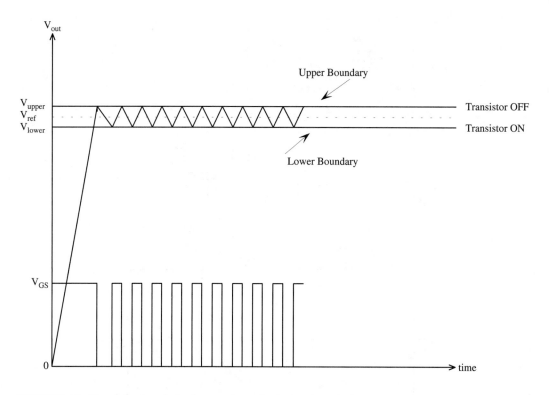

FIGURE 7.27 Time behavior of a buck converter under hysteresis control.

When the output voltage V_{out} is inside the dead band, the switch is OFF. Regardless of where the voltage starts, switching starts as soon as a boundary is encountered. In Fig. 7.27 the converter start-up process is illustrated. The transistor turns ON initially because the output voltage V_{out} is below the turn-on boundary. The output voltage rises from 0 to V_{upper} at a rate limited only by the inductor (L), the capacitor (C), and the load. The transistor then turns OFF as the voltage crosses the upper boundary and remains OFF until the output voltage falls below and crosses the lower boundary where the transistor is turned ON at V_{lower}. Once the voltage is between the boundaries, switching actions will keep it within the control boundaries under all conditions.

The operation of the system becomes independent of the line, the load, the inductor, and the capacitor values. The system will stay close to the desired output voltage V_{ref} even if the component values or the load changes drastically. A major drawback of this, however, is that the controller gives rise to an overvoltage during start-ups. But this problem can be solved by choosing the correct inductor and capacitor values that will allow the system to rise exponentially and settle somewhat close to the desired output voltage, while maintaining the desired ripple voltage. Initial voltage overshoots during start-ups can sometimes damage loads that are sensitive to high voltages or they may trigger the overvoltage protection circuits. Thus, it is essential to prevent such occurrences from happening.

Design Procedure

In the steady-state mode, the output voltage of the converter depends on the input voltage (V_{in}), the switching frequency (f_s), and the on duration of the switching period (t_{on}). This output voltage is given in Eq.(7.4):

$$V_{out} = t_{on} f_s V_{in} = DV_{in} \qquad (7.4)$$

Modulation Strategies

The product $t_{on}f_s$ is defined as the duty ratio D. The output voltage V_{out} is regulated by changing D while f_s is kept constant. This method is widely used in DC-DC converters. Another approach to regulating V_{out} is to vary f_s and keep D constant. However, this method is undesirable because it is difficult to filter the ripples in the input and output signals of the converter.

The hysteresis control of the buck converter is a fixed boundary control. V_{out} is regulated by the switching action of the switch as V_{out} crosses the upper or the lower boundary of the dead band of the hysteresis. In the hysteresis control of converters, f_s as well as D are not fixed values, but they change with the converter conditions. For a given set of converter parameters, both f_s and D are determined by the boundaries of the hysteresis. The frequency f_s and the duty ratio D are not used as control parameters in the design of hysteresis controllers. However, using a hysteresis controller, the basic operating principles of the buck converter do not change. The output voltage ripples of the converter depend on the dead band of the controller. As the dead band increases or decreases, the output voltage ripples will increase or decrease as well. The voltage ripple specifications can be guaranteed by setting the dead band of the controller at 50% of the ripple specifications with a suitable inductor value.

A properly designed hysteresis controller has excellent steady-state and dynamic properties. Its response quickly changes from its starting point to its desired operating value. Fixed boundary controllers have some important operating advantages. These controllers are stable under extreme disturbance conditions and can be chosen to directly guarantee the ripple specifications or other operating issues in a converter. The implementation of the hysteresis controller is illustrated by a design example.

Design Example

A DC-DC buck converter with voltage-based hysteresis control is designed to verify the method discussed in this section. Let $V_{out} = 5$ V and let the load vary between 1 and 5 Ω. The input voltage V_{in} also varies between 16 and 24 V with a nominal value of 20 V. The maximum ripple voltage is set at ±1%. A nominal switching frequency of $f_s = 100$ kHz is chosen. The output voltage ripple is then

$$\frac{\Delta V_{out}}{V_{out}} = 2\% \tag{7.5}$$

$$\Delta V_{out} = 0.02 \times 5 = 0.1 \text{ V}$$

The dead band is chosen to be 50% of the output voltage ripple to account for the small increase in the ripple level due to the natural response of the RLC circuit of the converter, after the switch is turned OFF, and to meet the required voltage ripple specification. The system will not meet the voltage ripple specification if the dead band is chosen to be equal to the output voltage ripple. The dead band is then

$$\Delta D_b = 0.5 \times 0.1 = 0.05 \text{ V} \tag{7.6}$$

Eq. (7.3) is solved for R_1:

$$R_1 = R_2 \left(\frac{V_o}{\Delta D_b} - 1 \right) \tag{7.7}$$

Let $R_2 = 100$ Ω and $V_o = 10$ V. Then R_1 is

$$R_1 = 100(10/0.05 - 1) = 19900 \text{ }\Omega$$

These resistor values produce a dead band that meets the output ripple specification. The maximum and minimum values of the load current are

$$I_{out,max} = V_{out}/R_{min} = 5/1 = 5 \text{ A} \tag{7.8}$$

and

$$I_{out,min} = V_{out}/R_{max} = 5/5 = 1\,A \quad (7.9)$$

Let the inductor current and the capacitor voltage swings be 10%. The inductor must limit the current swing at maximum load. The total current swing is as follows. Since

$$\frac{\Delta I_L}{I_{out,max}} = 10\%$$

using Eq. (7.8), $I_{out,max} = 5\,A$, then

$$\Delta I_L = 0.1 \times 5 = 0.5\,A \quad (7.10)$$

The capacitor voltage swing is

$$\frac{\Delta V_c}{\Delta V_{out}} = 10\%$$

$$\Delta V_c = 0.1 \times 0.1 = 0.01\,V \quad (7.11)$$

The equivalent series resistance (ESR) of the capacitor C must be included in the calculations. Since the waveform of the output ripple voltage is approximately sinusoidal, the output ripple voltage is then

$$\Delta V_{out} = \sqrt{(\Delta V_c)^2 + (\Delta V_{RC})^2} \quad (7.12)$$

where ΔV_{RC} is the voltage ripple across the capacitor resistance R_C. Note that ΔV_{RC} is usually much greater than ΔV_c. Thus a close approximation of the peak-to-peak output voltage ripple is obtained as follows:

$$\Delta V_{out} \cong \Delta V_{RC}$$
$$\cong \Delta I_c R_c$$
$$\cong (\Delta I_L - \Delta I_R) R_c \quad (7.13)$$

With the result from Eq. (7.10), the inductor L is

$$L = \frac{V_{out}(V_{in} - V_{out})}{V_{in} f_s \Delta I_L} = \frac{5(20-5)}{20(100000)0.5} = 75\,\mu H \quad (7.14)$$

The value of the capacitor is

$$C = \frac{\Delta I_L}{8 f_s \Delta V_c} = \frac{0.5}{8(100000)0.01} = 62.5\,\mu F \cong 65\,\mu F \quad (7.15)$$

The value of ESR of the capacitor can be determined from Eq. (7.13):

$$R_C = \frac{\Delta V_{out}}{\Delta I_L - \Delta I_R} = \frac{0.1}{0.5 - 0.1} = 0.25\,\Omega \quad (7.16)$$

Modulation Strategies

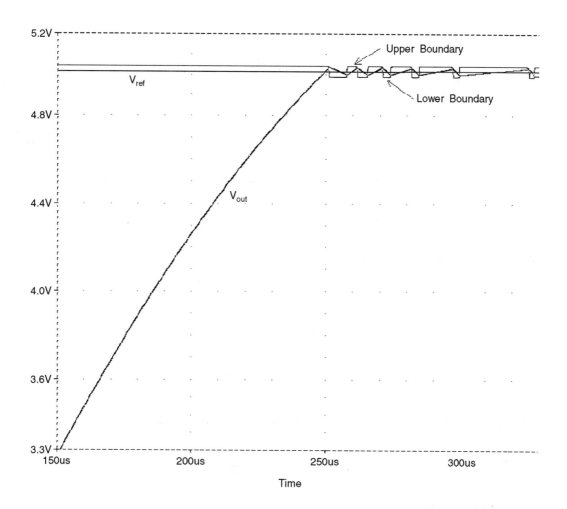

FIGURE 7.28 The output voltage of the buck converter using feedback hysteresis control.

The output waveform of the buck converter at 1 A load is shown in Fig. 7.28. As can be seen in Fig. 7.28, there are no excessive voltage overshoots during the start-up of the converter and V_{out} is kept within the specified boundaries of the hysteresis.

Transient overshoots, undershoots, and recovery times to step changes in the load and in the input are important performance parameters in buck converters. Since the current in the inductor cannot change instantaneously, the transient response is inherently inferior to that of the linear converters. The recovery time to step changes in the line and in the load is controlled by the characteristic of the feedback loop of the controller. Transient overshoots and undershoots resulting from step load changes can be analyzed and calculated in the following manner. The AC output impedance is

$$Z_{out} = \frac{V_{in} - V_{out}}{\Delta I_{load}} \tag{7.17}$$

Since

$$V_L = -L \frac{di_L}{dt} \tag{7.18}$$

and

$$I_{out} = C \frac{dv}{dt} \tag{7.19}$$

thus,

$$Z_{out} = \frac{L\Delta I_{out}}{(V_{in} - V_{out})C} \tag{7.20}$$

As a result, for an increasing current, the change in the output voltage will be as follows:

$$\Delta V_{out,min} = \Delta I_{out} Z_{out} = \frac{L\Delta I_{out}^2}{(V_{in} - V_{out})C} \tag{7.21}$$

and for a decreasing current, the change in the output voltage is

$$\Delta V_{out,max} = \frac{L\Delta I_{out}^2}{V_{out}C} \tag{7.22}$$

The transient undershoot for a load change from 1 to 5 A can be determined from Eq. (7.21).

$$\Delta V_{out,min} = \frac{L\Delta I_{out}^2}{(V_{in} - V_{out})C} = \frac{0.000075(5-1)^2}{(20-5)0.000065} = 1.231\,\text{V} \tag{7.23}$$

The transient overshoot for the load change from 5 to 1 A can be determined using Eq. (7.22).

$$\Delta V_{out,max} = \frac{L\Delta I_{out}^2}{V_{out}C} = \frac{0.000075(5-1)^2}{5(0.000065)} = 3.692\,\text{V} \tag{7.24}$$

Simulation Results

A DC-DC buck converter with a hysteresis feedback control circuit is constructed and simulated using the PSpice program. The results of the simulation were then compared to the theoretical results obtained in Eqs. (7.23) and (7.24) above. Figure 7.29 shows the results from the simulation. The transient behavior of the output voltage and the current of the converter can be obtained from Fig. 7.29. The maximum transient overshoot voltage for a load change from 5 to 1 A is found to be

$$\Delta V_{out,max} = 6.3463 - 5 = 1.3463\,\text{V} \tag{7.25}$$

and the minimum transient undershoot voltage for a load change from 1 to 5 A is

$$\Delta V_{out,min} = 5.0 - 3.4724 = 1.5276\,\text{V} \tag{7.26}$$

These results show that the controller has significantly reduced the overshoot transient. The overshoot value from the simulation is 36.47% of the theoretical value obtained using Eq. (7.24). However, a small increase in the undershoot transient is observed. The undershoot value from the simulation increased by 24.1% compared to the value obtained from Eq. (7.23). One can also see that the system has gone from a continuous-current mode to a discontinuous-current mode and back to a continuous-current mode. The switch is turned OFF when the output voltage crosses the upper boundary. The system then degrades linearly until it is back into a continuous-current mode where the switch is turned ON again and it continues its switching operation within the continuous-current mode region.

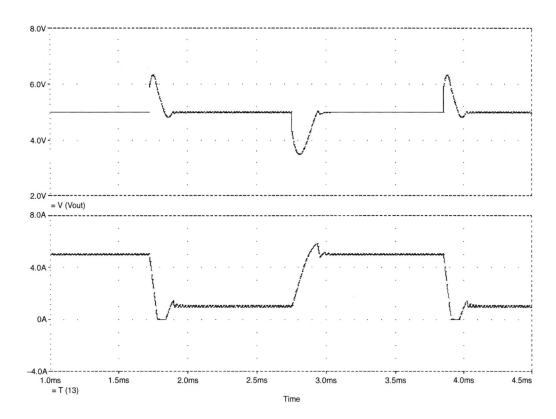

FIGURE 7.29 Simulation results, showing the transient analysis of the output voltage and current of the hysteresis feedback control for DC-DC buck converters.

Experimental Results

A DC-DC buck converter with a hysteresis feedback control circuit was physically built to verify the method under variable input and load conditions. The converter was built using an International Rectifier IRF540 power MOSFET, an International Rectifier Schottky MBR1045 diode, an inductor $L = 75$ μH, and a filter capacitor $C = 65$ μF. The DC input voltage, V_{in}, was set between 16 and 24 V, the nominal output voltage, V_{out}, was 5 V and the load resistance R_{load} was set between 1 and 5 Ω. The hysteresis feedback control circuit was built using a Motorola MC34081 operational amplifier as a comparator with a slew rate of 25 V/μS and resistors $R_1 = 19900$ Ω, $R_2 = 100$ Ω. The resulting waveforms were observed using the Lab-View program on a Pentium II PC. The output voltage waveform of the converter for a load change from 1 to 5 A, and vice versa, is shown in Fig. 7.30 with the input voltage V_{in} fixed at a nominal value of 20 V.

As expected, the transient undershoot for a load change from 1 to 5 A is found to be

$$\Delta V_{out,min} = 5.05 - 4.55 = 0.5 \text{ V} \quad (7.27)$$

which is 32.73% of that obtained from simulation and 40.62% of the theoretical value determined by Eq. (7.23). The transient overshoot for a load change from 5 to 1 A is found to be

$$\Delta V_{out,max} = 5.75 - 5.05 = 0.7 \text{ V} \quad (7.28)$$

which is 51.99% of that obtained from simulation and 18.96% of the theoretical value obtained using Eq. (7.24). The recovery time to the steady-state value of 5 V after an overshoot or undershoot is somewhat

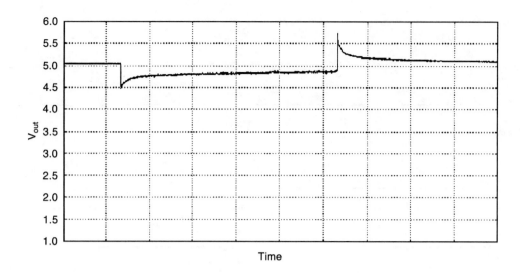

FIGURE 7.30 The transient response of the output voltage as load changes from 1 to 5 A, and vice versa.

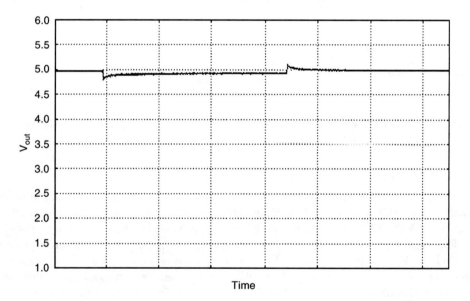

FIGURE 7.31 The transient response of the output voltage for a load change from 2 to 3 A, and vice-versa.

longer as compared to that from the simulation. This is due to the natural response of the *RLC* circuit in the system. The response decays gradually toward the steady-state value.

Figure 7.31 shows another transient response for a load change from 2 to 3 A, and vice versa. Here the recovery time to the steady-state value of 5 V is much shorter compared to the result shown in Fig. 7.30. Because the overshoot and the undershoot transients in this case are minimal, the resulting response time is much faster. This is mainly because the switching frequency is dependent on the natural response of the *RLC* circuit. The switching will only occur when the transient is within the boundary limits. Thus, one of the disadvantages of this type of feedback controller is a slower response time.

The waveform of the start-up transient response of the converter at 1-A load is shown in Figs. 7.32 and 7.33. These figures indicate that there are no excessive voltage overshoots during the start-up of the converter. Furthermore, the output ripple voltage of the converter is within the desired specification.

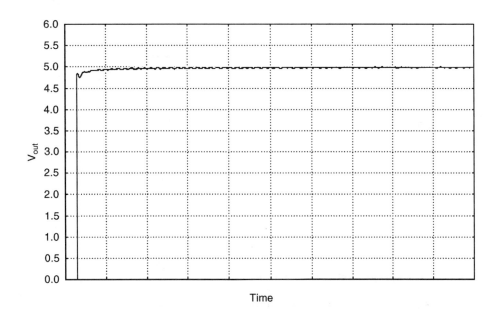

FIGURE 7.32 The start-up transient response at 1-A load.

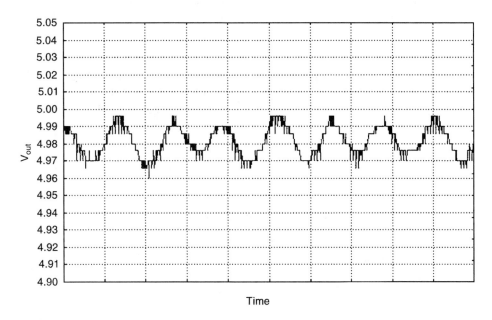

FIGURE 7.33 Zoom-in for the start-up transient response at 1-A load in Fig. 7.32.

From Fig. 7.33, one finds the ripple voltage to be

$$\Delta V_{\text{out}} = 4.995 - 4.97 = 0.025 \text{ V} = 25 \text{ mV} \tag{7.29}$$

which is five times less than the desired specification. The waveform of the start-up transient response of the converter at 5 A load is shown in Figs. 7.34 and 7.35. As expected, here also there are no excessive voltage overshoots during the start-up.

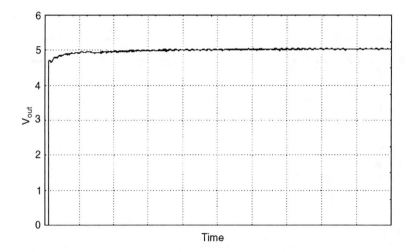

FIGURE 7.34 The start-up transient response at 5-A load.

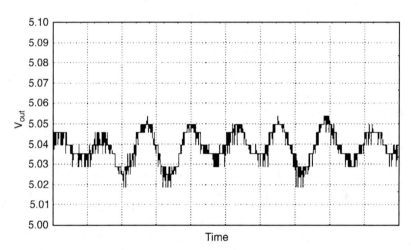

FIGURE 7.35 Zoom-in for the start-up transient response at 5-A load in Fig. 7.34.

The output voltage ripple of the buck converter is still within the desired specification. From Fig. 7.35, the ripple voltage is determined to be

$$\Delta V_{out} = 5.05 - 5.02 = 0.03 = 30 \text{ mV} \qquad (7.30)$$

which is within the desired specification. The waveform of the output voltage due to changes in the input voltage is shown in Fig. 7.36. The input voltage was sequentially changed from the nominal value of 20 V to 16 V to 24 V to 16 V and back to 20 V again. The load was fixed at 1 A.

As can be seen from Fig. 7.36, there is a small discrepancy in the output voltage due to changes in the input voltage. The output voltage is no longer maintained at a steady-state value of 5 V. Moreover, at the switching times between different levels of the input voltage, the output seems to decrease suddenly to zero and then returns back to a value close to 5 V. However, these sudden and sharp drops in the output voltage were not actually observed on the scope used for measurements. These sharp voltage drops seem to be caused by the slow response of the data acquisition card used in the experiment. Figure 7.37 shows another output voltage transient response to the same input voltages as the previous case except now the

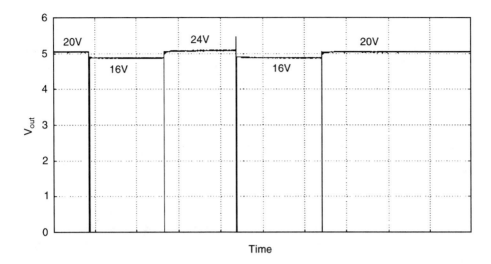

FIGURE 7.36 The output voltage response to various step changes in the input voltage at 1-A load.

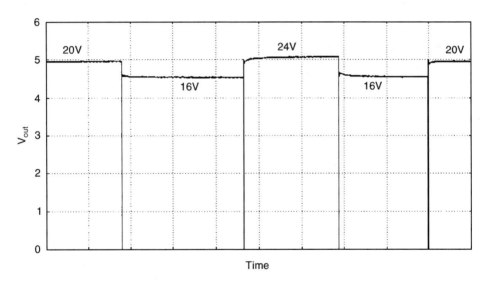

FIGURE 7.37 Output voltage response to various step change of input voltage at 2.5-A load.

load was fixed at 2.5 A. As Fig. 7.37 shows, there is an increase in the discrepancy of the output voltage. As the load increases and the input voltage decreases, the output voltage drops proportionally. But, on the other hand, as the input voltage increases, the output voltage maintains its steady-state value close to 5 V.

Conclusions

A voltage-based hysteresis feedback control circuit for DC-DC buck converters has been discussed in this section. Details of the design and analysis of the control circuit were presented. The basic concept of the controller was verified using simulation and actual experimental results. Results have shown that the buck converter with hysteresis control has a good load regulation. Line regulation is good over a wide range of load values. However, line regulation is good only over a limited range of input voltage values. The best values of line regulation were obtained at low-load resistance values. The buck converter with hysteresis control is simple to design and implement and it is suitable for practical applications. The hysteresis feedback loop decreases the overshoots in the buck converter. Results have also shown that the switching frequency of the buck converter with hysteresis control is dependent on the RLC circuit of the converter which causes a slow response time. A study of the dynamic behavior of the hysteresis controller to improve its response time is recommended for future research.

References

1. R. E. Tarter, *Solid-State Power Conversion Handbook*, Wiley, New York, 1993, 484–491.
2. S. S. Kelkar and F. C. Y. Lee, A fast time domain digital simulation technique for power converters: application of a buck converter with feedforward compensation, *IEEE Trans. Power Electron.*, PE-1, 21–31, January 1986.
3. D. W. Hart, *Introduction to Power Electronics*, Prentice-Hall, Englewood Cliffs, NJ, 1997, 194.
4. P. T. Krein, *Elements of Power Electronics*, Oxford University Press, New York, 1998, 666–674.
5. E. Vosicher and E. Lougee, Hysteretic controller fits process needs, *PCIM Power Electron. Syst. Mag.*, January 2000.
6. L. Hodson and R. Nowakowski, Hysteretic controller IC enables PC power supply to meet advanced CPU requirements, *PCIM Power Electron. Syst. Mag.*, July 2000.

7.8 Space-Vector Pulse Width Modulation

Hamid A. Toliyat and Tahmid Ur Rahman

One of the most preferred pulse width modulation (PWM) strategies today is space-vector modulation (SVPWM). This kind of scheme in voltage source inverter (VSI) drives offers improved bus voltage utilization and less commutation losses. Three-phase inverter voltage control by space-vector modulation includes switching between the two active and zero voltage vectors so that the time interval times the voltages in the chosen sectors equals the command voltage times the time period within each switching cycle. During the switching cycle the reference voltage is assumed to be constant as the time period would be very low. By simple digital calculation of the switching time one can easily implement the SVPWM scheme. However, the switching sequence may not be unique.

How the SVPWM Works

For a three-phase voltage source inverter as depicted in Fig. 7.38, each pole voltage may assume one of the two values depending upon whether the upper switch or the lower switch is on. Therefore, only eight combinations of switches are possible; these are shown in Fig. 7.39. Of these, two of them have zero states. Zero states occur when either the upper three or the lower three switches are conducting simultaneously.

The switches are termed as SA1, SA2 for pole A, SB1 and SB2 for pole B, and SC1 and SC2 for pole C. Different states are defined as follows:

$$A = 0 \text{ if SA1 off and SA2 on}$$
$$1 \text{ if SA1 on and SA2 off}$$
$$B = 0 \text{ if SB1 off and SB2 on}$$
$$1 \text{ if SB1 on and SB2 off}$$
$$C = 0 \text{ if SC1 off and SC2 on}$$
$$1 \text{ if SC1 on and SC2 off}$$

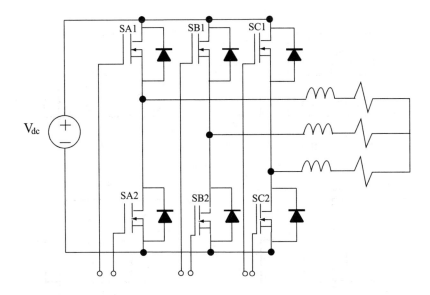

FIGURE 7.38 Six-switch voltage source inverter.

The instantaneous values of the line-to-line voltages of the inverter can be obtained from the above logic relations given by

$$\begin{aligned} V_{AB} &= V_{dc}(A-B) \\ V_{BC} &= V_{dc}(B-C) \\ V_{CA} &= V_{dc}(C-A) \end{aligned} \quad (7.31)$$

where V_{dc} is the DC bus voltage and V_{AB}, V_{BC}, V_{CA} are the line-to-line voltages. The line-to-neutral voltages are given by

$$\begin{aligned} V_A &= \frac{1}{3}(V_{AB} - V_{CA}) \\ V_B &= \frac{1}{3}(V_{BC} - V_{AB}) \\ V_C &= \frac{1}{3}(V_{CA} - V_{BC}) \end{aligned} \quad (7.32)$$

Replacing the values of line-to-line voltages in the previous set of equations yields the line-to-neutral voltages of the inverter:

$$\begin{aligned} V_A &= \frac{V_{dc}}{3}(2A - B - C) \\ V_B &= \frac{V_{dc}}{3}(2B - C - A) \\ V_C &= \frac{V_{dc}}{3}(2C - A - B) \end{aligned} \quad (7.33)$$

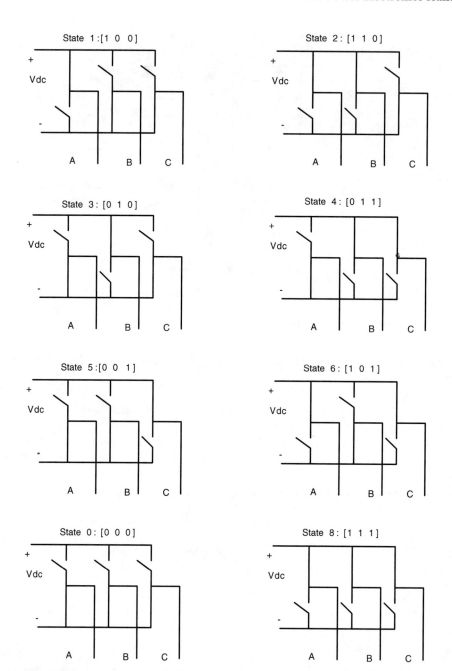

FIGURE 7.39 Possible switching pattern of the inverter.

For state 1, the values are

$$V_A = \frac{2}{3}V_{dc}$$
$$V_B = -\frac{1}{3}V_{dc} \quad (7.34)$$
$$V_C = -\frac{1}{3}V_{dc}$$

Modulation Strategies

By using the following transformation for a balanced three-phase system, the space vector for sector 1 with α and β components is given by

$$\begin{bmatrix} V_{s\alpha} \\ V_{s\beta} \end{bmatrix} = \sqrt{\frac{2}{3}} \begin{bmatrix} 1 & \cos 120° & \cos 240° \\ 0 & \sin 120° & \cos 240° \end{bmatrix} \begin{bmatrix} V_A \\ V_B \\ V_C \end{bmatrix} \quad (7.35)$$

Therefore, for the first sector space vector V_{s1} will be given by

$$\overline{V_{s1}} = \sqrt{\frac{2}{3}} V_{dc} e^{j0} \quad (7.36)$$

All the other six nonzero states are given by

$$\overline{V_{sk}} = \sqrt{\frac{2}{3}} V_{dc} e^{j(k-1)60°} \quad (7.37)$$

where the values of k vary from 1 to 6.

This divides the plane into six equal regions within a regular hexagon. These voltage vectors are of equal magnitude and mutually phase-displaced by 60°. Figure 7.40 shows the realizable voltage space vectors for a three-phase VSI. Whenever the reference vector is in a sector, the switches work according to the time interval T_m and T_{m+1} set by the projection of the vector on the adjacent sides as shown in Fig. 7.41. The [111] and the [000] states are defined as the zero states and they lie on the origin. Suppose it is necessary to generate a space-vector modulator for the following voltage system:

$$\begin{aligned} V_{an} &= V_1 \cos(\omega_m t - \gamma) \\ V_{bn} &= V_1 \cos(\omega_m t + 120 - \gamma) \\ V_{cn} &= V_1 \cos(\omega_m t - 120 - \gamma) \end{aligned} \quad (7.38)$$

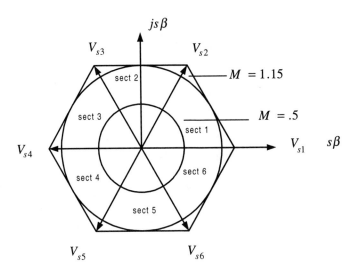

FIGURE 7.40 Possible space vectors.

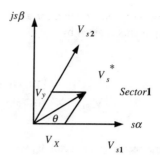

FIGURE 7.41 Time interval calculation for the space vector.

The system of voltages can be resolved into two components. These components in the stationary α and β axes are given as below:

$$V_{s\alpha} = V_1 \sin(\omega_m t - \gamma)$$
$$V_{s\beta} = -V_1 \cos(\omega_m t - \gamma) \tag{7.39}$$

These two voltages can be combined as

$$\overline{V}_s = -jV_1 e^{j(\omega_m t - \gamma)} = V_1 e^{j(\omega_m t - 90)}$$
$$= \sqrt{\frac{2}{3}} M V_{dc} e^{j(\omega_m t - 90)} \quad \text{assuming } \gamma \text{ to be zero} \tag{7.40}$$

where M is the modulation index. The voltage vector refers to a circular trajectory within the hexagon with an angular frequency of ω_m. The maximum voltage that can be achieved is proportional to the radius of the largest circle inscribed within the hexagon.

Implementation

First the position of the rotating vector is computed by taking the arctangent of the ratio between its two different quadrature axes components. The switching time interval calculation is the tricky part for this scheme. If the voltage vector lies in the first sector, the time intervals can be expressed as depicted in Fig. 7.41.

To synthesize a reference voltage vector $\overline{V}_{ref}$ to its adjacent states and to obtain a minimum switching frequency for that the total cycle T_{cycle} should be divided into three segments T_m, T_{m+1}, T_0. Using simple geometry in Fig. 7.41 yields

$$T_m + T_{m+1} + T_0 = T_{cycle}$$
$$\overline{V}_{ref} \times T_{cycle} = T_m \times \overline{V}_{sm} + T_{m+1} \times \overline{V}_{s(m+1)} \tag{7.41}$$

For a voltage vector residing in the first sector the equation can be expressed as

$$\overline{V}_{ref} \times T_{cycle} = T_1 \times \overline{V}_{s1} + T_2 \times \overline{V}_{s2} \tag{7.42}$$

By taking the components of the reference voltages in the quadrature axis,

$$\left|\overline{V}_{ref}\right| \sin\theta = \frac{T_2}{T_{cycle}} \times \left|\overline{V}_{s2}\right| \sin 60°$$

or

$$T_2 = T_{cycle} \times M \times \frac{\sin\theta}{\sin 60°}$$

where

$$M = \frac{|\overline{V_{ref}}|}{|\overline{V_{s2}}|} \tag{7.43}$$

Similarly it can be shown that

$$T_1 = T_{cycle} \times M \times \frac{\sin(60° - \theta)}{\sin 60°}$$

The rest of the cycle can be divided between and T_0 and T_8. This can be expressed as

$$T_{cycle} - T_1 - T_2 = T_0 \tag{7.44}$$

Switching Signals

As has been mentioned previously, the switching sequence is not unique. Of all these, the most prominent uses minimum inverter switching frequency, which is obtained by transitioning from one inverter state to another only by switching one inverter pole. The total zero time is divided between the two zero states. Figure 7.42 clearly demonstrated the switching in Sector 1. Here, the cycle begins in State 0, i.e., [000], with each inverter pole being successively toggled until State 8, [111], is obtained. The pattern is then reversed to complete the modulation cycle. Figure 7.42 shows the times from the start of each modulation cycle at which the inverter poles are toggled, T_{Aon}, T_{Bon}, and T_{Con}, respectively. Taking the variations from one sector to another into consideration, it is possible to tabulate these times as functions of both the active and zero state times. It can be easily seen that from one state to the other only one inverter pole is toggled.

The other type of switching is a bus-clamped one. Under this scheme, the switching is done in one sector using two inverter poles only. One of the poles remains clamped to the higher node. This makes switching easier. This is also called a 60° bus-clamped scheme because the pole is clamped for 60° (see Fig. 7.43). Now, one can alternate between [111] and [000] for the zero states in adjacent sectors. It is termed odd 60° bus-clamped switching if state 8 or [111] is used in odd sectors and vice versa.

For example, in case of a space vector in the first sector, the switching will be done according to the sequence [111], [110], [100], [110], [111]. So the upper switch of A is clamped to the positive terminal. The next sector would have a switching pattern that would look like this: [000], [010], [110], [010], [000]. Here the lower switch of C will be clamped to the negative terminal and all of the inverter legs will toggle when there is a change in sector. It is worth noting that the even sector only uses the [000] states.

In implementing space-vector modulation on a DSP or microcontroller, one must limit the space vector voltage. The required voltage can sometimes be over the range of the inverter. This may occur in cases where one needs to use current regulation from a current controller. There are two ways to limit the current. One is to limit it by the radius of the circle inscribed within the hexagon and the other one by limiting the interval time within T_{cycle}.

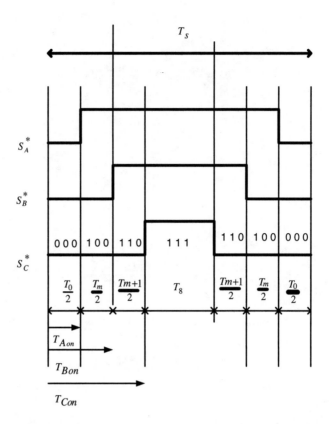

FIGURE 7.42 Switching signals for the SVPWM.

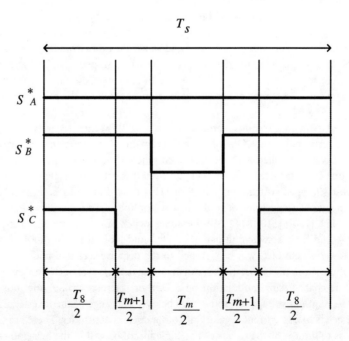

FIGURE 7.43 Switching signals for the bus-clamped SVPWM.

Modulation Strategies

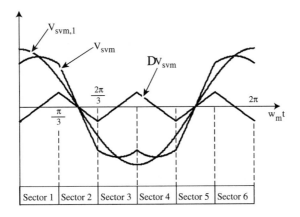

FIGURE 7.44 The modulating function of space-vector modulation with harmonic contents.

The limiting conditions can be given by the following equations:

$$\sqrt{(V_{s\alpha})^2 + (V_{s\beta})^2} > |\overline{V_{s1}}| \tag{7.45}$$

$$T_m + T_{m+1} < T_{cycle} \tag{7.46}$$

As mentioned previously, space-vector modulation provides 15% more bus utilization. The space-vector modulating function can be expressed as a sinusoidal and a triangular distorting waveform with a frequency three times the fundamental. This results in flattening of the peaks of the modulation signal, and, hence, a higher maximum voltage is achieved. This is shown in Fig. 7.44.

References

S. Ogasawara, H. Akagi, and A. Nabae, A novel PWM scheme of voltage source inverter based on space vector theory, *European Power Electron. Conf. Rec.*, Aachen, Germany, 1197–1202, 1989.

A. M. Tryzynadlowski, *The Field Orientation Principle in Control of Induction Motors*, Kluwer Academic Publishers, Dordrecht, The Netherlands, 1994.

H. Van Der Broeck, H. Skudelny, and G. Stanke, Analysis and realization of a pulse width modulator based on voltage space vectors, *IEEE-IAS Conf. Rec.*, 244–251, 1986.

8
Sliding-Mode Control of Switched-Mode Power Supplies

Giorgio Spiazzi
University of Padova

Paolo Mattavelli
University of Padova

8.1 Introduction 8-2
8.2 Introduction to Sliding-Mode Control 8-2
8.3 Basics of Sliding-Mode Theory 8-5
   Existence Condition • Hitting Conditions • System Description in Sliding Mode: Equivalent Control • Stability
8.4 Application of Sliding-Mode Control to DC-DC Converters—Basic Principle 8-8
8.5 Sliding-Mode Control of Buck DC-DC Converters 8-9
   Phase-Plane Description • Selection of the Sliding Line • Existence Condition • Current Limitation
8.6 Extension to Boost and Buck–Boost DC-DC Converters 8-14
   Stability Analysis
8.7 Extension to Cúk and SEPIC DC-DC Converters 8-18
   Existence Condition • Hitting Condition • Stability Condition
8.8 General-Purpose Sliding-Mode Control Implementation 8-20
8.9 Conclusions 8-22

Switch-mode power supplies represent a particular class of variable structure systems (VSS), and they can take advantage of nonlinear control techniques developed for this class of system. Sliding-mode control, which is derived from variable structure system theory [1, 2], extends the properties of hysteresis control to multivariable environments, resulting in stability even for large supply and load variations, good dynamic response, and simple implementation.

Some basic principles of sliding-mode control are first reviewed. Then the application of the sliding-mode control technique to DC-DC converters is described. The application to buck converter is discussed in detail, and some guidelines for the extension of this control technique to boost, buck–boost, Cúk, and SEPIC converters are given. Finally, to overcome some inherent drawbacks of sliding-mode control, improvements like current limitation, constant switching frequency, and output voltage steady-state error cancellation are described.

8.1 Introduction

Switch-mode power supplies (SMPS) are nonlinear and time-varying systems, and thus the design of a high-performance control is usually a challenging issue. In fact, control should ensure system stability in any operating condition and good static and dynamic performances in terms of rejection of input voltage disturbances and load changes. These characteristics, of course, should be maintained in spite of large input voltage, output current, and even parameter variations (robustness).

A classical control approach relies on the *state space averaging* method, which derives an equivalent model by circuit-averaging all the system variables in a switching period [3–5]. On the assumptions that the switching frequency is much greater than the natural frequency of system variables, low-frequency dynamics is preserved while high-frequency behavior is lost. From the average model, a suitable small-signal model is then derived by perturbation and linearization around a precise operating point. Finally, the small-signal model is used to derive all the necessary converter transfer functions to design a linear control system by using classical control techniques. The design procedure is well known, but it is generally not easy to account for the wide variation of system parameters, because of the strong dependence of small-signal model parameters on the converter operating point. Multiloop control techniques, such as current-mode control, have greatly improved power converter dynamic behavior, but the control design remains difficult especially for high-order topologies, such as those based on Cúk and SEPIC schemes.

The sliding-mode approach for variable structure systems (VSS) [1, 2] offers an alternative way to implement a control action that exploits the inherent variable structure nature of SMPS. In particular, the converter switches are driven as a function of the instantaneous values of the state variables to force the system trajectory to stay on a suitable selected surface on the phase space. This control technique offers several advantages in SMPS applications [6–19]: stability even for large supply and load variations, robustness, good dynamic response, and simple implementation. Its capabilities emerge especially in application to high-order converters, yielding improved performances as compared with classical control techniques.

In this chapter, some basic principles of sliding-mode control are reviewed in a tutorial manner and its applications to DC-DC converters are investigated. The application to buck converters is first discussed in details, and then guidelines for the extension of this control technique to boost, buck–boost, Cúk, and SEPIC converters are given. Finally, improvements like current limitation, constant switching frequency, and output voltage steady-state error cancellation are discussed.

8.2 Introduction to Sliding-Mode Control

Sliding-mode control is a control technique based on VSS, defined as systems where the circuit topology is intentionally changed, following certain rules, to improve the system behavior in terms of speed of response, stability, and robustness. A VSS is based on a defined number of independent subtopologies, which are defined by the status of nonlinear elements (switches); the global dynamics of the system is, however, substantially different from that of each single subtopology. The theory of VSS [1, 2] provides a systematic procedure for the analysis of these systems and for the selection and design of the control rules. To introduce sliding-mode control, a simple example of a second-order system is analyzed. Two different substructures are introduced and a combined action, which defines a sliding mode, is presented. The first substructure, which is referred as substructure I, is given by the following equations:

$$\begin{cases} \dot{x}_1 = x_2 \\ \dot{x}_2 = -K \cdot x_1 \end{cases} \quad (8.1)$$

where the eigenvalues are complex with zero real part; thus, for this substructure the phase trajectories are circles, as shown in Fig. 8.1 and the system is marginally stable. The second substructure, which is

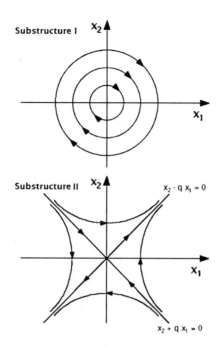

FIGURE 8.1 Phase-plane description corresponding to substructures I and II.

referred as substructure II, is given by

$$\begin{cases} \dot{x}_1 = x_2 \\ \dot{x}_2 = +K \cdot x_1 \end{cases} \quad (8.2)$$

In this case the eigenvalues are real and with opposite sign; the corresponding phase trajectories are shown in Fig. 8.1 and the system is unstable. Only one phase trajectory, namely, $x_2 = -qx_1$ ($q = \sqrt{K}$), converges toward the origin, whereas all other trajectories are divergent.

Divide the phase-plane in two regions, as shown in Fig. 8.2; accordingly, at each region is associated one of the two substructures as follows:

Region I: $x_1 \cdot (x_2 + cx_1) < 0 \Rightarrow$ Substructure I
Region II: $x_1 \cdot (x_2 + cx_1) > 0 \Rightarrow$ Substructure II

where c is lower than q. The switching boundaries are the x_2 axis and the line $x_2 + cx_1 = 0$. The system structure changes whenever the system representative point (RP) enters a region defined by the switching boundaries. The important property of the phase trajectories of both substructures is that, in the vicinity of the switching line $x_2 + cx_1 = 0$, they converge to the switching line. The immediate consequence of this property is that, once the RP hits the switching line, the control law ensures that the RP does not move away from the switching line. Figure 8.2a shows a typical overall trajectory starting from an arbitrary initial condition P_0 (x_{10}, x_{20}): after the intervals corresponding to trajectories $P_0 - P_1$ (substructure I) and $P_1 - P_2$ (substructure II), the final state evolution lies on the switching line (in the hypothesis of ideal infinite frequency commutations between the two substructures).

This motion of the system RP along a trajectory, on which the structure of the system changes and which is not part of any of the substructure trajectories, is called the *sliding mode*, and the switching line $x_2 + cx_1 = 0$ is called the *sliding line*. When sliding mode exists, the resultant system performance is completely different from that dictated by any of the substructures of the VSS and can be, under particular conditions, made independent of the properties of the substructures employed and dependent only on

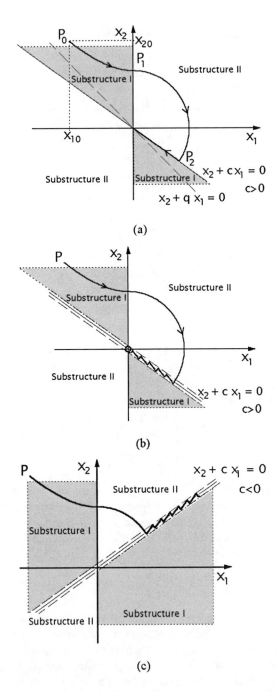

FIGURE 8.2 Sliding regime in VSS. (a) Ideal switching line; (b) switching line with hysteresis; (c) unstable sliding mode.

the control law (in this example the boundary $x_2 + cx_1 = 0$). In this case, for example, the dynamic is of the first order with a time constant equal to $1/c$.

The independence of the closed-loop dynamics on the parameters of each substructure is not usually true for more complex systems, but even in these cases it has been proved that the sliding-mode control maintains good robustness compared with other control techniques. For higher-order systems, the control rule can be written in the following way:

$$\sigma = f(x_1, \ldots, x_N) = \sum_{i=1}^{N} c_i x_i = 0 \qquad (8.3)$$

where N is the system order and x_i are the state variables. Note that the choice of using a linear combination of state variable in Eq. (8.3) is only one possible solution, which results in a particularly simple implementation in SMPS applications.

When the switching boundary is not ideal, i.e., the commutation frequency between the two substructures is finite, then the overall system trajectory is as shown in Fig. 8.2b. Of course, the width of the hysteresis around the switching line determines the switching frequency between the two substructures.

Following this simple example and looking at the Figs. 8.1 and 8.2, it is easy to understand that the conditions for realizing a sliding-mode control are:

- *Existence condition*: The trajectories of the two substructures are directed toward the sliding line when they are close to it.
- *Hitting condition*: Whatever the initial conditions, the system trajectories must reach the sliding line.
- *Stability condition*: The evolution of the system under sliding mode should be directed to a stable point. In Fig. 8.2b the system in sliding mode goes to the origin of the system, that is, a stable point. But if the sliding line were the following:

Region I: $x_1 \cdot (x_2 + cx_1) < 0 \Rightarrow$ Substructure I
Region II: $x_1 \cdot (x_2 + cx_1) > 0 \Rightarrow$ Substructure II

where $c < 0$, then the system trajectories would have been as shown in Fig. 8.2c. In this case, the resulting state trajectory still follows the sliding line, but it goes to infinity and the system is therefore unstable.

The approach to more complex systems cannot be expressed only with graphical considerations, and a mathematical approach should be introduced, as reported below.

8.3 Basics of Sliding-Mode Theory

Consider the following general system with scalar control [1, 2]:

$$\dot{\mathbf{x}} = \mathbf{f}(\mathbf{x}, t, u) \qquad (8.4)$$

where $\mathbf{x}$ is a column vector and $\mathbf{f}$ is a function vector, both of dimension N, and u is an element that can influence the system motion (control input). Consider that the function vector $\mathbf{f}$ is discontinuous on a surface $\sigma(\mathbf{x}, t) = 0$. Thus, one can write:

$$\mathbf{f}(\mathbf{x}, t, u) = \begin{cases} \mathbf{f}^+(\mathbf{x}, t, u^+) & \text{for} \quad \sigma \to 0^+ \\ \mathbf{f}^-(\mathbf{x}, t, u^-) & \text{for} \quad \sigma \to 0^- \end{cases} \qquad (8.5)$$

where the scalar discontinuous input u is given by

$$u = \begin{cases} u^+ & \text{for} \quad \sigma(\mathbf{x}) > 0 \\ u^- & \text{for} \quad \sigma(\mathbf{x}) < 0 \end{cases} \qquad (8.6)$$

The system is in sliding mode if its representative point moves on the sliding surface $\sigma(\mathbf{x}, t) = 0$.

Existence Condition

For a sliding mode to exist, the phase trajectories of the two substructures corresponding to the two different values of the vector function $\mathbf{f}$ must be directed toward the sliding surface $\sigma(\mathbf{x}, t) = 0$ in a small region close to the surface itself. In other words, approaching the sliding surface from points where $\sigma < 0$, the corresponding state velocity vector $\mathbf{f}^-$ must be directed toward the sliding surface, and the same must happen when points above the surface ($\sigma > 0$) are considered, for which the corresponding state velocity vector is $\mathbf{f}^+$. Indicating with subscript $\mathbf{N}$ the components of state velocity vectors $\mathbf{f}^+$ and $\mathbf{f}^-$ orthogonal to the sliding surface one can write:

$$\begin{array}{c} \lim_{\sigma \to 0^+} \mathbf{f}_N^+ < 0 \\ \lim_{\sigma \to 0^-} \mathbf{f}_N^- > 0 \end{array} \Rightarrow \begin{array}{c} \lim_{\sigma \to 0^+} \nabla \sigma \cdot \mathbf{f}^+ < 0 \\ \lim_{\sigma \to 0^-} \nabla \sigma \cdot \mathbf{f}^- > 0 \end{array} \quad (8.7)$$

where $\nabla \sigma$ is the gradient of surface σ. Since

$$\frac{d\sigma}{dt} = \sum_{i=1}^{N} \frac{\partial \sigma}{\partial x_i} \frac{dx_i}{dt} = \nabla \sigma \cdot \mathbf{f} \quad (8.8)$$

the existence condition of the sliding mode becomes

$$\begin{array}{c} \lim_{\sigma \to 0^+} \dfrac{d\sigma}{dt} < 0 \\ \lim_{\sigma \to 0^-} \dfrac{d\sigma}{dt} > 0 \end{array} \Rightarrow \lim_{\sigma \to 0} \sigma \dfrac{d\sigma}{dt} < 0 \quad (8.9)$$

When the inequality given by Eq. (8.9) holds in the entire state space and not only in an infinitesimal region around the sliding surface, then this condition is also a sufficient condition that the system will reach the sliding surface.

Hitting Conditions

Let $[\mathbf{x}^+]$ and $[\mathbf{x}^-]$ be the steady-state RPs corresponding to the inputs u^+ and u^- Eq. (8.6), respectively. Then a simple sufficient condition, that will be used later in the application of the sliding mode control to switch-mode power supplies, for reaching the sliding surface is given by:

$$[\mathbf{x}^+] \in \sigma(\mathbf{x}) < 0, \quad [\mathbf{x}^-] \in \sigma(\mathbf{x}) > 0 \quad (8.10)$$

In other words, if the steady-state point for one substructure belongs to the region of the phase space reserved to the other substructure, then sooner or later the system RP will hit the sliding surface.

System Description in Sliding Mode: Equivalent Control

The next focus of interest in the analysis of VSS is the behavior of the system operating in sliding regime. Consider here a particular class of systems that are linear with the control input, i.e.,

$$\dot{\mathbf{x}} = \mathbf{f}(\mathbf{x}, t) + \mathbf{B}(\mathbf{x}, t)u \quad (8.11)$$

where $\mathbf{x} \in \Re^N$, $\mathbf{f}$ and $\mathbf{B} \in \Im^N$, $u \in \Re^1$.

The scalar control input u is discontinuous on the sliding surface $\sigma(\mathbf{x}, t) = 0$, as shown in Eq. (8.6), whereas $\mathbf{f}$ and $\mathbf{B}$ are continuous function vectors. Under sliding mode control, the system trajectories stay on the sliding surface, hence:

$$\sigma(\mathbf{x}, t) = 0 \;\Rightarrow\; \dot{\sigma}(\mathbf{x}, t) = 0 \tag{8.12}$$

$$\dot{\sigma}(\mathbf{x}, t) = \frac{d\sigma}{dt} = \sum_{i=1}^{N} \frac{\partial \sigma}{\partial x_i} \frac{dx_i}{dt} = \nabla \sigma \cdot \dot{\mathbf{x}} = \mathbf{G}\dot{\mathbf{x}} \tag{8.13}$$

where $\mathbf{G}$ is a 1 by N matrix whose elements are the derivatives of the sliding surface with respect to the state variables. By using Eqs. (8.11) and (8.13),

$$\mathbf{G}\dot{\mathbf{x}} = \mathbf{G}\mathbf{f}(\mathbf{x}, t) + \mathbf{G}\mathbf{B}(\mathbf{x}, t)u_{eq} = 0 \tag{8.14}$$

where the control input u was substituted by an equivalent control u_{eq} that represents an equivalent continuous control input that maintains the system evolution on the sliding surface. On the assumption that $[\mathbf{GB}]^{-1}$ exists, from Eq. (8.14) one can derive the expression for the equivalent control:

$$u_{eq} = -(\mathbf{GB})^{-1}\mathbf{G}\mathbf{f}(\mathbf{x}, t) \tag{8.15}$$

Finally, by substituting this expression into Eq. (8.11),

$$\dot{\mathbf{x}} = [\mathbf{I} - \mathbf{B}(\mathbf{GB})^{-1}\mathbf{G}]\mathbf{f}(\mathbf{x}, t) \tag{8.16}$$

Equation (8.16) describes the system motion under sliding-mode control. It is important to note that the matrix $\mathbf{I} - \mathbf{B}(\mathbf{GB})^{-1}\mathbf{G}$ is less than full rank. This is because, under sliding regime, the system motion is constrained to be on the sliding surface. As a consequence, the equivalent system described by Eq. (8.16) is of order $N - 1$. This equivalent control description of a VSS in sliding regime is valid, of course, also for multiple control inputs. For details, see Refs. 1 and 2.

Stability

Analyzing the system behavior in the phase-plane for the second-order system, it was found that the system stability is guaranteed if its trajectory, in sliding regime, is directed toward a stable operating point. For higher-order systems, a direct view of the phase space is not feasible and one must prove system stability through mathematical tools. First consider a simple linear system with scalar control in the following canonical form:

$$\begin{cases} \dot{x}_i = x_{i+1} & i = 1, 2, \ldots, N-1 \\ \dot{x}_N = \displaystyle\sum_{j=1}^{N} a_{Nj} x_j + bu \end{cases} \tag{8.17}$$

and

$$\sigma(\mathbf{x}, t) = \sum_{i=1}^{N} c_i x_i = \sum_{i=1}^{N} c_i \frac{d^{i-1} x_1}{dt^{i-1}} = 0 \tag{8.18}$$

The latter equation completely defines the system dynamic in sliding regime. Moreover, in this case the system dynamic in sliding mode depends only on the sliding surface coefficients c_i, leading to a system behavior that is completely different from those given by the substructures defined by the two control input values u^+ and u^-. This is a highly desirable situation because the system dynamic can be directly determined by a proper c_i selection. Unfortunately, in the application to DC-DC converters this is possible only for the buck topology, whereas for other converters, state derivatives are not only difficult to measure, but also discontinuous. Therefore, we are obliged to select system states that are measurable, physical, and continuous variables. In this general case, the system stability in sliding mode can be analyzed by using the equivalent control method Eq. (8.16).

8.4 Application of Sliding-Mode Control to DC-DC Converters—Basic Principle

The general sliding-mode control scheme of DC-DC converters is shown in Fig. 8.3. U_g and u_o are input and output voltages, respectively, while i_{Li} and u_{Cj} ($i = 1 \div r, j = r + 1 \div N - 1$) are the internal state variables of the converter (inductor currents and capacitor voltages). Switch S accounts for the system nonlinearity and indicates that the converter may assume only two linear subtopologies, each associated to one switch status. All DC-DC converters having this property (including all single-switch topologies, plus push-pull, half and two-level full-bridge converters) are represented by the equivalent scheme of Fig. 8.3. The above condition also implies that the sliding-mode control presented here is valid only for *continuous conduction mode* (CCM) of operation.

In the scheme of Fig. 8.3, according to the general sliding-mode control theory, all state variables are sensed, and the corresponding errors (defined by difference to the steady-state values) are multiplied by proper gains c_i and added together to form the sliding function σ. Then, hysteretic block HC maintains this function near zero, so that

$$\sigma = \sum_{i=1}^{N} c_i \varepsilon_i = 0 \qquad (8.19)$$

Observe that Eq. (8.19) represents a hyperplane in the state error space, passing through the origin. Each of the two regions separated by this plane is associated, by block HC, to one converter sub structure. If one assumes (*existence condition* of the sliding mode) that the state trajectories near the surface, in both regions,

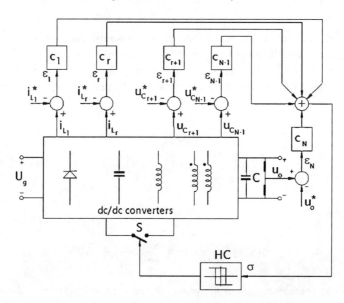

FIGURE 8.3 Principle scheme of a SM controller applied to DC-DC converters.

are directed toward the sliding plane, the system state can be enforced to remain near (lie on) the sliding plane by proper operation of the converter switch(es).

Sliding-mode controller design requires only a proper selection of the sliding surface Eq. (8.19), i.e., of coefficients c_i, to ensure existence, hitting, and stability conditions. From a practical point of view, selection of the sliding surface is not difficult if second-order converters are considered. In this case, in fact, the above conditions can be verified by simple graphical techniques. Instead, for higher-order converters, like Cúk and SEPIC, the more general approach outlined in Section 8.5 must be used.

One of the major problem of the general scheme of Fig. 8.3 is that inductor current and capacitor voltage references are difficult to evaluate, because they generally depend on load power demand, supply voltage, and load voltage. This is true for all basic topologies, except the buck converter, whose dynamic equations can be expressed in canonical form Eq. (8.17). Thus, for all converters, except the buck topology, some provisions are needed for the estimation of such references, strongly affecting the closed-loop dynamics, as discussed in the following sections.

8.5 Sliding-Mode Control of Buck DC-DC Converters

It was already mentioned that one of the most important features of the sliding-mode regimes in VSS is the ability to achieve responses that are independent of system parameters, the only constraint being the canonical form description of the system. From this point of view, the buck DC-DC converter is particularly suitable for the application of the sliding-mode control, because its controllable states (output voltage and its derivative) are all continuous and accessible for measurement.

Phase-Plane Description

The basic buck DC-DC converter topology is shown in Fig. 8.4.

In this case it is more convenient to use a system description, which involves the output error and its derivative, i.e.,

$$\begin{cases} x_1 = u_o - U_o^* \\ x_2 = \dfrac{dx_1}{dt} = \dfrac{du_o}{dt} = \dfrac{i_C}{C} \end{cases} \qquad (8.20)$$

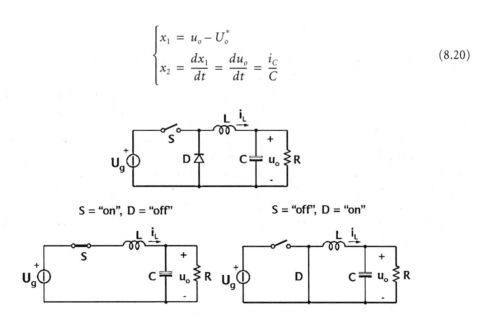

FIGURE 8.4 Buck DC-DC converter topology and related substructures corresponding to two different switch positions.

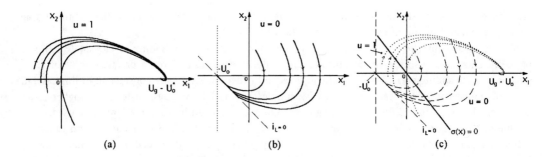

FIGURE 8.5 (a) Phase trajectories of the substructure corresponding to $u = 1$; (b) Phase trajectories of the substructure corresponding to $u = 0$; (c) Subsystem trajectories and sliding line in the phase-plane of the buck converter.

The system equations, in terms of state variables x_1 and x_2 and considering a continuous conduction mode (CCM) operation can be written as

$$\begin{cases} \dot{x}_1 = x_2 \\ \dot{x}_2 = -\dfrac{x_1}{LC} - \dfrac{x_2}{RC} + \dfrac{U_g}{LC} u - \dfrac{U_o^*}{LC} \end{cases} \quad (8.21)$$

where u is the discontinuous input, which can assume the values 0 (switch OFF) or 1 (switch ON). In state-space form:

$$\dot{\mathbf{x}} = \mathbf{A}\mathbf{x} + \mathbf{B}u + \mathbf{D}$$

$$\mathbf{A} = \begin{bmatrix} 0 & 1 \\ -\dfrac{1}{LC} & -\dfrac{1}{RC} \end{bmatrix}, \quad \mathbf{B} = \begin{bmatrix} 0 \\ \dfrac{U_g}{LC} \end{bmatrix}, \quad \mathbf{D} = \begin{bmatrix} 0 \\ -\dfrac{U_o^*}{LC} \end{bmatrix} \quad (8.22)$$

Practically, the damping factor of this second-order system is less than 1, resulting in complex conjugate eigenvalues with negative real part. The phase trajectories corresponding to the substructure $u = 1$ are shown in Fig. 8.5a for different values of the initial conditions. The equilibrium point for this substructure is $x_{1eq} = U_g - U_o^*$ and $x_{2eq} = 0$. Instead, with $u = 0$ the corresponding phase trajectories are reported in Fig. 8.5b and the equilibrium point for this second substructure is $x_{1eq} = -U_o^*$ and $x_{2eq} = 0$.

Note that the real structure of Fig. 8.5b has a physical limitation due to the rectifying characteristic of the freewheeling diode. In fact, when the switch S is OFF, the inductor current can assume only non-negative values. In particular, when i_L goes to zero it remains zero and the output capacitor discharge goes exponentially to zero. This situation corresponds to the discontinuous-conduction mode (DCM) and it poses a constraint on the state variables. In other words, part of the phase-plane does not correspond to possible physical states of the system and so need not be analyzed. The boundary of this region can be derived from the constraint $i_L = 0$ and is given by the equation:

$$x_2 = -\dfrac{1}{RC} x_1 - \dfrac{U_o^*}{RC} \quad (8.23)$$

which corresponds to the straight line with a negative slope equal to $-1/RC$ and passing through the point $(-U_o^*, 0)$ shown in dashed line in Fig. 8.5b. In the same figure, the line $x_1 = -U_o^*$ is also reported, which defines another not physically accessible region of the phase-plane, i.e., the region in which $u_o < 0$.

Selection of the Sliding Line

It is convenient to select the sliding surface as a linear combination of the state variables because the results are very simple to implement in the real control system and because it allows the use of the equivalent control method to describe the system dynamic in sliding mode. Thus, we can write:

$$\sigma(\mathbf{x}) = c_1 x_1 + x_2 = \mathbf{C}^T \mathbf{x} = 0 \tag{8.24}$$

where $\mathbf{C}^T = [c_1, 1]$ is the vector of sliding surface coefficients which corresponds to $\mathbf{G}$ in Eq. (8.13), and coefficient c_2 was set to 1 without loss of generality.

As shown in Fig. 8.5c, this equation describes a line in the phase-plane passing through the origin, which represents the stable operating point for this converter (zero output voltage error and its derivative). By using Eq. (8.21), Eq. (8.24) becomes

$$\sigma(\mathbf{x}) = c_1 x_1 + \dot{x}_1 = 0 \tag{8.25}$$

which completely describes the system dynamic in sliding mode. Thus, if existence and reaching conditions of the sliding mode are satisfied, a stable system is obtained by choosing a positive value for c_1. Figure 8.5c reveals the great potentialities of the phase-plane representation for second-order systems. In fact, a direct inspection of Fig. 8.5c shows that if we choose the following control law:

$$u = \begin{cases} 0 & \text{for } \sigma(\mathbf{x}) > 0 \\ 1 & \text{for } \sigma(\mathbf{x}) < 0 \end{cases} \tag{8.26}$$

then both existence and reaching conditions are satisfied, at least in a small region around the system equilibrium point. In fact, we can easily see that, using this control law, for both sides of the sliding line the phase trajectories of the corresponding substructures are directed toward the sliding line (at least in a small region around the origin). Moreover, the equilibrium point for the substructure corresponding to $u = 0$ belongs to the region of the phase-plane relative to the other substructure, and vice versa, thus ensuring the reachability of the sliding line from any allowed initial state condition. From Eq. (8.5) it is easy to see that the output voltage dynamics in sliding mode is simply given by a first-order system with time constant equal to $1/c_1$. Typical waveforms with $c_1 = 0.8/RC$ are reported in Fig. 8.6.

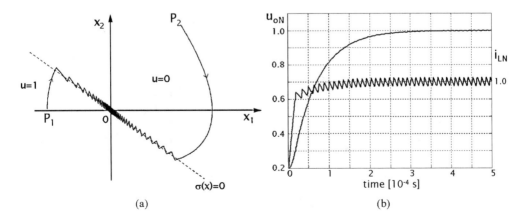

FIGURE 8.6 (a) Phase trajectories for two different initial conditions ($c_1 = 0.8/RC$); (b) Time responses of normalized output voltage u_{oN} and normalized inductor current i_{LN} ($c_1 = 0.8/RC$) (initial conditions in P_1).

Existence Condition

Let us analyze more precisely the existence of the sliding regime for the buck converter. From the sliding-mode theory, the conditions for the sliding regime to exist are (see Eq. 8.9):

$$\dot{\sigma}(\mathbf{x}) = \begin{cases} \mathbf{C}^T\mathbf{A}\mathbf{x} + \mathbf{C}^T\mathbf{B}\mathbf{u}^+ + \mathbf{C}^T\mathbf{D} < 0 & \text{for} \quad 0 < \sigma(\mathbf{x}) < \xi \\ \mathbf{C}^T\mathbf{A}\mathbf{x} + \mathbf{C}^T\mathbf{B}\mathbf{u}^- + \mathbf{C}^T\mathbf{D} > 0 & \text{for} \quad -\xi < \sigma(\mathbf{x}) < 0 \end{cases} \quad (8.27)$$

where ξ is an arbitrary small positive quantity. Using Eqs. (8.22) and (8.24) these inequalities become

$$\begin{cases} \lambda_1(\mathbf{x}) = \left(c_1 - \dfrac{1}{RC}\right)x_2 - \dfrac{1}{LC}x_1 - \dfrac{U_o^*}{LC} < 0 & \text{for} \quad 0 < \sigma(\mathbf{x}) < \xi \\ \lambda_2(\mathbf{x}) = \left(c_1 - \dfrac{1}{RC}\right)x_2 - \dfrac{1}{LC}x_1 + \dfrac{U_g - U_o^*}{LC} > 0 & \text{for} \quad -\xi < \sigma(\mathbf{x}) < 0 \end{cases}$$

Equations $\lambda_1(\mathbf{x}) = 0$ and $\lambda_2(\mathbf{x}) = 0$ define two lines in the phase-plane with the same slope passing through points $(-U_o^*, 0)$ and $(U_g - U_o^*, 0)$, respectively. The regions of existence of the sliding mode are depicted in Fig. 8.7 for two different situations: (1) $c_1 > 1/RC$, and (2) $c_1 < 1/RC$. As we can see, the increase of c_1 value causes a reduction of sliding-mode existence region. Remember that the sliding line coefficient c_1 determines also the system dynamic response in sliding mode, since the system dynamic response results are of first order with a time constant $\tau = 1/c_1$. Thus, high response speeds, i.e., $\tau < RC$, limit the existence region of the sliding mode. This can cause overshoots and ringing during transients.

To better understand this concept, let us take a look to some simulation results. Figure 8.7 shows the phase trajectories of a buck converter with sliding-mode control for two different c_1 values where the initial condition is in $(-U_o^*, 0)$: when the slope of the sliding line becomes too high, as shown in Fig. 8.7a, the system RP hits first the sliding line at a point outside the region of the existence of the sliding mode. As a consequence, the switch remains in a fixed position (open in this case) until the RP hits the sliding line again, now in a region where the existence condition is satisfied.

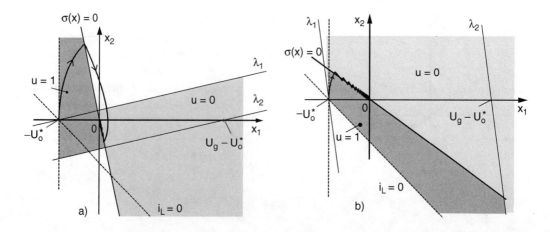

FIGURE 8.7 Regions of existence of the sliding mode in the phase-plane: (a) $c_1 > 1/RC$; (b) $c_1 < 1/RC$.

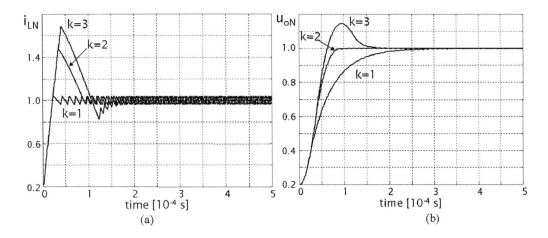

FIGURE 8.8 Time responses of normalized inductor current i_{LN} (a) and normalized output voltage u_{oN} (b) at different c_1 values ($k = c_1 RC$).

The time responses of the normalized inductor current i_{LN} and output voltage u_{oN} for different c_1 values are reported in Fig. 8.8a and b respectively ($i_{LN} = i_L/I_o$, $u_{oN} = u_o/U_o$). Note that with $c_1 = 1/RC$ neither the inductor current nor the output voltage has overshot during start-up.

Current Limitation

As we have seen from Fig. 8.8b, a fast output voltage dynamic calls for overshoots in the inductor current i_L. In fact, the first part of the transient response depends on the system parameters, and only when the system RP hits the sliding line at a point belonging to the existence region is the system dynamic dictated by the sliding equation (for the buck converter it is actually independent of the converter parameters and dependent only on the sliding coefficient c_1). The large inductor current could not be tolerated by the converter devices for two reasons: it can cause the inductor core to saturate with consequent even high-peak current value or can be simply greater than the maximum allowed switch current. Thus, it is convenient to introduce into the controller a protection circuit that prevents the inductor current from reaching dangerous values. This feature can be easily incorporated into the sliding-mode controller by a suitable modification of the sliding line. For example, in the case of buck converters, to keep constant the inductor current we have to force the system RP on the line:

$$x_2 = -\frac{1}{RC}x_1 + \frac{I_{L\max}}{C} - \frac{U_o^*}{RC} \tag{8.28}$$

Thus, the global sliding line consists of two pieces:

$$\sigma'(\mathbf{x}) = \begin{cases} \dfrac{x_1}{RC} + x_2 - \dfrac{1}{C}\left(I_{L\max} - \dfrac{U_o^*}{R}\right) & \text{for} \quad i_L > I_{L\max} \\ c_1 x_1 + x_2 & \text{for} \quad i_L < I_{L\max} \end{cases} \tag{8.29}$$

The phase plane trajectories for a buck converter with inductor current limitation and with $c_1 = 2/RC$ are shown in Fig. 8.9, and the corresponding normalized inductor current transient behavior is shown in Fig. 8.10. It is interesting to note that Eq. (8.29) gives an explanation of why the fastest response

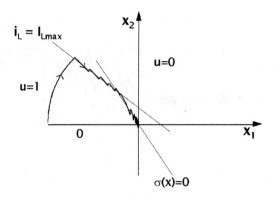

FIGURE 8.9 Phase trajectories for a buck converter with inductor current limitation ($c_1 = 2/RC$).

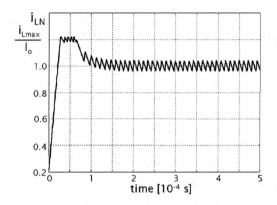

FIGURE 8.10 Time response of normalized inductor current i_{LN} of a buck converter with current limitation ($c_1 = 2/RC$).

without overshoots is obtained for $c_1 = 1/RC$. In fact, if $c_1 = 1/RC$ and $I_{Lmax} = U_o^*/R$ the two pieces of the sliding line σ' become a single line and thus the inductor current reaches its steady-state value U_o^*/R without overshoot.

8.6 Extension to Boost and Buck–Boost DC-DC Converters

For boost as well as buck–boost DC-DC converters, the derivative of the output voltage turns out to be a discontinuous variable, and we cannot express the system in canonical form as was done for the buck converter. Following the general scheme of Fig. 8.3, the inductor current and output voltage errors are chosen as state variables, i.e.,

$$\begin{cases} x_1 = i - I^* \\ x_2 = u_o - U_o^* \end{cases} \tag{8.30}$$

where the current reference I^* depends on the converter operating point (output power and input voltage). Choosing the same control law Eq. (8.26) of the buck converter, together with the following sliding line:

$$\sigma(\mathbf{x}) = x_1 + gx_2 = \mathbf{C}^T\mathbf{x} = 0 \tag{8.31}$$

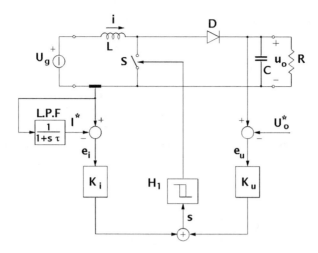

FIGURE 8.11 Boost converter with sliding-mode control.

it can be easily seen [13] that both existence and reaching conditions are satisfied (the former at least in a small region enclosing the origin) as long as

$$g < \frac{RC}{L} \frac{U_g}{U_o^*} \tag{8.32}$$

for both converters. However, current reference signal I^* is not usually available in practice, and some alternative techniques for its estimation are needed. One possible solution is to derive this reference signal directly from the inductor current by using a low-pass filter, as shown in Fig. 8.11. This estimation clearly affects the dynamic behavior of the sliding-mode control. To understand the closed-loop dynamics of this approach, we need to include the additional state variable introduced by the low-pass filter, i.e.,

$$\frac{di^*}{dt} = -\frac{1}{\tau}i^* + \frac{1}{\tau}i \tag{8.33}$$

Taking into account the boost converter, we can represent the overall system, choosing as state variables:

$$\begin{cases} x_1 = i \\ x_2 = u_o - U_o^* \\ x_3 = i^* \end{cases} \tag{8.34}$$

$$\dot{x} = Ax + B\bar{u} + D \tag{8.35}$$

$$A = \begin{bmatrix} 0 & 0 & 0 \\ 0 & -\dfrac{1}{RC} & 0 \\ \dfrac{1}{\tau} & 0 & -\dfrac{1}{\tau} \end{bmatrix}, \quad B = \begin{bmatrix} -\dfrac{u_o}{L} \\ \dfrac{i}{C} \\ 0 \end{bmatrix}, \quad D = \begin{bmatrix} \dfrac{U_g}{L} \\ -\dfrac{U_o^*}{RC} \\ 0 \end{bmatrix}$$

The sliding line becomes a sliding surface in the phase space:

$$\sigma(\mathbf{x}) = x_1 + g x_2 - x_3 = \mathbf{C}^T \mathbf{x} = 0 \qquad (8.36)$$

where $\mathbf{C}^T = [1, g, -1]$ is the vector of the sliding surface coefficients and $x_1 - x_3$ represents now the inductor current error. Fortunately, the existence conditions analysis for the system Eq. (8.35) leads to the same constraint (8.32), which can be derived without accounting for the low-pass filter dynamic [13]. However, unlike the buck converter, Eq. (8.32) does not directly give information on the system stability and on the possible values of filter time constant τ.

Stability Analysis

In the following, a procedure similar to the equivalent control method is used to derive a suitable small signal model for the system Eq. (8.35) in sliding mode. The starting point is the *small-signal state space averaged model* of the boost converter [3]:

$$\dot{\hat{\mathbf{x}}} = \mathbf{A}\hat{\mathbf{x}} + \mathbf{B}\hat{u}_g + \mathbf{C}\hat{d} \qquad (8.37)$$

$$\mathbf{A} = \begin{bmatrix} 0 & -\dfrac{D'}{L} & 0 \\ \dfrac{D'}{C} & -\dfrac{1}{RC} & 0 \\ \dfrac{1}{\tau} & 0 & -\dfrac{1}{\tau} \end{bmatrix}, \quad \mathbf{B} = \begin{bmatrix} \dfrac{1}{L} \\ 0 \\ 0 \end{bmatrix}, \quad \mathbf{D} = \dfrac{U_g}{D'} \begin{bmatrix} \dfrac{1}{L} \\ -\dfrac{1}{D'RC} \\ 0 \end{bmatrix}$$

where $\hat{\mathbf{x}} = [\hat{x}_1, \hat{x}_2, \hat{x}_3]^T = [\hat{i}, \hat{u}_o, \hat{i}^*]^T$ and $D' = 1 - D$. In Eq. (8.37), the dynamic equation of the low-pass filter Eq. (8.33) was added to the original boost equations. From the sliding surface definition we can write:

$$\sigma(\mathbf{x}) = (i - i^*) + g(u_o - U_o^*) = \hat{i} - \hat{i}^* + g\hat{u}_o = \mathbf{C}^T \hat{\mathbf{x}} \qquad (8.38)$$

where $\mathbf{C}^T = [1, g, -1]$ and the steady-state values $\mathbf{X}$ of the state variables coincide with the corresponding reference values $\mathbf{X}^*$. Now, if the system is in sliding regime, we can write

$$\sigma(\mathbf{x}) = 0 \Rightarrow \dot{\sigma}(\mathbf{x}) = \mathbf{C}^T \dot{\hat{\mathbf{x}}} = 0 \qquad (8.39)$$

From Eqs. (8.37) and (8.39) we can derive an expression for the duty-cycle perturbation as a function of the state variables and the input, which, substituted into Eq. (8.37), yields:

$$\dot{\hat{\mathbf{x}}} = \mathbf{A}' \hat{\mathbf{x}} + \mathbf{B}' \hat{u}_g \qquad (8.40)$$

In Eq. (8.40), which represents a third-order system, one equation (for example, the last one corresponding to the variable x_3) is redundant and can be eliminated by using the equation $\sigma = 0$. The result is the following second-order system:

$$\dot{\hat{\mathbf{x}}} = \mathbf{A}_T \hat{\mathbf{x}} + \mathbf{B}_T \hat{u}_g, \qquad \hat{\mathbf{x}} = [\hat{x}_1, \hat{x}_2]^T \qquad (8.41)$$

$$\mathbf{A}_T = \frac{1}{k} \begin{bmatrix} -\dfrac{gD'}{C} & g\left(\dfrac{2}{RC} - \dfrac{1}{\tau}\right) \\ \dfrac{D'}{C} & \dfrac{1}{RC}\left(\dfrac{gL}{D'\tau} - 2\right) \end{bmatrix}, \qquad \mathbf{B}_T = \frac{1}{kD'RC}\begin{pmatrix} -g \\ 1 \end{pmatrix} \quad \text{where} \quad k = 1 - \frac{gL}{D'RC}$$

Equation (8.41) completely describe the system behavior under sliding mode control. Moreover, they can be used to derive closed-loop transfer functions like output impedance and audiosusceptibility, which allows meaningful comparison with other control techniques. As far as system stability is concerned, by imposing positive values for the coefficients of the characteristic polynomial we obtain

$$0 < g < g_{\text{crit}} = \frac{RCD'}{L} \qquad (8.42)$$

and

$$\tau > \tau_{\text{crit}} = \frac{L}{D'^2 R} \cdot \frac{1}{1 + \dfrac{2}{RD'g}} \qquad (8.43)$$

It is interesting to note that constraint (8.42) coincides with the existence condition given by (8.32).

As an example of application of the discussed analysis, Fig. 8.12 reports the converter audiosusceptibility and output impedance predicted by the model and experimentally measured in a boost converter propotype [13] with the following parameters: $U_g = 24$ V, $U_o = 48$ V, $P_o = 50$ W, $f_s = 50$ kHz, $L = 570$ μH, $C = 22$ μF, $\tau = 0.4$ ms, $g = 0.35$. With this value of sliding coefficient g, the stability analysis shows that the filter time constant τ must be greater than 38 μs for stable operation. Moreover, the chosen value of 400 μs guarantees no output voltage overshoots during transient conditions, as depicted in Fig. 8.13, which reports simulated waveforms of the startup of the boost converter where the output voltage was precharged at the input voltage value.

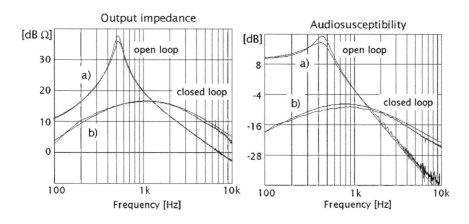

FIGURE 8.12 Comparison between model forecast and experimental results: (a) open loop; (b) closed loop.

TABLE 8.1 Values of τ_{crit} and g_{crit} for Boost and Buck–Boost Topologies

	Boost	Buck–Boost
g_{crit}	$\dfrac{RCD'}{L}$	$\dfrac{RC}{L}\dfrac{D'}{D}$
τ_{crit}	$\dfrac{L}{D'^2 R} \cdot \dfrac{1}{1 + \dfrac{2}{RD'g}}$	$\dfrac{L}{\dfrac{D'^2}{D}R + (2-D')\dfrac{L}{RC}}$

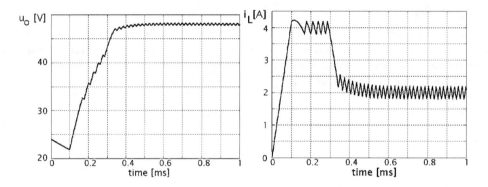

FIGURE 8.13 Output voltage and inductor current at startup of a boost converter (output capacitor precharged at input voltage).

The same stability analysis can be applied also to buck–boost converters. As a result, the critical value for the low-pass filter time constant τ_{crit} and g_{crit} are given in Table 8.1.

8.7 Extension to Cúk and SEPIC DC-DC Converters

As a general approach to high-order converters like Cúk and SEPIC, a sliding function can be built as a linear combination of all state variable errors x_i, i.e., $\sigma(\mathbf{x}) = \sum_{i=1}^{N} c_i x_i$, as depicted in Fig. 8.3. This general approach, although interesting in theory, is not practical. In fact, it requires sensing of too many state variables with an unacceptable increase of complexity as compared with such standard control techniques as current-mode control. However, for Cúk and SEPIC converters a reduced-order sliding-mode control can be used with satisfactory performances with respect to standard control techniques [9–11]. In this case, some sliding coefficients are set to zero. In particular, for Cúk and SEPIC converters, the sensing of only output voltage and input inductor current was proposed and the current reference signal was obtained, as for the boost and buck–boost converter, by using a low-pass filter. Taking the SEPIC converter as an example, the resulting scheme with reduced-order implementation is reported in Fig. 8.14. Of course, the same scheme can be applied to Cúk converters as well.

To give design criteria for selection of sliding-mode controller parameters, the system must be represented in a suitable mathematical form. To this purpose, the converter equations related to the two subtopologies corresponding to the switch status are written as

$$\dot{\mathbf{v}} = \mathbf{A}_{on}\mathbf{v} + \mathbf{F}_{on} \quad \text{switch on,} \tag{8.44}$$

$$\dot{\mathbf{v}} = \mathbf{A}_{off}\mathbf{v} + \mathbf{F}_{off} \quad \text{switch off,} \tag{8.45}$$

Sliding-Mode Control of Switched-Mode Power Supplies

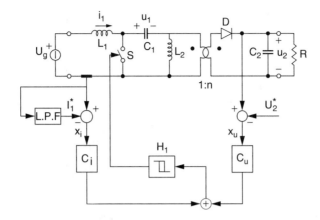

FIGURE 8.14 SEPIC converter with sliding-mode control.

where $\mathbf{v} = [i_1, i_2, u_1, u_2]^T$ is the state variable vector. These equations are combined in the following form (VSS)

$$\dot{\mathbf{v}} = \mathbf{A}\mathbf{v} + \mathbf{B}u + \mathbf{F}, \quad (8.46)$$

where u is the discontinuous variable corresponding to the switch status and matrices $\mathbf{A}$, $\mathbf{B}$, $\mathbf{F}$ are given by

$$\mathbf{A} = \mathbf{A}_{\text{off}}, \quad \mathbf{F} = \mathbf{F}_{\text{off}}, \quad (8.47)$$

$$\mathbf{B} = (\mathbf{A}_{\text{on}} - \mathbf{A}_{\text{off}})\mathbf{v} + (\mathbf{F}_{\text{on}} - \mathbf{F}_{\text{off}}) \quad (8.48)$$

It is convenient to write the system Eq. (8.47) in terms of state variables error x_i, where $\mathbf{x} = \mathbf{v} - \mathbf{V}^*$, being $\mathbf{V}^* = [I_1^*, I_2^*, U_1^*, U_2^*]^T$ the vector of state variable references. Accordingly, system equations become

$$\dot{x} = Ax + Bu + G, \quad (8.49)$$

where $G = A\mathbf{V}^* + F$.

Existence Condition

Assuming that the switch is kept on when σ is negative and off when σ is positive, we may express the existence condition in the form (see Eq. 8.9):

$$\begin{aligned}\frac{\partial \sigma}{\partial t} &= C^T A x + C^T G < 0 & 0 < \sigma < \xi \\ \frac{\partial \sigma}{\partial t} &= C^T A x + C^T B + C^T G > 0 & -\xi < \sigma < 0,\end{aligned} \quad (8.50)$$

where ξ is an arbitrary small positive quantity. Inequalities (8.50) are useful only if state variable errors x_i are bounded; otherwise, (8.50) must be analyzed under small-signal assumption. In this latter case, satisfying (8.50) means enforcing the existence condition in a small volume around the operating point, and this is equivalent to ensuring the stability condition as demonstrated in Ref. 13.

Hitting Condition

If sliding mode exists, a sufficient hitting condition is

$$C^T A_4 \leq 0 \quad (8.51)$$

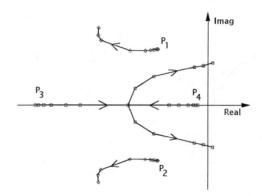

FIGURE 8.15 Root locus of closed-loop system for variation of low-pass filter time constant of the SEPIC converter [13].

where A_4 is the fourth column of matrix $\mathbf{A}$ [1, 2]. This yields the following constraint

$$-\frac{c_i}{nL_1} - \frac{c_u}{R_L C_2} \leq 0 \tag{8.52}$$

where c_i and c_u are the sliding line coefficients for the input current and output voltage errors, respectively, and n is the ratio between the second and the primary transformer windings.

Stability Condition

This issue must be addressed taking into account the effect of the time constant τ of the low-pass filter needed to extract the inductor current reference signal. To this purpose the small-signal analysis carried out for the sliding-mode control of boost converters can be generalized [13]. This can be used to derive useful design hints for the selection of sliding surface coefficients and filter time constants. As an example, from the small signal model in sliding mode, similar to (8.41), the root locus of closed-loop system eigenvalues can be plotted as a function of the low-pass filter time constant as shown in Fig. 8.15 in the case of a SEPIC converter [11]. Note that the system become unstable for low value of time constant τ. A similar analysis can be applied to Cúk converters as well.

8.8 General-Purpose Sliding-Mode Control Implementation

Compared to the current control, the sliding-mode approach has some aspects that still must be improved. The first problem arises from the fact that the switching frequency depends on the rate of change of function σ and on the amplitude of the hysteresis band. Since σ is a linear combination of state-variable errors, it depends on actual converter currents and voltages, and its behavior may be difficult to predict. This can be unacceptable if the range of variation becomes too high. One possible solution of the problem related to the switching frequency variations is the implementation of a variable hysteresis band, for example, using a PLL (phase locked loop). Another simple approach is to inject a suitable constant-frequency signal w into the sliding function as shown in Fig. 8.16 [10]. If, in the steady state, the amplitude of w is predominant in σ_f, then a commutation occurs at any cycle of w. This also allows converter synchronization to an external trigger. Instead, under dynamic conditions, error terms x_i and x_u increase, w is overridden, and the system retains the excellent dynamic response of the sliding mode. Simulated waveforms of ramp w, and σ_{PI}, σ_f signals are reported in Fig. 8.17.

The selection of the ramp signal w amplitude is worthy of further discussion. In fact, it should be selected by taking into account the slope of function σ_{PI} and the hysteresis band amplitude, so that function σ_f hits the lower part of the hysteresis band at the end of the ramp, causing the commutation. From the analysis of the waveform shown in Fig. 8.17, we can find that the slope S_e of the external ramp

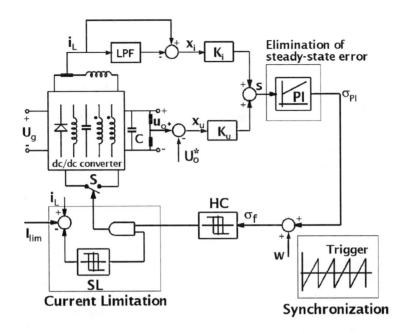

FIGURE 8.16 Reduced-order sliding-mode controller with inductor current limitation, constant switching frequency, and no output voltage steady-state error.

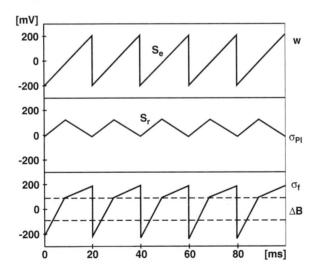

FIGURE 8.17 Simulated waveforms of ramp w, and σ_{PI}, σ_f signals.

must satisfy the following inequality

$$S_e > \frac{\Delta B}{\delta T_s} - S_r \qquad (8.53)$$

where ΔB represents the hysteresis band amplitude and S_r is the slope of function σ_{PI} during the switch on-time. Note that, in the presence of an external ramp, signal σ_{PI} must have a nonzero average value to accommodate the desired converter duty cycle (see Fig. 8.17). Of course, a triangular disturbing signal w is not the only waveform that can used. A pulse signal has been used alternatively as reported in Ref. 12.

The second problem derives from a possible steady state error on the output voltage. In fact, when the inductor current reference is evaluated using a low-pass filter, then the current error leads naturally to zero average value in steady state. Thus, if the sliding function, due to the hysteretic control or due to the added ramp signal w, has nonzero average value, a steady-state output voltage error necessarily appears. This problem can be solved by introducing a PI action on sliding function to eliminate its DC value (see Fig. 8.16). In practice, the integral action of this regulator is enabled only when the system is on the sliding surface; in this way, the system behavior during large transients, when σ can have values far from zero, is not affected, thus maintaining the large-signal dynamic characteristics of sliding-mode control. A general-purpose sliding-mode controller scheme that includes the aforementioned improvements, together with reduced-order implementation in the case of Cúk and SEPIC converters and a possible implementation of current limitation by means of another hysteretic comparator and an AND port, is reported in Fig. 8.16. Experimental results of this scheme are reported in Refs. 10 and 11.

8.9 Conclusions

Control techniques of VSS find a natural application to SMPS, since they inherently show variable structure properties as a result of the conversion process and switch modulation. Thus, the sliding-mode control represents a powerful tool to enhance performance of power converters. Sliding-mode control is able to ensure system stability even for large supply and load variations, good dynamic response, and simple implementation, even for high-order converters. These features make this control technique a valid alternative to standard control approaches.

The application of the sliding-mode control technique to DC-DC converters is the focus of this chapter. The application to buck converters is discussed in detail, whereas for the extension of this control technique to boost, buck–boost, Cúk, and SEPIC converters only some design guidelines are given. Finally, such improvements as current limitation, constant switching frequency, and output voltage steady-state error cancellation are outlined. This control approach can also be effectively used in other applications, not discussed here, such as inverters [14, 18], power factor controllers [16], and AC power supplies [17].

References

1. V. I. Utkin, *Sliding Modes and Their Application in Variable Structure Systems*, MIR Publishers, Moscow, 1978.
2. U. Itkis, *Control Systems of Variable Structure*, John Wiley & Sons, New York, 1976.
3. R. D. Middlebrook and S. Cuk, *Advances in Switched-Mode Power Conversion*, Vol. I and II, TESLAco, Pasadena, CA, 1983, 73–89.
4. R. Redl and N. Sokal, Current-mode control, five different types, used with the three basic classes of power converters: small-signal AC and large-signal DC characterization, stability requirements, and implementation of practical circuits, in *IEEE-PESC*, 1985, 771–785.
5. J. G. Kassakian, M. F. Schlecht, and G. C. Verghese, *Principles of Power Electronics*, Addison-Wesley, Reading, MA, 1991.
6. R. Venkataramanan, A. Sabanovic, and S. Cúk, Sliding-mode control of DC-to-DC converters, in *IECON Conf. Proc.*, 1985, 251–258.
7. B. Nicolas, M. Fadel, and Y. Chéron, Robust control of switched power converters via sliding mode, *ETEP*, 6(6), 413–418, 1996.
8. B. Nicolas, M. Fadel, and Y. Chéron, Sliding mode control of DC-to-DC converters with input filter based on Lyapunov-function approach, in *EPE Conf. Proc.*, 1995, 1.338–1.343.
9. L. Malesani, L. Rossetto, G. Spiazzi, and P. Tenti, Performance optimization of Cúk converters by sliding-mode control, *IEEE Trans. Power Electron.*, 10(3), 302–309, 1995.
10. P. Mattavelli, L. Rossetto, G. Spiazzi, and P. Tenti, General-purpose sliding-mode controller for DC/DC converter applications, in *Proc. of IEEE Power Electronics Specialists Conf. (PESC)*, Seattle, June 1993, 609–615.

11. P. Mattavelli, L. Rossetto, G. Spiazzi, and P. Tenti, Sliding mode control of SEPIC converters, in *Proc. of European Space Power Conf. (ESPC)*, Graz, August 1993, 173–178.
12. J. Fernando Silva and Sonia S. Paulo, Fixed frequency sliding mode modulator for current mode PWM inverters, in *Power Electronics Specialists Conf. Proc. (PESC)*, 1993, 623–629.
13. P. Mattavelli, L. Rossetto, and G. Spiazzi, Small-signal analysis of DC-DC converters with sliding mode control, *IEEE Trans. Power Electron.*, 12(1), 79–86, 1997.
14. N. Sabanovic, A. Sabanovic, and K. Ohnishi, Sliding mode control of three-phase switching converters, in *Proc. of Int. Conf. on Industrial Electronics, Control and Instrumentation (IECON)*, San Diego, 1992, 319–325.
15. N. Sabanovic-Behlilovic, A. Sabanovic, and T. Ninomiya, PWM in three-phase switching converters-sliding mode solution, in *Power Electronics Specialists Conference, PESC '94 Record, 25th Annual IEEE*, Vol. 1, 1994, 560–565.
16. L. Rossetto, G. Spiazzi, B. Fabiano, and C. Licitra, Fast-response high-quality rectifier with sliding mode control, *IEEE Trans. Power Electron.*, 9(2), 146–152, 1994.
17. L. Malesani, L. Rossetto, G. Spiazzi, and A. Zuccato, An AC power supply with sliding-mode control, *IEEE Ind. Appl. Mag.*, 2(5), 32–38, 1996.
18. H. Pinheiro, A. S. Martins, and J. R. Pinheiro, A sliding mode controller in single phase voltage source inverters, in *IEEE IECON '94*, 1994, 394–398.
19. H. Sira-Ramirez and M. Rios-Bolivar, Sliding mode control of DC-to-DC power converters via extended linearization circuits and systems I: fundamental theory and applications, *IEEE Transactions on Circuits and Systems* 41(10), 652–661, 1994.

Applications and Systems Considerations

9 **DC Motor Drives** *Ralph Staus* 9-1
   DC Motor Basics • DC Speed Control • DC Drive Basics • Transistor PWM DC Drives • SCR DC Drives

10 **AC Machines Controlled as DC Machines (Brushless DC Machines/Electronics)** *Hamid A. Toliyat, Tilak Gopalarathnam* 10-1
   Introduction • Machine Construction • Motor Characteristics • Power Electronic Converter • Position Sensing • Pulsating Torque Components • Torque-Speed Characteristics • Applications

11 **Control of Induction Machine Drives** *Daniel Logue, Philip T. Krein* 11-1
   Introduction • Scalar Induction Machine Control • Vector Control of Induction Machines • Summary

12 **Permanent Magnet Synchronous Machine Drives** *Patrick L. Chapman* 12-1
   Introduction • Construction of PMSM Drive Systems • Simulation and Model • Controlling the PMSM • Advanced Topics in PMSM Drives

13 **Switched Reluctance Machines** *Iqbal Husain* 13-1
   Introduction • SRM Configuration • Basic Principle of Operation • Design • Converter Topologies • Control Strategies • Sensorless Control • Applications

14 **Step Motor Drives** *Ronald H. Brown* 14-1
   Introduction • Types and Operation of Step Motors • Step Motor Models • Control of Step Motors

15 **Servo Drives** *Sándor Halász* 15-1
   DC Drives • Induction Motor Drives

16 **Uninterruptible Power Supplies** *Laura Steffek, John Hecklesmiller, Dave Layden, Brian Young* 16-1
   UPS Functions • Static UPS Topologies • Rotary UPSs • Alternate AC and DC Sources

17 **Power Quality and Utility Interface Issues** *Wayne Galli, Timothy L. Skvarenina, Badrul H. Chowdhury, Hirofumi Akagi, Rajapandian Ayyanar, Amit Kumar Jain* 17-1
   Overview • Power Quality Considerations • Passive Harmonic Filters • Active Filters for Power Conditioning • Unity Power Factor Rectification

18 **Photovoltaic Cells and Systems** *Roger Messenger* 18-1
   Introduction • Solar Cell Fundamentals • Utility Interactive PV Applications • Stand-Alone PV Systems

19 **Flexible, Reliable, and Intelligent Electrical Energy Delivery Systems**
   Alexander Domijan, Jr., Zhidong Song 19-1
   Introduction • The Concept of FRIENDS • Development of FRIENDS • The Advanced
   Power Electronic Technologies within QCCs • Significance of FRIENDS • Realization
   of FRIENDS • Conclusions

20 **Unified Power Flow Controllers**   *Ali Feliachi, Azra Hasanovic, Karl Schoder* 20-1
   Introduction • Power Flow on a Transmission Line • UPFC Description and Operation •
   UPFC Modeling • Control Design • Case Study • Conclusion

21 **More-Electric Vehicles**   *Ali Emadi, Mehrdad Ehsani* 21-1
   Aircraft • Terrestrial Vehicles

22 **Principles of Magnetics**   *Roman Stemprok* 22-1
   Introduction • Nature of a Magnetic Field • Electromagnetism • Magnetic
   Flux Density • Magnetic Circuits • Magnetic Field Intensity • Maxwell's
   Equations • Inductance • Practical Considerations

23 **Computer Simulation of Power Electronics**   *Michael Giesselmann* 23-1
   Introduction • Code Qualification and Model Validation • Basic Concepts—Simulation
   of a Buck Converter • Advanced Techniques—Simulation of a Full-Bridge (H-Bridge)
   Converter • Conclusions

9
DC Motor Drives

Ralph Staus
Pennsylvania State University

9.1 DC Motor Basics 9-1
9.2 DC Speed Control 9-2
9.3 DC Drive Basics 9-3
9.4 Transistor PWM DC Drives 9-4
9.5 SCR DC Drives 9-5

9.1 DC Motor Basics

The DC motor consists of two basic parts: a stationary magnetic field and a current-carrying coil on the armature. The force produced by the interaction of these two components produces a torque that causes the armature to rotate. The stationary magnetic field is produced by a permanent magnet for many small DC motors. Large and extended speed range motors use an electromagnet to produce the stationary field permitting the drive to control the field strength. The armature coils consists of a series of individual coils connected to the DC power source through a commutator and brushes. As the armature rotates, the commutator switches successive coils into the circuit to keep the armature coil and magnetic poles in the same relative position.

The field flux ϕ (Eq. 9.1) is a function of the field current I_f and a proportionality constant k_f. The torque T produced (Eq. 9.2) is related to the field flux and armature current i_a by the proportionality constant k_t. The speed of a DC motor is controlled by the torque produced by the motor and the torque required by the load. When the motor torque exceeds the load requirement, the rotational speed of the motor increases.

$$\phi = k_f I_f \quad (9.1)$$

$$T = k_t \phi i_a \quad (9.2)$$

The armature current is in response to the applied voltage V (Eq. 9.3) and is opposed by a countervoltage (e_a) produced by the armature coil rotating through the stationary magnetic field and the armature resistance (R_a). The countervoltage (Eq. 9.4) produced by the armature is proportional to the strength of the stationary field and the rotational speed (S). Below the rated base speed of the motor the current in the stationary magnetic field coil is kept constant. Therefore, at values below base speed the DC motor speed is a function of the applied voltage and resistive loss in the armature.

$$V = e_a + R_a i_a \quad (9.3)$$

$$e_a = k_e \phi S \quad (9.4)$$

Some DC motors have an additional winding in series with the armature to increase field strength in proportion to the armature current. With this series field, the voltage at the motor terminals is proportional to the speed of the motor without the reduction due to armature resistance. The improved performance permits open-loop speed control providing a reasonable steady-state response.

These equations for DC motor speed control are valid for steady-state analysis but do not account for the inertia of the motor or the inductance of the armature winding. To provide for a quick response, the DC drive must provide additional torque (current) to overcome the motor inertia and additional voltage to overcome the armature inductance.

DC motors are often used at greater than the base speed of the motor. Base speed is the rotational speed where the motor produces rated horsepower at rated torque. Below this speed, the motor may be operated at rated torque without overloading. Above base speed, the motor torque must be limited to prevent exceeding the horsepower limit.

9.2 DC Speed Control

The first applications of variable-speed control for large (10 to 1000 Hp) motors are known as the Ward Leonard Motor-Generator (M-G) systems (Fig. 9.1). In these systems a large constant-speed AC motor is mechanically coupled to a DC generator. The voltage produced by a DC generator is a function of rotational speed and the strength of the magnetic field. Controlling the field current of the generator with a rheostat can efficiently control the voltage produced by the generator operating at a constant speed. A rheostat in the generator field circuit controlled the field current of a few amps while the DC generator produced current in hundreds of amps at the controlled voltage. The speed of the connected motor is a function of the voltage supplied by the generator and the field current of the motor. When the same DC generator powers several DC motors, rheostats in each of the motor field circuits allow individual motor speed control.

State-of-the-art steel industry bar mills through the 1950s used M-G systems. A typical mill installed in 1952 used a 7500 hp synchronous AC motor coupled to three DC generators. One generator was electrically connected to the reversing mill, the second connected to the two-high mill, and the third connected to the bar mill so the voltage applied to each could be separately controlled. The reversing mill and the two-high mill motors were operated with full field current to enable the use of the full torque capability of the motors. To enable the use of small rheostats to control significant generator field current that would be varied continuously, a small M-G was applied with the main generator field as its load. The use of armature contactors for direction and field current rheostats for generator voltage control provided the full range of required voltages.

The bar mill consisted of five motors each requiring individual speed control. The product (steel bar) exit speed of each reduction stand must match the entering speed of the subsequent stand. The exit speed of each stand is a function of the entering speed and the bar reduction. In the mill, the speed of the product exiting a stand increased if the reduction rolls were moved closer together by the operator on

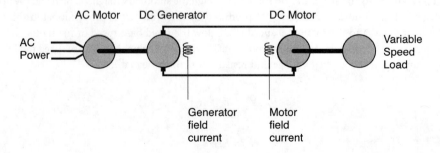

FIGURE 9.1 M-G system.

the production floor. As the floor operator adjusted the roll gap to produce the proper size, the field currents required adjustment. A second operator would monitor the size of the loop formed between the stands and adjust the rheostats to compensate for the roll gap changes.

With the introduction of semiconductors, systems were installed to take advantage of the diode. In the 1960s, large DC motors were replacing steam power turbines at many mills. Rather than installing a M-G set, a diode bank and an associated transformer provided a constant DC power source. The DC motor and gear reducers were selected to operate above base speed in the constant horsepower range. The goal was to increase efficiency and reduce maintenance costs. In the M-G set, both the AC motor and the DC generator were less then 100% efficient and each required maintenance of brushes and bearings. The solid-state constant DC voltage source increased efficiency and reduced maintenance costs. Speed regulation still used rheostats and the motor speed control was open loop. One significant disadvantage of the diode DC voltage source compared with the M-G source was the inability to regenerate power. Motor speed could not be reduced rapidly without having a bank of resistors to dissipate the excess energy. The M-G set would convert from a consumer of electric power to a supplier when the load speed was faster than the desired.

Transistor developments in the 1960s and 1970s brought further gains to large DC motor control. With the introduction of transistorized field current regulators, the motor speed was controlled electronically with a rapid closed-loop response. A small DC generator, referred to as a tachometer, provided a voltage proportional to the motor speed. The transistor field current regulator compared the tachometer feedback to the speed set point and adjusted the field current to control the motor speed.

The use of rheostats and early transistor regulators were inefficient for speed control when applied in the armature circuit. With the development of new systems using pulse width modulated (PWM) voltage regulators, the armature voltage for small motors could be controlled. Systems up to several horsepower were successfully applied to DC motors using permanent magnet fields. The power dissipated by the transistor when gated-on is minimal and the power while gated-off is zero. However, the power the transistor is required to dissipate while turning on and off limited the size of these systems. The application of PWM transistors as field current regulators to large DC motors of hundreds of horsepower provided the efficiency and closed-loop response of transistorized control to large systems.

The introduction of thyristor or silicon-controlled rectifiers (SCRs) provided for the revolution in control of large DC motors. SCRs were applied in both the field current regulators and the armature voltage regulators. With the SCR came increased efficiency, the ability to regenerate power, and high power capability. DC drives and motors from a few horsepower to hundreds of horsepower could be operated and controlled efficiently in all four quadrants when powered by three-phase line voltage. In addition, systems were developed for DC-to-DC control for the operation of battery-powered equipment.

The latest development in the DC motor control is the introduction of the insulated gate bipolar transistor (IGBT) to large power systems. These transistors have a fast switching time that reduces the power the transistor is required to dissipate. With the introduction of the IGBT came a movement to apply the concepts and lessons learned to large AC systems. Predictions were made that DC systems were obsolete and would soon be replaced. However, DC systems have proved to be dependable and cost-effective. They are still being applied when the application is suitable for economic reasons.

9.3 DC Drive Basics

Speed regulation in analogue drives is accomplished with a proportional–integral–derivative (PID) regulator for the voltage output. The input to the equation was the speed error and the output is used to drive the voltage regulator. A faster response is obtained by using the output of the PID equation as the input to the current regulator. A second PID regulator uses the output of the current regulator as an input to control the voltage. When additional speed is required, additional current is called for. The current regulator call for more voltage force increased current. The increased current provides additional torque resulting in increased speed.

In analog drives, a field current regulator provides the current for full field except when speed greater than base speed is required. The regulator monitors armature voltage and reduces the field current when the armature voltage is driven above rated values. Since the armature current is proportional to the motor torque and the current regulator is clamped at the rated value, an armature voltage above the rated value indicates the speed regulator output for a speed above the base value. Motor speed is proportional to armature voltage and inversely proportional to field current (see Eq. 9.5). Therefore:

- Reducing the field current reduces the field flux (Eq. 9.1).
- Reducing the field flux reduces the armature counter emf (Eq. 9.4).
- Reducing the armature counter emf increases the armature current (Eq. 9.3).
- Increasing the armature current increases the motor torque (Eq. 9.2).
- Increasing the motor torque increases the motor speed.
- Increasing the motor speed causes the regulator to decrease the armature voltage.

The system stabilizes with the armature voltage at base value and the field reduced within the range referred to as "weak field."

Modern digital microprocessor drives use a lookup table to determine the proper field current for the motor speed. The velocity feedback is fed to the field current regulator along with the armature voltage to determine the correct value of field current.

9.4 Transistor PWM DC Drives

The transistor PWM drive (Fig. 9.2) provides an efficient control for small motors nominally less then 5 hp. These drives operate at voltages and current levels limited by the transistors selected. The incoming AC voltage is rectified and filtered. The transistors are switched on/off to provide an average DC voltage (Fig. 9.3) to the motor. A switching frequency of 4 to 10 kHz prevents the motor speed from responding to individual cycles of the switching supply. The use of permanent magnet motors eliminated the need for field current regulators. A dedicated transformer determines the DC voltage supplied to the transistor regulator. The transformer is selected to provide the most efficient level of filtered DC and is an integral part of the drive system.

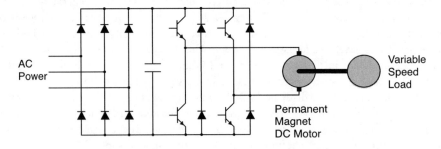

FIGURE 9.2 Transistor PWM system.

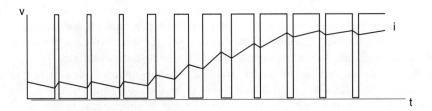

FIGURE 9.3 Transistor PWM output.

9.5 SCR DC Drives

DC drives for large DC motor use thyristors, commonly called SCRs, to convert incoming three-phase AC voltage to a regulated DC value. The use of 12 SCRs allows operation of the drive in all four quadrants of operation. The four quadrants are:

1. Forward motoring
2. Forward braking, regenerating power
3. Reverse motoring
4. Reverse braking, regenerating power

During operation in each of these quadrants, the SCRs are gated on at a phase angle to provide the DC voltage required. The angle is defined as the time in degrees from when the AC phase becomes the most positive or negative. The connection of the SCRs from the incoming AC power source to the drive output is shown in Fig. 9.4. The AC waveform in Fig. 9.5 starts with the SCRs from A phase to DC+ and B phase to DC− gated on. At 40° after the crossing point of the AC voltage, the SCR connects C phase to DC−. When this SCR begins to conduct, the B phase to DC− turn off. At 60° later the B phase to DC+ SCR is gated on and the motor is now connected from phase B to C. SCRs then connect B-A, C-A, C-B, A-B

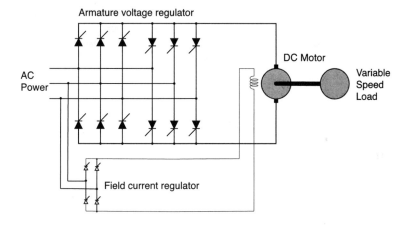

FIGURE 9.4 SCR system.

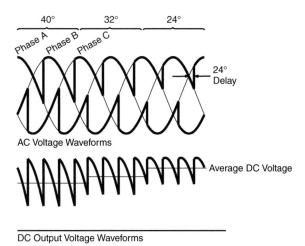

FIGURE 9.5 SCR system output.

before restarting the sequence with A-C. The resultant DC output waveform is shown in Fig. 9.5 as is its relation to the AC waveform. The resultant average DC is a function of the phase angle. The average DC output for 40°, 32°, and 24° firing phase angles is illustrated. The DC drive would function as a rectifier if the SCRs were gated on with a phase angle of 0°. The SCR provides unique advantages for the DC drive. First, the voltage across the SCR during the turn-on time is significantly less than the voltage switched in PWM drives. Second, the SCR does not switch off with current flowing. When the subsequent SCR turns on, the SCR is reverse-biased and current flow stops. The power consumed by the switching device is a function of the switched voltage, the switched current, and the switching time. The disadvantage of the SCR drive is the distortion of the AC power waveform due to line notching. The SCR proved to be well suited for operation at 480 V AC and from 10 to 1000 A.

Field current regulators require only single-phase power and only four SCRs. The field winding in the motor has significant inductance and the additional voltage ripple from using a single-phase source does not induce significant ripple in the field flux. In addition, the field is never operated with a negative current, eliminating the need for the complementary set required in the armature circuit when reversing the direction of the motor.

The disadvantage to SCR drives is the distortion to the incoming power waveform and the currents drawn during switching. As the drive switches from phase A to B, the drives turn on the SCR for phase B while the phase A SCR is conducting. Since the phase A SCR has a finite time requirement to stop conduction, this effectively shorts the two phases together. Additional inductance must be added to reduce the current during this time. The quantity of inductance required from isolation transformers or line reactors is based on the motor, drive, and power source parameters. The commutating inductance has been neglected in Fig. 9.5 for simplicity.

References

DC Drive Manual 1395-5.6, Allen-Bradley, 1989.
DC-300 Adjustable Speed Drives, General Electric, 1988.
Drive Basics, GE Industrial Control Systems, 2000.
Huffman, H. H., Introduction to Solid-State Adjustable Speed Drives, *IEEE Trans. Ind. Appl.*, 671, 1990.
IEEE Guide for Harmonic and Reactive Compensation of Static Power Converters, ANSI/IEEE Std 519-1981.
Mohan, N., Undeland, T., and Robbins, W., *Power Electronics: Converters, Applications, and Design*, Wiley, New York, 1989.
SCR Drives, General Electric Company, Erie, 1975.
Sen, P., *Thyristor DC Drives*, Wiley, New York, 1981.
Thyristor Regulated DC Motor Control Theory, DGI-2.0, Allen-Bradley, 1982.

10
AC Machines Controlled as DC Machines (Brushless DC Machines/Electronics)

Hamid A. Toliyat
Texas A&M University

Tilak Gopalarathnam
Texas A&M University

10.1 Introduction 10-1
10.2 Machine Construction...10-2
   Permanent Magnets • Stator Windings
10.3 Motor Characteristics 10-4
   Mathematical Model
10.4 Power Electronic Converter..10-7
   Unipolar Excitation • Fault-Tolerant Configuration • Current Source Inverter
10.5 Position Sensing 10-9
   Position Sensorless Control
10.6 Pulsating Torque Components.....................................10-11
10.7 Torque-Speed Characteristics.......................................10-11
10.8 Applications..10-15

10.1 Introduction

Brushless DC (BLDC) motors are synchronous motors with permanent magnets on the rotor and armature windings on the stator. Hence, from a construction point of view, they are the inside-out version of DC motors, which have permanent magnets or field windings on the stator and armature windings on the rotor. A typical BLDC motor with 12 stator slots and four poles on the rotor is shown in Fig. 10.1.

The most obvious advantage of the brushless configuration is the removal of the brushes, which eliminates brush maintenance and the sparking associated with them. Having the armature windings on the stator helps the conduction of heat from the windings. Because there are no windings on the rotor, electrical losses in the rotor are minimal. The BLDC motor compares favorably with induction motors in the fractional horsepower range. The former will have better efficiency and better power factor and, therefore, a greater output power for the same frame, because the field excitation is contributed by the permanent magnets and does not have to be supplied by the armature current.

These advantages of the BLDC motor come at the expense of increased complexity in the electronic controller and the need for shaft position sensing. Permanent magnet (PM) excitation is more viable in smaller motors, usually below 20 kW. In larger motors, the cost and weight of the magnets become

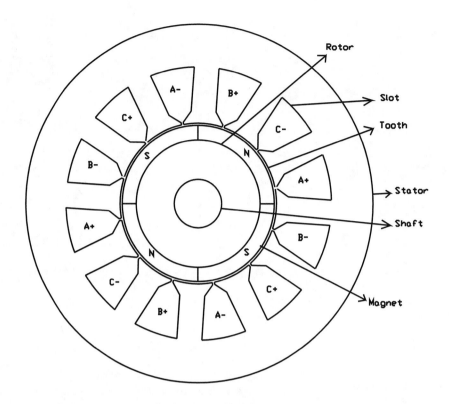

FIGURE 10.1 Three-phase BLDC motor with four poles on the rotor and 12 stator slots.

excessive, and it would make more sense to opt for excitation by electromagnetic or induction means. However, with the development of high-field PM materials, PM motors with ratings of a few megawatts have been built.

10.2 Machine Construction

BLDC motors are predominantly surface-magnet machines with wide magnet pole-arcs and concentrated stator windings. The design is based on a square waveform distribution of the air-gap flux density waveform as well as the winding density of the stator phases in order to match the operational characteristics of the self-controlled inverter [1].

Permanent Magnets

BLDC motors obtain life-long field excitation from permanent magnets mounted on the rotor surface. Advances in permanent magnet manufacturing and technology are primarily responsible for lowering the cost and increasing the applications of BLDC motors. Ferrite or ceramic magnets are the most popular choices for low-cost motors. These magnets are now available with a remanence of 0.38 T and an almost straight demagnetization characteristic throughout the second quadrant. For special applications, magnetic materials with high-energy products such as neodymium-iron-boron (Nd-Fe-B) are used. The high remanence and coercivity permit marked reductions in motor frame size for the same output compared with motors using ferrite magnets. However, the size reduction is at the expense of increased cost of the magnets.

The primary considerations while choosing the magnetic material for a motor are the torque per unit volume of the motor, the operating temperature range, and the severity of the operational duty of the magnet [2]. For maximum power density, the product of the electric and magnetic loadings of the motor

AC Machines Controlled as DC Machines

must be as high as possible. A high electric loading necessitates a long magnet length in the direction of magnetization and a high coercivity. A high power density also requires the largest possible magnet volume. Exposure to high temperatures tends to deteriorate the remanent flux density and coercive force of permanent magnets. Hence, the highest operating temperature must be considered while choosing the magnets. Magnets can also be demagnetized by fault currents such as short-circuit currents produced by inverter faults. Hence, protective measures are usually taken in the inverter and control electronics to limit the magnitude of the armature currents to a safe value.

The magnets are constructed in the form of arcs, radially magnetized, and glued onto the surface of the rotor with adjacent rotor poles of opposite magnetic polarity as shown in Fig. 10.1. The number of rotor poles is inversely proportional to the maximum speed of rotation, and is frequently chosen to meet manufacturing constraints. Most BLDC motors have four, six, or eight poles, with four the most popular choice.

Stator Windings

BLDC motors are often assumed to have three phases, but this is not always the case. Small motors for applications such as light-duty cooling fans have minimal performance requirements, and it is cost-effective to build them with just one or two phases. On the other hand, it is preferable to use a high phase number for large drives with megawatt ratings. This reduces the power-handling capacity of a single phase, and also incorporates some degree of fault tolerance. Machines with as many as 15 phases have been built for ship propulsion. Although these are special-purpose designs, motors with four and five phases are quite common.

The number of stator slots is chosen depending on the rotor poles, phase number, and the winding configuration. In general, a fractional slots/pole design is preferred to minimize cogging torque [3]. The motor of Fig. 10.2 has six slots, which is not a multiple of the number of poles, and is hence a fractional slots/pole design. The windings could be lap-wound or concentric-wound, and the coil span could be full-pitch or short-pitch, depending on the crest width of the back-emf desired. There are virtually infinite

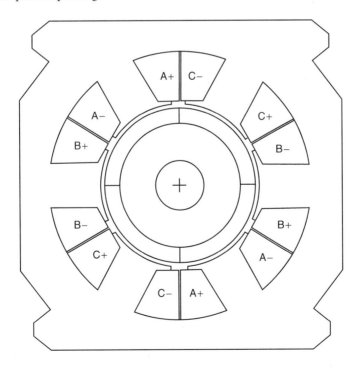

FIGURE 10.2 Three-phase BLDC motor with six slots and four poles.

combinations of the above design factors, and it is up to the ingenuity of the designer to select one that is best suited to the inverter characteristics and meets design specifications.

10.3 Motor Characteristics

The air-gap flux-density waveform is essentially a square wave, but fringing causes the corners to be somewhat rounded. As the rotor rotates, the waveform of the voltage induced in each phase with respect to time is an exact replica of the air-gap flux-density waveform with respect to rotor position. Because of fringing, the back-emf waveform takes on a trapezoidal shape. The shape of the back-emf waveform distinguishes the BLDC motor from the permanent magnet synchronous motor (PMSM), which has a sinusoidal back-emf waveform. This has given rise to the terminology "trapezoidal motor" and "sinusoidal motor" for describing these two permanent magnet AC (PMAC) machines.

The back-emf voltages induced in each phase are similar in shape and are displaced by 120° electrical with respect to each other in a three-phase machine. By injecting rectangular current pulses in each phase that coincides with the crest of the back-emf waveform in that phase, it is possible to obtain an almost constant torque from the BLDC motor. The crest of each back-emf half-cycle waveform should be as broad as possible (≥120° electrical) to obtain smooth output torque. This condition is satisfied by the 12-slot motor of Fig. 10.1 because it has full-pitched coils, but not by the six-slot motor of Fig. 10.2 because the coil spans are shorter than the pole arcs. The two back-emf waveforms calculated using the finite-element method are plotted in Fig. 10.3 and it can be seen that the six-slot motor has a smaller crest width, and is hence not suitable for 120° bipolar excitation. However, it can be used with other excitation waveforms as discussed in the section on unipolar excitation.

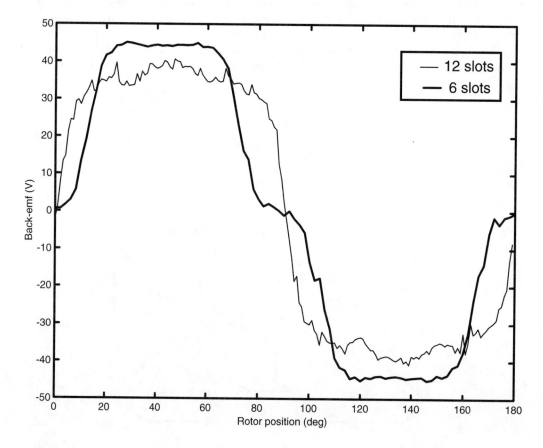

FIGURE 10.3 Back-emf waveforms of the 12-slot and the 6-slot motors.

AC Machines Controlled as DC Machines

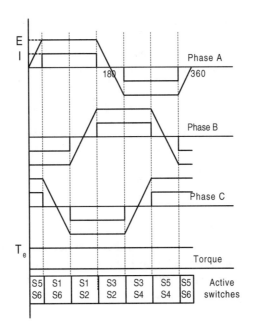

FIGURE 10.4 Back-emf and phase current waveforms for three-phase BLDC motor with 120° bipolar currents.

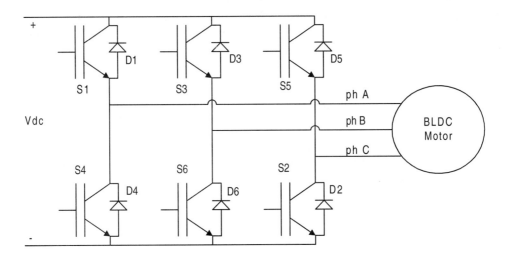

FIGURE 10.5 Schematic of IGBT-based inverter for three-phase BLDC motor.

The ideal back-emf voltage and 120° phase current waveforms for a three-phase BLDC motor are shown in Fig. 10.4. The inverter switches that are active during each 60° interval are also shown corresponding to the inverter circuit of Fig. 10.5. The simplicity of this scheme arises from the fact that during any conduction interval, there is only one current flowing through two phases of the machine, which can be sensed using a single current sensor in the DC link. Because there are only two inverter switches active at any time, this is also called the two-switch conduction scheme, as opposed to the three-switch conduction scheme used in PMSM motor drives.

The amplitude of the phase back-emf is proportional to the rotor speed, and is given by

$$E = k\phi\omega_m \tag{10.1}$$

where k is a constant that depends on the number of turns in each phase, ϕ is the permanent magnet flux, and ω_m is the mechanical speed.

During any 120° interval, the instantaneous power being converted from electrical to mechanical is the sum of the contributions from two phases in series, and is given by

$$P_o = \omega_m T_e = 2EI \tag{10.2}$$

where T_e is the output torque and I is the amplitude of the phase current. From Eqs. (10.1) and (10.2), the expression for output torque can be written as

$$T_e = 2k\phi I = k_t I \tag{10.3}$$

where k_t is the torque constant.

The similarity between the BLDC motor and the commutator DC motor can be seen from Eqs. (10.1) and (10.3). It is because of this similarity in control characteristics that the trapezoidal PMAC motor is widely known as the BLDC motor, although this term is a misnomer as it is actually a synchronous AC motor. But it is also not a rotating field machine in the AC sense, because the armature mmf rotates in discrete steps of 60° electrical as opposed to a smooth rotation in other AC machines.

Mathematical Model

Because of the nonsinusoidal nature of the back-emf and current waveforms, transformation of the machine equations to the d-q model is cumbersome, and it is easier to use the phase-variable approach for modeling and simulation. The back-emf can be represented as a Fourier series or by using piecewise linear curves [4]. The circuit equations of the three windings in phase variables can be written as [4]

$$\begin{bmatrix} v_a \\ v_b \\ v_c \end{bmatrix} = \begin{bmatrix} R & 0 & 0 \\ 0 & R & 0 \\ 0 & 0 & R \end{bmatrix} \cdot \begin{bmatrix} i_a \\ i_b \\ i_c \end{bmatrix} + \begin{bmatrix} L-M & 0 & 0 \\ 0 & L-M & 0 \\ 0 & 0 & L-M \end{bmatrix} \frac{d}{dt} \begin{bmatrix} i_a \\ i_b \\ i_c \end{bmatrix} + \begin{bmatrix} e_a \\ e_b \\ e_c \end{bmatrix} \tag{10.4}$$

where v_a, v_b, v_c are the phase voltages, i_a, i_b, i_c are the phase currents, e_a, e_b, e_c are the phase back-emf voltages, R is the phase resistance, L is the self-inductance of each phase, and M is the mutual inductance between any two phases.

The electromagnetic torque is given by

$$T_e = (e_a i_a + e_b i_b + e_c i_c)/\omega_m \tag{10.5}$$

where ω_m is the mechanical speed of the rotor.

The equation of motion is

$$\frac{d}{dt} w_m = (T_e - T_L - B w_m)/J \tag{10.6}$$

where T_L is the load torque, B is the damping constant, and J is the moment of inertia of the drive.

The electrical frequency is related to the mechanical speed by

$$\omega_e = (P/2)\omega_m \tag{10.7}$$

where P is the number of rotor poles.

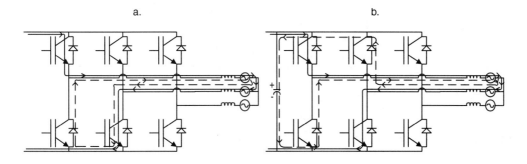

FIGURE 10.6 Illustration of soft chopping (a) and hard chopping (b) for current regulation.

10.4 Power Electronic Converter

BLDC motor drives require variable-frequency, variable-amplitude excitation that is usually provided by a three-phase, full-bridge inverter as shown in Fig. 10.5. The switches could be BJTs, MOSFETSs, IGBTs, or MCTs. The decreasing cost and drastic improvement in performance of these semiconductor devices have accelerated the applications of BLDC motor drives. The inverter is usually responsible for both the electronic commutation and current regulation [5]. The position information obtained from the position sensors is used to open and close the six inverter switches. For the given phase current waveforms, there are only two inverter switches—one upper and one lower that conduct at any instant, each for 120° electrical. If the motor windings are star-connected and the star point is isolated, the inverter input current flows through two of the three phases in series at all times. Hysteresis or pulse-width-modulated (PWM) current controllers are typically used to regulate the actual machine currents to the rectangular current reference waveforms shown in Fig. 10.4. Either soft chopping or hard chopping could be employed for this purpose. The flow of currents during one 60° interval when switches S_1 and S_6 are active is shown in Fig. 10.6a for soft chopping and Fig. 10.6b for hard chopping. When S_1 and S_6 are in their on state, the current builds up in the path shown by the solid lines. In soft chopping, the current regulator commands the turn-off of switch S_1 once the current crosses the threshold. The current then decays through diode D_4 and switch S_6 as shown by the dashed lines. Alternatively, S_6 could be turned off, and the current would then decay in the loop formed by S_1 and D_3. The fall time of the current can be made smaller by hard chopping, in which both the active switches are turned off. The current then freewheels through D_4, D_3, and the DC link capacitor, feeding energy back to the source. The freewheeling diodes thus provide important paths for the currents to circulate when the switches are turned off and during the commutation intervals.

The discussion thus far has concentrated on the operation of the BLDC machine as a motor. It can, however, operate equally well as a generator. The polarity of the torque can be reversed by simply reversing the polarity of the phase current waveforms with respect to the back-emfs. This can be used to advantage for regenerative braking operation, in vehicle propulsion, for example. Special arrangements may need to be made in the power converter to accept the energy returned by the machine, as conventional diode bridge rectifiers are incapable of feeding energy back to the AC supply. The situation is considerably simplified if the source is a battery, as in automotive applications.

Unipolar Excitation

Unipolar current conduction limits the phases to only one direction of current, and the commutation frequency is half that of a bipolar or full-wave drive. The unipolar motor needs fewer electronic parts and uses a simpler circuit than the bipolar motor. For these reasons, unipolar-driven motors are widely used in low-cost instruments. A typical application of BLDC motors of this class can be found in disk memory apparatus [6]. Unipolar excitation results in an inefficient winding utilization compared with bipolar excitation, but they have the following advantages over bipolar circuits [7]:

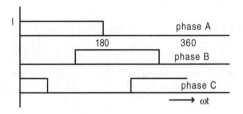

FIGURE 10.7 180° unipolar current waveforms.

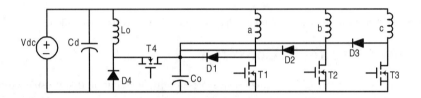

FIGURE 10.8 Schematic of C-dump topology for unipolar three-phase BLDC motor.

1. There is only one device in series with each phase, minimizing conduction losses.
2. The risk of shoot-through faults is eliminated.
3. Switching of devices connected to the supply rails, which generally requires some isolation circuitry, can be avoided.

Another factor that has to be considered before choosing unipolar excitation is that the motor neutral has to be available because the phase currents are no longer balanced. The main issue in unipolar BLDC motor drives is ripple torque. The reference case of the 12-slot motor with 120° bipolar currents gives a ripple torque of 13%. This value represents the ripple component caused by the nonideal back-emf alone, without considering inverter effects. Exciting the same motor with 120° unipolar currents, for example, would produce a torque ripple of 23.7%. However, by exciting the six-slot motor with the 180° unipolar current waveforms shown in Fig. 10.7, the torque ripple reduces to 8.5% [8]. It is thus important to match the motor characteristics to those of the inverter. Increasing the number of phases can also reduce the torque pulsation, but the cost of the drive increases. The simplest unipolar converter has a single switch in series with each motor winding, whereas a reverse-parallel diode provides a freewheeling path at turn-off. This drive has no regenerative control, but four-quadrant operation is possible by using topologies with more than one switch per phase but fewer than two switches per phase [9]. One such topology that has been used for switched reluctance drives is the C-dump converter shown in Fig. 10.8.

Fault-Tolerant Configuration

In applications requiring high reliability such as aerospace and defense, the inverter may be configured as a separate H-bridge supplying each phase of the motor as shown in Fig. 10.9. This doubles the number of power devices, but ensures complete electrical isolation between phases so that remedial strategies can be adopted to continue operation even with the failure of a power device or winding [10]. It is also important to design the machine to minimize the occurrence of a fault by winding each coil around a single tooth. High phase numbers are also used so that the healthy phases can partially compensate for the loss of torque resulting from the failure of one or more phases.

Current Source Inverter

As an alternative to the voltage source inverter (VSI), a current source inverter may be used to drive the BLDC motor. A load-commutated inverter as shown in Fig. 10.10 uses thyristors as the switching devices, and is cheaper than a VSI of similar rating [11]. It replaces the DC-link electrolytic capacitor by an inductor.

AC Machines Controlled as DC Machines

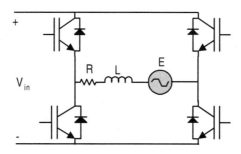

FIGURE 10.9 H-bridge configuration supplying one-phase winding.

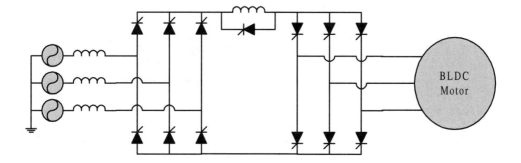

FIGURE 10.10 Load commutated inverter for BLDC motor drive.

Because of its inherent soft switching, switching losses are much lower than in a VSI. Another feature of the LCI drive is its inherent regenerating capability, by operating the line converter in the inverter mode. The thyristors in the load converter are commutated by the back-emf voltages of the motor. At low speed, when the back-emf magnitude is insufficient to commutate them, the line converter is operated in an inverter mode, forcing the link current to become zero and thus turning off the conducting thyristors of the load converter.

10.5 Position Sensing

The stator excitation for BLDC motors needs to be synchronized with rotor speed and position to produce constant torque. The controller has to keep track of the rotor angular position and switch the excitation among the motor phases appropriately. It performs the role of the mechanical commutator in the case of a DC machine, because of which the BLDC motor is also called the electronically commutated motor (ECM). The rotor position needs to be detected at six discrete points in each electrical cycle, i.e., at 60° electrical intervals for the commutation. The most common method of sensing the rotor position is by means of a Hall effect position sensor. A Hall effect position sensor consists of a set of Hall switches and a set of trigger magnets. The Hall switch is a semiconductor switch based on the Hall effect that opens or closes when the magnetic field is higher or lower than a certain threshold value. A signal conditioning circuit integrated with the Hall switch provides a TTL-compatible pulse with sharp edges and high noise immunity for connection to the controller. For a three-phase BLDC motor, three hall switches spaced 120° electrical apart are mounted on the stator frame. The trigger magnets can be a separate set of magnets aligned with the rotor magnets and mounted on the shaft in close proximity to the hall switches. The rotor magnets can also be used as the trigger magnets, with the hall switches mounted close enough to be energized by the leakage flux at the appropriate rotor positions. The digital signals from the Hall sensors are then decoded to obtain the three-phase switching sequence for the inverter. In the block diagram of a BLDC motor drive shown in Fig. 10.11, this function is performed by the controller, which

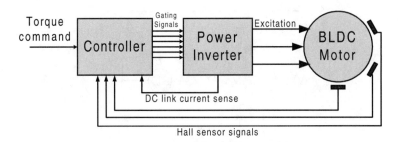

FIGURE 10.11 BLDC motor drive schematic.

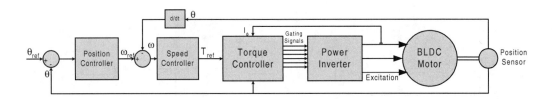

FIGURE 10.12 Position servo using the BLDC motor.

also processes the signal from the DC link current sensor. Based on these two inputs, gating signals are provided to the six inverter switches. High-resolution encoders or resolvers can also be used to provide position feedback for applications in which their cost is justified by the improved performance. For applications requiring speed or position control, the speed and position control loops can be built around the inner current control loop as shown in Fig. 10.12. The ability to operate with just three Hall sensors gives the trapezoidal brushless permanent magnet motor an edge over its sinusoidal counterpart in low-cost applications. It should be mentioned that PMSM motors are also sometimes operated with rectangular currents to minimize the cost of the position sensor, although the output torque waveform is far from ideal because of the mismatch between the motor and the inverter.

Position Sensorless Control

The mounting of hall sensors is a potentially adverse economic and reliability factor, which makes its elimination attractive for the appliance industry [12]. This has given rise to control schemes that eliminate the use of shaft position sensors. In these control methods, the rotor position is derived indirectly from the motor voltage or current waveform. The trapezoidal motor is especially amenable to position sensor elimination because of the availability of an unexcited phase in each 60° electrical conduction interval as can be seen in Fig. 10.4. The switching signals for the inverter can be derived by detecting the zero-crossing of the phase back-emf and introducing a speed-dependent time delay [13]. The terminal voltages are sensed and low-pass-filtered to eliminate the higher harmonics. A different algorithm has to be used for starting, since the generated back-emf is zero at standstill. Field-oriented control at high speeds using these methods is also problematic because of the speed-dependent phase shifts introduced by the capacitors in the low-pass filters. Another method that has a narrow speed range uses phase-locked loop circuitry to lock on to the back-emf of the inactive phase in every 60° interval. A wider speed range is obtained by using the third harmonic of the back-emf to obtain the switching signals [14]. The third harmonic component is obtained by summing the terminal voltages. This signal is also easier to filter, and can be integrated to obtain the third harmonic flux linkage. The zero crossings of the third harmonic of the flux linkage correspond to the commutation instants of the BLDC motor. Starting techniques for sensorless schemes are generally open loop or rely on bringing the rotor to an initial known position. Open-loop starting is accomplished by providing a slowly rotating stator field that gradually increases in magnitude or frequency until the rotor starts rotating. However, the direction of rotation cannot be

AC Machines Controlled as DC Machines

controlled using this method. This disadvantage can be overcome by exciting one phase to bring the rotor to a known initial position before applying the rotating stator field. There are several commercially available integrated circuits that use the back-emf sensing technique for sensorless control. The availability of fast digital signal processor (DSP) controllers has enabled the implementation of many computationally intensive algorithms for rotor position sensing.

10.6 Pulsating Torque Components

One of the drawbacks of the BLDC motor drive is its relatively high torque pulsation. Its pulsating torque components can be classified as cogging and ripple torques, which are produced by essentially different phenomena. Cogging torque is produced by the reluctance variation caused by the stator slot openings as the rotor rotates. It is space dependent, and exists even in the absence of any armature current. It is well known that skewing of the stator slots or rotor magnets by one slot pitch reduces cogging to a fraction of 1% of the rated torque [15]. Cogging can also be minimized without skewing by choosing a fractional slots/pole motor design [3], or by an appropriate choice of the magnet width relative to the slot pitch [16]. But any technique used to reduce the cogging torque generally results in more ripple torque due to a departure from the ideal trapezoidal induced emf. Ripple torque is a consequence of the interaction of armature currents with the machine back-emf waveforms. There are three main components of ripple torque, one motor-related and the other two inverter-related. The motor-related component is produced by the non-idealities in the back-emf waveform. It is desirable to minimize this component by designing the machine so that the crest of the back-emf waveform is as wide and flat as possible. The inverter-related components of ripple torque appear because of a departure from the ideal rectangular current profiles due to the finite inductance of the machine windings. The first inverter-related component is caused by the high-frequency current ripple that is present because of the current hysteresis or PWM control of the inverter. This component is usually filtered out by the load inertia, and so is generally not a problem. The second component is the commutation torque ripple, which occurs at every commutation instant. It develops because the sum of the currents in the off-going and oncoming phases is almost never constant during the commutation intervals. This is illustrated in Fig. 10.13, where the current is commutated from phase "b" to phase "c." The rate at which the current builds up in phase "c" is greater than the rate at which it decays in phase "b," which causes a current spike in phase "a," and a corresponding spike in the torque waveform. Commutation torque ripple appears as spikes or dips depending on the rotor speed and source voltage [17]. Control techniques to minimize this component generally involve using a current sensor in each phase [18], but it is more cost-effective to use a single current sensor in the DC link. A single current sensor would also be unable to detect transient overcurrents in the inverter switches, and so the current control scheme has to be modified to protect the switches [19]. Figure 10.14 shows the waveforms of the phase current and the electromagnetic torque under hysteresis current control. All three ripple torque components are visible here. The high-frequency ripple component is caused by the corresponding ripple in the phase current. The back-emf-related component has a frequency that is six times the electrical frequency, corresponding to the six conduction intervals in each cycle. The commutation torque ripple appears as spikes in the torque waveform at every commutation instant. At high speeds, these ripples may be filtered out by the load inertia, but at low speeds, they can affect the performance of the drive severely. This makes the BLDC motor drive unsuitable for high-performance positioning applications, where accuracy and repeatability would be compromised by torque pulsations.

10.7 Torque-Speed Characteristics

BLDC motors are ideally suited for constant-torque applications, as the field excitation is fixed and the torque is proportional to the armature current. However, operation beyond the base speed in the constant power region as shown in Fig. 10.15 is also desirable in many cases, and for this, fixed field excitation is a disadvantage, as discussed below. Current control is based on a positive voltage difference between the

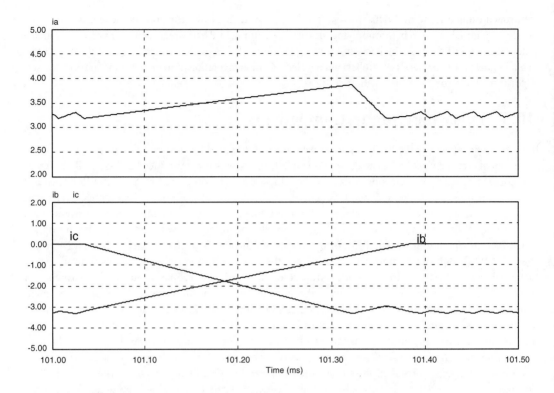

FIGURE 10.13 Phase currents during a commutation interval illustrating the source of commutation torque ripple.

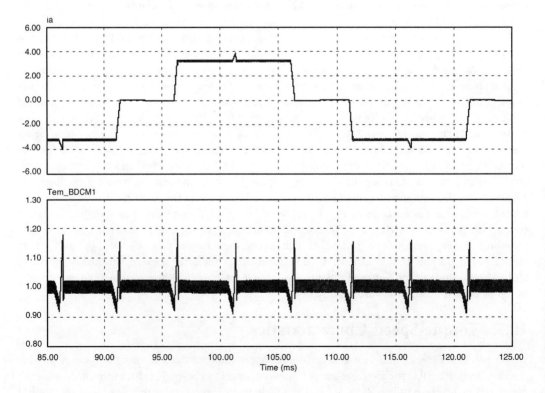

FIGURE 10.14 Phase current and torque waveforms with hysteresis current control (low speed).

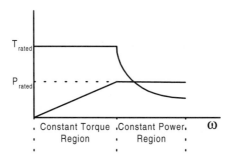

FIGURE 10.15 Constant torque and constant power regions in a variable-speed drive.

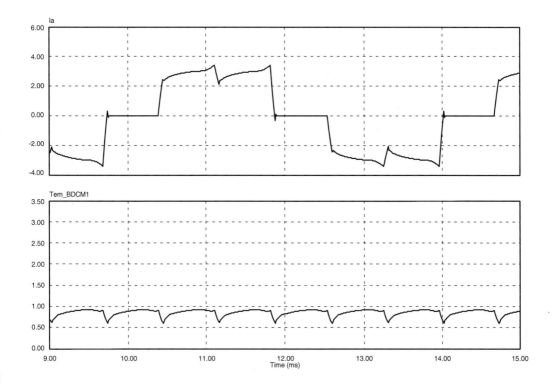

FIGURE 10.16 Phase current and torque waveforms under saturated regulator operation (high speed).

supply voltage and the line-line back-emf. The back-emf amplitude of the BLDC motor is proportional to the rotor speed. Hence, as the speed increases, a point is reached where the sum of the two back-emfs of the conducting motor phases approaches the amplitude of the DC link voltage. The reactance of the phases also increases with speed, and the inverter gradually loses its ability to force the commanded currents into the motor phases, and the current regulators are said to have "saturated." The phase current waveforms under these conditions differ considerably from their ideal rectangular waveshapes, as shown in Fig. 10.16. The inverter switches in this case are in their on state during the entire 120° interval that they are active. The adverse effect of the current waveforms on the torque pulsation can also be seen in the figure. As the speed is increased further, the phase currents and motor torque fall off quite abruptly. The operating envelope can be extended by "field-weakening," which is implemented by advancing the phase angle of the currents relative to the back-emfs. The phase advance angle α is illustrated in Fig. 10.17 as the angle between the back-emf of phase A and the gate signal of switch S_1. By allowing each phase to start conducting before its back-emf reaches its peak value, the current is given a time interval to build up to

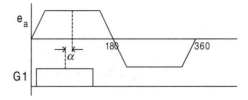

FIGURE 10.17 Illustration of phase advance angle.

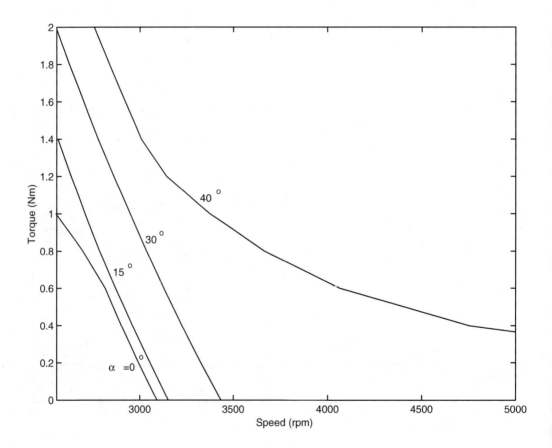

FIGURE 10.18 Field-weakening characteristics of BLDC motor as a function of advance angle.

its commanded value. Under these conditions, a current component is produced in the negative d-axis of the rotor which weakens the air-gap field, and hence the back-emf. This makes it possible to produce a torque-forming current component [20]. The effect of phase advancing on the torque-speed characteristic is shown in Fig. 10.18. At a given torque, the maximum speed of the motor increases as α is increased. However, the extended speed is accompanied by significant increases in torque pulsation. The torque-speed envelope achieved with phase advance is significantly larger with 180° conduction than with 120° conduction. However, if the excitation phase angle is fixed at 0° for all speeds, then the saturated-regulator performance of 120° conduction is better [21]. Implementation of adjustable excitation angle complicates the position-sensing scheme as a higher resolution is required than is provided by three Hall effect sensors. The maximum torque and speed range can also be increased by motor design changes, such as the use of a rotor with inset magnets instead of one with projecting magnets [22].

10.8 Applications

The many advantages of BLDC motors, combined with their rapidly decreasing cost, have led to their widespread applications in many variable-speed drives. Their high power density makes them ideal candidates for applications such as robotic actuators, computer disk drives, and office equipment. With their high efficiency, high power factor, and maintenance-free operation, domestic appliances and heating, ventilating, and air conditioning (HVAC) equipment are now increasingly employing BLDC motors in preference to DC and induction motors. They are also being developed for automotive applications such as electric power steering, power accessories, and active suspension, in addition to vehicle propulsion.

References

1. B.T. Ooi, P. Brissoneau, and L. Brugel, Optimal winding design of a permanent magnet motor for self-controlled inverter operation, *Electric Mach. Electromech.*, 6, 381–389, 1981.
2. T.J.E. Miller, *Brushless PM and Reluctance Motor Drives*, Clarendon Press, Oxford, U.K., 1989.
3. J.R. Hendershot, Jr. and T.J.E. Miller, *Design of Brushless Permanent Magnet Motors*, Magna Physics Publishing and Clarendon Press, Oxford, U.K., 1994.
4. P. Pillay and R. Krishnan, Modeling, simulation, and analysis of permanent-magnet drives. Part II: The brushless DC motor drive, *IEEE Trans. Ind. Appl.*, 5(2), 274–279, March/April 1989.
5. T.M. Jahns, Motion control with permanent-magnet AC machines, *Proc. IEEE*, 82(8), 1241–1252, Aug. 1994.
6. T. Kenjo and S. Nagamori, *Permanent-Magnet and Brushless DC Motors*, Clarendon Press, Oxford, U.K., 1985.
7. P.P. Acarnley, A.G. Jack, and P.T. Jowett, Power circuits for small permanent-magnet brushless dc drives, presented at Third International Conference on Power Electronics and Variable Speed Drives, 1988, 237–240.
8. T. Gopalarathnam, S. Waikar, H.A. Toliyat, M.S. Arefeen, and J.C. Moreira, Development of low-cost multi-phase brushless DC (BLDC) motors with unipolar current excitations, in *Proc. IEEE-IAS Annu. Meeting*, 1999, 173–179.
9. R. Krishnan and S. Lee, PM brushless DC motor drive with a new power converter topology, *Conf. Rec. IEEE-IAS Annual Meeting*, 1995, 380–387.
10. T. Gopalarathnam, H.A. Toliyat, and J.C. Moreira, Multi-phase fault-tolerant brushless DC motor drives, *Conf. Rec. IEEE-IAS Annual Meeting*, 2000.
11. H.A. Toliyat, N. Sultana, D.S. Shet, and J.C. Moreira, Brushless permanent magnet (BPM) motor drive system using load-commutated inverter, *IEEE Trans. Power Electron.*, 14(5), 831–837, Sept. 1999.
12. D.M. Erdman, H.B. Harms, and J.L. Oldenkamp, Electronically commutated DC motors for the appliance industry, *Conf. Rec. IEEE-IAS Annual Meeting*, 1984, 1339–1345.
13. K. Iizuka et al., Microcomputer control for sensorless brushless motor, *IEEE Trans. Ind. Appl.*, IA-27, 595–601, May/June 1985.
14. J. Moreira, Indirect sensing for rotor flux position of permanent magnet AC motors operating over a wide speed range, *IEEE Trans. Ind. Appl.*, 32(6), 1394–1401, Nov.–Dec. 1996.
15. T.M. Jahns and W.L. Soong, Pulsating torque minimization techniques for permanent magnet AC motor drives—a review, *IEEE Trans. Ind. Electron.*, 43(2), 321–330, Apr. 1996.
16. T. Li and G. Slemon, Reduction of cogging torque in permanent magnet motors, *IEEE Trans. Magnetics*, 24(6), 2901–2903, Nov. 1988.
17. R. Carlson, M. Lajoie-Mazenc, and J.C. Fagundes, Analysis of torque ripple due to phase commutation in brushless DC machines, *IEEE Trans. Ind. Appl.*, 28(3), 632–638, May/June 1992.
18. J. Cros, J.M. Vinassa, S. Clenet, S. Astier, and M. Lajoie-Mazenc, A novel current control strategy in trapezoidal EMF actuators to minimize torque ripples due to phase commutations, *Proc. of the European Conference on Power Electronics and Applications (EPE)*, 4, 266–271, 1993.

19. L. Schulting and H.-Ch. Skudelny, A control method for permanent magnet synchronous motors with trapezoidal electromotive force, *Proc. EPE*, 4, 117–122, 1991.
20. G. Schaefer, Field weakening of brushless permanent magnet servomotors with rectangular current, *Proc. EPE*, 3, 429–434, 1991.
21. T.M. Jahns, Torque production in permanent-magnet synchronous motor drives with rectangular current excitation, *IEEE Trans. Ind. Appl.*, IA-20(4), 803–813, July/August 1984.
22. T. Sebastian and G.R. Slemon, Operating limits of inverter-driven permanent magnet motor drives, *IEEE Trans. Ind. Appl.*, IA-23(2), 327–333, Mar./Apr. 1987.

11
Control of Induction Machine Drives

Daniel Logue
University of Illinois at Urbana-Champaign

Philip T. Krein
University of Illinois at Urbana-Champaign

11.1 Introduction 11-1
11.2 Scalar Induction Machine Control 11-2
11.3 Vector Control of Induction Machines 11-4
    Vector Formulation of the Induction Machine • Induction Machine Dynamic Model • Field-Oriented Control of the Induction Machine • Direct Torque Control of the Induction Machine
11.4 Summary 11-17

11.1 Introduction

Induction machines have become the staple for electromechanical energy conversion in today's industry; they are used more often than all other types of motors combined. Several factors have made them the machine of choice for industrial applications vs. DC machines, including their ruggedness, reliability, and low maintenance [1, 2]. The cage-induction machine is simple to manufacture, with no rotor windings or commutator for external rotor connection. There are no brushes to replace because of wear, and no brush arcing to prevent the machine from being used in volatile environments. The induction machine has a higher power density, greater maximum speed, and lower rotor inertia than the DC machine.

The induction machine has one significant disadvantage with regard to torque control as compared with the DC machine. The torque production of a given machine is related to the cross-product of the stator and rotor flux-linkage vectors [3–5]. If the rotor and stator flux linkages are held orthogonal to one another, the electrical torque of the machine can be controlled by adjusting either the rotor or stator flux-linkage and holding the other constant. The field and armature windings in a DC machine are held orthogonal by a mechanical commutator, making torque control relatively simple. With an induction machine, the stator and rotor windings are not fixed orthogonal to one another. The induction machine is singly excited, with the rotor field induced by the stator field, further complicating torque control. Until a few years ago, the induction machine was mainly used for constant-speed applications. With recent improvements in semiconductor technology and power electronics, the induction machine is seeing wider use in variable-speed applications [6].

This chapter discusses how these challenges related to the induction machine are overcome to effect torque and speed control comparable with that of the DC machine. The first section involves what is termed volts-per-hertz, or scalar, control. This control method is derived from the steady-state machine model and is satisfactory for many low-performance industrial and commercial applications. The rest of the chapter will present vector-controlled methods applied to the induction machine [7]. These methods are aimed at bringing about independent control of the machine torque- and flux-producing stator currents. Developed using the dynamic machine model, vector-controlled induction machines exhibit far better dynamic performance than those with scalar control [8].

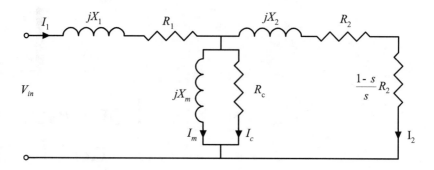

FIGURE 11.1 Induction machine steady-state model.

11.2 Scalar Induction Machine Control

Induction machine scalar control is derived using the induction machine steady-state model shown in Fig. 11.1 [1]. The phasor form of the machine voltages and currents is indicated by capital letters. The stator series resistance and leakage reactance are R_1 and X_1, respectively. The referred rotor series resistance and leakage reactance are R_2 and X_2, respectively. The magnetizing reactance is X_m; the core loss due to eddy currents and the hysteresis of the iron core is represented by the shunt resistance R_c. The machine slip s is defined as [1]

$$s = \frac{\omega_e - \omega_r}{\omega_e} \tag{11.1}$$

where ω_e is the synchronous, or excitation frequency, and ω_r is the machine shaft speed, both in electrical radians-per-second. The power supplied to the machine shaft can be expressed as

$$P_{shaft} = \frac{1-s}{s} R_2 i_2^2 \tag{11.2}$$

Solving for I_2 and using Eq. (11.2), the shaft torque can be expressed as

$$T_e = \frac{3|V_{in}|^2 R_2 s}{\omega_e[(sR_1 + R_2)^2 + s^2(X_1 + X_2)^2]} \tag{11.3}$$

where the numeral 3 in the numerator is used to include the torque from all three phases. This expression makes clear that induction machine torque control is possible by varying the magnitude of the applied stator voltage. The normalized torque vs. slip curves for a typical induction machine corresponding to various stator voltage magnitudes are shown in Fig. 11.2. Speed control is accomplished by adjusting the input voltage until the machine torque for a given slip matches the load torque. However, the developed torque decreases as the square of the input voltage, but the rotor current decreases linearly with the input voltage. This operation is inefficient and requires that the load torque decrease with decreasing machine speed to prevent overheating [1, 2]. In addition, the breakdown torque of the machine decreases as the square of the input voltage. Fans and pumps are appropriate loads for this type of speed control because the torque required to drive them varies linearly or quadratically with their speed.

Linearization of Eq. (11.3) with respect to machine slip yields

$$T_e = \frac{3|V_{in}|^2 s}{\omega_e R_2} = \frac{3|V_{in}|^2 (\omega_e - \omega_r)}{\omega_e^2 R_2} \tag{11.4}$$

The characteristic torque curve can be shifted along the speed axis by changing ω_e with the capability

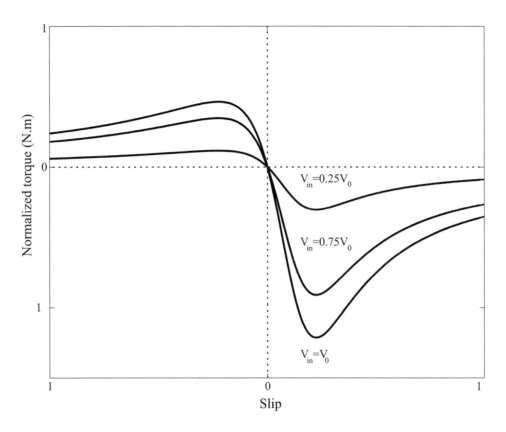

FIGURE 11.2 Normalized torque–slip curves with varying input voltage magnitudes for a typical induction machine.

for developing rated torque throughout the entire speed range given a constant stator voltage magnitude. An inverter is needed to drive the induction machine to implement frequency control.

One remaining complication is the fact that the magnetizing reactance changes linearly with excitation frequency. Therefore, with constant input voltage, the input current increases as the input frequency decreases. In addition, the stator flux magnitude increases as well, possibly saturating the machine. To prevent this from happening, the input voltage must be varied in proportion to the excitation frequency. From Eq. (11.4), if the input voltage and frequency are proportional with proportionality constant k_f, the electrical torque developed by the machine can be expressed as

$$T_e = \frac{3k_f^2}{R_2}(\omega_e - \omega_r) \tag{11.5}$$

and demonstrates that the torque response of the machine is uniform throughout the full speed range.

The block diagram for the scalar-controlled induction drive is shown in Fig. 11.3. The inverter DC-link voltage is obtained through rectification of the AC line voltage. The drive uses a simple pulse-width-modulated (PWM) inverter whose time-average output voltages follow a reference-balanced three-phase set, the frequency and amplitude of which are provided by the speed controller. The drive shown here uses an active speed controller based on a proportional integral derivative (PID), or other type of controller. The input to the speed controller is the error between a user-specified reference speed and the shaft speed of the machine. An encoder or other speed-sensing device is required to ascertain the shaft speed. The drive can be operated in the open-loop configuration as well; however, the speed accuracy will be reduced significantly.

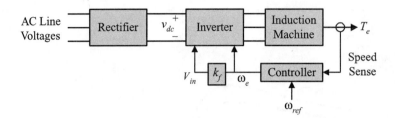

FIGURE 11.3 Block diagram of scalar induction machine drive.

Practical scalar-controlled drives have additional functionality, some of which is added for the convenience of the user. In a practical drive, the relationship between the input voltage magnitude and frequency takes the form

$$|V_{in}| = k_f \omega_e + V_{offset} \tag{11.6}$$

where V_{offset} is a constant. The purpose of this offset voltage is to overcome the voltage drop created by the stator series resistance. The relationship (11.6) is usually a piecewise linear function with several breakpoints in a standard scalar-controlled drive. This allows the user to tailor the drive response characteristic to a given application.

11.3 Vector Control of Induction Machines

The derivation of the vector-controlled (VC) method and its application to the induction machine is considered in this section. The vector description of the machine will be derived in the first subsection, followed by the dynamic model description in the second subsection. Field-oriented control (FOC) of the induction machine will be presented in the third subsection and the direct torque control (DTC) method will be described in the last subsection.

Vector Formulation of the Induction Machine

The stator and rotor windings for the three-phase induction machine are shown in Fig. 11.4 [3]. The windings are sinusoidally distributed, but are indicated on the figure as point windings. If N_0 is the number of turns for each winding, then the winding density distributions as functions of θ are given by

$$\begin{aligned} N_a(\theta) &= N_0 \cos(\theta) \\ N_b(\theta) &= N_0 \cos\left(\theta - \frac{2\pi}{3}\right) \\ N_c(\theta) &= N_0 \cos\left(\theta + \frac{2\pi}{3}\right) \end{aligned} \tag{11.7}$$

where θ is the angle around the stator referenced from phase *as*-axis. The magnemotive force (MMF) distributions corresponding to (11.7) are [5]

$$\begin{aligned} F_{as}(t, \theta) &= \frac{N_0}{2} i_{as}(t) \cos(\theta) \\ F_{bs}(t, \theta) &= \frac{N_0}{2} i_{bs}(t) \cos\left(\theta - \frac{2\pi}{3}\right) \\ F_{cs}(t, \theta) &= \frac{N_0}{2} i_{cs}(t) \cos\left(\theta + \frac{2\pi}{3}\right) \end{aligned} \tag{11.8}$$

Control of Induction Machine Drives

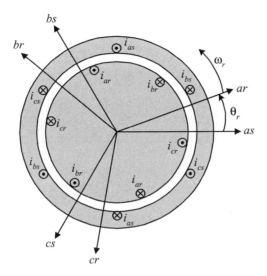

FIGURE 11.4 Induction machine stator and rotor windings.

These scalar equations can be represented by dot products between the following MMF vectors

$$\vec{F}_{as}(t) = \frac{N_0}{2} i_{as}(t) \hat{e}_{as}$$

$$\vec{F}_{bs}(t) = \frac{N_0}{2} i_{bs}(t) \hat{e}_{bs} \quad (11.9)$$

$$\vec{F}_{cs}(t) = \frac{N_0}{2} i_{cs}(t) \hat{e}_{cs}$$

and the unit vector whose angle with the *as*-axis is θ. The vectors $\hat{e}_{as}$, $\hat{e}_{bs}$, and $\hat{e}_{cs}$ represent unit vectors along the respective winding axes. All the machine quantities, including the phase currents and voltages, and flux linkages can be expressed in this vector form.

The vectors along the three axes *as*, *bs*, and *cs* do not form an independent basis set. It is convenient to transform this basis set to one that is orthogonal, the so-called *dq*-transformation, originally proposed by R. H. Park for application to the synchronous machine [3, 9]. Figure 11.5 illustrates the relationship between the degenerate *abc* and orthogonal *qd0* vector sets. If ϕ is the angle between i_{qs} and i_{as}, then the transformation relating the two coordinate systems can be expressed as

$$\mathbf{i}_{qd0s} = \mathbf{W}(\phi)\mathbf{i}_{abcs} = \frac{2}{3}\begin{bmatrix} \cos\phi & \cos\left(\phi - \frac{2\pi}{3}\right) & \cos\left(\phi + \frac{2\pi}{3}\right) \\ \sin\phi & \sin\left(\phi - \frac{2\pi}{3}\right) & \sin\left(\phi + \frac{2\pi}{3}\right) \\ \frac{1}{2} & \frac{1}{2} & \frac{1}{2} \end{bmatrix} \mathbf{i}_{abcs} \quad (11.10)$$

where $\mathbf{i}_{qd0s} = [i_{qs} \; i_{ds} \; i_{0s}]^T$ and $\mathbf{i}_{abcs} = [i_{as} \; i_{bs} \; i_{cs}]^T$. The variable i_{0s} is called the zero-sequence component and is obtained using the last row in the matrix **W** [3]. This last row is included to make the matrix invertible, providing a one-to-one transformation between the two coordinate systems. This row is not needed if the transformation acts on a balance set of variables, because the zero-sequence component is

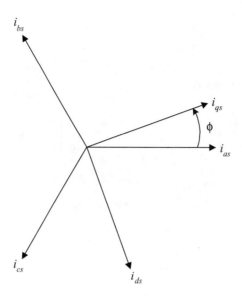

FIGURE 11.5 Illustration for reference frame transformation.

equal to zero. The zero-sequence component carries information about the neutral point of the *abc* variables being transformed. If the set is not balanced, this neutral point is not necessarily zero.

The constant multiplying the matrix of (11.10) is, in general, arbitrary. With this constant equal to $\frac{2}{3}$ as it is in (11.10), the result is the power invariant transformation. By using this transformation, the calculated power in the *abc* coordinate system is equal to that computed in the *qd*0 system [3].

If the angle $\phi = 0$, the result is a transformation from the stationary *abc* system to the stationary *qd*0 system. However, transformation to a reference frame rotating at an arbitrary speed ω is possible by defining

$$\phi(t) = \int_0^t \omega \, d\tau \qquad (11.11)$$

As will be seen later, the rotor flux–oriented vector control method makes use of this concept, transforming the machine variables to the synchronous reference frame where they are constants in steady state [4].

To understand this concept intuitively, consider the balanced set of stator MMF vectors of a typical induction machine given in (11.9). It is not difficult to show that the sum of these vectors produces a resultant MMF vector that rotates at the frequency of the stator currents. The length of the vector is dependent upon the magnitude of the MMF vectors. Observing the system from the synchronous reference frame effectively removes the rotational motion, resulting in only the magnitude of the vector being of consequence. If the magnitudes of the MMF vectors are constant, then the synchronous variables will be constant. Transients in the magnitudes of the stationary variables result in transients in the synchronous variables. This is true for currents, voltages, and other variables associated with the machine.

Induction Machine Dynamic Model

The six-state induction machine model in the arbitrary reference frame is presented in this section. This dynamic model will be used to derive the FOC and DTC methods. As will be seen, the derivations of these control methods will be simpler if they are performed in a specific coordinate reference frame. An additional advantage is that transforming to the *qd*0 coordinate system in any reference frame removes

Control of Induction Machine Drives

TABLE 11.1 Induction Machine Nomenclature

Induction Machine Parameter or Variable	Symbol
Stator voltages (V)	v_{qs}, v_{ds}
Stator currents (A)	i_{qs}, i_{ds}
Stator flux-linkages (Wb)	$\lambda_{qs}, \lambda_{ds}$
Rotor voltages (V)	v_{qr}, v_{dr}
Rotor currents (A)	i_{qr}, i_{dr}
Rotor flux-linkages (Wb)	$\lambda_{qr}, \lambda_{dr}$
Reference frame speed (rad/s)	ω
Rotor speed (rad/s)	ω_r
Stator series resistance (Ω)	r_s
Stator leakage inductance (H)	L_{ls}
Rotor series resistance (Ω)	r_r
Rotor leakage inductance (H)	L_{lr}
Magnetizing inductance (H)	L_m
Number of machine poles	P
Developed electrical torque (N·m)	T_e
Machine load torque (N·m)	T_{load}
Torque due to windage and friction losses (N·m)	T_{loss}

the time-varying inductances associated with the induction machine [10]. The machine model in a given reference frame is obtained by substituting the appropriate frequency for ω in the model equations.

The state equations for the six-state induction motor model in the arbitrary reference frame are given in Eqs. (11.12) through (11.22) [3, 4]. The induction machine nomenclature is provided in Table 11.1. The derivative operator is denoted by p, and the rotor quantities are referred to the stator. The state equations are

$$v_{qs} = r_s i_{qs} + p\lambda_{qs} + \omega\lambda_{ds} \tag{11.12}$$

$$v_{ds} = r_s i_{ds} + p\lambda_{ds} - \omega\lambda_{qs} \tag{11.13}$$

$$v_{qr} = 0 = r_r i_{qr} + p\lambda_{qr} + (\omega - \omega_r)\lambda_{dr} \tag{11.14}$$

$$v_{dr} = 0 = r_r i_{dr} + p\lambda_{dr} - (\omega - \omega_r)\lambda_{qr} \tag{11.15}$$

$$p\omega_r = \frac{P}{2J}(T_e - T_{load} - T_{loss}) \tag{11.16}$$

$$p\theta_r = \omega_r \tag{11.17}$$

where the stator and rotor flux linkages are given by

$$\lambda_{ds} = L_{ls} i_{ds} + L_m (i_{ds} + i_{dr}) \tag{11.18}$$

$$\lambda_{qs} = L_{ls} i_{qs} + L_m (i_{qs} + i_{qr}) \tag{11.19}$$

$$\lambda_{dr} = L_{lr} i_{dr} + L_m (i_{ds} + i_{dr}) \tag{11.20}$$

$$\lambda_{qr} = L_{lr} i_{qr} + L_m (i_{qs} + i_{qr}) \tag{11.21}$$

The electrical torque developed by the machine is [4, 5]

$$T_e = \frac{3PL_m}{4L_r}(\lambda_{dr}i_{qs} - \lambda_{qr}i_{ds}) = \frac{3PL_m}{L_r L_s'}(\lambda_{qs}\lambda_{dr} - \lambda_{qr}\lambda_{ds}) \qquad (11.22)$$

where the stator transient reactance is defined as $L_s' = L_s - L_m^2/L_r$, where $L_r = L_{lr} + L_m$ and $L_s = L_{ls} + L_m$. It is important to note that in Eqs. (11.14) and (11.15), the shaft speed ω_r is expressed in electrical radians-per-second, that is, scaled by the number of machine pole pairs.

Field-Oriented Control of the Induction Machine

Field-oriented control is probably the most common control method used for high-performance induction machine applications. Rotor flux orientation (RFO) in the synchronous reference frame is considered here [4]. There are other orientation possibilities, but rotor flux orientation is the most prominent, and so will be presented in detail.

The RFO control method involves making the induction machine behave similarly to a DC machine. The rotor flux is aligned entirely along the d-axis. The stator currents are split into two components: a field-producing component that induces the rotor flux and a torque-producing component that is orthogonal to the rotor field. This is analogous to the DC machine where the field flux is along one direction, and the commutator ensures an orthogonal armature current vector. This task is greatly simplified through transformation of the machine variables to the synchronously rotating reference frame.

Under FOC, the q-axis rotor flux linkage is zero in the synchronous reference frame, by using Eq. (11.22), the electric torque of the induction machine can be expressed as

$$T_e = \frac{3PL_m}{4L_r}\lambda_{dr}^e i_{qs}^e \qquad (11.23)$$

where the e superscript indicates evaluation in the synchronous reference frame. This torque equation is very similar to that of the DC machine. If either the flux linkage λ_{dr}^e or current i_{qs}^e is held constant, then the torque can be controlled by changing the other. Assuming the inverter driving the induction machine is current sourced, the stator currents can be controlled almost instantaneously. However, by setting $\lambda_{qr}^e = 0$ in Eq. (11.15) and substituting the result in Eq. (11.20), it can be shown that the d-axis rotor flux linkage is governed by

$$\lambda_{dr}^e = \frac{L_m r_r}{(L_{lr} + L_m)p + 1} i_{ds}^e = \frac{L_m}{\tau_r p + 1} i_{ds}^e \qquad (11.24)$$

where τ_r is termed the rotor time constant. Equation (11.24) dictates that the rotor flux cannot be changed arbitrarily fast. Therefore, the best dynamic torque response will result if the rotor flux linkage is held constant, and the electrical torque is controlled by changing i_{qs}^e. Assuming a current-sourced inverter, this control configuration allows torque control for which the response is limited only by the response time of the inverter driving the machine.

Implementation of RFO control requires that the machine variables be transformed to the synchronous reference frame. To accomplish this task, the synchronous reference frame speed must be calculated in some manner. There are two common methods of finding the synchronous speed. In indirect FOC, the synchronous speed is obtained by using a rotor speed measurement and a corresponding slip calculation [4, 11]. Direct FOC uses air-gap flux measurement or other machine-related quantities to compute the synchronous speed. The indirect method is the most common and will be presented here.

Control of Induction Machine Drives

In indirect FOC, the synchronous reference frame speed must be found, and this value integrated to obtain the angle used in the reference frame transformation $W(\phi)$. Rewriting Eq. (11.14) with $\lambda_{qr}^e = 0$ yields

$$\omega_e - \omega_r = -\frac{\lambda_{dr}^e}{r_r i_{qr}^e} \tag{11.25}$$

Again, with $\lambda_{qr}^e = 0$, rewrite Eq. (11.21) as

$$i_{qr}^e = -\frac{L_m i_{qs}^e}{L_{lr} + L_m} \tag{11.26}$$

Substitution of Eq. (11.26) into Eq. (11.25) yields the desired expression for ω_e

$$\omega_e = \omega_r + \frac{(L_{lr} + L_m)\lambda_{dr}^e}{L_m r_r i_{qs}^e} = \omega_r + \frac{\tau_r}{L_m} \frac{\lambda_{dr}^e}{i_{qs}^e} \tag{11.27}$$

This expression provides the needed synchronous speed in terms of the rotor flux, which is specified by the controller, and the q-axis stator current that is adjusted for torque control. The rotor flux time constant τ_r is required for the slip calculation, and in many cases must be estimated online because of its dependence on temperature and other factors [12, 13]. The d-axis stator current needed to produce a given rotor flux can be computed using Eq. (11.24). The angle ϕ used for the reference frame transformation is calculated via

$$\phi(t) = \int_0^t \omega_e \, d\tau + \phi(0) \tag{11.28}$$

The block diagram for the FOC drive is shown in Fig. 11.6. The current i_{qs}^{e*} is used for torque control, while the current i_{ds}^{e*} is calculated using the reference rotor flux λ_{dr}^{e*}. Also present in the diagram is an optional speed controller (connected via the dotted lines) that uses the error between a reference value and the actual

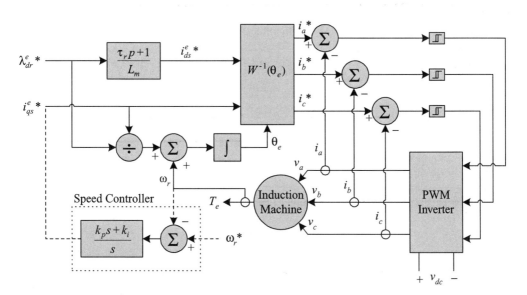

FIGURE 11.6 Block diagram of the indirect FOC drive.

machine speed to control the q-axis stator current. The machine reference currents in the stationary reference frame i_a^*, i_b^*, and i_c^* are computed using the transformation $W^{-1}(\phi)$. The inverter phase voltages are determined using hysteretic controllers [14]. Other methods include ramp comparison and predictive controllers. The shaft speed of the induction machine is obtained using a shaft encoder or similar device.

In the above setup, the inverter voltages were dynamically controlled using the stator current error. The stator voltages required to produce the currents i_a^*, i_b^*, and i_c^*, can also be computed directly using the induction machine model. The stator voltage Eqs. (11.12) and (11.13) must first be "decoupled" to control the armature currents independently. This is because these equations contain stator flux linkage terms that are dependent upon the rotor currents. The decoupling is accomplished by first substituting Eqs. (11.20) and (11.21) into Eqs. (11.18) and (11.19), respectively. The resulting forms of Eqs. (11.18) and (11.19) are then substituted into the stator voltage Eqs. (11.12) and (11.13) to yield [4]

$$v_{qs}^e = (r_s + L_s' p) i_{qs}^e + \omega_e \left(L_s' i_{ds}^e + \frac{L_m}{L_r \lambda_{dr}^e} \right) \tag{11.29}$$

$$v_{ds}^e = (r_s + L_s' p) i_{ds}^e - \omega_e L_s' i_{qs}^e + \frac{L_m}{L_r} p \lambda_{dr}^e \tag{11.30}$$

The decoupled voltage equations allow a voltage-sourced inverter to be used directly for FOC. Note that this is not the only method of performing the decoupling, that PID or other controllers can be used to generate the cross-coupled terms in the voltage equations. However, this technique requires estimation of the torque and rotor flux linkage.

Figures 11.7 and 11.8 display the response of a typical induction machine under FOC. The top plot in Fig. 11.7 shows the machine speed reference (dotted line) and the shaft speed (solid line). Initially, the

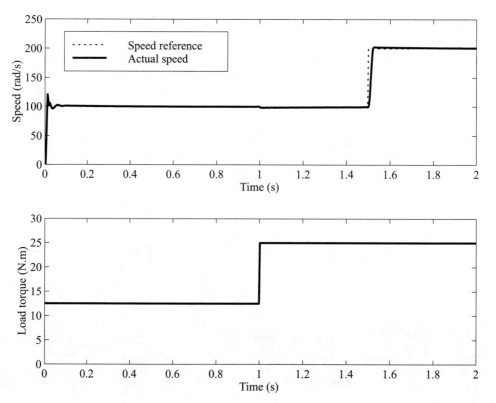

FIGURE 11.7 Induction machine speed reference, actual speed, and load torque.

Control of Induction Machine Drives

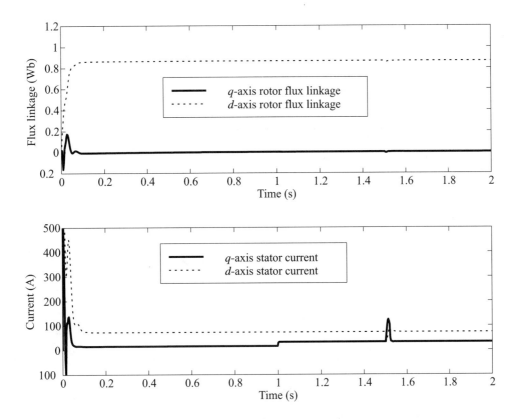

FIGURE 11.8 Rotor flux-linkage and stator currents for the FOC induction machine example.

speed reference is equal to 100 rad/s, and at $t = 1.5$ s, the reference is stepped to 200 rad/s. The machine load is shown in the lower plot. The initial load is 12 N·m and is stepped to 25 N·m at $t = 1$ s. These plots demonstrate that the FOC induction machine has a fast dynamic response and good disturbance rejection.

The rotor dq flux linkages are shown in the top plot of Fig. 11.8. The q-axis flux linkage settles to zero shortly after startup, and the d-axis flux linkage settles to the reference value. This plot verifies that the rotor flux is oriented along one axis in the synchronous reference frame. The synchronous frame stator currents are given in the lower plot of Fig. 11.8. The d-axis current settles to a constant value corresponding the constant rotor flux linkage value. The q-axis current is stepped at $t = 1$ s to satisfy the load torque and experiences a transient at $t = 1.5$ s to increase the machine speed.

Direct Torque Control of the Induction Machine

Whereas the FOC method maintains orthogonality between the rotor flux linkage and the stator torque-producing current, the DTC method directly controls the stator flux linkage to effect torque control [15–18]. The DTC method operates in the stationary reference frame and acts directly on the inverter switches to produce the necessary stator voltages. Hysteretic controllers are used to constrain the electrical torque and stator flux magnitude within certain bounds.

Space Vector Modulation

A DTC drive is constructed using a three-phase switch matrix as shown in Fig. 11.9. The DC input voltage is denoted v_{dc} and the each of the switches has an associated switching function, given by

$$q_{1i} = \begin{cases} 1 & q_{1i} \text{ on,} \\ 0 & q_{1i} \text{ off,} \end{cases} \quad i \in [1, 2, 3] \qquad (11.31)$$

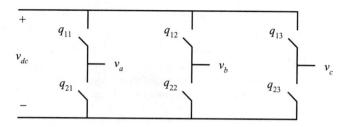

FIGURE 11.9 Switch matrix for the three-phase inverter.

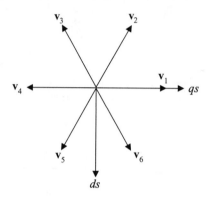

FIGURE 11.10 Voltage star for the three-phase inverter switch matrix.

where $q_{1i} = 1 - q_{2i}$. The result is eight inverter configurations enumerated by $[q_{11}\ q_{12}\ q_{13}]$. For example, in configuration [1 0 0], phase *a* is connected to the positive side of the DC bus, and phases *b* and *c* are connected to the negative side. These eight inverter configurations yield eight equivalent voltage vectors in the *dq*0 coordinate system as displayed in Fig. 11.10.

The diagram in Fig. 11.10 is called the *voltage star* for the three-phase inverter [15]. It is arrived at from the various configurations in Fig. 11.9 and using the transformation (11.10) with $\phi = 0°$. The voltage vector $\mathbf{v}_1$ is calculated using the coordinate transformation and the values $v_a = \frac{2}{3}v_{dc}$ and $v_b = v_c = -\frac{1}{3}v_{dc}$ and is equal to $\mathbf{v}_1 = \frac{2}{3}v_{dc}\hat{e}_q + 0\hat{e}_d$. The set of voltage vectors in polar coordinates provided by the inverter are collectively given by [5, 19]

$$\mathbf{v}_k = \begin{cases} \frac{2}{3}v_{dc}\hat{r}e^{[j(k-1)\pi/3]\hat{\theta}} & k = 1, \ldots, 6 \\ 0 & k = 0, 7 \end{cases} \quad (11.32)$$

where $(\hat{r}, \hat{\theta})$ is the polar coordinate representation of $(\hat{e}_q, \hat{e}_d)$. The vectors $\mathbf{v}_0$ and $\mathbf{v}_7$ correspond to the case where $q_{11} = q_{12} = q_{13} = 0$.

Direct Torque Control Concept

From Eqs. (11.12) and (11.13), the stator flux linkages in the stationary reference frame are computed via [5, 15]

$$\lambda_{qs}^s(t) = \int_0^t (v_{qs}^s(\tau) - r_s i_{qs}^s(\tau))\,d\tau \quad (11.33)$$

$$\lambda_{ds}^s(t) = \int_0^t (v_{ds}^s(\tau) - r_s i_{ds}^s(\tau))\,d\tau \quad (11.34)$$

where the superscripted s indicates evaluation in the stationary reference frame. If the stator resistance r_s is small, as is usually the case, the stator flux linkages can be approximated as the time integral of the stator voltages. This approximation coupled with the space vector development above provides the means for directly controlling the stator flux linkage vector by manipulation of the stator voltages.

The electrical torque developed by the three-phase induction machine (11.22) can be written as the cross product [3, 4]

$$T_e = \frac{3PL_m}{4L'_m} \boldsymbol{\lambda}_s^s \times \boldsymbol{\lambda}_r^s = \frac{3PL_m}{4L'_m} |\boldsymbol{\lambda}_s^s||\boldsymbol{\lambda}_r^s| \sin(\rho) \qquad (11.35)$$

where $\boldsymbol{\lambda}_s^s = [\lambda_{ds}^s \; \lambda_{qs}^s]^T$, $\boldsymbol{\lambda}_r^s = [\lambda_{dr}^s \; \lambda_{qr}^s]^T$, and ρ is the angle between the stator and rotor flux linkages. It is clear from Eq. (11.35), if the stator and rotor flux linkage magnitudes are held constant, then the machine torque can be controlled by changing the angle ρ. The angle ρ cannot be changed directly, but can be indirectly modified by changing the stator flux linkage angle rapidly. This is because the stator flux time constant is typically much faster than the rotor flux time constant. If the stator flux linkage is changed quickly, the rotor flux will lag behind, resulting in a change in ρ.

Hysteretic comparators are used to control the inverter switches to adjust the magnitude and angle of the stator flux linkage. This is because the voltages cannot be controlled through continuous ranges, only two discrete levels: each phase can only be connected to either the positive or negative DC bus voltage. The two-level hysteretic control function is defined as

$$g(t, x, \varepsilon, X_0) = \begin{cases} G_1 & x(t) > X_0 + \varepsilon \\ g(t) & X_0 + \varepsilon \geq x(t) \geq X_0 - \varepsilon \\ G_2 & x(t) < X_0 - \varepsilon \end{cases} \qquad (11.36)$$

The two quantities to be controlled are the electrical torque via the angle ρ, and the stator flux linkage magnitude. The hysteretic controllers are used to maintain these two quantities within the ranges $\lambda_{\text{ref}} + \Delta\lambda \geq |\boldsymbol{\lambda}_s^s| \geq \lambda_{\text{ref}} - \Delta\lambda$ and $T_{\text{ref}} + \Delta T \geq T_e \geq T_{\text{ref}} - \Delta T$. The comparators provide the necessary inverter switch configurations to ensure that the torque and stator flux linkage magnitude stay within these limits. A rule set relating the torque and stator flux linkage error to the set of inverter configurations must be developed. To simplify this task, the coordinate frame is separated into sectors as shown in Fig. 11.11. There are six sectors corresponding to the six active inverter states.

To understand why the sectors are used, consider the situation where the stator flux linkage vector is in Sector 1, and its magnitude and angle γ must be increased. If the stator voltage vector $\mathbf{v}_2$ is used, the flux linkage magnitude and the angle γ will both increase no matter where the flux linkage vector resides in Sector 1. The vector $\mathbf{v}_1$ cannot be reliably used to accomplish this goal, because if the stator flux linkage vector is ahead of $\mathbf{v}_1$, the angle γ will decrease, resulting in a torque decrease. All of the appropriate controller responses can be worked out this way to form the lookup table shown in Table 11.2 [4]. Given the sector number in which the stator flux linkage resides and the outputs of the flux linkage magnitude and torque hysteretic comparators, the table provides the required inverter voltage vector. The flux linkage comparator is two level, and the torque comparator is three level. If the estimated torque is within the specified bounds of the comparator, a zero voltage vector is selected. In this case, the zero voltage vector that requires the fewest inverter switches changing state is used.

To implement the control described above, the sector number must be determined. The most straightforward way of accomplishing this is to find the stator flux linkage angle γ trigonometrically:

$$\gamma = \tan^{-1}\left(\frac{\lambda_{sq}^s}{\lambda_{sd}^s}\right) \qquad (11.37)$$

TABLE 11.2 Optimum Lookup Table for DTC Inverter Control

$|\lambda_s^s|$	T_e	Sector 1	Sector 2	Sector 3	Sector 4	Sector 5	Sector 6
Increase	Increase	$\mathbf{v}_2$	$\mathbf{v}_3$	$\mathbf{v}_4$	$\mathbf{v}_5$	$\mathbf{v}_6$	$\mathbf{v}_1$
Increase	Within limits	$\mathbf{v}_7$	$\mathbf{v}_0$	$\mathbf{v}_7$	$\mathbf{v}_0$	$\mathbf{v}_7$	$\mathbf{v}_0$
Increase	Decrease	$\mathbf{v}_6$	$\mathbf{v}_1$	$\mathbf{v}_2$	$\mathbf{v}_3$	$\mathbf{v}_4$	$\mathbf{v}_5$
Decrease	Increase	$\mathbf{v}_3$	$\mathbf{v}_4$	$\mathbf{v}_5$	$\mathbf{v}_6$	$\mathbf{v}_1$	$\mathbf{v}_2$
Decrease	Within limits	$\mathbf{v}_0$	$\mathbf{v}_7$	$\mathbf{v}_0$	$\mathbf{v}_7$	$\mathbf{v}_0$	$\mathbf{v}_7$
Decrease	Decrease	$\mathbf{v}_5$	$\mathbf{v}_6$	$\mathbf{v}_1$	$\mathbf{v}_2$	$\mathbf{v}_3$	$\mathbf{v}_4$

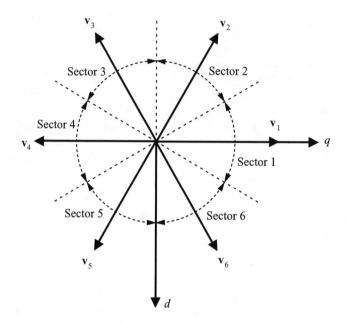

FIGURE 11.11 Sector diagram for the DTC control method.

Given the flux linkage angle, the sector number is easily found. In practice, this method is not used because of the computational burden that the trigonometric inverse places upon the controller. Practical controllers rely on the signs of the flux linkage components to determine the sector number.

One of the major disadvantages of the DTC method is the required accurate estimation of the stator flux linkage and the developed electrical torque. The stator flux linkage is estimated using Eqs. (11.33) and (11.34) with, perhaps, current feedback for correction. The torque is usually estimated using

$$T_e = \frac{3P}{4} \boldsymbol{\lambda}_s^s \times \mathbf{i}_s^s \tag{11.38}$$

There are several variants, but these are the most common ways of performing the estimates. The speed of the estimates must be quite fast, with common values for the sampling time in the neighborhood of 25 μs [15]. If the sampling time is not fast enough, excursion outside the limits imposed by the torque and flux comparators will occur.

The block diagram of the DTC induction machine drive is shown in Fig. 11.12. The error between the reference torque and the estimated torque is fed to a three-level hysteretic comparator, and the speed error is given to a two-level comparator. The outputs of the comparators are supplied to a vector lookup table that makes uses of the relationships in Table 11.2. The optimal switch states are supplied to the PWM inverter that drives the induction machine. The machine torque, stator flux vector, and stator flux sector are estimated online from the machine phase b and c voltages and currents.

Control of Induction Machine Drives

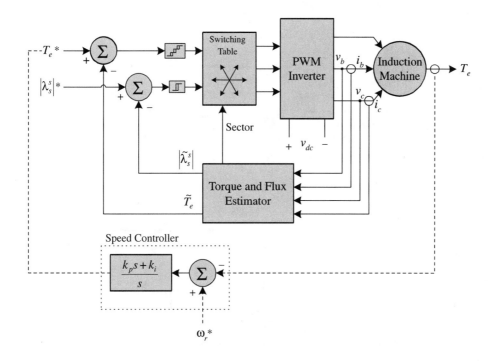

FIGURE 11.12 Block diagram of the DTC induction machine drive.

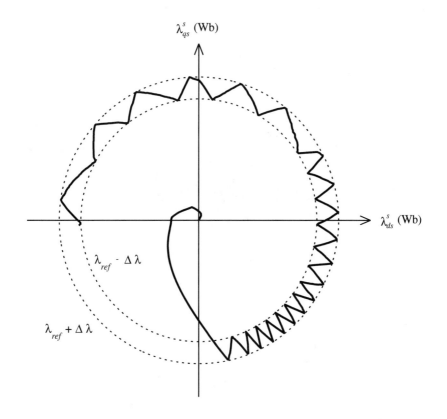

FIGURE 11.13 Stator flux linkage under DTC.

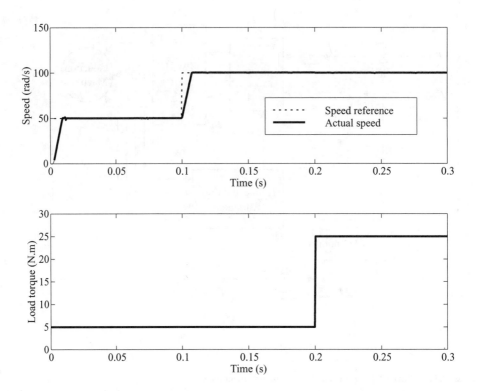

FIGURE 11.14 Speed reference, actual shaft speed, and load torque for the DTC drive example.

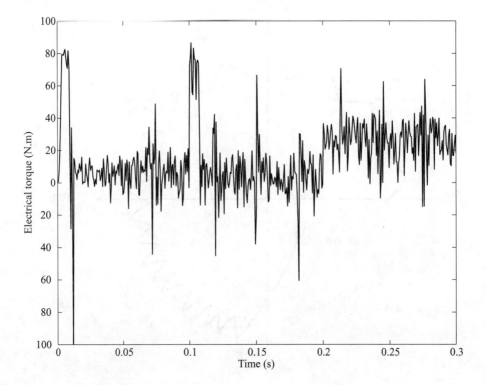

FIGURE 11.15 The electrical torque for the DTC drive.

Control of Induction Machine Drives

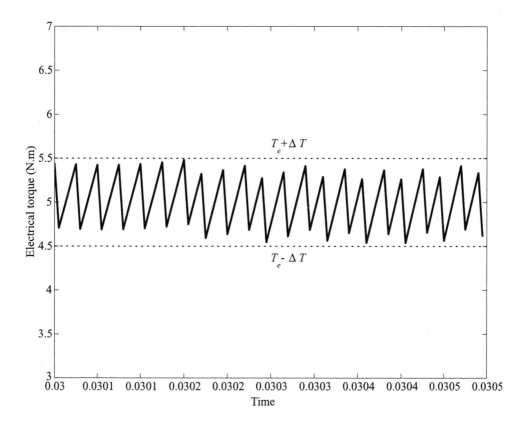

FIGURE 11.16 Electrical torque of the DTC drive.

The stator flux linkage vector is shown in Fig. 11.13 during start-up of a typical DTC drive. The flux linkage limits $\lambda_{ref} \pm \Delta\lambda$ limits of the hysteretic comparator are shown as dashed lines. The flux linkage vector circles the origin with its magnitude confined within these boundaries. The speed of the rotation of the flux linkage vector is determined by the estimated torque error of the machine. Since the machine requires knowledge of the stator flux linkage sector, start-up of the machine is not as simple as for the FOC drive. Typically, the machine is excited with a small DC current to establish the sector number needed for the controller.

An example of the operation of typical DTC drive will now be considered. The reference speed, actual speed, and load torque for the machine are displayed in Fig. 11.14. The initial speed reference is 50 rad/s, and is stepped to 100 rad/s at $t = 0.1$ s. The load torque is initially 5 N·m, and is stepped to 25 N·m at $t = 0.2$ s. Note that the drive has excellent response and disturbance rejection, typically better than that of the RFO controlled drive. The electrical torque as a function of time for the DTC drive is shown in Fig. 11.15. Due to the lower switching frequency of the drive, the torque ripple of the DTC drive is considerably greater than for the FOC drive. A close-up of the torque ripple is shown in Fig. 11.16. The minimum and maximum boundaries of the hysteretic comparator are shown as dashed lines.

11.4 Summary

The evolution of induction machine control began with the development of the scalar-controlled method allowing variable speed control. However, the scalar-controlled induction machine failed to match the dynamic performance of a comparable DC drive. The next step was the introduction of the vector-controlled methods. The goal of these methods is to make the induction machine emulate the DC machine by transforming the stator currents to a specific coordinate system where one coordinate is related to the torque production and the other to rotor flux. The FOC methods provide excellent dynamic response, matching that of the DC machine. The main disadvantage of such controls is the computational overhead required in the coordinate transformation.

The latest development in induction machine control is the DTC method. DTC does not rely on coordinate transformation, but rather controls the stator flux linkage in the stationary reference frame. Despite its control simplicity, the DTC method provides possibly the best dynamic response of any of the methods. The average switching frequency of the drive is lower as well, reducing switching loss as compared with the FOC drive. Since the control basis is the stator flux linkage, the DTC drive is capable of advanced functions such as performing "flying starts" and flux braking [15, 18]. Its main disadvantage lies in the need for accurate estimation of the machine electrical torque and stator flux linkage. At low speeds, loss of flux control can occur [20]. An additional drawback of the DTC drive is a greater torque ripple stemming from the low switching frequency.

Both drive types rely on knowledge of the machine parameters for control and observation; therefore, initial commissioning is usually required by both the FOC and DTC drives before start-up. The goal of the commissioning stage is to use low-level excitation to obtain estimates of the machine parameters. After the commissioning stage, and during normal operation, online parameter estimation is typically employed.

References

1. V. Del Toro, *Basic Electric Machines*, Prentice-Hall, Englewood Cliffs, NJ, 1990.
2. G. K. Dubey, *Power Semiconductor Controlled Drives*, Prentice-Hall, Englewood Cliffs, NJ, 1989.
3. P. C. Krause, O. Wasynczuk, and S. D. Sudhoff, *Analysis of Electric Machinery*, IEEE Press, New York, 1995.
4. D. W. Novotny and T. A. Lipo, *Vector Control and Dynamics of AC Drives*, Oxford University Press, New York, 1996.
5. P. Vas, *Sensorless Vector and Direct Torque Control*, Oxford University Press, New York, 1998.
6. T. Lipo, Recent progress in the development of solid-state AC motor drives, *IEEE Trans. Power Electron.*, 3(2), 105–117, April 1988.
7. W. Leonhard, Adjustable-speed AC drives, *Proc. IEEE*, 76(4), 455–471, April 1988.
8. G. O. Garcia, R. M. Stephan, and E. H. Watanabe, Comparing the indirect field-oriented control with a scalar method, *IEEE Trans. Ind. Appl.*, 41(2), 201–207, April 1994.
9. R. H. Park, Two-reaction theory of synchronous machines: generalized method of analysis-part I, *AIEE Trans.*, 48, 716–730, 1929.
10. H. C. Stanley, An analysis of the induction machine, *AIEE Trans.*, 57, 751–755, 1938.
11. N. P. Rubin, R. G. Harley, and G. Diana, Evaluation of various slip estimation techniques for an induction machine operating under field-oriented control conditions, *IEEE Trans. Ind. Appl.*, 28(6), 1367–1375, December 1992.
12. W. Leonhard, *Control of Electrical Drives*, Springer-Verlag, New York, 1996.
13. J. C. Moreira and T. A. Lipo, A new method for rotor time constant tuning in indirect field oriented control, in *Power Electronics Spec. Conf.*, 573–580, 1990.
14. D. M. Brod and D. W. Novotny, Current control of VSI-PWM inverters, *IEEE Trans. Ind. Appl.*, IA-21(4), 562–570, 1985.
15. P. Tiitinen, P. Pohjalainen, and J. Lalu, The next generation motor control method: direct torque control, DTC, in *Proc. EPE Chapter Symp. Electric Drive Design Appl.*, 1–7, October 1994.
16. M. Depenbrock, Direct self-control (DSC) of inverter fed induction machine, *IEEE Trans. Power Electron*, 3(4), 420–429, 1988.
17. J. Kang and S. Sul, New direct torque control of induction motor for minimum torque ripple and constant switching frequency, *IEEE Trans. Ind. Appl.*, 35(5), 1076–1081, 1999.

18. J. N. Nash, Direct torque control, induction motor vector control without an encoder, *IEEE Trans. Ind. Appl.*, 33(2), 333–341, 1997.
19. T. G. Habetler, F. Profumo, M. Pastorelli, and L. M. Tolbert, Direct torque control of induction machines using space vector modulation, *IEEE Trans. Ind. Appl.*, 28(5), 1045–1053, 1992.
20. D. Telford, M. W. Dunnigan, and B. W. Williams, A comparison of vector control and direct torque control of an induction machine, in *Power Electronics Spec. Conf.*, 1, 421–426, 2000.

12
Permanent-Magnet Synchronous Machine Drives

Patrick L. Chapman
University of Illinois at Urbana-Champaign

12.1 Introduction 12-1
12.2 Construction of PMSM Drive Systems 12-2
12.3 Simulation and Model 12-3
12.4 Controlling the PMSM 12-6
    Current-Based Drives • Voltage-Based Drives
12.5 Advanced Topics in PMSM Drives 12-9

12.1 Introduction

The permanent-magnet synchronous machine (PMSM) drive has emerged as a top competitor for a full range of motion control applications [1–3]. For example, the PMSM is widely used in machine tools, robotics, actuators, and is being considered in high-power applications such as vehicular propulsion and industrial drives. It is also becoming viable for commercial/residential applications. The PMSM is known for having high efficiency, low torque ripple, superior dynamic performance, and high power density. These drives often are the best choice for high-performance applications and are expected to see expanded use as manufacturing costs decrease. The purpose of this chapter is to introduce the PMSM and the application of power electronics technology to its control.

The PMSM is sometimes referred to as a permanent-magnet AC (PMAC) machine or simply as a PM machine. In some instances it is referred to as a brushless DC (BDC) machine because by appropriate control it can be made to have input/output characteristics much like a separately excited brush-type DC machine. It can also take on similarity with the DC machine when Hall effect sensors are utilized for position sensing, whereby electronic, instead of brush, commutation take place. The BDC machine, which is discussed in detail in Chapter 10, is a special case of the more general PMSM drive. The PMSM is a synchronous machine in the sense that it has a multiphase stator and the stator electrical frequency is directly proportional to the rotor speed in the steady state. However, it differs from a traditional synchronous machine in that it has permanent magnets in place of the field winding and otherwise has no rotor conductors.

The use of permanent magnets in the rotor facilitates efficiency, eliminates the need for slip rings, and eliminates the electrical rotor dynamics that complicate control (particularly vector control). The permanent magnets have the drawback of adding significant capital cost to the drive, although the long-term cost can be less through improved efficiency. The PMSM also has the drawback of requiring rotor position feedback by either direct means or by a suitable estimation system. Since many other high-performance drives utilize position feedback, this is not necessarily a disadvantage. Another disadvantageous aspect of the PMSM is cogging torque, which is the parasitic tendency of the rotor to align at discrete positions due to the interaction of the magnets and the stator teeth. Cogging torque is particularly troublesome at low speed, but can be virtually eliminated either by appropriate design of the machine or by electronic mitigation.

12.2 Construction of PMSM Drive Systems

As stated, the PMSM consists of a multiphase stator and a rotor with permanent magnets. The machines can have either radially or axially oriented flux. Some common radial-flux rotor configurations are depicted in Fig. 12.1. The magnets can be either mounted on the rotor surface (Fig. 12.1a) or buried in the rotor iron (Fig. 12.1b). The surface-mounted variety is popular because of the simplicity of construction and control, and virtual absence of reluctance torque since the stator inductance is essentially independent of rotor position. The buried magnet (or "interior magnet") variety of rotors has significant reluctance torque due to position-variant stator inductance that complicates analysis and control issues. However, the magnetic saliency can be used advantageously for operation above base speed. There are also variations of the stator design that are possible, particularly in regard to slot skewing and tooth shape. There is a wide variety of motor designs, each of which has its own performance and cost considerations.

There are several different magnet materials that are commonly used. Ferrite is an inexpensive but less magnetically powerful material that is frequently used. The rare earth magnets neodymium-iron-boron (NdFeB) and samarium-cobalt (SmCo) magnets are stronger magnetically and more resistant to temperature. SmCo magnets are particularly resistant to temperature but are comparably very expensive. Sintered NdFeB magnets have a stronger residual field and lower cost than SmCo magnets, but are less temperature resistant. Bonded NdFeB magnets are not quite as strong as SmCo, but are less expensive and are more easily shaped. Ferrite magnets are very common for lower-performance motors. Both radial and parallel magnetization are commonly used, depending on application. The particular choice of magnets and other design factors is important, but does not directly influence the basic principles of power converter control.

The multiphase stator is much like the stator of any other AC machine. Frequently, the slot design is distinctive in that measures are taken to reduce cogging torque. Use of tooth "shoes" and slot skewing is prevalent. Although distributed windings are common, lumped windings are also used when it is desired to have an approximately "trapezoidal" back emf. Advances are being made in the area of slotless (i.e., "toothless") PMSM design as well [4].

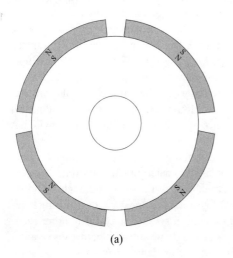 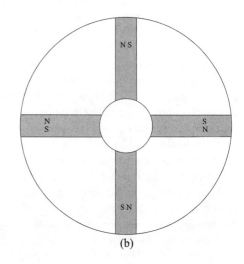

(a)   (b)

FIGURE 12.1 Typical PMSM rotor configurations: (a) surface-mount; (b) buried.

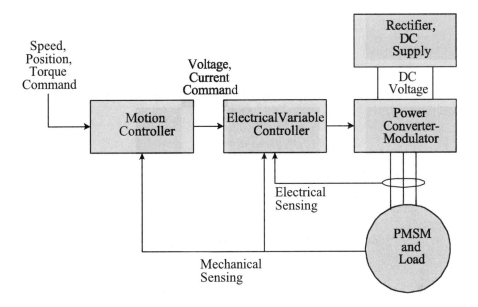

FIGURE 12.2 Diagram of conceptual drive system.

The aspects of motor construction that most significantly influence power converter design are the shape of the back emf, the cogging torque, the magnetic saliency (surface-mount or buried magnets), and the power requirements. Any of the standard inverter topologies discussed in Chapters 5 and 6 can be used to drive the machine. A conceptual drive system is pictured in Fig. 12.2. There, a speed, position, or torque command is input to the drive system. The motion controller implements feedback control based on mechanical sensors (or estimators). The controller outputs commands for the electrical variables to obey. The electrical control block converts its input commands into commands for the power converter/modulator block and sometimes utilizes feedback of voltage or current. The power converter block imposes the desired electrical signals onto the PMSM machine with the connected load.

12.3 Simulation and Model

When designing a PMSM drive, it is useful to compose a computer simulation before building a prototype. Such a model can also be used to develop the control. A suitable model of the PMSM is set forth in this section. Much of the detail of development is omitted because it is not the purpose of this chapter to provide derivations, but simply to provide the reader with useful formulas for designing the electronics for PMSM drive systems. Full development of PMSM drive models is available from a number of references [1, 2].

If there are N phases, then there are N stator voltages, currents, and flux linkages. Let the set of stator voltages be represented compactly as

$$\mathbf{v} = [v_1 \quad v_2 \quad \cdots \quad v_N]^T \tag{12.1}$$

where v_x is the voltage across the xth phase. The same relationship holds for the vectors of current ($\mathbf{i}$), and flux linkage (λ). For the special and common case of three-phase machines, the letters a, b, and c are used in place of 1, 2, and 3, respectively, in Eq. (12.1).

Since eddy current and hysteresis losses are generally small, it will suffice to attribute all stator losses to the winding resistance, r. Then, applying Faraday's and Ohm's laws, the stator voltage equation may be written as

$$\mathbf{v} = r\mathbf{i} + \frac{d}{dt}\lambda \quad (12.2)$$

Regarding the machine as balanced, symmetrical, and magnetically linear, the flux linkage equation may be written as

$$\lambda = \mathbf{L}\mathbf{i} + \lambda_{pm} \quad (12.3)$$

where $\mathbf{L}$ is a symmetric $N \times N$ matrix of the appropriate self- and mutual inductances and λ_{pm} is an $N \times 1$ vector of stator flux linkages due to the permanent magnet. The inductance matrix is constant for machines with surface-mounted magnets, but has rotor position–dependent terms for machines with buried magnets.

The torque equation can be derived from coenergy relationships:

$$T_e = \frac{P}{2}\frac{\partial}{\partial \theta_r}\left(\frac{1}{2}\mathbf{i}^T \mathbf{L}\mathbf{i} + \mathbf{i}^T \lambda_{pm}\right) + T_{cog} \quad (12.4)$$

where θ_r is the electrical rotor position in radians, and P is the number of poles. Mechanical rotor position is $\theta_{rm} = 2\theta_r/P$. The cogging torque is represented as T_{cog}.

Equations (12.2) to (12.4) represent a simulation model of the machine, provided that the resistance, r, the inductance matrix, $\mathbf{L}$, the cogging torque, T_{cog}, and the permanent magnet flux linkage vector, λ_{pm}, are known. The parameters can be determined from direct measurement or by calculation from motor geometry (i.e., finite-element analysis). The mechanical dynamics of the system, which are not discussed here since they can widely vary, must be simulated to determine position and speed.

The model set forth is general for any number of phases and for the buried or surface-mounted magnet cases. For a surface-mounted magnet machine, the air gap is effectively very wide and uniform since the magnet material has a relative permeability near 1. This results in stator inductance, which is generally not dependent upon rotor position. In both the surface-mount and buried magnet cases, λ_{pm} is a function of rotor position. Therefore, the torque equation for the surface-mounted case is

$$T_{e(SM)} = \frac{P}{2}\mathbf{i}^T \frac{\partial}{\partial \theta_r}\lambda_{pm} + T_{cog} \quad (12.5)$$

and the torque equation for a machine with buried magnets is

$$T_{e(BM)} = \frac{P}{2}\mathbf{i}^T \left(\frac{1}{2}\left(\frac{\partial}{\partial \theta_r}\mathbf{L}\right)\mathbf{i} + \frac{\partial}{\partial \theta_r}\lambda_{pm}\right) + T_{cog} \quad (12.6)$$

where it is noted that the current, $\mathbf{i}$, is not explicitly dependent on rotor position.

The cogging torque may be represented as

$$T_{cog} = \sum_{z \in Z} T_q^z \cos(zN_t \theta_r) + T_d^z \sin(zN_t \theta_r) \quad (12.7)$$

where Z is the set of natural numbers such that the Fourier series constants T_q^z and T_d^z are negligible and the constant, N_t, is the number of stator teeth. The cogging torque is frequently ignored in designing the motor drive electronics or it is sufficiently negligible because of special machine design efforts. If cogging torque is neglected, then the constants T_q^z and T_d^z are zero.

The power into the machine is simply the sum of the power into each phase:

$$P_{in} = \mathbf{v}^T \mathbf{i} \tag{12.8}$$

and the power output of the machine is

$$P_{out} = T_e \omega_{rm} \tag{12.9}$$

where ω_{rm} is the mechanical rotor speed. In Eq. (12.9), the frictional and windage dynamics are assumed to be negligible or to be accounted for in the mechanical system model.

As a common special case of the model in Eq. (12.9), the analysis is restricted to three-phase machines ($N = 3$). Frequently, the back emf of the machine has negligible harmonics, and thus it can be treated as if it is purely sinusoidal. As is common in buried magnet machine analysis, the rotor position variance of the stator inductance can be taken as sinusoidal. Furthermore, the cogging torque can be made small by utilizing certain design techniques. With these assumptions, a transformation of machine variables into the rotor reference frame can be made that facilitates vector control of the PMSM.

If the back emf is sinusoidal, then the flux linkage due the permanent magnets is as well. That is, λ_{pm} may be expressed as

$$\boldsymbol{\lambda}_{pm} = \lambda_m \left[\sin(\theta_r) \quad \sin\left(\theta_r - \frac{2\pi}{3}\right) \quad \sin\left(\theta_r + \frac{2\pi}{3}\right) \right]^T \tag{12.10}$$

where λ_m is a constant equal to the peak strength of the flux linkage due to the magnets. Note that Eq. (12.10) implies a certain interpretation of the measured rotor position. Specifically, it implies that the magnet flux linking the first phase is zero when $\theta_r = 0$. Then, the back emf due to the permanent magnets may be stated as

$$\mathbf{e}_{pm} = \omega_r \lambda_m \left[\cos(\theta_r) \quad \cos\left(\theta_r - \frac{2\pi}{3}\right) \quad \cos\left(\theta_r + \frac{2\pi}{3}\right) \right]^T \tag{12.11}$$

where ω_r is the electrical rotor speed and equals $P/2$ times its mechanical counterpart, ω_{rm}. Equation (12.11) is a useful expression for determining the constant λ_m experimentally.

The rotor position–dependent terms can be eliminated by transforming the variables into a reference frame fixed in the rotor. Only the results of this long process are given here. The transformation is applied as

$$\mathbf{v}_{qd0} = \mathbf{K}\mathbf{v} \tag{12.12}$$

where

$$\mathbf{v}_{qd0} = [v_q \quad v_d \quad v_0]^T \tag{12.13}$$

and [Ref. 1]

$$\mathbf{K} = \frac{2}{3} \begin{bmatrix} \cos(\theta_r) & \cos\left(\theta_r - \frac{2\pi}{3}\right) & \cos\left(\theta_r + \frac{2\pi}{3}\right) \\ \sin(\theta_r) & \sin\left(\theta_r - \frac{2\pi}{3}\right) & \sin\left(\theta_r + \frac{2\pi}{3}\right) \\ \frac{1}{2} & \frac{1}{2} & \frac{1}{2} \end{bmatrix} \tag{12.14}$$

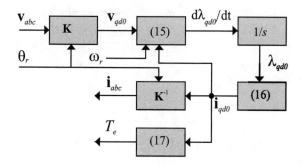

FIGURE 12.3 Diagram of simulation model.

Similar relationships hold for flux linkage and current by replacing v with λ and i, respectively.

After transforming the equations into the rotor reference frame with Eq. (12.14), the following relationships hold:

$$v_q = ri_q + \omega_r \lambda_d + \frac{d}{dt}\lambda_q$$

$$v_d = ri_d - \omega_r \lambda_q + \frac{d}{dt}\lambda_d \qquad (12.15)$$

$$v_0 = ri_0 + \frac{d}{dt}\lambda_0$$

$$\lambda_q = L_q i_q$$
$$\lambda_d = L_d i_d + \lambda_m \qquad (12.16)$$
$$\lambda_0 = L_0 i_0$$

$$T_e = \frac{3}{2}\frac{P}{2}\lambda_m i_q + (L_d - L_q)i_d i_q \qquad (12.17)$$

where L_q, L_d, and L_0 are constant inductances, which are obtainable from measurement from finite-element analysis. In general, the inductance values are not very high, since the flux due to stator current is predominantly leakage, due to the effectively wide air gap.

It is important to note that for the surface-mounted machine, $L_d = L_q$ because the stator inductances are independent of rotor position. This eliminates the second term in Eq. (12.17). For buried magnet motors, by the convention given $L_q > L_d$ in general because the high-permeability paths are aligned with the q-axis. Also, it is common to wye-connect the three phases. In that case, $i_0 = 0$ and therefore both $\lambda_0 = 0$ and $v_0 = 0$ thereby eliminating the zero-sequence components completely. Under constant torque and speed conditions, each current, flux linkage, and voltage are also constant in Eqs. (12.15) to (12.17). Equations (12.15) to (12.17) represent a standard model for a PMSM with sinusoidal flux linkages for either the surface-mounted or buried magnet cases. A diagram of the simulation structure is shown in Fig. 12.3, where numbers in parentheses refer to equation numbers.

12.4 Controlling the PMSM

There are many ways to use power converters to fix the voltage or current in a PMSM drive to desired set points. Chapter 7 details the modulation methods for inverter control. Among them are voltage control methods (such as six-step, sinusoidal, and space-vector modulation) and current control methods (such as hysteresis and delta modulation). There are also methods that augment the basic modulation strategies

Permanent-Magnet Synchronous Machine Drives

to ensure that the desired voltage or current is reached in the steady state (such as a synchronous current regulator). Because these topics are covered elsewhere in the book, the details are omitted in this chapter. In this section, the appropriate voltage and current commands are discussed. These commands become the inputs to the power electronics converter with the given modulation strategy. The specific topology of the inverter also can vary widely, such as the standard fully controlled bridge, H-bridge, ARCP converters, multilevel converters, etc.

Current-Based Drives

Current-based PMSM drives are typically used in motion control systems where a commanded torque is generated. It is not attempted in this chapter to develop the control loops associated with the mechanical apparatus. Rather, the focus is on the electrical aspects of the drive with specific emphasis on developing the appropriate torque. In particular, the current command can be derived directly from the torque equation.

Let the commanded torque be designated as T_e^*. By using the machine equations that were transformed into the rotor reference frame, other useful formulas can be developed. For example, the surface-mounted three-phase machine current command is simply

$$i_q^* = \frac{T_e^*}{\frac{3}{2}\frac{P}{2}\lambda_m} \tag{12.18}$$

where the actual phase currents would then be calculated from applying the inverse transformation of Eq. (12.14) to the vector $[i_q^* \ 0 \ 0]^T$. Note that for a constant torque command, i_q^* is constant. This relationship can also be used for the buried magnet case, but is not optimal for efficiency. If machines with nonsinusoidal back emf are considered, then the reader is referred to Refs. 5 and 6.

The buried magnet case is significantly more complicated. However, the current command that minimizes stator rms current for a given torque is

$$i_q^* = \frac{T_e^*}{\frac{3}{2}\frac{P}{2}\lambda_m + (L_d - L_q)i_d^*} \tag{12.19}$$

where the commanded d-axis current is obtain from solving

$$i_d^* \left(\frac{3}{2}\frac{P}{2}\lambda_m + (L_d - L_q)i_d^*\right)^2 + T_e^*(L_d - L_q) = 0 \tag{12.20}$$

numerically. The result could be stored as a table vs. torque command for application purposes.

As the speed of the machine increases, so does the back emf. At some speed there is no longer enough voltage available for the inverter to supply the current commanded by Eqs. (12.18) through (12.20). To increase the speed beyond this point, field (or "flux") weakening must be employed [7–10]. That is, the current must be adjusted to reduce the back emf and maintain the desired power. At rated torque and speed, the rms current is at its limit. Field weakening can be employed to increase the motor speed while holding the rms current constant and de-rating the torque, thereby maintaining a large constant power range (Fig. 12.4). There are a variety of methods to accomplish this task, each with its own restrictions, advantages, and complexity. The method is more effective for the buried magnet machine since the d-axis current can produce torque. As an example, the example for the qd model above is continued.

The peak phase voltage required may be expressed as

$$v_s = \sqrt{v_q^2 + v_d^2} \tag{12.21}$$

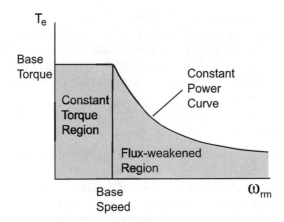

FIGURE 12.4 Torque–speed curve of a flux-weakened PMSM drive.

which cannot be in excess of the voltage that the inverter can supply.

The transformation to the rotor reference frame has the convenience that when steady state is reached, the state variables are constant. Therefore, the qd voltages can be written as

$$v_q = ri_q^* + \omega_r(L_d i_d^* + \lambda_m)$$
$$v_d = ri_d^* - \omega_r L_q i_q^* \tag{12.22}$$

from which the required phase voltage, Eq. (12.23), may be calculated. For the surface-mounted case, this constrained problem can be solved explicitly and the current commands above base speed can be written functionally. However, it cannot be solved explicitly for the buried magnet case. In general, it is perhaps most appropriate in this approach is to build a data set of current commands vs. torque and speed regardless of the structure of the machine.

This method of building the current commands relies heavily on machine parameterization. Advanced methods utilize closed-loop methods to eliminate or drastically reduce the dependence on parameters. However, they are too numerous to cover adequately here, and the reader is referred to the literature. Indeed, advanced control methods for PMSM drives is still an active research area.

Once the current commands have been developed for a given commanded torque, they can be directly issued to a current-regulated inverter. For example, either hysteresis or delta modulation could be used to achieve a good approximation to the desired current waveform. These techniques will cause the current to rise very quickly to the desired value, resulting in high-bandwidth torque control, limited only by the stator time constant and available voltage. If the command is sinusoidal, there are methods to ensure that the exact fundamental component is achieved in the steady state, such as the synchronous current regulator [11]. If nonsinusoidal current is used in case of nonsinusoidal back emf, a synchronous current regulator cast in multiple reference frames can be utilized to achieve steady-state convergence [12].

Voltage-Based Drives

It is not uncommon to utilize a voltage-based control rather than a direct current control. Voltage control PWM techniques can have preferable performance over current-based techniques like hysteresis or delta modulation. Because the torque is not an explicit function of voltage as it is for current, voltage control is not quite as straightforward.

One approach to voltage control is to begin with a torque command and synthesize a current command, as was done above. Then, use a current control loop to drive the phase currents to the appropriate values, thus achieving the desired torque. For example, if the current commands are formulated in the rotor reference frame as i_q^* and i_d^*:

$$v_q^* = \omega_r(L_d i_d + \lambda_m) + (K_p + K_i/s)(i_q^* - i_q)$$
$$v_q^* = -\omega_r L_q i_q + (K_p + K_i/s)(i_d^* - i_d)$$
(12.23)

where K_p and K_i are the proportional and integral control constants, respectively. The Laplace operator is s. Assuming that the voltage commanded is the voltage obtained, the transfer function between the commanded and actual current is

$$\frac{i_q}{i_q^*} = \frac{\frac{K_p}{L_q}\left(s + \frac{K_i}{K_p}\right)}{s^2 + \frac{r + K_p}{L_q}s + \frac{K_i}{L_q}}$$
(12.24)

where the same relationship holds if q is replaced with d. By utilizing this expression, the poles and zero of the current control can be arbitrarily set. In the steady state, the desired current, and therefore torque, is obtained. This method is common despite the inherent parameter dependence of pole placement.

Another example of PMSM control by voltage-based operation is a direct speed control. For example, the duty-cycle command given to the voltage source inverter could be formulated as

$$d = (K_p + K_i/s)(\omega_{rm}^* - \omega_{rm})$$
(12.25)

where ω_{rm}^* is the commanded speed. This control is simple, requires little in the area of sensors, and will ensure steady-state realization of the desired speed. It can, however, suffer from overmodulation and result in a comparatively sluggish control. By itself, Eq. (12.25) generally does not result in currents that obtain maximize efficiency. An additional control would need to be added to control the phase angle of the voltage.

Several variations on both current- and voltage-based control can be found in the literature. The methods presented only represent an introduction to these topics, and the reader is encouraged to investigate several other alternatives [13].

12.5 Advanced Topics in PMSM Drives

There are a large number of possible avenues for advancement of control of PMSM drives. A major research area is that of sensorless control [14–16]. This typically means that measures have been taken to eliminate the expensive rotor position sensor, but not necessarily current or voltage sensors. By employing estimation methods, a satisfactory measure of rotor position can often be obtained at minimal cost and added complexity.

Operation of the machine above base speed is of particular interest in vehicular propulsion applications. For this reason, many engineers have investigated field-weakening controls that extend the speed range of machine by effectively reducing the magnet flux [7–10]. This was briefly mentioned above, yet is an extensive topic with many challenges.

Many of the control systems set forth have parameter-dependent formulations. With time, wear, and temperature, the machine parameters can vary significantly, negatively affecting system performance. Therefore, the topic of dynamic parameter estimation is also an active research area.

Another area of active research is that of active vibration suppression. By prudent operation of the power converter, it is possible to manipulate the torque to mitigate ripple. There is also exploratory research in the area of mitigating nontorsional vibration as well. Adaptive controls, neural nets, and a host of other advanced topics in automatic control can also be applied to PMSM drives.

In short, the PMSM motor drive is a effective in high-performance drive applications. Rapid torque response is readily achieved as vector control of PMSM drives is much more straightforward than it is for induction motor drives. The capital cost can be high, but the high efficiency can at least partially offset this in high-power applications. For these reasons, the PMSM drive is expected to continue to be competitive for the near future.

References

1. P. C. Krause, O. Wasynczuk, and S. D. Sudhoff, *Analysis of Electric Machinery*, IEEE Press, Piscataway, NJ, 1996.
2. P. Pillay and R. Krishnan, Modeling of permanent magnet motor drives, *IEEE Trans. Ind. Electron.*, 35(4), 537–541, 1988.
3. B. K. Bose, Ed., *Power Electronics and Variable Frequency Drives*, IEEE Press, Piscataway, NJ, 1997, chap. 6.
4. D. C. Hanselman, *Brushless Permanent-Magnet Motor Design*, McGraw-Hill, New York, 1994.
5. P. L. Chapman, S. D. Sudhoff, and C. A. Whitcomb, Optimal current control strategies for surface-mounted permanent-magnet synchronous machine drives, *IEEE Trans. Energy Conversion*, 14(4), 1043–1050, 1999.
6. D. C. Hanselman, Minimum torque ripple, maximum efficiency excitation of brushless permanent magnet motors, *IEEE Trans. Ind. Electron.*, 41(3), 292–300, 1994.
7. S. D. Sudhoff, K. A. Corzine, and H. J. Hegner, A flux-weakening strategy for current-regulated surface-mounted permanent-magnet machine drives, *IEEE Trans. Energy Conversion*, 10(3), 431–437, 1995.
8. S. R. Macminn and T. M. Jahns, Control techniques for improved high-speed performance of interior PM synchronous motor drives, *IEEE Trans. Ind. Appl.*, 27(5), 997–1004, 1991.
9. A. K. Adnanes and T. M. Undeland, Optimum torque performance in PMSM drives above rated speed, in *Conference Record of the 1991 Industry Applications Society Annual Meeting*, 1991, 169–175.
10. M. F. Rahman, L. Zhong, and K. W. Lim, A DSP based instantaneous torque control strategy for interior permanent magnet synchronous motor drive with wide speed range and reduced torque ripples, in *Conference Record of the Thirty-First Annual Industry Applications Society Meeting*, 1996, 518–524.
11. T. M. Rowan and R. J. Kerkman, A new synchronous current regulator and an analysis of current-regulated drives, *IEEE Trans. Ind. Appl.*, IA-22(4), 678–690, 1986.
12. P. L. Chapman and S. D. Sudhoff, A multiple reference frame synchronous estimator/regulator, *IEEE Trans. Energy Conversion*, 15(2), 197–202, 2000.
13. W. Leohnard, *Control of Electric Drives*, 2nd ed., Springer-Verlag, New York, 1997.
14. K. A. Corzine and S. D. Sudhoff, Hybrid observer for high performance brushless DC motor drives, *IEEE Trans. Energy Conversion*, 11(2), 318–323, 1996.
15. D. Jung and I. Ha, Low-cost sensorless control of brushless DC motors using a frequency-independent phase shifter, *IEEE Trans. Power Electron.*, 15(4), 742–752, 2000.
16. J. P. Johnson, M. Ehsani, and Y. Guzelgunler, Review of sensorless methods for brushless DC, in *Conference Record of the 1999 Industry Applications Society Annual Meeting*, 1999, 143–150.

13
Switched Reluctance Machines

Iqbal Husain
The University of Akron

13.1 Introduction 13-1
      Advantages • Disadvantages
13.2 SRM Configuration... 13-2
13.3 Basic Principle of Operation.. 13-4
      Voltage Balance Equation • Energy Conversion • Torque
      Production • Torque–Speed Characteristics
13.4 Design 13-9
13.5 Converter Topologies.. 13-11
13.6 Control Strategies.. 13-14
      Control Parameters • Advance Angle Calculation •
      Voltage-Controlled Drive • Current-Controlled
      Drive • Advanced Control Strategies
13.7 Sensorless Control... 13-16
13.8 Applications... 13-19

13.1 Introduction

The concept of switched reluctance machines (SRMs) was established as early in 1838 by Davidson and was used to propel a locomotive on the Glasgow–Edinburgh railway near Falkirk [1]. However, the full potential of the motor could not be utilized with the mechanical switches available in those days. The advent of fast-acting power semiconductor switches revived the interest in SRMs in the 1970s when Professor Lawrenson's group established the fundamental design and operating principles of the machine [2]. The rejuvenated interest of researchers supplemented by the developments of computer-aided electromagnetic design prompted a tremendous growth in the technology over the next three decades. SRM technology is now slowly penetrating into the industry with the promise of providing an efficient drive system at a lower cost.

Advantages

The SRM possess a few unique features that makes it a vigorous competitor to existing AC and DC motors in various adjustable-speed drive and servo applications. The advantages of an SRM can be summarized as follows:

- Machine construction is simple and low-cost because of the absence of rotor winding and permanent magnets.
- There are no shoot-through faults between the DC buses in the SRM drive converter because each rotor winding is connected in series with converter switching elements.

- Bidirectional currents are not necessary, which facilitates the reduction of the number of power switches in certain applications.
- The bulk of the losses appear in the stator, which is relatively easier to cool.
- The torque–speed characteristics of the motor can be tailored to the application requirement more easily during the design stage than in the case of induction and PM machines.
- The starting torque can be very high without the problem of excessive in-rush current due to its higher self-inductance.
- The open-circuit voltage and short-circuit current at faults are zero or very small.
- The maximum permissible rotor temperature is higher, since there are no permanent magnets.
- There is a low rotor inertia and a high torque/inertia ratio.
- Extremely high speeds with a wide constant power region are possible.
- There are independent stator phases, which does not prevent drive operation in the case of loss of one or more phases.

Disadvantages

The SRM also comes with a few disadvantages among which torque ripple and acoustic noise are the most critical. The double saliency construction and the discrete nature of torque production by the independent phases lead to higher torque ripple compared with other machines. The higher torque ripple also causes the ripple current in the DC supply to be quite large, necessitating a large filter capacitor. The doubly salient structure of the SRM also causes higher acoustic noise compared with other machines. The main source of acoustic noise is the radial magnetic force induced resonant vibration with the circumferential mode shapes of the stator.

The absence of permanent magnets imposes the burden of excitation on the stator windings and converter, which increases the converter kVA requirement. Compared with PM brushless machines, the per-unit stator copper losses will be higher, reducing the efficiency and torque per ampere. However, the maximum speed at constant power is not limited by the fixed magnet flux as in the PM machine, and, hence, an extended constant power region of operation is possible in SRMs. The control can be simpler than the field-oriented control of induction machines, although for torque ripple minimization significant computations may be required for an SRM drive.

13.2 SRM Configuration

The SRM is a doubly-salient, singly-excited reluctance machine with independent phase windings on the stator, usually made of magnetic steel laminations. The rotor is a simple stack of laminations, without any windings or magnets. The cross-sectional diagrams of a four-phase, 8-6 SRM and a three-phase, 12-8 SRM are shown in Fig. 13.1. The three-phase, 12-8 machine is a two-repetition version of the basic 6-4 structure within the single stator geometry. The two-repetition machine can alternately be labeled as a 4-poles/phase machine, compared with the 6-4 structure with two poles/phase. The stator windings on diametrically opposite poles are connected either in series or in parallel to form one phase of the motor. When a stator phase is energized, the most adjacent rotor pole-pair is attracted toward the energized stator to minimize the reluctance of the magnetic path. Therefore, it is possible to develop constant torque in either direction of rotation by energizing consecutive phases in succession.

The aligned position of a phase is defined to be the situation when the stator and rotor poles of the phase are perfectly aligned with each other attaining the minimum reluctance position. The unsaturated phase inductance is maximum (L_a) in this position. The phase inductance decreases gradually as the rotor poles move away from the aligned position in either direction. When the rotor poles are symmetrically misaligned with the stator poles of a phase, the position is said to be the unaligned position. The phase has the minimum inductance (L_u) in this position. Although the concept of inductance is not valid for a highly saturating machine like SRM, the unsaturated aligned and unaligned inductances are two key reference positions for the controller.

Switched Reluctance Machines

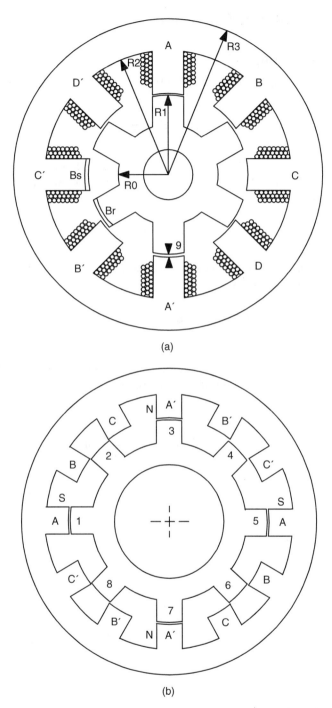

FIGURE 13.1 Cross sections of two SR machines: (a) four-phase, 8-6 structure; (b) three-phase, 12-8, two-repetition structure.

Several other combinations of the number of stator and rotor poles exist, such as 10-4, 12-8, etc. A 4-2 or a 2-2 configuration is also possible, but they have the disadvantage that, if the stator and rotor poles are aligned exactly, then it would be impossible to develop a starting torque. The configurations with higher number of stator/rotor pole combinations have less torque ripple and do not have the problem of starting torque.

13.3 Basic Principle of Operation

Voltage Balance Equation

The general equation governing the flow of stator current in one phase of an SRM can be written as

$$V_{ph} = iR + \frac{d\lambda}{dt} \tag{13.1}$$

where V_{ph} is the DC bus voltage, i is the instantaneous phase current, R is the winding resistance, and λ is the flux linking the coil. The SRM is always driven into saturation to maximize the utilization of the magnetic circuit, and, hence, the flux-linkage λ is a nonlinear function of stator current and rotor position

$$\lambda = \lambda(i, \theta) \tag{13.2}$$

The electromagnetic profile of an SRM is defined by the λ–i–θ characteristics shown in Fig. 13.2. The stator phase voltage can be expressed as

$$V_{ph} = iR + \frac{\partial \lambda}{\partial i}\frac{di}{dt} + \frac{\partial \lambda}{\partial \theta}\frac{d\theta}{dt} = iR + L_{inc}\frac{di}{dt} + k_v\omega \tag{13.3}$$

where L_{inc} is the incremental inductance, k_v is the current-dependent back-emf coefficient, and $\omega = d\theta/dt$ is the rotor angular speed. Assuming magnetic linearity (where $\lambda = L(\theta)i$), the voltage expression can be simplified as

$$V_{ph} = iR + L(\theta)\frac{di}{dt} + i\frac{dL(\theta)}{dt}\omega \tag{13.4}$$

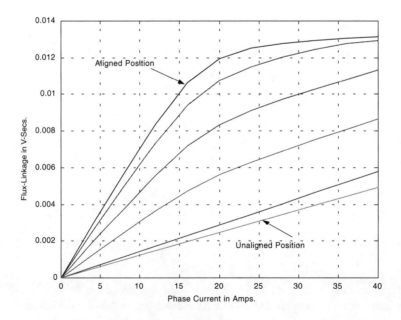

FIGURE 13.2 Flux–angle–current characteristics of a four-phase SRM.

The last term in Eq. (13.4) is the "back-emf" or "motional-emf" and has the same effect on SRM as the back-emf has on DC motors or electronically commutated motors. However, the back-emf in SRM is generated in a different way from the DC machines or ECMs where it is caused by a rotating magnetic field. In an SRM, there is no rotor field and back-emf depends on the instantaneous rate of change of phase flux linkage.

In the linear case, which is always valid for lower levels of phase current, the per phase equivalent circuit of an SRM consists of a resistance, an inductance, and a back-emf component. The back-emf vanishes when there is no phase current or when the phase inductance is constant relative to the rotor position. Depending on the magnitude of current and rotor angular position, the equivalent circuit changes its structure from being primarily an R-L circuit to primarily a back-emf dependent circuit.

Energy Conversion

The energy conversion process in an SRM can be evaluated using the power balance relationship. Multiplying Eq. (13.4) by i on both sides, the instantaneous input power can be expressed as

$$P = V_{ph}i = i^2 R + \left(Li\frac{di}{dt} + \frac{1}{2}i^2\frac{dL}{d\theta}\omega\right) + \frac{1}{2}i^2\frac{dL}{d\theta}\omega = i^2 R + \frac{d}{dt}\left(\frac{1}{2}Li^2\right) + \frac{1}{2}i^2\frac{dL}{d\theta}\omega \quad (13.5)$$

The first term represents the stator winding loss, the second term denotes the rate of change of magnetic stored energy, while the third term is the mechanical output power. The rate of change of magnetic stored energy always exceeds the electromechanical energy conversion term. The most effective use of the energy supplied is when the current is maintained constant during the positive $dL/d\theta$ slope. The magnetic stored energy is not necessarily lost, but can be retrieved by the electrical source if an appropriate converter topology is used. In the case of a linear SRM, the energy conversion effectiveness can be at most 50% as shown in the energy division diagram of Fig. 13.3a. The drawback of lower effectiveness is the increase in converter volt-amp rating for a given power conversion of the SRM. The division of input energy increases in favor of energy conversion if the motor operates under magnetic saturation. The energy division under saturation is shown in Fig. 13.3b. This is the primary reason for operating the SRM always under saturation. The term *energy ratio* instead of efficiency is often used for SRM, because of the unique situation of the energy conversion process. The energy ratio is defined as

$$ER = \frac{W}{W + R} \quad (13.6)$$

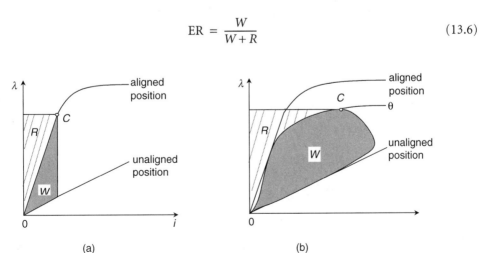

FIGURE 13.3 Energy partitioning during one complete working stroke. (a) Linear case. (b) Typical practical case. W = energy converted into mechanical work. R = energy returned to the DC supply.

where W is the energy converted into mechanical work and R is the energy returned to the source using a regenerative converter. The term energy ratio is analogous to the term power factor used for AC machines.

Torque Production

The torque is produced in the SRM by the tendency of the rotor to attain the minimum reluctance position when a stator phase is excited. The general expression for instantaneous torque for such a device that operates under the reluctance principle is

$$T_{ph}(\theta, i) = \left.\frac{\partial W'(\theta, i)}{\partial \theta}\right|_{i=\text{constant}} \qquad (13.7)$$

where W' is the coenergy defined as

$$W' = \int_0^i \lambda(\theta, i)\,di$$

Obviously, the instantaneous torque is not constant. The total instantaneous torque of the machine is given by the sum of the individual phase torques.

$$T_{inst}(\theta, i) = \sum_{\text{phases}} T_{ph}(\theta, i). \qquad (13.8)$$

The SRM electromechanical properties are defined by the static T–i–θ characteristics of a phase, an example of which is shown in Fig. 13.4. The average torque is a more important parameter from the user's point of view and can be derived mathematically by integrating Eq. (13.8).

$$T_{avg} = \frac{1}{T}\int_0^T T_{inst}\,dt \qquad (13.9)$$

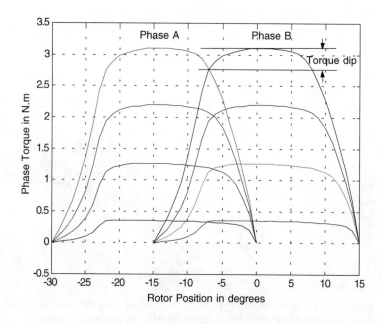

FIGURE 13.4 Torque–angle–current characteristics of a 4-phase SRM for four constant current levels.

Switched Reluctance Machines

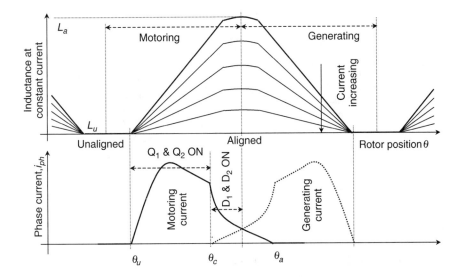

FIGURE 13.5 Phase currents for motoring and generating modes with respect to rotor position and idealized inductance profiles.

The average torque is also an important parameter during the design process.

When magnetic saturation can be neglected, the instantaneous phase torque expression becomes

$$T_{ph}(\theta, i) = \frac{1}{2} i^2 \frac{dL(\theta)}{d\theta} \qquad (13.10)$$

The linear torque expression also follows from the energy conversion term (last term) in Eq. (13.5).

The phase current needs to be synchronized with the rotor position for effective torque production. For positive or motoring torque, the phase current is switched such that rotor is moving from the unaligned position toward the aligned position. The linear SRM model is very insightful in understanding these situations. Equation (13.10) clearly shows that for motoring torque, the phase current must coincide with the rising inductance region. On the other hand, the phase current must coincide with the decreasing inductance region for braking or generating torque. The phase currents for motoring and generating modes of operation are shown in Fig. 13.5 with respect to the phase inductance profiles. The torque expression also shows that the direction of current is immaterial in torque production. The optimum performance of the drive system depends on the appropriate positioning of phase currents relative to the rotor angular position. Therefore, a rotor position transducer is essential to provide the position feedback signal to the controller.

Torque–Speed Characteristics

The torque–speed plane of an SRM drive can be divided into three regions as shown in Fig. 13.6. The constant torque region is the region below the base speed ω_b, which is defined as the highest speed when maximum rated current can be applied to the motor at rated voltage with fixed firing angles. In other words, ω_b is the lowest possible speed for the motor to operate at its rated power.

Region 1

In the low-speed region of operation, the current rises almost instantaneously after turn-on, since the back-emf is small. The current can be set at any desired level by means of regulators, such as hysteresis controller or voltage PWM controller.

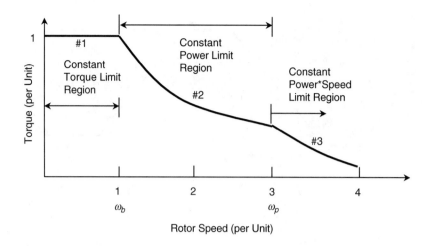

FIGURE 13.6 Torque–speed characteristics of an SRM drive.

As the motor speed increases, the back-emf soon becomes comparable to the DC bus voltage and it is necessary to phase advance-the turn-on angle so that the current can rise up to the desired level against a lower back-emf. Maximum current can still be forced into the motor by PWM or chopping control to maintain the maximum torque production. The phase excitation pulses are also needed to be turned off a certain time before the rotor passes alignment to allow the freewheeling current to decay so that no braking torque is produced.

Region 2

When the back-emf exceeds the DC bus voltage in high-speed operation, the current starts to decrease once pole overlap begins and PWM or chopping control is no longer possible. The natural characteristic of the SRM, when operated with fixed supply voltage and fixed conduction angle θ_{dwell} (also known as the dwell angle), is that the phase excitation time falls off inversely with speed and so does the current. Since the torque is roughly proportional to the square of the current, the natural torque–speed characteristic can be defined by $T \propto 1/\omega^2$. Increasing the conduction angle can increase the effective amps delivered to the phase. The torque production is maintained at a level high enough in this region by adjusting the conduction angle θ_{dwell} with the single-pulse mode of operation. The controller maintains the torque inversely proportional to the speed; hence, this region is called the constant power region. The conduction angle is increased by advancing the turn-on angle until the θ_{dwell} reaches its upper limit at speed ω_p.

The medium speed range through which constant power operation can be maintained is quite wide and very high maximum speeds can be achieved.

Region 3

The θ_{dwell} upper limit is reached when it occupies half the rotor pole-pitch, i.e., half the electrical cycle. θ_{dwell} cannot be increased further because otherwise the flux would not return to zero and the current conduction would become continuous. The torque in this region is governed by the natural characteristics, falling off as $1/\omega^2$.

The torque–speed characteristics of the SRM are similar to those of a DC series motor, which is not surprising considering that the back-emf is proportional to current, and the torque is proportional to the square of the current.

13.4 Design

The fundamental design rules governing the choice of phase numbers, pole numbers, and pole arcs were discussed in detail by Lawrenson et al. [2] and also by Miller [3]. From a designer's point of view, the objectives are to minimize the core losses, to have good starting capability, to minimize the unwanted effects due to varying flux distributions and saturation, and to eliminate mutual coupling. The choice of the number of phases and poles is open, but a number of factors need to be evaluated prior to making a selection.

The fundamental switching frequency is given by

$$f = \frac{N}{60} N_r \quad \text{Hz} \qquad (13.11)$$

where N is the motor speed in rev/m and N_r is the number of rotor poles. The "step angle" or "stroke" of an SRM is given by

$$\text{Step Angle } \varepsilon = \frac{2\pi}{N_{ph} \cdot N_{rep} \cdot N_r} \qquad (13.12)$$

The stoke angle is an important design parameter related to the frequency of control per rotor revolution. N_{rep} represents the multiplicity of the basic SRM configuration, which can also be stated as the number of pole pairs per phase. N_{ph} is the number of phases. N_{ph} and N_{rep} together set the number of stator poles.

The regular choice of the number of rotor poles in an SRM is

$$N_r = N_s \pm km \qquad (13.13)$$

where k is an integer such that $k \bmod q \neq 0$ and N_s is the number of stator poles. Some combinations of parameters allowed by Eq. (13.13) are not feasible, since sufficient space must exist between the poles for the windings. The most common choice for the selection of stator and rotor pole number for Eq. (13.13) is $km = 2$ with the negative sign.

The torque is produced during the partial overlap region between the stator and rotor pole arcs, and, hence, we must have

$$\min(\beta_r, \beta_s) > \varepsilon$$

where β_r and β_s are the rotor and stator pole arcs, respectively. Practical designs must insure that the rotor interpolar gap is greater than the stator pole arc so that the minimum possible unaligned inductance can be obtained to get the largest possible phase inductance variation between the aligned and unaligned rotor positions. The consideration leads to the second constraint:

$$\frac{2\pi}{N_r} - \beta_r > \beta_s$$

The above constraint prevents simultaneous overlap of one stator pole by two rotor poles.

The angular rate of change of phase flux can be doubled by doubling N_{rep} (while other parameters are held constant with all the coils of each phase connected in series), since this does not affect the maximum and minimum inductances. However, the torque remains unaffected, since the number of turns needs to be halved to keep the back-emf the same when N_{rep} is doubled. Further consideration of the rate of change of flux linkage, available coil area, saliency ratio, split ratio (ratio between rotor radius and motor outside radius), variation in the magnetic circuit reluctance, saturation behavior, and the iron loss due to the increase of the repetition modifies this simplistic conclusion. The advantages of increasing N_{rep}

are greater fault-tolerance and shorter flux-paths leading to lower core losses compared to the single-repetition machines. The contribution to the mean torque can also be increased in multiple repetition machines if the pole width is made more than 50% of that for a single-repetition machine. For $N_{rep} = 2$, the stator pole width needs to be approximately 70% of that of the single-repetition machine with the optimization criterion of maximizing the co-energy both under high and low current conditions [5]. This gives about 40% more thermally limited torque and horizontal force for the same copper loss and total volume.

The highest possible saliency ratio (the ratio between the maximum and minimum unsaturated inductance levels) is desired to achieve the highest possible torque per ampere, but as the rotor and stator pole arcs are decreased the torque ripple tends to increase. The torque ripple adversely affects the dynamic performance of an SRM drive. For many applications, it is desirable to minimize the torque ripple, which can be partially achieved through appropriate design. The torque dip observed in the T–i–θ characteristics of an SRM (see Fig. 13.4) is an indirect measure of the torque ripple expected in the drive system. The torque dip is the difference between the peak torque of a phase and the torque at an angle where two overlapping phases produce equal torque at equal levels of current. The smaller the torque dip, the less will be the torque ripple. The T–i–θ characteristics of the SRM depend on the stator-rotor pole overlap angle, pole geometry, material properties, number of poles and number of phases. A design trade-off needs to be considered to achieve the desired goals. The T–i–θ characteristics must be studied through finite element analysis during the design stage to evaluate both the peak torque and torque dip values.

Increasing the number of strokes per revolution can alleviate the problem of torque dips and hence the torque ripple. One way of achieving this is with a larger N_r, but the method has an associated penalty in the saliency ratio [2, 3]. The decrease in saliency ratio with increasing N_r will increase the controller volt-amps and decrease the torque output. The higher switching frequency will also increase the core losses. Increasing the phase numbers with a much smaller penalty in the saliency ratio is a better approach for reducing the torque dips. The average torque of the machine will also increase because of the smaller torque dips. The higher number of phases will increase the overlap between phase torques in the regions of commutation. The torque ripple can then be minimized through a controller algorithm that profiles the overlapping phase currents of adjacent phases during commutation. The SRMs with three or lower number of phases suffer more from the problem of torque dips near the commutation region. The four- or five-phase machines can deliver uniform torque without exceedingly boosting the current in rotor positions of low phase-torque per ampere. The cost and complexity of the drive increases with higher phase numbers, since additional switches are required for the power converter. In general, two- or three-phase machines are used in high-speed applications, while four-phase machines are chosen where torque-ripple is a concern.

The inductance overlap ratio K_L, which is the ratio of unsaturated inductance overlap of two adjacent phases to the angle over which the inductance is changing [1, 4], can be utilized during the design phase to analyze the torque characteristics. The ratio gives a direct measure of the torque overlap of adjacent phases. The higher the K_L, the lower will be the torque dip and the higher will be the mean torque as well. Mathematically, the inductance ratio is defined as

$$K_L = 1 - \frac{\varepsilon}{\min(\beta_s, \beta_r)} \tag{13.14}$$

The torque overlap can be increased by widening the stator and rotor poles. Figures 13.7 and 13.8 plot K_L vs. β_s (assuming $\beta_s \leq \beta_r$) and K_L vs. N_{ph} (assuming β_s for $N_{rep} = 2$ to be 70% of that of β_s for $N_{rep} = 1$). Figure 13.7 shows that high values of K_L are achievable at relatively low values of β_s for a machine with more phases and/or repetitions. The same machines have better starting capabilities. Also, the rate of change of K_L with respect to β_s is much higher near the minimum possible values for β_s. Additionally, a relatively higher stator pole width will reduce the available window area for winding and increase the copper losses. Therefore, the β_s should not be increased significantly from its minimum possible value. Figure 13.8 shows that the improvement on the problem of torque dip is noticeable in the lower range of N_{ph}.

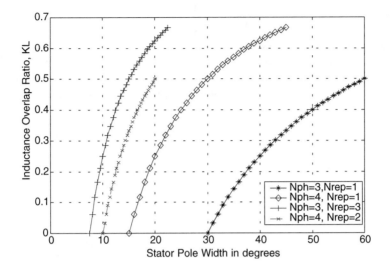

FIGURE 13.7 Inductance overlap ratio vs. stator pole width.

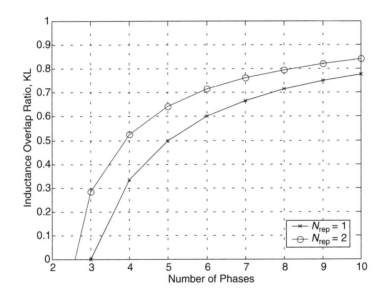

FIGURE 13.8 Inductance overlap ratio vs. number of phases.

13.5 Converter Topologies

The torque developed in an SRM is independent of the direction of current flow. Therefore, unipolar converters are sufficient to serve as the power converter circuit for the SRM, unlike induction motors or synchronous motors, which require bidirectional currents. This unique feature of the SR motor, together with the fact that the stator phases are electrically isolated from one another, has generated a wide variety of power circuit configurations. The type of converter required for a particular SRM drive is intimately related to motor construction and the number of phases. The choice also depends on the specific application.

The most flexible and the most versatile four-quadrant SRM converter is the bridge converter shown in Fig. 13.9a, which requires two switches and two diodes per phase. The switches and the diodes must be rated to withstand the supply voltage plus the factor of safety. During the magnetization period, both the switches are turned on and the energy is transferred from the source to the motor. Chopping or PWM, if necessary, can be accomplished by switching either or both the switches during the conduction period according to the control strategy. At commutation both switches are turned off and the motor phase is quickly defluxed through the freewheeling diodes. The main advantage of this converter is the independent control of each phase, which is particularly important when phase overlap is desired. The only

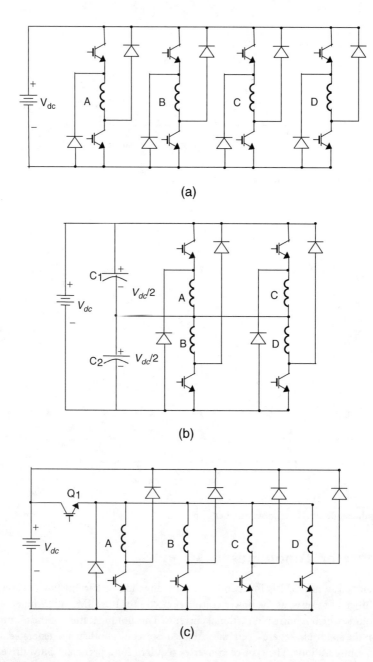

FIGURE 13.9 Converter topologies for SRM: (a) classic bridge power converter; (b) split-capacitor converter; (c) Miller converter.

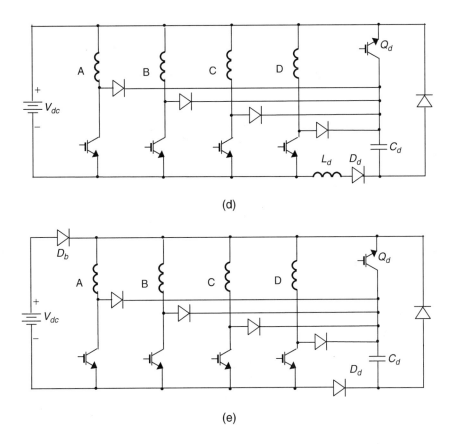

FIGURE 13.9 (Continued) Converter topologies for SRM: (d) energy-efficient converter I; and (e) energy-efficient converter II.

disadvantage is the requirement of two switches per phase. This converter is especially suitable for high-voltage, high-power drives.

The split-capacitor converter shown in Fig. 13.9b has only one switch per phase but requires a split DC supply [6]. The phases are energized through either the upper or the lower DC bus rail and the midpoint of the two capacitors. Therefore, only one half the DC bus voltage can be applied for torque production. To maintain power flow balance between the two supply capacitors, the switching device and the freewheeling diode are transposed for each phase winding, which means that the motor must have an even number of phases. Also, the power devices must be rated to withstand the full DC supply voltage.

The high cost of power semiconductor devices and the need to reduce the switching losses motivated researchers to develop converters with a reduced number of switches. The initial attempt was to solve the problem from the motor side, and the outcome was the bifilar topology, which requires the motor to be wound with a bifilar winding [6]. The complexity in adding the extra winding and the associated increase in motor losses made the topology unpopular.

In low-speed applications, where PWM current control is desirable over the entire range of operation, the bridge converter can be reduced to the circuit shown in Fig. 13.9c developed by Miller. In this converter, chopping is performed by one switch common to all phases. The circuit requires $(n + 1)$ switches for an n-phase motor. The main limitation of this circuit is that at higher speeds the off-going phase cannot be deenergized fast enough because the control switch Q_1 keeps turning on intermittently, disabling forced demagnetization. A class of power converter circuits with fewer than two switches per phase for SR motors having four or more phases has been developed by Pollock and Williams [7].

The energy-efficient C-dump converters I and II are two regenerative converter topologies with a reduced number of switches [8]. The converters are shown in Figs. 13.9d and e. The topologies were derived from the C-dump converter proposed earlier by Ehsani et al. [9]. The energy efficient converter topologies eliminate all the disadvantages of the C-dump converter without sacrificing its attractive features, and also provide some additional advantages. The attractive features of the converters are: lower number of power devices, full regenerative capability, freewheeling in chopping or PWM mode, simple control strategy, and faster demagnetization during commutation. Energy-efficient C-dump converter I has only one-switch forward voltage drop, whereas energy-efficient C-dump converter II has one-switch plus one diode forward voltage drop in the phase magnetization paths. The dump component energy requirements are much lower in these converters compared with those of the C-dump converter.

Converters with a reduced number of switches are typically less fault tolerant compared with the bridge converter. The ability to survive component or motor phase failure should be a prime consideration for high-reliability applications. On the other hand, in low-voltage applications, the voltage drop in two switches can be a significant percentage of the total bus voltage, which may not be affordable. Other factors to be considered in selecting a drive circuit include cost, complexity in control, number of passive components, number of floating drivers required, etc. The drive converter must be chosen to serve the particular needs of an application.

13.6 Control Strategies

Appropriate positioning of the phase excitation pulses relative to the rotor position is the key to obtaining effective performance from an SRM drive system. The turn-on time, the total conduction period, and the magnitude of the phase current determine torque, efficiency, and other performance parameters. The type of control to be employed depends on the operating speed of the SRM.

Control Parameters

The control parameters for an SRM drive are the turn-on angle (θ_{on}), turn-off angle (θ_{off}), and the phase current. The conduction angle is defined as $\theta_{dwell} = \theta_{off} - \theta_{on}$. The complexity of determination of the control parameters depends on the chosen control method for a particular application. The current command can be generated for one or more phases depending on the controller. In voltage-controlled drives, the current is indirectly regulated by controlling the phase voltage.

At low speeds, the current rises almost instantaneously after turn-on because of the negligible back-emf, and the current must be limited by either controlling the average voltage or regulating the current level. The type of control used has a marked effect on the performance of the drive. As the speed increases, the back-emf increases as explained before and opposes the applied bus voltage. Phase advancing is necessary to establish the phase current at the onset of rotor and stator pole overlap region. Voltage PWM or chopping control is used to force maximum current into the motor to maintain the desired torque level. Also, the phase excitation is turned off early enough so that the phase current decays completely to zero before the negative torque-producing region is reached.

At higher-speeds, the SRM enters the single-pulse mode of operation, and control is achieved by advancing the turn-on angle and adjusting the conduction angle. At very high speeds, the back-emf will exceed the applied bus voltage once the current magnitude is high and the rotor position is appropriate, which causes the current to decrease after reaching a peak even though a positive bus voltage is applied during the positive $dL/d\theta$. The control algorithm outputs θ_{dwell} and θ_{on} according to speed. At the end of θ_{dwell} the phase switches are turned off so that negative voltage is applied across the phase to commutate the phase as quickly as possible. The back-emf reverses polarity beyond the aligned position and may cause the current to increase in this region if the current does not decay to insignificant levels. Therefore, the phase commutation must precede the aligned position by several degrees so that the current decays before the negative $dL/d\theta$ region is reached.

In the high-speed range of operation, when the back-emf exceeds the DC bus voltage, the conduction window becomes too limited for current or voltage control and all the chopping or PWM has to be disabled. In this range, θ_{dwell} and θ_{adv} are the only control parameters and control is accomplished based on the assumption that approximately θ_{dwell} regulates torque and θ_{adv} determines efficiency.

Advance Angle Calculation

Ideally, the turn-on angle is advanced such that the reference current level i^* is reached just at the onset of pole overlap. In the unaligned position, phase inductance is almost constant, and, hence, during turn-on back-emf can be neglected. Also, assuming that the resistive drop is small, Eq. (13.4) can be written as

$$V_{ph} = L(\theta) \frac{\Delta i}{\Delta \theta} \omega \qquad (13.15)$$

Now, $\Delta i = i^*$ and $\Delta\theta = \theta_{overlap} - \theta_{on} = \theta_{adv}$, where $\theta_{overlap}$ is the position where pole overlap begins, θ_{on} is the turn-on angle and θ_{adv} is the required phase turn-on advance angle. Therefore, we have

$$\theta_{adv} = L_u \omega \frac{i^*}{V_{dc}} \qquad (13.16)$$

MacMinn and Sember [10] have described a more sophisticated method of controlling the advance angle, which accounts for the errors due to neglecting the back-emf and the resistive drop in the calculation of θ_{adv}.

Voltage-Controlled Drive

In low-performance drives, where precise torque control is not a critical issue, fixed-frequency PWM voltage control with variable duty cycle provides the simplest means of control of the SRM drive. The early proponents of SRM drives were driven by the fact that a highly efficient variable-speed drive having a wide speed range can be achieved with this motor by optimum use of the simple voltage feeding mode with closed loop position control only. The block diagram of the voltage-controlled drive is shown in Fig. 13.10. The angle controller generates the turn-on and turn-off angles for a phase depending on the rotor speed, which simultaneously determines the conduction period, θ_{dwell}. The duty cycle is adjusted according to the voltage command signal. The electronic commutator generates the gating signals based on the control inputs and the instantaneous rotor position. A speed feedback loop can be added on the outside as shown when precision speed control is desired. The drive usually incorporates a current sensor typically placed on the lower leg of the DC-link for overcurrent protection. A current feedback loop can also be added, which will further modulate the duty cycle and compound the torque–speed characteristics just like the armature voltage control of a DC motor.

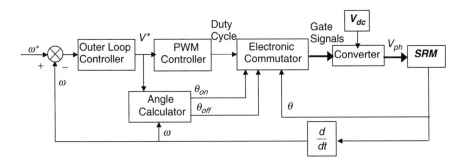

FIGURE 13.10 Voltage-controlled drive.

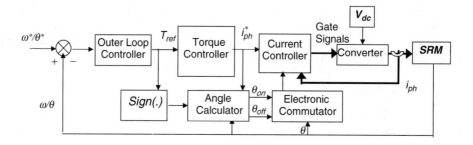

FIGURE 13.11 Current-controlled drive.

Current-Controlled Drive

In torque-controlled drives, such as in high-performance servo applications, the torque command is executed by regulating the current in the inner loop as shown in Fig. 13.11. The reference current i^* for a given operating point is determined from the load characteristics, the speed, and the control strategy. A wide-bandwidth current transducer provides the current feedback information to the controller from each of the motor phases. This mode of control allows rapid resetting of the current level and is used where fast motor response is desired. For loads whose torque increases monotonically with speed such as in fans or blowers, speed feedback can be introduced in the outer loop for accurate speed control.

The simpler control strategy is to generate one current command to be used by all the phases in succession. The electronic commutator (Fig. 13.11) selects the appropriate phase for current regulation based on θ_{on}, θ_{off}, and the instantaneous rotor position. The current controller generates the gating signal for the phases based on the information coming from the electronic commutator. The current in the commuated phase is quickly brought to zero applying negative V_{dc}, while the incoming phase assumes the responsibility of torque production based on the commanded current. The phase transition in these drives is not very smooth, which tends to increase the torque ripple of the drive.

Advanced Control Strategies

A higher performance index, such as torque/ampere maximization, efficiency maximization, or torque ripple minimization can be required in certain applications. For example, in direct drive or traction applications, the efficiency over a wide speed range is critical. For such applications as electric power steering in automobiles, the torque ripple is a critical issue. Typically, the torque/ampere maximization will go hand in hand with efficiency maximization, whereas torque ripple minimization will require the sacrifice of efficiency up to a certain extent.

The high-performance drives will typically be current-controlled drives with sophistication added to the controller as discussed earlier. For efficiency maximization, the key issue is the accurate determination of θ_{on} and θ_{off}, which may require modeling of the SRM and online parameter identification [11]. The modeling issue is equally important for torque ripple minimization, where the overlapping phase currents are carefully controlled during commutation [12, 13]. In these sophisticated drives, the electronic commutator works in conjunction with the torque controller to generate the gating signals. The torque controller will include either a model or tables describing the characteristics of the SRM.

13.7 Sensorless Control

A motor can be a very good sensor of the motion when its voltages and currents possess sufficient information to determine its position and velocity. Compared with other motors, a better position and motion estimator for an SRM is possible because of its double saliency construction. The position estimation is possible even at zero speed, since its inductance/flux varies in accordance with the rotor position. All the published methods of sensorless control of SRM exploit this fundamental idea to extract rotor position information by measuring either inductance or flux. A wide variety of methods have been proposed to eliminate the rotor position transducer over the past decade, which evolved with their own merits and demerits.

The various methods of indirect-position sensing reported so far in the literature fall under one of the following divisions:

- Phase current monitoring–based estimation [14].
- Phase flux measurement–based position estimation [15].
- Observer-based position estimation (Luenberger observer, sliding-mode observer, Kalman filter, etc.) [16, 17].
- Position estimation by pulse injection into an idle phase [18–21].
- Mutual voltage sensing–based system [22].
- Artificial intelligence–based position estimation (artificial neural network, fuzzy logic, etc.) [23, 24].

The methods of indirect position sensing can be broadly classified into two categories: (1) Nonintrusive methods, where position information is obtained from terminal measurements of voltages and currents and associated computations, and (2) intrusive (or active probing) methods, where low-level, high-frequency signals are injected into an idle phase to determine the position. Further classification of the broad categories is shown in Fig. 13.12.

Nonintrusive methods: The nonintrusive methods rely on the machine characteristics for estimating the rotor position. The terminal measurements of phase voltage or mutual voltage and current are used as inputs for an estimator to obtain the rotor position. The features of the nonintrusive methods are:

- Based on machine characteristics.
- Flux of jth phase is obtained from the integration of flux-inducing voltage,

$$\lambda_j = (v_j - i_j r_j) dt \qquad (13.17)$$

- The position information is obtained from flux and phase current information using the machine characteristics:

$$\lambda(i, \theta) \Rightarrow \theta$$

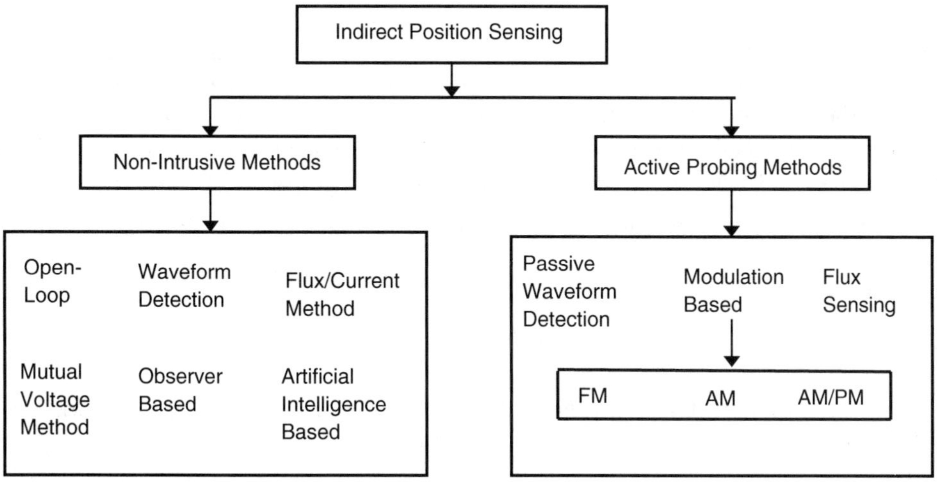

FIGURE 13.12 Classification of indirect rotor-position sensing methods for SRM.

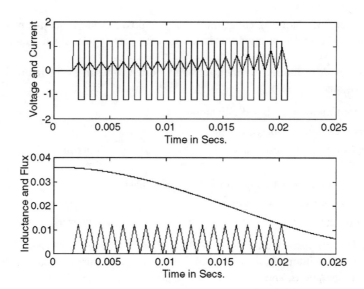

FIGURE 13.13 Rotor position sensing period in a diagnostic phase: (Top trace) applied voltage/20 and phase current; (bottom trace) phase inductance and flux.

- Advantages:
    - —Only terminal measurements of phase voltages and currents required.
    - —No inherent speed limitation.
    - —No diagnostic pulse or extra circuitry needed.
    - —Ease of implementation.
- Disadvantages:
    - —Mathematical computation may impose some burden on the microprocessor or DSP involved.
    - —Performance and accuracy depend highly on the machine model and the performance of the digital integrator.

Intrusive or active probing methods: In active probing methods, an idle or unexcited phase is injected with high-frequency diagnostic signals to obtain the unsaturated phase inductance characteristics, which are related to the rotor position information. The simplicity of these methods is a definite advantage, although inherent speed limitation and generation of negative torque in the sensing phases could be a drawback in some cases. Figure 13.13 shows the sample diagnostic pulses and their variation with rotor position for the active probing method. The appropriate idle phase available at lower to medium speeds is chosen as the diagnostic phase. In the case of a four-phase SRM, the phase diagonally opposite to the active phase has the highest sensitivity of phase current variation with rotor position, and, hence, is chosen as the diagnostic phase. The features of the active probing methods are:

- The position-dependent inductance profile is obtained from measured values of diagnostic pulses as

$$L(\theta) = \frac{V \cdot \Delta t}{\Delta i} \qquad (13.18)$$

- The position information is obtained either from a table or simply through comparison with a threshold value.

$$\theta = f^{-1}\{L(\theta)\} \qquad (13.19)$$

- Advantages:
  —Well suited for low-speed application.
  —Diagnostic pulse can be injected using the same power converter.
  —Self-starting can be implemented using the same probing pulses.
- Disadvantages:
  —Inherent speed limitations due to the pulse injection.
  —Unnecessary negative torque generation.
  —Needs extra circuitry in some cases, which adds cost and complexity to the overall drive system.

Continuous rotor position information can be obtained from indirect position sensing schemes by a mapping of inductance, flux, or current waveforms to rotor position. Alternatively, the task can be simplified in less sophisticated algorithms by threshold comparison of the indirectly measured position information to effectuate commutation. Phase advancing and retardation is possible by changing the threshold level appropriately.

13.8 Applications

The simple motor structure and inexpensive power electronic requirement have made the SRM an attractive alternative to both AC and DC machines in adjustable-speed drives. An example of SRM application is in heating, ventilation, and air conditioning (HVAC) systems. Here, the high-speed capability, low cost, ruggedness, and fault tolerance features of the SRM drive can produce low-cost adjustable-speed HVAC products for the consumer market. The appliance industry can also benefit from the low-cost and ruggedness of SRM drives. The Maytag Neptune washer uses a three-phase SRM drive for its high-end products. Ametek-Lamb Electric is also commercially manufacturing a heavy-duty vacuum/blower using SRM drives.

SRM drives have great potential for use within the various aspects of conventional automobiles. In this application, the SRM drive can be very simple, inexpensive, compatible with the low battery voltage in the car, very rugged, and fault-tolerant. The absence of temperature sensitivity in the harsh environment under the hood is a definite plus for SRM drives in comparison with permanent magnet motor drives. Example applications of SRMs within an automobile are for the electric power steering and antilock braking systems. The SRM drive is also a strong candidate for the main propulsion drive of an electric or hybrid vehicle. The wide constant power range of SRM drives is especially suitable for such applications. Electric Motorbike, Inc. has developed the Lectra motorcycle based on an SRM drive.

The SRM is also suitable for many industrial and manufacturing applications. For example, high-speed adjustable-speed pumping of fluids for a variety of petrochemical, food processing, and other applications can be done with large horsepower SRM drives. The fact that the SRM can operate under partial converter and motor failure means that these pumps can be placed in critical links in the process control loop. Other high-technology applications for which the SRM drives are being considered include robotic prime movers, aerospace starter/generators, fuel pumps, and servo systems.

Although sophisticated controllers are being designed for SRM drives, it must be realized that the simple PWM voltage control provides a highly efficient variable-speed drive that can be operated over a wide speed range. This mode of operation is usually adequate for a number of motion control applications.

References

1. J. V. Byrne et al., A high performance variable reluctance motor drive: a new brushless servo, in *Motorcon Proc.*, 1985, 147–160.
2. P. J. Lawrenson, J. M. Stephenson, P. T. Blenkinsop, J. Corda, and N. N. Fulton, Variable-speed switched reluctance motors, *IEE Proc. B*, 127(4), 253–265, 1980.

3. T. J. E. Miller, *Switched Reluctance Motors and Their Control*, Magna Physics Publishing, Hillsboro, OH; and Oxford Science Publications, Oxford, U.K., 1993.
4. M. N. Anwar, I. Husain, and A. V. Radun, A comprehensive design methodology for switched reluctance machines, in *IEEE-IAS Annual Conf.*, Rome, Italy, Oct. 2000.
5. H. C. Lovatt and J. M. Stephenson, Influence of the number of poles per phase in switched reluctance motors, *IEE Proc. B*, 139(4), 307–314, 1992.
6. R. M. Davis, W. F. Ray, and R. J. Blake, Inverter drive for switched reluctance motor: circuits and component ratings, in *IEE Proc.*, 128B(2), 126–136, 1981.
7. C. Pollock and B. W. Williams, Power converter circuits for switched reluctance motors with the minimum number of switches, *IEE Proc.*, 137B(6), 373–384, 1990.
8. S. Mir, I. Husain, and M. Elbuluk, Energy-efficient C-dump converters for switched reluctance motors, *IEEE Trans. Power Electronics*, 12(5), 912–921, 1997.
9. M. Ehsani, J. T. Bass, T. J. E. Miller, and R. L. Steigerwald, Development of a unipolar converter for variable reluctance motor drives, *IEEE Trans. Ind. Appl.*, IA-23(3), 545–553, 1987.
10. S. R. MacMinn and J. W. Sember, Control of a switched-reluctance aircraft engine starter-generator over a very wide speed range, in *IECEC Conf. Rec.*, Aug. 1989, 631–638.
11. S. Mir, I. Husain, and M. Elbuluk, Switched reluctance motor modeling with on-line parameter adaptation, *IEEE Trans. Ind. Appl.*, 34(4), 776–783, 1998.
12. K. Russa, I. Husain, and M. Elbuluk, Torque ripple minimization in switched reluctance machines over a wide speed range, *IEEE Trans. Ind. Appl.*, 34(5), 1105–1112, 1998.
13. S. Mir, M. Elbuluk, and I. Husain, Torque ripple minimization in switched reluctance motors using an adaptive fuzzy control, *IEEE Trans. Ind. Appl.*, 35(2), 461–468, 1999.
14. P. P. Acarnley, R. J. Hill, and C. W. Hooper, Detection of rotor position in stepping and switched reluctance motors by monitoring of current waveforms, *IEEE Trans. Ind. Electron.*, IE-32(3), 215–222, 1985.
15. J. P. Lyons, S. R. MacMinn, and M. A. Preston, Flux/Current methods for SRM rotor position estimation, in *IEEE-IAS Annu. Meet.*, Dearborn, MI, 1991, 482–487.
16. A. Lumsdaine and J. H. Lang, State observers for variable-reluctance motors, *IEEE Trans. Ind. Electron.*, IE-37(2), 133–142, 1990.
17. I. Husain, S. Sodhi, and M. Ehsani, Sliding mode observer based control for switched reluctance motors, in *IEEE-IAS Conf. Rec. 94*, Denver, CO, 1994, 635–643.
18. W. D. Harris and J. H. Lang, A simple motion estimator for variable-reluctance motors, *IEEE Trans. Ind. Appl.*, 26(2), 237–243, 1990.
19. S. R. MacMinn, W. J. Rzesos, P. M. Szczesny, and T. M. Jahns, Application of sensor integration techniques to switched reluctance motor drives, *IEEE Trans. Ind. Appl.*, 28(6), 1339–1344, 1992.
20. M. Ehsani, I. Husain, and A. Kulkarni, Elimination of discrete position sensor and current sensor in switched reluctance motor drives, *IEEE Trans. Ind. Appl.*, 28(1), 128–135, 1992.
21. M. Ehsani, I. Husain, S. Mahajan, and K. R. Ramani, New modulation techniques for rotor position sensing in switched reluctance motors, *IEEE Trans. Ind. Appl.*, 30(1), 85–91, 1994.
22. I. Husain and M. Ehsani, Rotor position sensing in switched reluctance motor drives by measuring mutually induced voltages, *IEEE Trans. Ind. Appl.*, 30(3), 665–672, 1994.
23. E. Mese and D. A. Torrey, Sensorless position estimation for variable-reluctance machines using artificial neural network, *IEEE-IAS Annu. Meeting Proc.*, 1, 540–547, 1997.
24. A. D. Cheok and N. Ertugal, Use off fuzzy logic for modeling, estimation, and prediction in switched reluctance motor drives, *IEEE Trans. Ind. Electron.*, 46(6), 1207–1224, 1999.

14
Step Motor Drives

Ronald H. Brown
Marquette University

14.1 Introduction 14-1
14.2 Types and Operation of Step Motors 14-2
   Variable Reluctance Step Motor • Drive Circuits for Variable Reluctance Step Motors • Permanent Magnet (Can-Stack) Step Motor • Hybrid Step Motor • Drive Circuitry for Permanent Magnet and Hybrid Step Motors
14.3 Step Motor Models 14-8
   Variable Reluctance Step Motor Model • Bifilar-Wound Hybrid Step Motor Model • Drive Circuit Modeling
14.4 Control of Step Motors 14-12
   Excitation of Step Motors • Open-Loop Control

14.1 Introduction

Step motors are used in many low-lost positioning applications due to their inherent ability to stop at discrete positions and follow position vs. time profiles while being controlled open loop. A step motor is a synchronous machine, but historically has been used almost exclusively in positioning and position tracking applications. Recently, however, some types of step motors have been applied in variable speed drives applications.

Step motors can be driven without feedback to stop at discrete angular positions, known as detent positions. The number of detent positions can be as low as 12 steps or detent positions per revolution to 400 or 500 or more steps per revolution. The location accuracy of the detent positions varies typically within 5% of the step size. The repeatability of the motor is also high, in that the rotor can start on one position, move away to other positions, and then return to within typically 3% of step size of the original position.

The motors with relatively few steps are typically used in higher-speed applications, where the motors with many steps per revolution are often used in high-torque, low-speed, direct drive applications or applications where many repeatable discrete positions are required. Step motors have been successfully applied in many applications such as computer peripherals (e.g., disk drives, pen plotters, and printers), office machines (e.g., copiers, scanners), automotive (e.g., seat positioning, speed control), aerospace (e.g., flap control, starter-generators), and industrial (e.g., robots, scanners, machine tools), to name a few.

Many step motor drives are driven with digital pulses, thus it is easy to interface and control step motors from computers or microcontrollers without digital to analog circuitry. For control purposes, the step motor and drive can be thought of as a digital to angular position converter.

It is perhaps easier to understand how a step motor works than any other rotating machine. However, the mathematical models of step motors are nonlinear, since the inductance and torque vary sinusoidally with position. The non-linear nature of the models requires that the engineer carefully design the controllers for the motors. In the following section the three types of step motors are discussed along with the operation and drive circuits of each. The mathematical models of each type are discussed in the second section. In the third section, the control of step motors is presented.

14.2 Types and Operation of Step Motors

The three common types of step motors are the variable reluctance, the permanent magnet, and the hybrid step motors. The variable reluctance and the hybrid step motors are double salient structures, i.e., teeth on both the stator and the rotor, with multiple windings or phases on the stator and no windings on the rotor (thus brushless machines). The variable reluctance step motors can have three, four, or even five phases, while the permanent magnet and the hybrid step motors usually have two phases.

Principle of Operation: A first understanding of the principle of operation can be easily seen from the variable reluctance step motor, but is common for all types. When a single winding or phase is energized, the motor generates a torque in the direction to align the rotor teeth with the teeth of the energized phase. The torque generated by current in a single phase, for example, phase A, is

$$T_e = -k_T i_A \sin(N_r \theta) \qquad (14.1)$$

where T_e is the generated torque, k_T is the torque constant, i_A is the current in phase A, N_r is number of electrical cycles per mechanical revolution, and θ is the mechanical rotor position. The cross-sectional view of a variable reluctance step motor is illustrated in Fig. 14.1, showing only the phase A winding. The rotor is shown in the aligned position, which is the detent position. The generated torque by the current in phase A vs. rotor position for this motor is shown in Fig. 14.2. If a load on the rotor were to displace the rotor in the position θ or clockwise direction, the generated torque would act in the direction to realign the rotor. From Fig. 14.2, we see that a displacement in the positive θ direction generates negative torque, which will push the rotor in the negative direction, thus trying to restore the detent position. If a load on the rotor were to displace the rotor in the negative θ or counterclockwise direction, the generated torque would also act in the direction to realign the rotor. From Fig. 14.2, we see that a displacement in the negative θ direction generates positive torque, which in turn is in the direction to restore the rotor to the detent position.

Now, suppose the current in phase A is deenergized and phase B is energized. It is apparent from Fig. 14.1 that the rotor will rotate 15° in the positive direction so that rotor teeth will align with the phase B stator teeth. Next, suppose that phase B is deenergized and phase C is energized. The rotor will rotate 15° additionally in the positive direction, aligning phase C stator teeth with rotor teeth. One more phase switching, deenergizing phase C and energizing phase A causes the rotor to rotate yet an additional

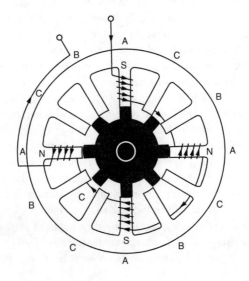

FIGURE 14.1 Cross-sectional view.

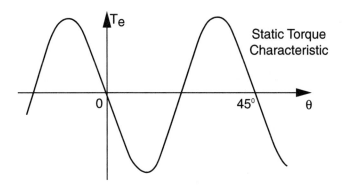

FIGURE 14.2 Static torque characteristic.

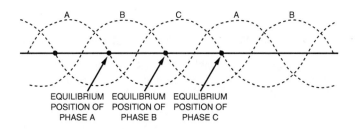

FIGURE 14.3 Static torque characteristics for all three phases.

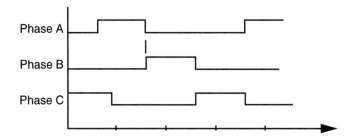

FIGURE 14.4 By exciting the motor windings during the positive portions of their torque curve, the motor can be made to produce nonzero average torque.

15° in the positive direction. Now the rotor has moved a total of 45° in the positive direction. This detent position is at 45° in Fig. 14.2. Detent positions occur where the torque vs. position curve crosses zero torque with negative slope.

The torque vs. position for all three phases of the motor (energized one at a time) are shown in Fig. 14.3. By exciting the motor phases in order during intervals of positive torque, as shown in Fig. 14.4, the motor can be made to run in the positive direction. Conversely, by exciting the motor phases in reverse order during intervals of negative torque, the motor can be made to run in the opposite direction.

Variable Reluctance Step Motor

As mentioned above, the variable reluctance step motors can have three, four, or five phases. The mode of operation discussed above is common to all variable reluctance step motors, and will not be repeated here. The motor discussed above is known as a 12/8 variable reluctance step motor, in that the stator has 12 teeth and the rotor has 8 teeth. This motor takes 15° steps and has 24 steps per revolution. The number of detent positions per revolution can be as low as 12 for a 6/4 motor and as high as 200 or 400 steps

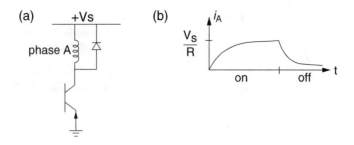

FIGURE 14.5 Drive circuit and phase current for one phase.

per revolution. The size of the motor can range from small fractional hp, with detent torque in the few ounce-inch range to 10 hp to more. The larger variable reluctance step motors are more commonly called switch reluctance motors and are usually used in variable speed applications.

Drive Circuits for Variable Reluctance Step Motors

The drive circuits are quite simple for the variable reluctance step motor. Figure 14.5a shows the simplest drive circuit for one phase (each phase requires its own identical drive circuit). The transistor acts as a switch, either off or on. When the switch is on, the current flows from the supply, through the phase winding, through the switch, to ground. When the switch is off, the current in the winding cannot drop to zero instantaneously due to the winding inductance. A path for the decay current is provided through the diode. The voltage across the winding when the switch is on is the supply voltage, V_s (neglecting the voltage drop across the switch). The current in the phase takes time to reach the value of V_s/R, where R is the phase resistance due to the phase inductance, L. Neglecting the back EMF, the phase current is approximately:

$$i_A \approx \frac{V_s}{R}(1 - e^{-(R/L)t}) \tag{14.2}$$

as shown in Fig. 14.5b. When the switch is off, the diode forms a short-circuit and the current flows through the diode, thus the voltage across the winding is zero (neglecting the voltage drop across the diode). Thus the current decays to zero with the L/R time constant. Once the current decays to zero, neglecting leakages, the winding is open circuited and no current flows.

The drive circuit in Fig. 14.6a can be used for higher-performance operation of the variable reluctance step motor. In this circuit, both transistors act as switches. When both switches are on, the current flows from the supply, through the top switch, through the phase winding, through the bottom switch, to ground. When both switches are off, the current in the winding cannot drop to zero instantaneously due to the winding inductance, so the current flows from ground through the lower diode, through the winding, through the top diode, and back into the supply. The voltage across the winding when both switches are on is the supply voltage (neglecting the voltage drop across the switches). When the switches are off, the voltage across the winding is the negative of the supply voltage (neglecting the voltage drop across the diodes). Thus the current decays toward $-V_s/R$ with the L/R time constant. Once the current decay reaches zero the diodes block and, neglecting leakages, the winding is open circuited and no current flows. The current decay time is much less with this circuit than with the circuit in Fig. 14.5. The current waveform of this operation of this circuit is shown in Fig. 14.6b.

Even greater performance of the variable reluctance step motor can be achieved with the circuit in Fig. 14.6a. V_s is set five to ten times larger than the motor rated voltage and the current is controlled by "chopping" one of the switches on and off. When the phase is energized, the current rises with the L/R time constant towards V_s/R as before, but now V_s/R is five to ten times larger than rated current, so the current reaches rated current much sooner. At this time, one of the two switches is then turned off,

Step Motor Drives

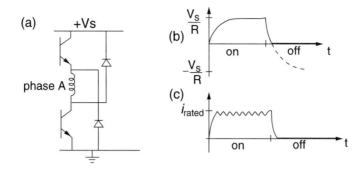

FIGURE 14.6 Drive circuit and current waveforms for higher-performance operation.

allowing the current to decay toward zero. A short time later, the switch is turned on again until current reaches rated current again. This process is repeated until the phase is to be deenergized. The current waveforms for this operation of the circuit is shown in Fig. 14.6c.

Permanent Magnet (Can-Stack) Step Motor

The permanent magnet step motor has a smooth, permanent magnet rotor. The rotor is constructed to have many pairs of magnetic poles. The windings are not wrapped around poles as in the variable reluctance motor, but around the circumference of the air gap. The stator poles are wrapped around the windings to form north and south magnetic poles to attract and repel the magnetic poles on the rotor. Two sets of windings and stator poles are required, with each set of stator poles offset by half a tooth pitch. This motor usually has a low number of steps or detent positions per revolution, and detent positions are less accurate than the other types of motors. The motor does have an unenergized detent torque.

The permanent magnet step motor has two phases, but can be wound in two different ways. If the motor is wound unifilar, that is, one winding per phase, bidirectional currents are required for proper operation. With bifilar windings, that is, two windings or a center tapped winding per phase, unidirectional currents can be used to run the motor.

Hybrid Step Motor

The hybrid step motor can be described as two two-phase (unidirectional) variable reluctance step motors put together with an axially mounted permanent magnet between the rotors. The magnetic flux paths are three-dimensional, aligned axially between the rotor halves and radially in the air gaps. The possible winding configurations are similar to the permanent magnet type motor, either unifilar, requiring bidirectional currents, or bifilar, requiring only unidirectional currents. As with variable reluctance motors, the number of steps per revolution typically range from 24 to 400. As with permanent magnet motors, there is some unenergized detent torque.

Drive Circuitry for Permanent Magnet and Hybrid Step Motors

The drive circuits for the permanent magnet and hybrid step motors are different from the drive circuits for the variable reluctance step motor. The drive circuit also depends on if the motor is wound unifilar or bifilar. Bifilar wound motors require fewer drive circuit components than for the unifilar wound motors, but at most only half the phase winding is energized at one time.

Figures 14.7 and 14.8 show partial drive circuits for unifilar wound step motors. The circuits shown are for only half or one winding of the motor. A second identical circuit is needed for the other motor winding. The drive circuit in Fig. 14.7, known as a half-H-bridge, requires both positive and negative voltage supplies. The transistors act as switches, connecting one end of the phase winding to either the

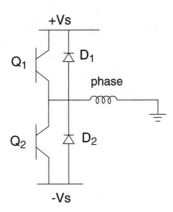

FIGURE 14.7 Half-H-bridge drive circuit.

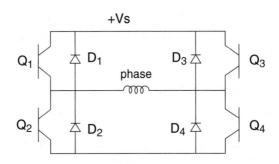

FIGURE 14.8 H-bridge drive circuit.

positive supply voltage, $+V_s$, or the negative supply voltage, $-V_s$. When switch Q_1 is on, the current flows from $+V_s$ through the switch, through the phase winding to ground. When switch Q_1 is off, the current cannot instantaneously drop to zero due to the winding inductance, thus the decay current flows through diode D_2 from $-V_s$ to the winding to ground. Once the current decays to zero, the diode blocks the current and, neglecting leakage, the winding is open circuited. When Q_1 is on, the phase voltage is $+V_s$ (neglecting the voltage drop across the switch). When Q_1 is off, the phase voltage is $-V_s$ (neglecting the voltage drop across the diode) until the current decays to zero, then the phase is open circuited. Similarly, when switch Q_2 is on, the current flows from ground through the phase winding in the opposite direction as before, through the switch to $-V_s$. When switch Q_2 is off the decay current flows through diode D_1 to $+V_s$ from the winding from ground.

The drive circuit in Fig. 14.8, known as an H-bridge, requires only a positive supply voltage. The four transistors act as switches, connecting each end of the phase winding to either the positive supply voltage, $+V_s$, or ground. When switches Q_1 and Q_4 are on, the current flows from $+V_s$ through Q_1, through the phase winding through Q_4 to ground, applying $+V_s$ across the winding. When switches Q_1 and Q_4 are off, the current in the winding cannot instantaneously drop to zero due to the winding inductance, thus the decay current flows through diodes D_2 and D_3, applying $-V_s$ across the winding until the current decays to zero, when the diode blocks the current and, neglecting leakage, the winding is open circuited. Similarly, when switches Q_2 and Q_3 are on, the current flows from $+V_s$ through Q_3, through the phase winding in the opposite direction as before, through Q_2 to ground, applying $-V_s$ across the winding. When switches Q_2 and Q_3 are off, the decay current flows through diodes D_1 and D_4, applying $+V_s$ across the winding until the current decays to zero, when the diode blocks the current and, neglecting leakage, the winding is open circuited.

Higher motor performance can be achieved from the circuit in Fig. 14.8 if the supply voltage is set at five to ten times the motor rated voltage and the phase currents are regulated by chopping either switch

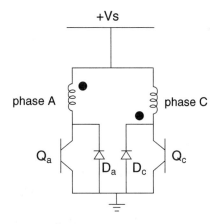

FIGURE 14.9 Inverse diode clamped drive circuit.

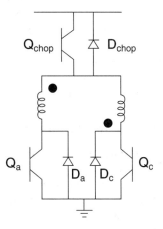

FIGURE 14.10 Drive circuit of Fig. 14.9 as a chopper drive.

Q_1 or Q_3. For example, when Q_1 and Q_4 are on, the phase current rises toward V_s/R with the L/R time constant, where R is the winding resistance and L is the winding inductance. When the phase current reaches rated current, Q_1 is off and the circuit path is through D_2, the phase winding, and Q_4, thus the applied voltage is zero (neglecting the diode and transistor voltage drops), and the current now starts to decay toward zero. A short time later, Q_1 is turned on again and the current builds towards rated current. Once the current reaches rated current, the cycle is repeated until the phase is to be deenergized.

Figures 14.9 and 14.10 show drive circuits for bifilar wound step motors. Both of these circuits are known as inverse diode clamped drive circuits. The circuits shown assume the center tapped winding configuration and are for only one bifilar winding of the motor. A second identical circuit is needed for the other motor winding. It is best to think of a bifilar wound step motor as having four phases, A, B, C, and D, but unlike with the variable reluctance step motor, phases A and C are inversely mutually coupled and phases B and D are inversely mutually coupled.

The drive circuit in Fig. 14.9 assumes that the supply voltage is set to motor rated voltage. The transistors in the circuit act as switches. When Q_a is on, the current flows from the supply through the phase A, through Q_a to ground. When Q_a is off, and since the current in the bifilar winding cannot decay to zero instantaneously, the mutual coupling in the bifilar winding couples the current to phase C, where the current flows from ground up through the diode, D_c, through phase C, into the supply. This applies $-V_s$ across the phase C, causing the current to decay quickly.

The drive circuit in Fig. 14.10 works the same way as the drive circuit in Fig. 14.9 when Q_{chop} is on. The addition of Q_{chop} and the additional diode allows the inverse diode clamp drive circuit to be a chopper drive. The supply voltage is set to five to ten times the motor rated voltage and when either leg of the circuit is on, the current is regulated using Q_{chop}. When Q_{chop} is off, the phase current drops to half of its original value, half of the conducting current couples to the opposite phase, and the current flows up through the clamp diode in the opposite phase, backward through the opposite phase, through the on phase, and through the on phase transistor.

14.3 Step Motor Models

When a constant current is passed through one phase of a step motor, the motor generates a torque. This torque is typically a sinusoidal function of rotor displacement from the detent position that causes the rotor to minimize this displacement. When the phases of the motor are excited so that the motor "runs," the generated torque is still a function of position and current, but the current becomes a varying quantity, dependent on time, position, velocity, and of course, the drive circuit and drive scheme. Selection of a motor, drive circuit, and drive scheme depends on predicting the performance and the dynamic torque–speed characteristics of a particular motor with a drive circuit and drive scheme. These performances of step motors can be predicted to within reasonable accuracy using mathematical models for both the motor and drive circuit. Ways to model the motor and drive circuit are presented in this section. As with modeling most physical systems, more accurate models produce more accurate results. Tradeoffs between accuracy and simplicity are also discussed with each model.

The variable reluctance step motor model needs to be modeled separately from the permanent magnet and hybrid step motor models, and separate models are needed for unifilar and bifilar windings.

Variable Reluctance Step Motor Model

Precise mathematical modeling of variable reluctance step motors requires knowledge of both the geometry of the machine and of the ferromagnetic material characteristics. These requirements are often relaxed and assumptions are made to simplify the model to a set of nonlinear differential equations.

Assumption 1. The ferromagnetic material does not saturate. This is a poor assumption for variable reluctance step motors in that the motors are usually operated with a high degree of saturation. This assumption is replaced after the "non-saturated" model is presented.

Assumption 2. The inductance for each phase varies sinusoidally around the circumference of the air gap, for example, the phase A inductance is $L_A(\theta) = L_0 + L_1 \cos(N_r\theta)$. This assumption required Assumption 1, otherwise L_A is a function of both θ and i_A.

The terminal voltage for phase A can be found using Faraday's law as:

$$V_A = R_A i_A + \frac{d\lambda_A}{dt} \tag{14.3}$$

where V_A is the terminal voltage, R_A is the winding resistance, i_A is the winding current, and λ_A is the phase flux linkages. Since $\lambda_A = L_A i_A$:

$$\frac{d\lambda_A}{dt} = L_A \frac{di_A}{dt} + i_A \frac{dL_A}{dt} = L_A \frac{di_A}{dt} - L_1 i_A \omega N_r \sin(N_r\theta) \tag{14.4}$$

where the first term is the magnetizing voltage and the second term is the speed voltage. Equation (14.6) can be rewritten as

$$\frac{di_A}{dt} = \frac{1}{L_A} V_A - \frac{R_A i_A}{L_A} + \frac{L_1 N_r}{L_A} i_A \omega \sin(N_r\theta) \tag{14.5}$$

Step Motor Drives

The differential equations for the remaining phases are the same as the above equations, replacing the subscripts with the appropriate phase letter. The inductances for the other phases, however, need to be shifted in position. For a three-phase motor, the inductances are

$$L_B(\theta) = L_0 + L_1 \cos\left(N_r\theta - \frac{2\pi}{3}\right)$$

$$L_C(\theta) = L_0 + L_1 \cos\left(N_r\theta - \frac{4\pi}{3}\right)$$
(14.6)

For a four-phase motor, the inductances are

$$L_B(\theta) = L_0 + L_1 \cos\left(N_r\theta - \frac{\pi}{2}\right)$$

$$L_C(\theta) = L_0 + L_1 \cos(N_r\theta - \pi)$$

$$L_D(\theta) = L_0 + L_1 \cos\left(N_r\theta - \frac{3\pi}{2}\right)$$
(14.7)

The mechanical equations can be found from Newton's law and conservation of energy. Newton's law states

$$J\frac{d\omega}{dt} = T_e - T_L - B\omega$$
(14.8)

where J is the rotor and load moment of inertia, ω is the rotor velocity in mechanical radians per second, T_e is the torque generated by the motor, T_L is the load torque, and B is the rotor and load viscous friction coefficient. Using conservation of energy, the torque generated by i_A for the variable reluctance step motor assuming no magnetic saturation is

$$T_A = -\frac{L_1 N_r}{2} \sin(N_r\theta) i_A^2$$
(14.9)

where T_A is the torque generated by the current in phase A.

Summarizing, the differential equations for the three-phase variable reluctance step motor are

$$\frac{di_A}{dt} = \frac{1}{L_A}V_A - \frac{R_A}{L_A}i_A + \frac{L_1 N_r}{L_A}i_A\omega \sin(N_r\theta)$$

$$\frac{di_B}{dt} = \frac{1}{L_B}V_B - \frac{R_B}{L_B}i_B + \frac{L_1 N_r}{L_B}i_B\omega \sin\left(N_r\theta - \frac{2\pi}{3}\right)$$

$$\frac{di_C}{dt} = \frac{1}{L_C}V_C - \frac{R_C}{L_C}i_C + \frac{L_1 N_r}{L_C}i_C\omega \sin\left(N_r\theta - \frac{4\pi}{3}\right)$$
(14.10)

$$\frac{d\omega}{dt} = \frac{T_e}{J} - \frac{T_L}{J} - \frac{B}{J}\omega$$

$$\frac{d\theta}{dt} = \omega$$

where

$$T_e = -\frac{L_1 N_r}{2}\left[\sin(N_r\theta)i_A^2 + \sin\left(N_r\theta - \frac{2\pi}{3}\right)i_B^2\right.$$
$$\left. + \sin\left(N_r\theta - \frac{4\pi}{3}\right)i_C^2\right] \quad (14.11)$$

The differential equations for the four-phase variable reluctance step motor are

$$\frac{di_A}{dt} = \frac{1}{L_A}V_A - \frac{R_A}{L_A}i_A + \frac{L_1 N_r}{L_A}i_A\omega \sin(N_r\theta)$$

$$\frac{di_B}{dt} = \frac{1}{L_B}V_B - \frac{R_B}{L_B}i_B + \frac{L_1 N_r}{L_B}i_B\omega \sin\left(N_r\theta - \frac{\pi}{2}\right)$$

$$\frac{di_C}{dt} = \frac{1}{L_C}V_C - \frac{R_C}{L_C}i_C + \frac{L_1 N_r}{L_C}i_C\omega \sin(N_r\theta - \pi) \quad (14.12)$$

$$\frac{di_D}{dt} = \frac{1}{L_D}V_D - \frac{R_D}{L_D}i_D + \frac{L_1 N_r}{L_D}i_D\omega \sin\left(N_r\theta - \frac{3\pi}{4}\right)$$

$$\frac{d\omega}{dt} = \frac{T_e}{J} - \frac{T_L}{J} - \frac{B}{J}\omega$$

$$\frac{d\theta}{dt} = \omega$$

where

$$T_e = -\frac{L_1 N_r}{2}\left[\sin(N_r\theta)i_A^2 + \sin\left(N_r\theta - \frac{\pi}{2}\right)i_B^2\right.$$
$$\left. + \sin(N_r\theta - \pi)i_C^2 + \sin\left(N_r\theta - \frac{3\pi}{2}\right)i_D^2\right] \quad (14.13)$$

The above model does not account for magnetic saturation of the ferromagnetic material used to construct the variable reluctance step motor. A common and effective way to account for the magnetic saturation is to replace the torque expressions with an expression that is linear, instead of quadratic, in phase current. For example, the torque due to the current in phase A is modeled as:

$$T_A = -k_T \sin(N_r\theta) i_A \quad (14.14)$$

Equation (14.11) is replaced by

$$T_e = -k_T\left[\sin(N_r\theta) i_A + \sin\left(N_r\theta - \frac{2\pi}{3}\right)i_B\right.$$
$$\left. + \sin\left(N_r\theta - \frac{4\pi}{3}\right)i_C\right] \quad (14.15)$$

and Eq. (14.13) is replaced by

$$T_e = -k_T \left[\sin(N_r\theta)i_A + \sin\left(N_r\theta - \frac{\pi}{2}\right)i_B \right.$$
$$\left. + \sin(N_r\theta - \pi)i_C + \sin\left(N_r\theta - \frac{3\pi}{2}\right)i_D \right] \quad (14.16)$$

where k_T is the torque constant, equal to the zero speed one phase on holding torque.

Bifilar-Wound Hybrid Step Motor Model

A two-phase bifilar-wound hybrid step motor is wound with two windings or one center-tapped winding per pole. A positive current in one center-tapped winding will cause the magnetic flux to align in one direction, while a positive current in the other half of the winding causes the flux to align in the reverse direction. In both cases, the current can be supplied through the center tap. Thus, this motor can be driven from a single, or unipolar, supply. For convenience, the bifilar-wound hybrid step motor is considered to have four phases, with each center-tapped winding consisting of two opposite phases. One center-tapped winding consists of phases A and C, the other consists of phases B and D. Nearly perfect flux coupling exists between phases A and C as well as between phases B and D, whereas practically no flux coupling exists between the two separate winding pairs. As a result, the flux linkage in the kth phase is due to current in the kth phase winding, the current in the other half of the winding pair, and the flux due to the permanent magnet. These relationships are

$$\lambda_A = \lambda_{AA} + \lambda_{AC} + \lambda_{AF}$$
$$\lambda_B = \lambda_{BB} + \lambda_{BD} + \lambda_{BF}$$
$$\lambda_C = -\lambda_A$$
$$\lambda_D = -\lambda_B \quad (14.17)$$

where λ_k is the total flux in winding k: λ_{kj} is the flux in winding k due to the j current winding and λ_{kf} is the flux in winding k due to the permanent magnet.

If the per phase inductances, L_k, are assumed to be equal, i.e., in the kth phase $L_k = L$ for all k, with the assumption of no saturation, and with eddy currents neglected, the flux linkage due to self-inductance is $\lambda_{kk} = L i_k$ and flux linkage in the opposite (jth) phase due to mutual inductance is $\lambda_{jk} = -\lambda_{kk}$ where k and j are phases on the same pole. With these relationships between flux linkage and current, Eq. (14.17) reduces to

$$\lambda_A = L(i_A - i_C) + \lambda_{AF}$$
$$\lambda_B = L(i_B - i_D) + \lambda_{BF} \quad (14.18)$$

where λ_{AF}, λ_{BF} are the flux linkages due to the permanent magnet given by $\lambda_{AF} = k_o \cos(\theta)$ and $\lambda_{BF} = k_o \sin(\theta)$; where k_o is the flux constant due to the permanent magnet; θ is the rotor angular position in electrical radians; and $\theta = N_r\theta_m$, where θ_m is the mechanical position or displacement of the rotor with respect to the detent position of phase A which is at $\theta = 0$ rad. N_r is the number of rotor teeth (one mechanical period = N_r electrical period).

The phase voltages, being the voltages measured on the motor terminals, are modeled as

$$V_k = R_k i_k + \dot{\lambda}_k \tag{14.19}$$

where $k = A, B, C, D$. R_A, R_B, R_C, and R_D represent the resistance of phases A, B, C, and D, respectively, which are not assumed equal, and the super-dot denotes derivative with respect to time.

From Eqs. (14.18) and (14.19), four differential equations corresponding to the four phases can be derived with the flux linkages as the state variables; however, two of these four state variables are dependent since the flux linkage in phase C is equal to the opposite of that in phase A and flux linkage in phase D is equal to the opposite of that of phase B, as stated in Eq. (14.17). Only two differential equations are necessary to model the flux linkages in the motor, one for each half of the motor.

Let the flux linkages in phase A and phase B be the two state variables. Flux linkages are used as state variables instead of currents because step discontinuities can occur in the current when a phase in switched, while the flux linkages are continuous in time. For phase A, replacing for i_C and i_A (for $k = A, C$) in Eq. (14.18), replacing for λ_C from Eq. (14.17), and rearranging terms yields

$$\begin{aligned}\dot{\lambda}_A &= \frac{1}{2}V_A\left[1 - \frac{R_A - R_C}{R_A + R_C}\right] - \frac{1}{2}V_C\left[1 + \frac{R_A - R_C}{R_A + R_C}\right] - \frac{R_A R_C(\lambda_A - \lambda_{AF})}{L(R_A + R_C)} \\ \dot{\lambda}_B &= \frac{1}{2}V_B\left[1 - \frac{R_B - R_D}{R_B + R_D}\right] - \frac{1}{2}V_D\left[1 + \frac{R_B - R_D}{R_B + R_D}\right] - \frac{R_B R_D(\lambda_B - \lambda_{BF})}{L(R_B + R_D)}\end{aligned} \tag{14.20}$$

The torque generated by the motor can be modeled by

$$T_E = \frac{k_T}{L}(-\lambda_A \sin(\theta) + \lambda_B \cos(\theta)) - k_m \sin(4\theta) \tag{14.21}$$

where

k_T = torque constant
k_m = maximum torque due to the permanent magnet

Drive Circuit Modeling

The easiest and most common way to model a motor and drive circuit is to model the model as presented above with all the resistances equal to the phase resistance, and the phase voltages as V_s (or $-V_s$ when appropriate) when a phase is on and either 0 or $-V_s$, whichever is appropriate when a phase is off. However, this models the unenergized phases as short circuited, which, after the current has decayed, would be better modeled as an open circuit. In general, this has the effect of underpredicting the performance of the motor, especially at higher speeds.

When more accurate motor and drive system models are indicated, the phase voltages should be set to V_s, and the transistors and diodes modeled as variable resistors. Conducting transistor and diode resistances can be set to zero and nonconducting transistor and diode resistances can be set to a large value.

14.4 Control of Step Motors

Successful application of a step motor to a positioning or position tracking application requires careful attention to the control of the step motor. In this section, techniques are discussed that show how to increase the torque, double the number of detent positions, and open-loop control characteristics. The drive circuits for the various types of step motors and winding configurations was discussed in Section 14.2.

Excitation of Step Motors

Although energizing one phase at a time is the simplest way to control the step motor, greater performance from the motor is possible by exciting two phases at a time, or by switching between one phase on and two phases on at a time. This latter switching scheme is known as half-stepping.

One Phase-On Excitation: By exciting the phases one at a time, the motor will move from detent position to detent position, for example, A-B-C-D-A-B-C ... for a four-phase motor.

Two Phase-On Excitation: When two phases are energized at a time, the torque curves for the individual phases add. The stable detent position is halfway between the detent positions of the motor when the phases are energized one at a time. By exciting the phases two at a time, the motor will move from the new detent position to new detent position, for example, AB-BC-CD-DA-AB-BC-CD ... for a four-phase motor, where AB is the position halfway between detent position A and detent position B. Exciting the four-phase motor with two phases on at a time produces $\sqrt{2}$ more torque but consumes twice the power of exciting the motor one phase on at a time. Exciting a three-phase motor with two phases on at a time produces no additional torque but also consumes twice the power of exciting the motor one phase on at a time.

Half-Stepping Excitation: Switching the excitation alternately between one phase on and two phases on is called half-stepping excitation. This mode of operation doubles the number of detent positions of the motor, in that all of the one phase on detent positions and all of the two phase on detent positions are available. For example, by exciting a four-phase motor using half-stepping, the rotor can be made to step from detent position to detent position such as A, AB-B-BC-C-CD-D-DA-A ....

Open-Loop Control

The most common way to control a step motor is open loop, that is, without position and/or velocity feedback. Once the motor and drive scheme have been chosen, the step command sequence must be chosen. The following examples illustrate some of the characteristics and hazards of open-loop controlled step motors.

The One-Step Move: Fig. 14.11a illustrates the position vs. time and Fig. 14.11b illustrates the velocity vs. time for a one-step move achieved by deenergizing phase A and energizing phase B. The characteristic position overshoot and ringing can be seen from these plots. Figure 14.11c shows the velocity vs. position or phase–plane plot of this one-step move. Observe that the peak overshoot is almost half a step. The peak velocity reaches over 500 steps per second.

A Six-Step Move at a Rate of 50 Steps per Second: Fig. 14.12a shows the position vs. time and Fig. 14.12b shows the velocity vs. time for a six-step move at a rate of 50 steps per second. The stair step in Fig. 14.12a is the commanded position, the straight line in Fig. 14.12b is commanded velocity. We can see the motor is moving in discrete steps, in that the position and velocity profiles are almost six individual single-step responses with overshoots close to half a step and peak velocities over 500 steps per second and close to −400 steps per second.

A Six-Step Move at a Rate of 500 Steps per Second: Fig. 14.13a shows the position vs. time and Fig. 14.13b shows the velocity vs. position for a six-step move at a rate of 500 steps per second. The stair step in Fig. 14.13a is the commanded position and the straight line in Fig. 14.13b is the commanded velocity. In these figures, we no longer see the individual step responses in that the next phase is energized before the rotor has reached the peak overshoot for the previously energized phase. The velocity profile shows that the motor is still not running at a constant speed, but the wild velocity oscillations are gone. The characteristic ringing is seen at the end of the move.

A Six-Step Move at a Rate of 1000 Steps per Second: Fig. 14.14a shows the position vs. time and Fig. 14.14b shows the velocity vs. position for a six-step move at a rate of 1000 steps per second. The stair step in Fig. 14.14a is the commanded position and the straight line in Fig. 14.14b is the commanded velocity. In these figures, we see that the rotor is unable to keep up with the commanded position. A step motor is a synchronous machine, and can generate nonzero average torque only at synchronous speed.

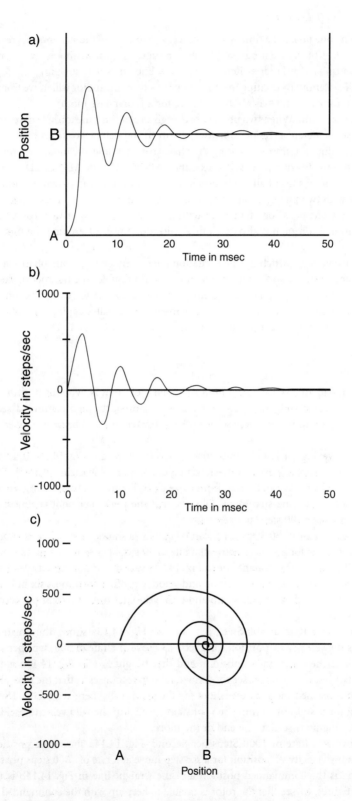

FIGURE 14.11 The one-step move: (a) Position vs. time; (b) velocity vs. time.

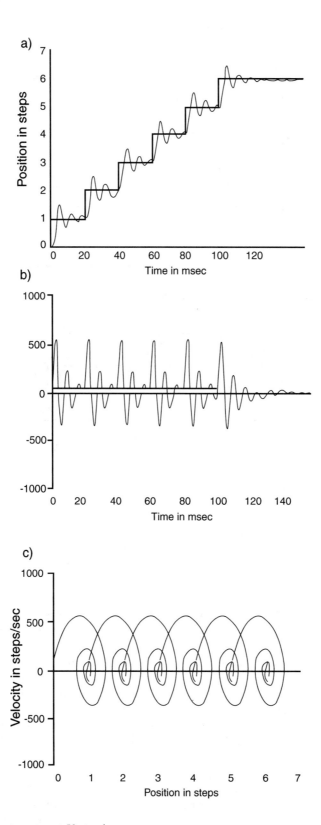

FIGURE 14.12 A six-step move at 50 steps/sec.

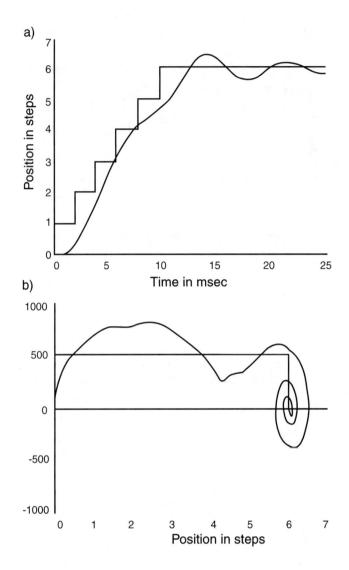

FIGURE 14.13 A six-step move at 500 steps/sec.

Typically, the motor will just vibrate when it loses synchronism. Here we see that the motor is attracted to the wrong detent position after the move is completed, which is four steps away from the desired position.

Error-Free Start-Stop Rate: From the above examples we see that the motor can run at speeds up to 500 steps per second, but cannot start at a speed of 1000 steps per second. The error-free start-stop rate for this motor is between 500 and 1000 steps per second. Speed greater than the error-free start-stop rate can be achieved by starting the motor at or below the error-free start-stop rate and accelerating the motor up to a higher speed. Near the end of the move, the motor must decelerate to a speed at or below the error-free start-stop rate or it may not be able to stop at the desired position, but somewhere beyond.

Low-Frequency Resonance: When a step motor is run at or near the natural frequency of its one-step response, synchronism with the commanded position can be lost due to low-frequency resonance.

Midfrequency Resonance: Fig. 14.15 illustrates another problem known as midfrequency resonance. In this position vs. time plot, the motor is started out at 400 steps per second, well below its error-free start-stop rate, and accelerated up to a speed of 1200 steps per second. At this speed, large oscillations

Step Motor Drives 14-17

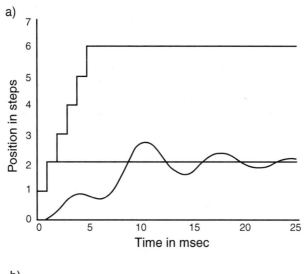

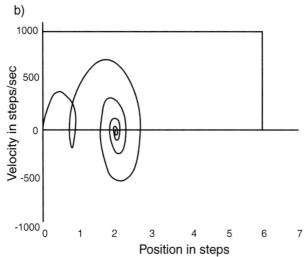

FIGURE 14.14 A six-step move at 1000 steps/sec.

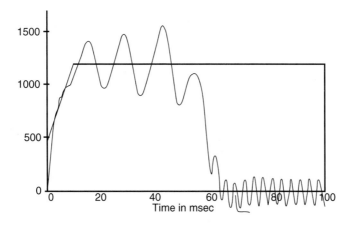

FIGURE 14.15 Midfrequency resonance.

occur in the velocity until, finally, the rotor loses synchronism with the commanded signal. This phenomenon can be avoided by accelerating through this speed range.

References

Kuo, B. C., Ed. 1979. *Incremental Motion Control*, Vols. 1 and 2, SRL Publishing Co., Champaign, IL.
Leenhouts, A. C. 1986. *The Art and Practice of Step Motor Control*, Intertec Communications, Ventura, CA.
Miller, T. J. E. 1993. *Switched Reluctance Motors and Their Control*, Oxford University Press, Oxford, U.K.

15
Servo Drives

Sándor Halász
Budapest University of Technology and Economics

15.1 DC Drives 15-2
15.2 Induction Motor Drives 15-3
   PM Synchronous Motor Drive

A significant and very special class of industrial drives are those that are used for position control. These drives are typically called servo drives and the intelligent control of these drives is often called motion control. Some of the application areas of servo drives are machine tool servos, robotic actuator drives, electric vehicles, computer disk drives and the like. The power level for these drives usually range below 20 to 30 kW; however, drives with slightly lower control quality usually have power levels below 50 to 60 kW.

Servo drives must meet several quality requirements, such as:

1. High dynamic response, which can be realized only with special control schemes and special motors with a high torque/inertia torque ratio
2. Smooth torque production in order to achieve smooth rotation and the elimination of position angle oscillations
3. High reliability with quick maintenance and repair
4. Robust control, i.e., the ability of the drives to tolerate wide swings in load inertia or motor parameters

As a result of these quality requirements, the price of servo drives can be several times that of common industrial drives of the same size.

The control scheme for servo drives usually consists of three subordinate loops as shown in Fig. 15.1. The first and most inner one is the current control loop. The Y_I transfer function of the current controller is generally chosen in such a manner that the current closed loop must have a cutoff angular frequency of $\omega_{0I} \geq 1000$ r/s. The second control loop, referred to as the speed loop, usually has a closed-loop control band width of $\omega_{0\omega} \geq 300$ r/s. The outer loop, or position control loop, must accurately follow the position reference. All the control loops, as a rule, are proportional integral (PI) controllers; the position controller is the only one that often employs proportional, sometimes proportional differential controller. But if a parameter of the system changes in some reasonable fashion, e.g., the inertia torque in a robotics application, this control system cannot achieve fast and accurate position control without overshoot. In this case, other types of control schemes have been used such as feed-forward, optimal, and sliding mode control.

In some applications the servo drives require only torque control for positioning, e.g., in robotics applications. In this case, the torque control loop becomes the outer loop. For most drives a proportionality exists between torque and motor current; therefore, in this case torque control means current control. All the control loops, at present, usually employ digital control. Only the current loop, at high operating frequencies, is sometimes implemented with analog circuits. About 10% of all servo drives are used in single applications. In machine tools applications, where there are several axes of control, all the servo

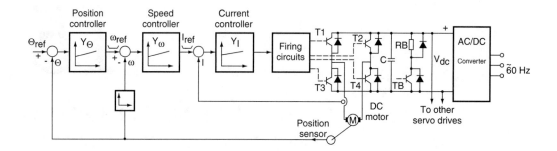

FIGURE 15.1 DC servo drive (with control scheme).

drives have a common DC supply, which is obtained from a standard AC supply through a common rectifier as shown in Fig. 15.1 When electric motors are used, they are either permanent magnet (PM) DC motors, permanent magnet synchronous motors or, rarely, induction motors. For low-power applications stepping motors can be used, but this type of motor exhibits a considerable amount of torque pulsation. The switched reluctance motor (SRM) can also be used. The permanent magnet for DC and synchronous motors is manufactured from various types of ferrite (strontium ferrite, hard ferrite, etc.), ceramic or samarium cobalt.

Servo motors normally come with different built-in sensors, e.g., encoder or resolver for position control and tachometer for velocity control. The motors generally are of a rugged design.

15.1 DC Drives

At present most servo drives are DC drives. The servo motor with permanent magnet excitation permits a 400 to 1000% torque overload. The torque limitation areas are shown in Fig. 15.2. Area I is the continuous operating area; area II is the intermittent operating area; and area III can be used only for accelerating and decelerating. These areas are limited by absolute maximum speed, an absolute commutation limit and the peak stall torque. The speed and the torque (current) control loops must take into account these limitations of the DC servo motors. DC servo motors have such low torque (size) ratings that normally their rated voltage must be less than 100 to 200 V. Therefore, DC servo drives when supplied from an AC source use a transformer for the creation of the supply with a reasonably rated value of voltage.

The motor supply circuits are shown in Fig. 15.1. The four-quadrant transistor chopper with a commutation frequency of 5 to 20 kHz ensures a very good dynamic control of the motor with very little, usually below 1 to 2 μs, dead time (the time between turn-off of one transistor and turn-on of the next). The transistors are either MOSFETs (metal oxide semiconductor field effect transistors) or IGBTs (insulated gate bipolar transistors).

The servo drives are normally unable to return the braking energy to the AC supply: the energy is lost in the DC circuit resistance, i.e., RB in Fig. 15.1. The resistor current is controlled by the transistor TB. During braking, DC current flows through the capacitor C and the DC voltage increases. When this voltage achieves its maximum permitted value, the transitor TB turns on, and DC current flows through resistance RB and then the DC voltage decreases to the minimum value when transistor TB turns off. Thus, during braking, the DC line voltage is maintained between a maximum and a minimum value.

The control scheme for the chopper transistors is presented in Fig. 15.3. A so-called overlapping control is commonly used. The T_C time-cycle is derived from the times αT_C and $(1 - \alpha)T_C$. The T_1 and T_4 transistors turn on during the time period $(1 - \alpha)T_C$, and the T_2 and T_3 transistors turn on during the time period αT_C; however, the turn-on and turn-off times of the odd and even transistors are shifted, i.e., overlapping, by $(1-2\alpha)T_C/2$ as shown in Fig. 15.3. If α is between 0 and 0.5 the motor voltage is positive and if α is between 0.5 and 1.0 the motor voltage becomes negative as illustrated in Fig. 15.3. A very important advantage of this control scheme is that the motor voltage and current waveforms repeat twice during one period (T_C) of the transistor control.

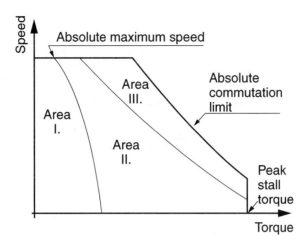

FIGURE 15.2 Limitations on the operating areas of the DC servo motors.

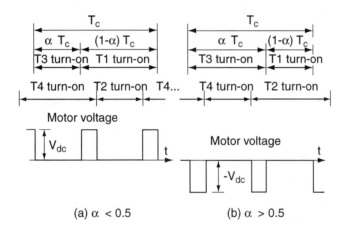

FIGURE 15.3 DC chopper transistor control.

As a result of the high-frequency control, the motor current (and torque) consists only of high-frequency harmonics with very low amplitudes, usually under 1% of the motor's rated current. This means that in both the transient and steady states the motor current and torque consist of virtually only a DC component and therefore there are no speed (or position) oscillations.

15.2 Induction Motor Drives

The induction servo motor with a squirrel cage rotor has very small rotor inertia torque, high reliability, and it is very economical. However, the control system for the induction motor is very complicated, expensive, and the quality of the control is sensitive to motor parameter changes. Therefore this motor is not widely used.

The typical supply circuits for the induction servo motor are shown in Fig. 15.4. The AC supply voltage feeds the diode rectifier, which creates the DC link. The DC link consists of the capacitor C, braking resistor RB, and transistor TB. The control of the DC voltage during the braking operation is performed in the same manner as that for DC drives. The voltage source inverter is usually constructed with IGBT transistors and very fast parallel diodes. In the last several years, the use of IGBT modules with six transistors and six diodes has been the preferred configuration.

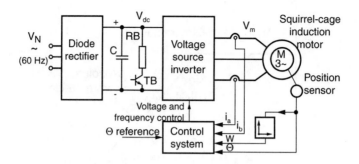

FIGURE 15.4 Servo drive with induction motor.

The drive does not need a transformer since high-voltage motors are available. If the AC phase voltage is V_N (rms value), then the DC link voltage will be $V_{dc} \cong 3/\pi \sqrt{6} V_N$ and the maximum possible motor phase voltage will be

$$V_m = \frac{2}{\pi}\frac{1}{\sqrt{2}} V_{dc} = 3\frac{\sqrt{12}}{\pi^2} V_N \cong 1.05 V_N \qquad (15.1)$$

Hence, if the rated voltage of the motor is equal to the AC supply voltage, then as a result of the voltage drop in both the rectifier and the inverter, the motor can operate with a rated flux between 0 Hz and the approximate frequency of the AC supply.

A position control system usually uses the indirect field oriented principle. The rotor flux is generated by the two phase currents as well as the speed, as shown in Fig. 15.4. The calculation is a function of the rotor time constant, which is dependent upon both the rotor resistance and rotor inductance. Variations in these parameters must be taken into consideration; however, the identification of the parameter changes is very complicated.

PM Synchronous Motor Drive

The permanent magnet synchronous motor is much more expensive than the squirrel cage induction motor, but the control system of the PM synchronous motor drive is much simpler than that used for the induction motor. When compared to DC motors, PM synchronous motors normally have less inertia torque and require less maintenance. As a result of these features, the PM synchronous servo drive has become one of the most popular types of servo drives. The converter circuits for PM synchronous motor drives are identical to those for induction motors as shown in Fig. 15.4. PM synchronous motors, such as induction servo motors, are usually manufactured for high voltage and therefore transformers are not required in their use.

There are two classes of PM synchronous motors:

1. Those with a square flux density distribution along the rotor air gap surface, as shown in Fig. 15.5a, which produces a trapezoidal back-emf (electromotive force) in the stator coil—the so-called trapezoid PM machines.
2. Those with a sinusoidal flux density distribution, as shown in Fig. 15.5b, which produces a sinusoidal back-emf—the so-called sinusoidal PM machines.

In the trapezoidal machines the angle β illustrated in Fig. 15.5a is the width of the magnet. In general, $\beta \cong 180°$. In Fig. 15.6 the trapezoidal machine with $\beta = 180°$ is presented and a two-pole machine is assumed. In steady-state the machine is rotated with constant synchronous speed, which is a function of the number of pole pairs p and the frequency of the stator supply f_1

$$\omega_1 = \frac{2\pi f_1}{p} = \text{const} \qquad (15.2)$$

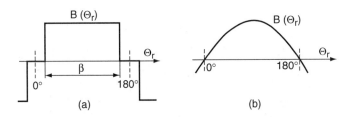

FIGURE 15.5 Two types of PM synchronous servo motors: (a) with square flux density distribution (trapezoidal PM machines); (b) with sinusoidal flux density distribution (sinusoidal PM machines).

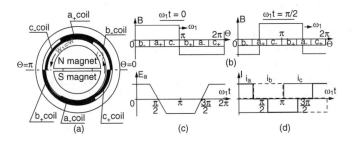

FIGURE 15.6 Trapezoidal PM machines with $\beta = 180°$: (a) motor construction; (b) flux density displacement at $\omega_1 t = 0$ and $\omega_1 t = 90$; (c) back-emf vs. time; (d) motor phase currents vs. time.

Consider Fig. 15.6a or b where the machine is expanded along the stator air gap surface. In the range $-60° \leq \omega_1 t \leq 60°$, the a phase conductors are located under the maximum flux density B, i.e., $a+$ is under $+B$ and $a-$ is under $-B$. Hence in this timeframe in the a phase the maximum value of the back-emf E_a is induced as shown in Fig. 15.6c. For $\omega_1 t \geq 60°$, E_a begins to decrease since the $a+$ conductors (or $a-$) are in the flux density of different directions. As shown in Fig. 15.6b at $\omega_1 t = \pi/2$ half of the a phase coil will be under a positive and the other half under a negative value of the flux density; therefore, at this time $E_a = 0$. As a result this analysis indicates that the back-emf-time function is a trapezoidal shape of the form shown in Fig. 15.6c.

Suppose that the drive control only permits stator current to flow in two phases at any time. With reference to Fig. 15.6a, positive current is supplied to phase a and negative current is supplied to phase c. The resulting stator phase currents are shown in Fig. 15.6d. This current distribution is achieved by the appropriate phase current commutations through the use of a position sensor signal (once for every 60°). The motor torque will be

$$T = cBI \tag{15.3}$$

where c is a motor constant. The torque will not have ripples if the current is constant and this constant current is ensured by DC current control just as it is for DC servo drives. But now under control are only the transistors that belong to the two current conducting phases. As a result of the high-frequency current control, the torque is essentially constant; however, during the phase current commutations, i.e., every $\omega_1 t = 60°$, current control is not possible. Hence, torque oscillations occur at a frequency of $6f_1$, which is a considerable disadvantage of trapezoidal machines.

In the sinusoidal PM machines the sinusoidal flux density distribution will produce a constant torque only if the phase currents are also sinusoidal. The sinusoidal values can be characterized by vectors as shown in Fig. 15.7. The $\overline{\Lambda}_p$ pole flux linkage vector and the $\overline{I}$ current vector will produce the torque

$$T = c_1 \overline{\Lambda}_p x \overline{I} = c_1 \Lambda_p I \sin(\Lambda_p I) \tag{15.4}$$

where c_1 is a constant. Therefore, if the angle between these two vectors is equal to 90°, as shown in Fig. 15.7, the torque is maximized. Current control is normally achieved as shown in Fig. 15.7b. The position

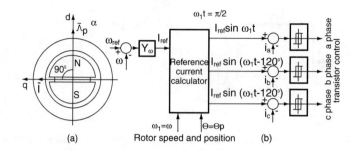

FIGURE 15.7 Sinusoidal PM machines: (a) pole flux and current vector orientation; (b) control schemes.

sensor signal requires the creation of three sinusoidal phase current reference signals, which generate the current vector with a 90° displacement from the pole flux vector. The three Schmidt triggers ensure two-point phase current control with the desired hysteresis. Because the phase current hysteresis is very small the motor torque ripples are very high frequency and have very small values. The important advantage of sinusoidal machine drives is that there are no torque oscillations with $6f_1$ frequency, as is the case with trapezoidal machines.

References

Jahns, T. M. 1994. Motion control with permanent-magnet AC machines, *Proc. IEEE*, 82(8), 1241–1255, August, special issue.

Kenjo, T. and Nagamori, S. 1985. *PM and Brushless DC Motors,* Clarendon Press, Oxford, U.K.

Miller, T. J. E. 1993. *Brushless Permanent-Magnet and Reluctance Motor Drive,* Clarendon Press, Oxford, U.K.

16
Uninterruptible Power Supplies

Laura Steffek
Best Power

John Hecklesmiller
Best Power

Dave Layden
Best Power

Brian Young
Best Power

16.1 UPS Functions 16-1
    Power Conditioning • System Integration
16.2 Static UPS Topologies 16-3
    Double-Conversion UPS • Line-Interactive UPS • Standby Power Supplies
16.3 The Rotary UPS 16-6
16.4 Alternate AC and DC Sources 16-7
    Batteries • DC Generators • Superconducting Magnetic Energy Storage • AC Generators

With the proliferation of electronic loads such as computers, the incidence of power quality-related problems is growing. As a result, the uninterruptible power supply (UPS) market has grown significantly in the last few years. What follows is an overview of UPS functions and descriptions of common types of UPSs and backup power sources.

16.1 UPS Functions

The primary purpose of a UPS is to provide conditioned, continuous power to its load. Another UPS function that is of growing importance in today's market is system integration, or the ability to communicate over a network to facilitate the monitoring and orderly shutdown of loads.

Power Conditioning

A UPS provides continuous, regulated power to its load, under all conditions of the utility power line. Unlike other types of power conditioning equipment, a UPS provides power during outages. Typically, a UPS will provide backup power for 10 or 15 min, although longer times are possible with large battery strings or a DC generator.

A UPS will also correct for high- and low-voltage events, known as surges and sags. This regulation is provided either electronically or by a tapped transformer or a ferroresonant transformer.

Normal mode, or line-to-line, transients are prevented from reaching the load. This is accomplished either with filter components, or in a double conversion UPS, by converting the AC to DC and then back to AC. There is quite some variation in the ability of UPS systems to protect the load from common mode, or line-to-ground, transients. Safety agency requirements preclude most forms of common mode transient protection. The best common mode transient suppression is achieved with an isolation transformer. Some UPSs have isolation transformers and some do not.

System Integration

The industrial electronics environment is very similar to the typical office LAN/WAN environment when it comes to using a UPS to provide power protection for industrial-grade PCs, PLCs and other equipment that make use of any form of microprocessor control.

The fact that a UPS only provides a finite amount of battery backup during an extended power outage should encourage us to take certain precautions to prevent the corruption and loss of data once the UPS reaches a point where it can no longer support the load equipment.

Certain methods may be used to communicate to the load equipment when a power outage has occurred and, in extreme cases, when a low battery condition exists. The load equipment should be configured to react to critical UPS conditions by saving data and preparing the system for a safe shutdown. Creating this "communication" between UPS and the load equipment is called UPS integration. There are several ways that the UPS can be integrated. These methods may be classified into three integration categories:

Basic
Enhanced
Network

No matter what integration methodology is utilized, four items are required to integrate the UPS. First, the UPS must have a communication port. Second, the equipment being protected must also have a communication port. Third, some medium (cabling) must be used to connect the two together. Finally, some form of software must be used to monitor the UPS and provide the appropriate actions relevant to specific UPS conditions.

Basic

The first and most common integration method communicates the status of the UPS via contact closures. Typically, normally open or normally closed relay contacts are used to signal two UPS conditions to the load equipment. These conditions are "AC Failure" and "Low Battery." An "AC Failure" should be signaled by the UPS whenever a power failure condition exists for more than 5 sec. The "Low Battery" signal exists when a minimum of 2 min of battery runtime remains to support the load. However, most UPS manufacturers allow this setpoint to be programmed by the user to allow more time to shutdown the system.

In most cases, the software to monitor the UPS is provided as a part of the computer's operating system. The UPS manufacturer typically provides the cable and appropriate setup information required to connect the two together.

Note that UPS manufacturers often substitute open-collector type circuits in place of actual relays, to provide the UPS signals. Users should pay close attention to this detail if they choose to build their own interface cable, since current is only allowed to pass in one direction through an open-collector circuit.

Enhanced

To provide more than just the basic UPS status information, many UPS manufacturers have chosen to offer RS232 and other forms of serial communication that allow real-time UPS data to be monitored by software running on the load equipment. Instead of knowing only that a power failure has occurred or that a low battery condition exists, the user may now know how much calculated runtime is available and the measured battery voltage at any given time. Other data values are typically available that represent the input and output voltage, percent of full load, UPS temperature, as well as many others.

Since the way this UPS data is presented is usually proprietary, the UPS manufacturer most often supplies the software to run on the protected load. Because the software is capable of monitoring real time data from the UPS, a GUI (graphical user interface) is typically used to portray the data using easy-to-read digital displays and historical graphs.

Network

The size and complexity of today's local and wide area networks has led to an increase in the use of network management tools to monitor and control network devices. The Simple Network Management Protocol (SNMP) has become the de facto standard for network management and is backed by many network management software products including SunNet Manager, HP-Open View, IBM's Netview/6000, and Novell's NMS.

Today, many UPS manufacturers offer software or a software/hardware combination that effectively makes the UPS a network peripheral. In some cases an internal or external network adapter is provided that through its own microprocessor and associated components effectively translates proprietary UPS data and commands into a format that is compatible with the SNMP standards set forth by a working group of the Internet Engineering Task Force (IETF). This group recently adopted a standard database of UPS-related information, called a Management Information Base (MIB), for all UPS products. The official IETF document that describes this MIB is RFC-1628, which is available on the Internet.

The SNMP-capable UPS provides three basic functions when communicating with a network management station. It responds to "get" requests by replying back to the management console with a value corresponding to the requested MIB variable. It responds to "set" requests by allowing the UPS configuration to be changed by the management console. And it broadcasts unsolicited alarm "traps" to the network management console alerting the network administrator to the existence of potential power problems.

The UPS in an industrial environment presents a new challenge to the integrator due to the existence of many different industrial network protocols. In some cases, many of the same protocols exist that are present in the office LAN environment, but they are often joined by such protocols as SP50 and PROFIBus, which are adaptations of Field Bus. Other industrial control protocols include FIP (Factory Instrumentation Protocol), MAP (Manufacturing Automation Protocol), and Echelon's LonTalk. UPS manufacturers have not yet built in direct connections to these industrial networks. In some cases, protocol adapters are available that translate the RS232 information from the UPS into the required network protocol. Future developments from UPS vendors may enhance and simplify the UPS connectivity in the industrial environment.

16.2 Static UPS Topologies

A static UPS is one that relies on power electronics, rather than a motor generator, to provide power to the load. Most UPSs today are of this type.

There are several basic UPS topologies, each of which has its advantages and disadvantages. The terms "on-line UPS" and "off-line UPS" have commonly been used to describe some UPS topologies. Unfortunately, UPS manufacturers have not been able to agree on the meanings of these terms, leading to confusion among users. Terms which are more descriptive of the differences between various topologies are double-conversion UPS, line-interactive UPS, and standby power supply.

Double-Conversion UPS

A double-conversion UPS (Fig. 16.1) first rectifies incoming AC line to a DC voltage, then inverts that DC voltage to provide an AC output. During normal operation, the rectifier is providing current to charge the batteries and also to the inverter. The inverter supports the load and provides regulation of the output voltage and frequency. In the event that line is lost or deviates from the specified input voltage

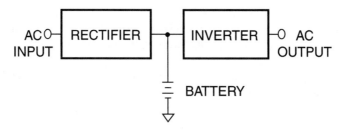

FIGURE 16.1 Double-conversion UPS. There is no disruption in output power when the UPS transfers from its line source to battery power, because the inverter is always operating.

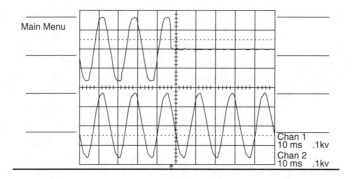

FIGURE 16.2 Typical double-conversion UPS response to a power disturbance. Top trace: AC input; bottom trace: AC output. (Courtesy of National Power Laboratory of Best Power, a unit of General Signal.)

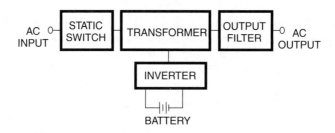

FIGURE 16.3 Typical line-interactive UPS.

and frequency tolerances, the inverter uses the batteries as an energy source and operates until the batteries are depleted or line is restored. (See Fig. 16.2 for typical response.)

Some double-conversion UPS have an automatic bypass switch. This switch connects the load to the AC source in the event of a UPS failure. It may also be used to help support a temporary overload that the inverter cannot support alone.

Traditionally, phase-controlled thyristor rectifiers have been used in double conversion UPSs. These rectifiers cause distortion of the input current and voltage waveforms. Distorted current waveforms can cause excess neutral currents in the building wiring, and distorted voltage waveforms can cause problems in other equipment on the same circuit. Some newer rectifier designs use pulse width modulation (PWM) techniques to reduce waveform distortion. These techniques can result in harmonic distortion levels of 5% or less.

Line-Interactive UPS

In normal operation (Fig.16.3), the AC input passes through a filter or transformer to the load. The inverter is normally not supporting the entire load, but may be used to buck or boost the line voltage, or even fill in "notches" of the incoming line voltage waveform on a subcycle basis. It is this ability of the inverter to interact with line that gives the line-interactive UPS its name. The inverter does not support the load unless there is a power outage, or the AC input falls outside the specified voltage and frequency tolerances. (See Fig. 16.4 for typical response.)

The key to a line-interactive unit is its ability to respond to line disturbances quickly. This is necessary to ensure that power is supplied continuously to the load. Some energy is stored in the magnetics and output filter, which can support the load for a short time. The static switch must open quickly and the inverter become active before that energy is lost to the load.

Voltage regulation during line operation may be achieved by phase-controlling the inverter, by using a tapped transformer, or by using a ferroresonant transformer.

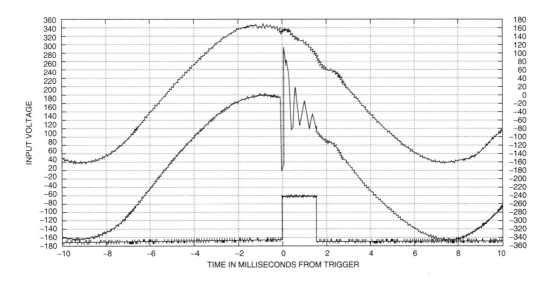

FIGURE 16.4 Typical response of a line-interactive UPS to a power disturbance. Top: AC output (right scale); middle: AC input (left scale); bottom: inverter active signal (no scale). (Courtesy of Best Power, a unit of General Signal.)

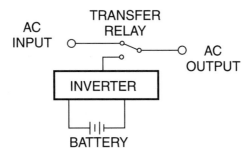

FIGURE 16.5 Standby power supply.

Line-interactive UPSs do not themselves cause harmonic distortions on the utility line. However, they may or may not pass harmonic load currents to the input. Ferroresonant-based units correct the current harmonic distortion of the load and present a near-sinusoidal current waveform to the utility line. Other line-interactive units provide little harmonic correction. This is of diminishing importance as computer power supplies are being redesigned to reduce the harmonic currents they cause, to meet the requirements of standards such as IEC 555-2.

Standby Power Supplies

Standby power supplies (SPS) (Fig. 16.5) are not properly called UPS because they do not provide continuous power to the load. A standby power supply is similar to a line interactive UPS in that the inverter is not normally supporting the load. However, when the load is transferred from line to inverter, an interruption in power occurs due to the break time of the transfer switch. Typically this switching device is an electromechanical relay and takes several milliseconds to open or close. The minimum operation on inverter is usually several seconds, as compared to the subcycle control possible with a line-interactive UPS. (See Fig. 16.6 for typical SPS response.)

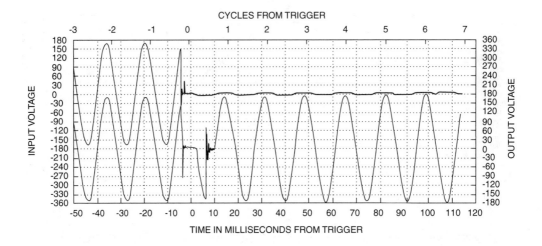

FIGURE 16.6 Standby power supply response to power disturbance. Top trace: AC input; bottom trace: AC output. (Courtesy of National Power Laboratory of Best Power, a unit of General Signal.)

Standby power supplies are typically low-cost products and provide minimum levels of voltage regulation and line conditioning. They usually provide square wave or stepped-square wave outputs on inverter, rather than the sinewave outputs provided by most double-conversion and line-interactive products. They are most appropriately used in less critical applications, where power interruptions several milliseconds in duration and voltage fluctuations can be tolerated.

16.3 The Rotary UPS

The earliest form of UPS is the rotating, or rotary, UPS. Motor and generator combinations have provided uninterruptible power since circa 1950. These early systems offered excellent isolation and fairly good overall performance. They consist of little more than a DC motor coupled to an AC generator. Rectified line normally powers the DC motor. Power switch-over to batteries occurs when the utility (line) fails. Due to the inertia of the rotating mass, switch-over times on the order of 0.3 sec are typical. In practicality, however, the decay in frequency is usually more of a problem than the decay in voltage. To remedy this problem, a supplemental flywheel increases the inherent ride-through to 1 sec or longer. Thus, large mechanical contactors are acceptable to make the power transfer from line to battery. However, modern systems use power semiconductors to do the switching. The mechanical coupling between motor and generator can be either direct, where the components share a common shaft, or by belt. Belt drives, while less efficient, do allow for different speeds between the motor and the generator. Rotary UPSs are currently available in sizes from 35 kVA up to 1000 kVA.

As have other forms of UPSs, the rotary UPS has continued to evolve. Most rotary UPSs today use AC induction motors instead of the DC motor, as AC motors do not require brush maintenance. A typical modern system rectifies and controls incoming AC to charge batteries. The batteries then power a simple three-phase inverter. This inverter, which requires no commutation or voltage-regulation circuitry, drives the induction motor. An added benefit is that this system requires no flywheel for energy storage, as there is no transfer time from line to battery power. Figure 16.7 depicts the block diagram of a typical rotary UPS.

Some of the newer rotating UPSs combine the motor and generator on one stator, and apply a DC field to the rotor. This scheme makes a very compact and cost-effective system. Other advances include the introduction of a "pole-writing" generator. In this topology, there are no pole windings as such. The poles of the generator write on a ferrite stator with varying position and frequency, depending on the speed of the rotor. Pickup coils read these poles and use them to produce the AC output, much as a tape recorder records a signal, then plays it back. This design can give as much as 15 sec of ride-through. Frequency and voltage stability are excellent.

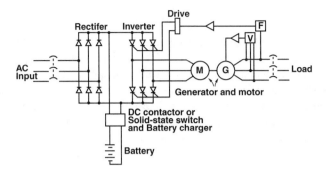

FIGURE 16.7 Typical rotary UPS.

A recent entry is a line-interactive rotary UPS. The line-interactive rotary UPS uses a normally free-spinning unloaded synchronous motor, with an additional motor or engine used for backup power. The synchronous motor has an overexcited field connected to a tapped line inductor. The motor acts as a synchronous capacitor, thereby providing power factor correction. When power fails, the mass of the synchronous motor powers the load until the engine comes up to speed and can assume the load.

Advantages of rotating UPSs include unmatched isolation, and the ability to use many different sources of energy. Single- or three-phase AC power, or power from a turbine or diesel engine can provide rotation. Reliability is excellent, with a demonstrable MTBF that exceeds 10^6 hours. High thermal inertia means that the UPS can sustain very heavy overloads for a short period of time. The units are efficient, with typical efficiencies running from 84 to 88%. Some newer designs can exceed 90% efficiency. For example, a 500 kVA Uniblock demonstrates an efficiency of over 94% at load. As with all rotary UPSs, the rotating mass offers a degree of frequency stability as well as immunity to small load fluctuations.

Disadvantages include an inherent difficulty in starting the system into high-surge loads. It is difficult to make a completely redundant system, and some maintenance, such as bearing replacement, will require shutdown. Rotary UPSs usually cannot start from the inverter, requiring a secondary motor to start rotation.

The rotary UPS is a very practicable system for any application requiring a medium- to high-powered premium UPS system that is reliable and cost-effective.

16.4 Alternate AC and DC Sources

Most UPSs use utility line for the AC source and batteries for the DC source. Batteries are a critical but often misunderstood component that bear further mention. Some installations use alternate power sources that are described below.

Batteries

Both flooded and valve-regulated lead acid (VRLA) batteries are commonly used in UPS applications. Wet cell batteries require maintenance of the electrolyte level and special precautions to prevent build-up of hydrogen gas. VRLA batteries have become increasingly popular in the last few years because of their relative ease of installation and maintenance.

All batteries require some maintenance. Battery terminals and connections should be checked for cleanliness and tightness. The batteries should be discharged periodically to test for battery capacity. End of life is usually defined as a 20% loss of the specified battery capacity at the desired discharge rate.

Battery life may be degraded by several factors, the greatest of which is battery temperature. Battery life is typically reduced by 50% for every 10°C increase in its temperature. Note that the battery temperature may be significantly higher than the ambient temperature of the room, especially if the battery is in an enclosure. Other factors affecting life include the charging method, the number of discharge cycles, the depth of discharge, the rate of discharge, and the ripple voltage across the battery terminals.

Battery storage life is also temperature dependent. Batteries experience a self-discharge at a rate that increases with temperature. This self-discharge is in addition to any current drain the UPS may have when it is off. UPS batteries should be charged upon receipt and every 6 months of storage after that, or more often if the storage temperature exceeds 25°C. If a battery is stored longer than this without being recharged, then a phenomenon called sulfation will occur. Sulfation is the formation of lead sulfate on the battery plates. This lead sulfate is an insulator and causes a loss of battery capacity. Many battery users have stored their batteries for long periods of time, only to find that at installation those batteries have no useful capacity at all. Most of the lost capacity can be recovered by exercising the batteries with repeated charge/discharge cycles, preferably at a high charge rate.

DC Generators

Direct current (DC) power generation has developed over the years to become a viable replacement for batteries in a variety of applications. These applications range from remote island power, such as railroad signal and switching, to uninterruptible power system backup and even lighting applications. Anywhere batteries are traditionally used, a DC generator can be installed to reduce the battery requirement or work in conjunction with alternate power sources such as solar.

To a great extent, the DC generator of today has changed from the days of maintenance-intensive brushes and commutators to highly efficient rectified systems. Now, instead of relying on the brush and commutator to perform the rectification process, AC alternators and diodes are used to produce near-"battery quality" DC power. Reduced maintenance requirements through elimination of brushes, commutators, and slip rings are a few of the obvious advantages of a solid-state rectified system. High frequency alternators and rectifier assemblies provide years of reliable service in less floor space than traditional AC generators or batteries.

Applications such as railroad switch and signal locations are examples of the versatility of DC power generation. Traditionally signal maintenance staff would replace a discharged battery with a recharged battery every few days. This was required to keep the trains rolling by properly signaling the track's availability. Even at sites where solar power sources were utilized, dark days could force increased signal maintenance due to low battery conditions. DC power generation has successfully demonstrated long term battery backup and cooperation with other alternate energy sources. Some railroad applications have adopted a "cycling duty" system to maintain signal integrity. By allowing the battery to discharge or the solar charger to operate, a DC generator can be used to automatically start and recharge the battery when required. This reduces the site maintenance requirement to about twice a year.

UPS backup applications for DC generation have also proven viable alternatives to large strings of batteries. By installing a "minimum" battery, most short-term power outages can be supported. For the long duration power outages, DC power generation can be used. DC power generation generally requires only a connection to the two battery terminals for the system to operate. Through these two battery connections the battery condition is monitored and the DC power generator will start and run automatically to provide long-term reliable DC power to the UPS inverter. Using the DC generator topology, oversized AC generators are not required. Generally the AC generator manufacturers recommend oversizing the generator to reduce the power factor induced by many UPS installations. With a DC generator, the induced power factor problem is not possible.

DC power generation has even grown into the lighting arena. By using a DC power source for floor lighting, HID bulb life is increased substantially. In some cases, this increase can be tenfold. Lighting manufacturers have doubled their lamp warranties due to DC power. In conjunction with these installations, DC power users have discovered the many utility rebates and rate credits available for peak-shaving with DC power generators. Generator run times as low as 50 hours per year can qualify for utility power reduction programs.

Telecommunications applications such as remote offices and cellular radio sites also enjoy the reliability of DC power generation. Even the information superhighway is powered by DC generators, bringing the benefits of the new technology to your doorstep.

Figure 16.8 shows two examples of DC generators available for operation with UPSs.

Uninterruptible Power Supplies

FIGURE 16.8 DC generators available for operation with UPSs. (Courtesy of Best Power Technology, Inc.)

Superconducting Magnetic Energy Storage

Superconducting magnetic energy storage (SMES) systems are relative newcomers to the field of backup power systems. SMES systems store DC energy in a superconducting magnetic coil. The niobium-titanium coil is cooled by liquid helium to 4.2 K or by superfluid helium to 1.8 K.

SMES units are used to provide large amounts of power for short durations. This is useful in industrial applications where even momentary power disturbances can cause expensive equipment downtime and production losses. Commercially available units store 0.3 to 1 kW-hour and are rated for 0.75 to 1.5 MW.

A complete SMES system is functionally the same as a traditional UPS with a more conventional DC storage element. AC line is fed to the load under normal operating conditions. A line fault detector monitors the AC line, and if line is unacceptable, disconnects line from the load. The magnetic storage element provides DC power to an inverter, which in turn supports the load. When acceptable line returns, the load is transferred back to the line source.

SMES has several advantages over batteries. The expected life of a SMES unit is claimed to be as long as 30 years, compared to 10 years or less for batteries. A SMES can be recharged completely in several minutes and the charge-discharge cycle can be repeated thousands of times without degrading the magnet.

AC Generators

What of extended autonomy, where utility may be out for hours at a time? For many applications where the AC line quality is unimportant, the AC generator is still a viable alternative for extended-run applications. With the potentially unlimited runtimes obtainable, the AC generator is certainly attractive. But with today's more sensitive loads, the AC generator may not be the best solution. It is well known in the industry that the AC generator suffers from poor regulation, and unless the unit is very large in comparison to the load, will also exhibit poor frequency stability. So, many users will run a hybrid combination of the AC generator and UPS to power critical loads for prolonged times.

A generator has its own set of maintenance issues. Aside from fuel, oil, and water requirements, the engine must be run periodically to maintain a degree of readiness. Any engine, especially a gasoline engine, must be run occasionally to keep moving parts lubricated, and fuel must be treated against the formation of varnish or bacterial growth, which can restrict fuel flow. Generators are usually kept outside, so shelters must be built and, in many areas, cold-weather starting packages must also be used.

Sometimes, compatibility problems between AC generators and UPSs occur. Double-conversion is usually the most trouble-free of the different topologies of UPSs when used with an AC generator. As the name implies, the power for the double-conversion UPS is converted twice—once from AC to DC (for the batteries) and then from DC back to AC. This scheme assures that no matter what is happening on the input, the output can be controlled precisely. A line-interactive or single-conversion UPS that typically passes line through to the output must incorporate design features to accommodate AC generator operation. The primary trade-off is the desensitization of the UPS to the fluctuating inputs. Often, generator outputs are far from sinusoidal and rather unstable, so the voltage window in which the UPS stays on utility power must be widened. The out-of-frequency window must also be widened, and the tracking capabilities of the phase-locked-loop must be increased. If these alterations are not taken, protracted inverter runs result, depleting the batteries and thus negating the purpose of the AC generator. Needless to say, the output reflects these widened windows. As most modern UPSs can be adjusted to function with an AC generator, the user must assess the impacts of the somewhat diminished performance. The vast majority of loads will function acceptably.

References

DeWinkel, C., Losleben, J., and Billman, J. 1993. Recent applications of superconductivity magnet energy storage, *Proc. Power Quality Conf.*, 462–469.

Griffith, D. 1989. *Uninterruptible Power Supplies: Power Conditioners for Critical Equipment*, Marcel Dekker, New York.

Platts, J. and St. Aubyn, J. D. 1992. *Uninterruptible Power Supplies*, Peter Peregrinus, Stevanage, Herts., U.K.

The Institute of Electrical and Electronic Engineers, 1992. *IEEE Recommended Practice for Powering and Grounding Sensitive Electronic Equipment* (IEEE Emerald Book), IEEE Std. 1100-1992, Institute of Electrical and Electronic Engineers, New York.

17
Power Quality and Utility Interface Issues

Wayne Galli
Southwest Power Pool

Timothy L. Skvarenina
Purdue University

Badrul H. Chowdhury
University of Missouri–Rolla

Hirofumi Akagi
Tokyo Institute of Technology

Rajapandian Ayyanar
Arizona State University

Amit Kumar Jain
University of Minnesota

17.1 Overview 17-1
Harmonics and IEEE 519 • Surge Voltages and C62.41 • Other Standards Addressing Utility Interface Issues

17.2 Power Quality Considerations 17-3
Harmonics • What Are Harmonics? • Harmonic Sequence • Where Do Harmonics Come From? • Effects of Harmonics on the System Voltage • Notching • Effects of Harmonics on Power System Components • Conductors • Three-Phase Neutral Conductors • Transformers • Effects of Harmonics on System Power Factor • Power Factor Correction Capacitors • IEEE Standard 519

17.3 Passive Harmonic Filters 17-20
Passive Filter Design • Appendix—IEEE Recommended Practices and Requirements for Harmonic Control in Electric Power Systems

17.4 Active Filters for Power Conditioning 17-30
Harmonic-Producing Loads • Theoretical Approach to Active Filters for Power Conditioning • Classification of Active Filters • Integrated Series Active Filters • Practical Applications of Active Filters for Power Conditioning

17.5 Unity Power Factor Rectification 17-49
Diode Bridge and Phase-Controlled Rectifiers • Standards for Limiting Harmonic Currents • Definitions of Some Common Terms • Passive Solutions • Active Solutions • Summary

17.1 Overview

Wayne Galli

In the traditional sense, when one thinks of power quality, visions of classical waveforms containing 3rd, 5th, and 7th, etc. harmonics appear. It is from this perspective that the IEEE in cooperation with utilities, industry, and academia began to attack the early problems deemed "power quality" in the 1980s. However, in the last 5 to 10 years, the term *power quality* has come to mean much more than simple power system harmonics. Because of the prolific growth of industries that operate with sensitive electronic equipment (e.g., the semiconductor industry), the term power quality has come to encompass a whole realm of anomalies that occur on a power system. Without much effort, one can find a working group or a standards committee at various IEEE meetings in which there is much lively debate on the issue of what power quality entails and how to define various power quality events. The IEEE Emerald Book (IEEE Standard 1100-1999, *IEEE Recommended Practice for Power and Grounding Electronic Equipment*) defines power quality as:

> The concept of powering and grounding electronic equipment in a manner that is suitable to the operation of that equipment and compatible with the premise wiring system and other connected equipment.

The increased use of power electronic devices at all levels of energy consumption has forced the issue of power quality out of abstract discussions of definitions down to the user level. For example, a voltage sag of only a few cycles could cause the variable frequency drives or programmable logic controllers (PLC) on a large rolling mill to "trip out" on low voltage, thereby causing lost production and costing the company money. The question that is of concern to everyone is, "What initiated the voltage sag?" It could have been a fault internal to the facility or one external to the facility (either within a neighboring facility or on the utility system). In the latter two cases, the owner of the rolling mill will place blame on the utility and, in some cases, seek recompense for lost product.

Another example is the semiconductor industry. This segment of industry has been increasingly active in the investigation of power quality issues, for it is in this volatile industry that millions of dollars can possibly be lost due to a simple voltage fluctuation that may only last two to four cycles. Issues of power quality have become such a concern in this industry that Semiconductor Equipment and Materials International (SEMI), the worldwide trade association for the semiconductor industry, has been working to produce power quality standards explicitly related to the manufacturing of equipment used within that industry.

Two excellent references on the definitions, causes, and potential corrections of power quality issues are Refs. 1 and 2. The purpose of this section, however, is to address some of the main concerns regarding the power quality related to interfacing with a utility. These issues are most directly addressed in IEEE 519-1992 [3] and IEEE/ANSI C62.41 [4].

Harmonics and IEEE 519

Harmonic generation is attributed to the application of nonlinear loads (i.e., loads that when supplied a sinusoidal voltage do not draw a sinusoidal current). These nonlinear loads not only have the potential to create problems within the facility that contains the nonlinear loads but also can (depending on the stiffness of the utility system supplying energy to the facility) adversely affect neighboring facilities. IEEE 519-1992 [3] specifically addresses the issues of steady-state limits on harmonics as seen at the *point of common coupling* (PCC). It should be noted that this standard is currently under revision and more information on available drafts can be found at http://standards.ieee.org.

The whole of IEEE 519 can essentially be summarized in several of its own tables. Namely, Tables 10.3 through 10.5 in Ref. 3 summarize the allowable harmonic current distortion for systems of 120 V to 69 kV, 69.001 kV to 161 kV, and greater than 161 kV, respectively. The allowable current distortion (defined in terms of the total harmonic distortion, THD) is a function of the stiffness of the system at the PCC, where the stiffness of the system at the PCC is defined by the ratio of the maximum short-circuit current at the PCC to the maximum demand load current (at fundamental frequency) at the PCC. Table 11.1 in Ref. 3 provides recommended harmonic voltage limits (again in terms of THD). Tables 10.3 and 11.1 are of primary interest to most facilities in the application of IEEE 519. The total harmonic distortion (for either voltage or current) is defined as the ratio of the rms of the harmonic content to the rms value of the fundamental quantity expressed in percent of the fundamental quantity. In general, IEEE 519 refers to this as the distortion factor (DF) and calculates it as the ratio of the square root of the sum of the squares of the rms amplitudes of all harmonics divided by the rms amplitude of the fundamental all times 100%. The PCC is, essentially, the point at which the utility ceases ownership of the equipment and the facility begins electrical maintenance (e.g., the secondary of a service entrance transformer for a small industrial customer or the meter base for a residential customer).

Surge Voltages and C62.41

Reference 4 is a *guide* (lowest level of standard) for characterizing the ability of equipment on low-voltage systems (<1000 V) to withstand voltage surges. This guide provides some practical basis for selecting appropriate test waveforms on equipment. The primary application is for residential, commercial, and industrial systems that are subject to lightning strikes because of their close proximity (electrically speaking) to unshielded overhead distribution lines. Certain network switching operations may also result in similar voltage transients being experienced.

Other Standards Addressing Utility Interface Issues

Many power quality standards are at present in existence and are under constant revision. The following standards either directly or indirectly address issues with the utility interface and can be applied accordingly: IEEE 1159 for the monitoring of power quality events, IEEE 1159.3 for the exchange of measured power quality data, IEEE P1433 for power quality definitions, IEEE P1531 for guidelines regarding harmonic filter design, IEEE P1564 for the development of sag indices, IEEE 493 (the *Gold Book*) for industrial and commercial power system reliability, IEEE 1346 for guidelines in evaluating component compatibility with power systems (this guideline is an attempt to better quantify the CBEMA and ITIC curves), C84.1 for voltage ratings of power systems and equipment, IEEE 446 (the *Orange Book*) for emergency and standby power systems (this standard contains the so-called power acceptability curves), IEEE 1100 (the *Emerald Book*), IEEE 1409 for development of guidelines for the application of power electronic devices/technologies for power quality improvement on distribution systems, and IEEE P1547 for the power quality issues associated with distributed generation resources.

As previously mentioned, the IEEE is not the only organization to continue investigation into the impacts of nonlinear loads on the utility system. Other organizations such as CIGRE, UL, NEMA, SEMI, IEC, and others all play a role in these investigations.

References

1. IEEE Standards Board, *IEEE Recommended Practice for Powering and Grounding Electronic Equipment*, IEEE Std. 1100-1999.
2. R. C. Dugan, M. F. McGranaghan, and H. W. Beaty, *Electrical Power Systems Quality*, McGraw-Hill, New York, 1996.
3. IEEE Standards Board, *IEEE Recommended Practices and Requirements for Harmonic Control in Electrical Power Systems*, IEEE Std. 519-1992.
4. IEEE Standards Board, *IEEE Recommended Practice on Surge Voltages in Low-Voltage AC Power Circuits*, IEEE Std. C62.41-1991.

17.2 Power Quality Considerations

Timothy L. Skvarenina

Harmonics

In the past, utilities had the responsibility to provide a single-frequency voltage waveform, and for the most part, customers' loads had little effect on the voltage waveform. Now, however, power electronics are used widely and create nonsinusoidal currents that contain many harmonic components. Harmonic currents cause problems in the power system and for other loads connected to the same portion of the power system. Because utility customers can now cause electrical problems for themselves and others, the Institute of Electrical and Electronic Engineers (IEEE) developed IEEE Standard 519, which places the responsibility of controlling harmonics on the user as well as the utility. This section describes harmonics, their cause, and their effects on the system voltage and components.

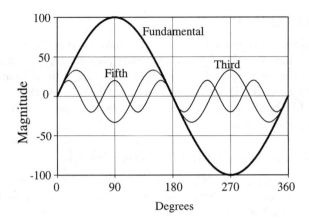

FIGURE 17.1 Fundamental, third, and fifth harmonics.

What Are Harmonics?

Ideally, the waveforms of all the voltages and currents in the power system would be single-frequency (60 Hz in North America) sine waves. The actual voltages and currents in the power system, however, are not purely sinusoidal, although in the steady state they do look the same from cycle to cycle; i.e., $f(t + T) = f(t)$, where T is the period of the waveform and t is any value of time. Such repeating functions can be viewed as a series of components, called harmonics, whose frequencies are integral multiples of the power system frequency. The second harmonic for a 60-Hz system is 120 Hz, the third harmonic is 180 Hz, etc. Typically, only odd harmonics are present in the power system.

Figure 17.1 shows one cycle of a sinusoid (labeled as the fundamental) with a peak value of 100. The fundamental is also know as the first harmonic, which would be the nominal frequency of the power system. Two other waveforms are shown on the figure—the third harmonic with a peak of 50 and the fifth harmonic with a peak of 20. Notice that the third harmonic completes three cycles during the one cycle of the fundamental and thus has a frequency three times that of the fundamental. Similarly, the fifth harmonic completes five cycles during one cycle of the fundamental and thus has a frequency five times that of the fundamental. Each of the harmonics shown in Fig. 17.1 can be expressed as a function of time:

$$V_1 = 100\sin(\omega t), \quad V_3 = 50\sin(3\omega t), \quad V_5 = 20\sin(5\omega t) \quad (17.1)$$

Equation 17.1 shows three harmonic components of voltage or current that could be added together in an infinite number of ways by varying the phase angles of the three components. Thus, an infinite number of waveforms could be produced from these three harmonic components. For example, suppose V_3 is shifted in time by 60° and then added to V_1 and V_5. In this case, all three waveforms have a positive peak at 90° and a negative peak at 270°. One half cycle of the resultant waveform is shown in Fig. 17.2, which is clearly beginning to look like a pulse. In this case, we have used the harmonic components to synthesize a waveform. Generally, we would have a nonsinusoidal voltage or current waveform and would like to know its harmonic content. The question, then, is how to find the harmonic components given a waveform that repeats itself every cycle.

Fourier, the mathematician, showed that it is possible to represent any periodic waveform by a series of harmonic components. Thus, any periodic current or voltage in the power system can be represented by a Fourier series. Furthermore, he showed that the series can be found, assuming the waveform can be expressed as a mathematical function. We will not go into the mathematics behind the solution of Fourier series here; however, we can use the results. In particular, if a waveform $f(t)$ is periodic, with period T, then it can be approximated as

$$f(t) = a_0 + a_1\sin(\omega t + \theta_1) + a_2\sin(2\omega t + \theta_2) + a_3\sin(3\omega t + \theta_3) + \cdots + a_n\sin(n\omega t + \theta_n) \quad (17.2)$$

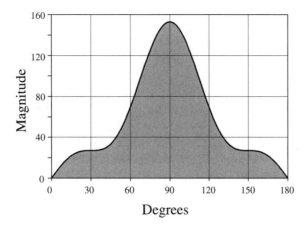

FIGURE 17.2 Pulse wave formed from the three harmonics in Eq. 17.1 with 60° shift for V_3.

where a_0 represents any DC (average) value of the waveform, a_1 through a_n are the Fourier amplitude coefficients, and θ_1 through θ_n are the Fourier phase coefficients. The amplitude coefficients are always zero or positive and the phase coefficients are all between 0 and 2π radians. As "n" gets larger, the approximation becomes more accurate.

For example, consider an alternating square wave of amplitude 100. The Fourier series can be shown to be

$$V_{\text{square}} = 100 \sum_{n=1}^{\infty} \frac{1}{2n-1} \sin[(2n-1)\omega t] \qquad (17.3)$$

Since the alternating waveform has zero average value, the coefficient a_0 is zero. Note also that only odd harmonics are included in the series given by Eq. (17.3), since $(2n - 1)$ will always be an odd number, and all of the phase coefficients are zero. Expanding the first five terms of Eq. (17.3) yields:

$$V_{\text{square}} = 100\left[\sin(\omega t) + \frac{1}{3}\sin(3\omega t) + \frac{1}{5}\sin(5\omega t) + \frac{1}{7}\sin(7\omega t) + \frac{1}{9}\sin(9\omega t)\right] \qquad (17.4)$$

Figure 17.3 shows one cycle for the waveform represented by the right-hand side of Eq. (17.3). Although only the first five terms of the Fourier series were used in Fig. 17.3, the resultant waveform already resembles a square wave. Harmonics have a number of effects on the power system as will be seen later, but for now we would like to have some way to indicate how large the harmonic content of a waveform is. One such figure of merit is the total harmonic distortion (THD).

Total harmonic distortion can be defined two ways. The first definition, in Eq. (17.5), shows the THD as a percentage of the fundamental component of the waveform, designated as THD_F. This is the IEEE definition of THD and is used widely in the United States.

$$\text{THD}_F = \frac{\sqrt{\sum_{h=2}^{\infty} V_{h\text{rms}}^2}}{V_{1\text{rms}}} \times 100\% \qquad (17.5)$$

In Eq. (17.5), $V_{1\text{rms}}$ is the rms of the fundamental component and $V_{h\text{rms}}$ is the amplitude of the harmonic component of order "h" (i.e., the "hth" harmonic). Although the symbol "V" is used in Eqs. (17.5) to (17.10), the equations apply to either current or voltage. The rms of a waveform composed of harmonics

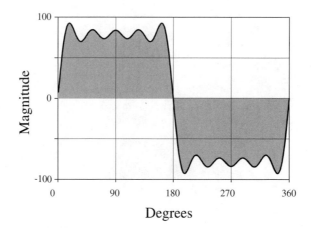

FIGURE 17.3 Approximation to a square wave using the first five terms of the Fourier series.

is independent of the phase angles of the Fourier series, and can be calculated from the rms values of all harmonics, including the fundamental:

$$V_{rms} = \sqrt{\sum_{h=1}^{\infty} V_{hrms}^2} \qquad (17.6)$$

Because the series in Eq. (17.6) has only one more term (the rms of the fundamental) than the series in the numerator of Eq. (17.5), we can also find the total rms in terms of percent THD_F.

$$V_{rms} = V_{1rms} \sqrt{1 + \left(\frac{\%THD_F}{100}\right)^2} \qquad (17.7)$$

In the opinion of some, Eq. (17.5) exaggerates the harmonic problem. Thus, another technique is also used to calculate THD. The alternate method, designated as THD_R, calculates THD as a percentage of the total rms instead of the rms of the fundamental. From Eq. (17.7), it is clear that the total rms will be larger than the rms of the fundamental, so such a calculation will yield a lower value for THD. This definition is used by the Canadian Standards Association and the IEC:

$$THD_R = \frac{\sqrt{\sum_{h=2}^{\infty} V_{hrms}^2}}{V_{rms}} \times 100\% \qquad (17.8)$$

The value for THD_R can be obtained from THD_F by multiplying by V_{1rms} and dividing by V_{rms}.

$$THD_R = THD_F \times \left(\frac{I_{1rms}}{I_{rms}}\right) \qquad (17.9)$$

Substituting Eq. (17.7) into Eq. (17.9) yields another expression for THD_R in terms of THD_F:

$$THD_R = \frac{THD_F}{\sqrt{1 + \left(\frac{\%THD_F}{100}\right)^2}} \qquad (17.10)$$

THD_R, as given by Eq. (17.8) and (17.10), will always be less than 100%. THD is very important because the IEEE Standard 519 specifies maximum values of THD for the utility voltage and the

Power Quality and Utility Interface Issues

customer's current. Having considered what harmonics are, we can now look at some of their properties. The next section deals with the phase sequence of various harmonics.

Harmonic Sequence

In a three-phase system, the rotation of the phasors is assumed to have an A-B-C sequence as shown in Fig. 17.4a. As the phasors rotate, phase A passes the *x*-axis, followed by phase B and then phase C. An A-B-C sequence is called the *positive sequence*. However, phase A could be followed by phase C and then phase B, as shown in Fig. 17.4b. A set of phasors whose sequence is reversed is called the *negative sequence*. Finally, if the waveforms in all three phases were identical, their phasors would be in line with each other as shown in Fig. 17.4c. Because there are no phase angles between the three phases, this set of phasors is call the *zero sequence*.

When negative and zero sequence currents and voltages are present along with the positive sequence, they can have serious effects on power equipment. Not all harmonics have the same sequence; in fact, the sequence depends on the number of the harmonic, as shown in Fig. 17.5. Figure 17.5a, b, and c show the fundamental component of a three-phase set of waveforms (voltage or current) as well as their second harmonics. In each case, the phase-angle relationship has been chosen so both the fundamental and the second harmonic cross through zero in the ascending direction at the same time.

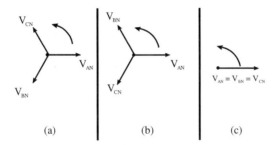

FIGURE 17.4 Positive (a), negative (b), and zero (c) sequences.

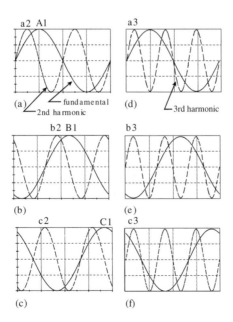

FIGURE 17.5 First, second, and third harmonics.

To establish the sequence of the fundamental components, label the positive peak values of the three phases A1, B1, and C1. Clearly, A1 occurs first, then B1, and finally C1. Thus, we can conclude that the fundamental component has an A-B-C, or positive, sequence. In fact, it was chosen to have a positive sequence. Given that the fundamental has a positive sequence, we can now look at other harmonics. In a similar manner, the first peak of each of the second harmonics are labeled a2, b2, and c2. In this case, a2 occurs first, but it is followed by c2 and then b2. The second harmonic thus has an A-C-B, or negative, sequence.

Now consider Fig. 17.5d, e, and f, which also show the same fundamental components, but instead of the second harmonic, the third harmonic is shown. Both the fundamental and third harmonics were chosen so they cross through zvoero together. When the peaks of the third harmonics are labeled as a3, b3, and c3, it is evident that all three occur at the same time. Since the third harmonics are concurrent, they have no phase order. Thus, they are said to have zero sequence. If the process in Fig. 17.5 was continued, the fourth harmonic would have a positive sequence, the fifth a negative sequence, the sixth a zero sequence, and so on.

All harmonics whose order is $3n$, where n is any positive integer, are zero sequence and are called *triplen* harmonics. Triplen harmonics cause serious problems in three-phase systems as discussed later in this section. First, however, consider what causes harmonics in the power system.

Where Do Harmonics Come From?

Electrical loads that have a nonlinear relationship between the applied voltages and their currents cause harmonic currents in the power system. Passive electric loads consisting of resistors, inductors, and capacitors are linear loads. If the voltage applied to them consists of a single-frequency sine wave, then the current through them will be a single-frequency sine wave as well. Power electronic equipment creates harmonic currents because of the switching elements that are inherent in their operation. For example, consider a simple switched-mode power supply used to provide DC power to devices such as desktop computers, televisions, and other single-phase electronic devices.

Figure 17.6 shows an elementary power supply in which a capacitor is fed from the power system through a full-wave, diode bridge rectifier. The instantaneous value of the AC source must be greater than the voltage across the capacitor for the diodes to conduct. When first energized, the capacitor charges to the peak of the AC waveform and, in the absence of a load, the capacitor remains charged and no further current is drawn from the source.

If there is a load, then the capacitor acts as a source for the load. After the capacitor is fully charged, the AC voltage waveform starts to decrease, and the diodes shut off. While the diode is off, the capacitor discharges current to the DC load, which causes its voltage, V_{dc}, to decrease. Thus, when the AC source becomes larger than V_{dc} during the next half-cycle, the capacitor draws a pulse of current to restore its charge.

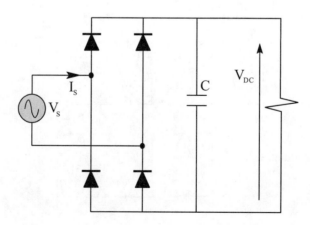

FIGURE 17.6 Simple single-phase switch-mode power supply.

Power Quality and Utility Interface Issues

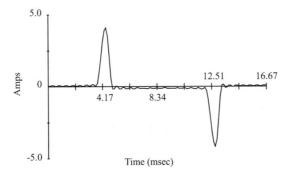

FIGURE 17.7 Input current to single-phase, full-wave rectifier.

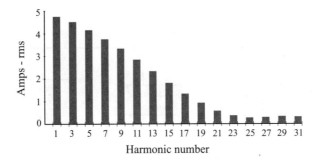

FIGURE 17.8 Harmonic spectrum of current for the circuit shown in Fig. 17.6.

Figure 17.7 shows the current of such a load (actually the input current to a variable-speed motor drive). Since the current has a repetitive waveform, it is composed of a series of harmonics. The harmonics can be found using a variety of test equipment with the capability to process a fast Fourier transform (FFT). This particular waveform has a large amount of harmonics, as shown by the harmonic spectrum (through the 31st harmonic) in Fig. 17.8. Note that the first several harmonics after the fundamental are almost as large as the fundamental. This waveform, as shown in Fig. 17.7, has a peak value of 4.25 A, but the rms of the waveform is only 1.03 A. This leads to another quantity that is an indicator of harmonic distortion. The *crest factor* (CF) is defined as the ratio of the peak value of the waveform divided by the rms value of the waveform:

$$\text{CF} = \frac{\text{peak of waveform}}{\text{rms of waveform}} \qquad (17.11)$$

For the current shown in Fig. 17.7, the crest factor is 4.25 divided by 1.03, or 4.12. For a sinusoidal current or voltage, the crest factor would be the square root of 2 (1.414). Waveforms whose crest factor are substantially different from 1.414 will have harmonic content. Note that the crest factor can also be lower than 1.414. A square wave, for example, would have a CF of 1.

As shown in Fig. 17.8, the third harmonic of a single-phase bridge rectifier is very large. Putting such loads on the three phases of a three-phase, wye-connected system could cause problems because the third harmonics add on the neutral conductor. The best way to handle these problems is to eliminate the triplen harmonics.

Whereas single-phase rectifiers require a large amount of triplen current, three-phase bridge rectifiers do not. Figure 17.9 shows the input current and harmonic content for a three-phase bridge rectifier (again, the input current to a variable-frequency motor drive). In this case, the phase current contains two pulses in each half-cycle, which results in the elimination of all the triplen harmonics. Examination of

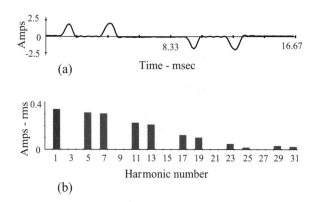

FIGURE 17.9 Line current and harmonic content for three-phase bridge rectifier.

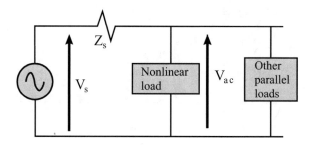

FIGURE 17.10 Simple single-phase power system.

the spectrum in Fig. 17.9 shows that the only harmonics that remain are those whose order numbers are of the form:

$$h = 6n \pm 1 \qquad (17.12)$$

where n is any positive integer, beginning with 1. Setting $n = 1$, indicates the 5th and 7th harmonics will be present, $n = 2$ yields the 11th and 13th harmonics, and so on.

Harmonic currents have many impacts on the power system, both on the components of the system as well as the voltage. The next section considers some of these effects.

Effects of Harmonics on the System Voltage

A simple circuit representing a single-phase power system is shown in Fig. 17.10. In North America, the utility generates a 60-Hz sinusoidal voltage, indicated by the ideal source. However, the load current flows through transmission lines, transformers, and distribution feeders, which all have impedance. The impedance of the system is represented in Fig. 17.10 by Z_s. Finally, the load for this system is considered to be a nonlinear load in parallel with other loads.

Harmonic currents drawn from the power system by nonlinear loads create harmonic voltages ($RI + j\omega_h LI$) across the system impedance, and their effect can be significant for higher-order harmonics because inductive reactance increases with frequency. The load voltage is the difference between the source voltage and the voltage drop across the system impedance. Since the voltage drop across the system impedance contains harmonic components, the load voltage may become distorted if the nonlinear loads are a large fraction of the system capacity.

Referring back to Fig. 17.6, note the current pulse drawn by the rectifier occurs only when the AC source voltage is near its peak. This means the voltage drop across the source impedance will be large when

Power Quality and Utility Interface Issues

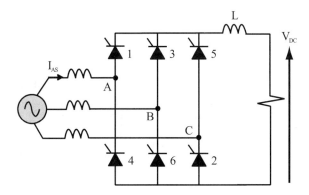

FIGURE 17.11 Three-phase bridge rectifier.

the source voltage is near its peak and essentially zero during the remainder of the half-cycle. Thus, the voltage delivered to the load will be "flattened" by the subtraction of the system impedance voltage drop. Unfortunately, some power electronic devices, such as the rectifier front-end of motor drives, are sensitive to the peak value of the AC voltage waveform, and may shut down or operate incorrectly when the incoming AC voltage is distorted. Voltage distortion affects the nonlinear load that created the harmonics and any other load that is connected in parallel with it. The interface between the loads and the power system is called the point of common coupling (PCC), and the PCC is where the harmonic content of system voltage and current must be controlled to comply with IEEE Standard 519. Although three-phase rectifiers do not cause triplen harmonic currents, they do cause another problem as a result of their operation.

Notching

A three-phase bridge rectifier is shown in Fig. 17.11, the details of which are described in Chapter 4 of this handbook. However, consider briefly how the diodes operate. Each diode in the top or bottom half conducts while one or two diodes in the oppositie half conduct. For example, diode 1 is connected to phase A and conducts during the period of time when diode 6 (phase B) and diode 2 (phase C) are conducting. Clearly, diode 1 should not conduct when diode 4 is conducting as that would constitute a short circuit. The inductor in series with the DC load tends to keep the current constant, so current must be passed from one diode to another. This transfer of the load current from one diode to another is called *commutation*. While diode 1 is conducting in the upper half of the bridge, the current in the lower half of the bridge will commutate from diode 6 to diode 2. Since the three-phase source has inductance as well, this transfer of current cannot occur instantaneously. Instead, the current in diode 2 must increase while the current in diode 6 decreases.

While it is conducting, a diode is essentially a short circuit, so during the commutation interval, two diodes in one side of the bridge are conducting. This results in two phases of the source being shorted together. For example, while the load current commutates from diode 6 to diode 2, points B and C are connected together, which means the voltage from B to ground and from C to ground is the same. The effect of commutation is to create a notch in the voltage waveform. Figure 17.12 shows the voltage from A to ground as calculated by a simulation of a three-phase bridge rectifier. The notching effect is evident, six times per cycle.

Notching is a repetitive event and the voltage waveform shown in Fig. 17.12 could be represented by a Fourier series. However, the order of the harmonics is extremely high, well above the range of many monitors normally used for making power quality measurements. Thus, notching is a special case somewhere between harmonics and transients. Devices connected in parallel with the bridge rectifier could be affected by notching, especially if the rectifier load is large relative to the size of the system from which it is fed.

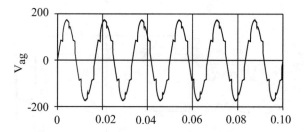

FIGURE 17.12 Voltage notching of the AC source voltage due to commutation of diodes in a three-phase rectifier.

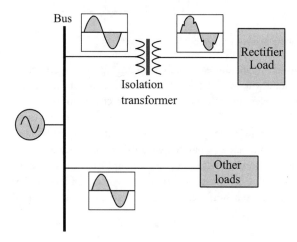

FIGURE 17.13 Use of an isolation transformer to keep notching from affecting other loads.

An isolation transformer can be used to supply the offending equipment and thus reduce the amount of notching seen by other loads. Figure 17.13 shows a rectifier load and other loads fed from a common bus with an isolation transformer between the rectifier load and the bus. The voltage on the secondary of the transformer is notched; however, the voltage on the primary side is relatively unaffected because the impedance of the isolation transformer tends to smooth out the notches. Thus, the other loads do not see the notching or at least see much smaller notches in the voltage waveform.

Effects of Harmonics on Power System Components

Harmonic currents from nonlinear loads can seriously affect electric power distribution equipment. Components that may be affected include transformers, conductors, circuit breakers, bus bars and connecting lugs, and electrical panels. Harmonic problems can occur in both single-phase and three-phase systems.

Conductors

Higher-order harmonic current components cause additional I^2R heating in every conductor through which they flow, because conductor resistance increases with frequency as a result of the skin effect. This means that as the frequency of a current increases, its ability to "soak" into a conductor is reduced, resulting in a higher current density at the edge of the conductor than at its center. A conductor can be carrying rated current (rms amps) and still overheat if the current contains significant higher-order harmonics. Because every conductor carrying the harmonic currents will have increased losses, there will be more heat to be dissipated in the system and the overall efficiency of the system will be reduced.

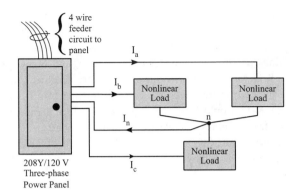

FIGURE 17.14 Three-phase power system with balanced, single-phase nonlinear loads.

Three-Phase Neutral Conductors

Triplen harmonics pose a problem for the neutral conductor in three-phase, wye-connected systems, such as the one shown in Fig. 17.14. Therein, a feeder circuit provides three-phase power to a circuit breaker panel board from which branch circuits provide power to outlets and lighting, including three single-phase loads connected via a four-wire branch circuit. When identical linear loads are placed on each of the three phases, the phase currents add to zero at point "n" and no current flows on the neutral wire. Again assuming linear loads, then even if the load are not identical, the current in the neutral could not be higher than the highest phase current.

If the loads in Fig. 17.14 are nonlinear, there will be harmonic currents in each phase. For balanced loads, the fundamentals and all non-triplen harmonic currents add to zero at the neutral point. If triplen harmonics are present in the phase currents, however, they will be in phase and add directly on the branch circuit and feeder neutrals. Since the neutral conductors carry the sum of the triplens from the three phases, the neutral current can actually exceed the current in the phase conductors. Since neutral conductors are not protected by circuit breakers, this can damage to the conductors.

To find the current in the neutral, we must recognize that all positive and negative sequence harmonics from the three phases will cancel out at the neutral point. The triplen harmonics, on the other hand, will *add* together at the neutral:

$$I_{N\text{rms}} = 3 \times (I_{3\text{rms}}^2 + I_{6\text{rms}}^2 + I_{9\text{rms}}^2 + \cdots)^{1/2} \qquad (17.13)$$

From Eq. (17.13), it is evident that the neutral current is three times the rms of all the triplens on one phase of the system.

Transformers

Current flow in the windings and flux in the ferromagnetic core cause real power losses in a transformer. Because of their higher frequencies, harmonic currents cause additional losses in every conductor through which they flow, including the conductors of the transformer coils. Harmonic currents in the windings create harmonic flux components in the core of the transformer, which cause additional hysteresis and eddy current losses in the steel. Hysteresis loss is proportional to the frequency of the magnetic flux, and eddy currents are proportional to frequency squared. Thus, harmonic currents can cause significant increases in the core loss of the transformer. These additional losses may result in transformer overheating and electrical insulation failure.

To provide for the effects of nonlinear loads, manufacturers build specially designed transformers, called "K-factor rated," that are capable of supplying rated output current to loads with a specific level of harmonic content. K-factor-rated transformers have larger conductors in the windings and thinner, low-loss steel laminations in the core to reduce the losses. A transformer with a K-factor of K-1 is rated only for single-frequency current; thus, if the load is nonlinear, the transformer cannot provide rated current without overheating. Transformers rated K-4, K-9, K-13, and higher are available to provide power to nonlinear loads. K-factors of K-4 or K-9 indicate the transformer can supply rated current to loads that would increase the eddy current loss of a K-1 transformer by a factor of 4 or 9, respectively. Transformers rated K-9 or K-13 would likely be required for office areas containing many desktop computers, copy machines, fax machines, and electronic lighting ballasts. A large variable-speed motor drive could require a transformer rated K-30 or higher. The K-factor of a load can be calculated, if the harmonic components are known, as follows:

$$K = \sum h^2 \left(\frac{I_{h,\mathrm{rms}}}{I_{\mathrm{tot,rms}}}\right)^2 \tag{17.14}$$

where h is the harmonic order number, $I_{h,\mathrm{rms}}$ is the rms of the harmonic current whose frequency is "h" times the fundamental frequency, and $I_{\mathrm{tot,rms}}$ is the rms of the total current.

Effects of Harmonics on System Power Factor

Earlier, Eq. (17.6) showed that the addition of harmonic currents to the fundamental component increases the total rms current. Because they affect the rms value of the current, harmonics will affect the power factor of the circuit. Consider the voltage and current waveforms shown in Fig. 17.15 in which current lags the voltage by an angle θ. The apparent power of the circuit would be found by multiplying the rms voltage magnitude by the rms current magnitude. Power factor, F_p, is then defined as the ratio of the real power to the apparent power:

$$F_p = \frac{P}{V_{\mathrm{rms}} I_{\mathrm{rms}}} = \cos(\theta) \tag{17.15}$$

For linear loads, the phase shift (time displacement) between voltage and current results in different values for real power and apparent power. Since the current can only lag or lead the voltage by 0 to 90°, the power factor will always be positive and less than or equal to 1.

Instead of a sinusoidal current, suppose the current and voltage shown by Fig. 17.16. The current is the quasi-square wave, consisting of the Fourier series shown in Eq. (17.4). The voltage is a sine wave,

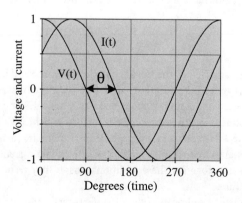

FIGURE 17.15 Voltage and current for a lagging load.

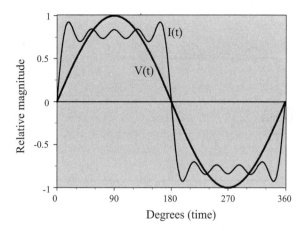

FIGURE 17.16 Sinusoidal voltage and quasi-square wave current.

which is in phase with the fundamental harmonic component of the current. The power can be found as a function of time by multiplying the voltage times the current at each time step.

Because the voltage consists of a single component, the power is a series of terms consisting of the voltage times each harmonic component of current. The first term of the series is of the form $\sin^2 \omega t$ since the voltage is in phase with the fundamental current component. Obviously, this term is always positive; therefore, it indicates real power (energy) being delivered to the load.

The remaining terms contain the product of the fundamental frequency voltage and one of the higher-order harmonic current components. Multiplying two sinusoidal waveforms of different frequencies creates a sinusoidal waveform, which has a zero average value. Thus, none of higher-order harmonic currents produces real power if the voltage is a single frequency. Substituting Eq. (17.7) for the total rms current into Eq. (17.15) yields a new expression for the power factor:

$$F_{p\,tot} = \frac{P}{V_{1rms}I_{rms}} = \frac{P}{V_{1rms}I_{1rms}\sqrt{1+\left(\frac{\%THD_F}{100}\right)^2}} \quad (17.16)$$

where the current $\%THD_F$ is used in the denominator of Eq. (17.16). Rewriting Eq. (17.16):

$$F_{p\,tot} = \frac{P}{V_{1rms}I_{1rms}} \times \frac{1}{\sqrt{1+\left(\frac{\%THD_F}{100}\right)^2}} \quad (17.17)$$

In Eqs. (17.16) and (17.17), the subscript "tot" indicates the *total power factor*, which is sometimes called the *true power factor*. The total power factor in Eq. (17.17) is the product of two components, the first of which is called the *displacement power factor*:

$$F_{p\,disp} = \frac{P}{V_{1rms}I_{1rms}} \quad (17.18)$$

The second component of the total power factor is the *distortion power factor*, which results from the harmonic components in the current:

$$F_{p\,disp} = \frac{1}{\sqrt{1+\left(\frac{\%THD_F}{100}\right)^2}} \quad (17.19)$$

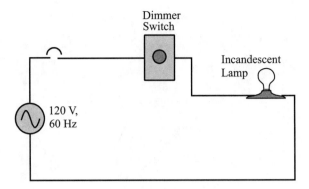

FIGURE 17.17 Simple incandescent lamp dimmer circuit.

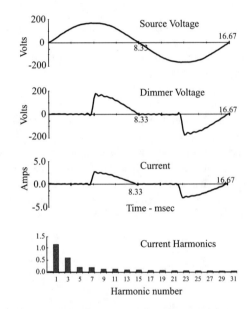

FIGURE 17.18 Source voltage, lamp voltage, lamp current, and current harmonic spectrum for the system shown in Fig. 17.17.

In the event that the voltage also has harmonic components, then the distortion power factor would be the product of two terms similar to the right side of Eq. (17.19), one for the voltage and one for the current. However, voltage distortion is normally very low compared with the current distortion.

Power electronic devices can cause unusual results with respect to power factor. The circuit shown in Fig. 17.17 consists of an incandescent lamp fed by a simple wall-mounted dimmer. Incandescent lamps operate at essentially unity power. In this case, the lamp voltage was set at 85 V rms by adjusting the dimmer switch. The voltage was observed at the source and at the lamp and the circuit current was measured, all with a harmonics analyzer.

The results are shown in Fig. 17.18, where the top waveform is the source voltage, with a 60-Hz component of 118 V and a 5th harmonic (300-Hz) component of 1.6 V. The second trace shows the lamp voltage, i.e, the dimmer output. At this voltage setting, the dimmer was conducting for approximately one half (90°) of each half-cycle, as shown. The bottom waveform shows the lamp current. Because the incandescent lamp is essentially a resistive load, the current waveform looks identical to the lamp voltage, except for the scale. The bar chart in Fig. 17.18 shows the harmonic spectrum of the current, which, except for the scale, was identical to the voltage spectrum.

Since the lamp voltage and current have the same shape and are in phase, the harmonic components of current and voltage are in phase. With no phase angle between each of the voltage and current components, the power factor of the lamp is unity, and that was found to be the case with the harmonic analyzer. Because the dimmer creates harmonic voltages to the lamp, each harmonic of voltage and its respective current delivers some power to the lamp at unity power factor.

Looking at the source results in a much different picture. The source voltage consists almost solely of a single frequency, whereas the current contains all of the harmonics shown in Fig. 17.18. Since the product of two sine waves of different frequencies is another sine wave, only the fundamental harmonic of current can deliver real power to the circuit. The harmonics analyzer showed that the fundamental component of the current lagged the source voltage by 28°. Taking the cosine of 28° results in a measured displacement power factor of 0.88 at the source.

THD_F for the current was found to be 60.7%, and the distortion power factor was calculated to be 0.855 from Eq. (17.19). The product of the distortion power factor and the displacement power factor yields the total power factor, 0.75 in this case. Thus, the incandescent lamp, a resistive load, appears to the power system as a 0.75 power factor lagging load. Low power factor results in higher losses in the system due to higher I^2R losses. In fact, both I and R increase in this case because the rms current is higher due to the harmonics and because skin effect causes higher resistance in the conductors. While a single lamp on a dimmer switch does not seriously affect the power system, very large nonlinear loads, such as a DC motor drive, could require the installation of harmonic filters to reduce the distortion power factor. Passive harmonic filters are briefly described here and in more detail in Section 17.3.

Power Factor Correction Capacitors

Many industrial loads are inductive, so capacitors are often used to improve the power factor. Although capacitors do not cause harmonics, they can resonate with the inductance of the power system. When resonant frequencies occur near harmonic frequencies, capacitors can amplify the harmonic currents created by nonlinear loads. Figure 17.19 shows a power circuit including power factor correction capacitors. The parallel combination of the system inductance and the power factor correction capacitors has a resonant frequency, f_r. The resonant frequency is given by

$$f_r = \frac{1}{\sqrt{LC}} \quad (17.20)$$

where L is the system inductance (X_s divided by 2π), and C is the capacitance.

Normally, we do not deal with inductance and capacitance, however. It is much more convenient to express the resonant frequency in other terms. In particular, we normally size power factor capacitors in kVAR. Figure 17.19 also shows a switch that can be closed to create a short circuit. If the switch is closed to short out the loads, the source voltage will be dropped across the system impedance, which in this

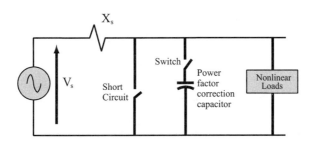

FIGURE 17.19 Circuit demonstrating how resonance can form with power factor correction.

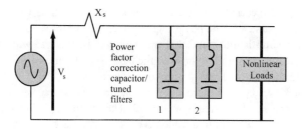

FIGURE 17.20 Use of harmonic filters.

case is considered to be inductive. Thus, the short-circuit kVA can be calculated as

$$\text{kVA}_{sc} = \frac{V^2}{X_s} \tag{17.21}$$

The utility normally provides the available short-circuit capacity upon request. Neglecting the voltage drop across X_s during normal operation, the total kVAR of the capacitance would be

$$\text{kVAR}_{cap} = \frac{V^2}{X_{cap}} \tag{17.22}$$

Since $X_s = 2\pi f L$ and $X_{cap} = 1/(2\pi f C)$, it can be shown from Eqs. (17.20) through (17.22) that

$$h_r \cong \sqrt{\frac{\text{kVA}_{sc}}{\text{kVAR}_{cap}}} \tag{17.23}$$

where h_r is the multiple of the system frequency at which the resonance occurs.

For example, if h_r is five, the resonant frequency is 300 Hz for a 60-Hz power system. Unfortunately, it is not uncommon for the value calculated by Eq. (17.23) to be near the 5th harmonic, which, as we have seen in Fig. 17.9, is the dominant harmonic for some three-phase bridge rectifiers. When the capacitors cause a resonance near one of the harmonics, the original harmonic current can be amplified by as much as a factor of 16, which can in turn cause excessive voltage drop and voltage distortion, damage to the capacitors, and lower power factor.

When harmonics cause serious voltage distortion, tuned filters can be used to reduce the amount of harmonic current drawn from the source. Figure 17.20 shows a circuit with two filters, each designed to reduce the effects of one particular harmonic. The inductance added in series with the capacitor should be chosen to create a series resonance frequency that is slightly below the frequency of the harmonic that is to be reduced. For example, if it was desired to reduce the 5th and 7th harmonics, then the filters would be designed to have resonance frequencies about 4.7 and 6.7 times the normal system frequency. This allows for tolerances in the actual values of the devices and causes the majority of the 5th and 7th harmonic currents to be diverted through the filters. A small portion of the harmonic current is still supplied by the source.

IEEE Standard 519

Recognizing the problems caused by nonlinear loads, the IEEE Standards board approved a revised and renamed Standard 519 in the fall of 1992. The 1981 version of the standard was titled, "Guide for Harmonic Control and Reactive Compensation of Static Power Converters." The 1981 version recommended specific limits for voltage THD from the utility, but did not recognize the possibility of customer load currents causing voltage distortion. The 1992 version was titled, "IEEE Recommended Practices and

TABLE 17.1 Harmonics Allowed by IEEE Standard 519

SCR	$h < 11$	11–15	17–21	23–33	$h > 33$	% THD
<20	4.0	2.0	1.5	0.6	0.3	5.0
20–50	7.0	3.5	2.5	1.0	0.5	8.0
50–100	10.0	4.5	4.0	1.5	0.7	12.0
100–1000	12.0	5.5	5.0	2.0	1.0	15.0
>1000	15.0	7.0	6.0	2.5	1.4	20.0

Requirements for Harmonic Control in Electrical Power Systems." The new version places the responsibility for ensuring power quality on both the utility and the customer.

As indicated by the title, IEEE Standard 519 is a "recommended practice," which means it is not a law or rule for all utility–customer interfaces, but it may be used as a design guideline for new installations. Utilities may also include the requirements from Standard 519 in service agreements with their customers, which could result in financial penalties for customers that do not comply. The standard makes the customer responsible for limiting the harmonic currents injected into the power system and the utility responsible for avoiding unacceptable voltage distortion.

IEEE Standard 519 defines harmonic current limits (shown in Table 17.1) for individual customers at the point of common coupling (PCC). Because voltage distortion is caused by the amount of harmonic currents in the system, larger customers are capable of causing more voltage distortion than smaller ones. Recognizing this, the standard allows a higher current THD for smaller customers' loads. The short-circuit ratio (SCR) is used to differentiate customer size.

When the load of Fig. 17.19 was shorted, the only impedance limiting the current was the system impedance. That current is called the available short-circuit current, and is generally high since the system impedance is much lower than the load impedance. SCR is defined as the "average maximum demand (load) current" for the facility divided by the available short-circuit current. The maximum load current drawn by a large customer would be a higher fraction of the available short-circuit current, so the large customer's SCR would be lower. The lower the SCR, the more stringent are the IEEE 519 limitations on harmonic currents.

IEEE Standard 519 also provides limits for specific ranges of frequencies, as shown in Table 17.1. Higher-order harmonics are constrained to have lower amplitudes for two reasons. First, higher-order harmonics cause greater voltage distortion than lower-order harmonics, even if they have the same amplitude, because the system inductive reactance is proportional to frequency. Second, interference with telecommunication equipment is more severe for higher-frequency harmonics. Note that Table 17.1 applies only to odd harmonics; even harmonics are limited to 25% of the values for the ranges they would occupy in Table 17.1.

The utility is required by Standard 519 to maintain acceptable levels of voltage distortion. Below 69 kV, individual harmonic components in the voltage should not exceed 3% of the fundamental, and the voltage THD must be less than 5%. Higher voltages have even lower limits, but those apply primarily to utility interconnections.

References

Bingham, R. P., Planning and performing a power quality survey, *Power Qual. Assurance*, 9(3), May/June 1998.

Dugan, R. C., M. F. McGranaghan, and H. W. Beaty, *Electrical Power Systems Quality*, McGraw-Hill, New York, 1996.

Grady, W. M., Harmonics and how they relate to power quality, in *Proceedings of the EPRI Power Quality and Opportunities Conference (PQA '93)*, San Diego, CA, November 1993.

Guth, B., Pay me now or pay me later—power monitors and conditioners provide valuable insurance when it comes time to install modern and expensive electrical systems and equipment, *Consulting-Specifying Eng.*, Vol. 3, September 1997.

Handbook of Power Signatures, 1993, Basic Measuring Instruments, Santa Clara, CA.

IEEE Standard 519-1992, IEEE Recommended Practices and Requirements for Harmonic Control in Electrical Power Systems.

IEEE Standard 1100-1992, IEEE Recommended Practice for Powering and Grounding of Sensitive Electronic Equipment.

IEEE Standard 1159-1995, IEEE Recommended Practice for Monitoring Electric Power Quality.

Waller, M., *Managing the Computer Power Environment—A Guide to Clean Power for Electronic Systems*, 1992, Prompt Publications, Indianapolis, IN.

17.3 Passive Harmonic Filters

Badrul H. Chowdhury

Currently in the United States, only 15 to 20% of the utility distribution loading consists of nonlinear loads. Loads, such as AC and DC adjustable speed drives (ASD), power rectifiers and inverters, arc furnaces, and discharge lighting (metal halide, fluorescent, etc.), and even saturated transformers, can be considered nonlinear devices. It is projected over the next 10 years that such nonlinear loads will comprise approximately 70 to 85% of the loading on utility distribution systems in the United States. These loads may generate enough harmonics to cause distorted current and voltage waveshapes.

The deleterious effects of harmonics are many. A significant impact is equipment overheating because of the presence of harmonics in addition to the fundamental. Harmonics can also create resonance conditions with power factor correction capacitors, resulting in higher than normal currents and voltages. This can lead to improper operation of protective devices, such as relays and fuses.

Harmonic frequency currents can cause additional rotating fields in AC motors. Depending on the frequency, the motor will rotate in the opposite direction (countertorque). In particular, the 5th harmonic, which is the most prevalent harmonic in three-phase power systems, is a negative sequence harmonic causing the motor to have a backward rotation, thus shortening the service life.

A typical current wave, as drawn by a three-phase AC motor drive, may look like the waveshape shown in Fig. 17.21. A Fourier analysis of the current would reveal the nature of the harmonics present. Three-phase ASDs generate primarily the 5th and 7th current harmonics and a lesser amount of 11th, 13th, and higher orders. The triplen harmonics (3rd, 9th, 15th, i.e., odd multiples of three) are conspicuously missing, as is usually the case in six-pulse converters, giving them an added advantage over single-phase converters. However, the triplen harmonics are additive in the neutral and can cause dangerous overheating.

In general, the characteristic harmonics generated by a converter is given by

$$h = pn \pm 1 \begin{cases} p = 6, 12, 18, \ldots \\ n = 1, 2, \ldots \end{cases} \tag{17.24}$$

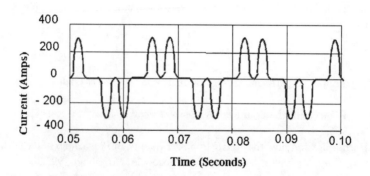

FIGURE 17.21 Typical current waveform of a three-phase adjustable-speed drive.

where h is the order of harmonics, n is any integer, and p is the number of pulses generated in each cycle (six for a three-phase converter).

To understand the impact of harmonics and to design remedies, one must quantify the amount of harmonics present. This is done by combining all of the harmonic frequency components (voltage or current) with the fundamental component (voltage or current) to form the total harmonic distortion, or THD. A commonly accepted definition of THD is as follows:

$$\text{THD}_I = \frac{\sqrt{I_2^2 + I_3^2 + I_4^2 + \cdots}}{I_1} \times 100\% = \frac{\sqrt{\sum_{h=2}^{\infty} I_h^2}}{I_1} \times 100\% \qquad (17.25)$$

where I_1 is the fundamental component of the current, I_2 is the second harmonic, I_3 the third harmonic, and so on. A similar equation can be written for voltage distortion.

Any THD values over 5% are significant enough for concern. Harmonic current distortion greater than 5% will contribute to the additional heating of a power transformer, so it must be derated for harmonics. It is not uncommon for THD levels in industrial plants to reach 25%. Normally, THD levels in office settings will be lower than in industrial plants, but office equipment is much more susceptible to variations in power quality. Odd-number harmonics (3rd, 5th, 7th, etc.) are of the greatest concern in the electrical distribution system. Even-number harmonics are usually mitigated because the harmonics swing equally in both the positive and negative direction. Pesky harmonics can be mitigated by the use of passive and active filters. Passive filters, consisting of tuned series L-C circuits, are the most popular. However, they require careful application, and may produce unwanted side effects, particularly in the presence of power factor correction capacitors.

The active filter concept uses power electronics to produce harmonic components that cancel the harmonic components from the nonlinear loads so that the current supplied from the source is sinusoidal. These filters are costly and relatively new.

Passive harmonic filters are constructed from passive elements (resistors, inductors, and capacitors) and thus the name. These filters are highly suited for use in three-phase, four-wire electrical power distribution systems. They should be applied as close as possible to the offending loads, preferably at the farthest three- to single-phase point of distribution. This will ensure maximum protection for the upstream system. Harmonics can be substantially reduced to as low as 30% by use of passive filters.

Passive filters can be categorized as parallel filters and series filters. A parallel filter is characterized as a series resonant and trap-type exhibiting a low impedance at its tuned frequency. Deployed close to the source of distortion, this filter keeps the harmonic currents out of the supply system. It also provides some smoothing of the load voltage. This is the most common type of filter.

The series filter is characterized as a parallel resonant and blocking type with high impedance at its tuned frequency. It is not very common because the load voltage can be distorted.

Series Passive Filter

This configuration is popular for single-phase applications for the purpose of minimizing the 3rd harmonic. Other specific tuned frequencies can also be filtered. Figure 17.22 shows the basic diagram of a series passive filter.

The advantages of a series filter are that it:

- Provides high impedance to tuned frequency;
- Does not introduce any system resonance;
- Does not import harmonics from other sources;
- Improves displacement power factor and true power factor.

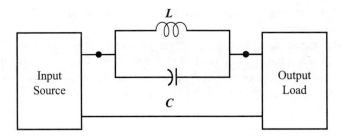

FIGURE 17.22 A series passive filter.

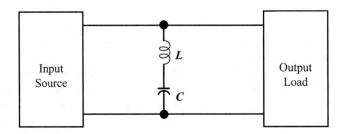

FIGURE 17.23 A shunt passive filter.

Some disadvantages are that it:

- Must handle the rated full load current;
- Is only minimally effective other than tuned harmonic frequencies;
- Can supply nonlinear loads only.

Shunt Passive Filter

The shunt passive filter is also capable of filtering specific tuned harmonic frequencies such as, 5th, 7th, 11th, etc. Figure 17.23 shows a commonly used diagram of a shunt filter. The advantages of a parallel filter are that it:

- Provides low impedance to tuned frequency;
- Supplies specific harmonic component to load rather than from AC source;
- Is only required to carry harmonic current and not the full load current;
- Improves displacement power factor and true power factor.

Some disadvantages are that:

- It only filters a single (tuned) harmonic frequency;
- It can create system resonance;
- It can import harmonics form other nonlinear loads;
- Multiple filters are required to satisfy typical desired harmonic limits.

Series Passive AC Input Reactor

The basic configuration is shown in Fig. 17.24. This type filters all harmonic frequencies, by varying amounts. The advantages of a series reactor are:

- Low cost;
- Higher true power factor;
- Small size;

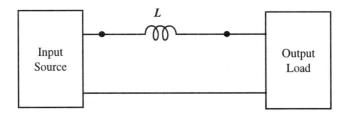

FIGURE 17.24 A series passive AC input reactor.

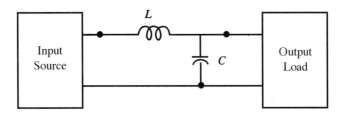

FIGURE 17.25 Low-pass filter.

- Filter does not create system resonance;
- It protects against power line disturbances.

Some disadvantages are that it:

- Must handle the rated full load current;
- Can only improve harmonic current distortion to 30 to 40% at best;
- Only slightly reduces displacement power factor.

Low-Pass (Broadband) Filter

The basic configuration is shown in Fig. 17.25. It is capable of eliminating all harmonic frequencies above the resonant frequency. The specific advantages of a low-pass filter are that it:

- Minimizes all harmonic frequencies;
- Supplies all harmonic frequencies as opposed to the AC source supplying those frequencies;
- Does not introduce any system resonance;
- Does not import harmonics from other sources;
- Improves true power factor.

Some of the disadvantages are that it:

- Must handle the rated full load current;
- Can supply nonlinear loads only.

Passive Filter Design

The filter design process involves a number of steps that will ensure lowest possible cost and proper performance under the THD limits. Figure 17.26 shows a flowchart of the entire process.

Characterizing Harmonic-Producing Loads

This is the first step in the process that will produce a summary of the level of harmonics being generated by nonlinear loads, such as AC adjustable speed drives, power rectifiers, arc furnaces, etc. Harmonic measurements must be used to characterize the level of harmonic generation for an existing nonlinear load.

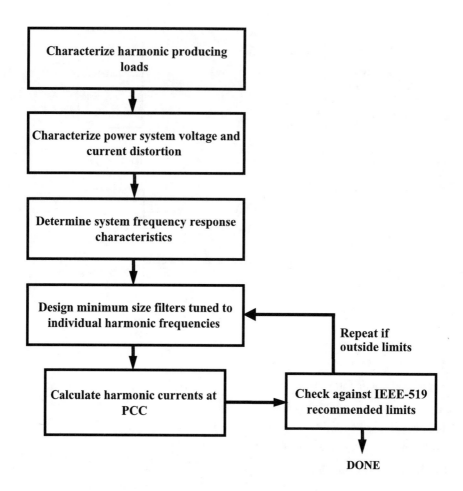

FIGURE 17.26 Flowchart for harmonic filter design.

Characterizing Power System Voltage and Current Distortion

In this step, a power system model is developed for analysis. The model is developed from one-line diagrams, manufacturer's data for various electrical equipment, the utility system characteristics, such as fault MVA, representative impedance, nominal voltage level, and the loading information. Figure 17.27 shows a sample representation of a utility system and an industrial plant supplied by a step-down transformer. The equivalent utility system can be represented as a simple impedance consisting of a resistance and an inductive reactance.

Determining System Frequency Response Characteristics

Switching transients created from regular utility operations as well as harmonics emanating from non-linear loads can both be magnified by power factor correction capacitors if resonant conditions exist. Therefore, it is necessary to perform simulations or frequency scans to determine the frequency response characteristics, looking from the low voltage bus. Simulations can be easily carried out by representing the system as a Thevenin's equivalent circuit. Such a circuit is shown in Fig. 17.28.

In the figure, L_{eq} and R_{eq} represent the combined inductance and resistance of the utility system and the step-down transformer.

$$Z_{in} = \frac{(R_{eq} + j\omega L_{eq})(-j/\omega C)}{R_{eq} + j\omega L_{eq} - j/\omega C} \quad (17.26)$$

Power Quality and Utility Interface Issues

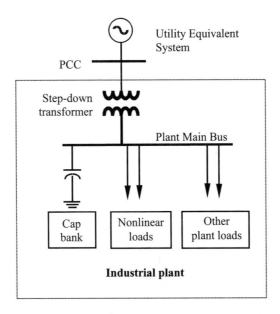

FIGURE 17.27 A typical representation of an industrial plant being supplied by a utility system.

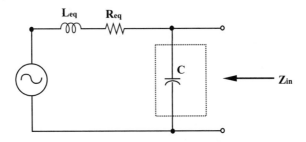

FIGURE 17.28 System equivalent circuit with reactive compensation at the load.

Parallel resonance occurs when the imaginary part of the denominator is equal to zero. That is,

$$\omega_0 L_{eq} - 1/\omega_0 C = 0 \qquad (17.27)$$

Solving for ω_0:

$$\omega_0 = \frac{1}{[L_{eq}C]^{1/2}} \qquad (17.28)$$

In hertz:

$$f_0 = \frac{1}{2\pi[L_{eq}C]^{1/2}} \qquad (17.29)$$

Frequency scan output consists of magnitude and phase angle for the driving point impedance. The effect of important system parameters, such as a capacitor, is evaluated and the potential for problem resonance conditions is determined. Figure 17.29 shows a typical output of a frequency scan simulation for studying the impact of power factor correction capacitors. Figure 17.30 depicts the proximity of the resonance points to some of the important harmonic characteristics, such as the 5th, the 7th, the 11th, and the 13th harmonics for varying levels of capacitive compensation.

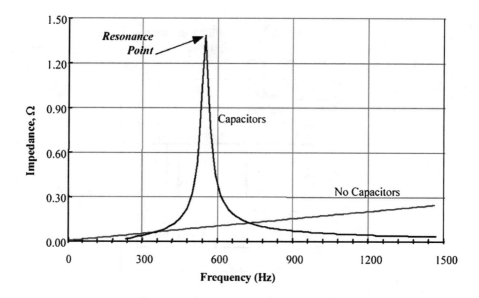

FIGURE 17.29 Frequency scan showing resonance point.

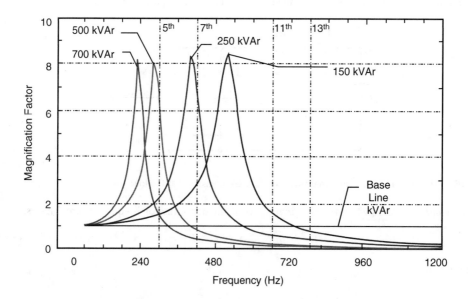

FIGURE 17.30 Resonance magnification due to low voltage reactive compensation.

Designing Minimum Size Filters Tuned to Individual Harmonic Frequencies

If not filtered, the harmonics generated by the industrial customer, downstream of the plant main bus, are returned upstream to the point of common coupling (PCC). If the short-circuit level available at the PCC is high enough, the voltage distortion will be significantly high and will thus affect other customers.

One of the most widely adopted solutions for reducing the impact of harmonics is the application of capacitor banks as tuned harmonic filters. This amounts to a very cost-effective solution since power factor correction capacitors are quite commonly installed in industrial facilities. To avoid resonance, one can use inductors in series with power factor capacitors to produce a harmonic filter. The inductor allows the parallel resonant frequency to be shifted somewhat. Figure 17.31a and b show a tuned delta-connected and a wye-connected filter bank, respectively. Without the series inductor, the bank simply becomes a

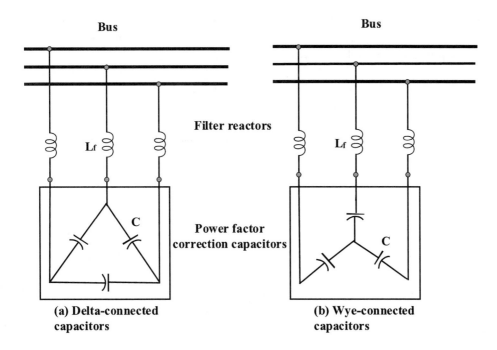

FIGURE 17.31 Application of capacitor banks as tuned harmonic filters.

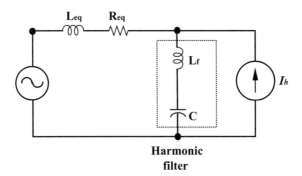

FIGURE 17.32 Equivalent circuit of the compensated system showing the harmonic filter equivalent.

power factor correction capacitor bank. This combination is referred to as *harmonic filter bank* or *detuned capacitor bank*.

The *series resonant frequency* or tuning frequency of the filter is selected to be about 3 to 10% below the lowest-order harmonic produced by the load. For typical six-pulse converters, this happens to be the 5th harmonic or a frequency of 300 Hz (on a 60-Hz system).

Typically, the tuning frequency of the filter is 282 Hz, corresponding to the 4.7th harmonic. In addition to shifting the parallel resonant frequency, the filter also supplies a portion of the harmonic current demanded by the load. Hence, the source current has less of the 5th harmonic content.

Figure 17.32 shows the equivalent circuit of the compensated system where

I_h = hth harmonic current source representing contributions from all harmonic sources at the plant main bus
L_f = inductance of the series reactor in the filter
C = capacitance of the harmonic filter

Depending on the severity of the individual harmonic level, harmonic filters should be tuned to a harmonic that causes the resonance condition. Some useful equations are given below.

Series resonant frequency of the filter element is

$$f_0 = \frac{1}{2\pi[L_f C]^{1/2}} \qquad (17.30)$$

where L_f = inductance of the tuning inductor and C = capacitance of the capacitor bank.

Solving for L_f:

$$L_f = \frac{1}{C[2\pi f_0]^2} \qquad (17.31)$$

Since

$$C = \frac{1}{2\pi f_{sys} X_{cap}} = \frac{1}{\omega_{sys} X_{cap}} \qquad (17.32)$$

where

$$X_{cap} = \frac{KV_{cap}^2}{MVA_{cap}} \qquad (17.33)$$

$$L_f = \frac{f_{sys} X_{cap}}{2\pi(f_0)^2} = \frac{f_{sys} kV_{cap}^2}{2\pi(f_0)^2 MVA_{cap}} \qquad (17.34)$$

Thus, the driving point impedance, as seen in Fig. 17.32, is

$$Z_{in} = \frac{(R_{eq} + j\omega L_{eq})(j\omega L_f - j/\omega C)}{R_{eq} + j\omega L_{eq} + j\omega L_f - j/\omega C} \qquad (17.35)$$

Parallel resonance occurs when the imaginary part of the denominator is equal to zero. That is,

$$\omega_0 L_{eq} + \omega_0 L_f - 1/\omega_0 C = 0 \qquad (17.36)$$

Solving for ω_0:

$$\omega_0 = \frac{1}{[(L_{eq} + L_f)C]^{1/2}} \qquad (17.37)$$

In hertz:

$$f_0 = \frac{1}{2\pi[(L_{eq} + L_f)C]^{1/2}} \qquad (17.38)$$

If the design is correct, then the frequency given by Eq. (17.38) should be farther away from the resonance point.

Calculating Harmonic Currents at PCC

To carry out simulations to estimate actual harmonic distortion levels, one needs to represent each harmonic-generating device, the system parameters, and the tuned filter characteristics. The output for these simulations consists of individual harmonic levels, bus voltage distortion and current distortion levels, and rms voltage and current levels. Harmonic levels are calculated at the PCC where the consumer's load connects to other loads in the power system.

Checking against IEEE-519 Recommended Limits

Current and voltage distortion levels as determined through simulations are compared with recommended limits outlined in IEEE Standard 519-1992. This standard is explained in the Appendix to this section. If harmonic voltage distortion levels are still not within acceptable limits, it is easy to change capacitor sizes and/or locations, or the size of the series reactor. These changes affect the frequency response characteristics of the industrial facility such that proximity to resonance points can be altered.

Further Information

Cameron, M., Trends in power factor correction with harmonic filtering, *IEEE Trans. Ind. Appl.*, 29, 60–65, 1993.

Dugan, R. C., McGranaghan, M. F., and Beaty, H. W., *Electrical Power Systems Quality*, McGraw-Hill, New York, 1996.

Fujita, H. and Akagi, H., A practical approach to harmonic compensation in power systems—series connection of passive and active filters, *IEEE Trans. Ind. Appl.*, 27(6), 1020–1025, 1991.

Grady, W. W., Samotyj, M. J., and Noyola, A. H., Survey of active power line conditioning methodologies, *IEEE Trans. Power Delivery*, 5(3), 1536–1542, 1990.

Grebe, T. E., Application of distribution system capacitor banks and their impact on power quality, *IEEE Trans. Ind. Appl.*, 32(3), 714–719, 1996.

IEEE Harmonics Working Group, IEEE Recommended Practices and Requirements for Harmonic Control in Electrical Power Systems, IEEE STD 519-1992, New York: IEEE, 1993.

IEEE P519A Task Force of the PES Harmonics Working Group, *Guide for Applying Harmonic Limits on Power Systems*, IEEE, New York, 1996.

Lemieux, G., Power system harmonic resonance—a documented case, *IEEE Trans. Ind. Appl.*, 26(3), 483–485, 1990.

Lowenstein, M. Z., Improving power factor in the presence of harmonics using low-voltage tuned filters, *IEEE Trans. Ind. Appl.*, 29, 528–535, 1993.

Makram, E. B. et al., Harmonic filter design using actual recorded data, *IEEE Trans. Ind. Appl.*, 29(6), 1176–1183, 1993.

Peng, F. Z., Akagi, H., and Nabae, A., Compensation characteristics of combined system of shunt passive and series active filters, in *IEEE/IAS Annual Meet. Conf. Record*, 1989, 959–966.

Phipps, J. K., A transfer function approach to harmonic filter design, *Ind. Appl. Mag.*, 3(2), 68–82, 1997.

Appendix—IEEE Recommended Practices and Requirements for Harmonic Control in Electric Power Systems

IEEE 519-1992 recommends limits on harmonic distortion according to two distinct criteria:

1. There is a limitation on the amount of harmonic current that a consumer can inject into a utility network.
2. A limitation is placed on the level of harmonic voltage that a utility can supply to a consumer.

$$\text{SCR} = \frac{\text{Short-circuit MVA}}{\text{Load MW}} = \frac{I_{SC}}{I_L}$$

IEEE 519 is limited to being a collection of *Recommended Practices* that serve as a guide to both suppliers and consumers of electrical energy with regard to excessive harmonic current injection or excessive voltage distortion. All of the current distortion values are given in terms relative to the maximum demand load current. The total distortion is in terms of total demand distortion (TDD) instead of the more common THD term. I_{SC} = maximum short circuit current at PCC; I_L = maximum demand load current (fundamental frequency) at point of common coupling; TDD = Total demand distortion in % of max demand and given by

$$\text{TDD} = \frac{\sqrt{\sum_{h=2}^{\infty} I_h^2}}{I_L} \times 100\%$$

17.4 Active Filters for Power Conditioning

Hirofumi Akagi

Much research has been performed on active filters for power conditioning and their practical applications since their basic principles of compensation were proposed around 1970 [Bird et al., 1969; Gyugyi and Strycula, 1976; Kawahira et al., 1983]. In particular, recent remarkable progress in the capacity and switching speed of power semiconductor devices such as insulated-gate bipolar transistors (IGBTs) has spurred interest in active filters for power conditioning. In addition, state-of-the-art power electronics technology has enabled active filters to be put into practical use. More than one thousand sets of active filters consisting of voltage-fed pulse width modulation (PWM) inverters using IGBTs or gate turn-off (GTO) thyristors are operating successfully in Japan.

Active filters for power conditioning provide the following functions:

- Reactive-power compensation,
- Harmonic compensation, harmonic isolation, harmonic damping, and harmonic termination,
- Negative sequence current/voltage compensation,
- Voltage regulation.

The term *active filters* is also used in the field of signal processing. In order to distinguish active filters in power processing from active filters in signal processing, the term *active power filters* often appears in many technical papers or literature. However, the author prefers *active filters for power conditioning* to active power filters, because the term active power filters is misleading to either active filters for power or filters for active power. Therefore, this section takes the term active filters for power conditioning or simply uses the term active filters as long as no confusion occurs.

Harmonic-Producing Loads

Identified Loads and Unidentified Loads

Nonlinear loads drawing nonsinusoidal currents from utilities are classified into identified and unidentified loads. High-power diode/thyristor rectifiers, cycloconverters, and arc furnaces are typically characterized as identified harmonic-producing loads because utilities identify the individual nonlinear loads installed by high-power consumers on power distribution systems in many cases. The utilities determine the point of common coupling with high-power consumers who install their own harmonic-producing loads on power distribution systems, and also can determine the amount of harmonic current injected from an individual consumer.

A "single" low-power diode rectifier produces a negligible amount of harmonic current. However, multiple low-power diode rectifiers can inject a large amount of harmonics into power distribution

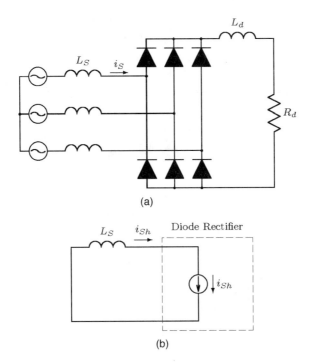

FIGURE 17.33 Diode rectifier with inductive load. (a) Power circuit; (b) equivalent circuit for harmonic on a per-phase base.

systems. A low-power diode rectifier used as a utility interface in an electric appliance is typically considered as an unidentified harmonic-producing load. Attention should be paid to unidentified harmonic-producing loads as well as identified harmonic-producing loads.

Harmonic Current Sources and Harmonic Voltage Sources

In many cases, a harmonic-producing load can be represented by either a harmonic current source or a harmonic voltage source from a practical point of view. Figure 17.33a shows a three-phase diode rectifier with a DC link inductor L_d. When attention is paid to voltage and current harmonics, the rectifier can be considered as a harmonic current source shown in Fig. 17.33b. The reason is that the load impedance is much larger than the supply impedance for harmonic frequency ω_h, as follows:

$$\sqrt{R_L^2 + (\omega_h L_d)^2} \gg \omega_h L_S$$

Here, L_S is the sum of supply inductance existing upstream of the point of common coupling (PCC) and leakage inductance of a rectifier transformer. Note that the rectifier transformer is disregarded from Fig. 17.33a. Figure 17.33b suggests that the supply harmonic current i_{Sh} is independent of L_S.

Figure 17.34a shows a three-phase diode rectifier with a DC link capacitor. The rectifier would be characterized as a harmonic voltage source shown in Fig. 17.34b if it is seen from its AC terminals. The reason is that the following relation exists:

$$\frac{1}{\omega_h C_d} \ll \omega_h L_S$$

This implies that i_{Sh} is strongly influenced by the inductance value of L_S.

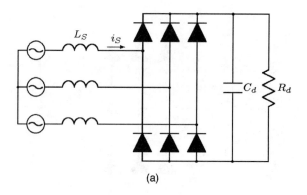

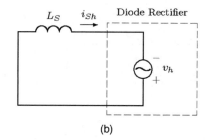

FIGURE 17.34 Diode rectifier with capacitive load. (a) Power circuit; (b) equivalent circuit for harmonic on a per-phase base.

Theoretical Approach to Active Filters for Power Conditioning
The Akagi-Nabae Theory

The theory of instantaneous power in three-phase circuits is referred to as the "Akagi-Nabae theory" [Akagi et al., 1983; 1984]. Figure 17.35 shows a three-phase three-wire system on the a-b-c coordinates, where no zero-sequence voltage is included in the three-phase three-wire system. Applying the theory to Fig. 17.35 can transform the three-phase voltages and currents on the a-b-c coordinates into the two-phase voltages and currents on the α-β coordinates, as follows:

$$\begin{bmatrix} e_\alpha \\ e_\beta \end{bmatrix} = \sqrt{\frac{2}{3}} \begin{bmatrix} 1 & -1/2 & -1/2 \\ 0 & \sqrt{3}/2 & -\sqrt{3}/2 \end{bmatrix} \begin{bmatrix} e_a \\ e_b \\ e_c \end{bmatrix} \quad (17.39)$$

$$\begin{bmatrix} i_\alpha \\ i_\beta \end{bmatrix} = \sqrt{\frac{2}{3}} \begin{bmatrix} 1 & -1/2 & -1/2 \\ 0 & \sqrt{3}/2 & -\sqrt{3}/2 \end{bmatrix} \begin{bmatrix} i_a \\ i_b \\ i_c \end{bmatrix} \quad (17.40)$$

As is well known, the instantaneous real power either on the a-b-c coordinates or on the α-β coordinates is defined by

$$p = e_a i_a + e_b i_b + e_c i_c = e_\alpha i_\alpha + e_\beta i_\beta \quad (17.41)$$

To avoid confusion, p is referred to as three-phase instantaneous real power. According to the theory, the three-phase instantaneous imaginary power, q, is defined by

$$q = e_\alpha i_\beta - e_\beta i_\alpha \quad (17.42)$$

Power Quality and Utility Interface Issues

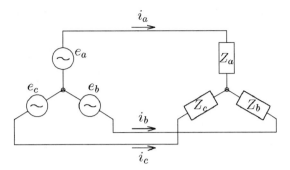

FIGURE 17.35 Three-phase three-wire system.

The combination of the above two equations bears the following basic formulation:

$$\begin{bmatrix} p \\ q \end{bmatrix} = \begin{bmatrix} e_\alpha & e_\beta \\ -e_\beta & e_\alpha \end{bmatrix} \begin{bmatrix} i_\alpha \\ i_\beta \end{bmatrix} \quad (17.43)$$

Here, $e_\alpha \cdot i_\alpha$ or $e_\beta \cdot i_\beta$ obviously means instantaneous power in the α-phase or the β-phase because either is defined by the product of the instantaneous voltage in one phase and the instantaneous current in the same phase. Therefore, p has a dimension of [W]. Conversely, neither $e_\alpha \cdot i_\beta$ nor $e_\beta \cdot i_\alpha$ means instantaneous power because either is defined by the product of the instantaneous voltage in one phase and the instantaneous current in the other phase. Accordingly, q is quite different from p in dimension and electric property although q looks similar in formulation to p. A common dimension for q should be introduced from both theoretical and practical points of view. A good candidate is [IW], that is, "imaginary watt."

Equation (17.43) is changed into the following equation:

$$\begin{bmatrix} i_\alpha \\ i_\beta \end{bmatrix} = \begin{bmatrix} e_\alpha & e_\beta \\ -e_\beta & e_\alpha \end{bmatrix}^{-1} \begin{bmatrix} p \\ q \end{bmatrix} \quad (17.44)$$

Note that the determinant with respect to e_α and e_β in Eq. (17.43) is not zero. The instantaneous currents on the α-β coordinates, i_α and i_β, are divided into two kinds of instantaneous current components, respectively:

$$\begin{bmatrix} i_\alpha \\ i_\beta \end{bmatrix} = \begin{bmatrix} e_\alpha & e_\beta \\ -e_\beta & e_\alpha \end{bmatrix}^{-1} \begin{bmatrix} p \\ 0 \end{bmatrix} + \begin{bmatrix} e_\alpha & e_\beta \\ -e_\beta & e_\alpha \end{bmatrix}^{-1} \begin{bmatrix} 0 \\ q \end{bmatrix}$$
$$= \begin{bmatrix} i_{\alpha p} \\ i_{\beta p} \end{bmatrix} + \begin{bmatrix} i_{\alpha q} \\ i_{\beta q} \end{bmatrix} \quad (17.45)$$

Let the instantaneous powers in the α-phase and the β-phase be p_α and p_β, respectively. They are given by the conventional definition as follows:

$$\begin{bmatrix} p_\alpha \\ p_\beta \end{bmatrix} = \begin{bmatrix} e_\alpha i_\alpha \\ e_\beta i_\beta \end{bmatrix} = \begin{bmatrix} e_\alpha i_{\alpha p} \\ e_\beta i_{\beta p} \end{bmatrix} + \begin{bmatrix} e_\alpha i_{\alpha q} \\ e_\beta i_{\beta q} \end{bmatrix} \quad (17.46)$$

The three-phase instantaneous real power, p, is given as follows, by using Eqs. (17.45) and (17.46):

$$p = p_\alpha + p_\beta = e_\alpha i_{\alpha p} + e_\beta i_{\beta p} + e_\alpha i_{\alpha q} + e_\beta i_{\beta q}$$
$$= \frac{e_\alpha^2}{e_\alpha^2 + e_\beta^2} p + \frac{e_\beta^2}{e_\alpha^2 + e_\beta^2} p + \frac{-e_\alpha e_\beta}{e_\alpha^2 + e_\beta^2} q + \frac{e_\alpha e_\beta}{e_\alpha^2 + e_\beta^2} q \qquad (17.47)$$

The sum of the third and fourth terms on the right-hand side in Eq. (17.47) is always zero. From Eqs. (17.46) and (17.47), the following equations are obtained:

$$p = e_\alpha i_{\alpha p} + e_\beta i_{\beta p} \equiv p_{\alpha p} + p_{\beta p} \qquad (17.48)$$

$$0 = e_\alpha i_{\alpha q} + e_\beta i_{\beta q} \equiv p_{\alpha q} + p_{\beta q} \qquad (17.49)$$

Inspection of Eqs. (17.48) and (17.49) leads to the following essential conclusions:

- The sum of the power components, $p_{\alpha p}$ and $p_{\beta p}$, coincides with the three-phase instantaneous real power, p, which is given by Eq. (17.41). Therefore, $p_{\alpha p}$ and $p_{\beta p}$ are referred to as the α-phase and β-phase instantaneous active powers.
- The other power components, $p_{\alpha q}$ and $p_{\beta q}$, cancel each other and make no contribution to the instantaneous power flow from the source to the load. Therefore, $p_{\alpha q}$ and $p_{\beta q}$ are referred to as the α-phase and β-phase instantaneous reactive powers.
- Thus, a shunt active filter without energy storage can achieve instantaneous compensation of the current components, $i_{\alpha q}$ and $i_{\beta q}$ or the power components, $p_{\alpha q}$ and $p_{\beta q}$. In other words, the Akagi–Nabae theory based on Eq. (17.43) exactly reveals what components the active filter without energy storage can eliminate from the α-phase and β-phase instantaneous currents, i_α and i_β or the α-phase and β-phase instantaneous real powers, p_α and p_β.

Energy Storage Capacity

Figure 17.36 shows a system configuration of a shunt active filter for harmonic compensation of a diode rectifier, where the main circuit of the active filter consists of a three-phase voltage-fed PWM inverter and a DC capacitor, C_d. The active filter is controlled to draw the compensating current, i_{AF}, from the utility, so that the compensating current cancels the harmonic current flowing on the AC side of the diode rectifier with a DC link inductor.

Referring to Eq. (17.44) yields the α-phase and β-phase compensating currents,

$$\begin{bmatrix} i_{AF\alpha} \\ i_{AF\beta} \end{bmatrix} = \begin{bmatrix} e_\alpha & e_\beta \\ -e_\beta & e_\alpha \end{bmatrix}^{-1} \begin{bmatrix} p_{AF} \\ q_{AF} \end{bmatrix} \qquad (17.50)$$

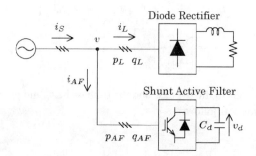

FIGURE 17.36 Shunt active filter.

Here, p_{AF} and q_{AF} are the three-phase instantaneous real and imaginary power on the AC side of the active filter, and they are usually extracted from p_L and q_L. Note that p_L and q_L are the three-phase instantaneous real and imaginary power on the AC side of a harmonic-producing load. For instance, when the active filter compensates for the harmonic current produced by the load, the following relationships exist:

$$p_{AF} = -\tilde{p}_L, \qquad q_{AF} = -\tilde{q}_L \qquad (17.51)$$

Here, $\tilde{p}_L$ and $\tilde{q}_L$ are AC components of p_L and q_L, respectively. Note that the DC components of p_L and q_L correspond to the fundamental current present in i_L and the AC components to the harmonic current. In general, two high-pass filters in the control circuit extract $\tilde{p}_L$ from p_L and $\tilde{q}_L$ from q_L.

The active filter draws p_{AF} from the utility, and delivers it to the DC capacitor if no loss is dissipated in the active filter. Thus, p_{AF} induces voltage fluctuation of the DC capacitor. When the amplitude of p_{AF} is assumed to be constant, the lower the frequency of the AC component, the larger the voltage fluctuation [Akagi et al., 1984; 1986]. If the period of the AC component is one hour, the DC capacitor has to absorb or release electric energy given by integration of p_{AF} with respect to time. Thus, the following relationship exists between the instantaneous voltage across the DC capacitor, v_d and p_{AF}:

$$\frac{1}{2}C_d v_d^2(t) = \frac{1}{2}C_d v_d^2(0) + \int_0^t p_{AF} \, dt \qquad (17.52)$$

This implies that the active filter needs an extremely large-capacity DC capacitor to suppress the voltage fluctuation coming from achieving "harmonic" compensation of $\tilde{p}_L$. Hence, the active filter is no longer a harmonic compensator, and thereby it should be referred to as a "DC capacitor-based energy storage system," although it is impractical at present. In this case, the main purpose of the voltage-fed PWM inverter is to perform an interface between the utility and the bulky DC capacitor.

The active filter seems to "draw" q_{AF} from the utility, as shown in Fig. 17.36. However, q_{AF} makes no contribution to energy transfer in the three-phase circuit. No energy storage, therefore, is required to the active filter, independent of q_{AF}, whenever $p_{AF} = 0$.

Classification of Active Filters

Various types of active filters have been proposed in technical literature [Moran, 1989; Grady et al., 1990; Akagi, 1994; Akagi and Fujita, 1995; 1997; Aredes et al., 1998]. Classification of active filters is made from different points of view [Akagi, 1996]. Active filters are divided into AC and DC filters. Active DC filters have been designed to compensate for current and/or voltage harmonics on the DC side of thyristor converters for high-voltage DC transmission systems [Watanabe, 1990; Zhang et al., 1993] and on the DC link of a PWM rectifier/inverter for traction systems. Emphasis, however, is put on active AC filter in the following because the term "active filters" refers to active AC filters in most cases.

Classification by Objectives: Who Is Responsible for Installing Active Filters?

The objective of "who is responsible for installing active filters" classifies them into the following two groups:

1. Active filters installed by *individual consumers* on their own premises in the vicinity of one or more identified harmonic-producing loads.
2. Active filters being installed by *electric power utilities* in substations and/or on distribution feeders.

Individual consumers should pay attention to current harmonics produced by their own harmonic-producing loads, and thereby the active filters installed by the individual consumers are aimed at compensating for current harmonics.

TABLE 17.2 Comparison of Shunt Active Filters and Series Active Filters

	Shunt Active Filter	Series Active Filter
System configuration	Fig. 17.36	Fig. 17.37
Power circuit of active filter	Voltage-fed PWM inverter *with* current minor loop	Voltage-fed PWM inverter *without* current minor loop
Active filter acts as	Current source: i_{AF}	Voltage source: v_{AF}
Harmonic-producing load suitable	Diode/thyristor rectifiers with *inductive* loads, and cycloconverters	Large-capacity diode rectifiers with *capacitive* loads
Additional function	Reactive power compensation	AC voltage regulation
Present situation	Commercial stage	Laboratory stage

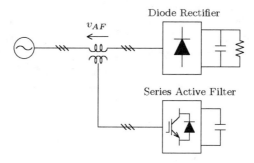

FIGURE 17.37 Series active filter.

Utilities should concern themselves with voltage harmonics, and therefore active filters will be installed by utilities in the near future for the purpose of compensating for voltage harmonics and/or of achieving "harmonic damping" throughout power distribution systems or "harmonic termination" of a radial power distribution feeder.

Classification by System Configuration

Shunt Active Filters and Series Active Filters

A stand-alone shunt active filter shown in Fig. 17.36 is one of the most fundamental system configurations. The active filter is controlled to draw a compensating current, i_{AF}, from the utility, so that it cancels current harmonics on the AC side of a general-purpose diode/thyristor rectifier [Akagi et al., 1990; Peng et al., 1990; Bhattacharya et al., 1998] or a PWM rectifier for traction systems [Krah and Holtz, 1994]. Generally, the shunt active filter is suitable for harmonic compensation of a current harmonic source such as diode/thyristor rectifier with a DC link inductor. The shunt active filter has the capability of damping harmonic resonance between an existing passive filter and the supply impedance.

Figure 17.37 shows a system configuration of a series active filter used alone. The series active filter is connected in series with the utility through a matching transformer, so that it is suitable for harmonic compensation of a voltage harmonic source such as a large-capacity diode rectifier with a DC link capacitor. The series active filter integrated into a diode rectifier with a DC common capacitor is discussed elsewhere. Table 17.2 shows comparisons between the shunt and series active filters. This concludes that the series active filter has a "dual" relationship in each item with the shunt active filter [Akagi, 1996; Peng, 1998].

Hybrid Active/Passive Filters

Figures 17.38 through 17.40 show three types of hybrid active/passive filters, the main purpose of which is to reduce initial costs and to improve efficiency. The shunt passive filter consists of one or more tuned LC filters and/or a high-pass filter. Table 17.3 shows comparisons among the three hybrid filters in which

Power Quality and Utility Interface Issues

TABLE 17.3 Comparison of Hybrid Active/Passive Filters

	Shunt Active Filter Plus Shunt Passive Filter	Series Active Filter Plus Shunt Passive Filter	Series Active Filter Connected in Series with Shunt Passive Filter
System configuration	Fig. 17.38	Fig. 17.39	Fig. 17.40
Power circuit of active filter	• Voltage-fed PWM inverter *with* current minor loop	• Voltage-fed PWM inverter *without* current minor loop	• Voltage-fed PWM inverter *with* or *without* current minor loop
Function of active filter	• Harmonic compensation	• Harmonic isolation	• Harmonic isolation or harmonic compensation
Advantages	• General shunt active filters applicable • Reactive power controllable	• Already existing shunt passive filters applicable • No harmonic current flowing through active filter	• Already existing shunt passive filters applicable • Easy protection of active filter
Problems or issues	• Share compensation in frequency domain between active filter and passive filter	• Difficult to protect active filter against overcurrent • No reactive power control	• No reactive power control
Present situation	• Commercial stage	• A few practical applications	• Commercial stage

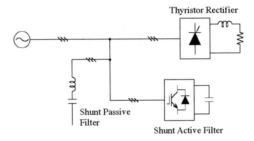

FIGURE 17.38 Combination of shunt active filter and shunt passive filter.

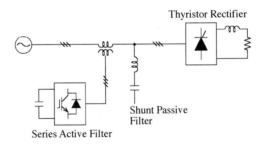

FIGURE 17.39 Combination of series active filter and shunt passive filter.

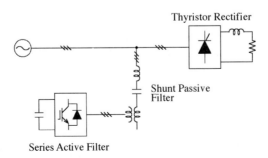

FIGURE 17.40 Series active filter connected in series with shunt passive filter.

the active filters are different in function from the passive filters. Note that the hybrid filters are applicable to any current harmonic source, although a harmonic-producing load is represented by a thyristor rectifier with a DC link inductor in Figs. 17.38 through 17.40.

Such a combination of a shunt active filter and a shunt passive filter as shown in Fig. 17.38 has already been applied to harmonic compensation of naturally-commutated twelve-pulse cycloconverters for steel mill drives [Takeda et al., 1987]. The passive filters absorbs 11th and 13th harmonic currents while the active filter compensates for 5th and 7th harmonic currents and achieves damping of harmonic resonance between the supply and the passive filter. One of the most important considerations in system design is to avoid competition for compensation between the passive filter and the active filter.

The hybrid active filters, shown in Fig. 17.39 [Peng et al., 1990; 1993; Kawaguchi et al., 1997] and in Fig. 17.40 [Fujita and Akagi, 1991; Balbo et al., 1994; van Zyl et al., 1995], are right now on the commercial stage, not only for harmonic compensation but also for harmonic isolation between supply and load, and for voltage regulation and imbalance compensation. They are considered prospective alternatives to pure active filters used alone. Other combined systems of active filters and passive filters or LC circuits have been proposed in Bhattacharya et al. [1997].

Classification by Power Circuit

There are two types of power circuits used for active filters: a voltage-fed PWM inverter [Akagi et al., 1986; Takeda et al., 1987] and a current-fed PWM inverter [Kawahira et al., 1983; van Schoor and van Wyk, 1987]. These are similar to the power circuits used for AC motor drives. They are, however, different in their behavior because active filters act as nonsinusoidal current or voltage sources. The author prefers the voltage-fed to the current-fed PWM inverter because the voltage-fed PWM inverter is higher in efficiency and lower in initial costs than the current-fed PWM inverter [Akagi, 1994]. In fact, almost all active filters that have been put into practical application in Japan have adopted the voltage-fed PWM inverter as the power circuit.

Classification by Control Strategy

The control strategy of active filters has a great impact not only on the compensation objective and required kVA rating of active filters, but also on the filtering characteristics in transient state as well as in steady state [Akagi et al., 1986].

Frequency-Domain and Time-Domain

There are mainly two kinds of control strategies for extracting current harmonics or voltage harmonics from the corresponding distorted current or voltage; one is based on the Fourier analysis in the frequency domain [Grady et al., 1990], and the other is based on the Akagi–Nabae theory in the time-domain. The concept of the Akagi–Nabae theory in the time-domain has been applied to the control strategy of almost all the active filters installed by individual high-power consumers over the last 10 years in Japan.

Harmonic Detection Methods

Three kinds of harmonic detection methods in the time-domain have been proposed for shunt active filters acting as a current source i_{AF}. Taking into account the polarity of the currents i_S, i_L and i_{AF} in Fig. 17.36 gives

load-current detection: $\quad i_{AF} = -i_{Lh}$

supply-current detection: $\quad i_{AF} = -K_S \cdot i_{Sh}$

voltage detection: $\quad i_{AF} = K_V \cdot v_h$

Note that load-current detection is based on feedforward control, while supply-current detection and voltage detection are based on feedback control with gains of K_S and K_V, respectively. Load-current detection and supply-current detection are suitable for shunt active filters installed in the vicinity of one or more harmonic-producing loads by individual consumers. Voltage detection is suitable for shunt active filters that will be dispersed on power distribution systems by utilities, because the shunt active filter based on voltage detection is controlled in such a way to present infinite impedance to the external circuit for the fundamental frequency, and to present a resistor with low resistance of $1/K_V$ [Ω] for harmonic frequencies [Akagi et al., 1999].

Supply-current detection is the most basic harmonic detection method for series active filters acting as a voltage source v_{AF}. Referring to Fig. 17.37 yields

$$\text{supply-current detection:} \quad v_{AF} = G \cdot i_{Sh}$$

The series active filter based on supply-current detection is controlled in such a way to present zero impedance to the external circuit for the fundamental frequency and to present a resistor with high resistance of G [Ω] for the harmonic frequencies. The series active filters shown in Fig. 17.37 [Fujita and Akagi, 1997] and Fig. 17.39 [Peng et al., 1990] are based on supply-current detection.

Integrated Series Active Filters

A small-rated series active filter integrated with a large-rated double-series diode rectifier has the following functions [Fujita and Akagi, 1997]:

- Harmonic compensation of the diode rectifier,
- Voltage regulation of the common DC bus,
- Damping of harmonic resonance between the communication capacitors connected across individual diodes and the leakage inductors including the AC line inductors,
- Reduction of current ripples flowing into the electrolytic capacitor on the common DC bus.

System Configuration

Figure 17.41 shows a harmonic current-free AC/DC power conversion system described below. It consists of a combination of a double-series diode rectifier of 5 kW and a series active filter with a peak voltage and current rating of 0.38 kVA. The AC terminals of a single-phase H-bridge voltage-fed PWM inverter are connected in "series" with a power line through a single-phase matching transformer, so that the combination of the matching transformers and the PWM inverters forms the "series" active filter. For small to medium-power systems, it is economically practical to replace the three single-phase inverters with a single three-phase inverter using six IGBTs. A small-rated high-pass filter for suppression of switching ripples is connected to the AC terminals of each inverter in the experimental system, although it is eliminated from Fig. 17.41 for the sake of simplicity.

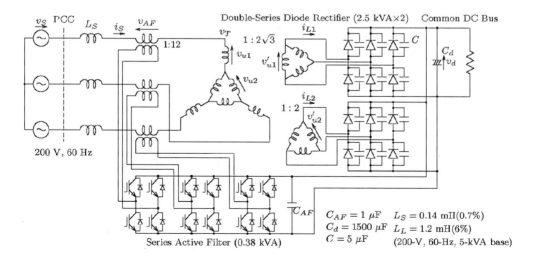

FIGURE 17.41 The harmonic current-free AC/DC power conversion system.

The primary windings of the Y-Δ and Δ-Δ connected transformers are connected in "series" with each other, so that the combination of the three-phase transformers and two three-phase diode rectifiers forms the "double-series" diode rectifier, which is characterized as a three-phase twelve-pulse rectifier. The DC terminals of the diode rectifier and the active filter form a common DC bus equipped with an electrolytic capacitor. This results not only in eliminating any electrolytic capacitor from the active filter, but also in reducing current ripples flowing into the electrolytic capacitor across the common DC bus.

Connecting only a commutation capacitor C in parallel with each diode plays an essential role in reducing the required peak voltage rating of the series active filter.

Operating Principle

Figure 17.42 shows an equivalent circuit for the power conversion system on a per-phase basis. The series active filter is represented as an AC voltage source v_{AF}, and the double-series diode rectifier as the series connection of a leakage inductor L_L of the transformers with an AC voltage source v_L. The reason for providing the AC voltage source to the equivalent model of the diode rectifier is that the electrolytic capacitor C_d is directly connected to the DC terminal of the diode rectifier, as shown in Fig. 17.41.

The active filter is controlled in such a way as to present zero impedance for the fundamental frequency and to act as a resistor with high resistance of K [Ω] for harmonic frequencies. The AC voltage of the active filter, which is applied to a power line through the matching transformer, is given by

$$v_{AF}{}^* = K \cdot i_{Sh} \tag{17.53}$$

where i_{Sh} is a supply harmonic current drawn from the utility. Note that v_{AF} and i_{Sh} are instantaneous values. Figure 17.43 shows an equivalent circuit with respect to current and voltage harmonics in Fig. 17.42. Referring to Fig. 17.43 enables derivation of the following basic equations:

$$I_{Sh} = \frac{V_{Sh} - V_{Lh}}{Z_S + Z_L + K} \tag{17.54}$$

$$V_{AF} = \frac{K}{Z_S + Z_L + K}(V_{Sh} - V_{Lh}) \tag{17.55}$$

where V_{AF} is equal to the harmonic voltage appearing across the resistor K in Fig. 17.43.

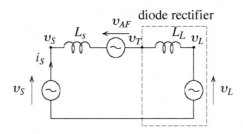

FIGURE 17.42 Single-phase equivalent circuit.

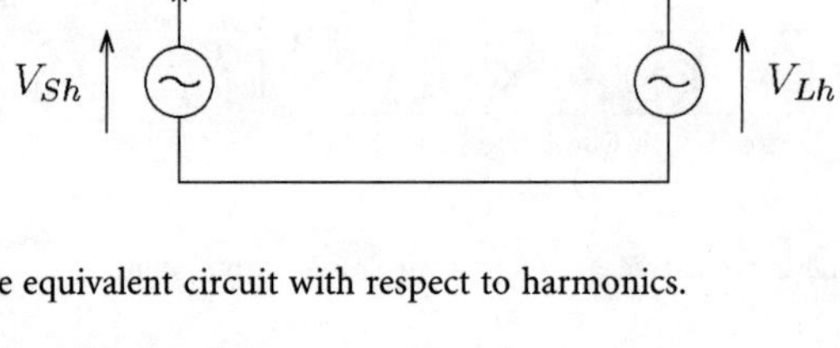

FIGURE 17.43 Single-phase equivalent circuit with respect to harmonics.

Power Quality and Utility Interface Issues

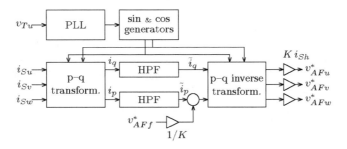

FIGURE 17.44 Control circuit for the series active filter.

If $K \gg Z_s + Z_L$, Eqs. (17.54) and (17.55) are changed into the following simple equations.

$$I_{Sh} \approx 0 \tag{17.56}$$

$$V_{AF} \approx V_{Sh} - V_{Lh} \tag{17.57}$$

Equation (17.56) implies that an almost purely sinusoidal current is drawn from the utility. As a result, each diode in the diode rectifier continues conducting during a half cycle. Eqution (17.57) suggests that the harmonic voltage V_{Lh}, which is produced by the diode rectifier, appears at the primary terminals of the transformers in Fig. 17.41, although it does not appear upstream of the active filter or at the utility–consumer point of common coupling (PCC).

Control Circuit

Figure 17.44 shows a block diagram of a control circuit based on hybrid analog/digital hardware. The concept of the Akagi–Nabae theory [Akagi, 1983; 1984] is applied to the control circuit implementation. The *p-q* transformation circuit executes the following calculation to convert the three-phase supply current i_{Sv}, i_{Sv}, and i_{Sw} into the instantaneous active current i_p and the instantaneous reactive current i_q.

$$\begin{bmatrix} i_p \\ i_q \end{bmatrix} = \sqrt{\frac{2}{3}} \begin{bmatrix} \cos\omega t & \sin\omega t \\ -\sin\omega t & \cos\omega t \end{bmatrix} \cdot \begin{bmatrix} 1 & -1/2 & -1/2 \\ 0 & \sqrt{3}/2 & -\sqrt{3}/2 \end{bmatrix} \begin{bmatrix} i_{Su} \\ i_{Sv} \\ i_{Sw} \end{bmatrix} \tag{17.58}$$

The fundamental components in i_{Su}, i_{Sv}, and i_{Sw} correspond to DC components in i_p and i_q, and harmonic components to AC components. Two first-order high-pass-filters (HPFs) with the same cut-off frequency of 10 Hz as each other extract the AC components $\tilde{i}_p$ and $\tilde{i}_q$ from i_p and i_q, respectively. Then, the *p-q* transformation/inverse transformation of the extracted AC components produces the following supply harmonic currents:

$$\begin{bmatrix} i_{Shu} \\ i_{Shv} \\ i_{Shw} \end{bmatrix} = \sqrt{\frac{2}{3}} \begin{bmatrix} 1 & 0 \\ -1/2 & \sqrt{3}/2 \\ -1/2 & -\sqrt{3}/2 \end{bmatrix} \cdot \begin{bmatrix} \cos\omega t & -\sin\omega t \\ \sin\omega t & \cos\omega t \end{bmatrix} \begin{bmatrix} \tilde{i}_p \\ \tilde{i}_q \end{bmatrix} \tag{17.59}$$

Each harmonic current is amplified by a gain of K, and then it is applied to the gate control circuit of the active filter as a voltage reference v^*_{AF} in order to regulate the common DC bus voltage, v^*_{AFf} is divided by the gain of K, and then it is added to $\tilde{i}_p$.

The PLL (phase locked loop) circuit produces phase information ωt which is a 12-bit digital signal of 60×2^{12} samples per second. Digital signals, sin ωt and cos ωt, are generated from the phase information, and then they are applied to the p-q (inverse) transformation circuits. Multifunction in the transformation circuits is achieved by means of eight multiplying D/A converters. Each voltage reference, v_{AF}^* is compared with two repetitive triangular waveforms of 10 kHz in order to generate the gate signals for the IGBTs. The two triangular waveforms have the same frequency, but one has polarity opposite to the other, so that the equivalent switching frequency of each inverter is 20 kHz, which is twice as high as that of the triangular waveforms.

Experimental Results

In the following experiment, the control gain of the active filter, K, is set to 27 Ω, which is equal to 3.3 p.u. on a 3ϕ 200-V, 15-A, 60-Hz basis. Equation (17.54) suggests that the higher the control gain, the better the performance of the active filter. An extremely high gain, however, may make the control system unstable, and thereby a trade-off between performance and stability exists in determining an optimal control gain. A constant load resistor is connected to the common DC bus, as shown in Fig. 17.41.

Figures 17.45 and 17.46 show experimental waveforms, where a 5-μF commutation capacitor is connected in parallel with each diode used for the double-series diode rectifier. Table 17.4 shows the THD of i_S and the ratio of each harmonic current with respect to the fundamental current contained in i_S. Before starting the active filter, the supply 11th and 13th harmonic currents in Fig. 17.45 are slightly magnified due to resonance between the commutation capacitors C and the AC line and leakage inductors, L_S and L_L. Nonnegligible amounts of 3rd, 5th, and 7th harmonic currents, which are so-called "non-characteristic current harmonics" for the three-phase twelve-pulse diode rectifier, are drawn from the utility.

Figure 17.46 shows experimental waveforms where the peak voltage of the series active filter is imposed on a limitation of ±12 V inside the control circuit based on hybrid analog/digital hardware. Note that

TABLE 17.4 Supply Current THD and Harmonics Expressed as the Harmonic-to-Fundamental Current Ratio [%], Where Commutation Capacitors of 5 μF are Connected

	THD	3rd	5th	7th	11th	13th
Before (Fig. 17.46)	16.8	5.4	2.5	2.2	12.3	9.5
After (Fig. 17.47)	1.6	0.7	0.2	0.4	0.8	1.0

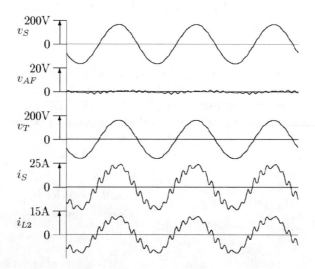

FIGURE 17.45 Experimental waveforms before starting the series active filter.

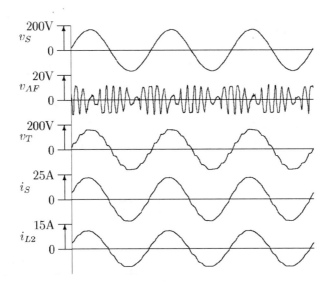

FIGURE 17.46 Experimental waveforms after starting the series active filter.

the limitation of ±12 V to the peak voltage is equivalent to the use of three single-phase matching transformers with turn ratios of 1:20 under the common DC-link voltage of 240 V. After starting the active filter, a sinusoidal current with a leading power factor of 0.96 is drawn because the active filter acts as a high resistor of 27 Ω, having the capability of compensating for both voltage harmonics V_{Sh} and V_{Lh}, as well as of damping the resonance. As shown in Fig. 17.46, the waveforms of i_S and v_T are not affected by the voltage limitation, although the peak voltage v_{AF} frequently reaches the saturation or limitation voltage of ±12 V.

The required peak voltage and current rating of the series active filter in Fig. 17.46 is given by

$$3 \times 12^V/\sqrt{2} \times 15^A = 0.38 \text{ kVA} \tag{17.60}$$

which is only 7.6% of the kVA-rating of the diode rectifier.

The harmonic current-free AC-to-DC power conversion system has both practical and economical advantages. Hence, it is expected to be used as a utility interface with large industrial inverter-based loads such as multiple adjustable speed drives and uninterruptible power supplies in the range of 1 to 10 MW.

Practical Applications of Active Filters for Power Conditioning

Present Status and Future Trends

Shunt active filters have been put into practical applications mainly for harmonic compensation, with or without reactive-power compensation. Table 17.5 shows ratings and application examples of shunt active filters classified by compensation objectives.

Applications of shunt active filters are expanding, not only into industry and electric power utilities but also into office buildings, hospitals, water supply utilities, and rolling stock. At present, voltage-fed PWM inverters using IGBT modules are usually employed as the power circuits of active filters in a range of 10 kVA to 2 MVA, and DC capacitors are used as the energy storage components.

Since a combined system of a series active filter and a shunt passive filter was proposed in 1988 [Peng et al., 1990], much research has been done on hybrid active filters and their practical applications [Bhattacharya et al., 1997; Aredes et al., 1998]. The reason is that hybrid active filters are attractive from both practical and economical points of view, in particular, for high-power applications. A hybrid active filter for harmonic damping has been installed at the Yamanashi test line for high-speed magnetically

TABLE 17.5 Shunt Active Filters on Commercial Base in Japan

Objective	Rating	Switching Devices	Applications
Harmonic compensation with or without reactive/negative-sequence current compensation	10 kVA ~ 2 MVA	IGBTs	Diode/thyristor rectifiers and cycloconverters for industrial loads
Voltage flicker compensation	5 MVA ~ 50 MVA	GTO thyristors	Arc furnaces
Voltage regulation	40 MVA ~ 60 MVA	GTO thyristors	Shinkansen (Japanese "bullet" trains)

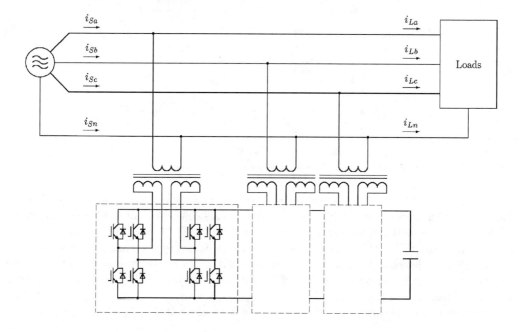

FIGURE 17.47 Shunt active filter for three-phase four-wire system.

levitated trains [Kawaguchi et al., 1997]. The hybrid filter consists of a combination of a 5-MVA series active filter and a 25-MVA shunt passive filter. The series active filter makes a great contribution to damping of harmonic resonance between the supply inductor and the shunt passive filter.

Shunt Active Filters for Three-Phase Four-Wire Systems

Figure 17.47 depicts the system configuration of a shunt active filter for a three-phase four-wire system. The 300-kVA active filter developed by Meidensha has been installed in a broadcasting station [Yoshida et al., 1998]. Electronic equipment for broadcasting requires single-phase 100-V AC power supply in Japan, and therefore the phase-neutral rms voltage is 100 V in Fig. 17.47. A single-phase diode rectifier is used as an AC-to-DC power converter in an electronic device for broadcasting. The single-phase diode rectifier generates an amount of third-harmonic current that flows back to the supply through the neutral line. Unfortunately, the third-harmonic currents injected from all of the diode rectifiers are in phase, thus contributing to a large amount of third-harmonic current flowing in the neutral line. The current harmonics, which mainly contain the 3rd, 5th, and 7th harmonic frequency components, may cause voltage harmonics at the secondary of a distribution transformer. The induced harmonic voltage may produce a serious effect on other harmonic-sensitive devices connected at the secondary of the transformer.

Figure 17.48 shows actually measured current waveform in Fig. 17.47. The load currents, i_{La}, i_{Lb}, and i_{Lc}, and the neutral current flowing on the load side, i_{Ln}, are distorted waveforms including a large amount of harmonic current, while the supply currents, i_{Sa}, i_{Sb}, and i_{Sc}, and the neutral current flowing on the supply side, i_{Sn}, are almost sinusoidal waveforms with the help of the active filter.

Power Quality and Utility Interface Issues

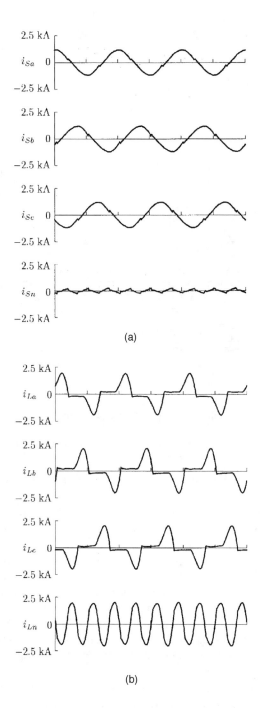

FIGURE 17.48 Actual current waveforms: (a) supply currents; (b) load currents.

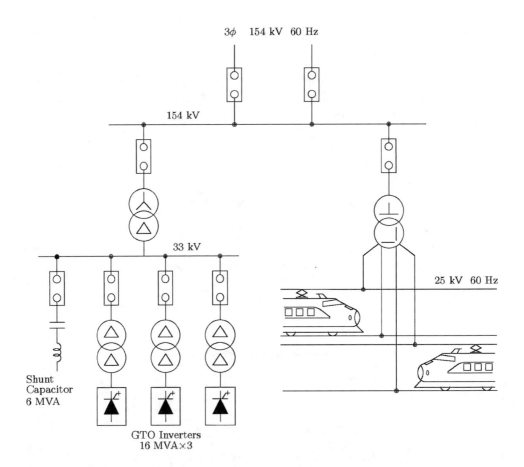

FIGURE 17.49 The 48-MVA shunt active filter installed in the Shintakatsuki substation.

The 48-MVA Shunt Active Filter for Compensation of Voltage Impact Drop, Variation, and Imbalance

Figure 17.49 shows a power system delivering electric power to the Japanese "bullet trains" on the Tokaido Shinkansen. Three shunt active filters for compensation of fluctuating reactive current/negative-sequence current have been installed in the Shintakatsuki substation by the Central Japan Railway Company [Iizuka et al., 1995]. The shunt active filters, manufactured by Toshiba, consist of voltage-fed PWM inverters using GTO thyristors, each of which is rated at 16 MVA. A high-speed train with maximum output power of 12 MW draws unbalanced varying active and reactive power from the Scott transformer, the primary of which is connected to the 154-kV utility grid. More than twenty high-speed trains pass per hour during the daytime. This causes voltage impact drop, variation, and imbalance at the terminals of the 154-kV utility system, accompanied by a serious deterioration in the power quality of other consumers connected to the same power system. The purpose of the shunt active filters with a total rating of 48 MVA is to compensate for voltage impact drop, voltage variation, and imbalance at the terminals of the 154-kV power system, and to improve the power quality. The concept of the instantaneous power theory in the time-domain has been applied to the control strategy for the shunt active filter.

Figure 17.50 shows voltage waveforms on the 154-kV bus and the voltage imbalance factor before and after compensation, measured at 14:20–14:30 on July 27, 1994. The shunt active filters are effective not only in compensating for the voltage impact drop and variation, but also in reducing the voltage imbalance factor from 3.6 to 1%. Here, the voltage imbalance factor is the ratio of the negative to positive sequence component in the three-phase voltages on the 154-kV bus. At present, several active filters in a range of 40 MVA to 60 MVA have been installed in substations along the Tokaido Shinkansen [Takeda et al., 1995].

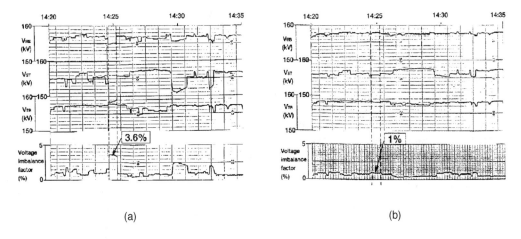

FIGURE 17.50 Installation effect: (a) before compensation; (b) after compensation.

Acknowledgment

The author would like to thank Meidensha Corporation and Toshiba Corporation for providing helpful and valuable information of the 300-kVA active filter and the 48-MVA active filter.

References

Akagi, H., New trends in active filters for power conditioning, *IEEE Trans. Ind. Appl.*, 32, 6, 1312–1322, 1996.

Akagi, H., Trends in active power line conditioners, *IEEE Trans. Power Electron.*, 9, 3, 263–268, 1994.

Akagi, H. and Fujita, H., A new power line conditioner for harmonic compensation in power systems, *IEEE Trans. Power Delivery*, 10(3), 1570–1575, 1995.

Akagi, H., Fujita, H., and Wada, K., A shunt active filter based on voltage detection for harmonic termination of a radial power distribution line, *IEEE Trans. Ind. Appl.*, 35(3), 638–645, 1999.

Akagi, H., Kanazawa, Y., and Nabae, A., Generalized theory of the instantaneous reactive power in three-phase circuits, in *Proceedings of the 1983 International Power Electronics Conference*, Tokyo, Japan, 1983, 1375–1386.

Akagi, H., Kanazawa, Y., and Nabae, A., Instantaneous reactive power compensators comprising switching devices without energy storage components, *IEEE Trans. Ind. Appl.*, 20(3), 625–630, 1984.

Akagi, H., Nabae, A., and Atoh, S., Control strategy of active power filters using multiple voltage-source PWM converters, *IEEE Trans. Ind. Appl.*, 22(3), 460–465, 1986.

Akagi, H., Tsukamoto, Y., and Nabae, A., Analysis and design of an active power filter using quad-series voltage-source PWM converters, *IEEE Trans Ind. Appl.*, 26(1), 93–98, 1990.

Aredes, M., Heumann, K., and Watanabe, E. H., A universal active power line conditioner, *IEEE Trans. Power Delivery*, 13, 2, 545–551, 1998.

Balbo, N., Penzo, R., Sella, D., Malesani, L., Mattavelli, P., and Zuccato, A., Simplified hybrid active filters for harmonic compensation in low voltage industrial applications, in *Proceedings of the 1994 IEEE/PES International Conference on Harmonics in Power Systems*, 1994, 263–269.

Bhattacharya, S., Cheng, P., and Divan, D., Hybrid solutions for improving passive filter performance in high power applications, *IEEE Trans. Ind. Appl.*, 33(3), 732–747, 1997.

Bhattacharya, S., Frank, T. M., Divan, D., and Banerjee, B., Active filter system implementation, *IEEE Ind. Appl. Mag.*, 4(5), 47–63, 1998.

Bird, B. M., Marsh, J. F., and McLellan, P. R., Harmonic reduction in multiple converters by triple-frequency current injection, *IEE Proc.*, 116(10), 1730–1734, 1969.

Fujita, H. and Akagi, H., An approach to harmonic current-free AC/DC power conversion for large industrial loads: the integration of a series active filter and a double-series diode rectifier, *IEEE Trans. Ind. Appl.*, 33(5), 1233–1240, 1997.

Fujita, H. and Akagi, H., A practical approach to harmonic compensation in power systems—series connection of passive and active filters, *IEEE Trans. Ind. Appl.*, 27(6), 1020–1025, 1991.

Fujita, H. and Akagi, H., The unified power quality conditioner: the integration of series- and shunt-active filters, *IEEE Trans. Power Electron.*, 13(2), 315–322, 1998.

Grady, W. M., Samotyj, M. J., and Noyola, A. H., Survey of active power line conditioning methodologies, *IEEE Trans. Power Delivery*, 5(3), 1536–1542, 1990.

Gyugyi, L. and Strycula, E. C., Active AC power filters, in *Proceedings of the 1976 IEEE/IAS Annual Meeting*, 1976, 529–535.

Iizuka, A., Kishida, M., Mochinaga, Y., Uzuka, T., Hirakawa, K., Aoyama, F., and Masuyama, T., Self-commutated static var generators at Shintakatsuki substation, in *Proceedings of the 1995 International Power Electronics Conference*, Yokohama, Japan, 1995, 609–614.

Kawaguchi, I., Ikeda, H., Ogihara, Y., Syogaki, M., and Morita, H., Novel active filter system composed of inverter bypass circuit for suppression of harmonic resonance at the Yamanashi maglev test line, in *Proceedings of the IEEE-IEEJ/IAS Power Conversion Conference*, 1997, 175–180.

Kawahira, H., Nakamura, T., Nakazawa, S., and Nomura, M., Active power filters, in *Proceedings of the 1983 International Power Electronics Conference*, Tokyo, Japan, 1983, 981–992.

Krah, J. O. and Holtz, J., Total compensation of line-side switching harmonics in converter-fed AC locomotives, in *Proceedings of the 1994 IEEE/IAS Annual Meeting*, 1994, 913–920.

Lee, T-N., Pereira, M., Renz, K., and Vaupel, G., Active damping of resonances in power systems, *IEEE Trans. Power Delivery*, 9(2), 1001–1008, 1994.

Moran, S., A line voltage regulator/conditioner for harmonic-sensitive load isolation, in *Proceedings of the 1989 IEEE/IAS Annual Meeting*, 1989, 947–951.

Peng, F. Z., Application issues of active power filters, *IEEE Ind. Appl. Mag.*, 4(5), 21–30, 1998.

Peng, F. Z., Akagi, H., and Nabae, A., Compensation characteristics of the combined system of shunt passive and series active filters, *IEEE Trans. Ind. Appl.*, 29(1), 144–152, 1993.

Peng, F. Z., Akagi, H., and Nabae, A., A new approach to harmonic compensation in power systems—a combined system of shunt passive and series active filters, *IEEE Trans. Ind. Appl.*, 26(6), 983–990, 1990.

Peng, F. Z., Akagi, H., and Nabae, A., A study of active power filters using quad-series voltage-source PWM converters for harmonic compensation, *IEEE Trans. Power Electron.*, 5(1), 9–15, 1990.

Takeda, M., Murakami, S., Iizuka, A., Kishida, M., Mochinaga, Y., Hase, S., and Mochinaga, H., Development of an SVG series for voltage control over three-phase unbalance caused by railway load, in *Proceedings of the 1995 International Power Electronics Conference*, Yokohama, Japan, 1995, 603–608.

Takeda, M., Ikeda, K., and Tominaga, Y., Harmonic current compensation with active filter, in *Proceedings of the 1987 IEEE/IAS Annual Meeting*, 1987, 808–815.

Van Schoor, G. and van Wyk, J., A study of a system of a current-fed converters as an active three-phase filter, in *Proceedings of the 1987 IEEE/PELS Power Electronics Specialist Conference*, 1987, 482–490.

Van Zyl, A., Enslin, J. H. R., and Spée, R., Converter based solution to power quality problems on radial distribution lines, in *Proceedings of the 1995 IEEE/IAS Annual Meeting*, 1995, 2573–2580.

Watanabe, E. H., Series active filter for the DC side of HVDC transmission systems, in *Proceedings of the 1990 International Power Electronics Conference*, Tokyo, Japan, 1990, 1024–1030.

Yoshida, T., Nakagawa, G., Kitamura, H., and Iwatani, K., Active filters (multi-functional harmonic suppressors) used to protect the quality of power supply from harmonics and reactive power generated in loads, *Meiden Rev.*, 262(5), 13–17, 1998 (in Japanese).

Zhang, W., Asplund, G., Åberg, A., Jonsson, U., and Lööf, O., Active DC filter for HVDC system—A test installation in the Konti-Skan at Lindome converter station, *IEEE Trans. Power Delivery*, 8(3), 1599–1605, 1993.

17.5 Unity Power Factor Rectification

Rajapandian Ayyanar and Amit Kumar Jain

The proliferation of power electronic converters with front-end rectifiers has resulted in numerous problems for the utility distribution network. The currents drawn by the rectifier systems are nonsinusoidal and have large harmonic components, which interfere with other loads connected to the utility. Phase displacement of fundamental current and voltage requires the source and distribution equipment to handle reactive power and therefore higher rms currents for a given real output power. This section introduces the problems associated with the rectifier systems and discusses briefly the standards that are being enforced to limit the harmonic content in the line currents to acceptable levels. Three approaches—passive filters, active current-shaping techniques, and active filters—are introduced. Of these, the active current-shaping techniques for both the single-phase and three-phase applications are discussed in detail.

Diode Bridge and Phase-Controlled Rectifiers

In a majority of power electronic applications, for example, in switch-mode power supplies (SMPS), the utility voltage is first converted to an unregulated DC voltage using a single-phase or three-phase diode bridge rectifier. In a typical SMPS, this DC-link voltage is then converted to the desired voltage levels with isolation, using high-frequency DC-DC converters. In adjustable-speed drives, the DC-link voltage is converted to either a variable magnitude DC voltage as in DC drives, or a variable-frequency, variable-magnitude voltage suitable for AC motors. To minimize the ripple in the DC-link voltage, large capacitors are used as shown in Fig. 17.51.

The currents drawn by these diode bridge rectifiers from the utility, as shown in Fig. 17.51a and b, are not sinusoidal. The DC-link capacitor is charged to a value close to the peak of the utility voltage, and draws current only when the utility voltage is near its peak value. Hence, the current drawn from the utility is discontinuous and rich in harmonics. Table 17.6 gives the harmonic spectrum of the current drawn by a typical single-phase diode bridge rectifier with a capacitive filter [1]. As seen, it has a large

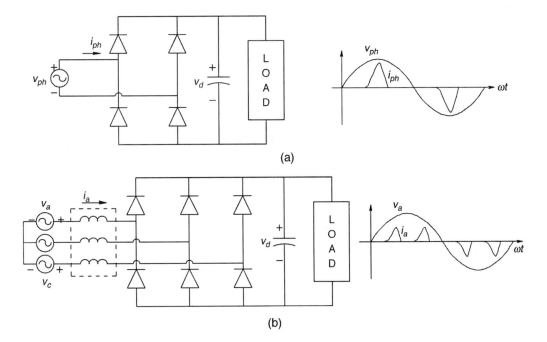

FIGURE 17.51 Diode bridge rectifiers: (a) single phase, (b) three phase.

TABLE 17.6 Typical Harmonic Spectrum of Line Current Drawn by Single-Phase Diode Bridge Rectifier

h	3	5	7	9	11	13	15	17
$\frac{I_h}{I_1}$ %	73.2	36.6	8.1	5.7	4.1	2.9	0.8	0.4

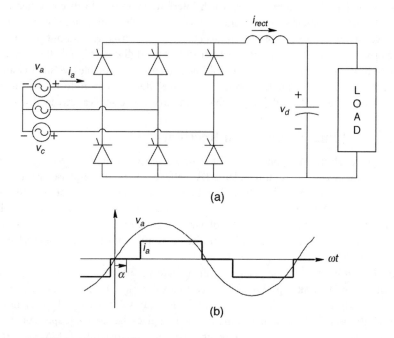

FIGURE 17.52 Three-phase phase-controlled rectifiers.

third harmonic component. For a given output power, the rms value of the line current for the case shown in Table 17.6 is about 30% higher than that of a sinusoidal current at unity power factor.

Figure 17.52a shows the schematic diagram of a three-phase phase-controlled rectifier. Unlike diode bridge rectifiers, the DC-link voltage here can be regulated by controlling the firing angle of the thyristors with respect to the utility voltage. Such phase-controlled three-phase rectifiers with inductive filter are commonly used in high-power applications. Figure 17.52b shows an input phase voltage and the corresponding phase current for a three-phase, phase-controlled rectifier, neglecting the source inductance. The DC side inductor is assumed to be large enough so that the inductor current is pure DC. As seen from Fig. 17.52b, the current drawn from the utility is a quasi-square wave, with its fundamental component displaced from the mains voltage by the firing angle α. The dominant harmonics in the input currents are the 5th and the 7th [1].

Standards for Limiting Harmonic Currents

Various national and international agencies have specified standards and guidelines or recommended practices for the maximum allowable harmonic currents that can be injected into the utility. The standards widely used are those formulated by the International Electrotechnical Commission (IEC), ANSI/IEEE Standards, and the VDE series of German standards [2, 3].

IEC 555 standards, released in 1982, underwent several amendments and eventually became part of the IEC 1000 family, which deals with electromagnetic compliance [4]. IEC 1000-3-2 sets the limits for harmonic emission—both absolute values as well as per-watt magnitude—for equipment with input current less than or equal to 16 A per phase. The standard refers to four different classes of equipment:

TABLE 17.7 IEC 1000-3-2 Limits for Class D Equipment

Harmonic Order n	Maximum Permissible Harmonic Current per Watt mA/W	Maximum Permissible Harmonic Current A
3	3.4	2.30
5	1.9	1.14
7	1.0	0.77
9	0.5	0.40
11	0.35	0.33
$13 \leq n \leq 39$	$\dfrac{3.85}{n}$	Refer Class A

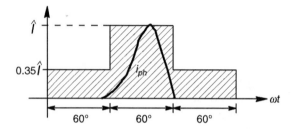

FIGURE 17.53 Envelop of current waveshapes covered in Class D.

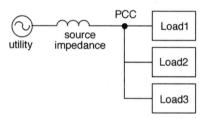

FIGURE 17.54 Point of common coupling.

Class A refers to balanced three-phase equipment, Class B refers to portable tools, Class C to lighting equipment, and Class D deals with equipment that draws currents having a special nonsinusoidal waveshape lying under the envelope shown in Fig. 17.53 with an active input power under 600 W. The diode bridge rectifiers discussed earlier belong to this class of equipment. Table 17.7 lists the 1000-3-2 limits for Class D equipment with maximum current less than or equal to 16 A[4]. IEC 1000-3-3 deals with limits on voltage flicker in low-voltage supply systems with current less than 16 A. IEC 1000-3-4 sets limits for both individual equipment as well as for the whole system installation. IEC 1000-3-2 has been adopted as EN 61000-3-2, and hence any equipment sold in Europe must comply with the standard.

Whereas the IEC standards limit the harmonic emission produced by the equipment, the IEEE standard 519 limits the harmonic currents and voltages at the point of common coupling (PCC) [5]. PCC refers to the point at which all the loads at a particular location are connected to the utility as shown in Fig. 17.54. Table 17.1 lists the harmonic current limits recommended by this standard for power systems below 69 kV. The short circuit ratio (SCR) shown in column 1 is defined as the ratio of maximum short-circuit current, I_s, to the maximum load current, I_L. As seen, stricter limits are applied to low SCR, since low SCR, implying high source impedance, results in higher voltage distortion for a given harmonic current magnitude.

Definitions of Some Common Terms

This section discusses definitions of some of the commonly used terms related to power factor. For a single-phase system, *power factor* is defined as

$$\text{Power factor} = \frac{\text{Average power}}{\text{Apparent power}} = \frac{\frac{1}{T}\int_0^T v_s(t) i_s(t)}{V_{\text{rms}} I_{\text{rms}}} \tag{17.61}$$

where $v_s(t)$ and $i_s(t)$ are the instantaneous phase voltage and current, and V_{rms} and I_{rms} are the corresponding rms quantities.

If the voltage is sinusoidal,

$$\text{Power factor} = \frac{V_{\text{rms}} I_{1,\text{rms}}}{V_{\text{rms}} I_{\text{rms}}} \cos\theta = \frac{I_{1,\text{rms}}}{I_{\text{rms}}} \cos\theta \tag{17.62}$$

In Eq. (17.62), $\cos\theta$ is called the *displacement power factor*, and θ is the phase angle between the voltage and the fundamental component of the current. The ratio $I_{1,\text{rms}}/I_{\text{rms}}$ is called the *distortion power factor* and it accounts for the additional losses due to harmonics. Hence, the power factor with a sinusoidal voltage waveform and an arbitrary current waveform is equal to the product of the distortion power factor and the displacement power factor.

The harmonic content in the current waveform is usually quantified by the *total harmonic distortion* (THD) defined as:

$$\text{THD} = \frac{\sqrt{I_{\text{rms}}^2 - I_{1,\text{rms}}^2}}{I_{1,\text{rms}}} \tag{17.63}$$

Passive Solutions

To achieve compliance with the above-mentioned harmonic standards, a simple but less effective approach is to insert a passive filter between the harmonic generating sources, namely, the diode bridge rectifier or the phase-controlled thyristor rectifier and the utility. The other approach, referred to as the active solution, is to use a high-frequency power electronic converter as a pre-regulator that will shape the current drawn from the utility to be sinusoidal. The passive filters are reliable, less complex, and usually less expensive compared with the active solutions, but are bulky, heavy, and cannot be used for applications that have to meet stringent harmonic distortion standards. Another drawback of passive filters as compared with active solutions is their inability to regulate the DC-link voltage.

Passive filters can broadly be classified into series filters, shunt filters, and a hybrid combination of the two. Series filters introduce an impedance in series with the utility to reduce harmonic currents. Shunt filters provide a low-impedance path for the harmonic currents generated by the rectifiers so that they are not reflected in the current drawn from the utility. Figure 17.55 shows a simple series-type filter, where a suitably designed inductor is inserted between the diode bridge and the mains. The waveform of the corresponding line current is shown in Fig. 17.55b. Comparing this current to that without a series filter as shown in Fig. 17.51a, it can be seen that the rms value for a given power is lower in the present case. Tuned filters to attenuate selected current harmonics, usually 3rd and 5th, are also employed to improve the power factor. The main drawback of such an approach is that if the input voltage has even a small component at the tuned frequency, the filter can resonate with the mains and draw large harmonic currents.

Active Solutions

An active power factor correction (PFC) circuit refers to the use of a power electronic converter, switching at high frequencies, to shape the input current to be sinusoidal and in phase with the input utility voltage. Using active PFC techniques, it is possible to achieve a power factor greater than 0.99 and a THD less

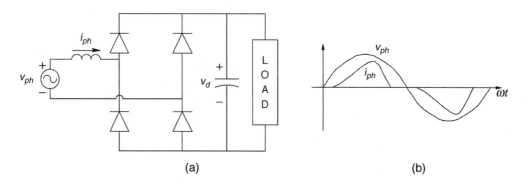

FIGURE 17.55 Passive solution for single-phase rectifier.

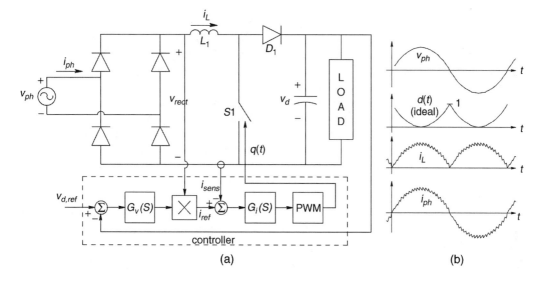

FIGURE 17.56 Single-phase boost power factor correction.

than 5%. Compared with the passive solutions, they are less bulky and can easily meet the standards on harmonic distortion. Another major advantage of the active approach is that the DC-link voltage can be regulated (although usually with slow dynamics), even when the input voltage varies over a wide range. Commonly used PFC circuits for single-phase and three-phase power factor correction applications are discussed below.

Single-Phase Active PFC Circuits

Boost Converter–Based PFC

The boost converter with a front-end diode bridge rectifier (without the large input filter capacitor) is perhaps the most popular single-phase PFC topology at present [6, 7]. Figure 17.56 shows the schematic diagram of the single-phase boost PFC circuit along with the controllers. The duty ratio of S_1 is modulated such that the waveshape of the inductor current closely tracks the rectified sinusoidal voltage obtained at the output of the diode bridge rectifier. The inductor L_1 is designed such that the operation of the converter is always in continuous conduction mode (CCM).

The controller involves a slow outer voltage loop and an inner fast current loop. The outer voltage loop produces a signal that depends on the error between the actual and the desired output voltages. The reference to the inner current loop is derived from the rectified sinusoidal input voltage, suitably scaled by multiplying it with the output signal from the voltage loop. For the current loop, either peak current mode control or average current control can be used. In average current mode control, the current controller amplifies the error between the actual inductor current and the current reference. This amplified error signal is then compared with a fixed frequency ramp to obtain the gate control signal $q(t)$ for the switch. In peak current mode control, the actual inductor current $i_L(t)$ is compared with the current reference, $i_{ref}(t)$. When $i_L(t)$ just exceeds the reference the switch is turned off, and is turned on again at start of the next switching cycle by a fixed-frequency clock signal. It is also possible to control the current using a hysteresis controller.

As in DC-DC converter applications, the output voltage of the boost converter needs to be above the maximum input voltage. Hence, for single-phase PFC applications using the boost converter, the DC-link voltage has to be higher than the peak of the maximum input utility voltage. For a nominal utility voltage of 110 V rms, the DC-link voltage needs to be close to 200 V DC. In a single-phase rectifier with unity power factor, the input power fluctuates at twice the fundamental frequency (for example, 120 Hz), whereas the output power is constant (within the 120-Hz cycle). Hence, energy storage elements are needed to account for the difference between the instantaneous values of the input power and the output power. The output capacitor of the boost converter thus needs to be large in single-phase PFC applications.

Figure 17.56b shows the waveforms corresponding to the input utility voltage, the idealized instantaneous switch duty ratio, the inductor current, and the current drawn from the utility. Usually small, low-pass filters are used to reduce the high-frequency switching component in the input current, and to meet EMI standards.

The design of the feedback controller for both the outer voltage loop and the inner current loop is discussed in detail in Refs. 6 and 7. The reference signal to the current loop, which is mainly a rectified sinusoidal voltage, contains a large second-order harmonic term, and substantial higher-order harmonics. Therefore, to obtain low THD for the input current, the current loop should be able to track the higher-order harmonics. Thus, the current loop should have a high bandwidth. This is especially important if the input frequency is higher, as, for example, 400 Hz in avionic applications.

The current through the filter capacitor has a large second harmonic component apart from the switching frequency component. This is reflected in the DC-link voltage as large second harmonic ripple voltage. The outer voltage loop therefore should be designed for low bandwidth (much less than 120 Hz for utility applications) so that the above-mentioned ripple voltage does not distort the reference signal applied to the current loop. Since voltage control is required to have a low bandwidth it does not compensate for 120 Hz and the dynamic response for step-load changes is very poor. Usually the boost PFC is used as a pre-regulator with main emphasis on drawing sinusoidal currents at unity power factor. The pre-regulator is followed by a second-stage DC-DC converter, which provides electrical isolation as well as good dynamic regulation of load voltage at any desired voltage levels.

Other CCM Topologies

Similar to the boost converter, several other basic DC-DC converter topologies can be used with a diode bridge rectifier for PFC applications. SEPIC and Cúk converters operating in CCM have the same advantage as the boost converter in that they draw continuous current from the utility with small, high-frequency ripple [8]. An additional advantage of SEPIC or Cúk converter over the boost converter for PFC applications is that the output voltage can be less than the peak of the input voltage. The trade-off is in the complex controller design.

For single-phase PFC applications, buck-derived converters are not practical because such a converter will not be able to draw any current from the input when the input voltage is instantaneously less than the output voltage. Figure 17.57 shows an interesting combination of a buck and a boost converter suitable for PFC applications that require the output voltage to be less than the peak input voltage [9]. When the main voltage is instantaneously less than the desired output voltage, S_1 is fully on, and the duty ratio of S_2 is controlled in response to the current reference. The operation is now in boost mode. When the input voltage is higher than the output, S_2 is fully off and the duty ratio of S_1 is controlled, and the operation is now in the buck mode. In the buck mode the input current is discontinuous, but its average value tracks the input voltage.

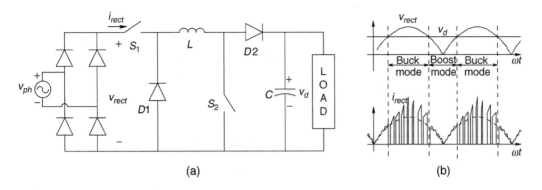

FIGURE 17.57 Buck–boost single-phase power factor correction.

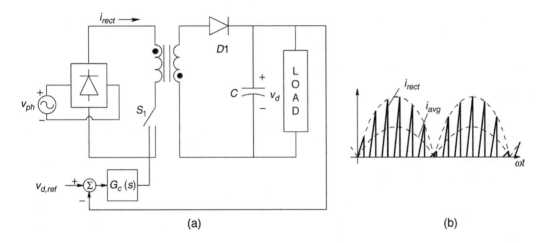

FIGURE 17.58 Single-phase PFC: flyback operation in discontinuous mode.

PFC Converters Utilizing DCM Operation

All the converters discussed so far for single-phase PFC applications operate in CCM. In these converters the duty ratio of the switch is modulated by a current loop to shape the input current within a line-frequency cycle, and a slow outer voltage loop is used to regulate the output DC voltage. There is a family of converters that, while operating in discontinuous conduction mode (DCM), inherently behave as a resistive load and thus draw sinusoidal current at unity power factor (UPF). Flyback, SEPIC, and Cúk converters belong to this family of converters. Their operation involves maintaining a constant duty ratio within a line-frequency cycle, without the need for a current loop. A slow output voltage loop adjusts the duty ratio to regulate the DC-link voltage.

Figure 17.58a shows the schematic diagram of a flyback power factor correction circuit operating in DCM. The design details, including the output voltage control loop, are discussed in detail in Ref. 10. The expression for the input current, $i_g(t)$ (average over a switching period), is as shown in Eq. (17.64) and is derived in Ref. 10.

$$i_g(t) = \frac{V_{pk}}{R_e} |\sin(\omega t)| \qquad (17.64)$$

where $R_e = 2n^2 L/D^2 T_s$, ω is the fundamental frequency in rad/s, n is the turns ratio of the flyback transformer, D is the duty ratio of the switch, and T_s is the switching period. The corresponding input current is shown in Fig. 17.58b, and as seen its average follows the input main voltage. Boost converters

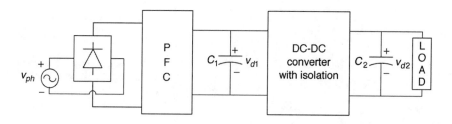

FIGURE 17.59 Two-stage solution for single-phase AC-to-DC converters with PFC.

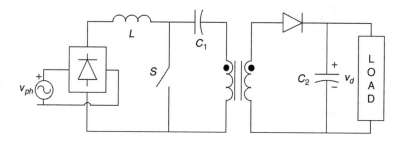

FIGURE 17.60 Single-phase AC-DC converter with PFC: isolated SEPIC.

can also be operated in DCM mode, but unlike the flyback converter it introduces harmonic distortion. THD in a boost DCM converter reduces with increasing output voltage.

The operation of Cúk and SEPIC converters in DCM mode as natural sinusoidal rectifiers is discussed in Refs. 11 and 12, respectively. The advantage of Cúk or SEPIC converters in DCM compared with the flyback or boost is that the input current does not return to zero in each switching period. DCM mode for Cúk and SEPIC is defined as operating condition when the sum of the two inductor currents becomes discontinuous. Also, the two inductors can be coupled, resulting in significant reduction in the high-frequency ripple component in the input current.

Single-Stage Solutions

The PFC circuits described above are used as pre-regulators with a UPF feature. They are usually followed by a DC-DC converter stage as shown in Fig. 17.59, which provides isolation and achieves good dynamic regulation for the load voltage. The voltage levels required by the modern loads are very low—for example, 2 V and below for some integrated circuits. The DC-link voltage at the output of the PFC stage is around 200 V. Conversion of this high DC voltage directly to 2 V as required by the loads is highly inefficient, mainly because of the large turns ratios required. There are two options. The first is to use a three-stage approach with a PFC pre-regulator, a DC-DC converter that converts the 200-V DC-link voltage to isolated distribution level voltage (around 28 V, for example), and another non-isolated DC-DC converter to provide the final well regulated DC voltage at 2 V. The second option is to use what are referred to as single-stage PFC circuits, which directly convert the main voltage to an isolated distribution-level DC voltage (28 V). This is then followed by another DC-DC converter to obtain the well-regulated low DC voltage. The second approach is more efficient and also results in higher overall power density.

Although the boost converter in CCM is a popular topology for non-isolated PFC solutions, it is not very attractive for the single-stage applications, mainly because it is difficult to achieve isolation using the boost configuration in CCM mode. Many DCM solutions have been suggested [13] to realize single-stage PFC circuits, and are useful for lower power applications. For applications that have strict restrictions on the high-frequency component in the input current, the isolated version of SEPIC converter as shown in Fig. 17.60 and operating in CCM is an attractive solution [8]. By suitable choice of turns ratio any desired output voltage can be obtained. Similar to DC-DC converters, interleaving of two or more converters can result in ripple reduction [14].

Power Quality and Utility Interface Issues   17-57

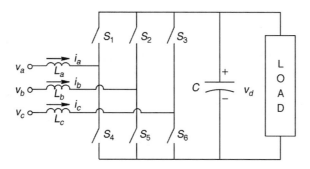

FIGURE 17.61 Three-phase six-switch boost PFC.

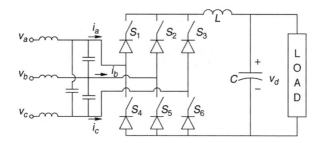

FIGURE 17.62 Three-phase six-switch buck PFC.

Three-Phase PFC Circuits

At higher power levels, for example, above a few kilowatts, three-phase rectifiers are normally preferred. As in the case of single-phase rectifiers, passive PFC solutions are acceptable for less-demanding applications. The passive solutions are typically bulkier. For applications with stringent restrictions on the allowable harmonic distortion active current shaping using high-frequency power electronic circuits is necessary.

Six-Switch PWM Converters

Perhaps the most effective, but expensive, solution is the six-switch boost PWM (pulse-width-modulated) converter [15] shown in Fig. 17.61. The design of the converter ensures that the operation is always in CCM. The control involves an outer voltage loop, which gives the reference to the inner current loop. A preferred method of current control for the inner loop is to use space-vector modulation (SVM) techniques, with digital implementation. The output voltage needs to be higher than the peak of the input line-to-line voltage [16]. A major advantage of the six-switch PWM boost rectifier is its bidirectional power flow capability. However, because of the higher cost and complex control requirement, its use is limited to higher power applications and applications requiring bidirectional power flow. Unlike single-phase converters, in all the three-phase PFC circuits, the power drawn from the input is not pulsating (at two or six times the fundamental frequency). Hence, the output capacitor can be designed based on just the hold-up time requirement. Figure 17.62 shows the schematic diagram of a six-switch buck-derived PWM PFC circuit. The advantage of the buck-derived converter is that the output voltage can be lower than the peak value of the input voltage. However, the input current is discontinuous, and hence large filters are needed to reduce the switching frequency component in the input current.

Single-Switch Three-Phase PFCs

The simplest active three-phase PFC circuit is the single-switch DCM boost converter shown in Fig. 17.63a [17]. The duty ratio S is kept constant within a line-frequency cycle, and is controlled by a slow, outer voltage loop. Because of DCM operation, all the three-phase currents are zero at the beginning of the

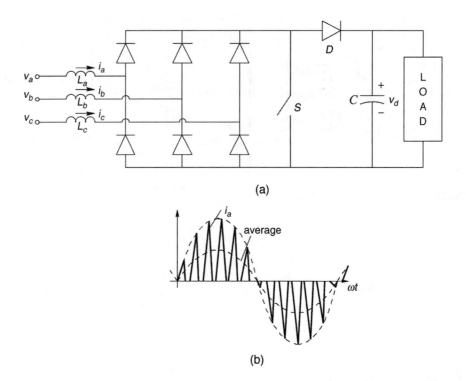

FIGURE 17.63 Single-switch three-phase DCM boost PFC.

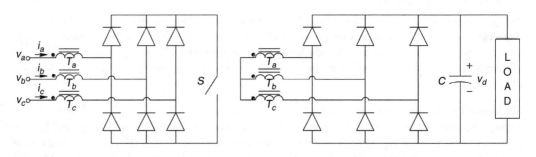

FIGURE 17.64 Single-switch three-phase DCM flyback PFC.

switching period, and change linearly at a rate proportional to the corresponding instantaneous phase voltages when S is turned on. Therefore, the input current peaks are proportional to the input voltages. However, during the off period, this linear relation does not hold, and therefore the current is distorted. Harmonic analysis shows that the input current exhibits a large fifth-order harmonic. The THD depends on the ratio of the output voltage to the peak of the input main voltage and reduces with increasing output voltage. The output voltage must be very high to achieve low THD. High peak currents and high filter rating at the input are other major disadvantages. Techniques to reduce the distortion with fairly low output voltage levels are discussed in Ref. 18.

Figure 17.64 shows a scheme where the three inductors of the earlier scheme (Fig. 17.63) are replaced by three flyback transformers [19]. The main advantage of this scheme is that during the on-interval the primary current is proportional to the input voltage and during the off-interval it is zero. Hence, on an average, the input current accurately follows the input voltage resulting in lower THD as compared with the boost scheme. Also, the turns ratio can be chosen appropriately to obtain any desired output voltage. Hence, this scheme is an attractive option especially at low power levels, around 1 kW.

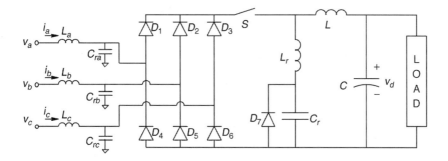

FIGURE 17.65 Single-switch three-phase buck-derived multiresonant ZSC PFC.

Figure 17.65 shows a representative topology from a class of multiresonant single-switch three-phase PFC circuits [20]. Low THD can be achieved with the additional advantage of soft-switching for the main switch. However, the VA rating of the switch is high for a given output power, and hence its application is limited to lower power levels. Control of output voltage is achieved by varying the switching frequency. The operation is designed to be in discontinuous voltage mode (DVM) with respect to the input capacitors, and the average voltage across the capacitors is proportional to the corresponding instantaneous phase voltages.

Use of Single-Phase PFCs for Three-Phase Applications

An attractive solution to achieve high-performance three-phase rectification is to use three single-phase PFC circuits connected in star or delta at the input side and in parallel at the output side. When non-isolated single-phase converters are used in such a scheme, precaution should be taken to avoid interaction among the three phases. Ref. 21 discusses in detail the operation of three nonisolated boost PFC circuits with extra circuitry to avoid phase interactions. The advantages include low overall switch ratings and low passive component ratings compared with other solutions of similar performance. Also, well-known single-phase PFC methods, as well as control ICs can be directly applied for the three-phase extension.

Figure 17.66 shows three isolated single-phase SEPIC converters operating in CCM, with delta connection at the input and parallel connection at the output [22]. Since the three single-phase PFCs are isolated, there is no problem of adverse interaction among the three phases, and hence it is an attractive single-stage solution.

Vienna Rectifier

A three-switch, three-level, three-phase PFC circuit referred to as Vienna rectifier [23] is shown in Fig. 17.67. The three switches are controlled usually by space-vector PWM such that the input currents accurately track the phase voltages. When a switch is turned on, the corresponding phase is connected to the midpoint O, resulting in an increase in the phase current. Turning off the switch results in the conduction of the associated diodes in the upper or lower bridge half, causing the phase currents to reduce. Because of the inclusion of the midpoint of the output voltage into the system function, the circuit has three-level characteristics. Hence, the high-frequency content in the mains current is reduced significantly as compared with that in the two-level converters. Another advantage is the relatively lower combined VA ratings for the controllable switches. The disadvantage is the slightly more complex control circuitry.

Buck-Derived Three-Phase PFC

Figure 17.68 shows the schematic diagram of two interleaved buck-derived three-phase PFC circuits with three switches each. The control is very simple, as it is based on conventional sinusoidal PWM, and retains good harmonic performance [24]. The output voltage can be lower than the peak of the input voltage. However, the input currents are discontinuous, making it necessary to use large low-pass filters at the input to attenuate the switching frequency content. As suggested in Fig. 17.68, two or more such

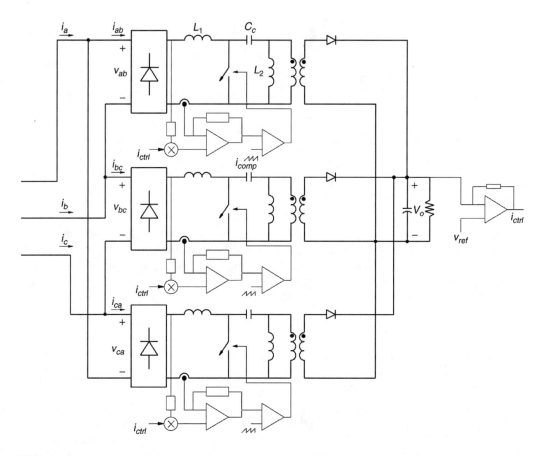

FIGURE 17.66 Three single-phase SEPIC converters delta-connected at input and paralled at the ouptput.

buck-derived PFC circuits can be interleaved, i.e., connected in parallel with their switching signals suitably sequenced. Interleaving leads to ripple cancellation, resulting in lower filter requirement.

Active Filters

Use of passive filters between the harmonic generating rectifier load and the utility is discussed briefly in the section on passive solutions. The main drawbacks mentioned were that passive filters are bulky, can result in over/undercompensation as the load varies, and can result in a low-impedance current path for the harmonic components present in the utility voltage. In this section, another approach to reduce the harmonic distortion, namely, the active filters, is discussed [25, 26]. Figure 17.69 shows a block diagram representation of the concept of active filters to eliminate the harmonic currents from entering the utility. The current drawn by the rectifier system (single-phase or three-phase) has a fundamental component i_{L_1} and a distortion component $i_{L_{distortion1}}$. The active filter senses the current drawn by the rectifier system and, by high-frequency current-mode control, delivers the distortion component $i_{L_{distortion1}}$. Therefore, the utility needs to deliver only the fundamental component of the current drawn by the rectifier system. Since the active filter needs to supply only the harmonic component of the current, no active power needs to be delivered to the load by the active filter. Hence, the active filter does not require a separate energy source. Figure 17.70 shows a typical implementation of an active filter for three-phase PFC applications. For high-power systems, where the use of active filter may be limited by the availability of high-voltage and high-current devices, hybrid filters, which are a combination of active filters and passive LC filters, are an effective approach to meet harmonic standards.

Power Quality and Utility Interface Issues

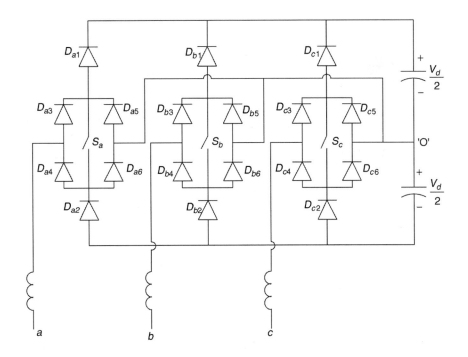

FIGURE 17.67 Vienna rectifier–I.

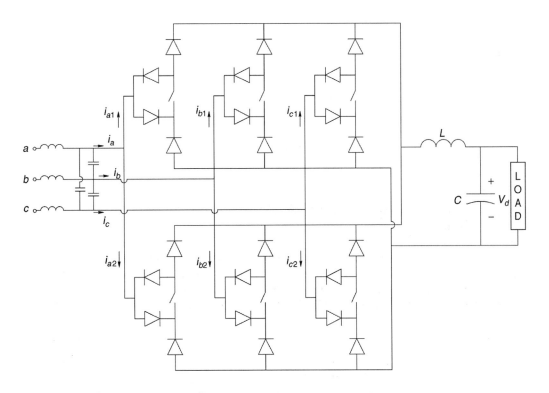

FIGURE 17.68 Three-phase interleaved PFC.

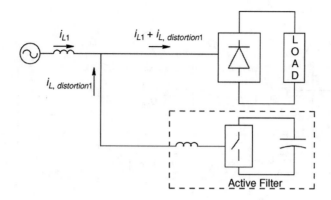

FIGURE 17.69 Active filter.

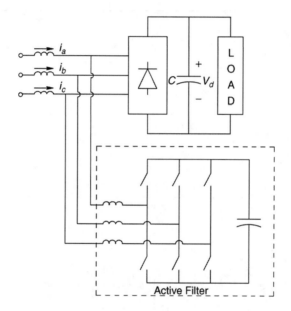

FIGURE 17.70 A three-phase active filter for power factor correction.

Summary

The input currents drawn by conventional diode bridge and phase-controlled thyristor rectifiers are far from sinusoidal and are rich in harmonics. The harmonic currents pollute the utility system and interfere with other loads connected to the system. Several standards that place restrictions on the amount of distortion in the line current were discussed briefly. Several techniques, such as passive filters, active current shaping using high-frequency power electronic converters, and the active filter approach were introduced. Various topologies were discussed for the active current-shaping technique, for both single-phase and three-phase applications. The relative merits of these topologies and the applications for which they are best suited were pointed out.

References

1. Mohan, N., Undeland, T., and Robbins, W. P., *Power Electronics: Converters, Applications, and Design*, 2nd ed., John Wiley & Sons, New York, 1995.
2. Redl, R., Tenti, P., and Van Wyk, J. D., Combatting the pollution of the power distribution systems by electronic equipment, in *Proc. IEEE Applied Power Electronics Conf.*, 1997, 42.

3. Key, T. S. and Lai, J. S., IEEE and international harmonic standards impact on power electronic equipment design, in *Proc. IECON*, 1997, 430.
4. IEC 1000-3, *Electromagnetic Compatibility Part 3: Limits*, March 1995.
5. IEEE-519, *IEEE Recommended Practices and Requirements for Harmonic Control in Electric Power Systems*, IEEE, New York, 1992.
6. Dixon, L. H., High power factor pre-regulators for off-line power supplies, presented at Unitrode Seminar SEM600, 1988.
7. Mohan, N., Design of Feedback Controller in Power Factor Correction (PFC) Circuits Using PSPICE: Explained by a Numerical Example, Tutorial.
8. Spiazzi, G. and Mattavelli, P., Design criteria for power factor pre-regulators based on SEPIC and Cúk converters in continuous conduction mode, in *Proc. IAS*, 1994, 1084.
9. Ridley, R., Kern, S., and Fuld, B., Analysis and design of a wide input range power factor correction circuit for three phase applications, in *Proc. IEEE Applied Power Electronics Conf.*, 1993, 299.
10. Erickson, R., Madigan, M., and Singer, S., Design of a simple high-power-factor rectifier based on the flyback converter, in *Proc. IEEE Applied Power Electronics Conf.*, 1990, 792.
11. Brkovic, M. and Cúk, S., Input current shaper using Cúk converter, in *Proc. International Telecommunications Energy Conf.*, 1992, 532.
12. Simonetti, D. S. L., Sebastian, J., Reis, F. S. D., and Uceda, J., Design criteria for SEPIC and Cúk converters as power factor preregulators in discontinuous conduction mode, in *Proc. IECON*, 1992, 283.
13. Chow, M. H. L., Siu, K. W., Tse, C. K., and Lee, Y.-S., A novel method for elimination of the line current harmonics in single-stage PFC switching regulators, *IEEE Trans. Power Electron.*, 13(1), 75, Jan 1998.
14. Miwa, B. A., Otten, D. M., and Schlecht, M. F., High efficiency power factor correction using interleaving techniques, in *Proc. IEEE Applied Power Electronics Conf.*, 1992, 557.
15. Green, A. W., Boys, J. T., and Gates, G. F., Three-phase voltage sourced reversible rectifier, *IEE Proc. Electric Power Appl.*, 135(6), 362, Nov. 1988.
16. Lee, F. C. and Boroyevich, D., Power factor correction circuits topologies and control, Seminar held in conjunction with the IEEE Applied Power Electronics Conference, 1993.
17. Prasad, A. R., Ziogas, P. D., and Manias, S., An active power factor correction technique for three phase diode rectifiers, in *Proc. IEEE PESC Conf.*, 1989, 58.
18. Sun, J., Frohleke, N., and Grotstollen, H., Harmonic reduction techniques for single switch three-phase rectifiers, in *Proc. IAS Conf.*, 1996, 1225.
19. Kolar, J. W., Ertl, H., and Zach, F. C., A novel three-phase single switch discontinuous mode AC-DC buck-boost converter with high quality input current waveforms and isolated output, *IEEE Trans. Power Electronics*, 160, March 1994.
20. Ismail, E. H., Oliveira, C. M., and Erickson, R. W., A low-distortion three-phase multi-resonant boost rectifier with zero current switching, *IEEE Trans. Power Electron.*, 718, July 1998.
21. Spiazzi, G. and Lee, F. C., Implementation of single-phase boost power factor correction circuits in three-phase applications, *IEEE Trans. Ind. Electron.*, 44, 365, June 1997.
22. Ayyanar, R., Mohan, N., and Sun, J., Single-stage three-phase power factor correction circuit using three isolated single phase SEPIC converters operating in CCM, in *Proc. IEEE PESC Conf.*, June 2000, 353.
23. Kolar, J. W., Drofenik, U., and Zach, F. C., Current handling capability of the neutral point of a three-phase/switch/level boost-type PWM (VIENNA) rectifier, in *Proc. IEEE PESC Conf.*, 1996, 1329.
24. Malesani, L. and Tenti, P., Three-phase AC/DC PWM converter with sinusoidal AC currents and minimum filter requirements, *IEEE Trans. Ind. Appl.*, IA-23, 71, 1987.
25. Akagi, H., New trends in active filters for power conditioning, *IEEE Trans. Ind. Appl.*, 32, 1312, 1996.
26. Rastogi, M., Mohan, N., and Edris, A-A., Filtering of harmonic currents and damping of resonances in power systems with a hybrid-active filter, in *Proc. IEEE APEC Conf.*, 1995, 607.

# 18
Photovoltaic Cells and Systems

Roger Messenger
Florida Atlantic University

18.1 Introduction 18-1
18.2 Solar Cell Fundamentals 18-1
   Conversion of Sunlight to Electricity • Cell Performance
18.3 Utility Interactive PV Applications 18-4
   The PCU • Simple UI PV System • UI PV System with Battery Backup
18.4 Stand-Alone PV Systems 18-7
   Systems with No Storage • Systems with Storage

18.1 Introduction

The ability of certain materials to convert sunlight to electricity was first discovered by Becquerel in 1839, when he discovered the photogalvanic effect. A number of other significant discoveries ultimately paved the way for the fabrication of the first solar cell in 1954 by Chapin, Fuller, and Pearson [1]. This cell had a conversion efficiency of 6%. Within 4 years, solar cells were used on the Vanguard I orbiting satellite. The high cost of boosting a payload into space readily justified the use of these cells, even though they were quite expensive.

Space applications eventually led to improved production efficiencies, higher conversion efficiencies, higher reliability, and lower cost for photovoltaic (PV) cells. By the 1980s, PV cells had been introduced to terrestrial applications where conventional electrical sources were expensive, and by the turn of the millennium, PV cells have become cost-effective in a wide range of utility interactive and stand-alone applications. Conversion efficiencies at the turn of the millennium for large-scale PV modules ranged from just under 10% for thin-film modules to over 30% for gallium arsenide (GaAs) concentrating cells.

This chapter presents a thumbnail sketch of the basic theory of solar cells and then focuses on several examples of PV applications in utility interactive systems and stand-alone systems. Each type of system generally requires a power electronics interface unit to enable the PV system to transfer solar electricity optimally to the desired load or storage system. The intent of this chapter is to present the *what* of PV systems. For the *why* and *how* of PV systems, the reader is referred to the references at the end of the chapter that provide detailed examples of specific PV system designs [1–3].

18.2 Solar Cell Fundamentals

Conversion of Sunlight to Electricity

To date, the most popular materials for direct conversion of sunlight to electricity have been crystalline silicon (Si), amorphous silicon (a-SiH), copper indium diselenide (CIS), cadmium telluride (CdTe), and gallium arsenide (GaAs). All of these semiconductor materials have band-gap energies between 1 and 2 eV. The band gap of a semiconductor is the energy required to excite an electron from the valence band to

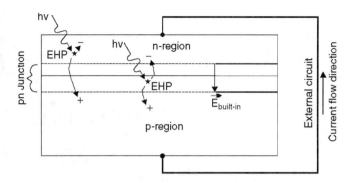

FIGURE 18.1 Effect of sunlight incident on the PV cell.

the conduction band of the semiconductor. Transferring the negative electron to the conduction band creates a positive hole in the valence band. Both charge carriers are then available for electrical conduction. Sunlight is a very convenient source of energy for creation of these electron–hole pairs (EHPs), since most of the energy in the solar spectrum is at levels higher than the band-gap energies of PV materials.

Once the EHP has been produced by an incident photon, the electron and hole must flow in opposite directions. Separation of electron and hole can be achieved by using a *pn*-junction. A *pn*-junction is composed of material that is rich in electrons on one side (the *n*-side) and rich in holes on the other side (the *p*-side). The *pn*-junction produces a built-in electric field, directed from the *n*-side to the *p*-side, that separates the photon-generated EHPs. The electrons are forced to the *n*-side and the holes are forced to the *p*-side by the junction electric field as long as the EHP is produced within or close to the *pn*-junction. If the EHP is generated too far from the junction, the electron and hole will recombine before they can be separated by the junction electric field.

Figure 18.1 shows photons ($h\nu$) entering a typical PV cell. Some of the photons will create EHPs close to the surface, some will create EHPs near or within the junction region, and some will penetrate beyond the junction. Generally, the highest-energy photons produce EHPs close to the surface, whereas the lowest-energy photons penetrate the deepest. This process of liberating an EHP results in the conversion of part of the energy of the incident photon to electricity. Any leftover energy is converted to heat.

If the EHP is produced near or within the *pn*-junction, the electron is swept into the *n*-region and the hole is swept into the *p*-region. The electrons (−) then diffuse toward the top of the cell and the holes (+) diffuse toward the bottom of the cell.

As the electrons reach the top surface, where there is a contact to an external circuit, they continue to flow into the external circuit. As the holes reach the bottom surface, where there is another contact to the external circuit, they recombine with electrons flowing in from the external circuit. For each electron that leaves the top, another enters the bottom. This completes the circuit, with electron flow in the external circuit and the flow of both electrons and holes within the PV cell. The challenge in the design of the PV cell is to absorb all incident photons close enough to the *pn*-junction so all electrons and holes generated will be collected. A further challenge in cell design is to minimize conversion of sunlight to heat and maximize conversion to electricity.

Because of the *pn*-junction, a voltage appears between the bottom and the top of the cell. This voltage is what forces the current through the external circuit. Depending upon the cell material, the voltage developed by the cell may range from very small up to about 1 V. Thus, to produce higher voltages, the cells must be connected in series. When cells are connected together, normally they are incorporated into PV *modules*, which often combine as many as 40 cells in series to produce voltages in the range of 20 V and currents of several amperes.

When voltages and currents beyond the capability of an individual module are desired, the modules can be connected into *arrays* that will produce higher voltages and higher currents. Although most cells produce only a few watts, and most modules produce 10 to 300 W, most arrays produce a few thousand watts. A few very large systems have been deployed that produce power in the megawatt range.

An important feature of all modern PV cells is that, over their lifetimes, they can produce up to ten times as much energy as was used in their fabrication and deployment.

Cell Performance

The ideal solar cell operates as a diode when in the dark, and operates almost as an ideal current source when operated under short-circuit conditions. The short-circuit current of the cell is close to directly proportional to the intensity of the sunlight incident on the cell. The current source nature of the cell means that if cells are connected in series to increase their overall voltage, the cells must be closely matched so each cell produces identical current under identical illumination conditions. If this is not the case, the voltage of the series combination will not be optimized. The I–V relationship for the ideal PV cell is given by

$$I = I_l - I_o(e^{qV/kT} - 1) \tag{18.1}$$

where I_l is the photon-generated current component, I_o is the cell reverse saturation current, and $kT/q = 25.7$ mV at a temperature of 25°C. More specifically, the photocurrent is related to sunlight intensity by the relationship:

$$I_l = I_l(G_o)\frac{G}{G_o} \tag{18.2}$$

where G is the sunlight intensity in W/m^2 and $G_o = 1000$ W/m^2. Note also from Eq. (18.1) that I_l is the short-circuit current of the PV cell, I_{SC}.

Figure 18.2a shows typical PV cell I–V characteristics as a function of incident sunlight, and Fig. 18.2b shows the temperature dependence of the output power of a cell. Note that as the temperature rises, the open circuit voltage of the cell, V_{OC}, decreases. For Si cells, the rate of decrease is 2.3 mV/°C/cell. Thus, a 36-cell module operating 25°C above ambient will lose $36 \times 2.3 \times 25 = 2070$ mV = 2.07 V. This is nearly a 10% loss in output voltage, which, when coupled with approximately temperature-independent current, results in a 10% power loss.

The departure of the I–V characteristic of a real cell from that of a perfect cell is measured by the *fill factor* (FF) of the cell. The assumption is that a perfect cell would have a rectangular characteristic, with

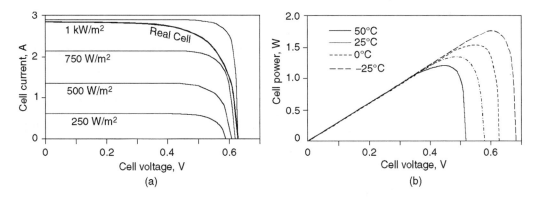

FIGURE 18.2 Dependence of PV cell characteristics on sunlight intensity and temperature. (a) Real and ideal PV cell I–V vs. sunlight intensity; (b) cell output power vs. cell temperature.

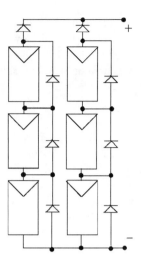

FIGURE 18.3 PV array showing modules with bypass and blocking diodes.

constant current up to the maximum cell voltage, and then constant voltage. The constant current would be the short-circuit current and the constant voltage would be the open-circuit voltage. The fill factor is thus defined as

$$\text{FF} = \frac{P_{\max}}{I_{\text{SC}} V_{\text{OC}}} \tag{18.3}$$

Since the current produced by a cell depends upon the total power incident upon the cell, if a cell is shaded even partially, it will not produce the same current as unshaded cells. At a certain point of shading, the polarity of the cell voltage reverses to enable the cell to carry the current generated by the unshaded cells in the module. When this happens, the cell dissipates power, and can overheat to the point of cell degradation. To protect the module against cell degradation, bypass diodes are normally incorporated into the module design to shunt current away from shaded cells, as shown in Fig. 18.3.

If the voltage of a module drops below the voltage of other modules connected in parallel, it is possible for the current produced by the higher-voltage modules to flow in the reverse direction of the lower-voltage module. To prevent reverse flow of current through a module, a blocking diode is sometimes used in series with the module, as shown in Fig. 18.3.

18.3 Utility Interactive PV Applications

Perhaps the simplest PV application, except for connecting the PV array output directly to the load, is the utility interactive (UI) system. In a simple UI system, the PV array output is connected to the input of a DC-to-AC inverter, known as a power conditioning unit (PCU), the output of which is connected directly to the utility. When battery storage and alternative generation means are incorporated in the UI system, however, the system is no longer quite as simple. In either case, the PCU must be designed to meet a wide range of utility concerns. If the heart of the UI PV system is the PV array, then the PCU may be considered to be the brains of the system.

The PCU

Although the basic UI PV system is quite straightforward, the PCU is a very sophisticated piece of power electronics equipment. The PCU must meet the stringent design requirements of IEEE Standard 929 [4] and the stringent performance requirements of UL 1741 [5].

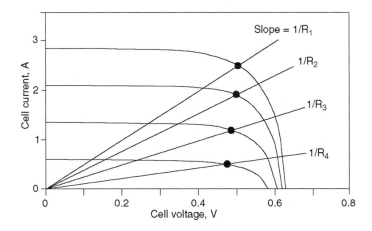

FIGURE 18.4 Cell I–V showing maximum power points and associated resistances.

Since PV arrays are still relatively costly, it is important for the PCU to extract maximum power from the array. This is done by incorporating maximum power-tracking circuitry into the PCU. Figure 18.4 shows the I–V characteristics of an array with the maximum power points indicated for each level of sunlight. The design challenge for the PCU is to vary the PCU input resistance, defined as the ratio of input voltage to input current, while sampling the PCU output power. When the PCU output power reaches a maximum level, the input resistance is fixed at the value that produces this level. Presumably when output power is a maximum, input power is also at a maximum, provided that PCU conversion losses remain at a constant percentage of the output power. The effective input resistance of the PCU can be varied by the use of a buck–boost DC-DC converter.

Nearly all modern PCUs use a pulse code modulation (PCM) scheme for generating an output waveform of appropriate amplitude and frequency. The PCU is generally designed to perform as a current source when it is connected to the utility, so the utility voltage can be used as a synchronization signal for PCU output frequency control. As long as the utility voltage is present at the proper amplitude and frequency, the PCU supplies power to the grid. However, if the utility voltage or frequency drifts outside prescribed limits for too long, the PCU is programmed to shut down its output to the utility. Although output is shut down, the PCU continues to monitor the utility voltage. The PCU reconnects to the utility after the PCU senses that the utility has remained within amplitude and frequency limits for a predetermined time.

IEEE 1741 prescribes limits for PCU output harmonics and general PCU power quality. It also requires the PCU to shut down under utility islanding conditions. Islanding occurs when the utility shuts down, leaving the PV source along with other PV sources connected to the disabled utility line. The trick here is for every UI PCU to be able to recognize that the presence of power on the utility line from other PCUs is not the same as power on the line from the utility. Several elegant software algorithms have been developed to prevent islanding [6].

The output power range of modern PCUs is from a few hundred watts to 100 kW. These units typically operate with efficiencies in excess of 90% for output powers between 10 and 100% of rated output power. They are capable of maximum power tracking over a wide range of incident sunlight. Many modern PCUs that can be used in a grid-independent mode have a sleep mode. In the sleep mode, the PCU sends out short pulses of AC voltage at regular intervals to sense for connected loads. If current is drawn when the voltage pulse is sent, the PCU recognizes that a load is connected and remains on until the load is no longer sensed. The sleep mode sensitivity is usually adjustable. If the PCU includes battery-charging capability from utility line input, then it normally also incorporates battery protection from overcharge or overdischarge. Additional features may include provision for code-required fusing and disconnects on the DC and AC sides of the PCU. A Web search for PV power conditioning units, or PV inverters, will yield information on a wide range of products. The important consideration in selection of a PCU for grid-connected applications is the UL 1741 listing.

It should also be noted that inverter output waveforms range from square to modified sine to pure sine. The cost of the inverter is generally increased as the quality of output waveform approaches pure sine. In the event that the PCU will operate in a stand-alone mode during utility failure, it is important to consider whether all loads to be supplied by the PCU will operate on the waveform generated by the PCU. For example, an electric igniter on a gas appliance may not operate when connected to a modified sine wave or to a square wave voltage waveform.

Simple UI PV System

Because the output of a UI PV system is connected directly to the electric grid, any excess output will be used by the grid and any deficiency in PV output will be made up by the grid. Hence, the sizing of the simple UI PV system is not necessarily related to the load at the UI PV installation site. The sizing is more likely to be governed either by available space for the array, available funding for the system, or a preference for PCU or modules.

The system shown schematically in Fig. 18.5 consists of matching the input voltage and current requirements of the PCU with the voltage and current output capabilities of the PV array and then connecting to the utility. The connection may be made either on the customer side or the utility side of the revenue meter. It should be noted that Fig. 18.5 is somewhat simplified, since actual installations may require DC and AC disconnect switches, fuses, ground-fault protection, and system grounding per the requirements of the National Electrical Code® (NEC) [7]. Wire size and insulation type are also specified by the NEC. Additional requirements may be placed upon the system by the local utility, local electrical inspector, or local fire inspector. All of these potential sources of system requirements should be consulted prior to any installation.

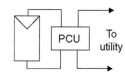

FIGURE 18.5 Simple UI PV system.

The PV output circuit is that part of the PV system that connects the PV array to the PCU. The NEC imposes a number of requirements on this DC circuit, including fusing, disconnects, and, in some cases, ground-fault protection, all of which add to the installation cost.

The desire to eliminate the PV output circuit has led to the development of the AC module. By mounting the PCU on the DC PV module, the PCU becomes a part of the module, so there is no PV output circuit—only the PCU output circuit. Since there is no particular economy of scale for PCUs, a popular match is between a 300 W module and a 300 W PCU, resulting in a 300 W AC module. When used in the United States, the AC module must meet all the requirements of IEEE 929 and UL 1741.

Although IEEE 929, UL 1741, and the NEC have solved the technical considerations for a UI PV installation, additional redundant requirements are still imposed in many jurisdictions. For example, at the time of this writing, one investor-owned utility requires the owner of a UI PV system to carry $1 million in liability insurance and to install an isolation transformer between the PV system and the grid connection.

UI PV System with Battery Backup

Figure 18.6 shows a simplified diagram of one way to configure a system with battery backup to provide for emergency power in the event of grid failure. Required fuses, disconnects, grounding, surge protection, and ground-fault protection are not shown. If a PV system is to provide emergency power in the event of grid failure, then the PV system must be designed to provide power to all designated emergency loads for a length of time required either by the owner or by a regulation. Sizing of the array and battery system will normally follow the procedure used for stand-alone systems. This requires identification of the energy requirements of each emergency load over the anticipated duration of a power outage. Since storage batteries are normally rated in ampere hours (Ah), the energy requirements of the load are normally converted to Ah at the voltage of the storage battery system.

For example, suppose it is determined that 10 kWh of battery storage will operate all the emergency loads over the anticipated duration of the power outage. Suppose also a PCU is chosen that has a nominal 48 VDC input. Dividing 10,000 Wh by 48 V results in a battery capacity of 208 Ah. However, if

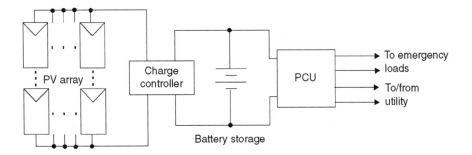

FIGURE 18.6 UI PV system with battery backup and emergency loads.

deep-discharge lead acid batteries are used, the batteries should not be allowed to discharge below 20% of their rated charge. Thus, the battery capacity needs to be 208 ÷ 0.8 = 260 Ah. If a 6-V, 130-Ah deep-discharge battery is available, then the system will require 16 of these batteries to store 260 Ah at 48 V. When batteries are connected in series–parallel combinations, it is important to use connecting cable lengths that will ensure equal charging and discharging rates for all batteries in the system. Furthermore, connecting cables must be fused close to the batteries in accordance with NEC requirements and a disconnect switch must be provided between batteries and PCU.

Sometimes the PCU in a UI system with battery backup will allow for charging of the batteries from the utility connection. In a system of this type, the utility is connected to the emergency loads when the utility is energized. In the event of utility failure, the PCU automatically switches to its battery input and powers the emergency loads from the battery storage, much the same as an uninterruptible power supply used in computer systems. If the utility remains down for a prolonged time, the PV system can be designed to provide the energy required by the emergency loads. This type of system is not UI in the purest sense, since it does not feed back power to the utility if the PV array is providing more power than is needed by the emergency load. Normally, however, the PV array is sized to meet the needs of the emergency loads over a prolonged utility outage. Array sizing will be discussed in Section 18.4.

If the PCU connection to the utility is bidirectional, then the system is interactive, with interesting control possibilities. Normally it would be desirable to size the PV system to meet the entire emergency load, since the emergency loads would then become essentially grid independent. When a significant PV array is incorporated into the system, it is desirable to optimize the utilization of the PV output. The control algorithm would be designed to incorporate utility charging of batteries only if the PV array has not brought the batteries to full charge by the end of daylight hours. In addition, as soon as the array has fully charged the batteries, any excess array output would be directed back to the utility. In many areas, this feeding of the array output to the utility will occur during utility peaking hours, thus increasing the value of the PV-generated electrical power. In fact, with adequate battery storage, the system can be controlled to feed power to the utility *only* during peaking times. During off-peak hours, assuming utility wholesale prices are low, the utility can then be used to charge the batteries with relatively low-cost electricity. Such a control strategy wins for everyone. A leveling effect is provided to the utility by the assist during peak hours and the battery charging during off-peak hours. The system owner experiences full utilization of the PV system output and the overall customer base benefits from incrementally cleaner air and a reduced need for siting of power plants and transmission lines.

18.4 Stand-Alone PV Systems

Systems with No Storage

Probably the simplest stand-alone PV system is the direct connection of array to either a fan or a pump. When enough sunlight is available to start the fan or pump, it starts and continues until insufficient sunlight is available to meet the fan or pump power requirements.

The next step is to incorporate some sort of power electronics into the system to provide a better match between PV system output and load requirements. For example, a small amount of energy storage can assist the PV system in overcoming the starting torque of the fan or pump. A maximum power tracker (MPT), as described earlier, can ensure that the PV output is matched to the load resistance. Providing an MPT allows a system with a fixed fan or pump size to provide maximum water or airflow for a given array size. Without the MPT, a larger array size will be needed to achieve the same performance from the fan or pump. Sizing of such a system involves the determination of minimum operating parameters for the pump or fan and then choosing the combination of array and MPT to meet these conditions.

Systems with Storage

If a system is to operate on cloudy days or after sundown, a means of energy storage will be required. Although exotic means such as fuel cells or hydrogen storage are possible, the most common storage means at the time of this writing is still the lead acid storage battery. For PV energy storage applications, the deep-discharge variety of lead acid battery is used, as opposed to the automotive type that provides high current for short time periods. Examples of systems that require storage include lighting systems, highway information and warning signs, cathodic protection systems, refrigeration systems, communication systems, and remote dwellings. If the grid connection is eliminated, the system shown in Fig. 18.6 is representative of a stand-alone system with battery storage. If the system has only DC loads, then the PCU is replaced with a discharge controller to disconnect the loads if the battery state of charge drops too low.

If the utility connection in Fig. 18.6 is replaced with an auxiliary fossil fuel or wind generator, the system then becomes a hybrid system, where an auxiliary power source is made available to augment the PV power. Hybrid systems are most cost-effective when the PV array must be sized to meet the demands of low sun periods with resulting significant excess PV energy during high sun periods. For example, in Fairbanks, AK, a one-axis tracking array that follows the sun from sunrise to sunset will be exposed to the equivalent of nearly 9 h of full sun in May, but will receive no sun at all in December. An alternative power source must then be provided for the winter months, unless no load is present.

Design of a system with storage involves determining the daily or weekly system load in Ah, determining the number of storage days required, determining system losses, determining the battery requirements, and determining the array requirements. The design also requires selecting appropriate charge controller, PCU or other power electronic equipment, switches, fuses, wires, and surge protectors.

The *connected* daily Ah load of the stand-alone system is the Ah that must be delivered to the load, including any losses in inversion of DC to AC to provide power to AC loads. Inverter losses are generally assumed to be 10% unless specific inverter (PCU) data indicate different loss values. The *corrected* Ah load of a PV system is the Ah that must be supplied to the batteries to overcome battery losses and wiring losses and still supply the connected load. Battery sizing and array sizing are based upon the corrected load of the system, which is typically about 12% higher than the connected Ah load.

The number of storage days, also known as days of autonomy, is determined by whether the load is considered to be critical or noncritical. Critical loads are defined as loads that are met 99% of the time by the PV system, whereas noncritical loads only require 95% system availability. Hence, critical loads require more days of autonomy, and the days of autonomy depend upon the specific geographic location. For example, in Albuquerque, NM, a critical load requires approximately 7 days of autonomy, whereas a critical load in Seattle, WA, requires approximately 16 days of autonomy. Noncritical loads in Albuquerque require only 2 days of autonomy, but in Seattle require 4 days of autonomy.

System losses can normally be considered to be 10% for battery charging and discharging losses, and 2 to 3% for wiring losses. Battery losses are due to conversion of electrical energy to heat energy while charging or discharging the batteries. In addition, the PV array is generally derated to account for elevated array temperatures and degradation of the array from dust and dirt.

Once the corrected load is known, battery sizing can be accomplished by using

$$\text{Ah} = \left(\frac{\text{Ah}}{\text{day}}\right)\left(\frac{\text{days}}{D_T D_{\text{ch}}(\text{disch})}\right) \tag{18.4}$$

where Ah is the required battery capacity, Ah/day is the system corrected load, days is the number of days of autonomy required, and D_T is a temperature correction factor that varies linearly from unity at 80°F to 0.72 at 32°F. D_{ch} is a discharge correction factor that is unity as long as the batteries discharge at less than the rated discharge rate. For faster discharge rates lasting more than 10 min, D_{ch} is the ratio of the rated discharge rate to the actual discharge rate. Finally, (disch) is the design depth of discharge, normally in the range of 0.5 to 0.8, with 0.8 representing the loss of 80% of rated charge.

Array sizing is also based on the corrected load of the system, since the array must supply the corrected load to the batteries at the selected battery voltage. Array sizing also depends upon any degradation of array output and upon available sunlight. In areas where the array is likely to become dusty or where it is likely to operate more than a few degrees above 25°C, the degradation factor may be as high as 15 to 20%. Sunlight is measured in terms of peak sun hours (PSH). The PSH for a location on a particular day is determined by measuring the daily kWh/m² incident upon the array. Monthly expected PSH are tabulated for many locations, based upon measurements over periods of 20 years or more. In addition to the PSH, standard deviations are also available for use when worst-case estimates need to be made. The Florida Solar Energy Center maintains a Web site with links to a wide range of PV information, including sunlight intensity data [8]. When the corrected load, system voltage, degradation factor, and PSH information are available, the array size can be determined from

$$I = \frac{\text{Ah}}{\text{PSH}_{min} \times \eta} \tag{18.5}$$

where I is the rated array current, Ah is the system corrected load, PSH_{min} is the minimum PSH for the PV system site, and η is the array degradation factor. Note that as PSH decreases for the winter months, η increases as the array temperature decreases. It is thus useful to estimate PSH and η for winter and summer and then use the smaller of the two products in Eq. (18.5). Modules are then connected in series to achieve the system design voltage and in parallel to obtain the system design current. System voltages typically are selected in multiples of 12, while system currents depend upon the specific loads.

The PSH for a PV array depends upon the month of the year as well as the tilt of the array. The PSH also depends upon whether the array is a fixed array or a single- or double-axis tracking array. The PSH tables typically list the available PSH for several different tilt angles. For best annual performance, the optimal tilt of a fixed PV array is approximately latitude −10°. If an array is to supply different loads for different seasons, then the array may be tilted to optimize seasonal performance.

After the batteries and array have been selected, the system power electronic components are selected. The charge controller is an important part of the system and can range from relatively simple design to relatively complex design. A simple controller monitors battery voltage and diverts the array current output when the battery voltage reaches the manufacturer's specified full-charge voltage. A better controller also monitors when the batteries are at the design discharge voltage and disconnects the batteries from the load at this point. An even better controller monitors the battery temperature and compensates the array disconnect voltage and the load disconnect voltage for temperature. A further controller improvement employs maximum power tracking to ensure maximum transfer of charge from array to batteries and provides for diversion of the PV output to an alternate load after the batteries are fully charged. Some controllers incorporate most of the features listed, but must be set either for charge control or discharge control. In a stand-alone system with AC loads, the PCU will normally incorporate the discharge protection feature. If no AC loads are present, a separate discharge controller may be needed.

The internal resistance of a battery system introduces a hysteresis effect into the charge and discharge process. When the batteries are charging, the terminal voltage becomes greater than the cell voltage as a result of the additional voltage drop across the internal battery resistance between the terminals and the cells. Hence, if the terminal voltage is sensed, it will drop when the charger is disconnected. This may cause the charger to be reconnected, if the terminal voltage drops below the full-charge value. Because of the hysteresis effect, charge controllers normally are designed to provide a three-stage charging process. Initially, the batteries are charged at constant current. When the batteries reach a predetermined terminal voltage, they are charged at constant voltage until the current drops to a prescribed level. At this point, the charge controller voltage decreases to complete the charging process at a lower current level.

During discharge, the battery terminal voltage is lower than the battery cell voltage, since the current through the battery internal resistance is now in the opposite direction. The discharge control must thus incorporate a means of avoiding oscillation on and off when the battery terminal voltage indicates that the battery is approaching minimum allowable charge.

The PCU in a stand-alone system does not need to comply with IEEE 929 or UL 1741. It thus can range from a relatively simple square wave inverter to a more-sophisticated, microprocessor-controlled PCM inverter that employs amplitude and/or frequency control of the output waveform as well as many other features described earlier in this chapter.

References

1. Markvart, T., Ed., *Solar Electricity,* John Wiley & Sons, Chichester, U.K., 1994.
2. Messenger, R. A. and Ventre, J., *Photovoltaic Systems Engineering,* CRC Press, Boca Raton, FL, 1999.
3. *Stand-Alone Photovoltaic Systems: A Handbook of Recommended Design Practices,* Sandia National Laboratories, Albuquerque, NM, 1995.
4. ANSI/IEEE P929, IEEE Recommended Practice for Utility Interface of Residential and Intermediate Photovoltaic (PV) Systems, IEEE Standards Coordinating Committee 21, Photovoltaics, Draft 10, February 1999.
5. UL 1741: 1999, Standard for Static Inverters and Charge Controllers for Use in Photovoltaic Power Systems, Underwriters Laboratories, Inc., Northbrook, IL, May 1999.
6. Kern, G. A., Bonn, R. H., Ginn, J., and Gonzalez, S., Results of Sandia National Laboratories grid-tied inverter testing, in *Proc. 2nd World Conference and Exhibition on PV Solar Energy Conversion,* Vienna, Austria, July, 1998.
7. *NFPA 70 National Electrical Code, 1999 Edition,* National Fire Protection Association, Quincy, MA, 1998.
8. http://alpha.fsec.ucf.edu/~pv (Florida Solar Energy Center PV home page).

19
Flexible, Reliable, and Intelligent Electrical Energy Delivery Systems

Alexander Domijan, Jr.
University of Florida

Zhidong Song
University of Florida

19.1 Introduction 19-1
19.2 The Concept of FRIENDS 19-2
   What Is FRIENDS? • Quality Control Centers • Intelligent Function of FRIENDS
19.3 Development of FRIENDS 19-5
19.4 The Advanced Power Electronic Technologies within QCCs 19-7
   Distribution Static Compensator • Dynamic Voltage Restorer • Solid-State Breaker
19.5 Significance of FRIENDS 19-9
19.6 Realization of FRIENDS 19-11
19.7 Conclusions 19-12

19.1 Introduction

There have been considerable changes in today's power systems in recent years, due, in large part, to the emergence of competition and deregulation in power industries. The goal of deregulation is to provide the customer with reduced costs. The global competition between utilities leads to utility cost-cutting, downsizing, and reducing maintenance on both transmission and distribution systems. Power systems that are hierarchically integrated are separated such that tasks normally carried out within traditional organizations have been open to competition whenever practical and profitable. This process, which is called "unbundling," consists of unbundling of vertically integrated utilities, unbundling of functions within a corporation, and service unbundling. As a consequence, the common structure in the deregulated power industry is the separation of the generation, transmission, and distribution business into separate entities. There is, however, a possible drop in service quality as a result of deregulation. Additionally, installation and penetration of distributed generators (DGs) and distributed energy storage systems (DESSs), allocated in the demand-side of power systems, have been increased. Unexpected power quality problems may occur, because these distributed facilities are operated independently and are uncommitted to the operating situation of power systems.

Power quality issues have become more important in the face of this open competition. In recent years there has been an increase in the number of system loads and controls that are sensitive to power quality, as well as an increase in the number of system loads that are themselves a source of poor power quality (such as harmonics caused by static drives). Loads such as computers, process controls, and communications equipment are more sensitive to power quality variations due to frequent system disturbances than is equipment applied in the past. On the other hand, there has been continuing growth in the application of power electronics devices, both to improve the overall power system efficiency and to facilitate controls. This is resulting in increasing power quality problems on power systems, for these loads themselves are major causes of the degradation of power quality. Yet awareness of power quality issues is increased by the end users who need a high quality of power for their equipment and are becoming better informed about such issues as voltage sags, swells, surges, harmonic distortions, interruptions, outages, undervoltages, overvoltages, electrical noise, frequency deviations, impulses, transients, and notches. They need to understand how the disturbances will affect sensitive loads and therefore develop appropriate specifications, or install appropriate power conditioning systems [1]. Harmonics, for example, can result in equipment heating and voltage stress, communication interference, and control malfunctions. Voltage sags of only a few cycles can cause tripping of drives, loss of computer data, or errors. As a matter of fact, utilities are being challenged to improve the quality of power delivered by every possible means.

Power quality, therefore, is the benchmark by which we may judge the success of our problem solution. Reliable and secure power service needs to be made available to customers. Customers, who need a high quality of power for their equipment and are becoming better informed about power quality issues, provide the principal motivation for the utility industry with unbundled services and greater reliance on competitive forces. They, then, would be free to select their desired level of power quality. There is opportunity for enhanced value of service through greater choice among customized and competitive services. Yet the utility industry itself, under the threat of competition, on the other hand, is likely to be forced to offer a variety of valued-added unbundled services and prices to customers to increase customer satisfaction. It may be assumed that the customer or market will play the lead over that of the energy supplier. Information-based electric suppliers, both with and without generation sources, will be able to attract large portions of market share from the traditional suppliers. Therefore, electric utilities must be prepared to develop a strategic methodology both to satisfy their existing customer base and to capture additional customers as competition unfolds. One way to accomplish these goals is via the business strategy of providing unbundled or differentiated services/products. The rapid advance in technologies such as power electronics, small-scale energy generators and storage, and computer control has made feasible distributed systems and the unbundling of power services.

Advanced technologies including power electronics, together with new coordination mechanisms, should be taken into account to ensure the quality of an unbundling service on reasonable levels. New system configurations and control schemes are desired under the above situations. Many technical problems need to be solved. The new concept of Flexible, Reliable, and Intelligent Electrical Energy Delivery Systems (FRIENDS) is such a methodology for electric power infrastructure development and information management systems to realize such an unbundled power quality service.

19.2 The Concept of FRIENDS

What Is FRIENDS?

Similar to the early days of the electric utility industry, a product, unbundled power quality services, is needed first to create a market. The unbundling of power quality services, defined as permitting several levels of power quality at a differentiated cost, is a business concept that achieves its potential in the presence of technologies that can provide specific customers with more than one level of power quality. Ideally, such technologies would provide a range of user choices, of readily calculable economic value, that could be reoptimized as the needs of the user evolve. This then allows expansion of the FRIENDS infrastructure (from revenues generated by the first products introduced) to reach a larger market for services rendered. This process is not likely to be implemented suddenly. Rather, a systematic approach, [2] as illustrated in Fig. 19.1, is the more likely scenario for this century. The enabling products are those

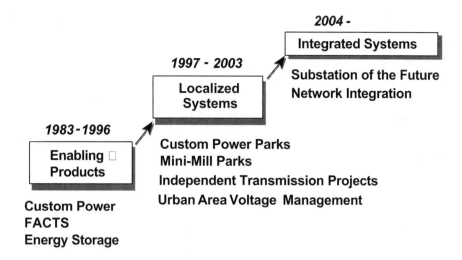

FIGURE 19.1 The evolution to FRIENDS.

that permit a flexible and reliable tailoring of electrical supply that are promoted by the concept of unbundling.

The result of the process will lead to fully integrated and distributed power systems—FRIENDS. Such power systems, with substations using quality control center (QCC) concepts, will give the distributor the ability to provide multimenu (custom) services to customers. The customers can select the supplier according to the power cost by automatically selecting the supplier and accounting for the power cost. For providing multiquality power supply, the QCC in FRIENDS can control and operate customer's switching facilities with a priority order, such as customer's switchboard, panel-board, and/or electric apparatus, via a powerful communication network and intelligent computers. The priority order here means the customer's reliability power supply and that the customer can decide this order. However, in an emergency case, the QCC will control those switches flexibly. Moreover, in the near future, utilization of DC power on the customer side may become general. It is expected that the QCC will have ability of supplying DC power directly to customers. In such a case, reverse power flow from customer to the QCC is also DC power so that the loss of an AC-DC converter can be expected. Multimenu services for each consumer can be realized by using information through data communication lines.

The FRIENDS systems can be operated without interrupting power supply by flexibly changing the system configuration after a disturbance occurs. Each customer, as a consequence, can select the quality level of power independently through QCCs. Multimenu services for each consumer can also be realized by using information through data communication lines. In addition, distributed generators (DGs) and distributed energy storage systems (DESSs) will be allocated on the demand side of FRIENDS. Energy-conserving measures can be expected with the full use of flexibility of the system configuration and advanced demand-side management (DSM). On the other hand, the custom power park concept [2] is geared toward a local system, but may be integrated in the overall QCC concept of FRIENDS. Furthermore, the unbundled services provided, whether they are of commodity or premium grades, may be implemented in a modular manner. Advanced power electronics provides the wherewithal in an unbundled business scenario to differentiate power quality levels provided to customers. The equipment may be individually customized via dynamic voltage restorers (DVR), distribution static compensators (DSTATCOM), solid state breakers (SSB), etc., depending on customer needs and resources available. Such an approach permits business strategies whereby customer retention and acquisition become possible in a competitive environment.

FRIENDS is a forward-looking type of power delivery system that takes into account deregulation, a varied and distributed power supply, and a delivery system that provides flexible, yet reliable, power to a diversified customer base. There are several possible conceptual implementations of FRIENDS. However, fundamentally the network is an integrated system with the ability to provide flexibility in reconfiguration and reliability so that uninterrupted service is obtained. To implement such a system requires advanced power electronic technologies to create the desired conditioned power and command-and-control technologies.

From the viewpoint of consumers, the characteristic of FRIENDS can be summarized as multiple menu service, which may be classified into three types: multiple rate service, multiple quality service, and multiple supplier service. Specifically, FRIENDS is directed to obtain the following technical features:

- Flexibility in reconfiguration of a power system
- Reliability of power supply
- Multimenu services to allow customers to select the type of power provided and other services offered by the electric utility
- Revenue-enhancement techniques
- Load leveling and energy conservation
- Enhancement of information services (utility/customer, and utility/utility)
- Effective demand-side management
- Flexibility in incorporating dispersed generation
- New automatic generation control methodologies
- Voltage regulation functions
- Fault current-limiting functions
- Prevention of sags, swells, and harmonics
- Instrumentation and revenue metering

In brief, the purpose of FRIENDS is to develop a desirable structure for future power delivery systems, where dispersed energy storage systems and dispersed generators are installed near the demand side, and to develop reliable and energy-conserving operation strategies of power systems, taking into account the ways of enhancing service to customers through intelligent functions. It will be necessary for the power industry to create technical product differentiation that provides for the effective unbundling of power quality services, of readily calculable value, as the needs of the customer evolve [2].

Quality Control Centers

FRIENDS can realize the most efficient use of the flow of electric energy generated by remote generators, dispersed generators, and dispersed storage devices in electric power networks. One of the most important features of FRIENDS is that new power improvement facilities referred to as quality control centers (QCCs) are installed very closely to the customers. A desirable operating algorithm of DGs and DESSs in QCCs have been investigated and the capacity required for the multiquality power supply have been evaluated [3–9]. QCCs are developed to cope with providing flexible services to customers in an uncertain electric utility business in near future delivery systems, with which multimenu electrical power quality services are realized in FRIENDS. A QCC is much like a switching facility with the capability to change flexibly the system configuration depending on the system states. The QCCs play a vital role in the operation of FRIENDS, such as:

- Flexible change in system configuration
- Multiple menu services for customers (unbundled services)
- Information management (information processing and data exchange centers)
- Monitoring and controlling power flow by interchange of information between QCCs so that the system can operate in the most effective and economical way

A QCC can be supplied with electrical power from several distribution substations through several power lines to enhance the supply reliability as shown in Fig. 19.2. It may cover a wide area or may be local and located at a section of the current distribution system, an underground floor of a large building, inside the area of a factory, and so forth, depending on its role. In a QCC, it is expected that electric

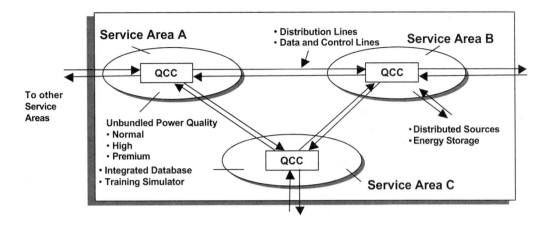

FIGURE 19.2 A FRIENDS network with quality control centers.

power with various levels of quality is produced for multiple-menu services, and a number of static open and closed switches are equipped to establish flexibility in the configuration of the delivery system. Distributed power generation resources and energy storage systems may be incorporated into the QCC for high reliability and energy conservation. Further, the QCC has roles as an information processing center and data communication center for controlling certain apparatus in the QCC and supplying many kinds of information. The operation and control of certain apparatus in the QCC, such as open and closed switches, are implemented from a global point of view by cooperation between computers used at a distribution substation or by individual consumers.

Intelligent Function of FRIENDS

The principal computer functions of FRIENDS include not only the current distribution automation functions, but many other functions to cope with the new system. They are typically the DSM functions, power quality selection/control functions, and information services.

The most important role among the computer functions is the high-speed and flexible switching operation of the distribution network that includes protective relaying functions. This function flexibly controls the static switching facilities (such as thyristors) in the system according to the situation of the system, to minimize loss in the normal state, to provide protective relaying scheme in the emergency state, and to conduct restorative operation in the restoration state. These functions are installed on every computer according to its control level, and achieved on the basis of the distributed computers.

Other important functions are related to the control of DGs and DESSs, or multiquality power control and load leveling. These functions are mainly installed on the computers at the QCC. The DSM functions are also a remarkable feature of FRIENDS. Moreover, through communication lines, control computers are connected to an off-line computer in which the integrated database for control and maintenance, rate calculation, and mapping (see Fig. 19.2). Therefore, only one database must be maintained for the distribution systems operation and management. This database can also be used for the operator-training simulator installed within FRIENDS. Training of mapping and management as well as systems operation may be a key issue for the operators and the manager in the future automated power system.

19.3 Development of FRIENDS

In the near future, power systems may change considerably as a result of the worldwide deregulation occurring in the electric power industry. Further, various customers' requirements with regard to their power supply will increase. Therefore, it is desired to establish new power distribution systems [10]. On the other hand, automatic operation technologies in power distribution systems such as SCADA, AM/FM-GIS

(Automatic Mapping/Facility Management Geographic Information Systems) have been developing rapidly. Meanwhile, power electronic technologies, communication technologies, DSM technologies, etc. have also been developing. Some DGs and energy storage systems have been installed in distribution systems. The complexity of the existing systems as well as the specific quality requirements of each customer are expected to increase, and will consequently require more advanced electric power delivery systems and advanced technologies. Therefore, it is possible to build new distribution systems that have many excellent functions with these new technologies.

The advantages of FRIENDS are increasingly apparent by progress made via international collaboration in recent years. They are as follows: multiquality electric power, multimenu services and rate information to customers, energy conservation effect with the full use of flexibility of the system configuration and advanced DSM, and powerful intelligent communication networks along with QCCs. So far, research in the following aspects of this field has attracted much attention:

- Optimal FRIENDS configuration for reliable power supply
- Concepts and configurations of QCCs for unbundled power quality services
- Optimal allocation of DGs
- Autonomous decentralized power distribution system (ADPDS)
- Utility and customer interface

It is necessary to develop an optimal configuration of FRIENDS from the viewpoint of power supply reliability, to realize the unbundled power quality services through DGs and DESSs. Network connection of FRIENDS must be flexibly changed to minimize distribution loss in a normal state, and to supply power without interruption to customers by utilizing DGs and DESSs in a fault state, through ON/OFF status of solid-state transfer switches (SSTS). To solve the network configuration problem in FRIENDS, from the viewpoint of power supply reliability, an effort has been made to determine power lines and DG installations, to minimize total installation cost under constraints of the maximum capacity of DGs, line current capacity, and uninterrupted power supply. Special algorithms have been employed to solve this problem and numerical results have been obtained [3–9]. It is expected that system configurations can be changed flexibly, to minimize distribution losses in normal states, to restore the system swiftly in an emergency state, and to repair facilities without power interruption in a maintenance state.

A QCC is a new facility that can improve the quality of power according to customers' requests and provide multiquality power to customers. A QCC may be connected to multiple power lines and can be supplied with commodity quality power from several substations. The connection of many QCCs will lead to the development of a new type of high-voltage power distribution network such that QCCs can provide power with various qualities to customers in a multimenu manner. To realize FRIENDS, detailed interior structures of QCCs for accomplishing its functions and algorithms for optimal selection of power quality and power provider are needed. The problems for development of a structure of QCCs are mainly focused on (1) a circuit configuration that makes it possible to deliver electric power with different power quality levels and (2) the control of the static switches that are the most fundamental components of a QCC. In general, QCCs will take various kinds of interior structures according to the features of customers connected to the QCC, conditions in the neighborhood of the QCC, location of the QCC, and installation cost of the QCC, and so on. Progress has been made on development of concepts and physical configurations of the QCC. The configuration of a QCC, which will be located between distribution substations and loads, has been investigated and a prototype model of generating end-use power demand proposed [11, 12]. This model can simulate the activity of a customer and the usage patterns of energy using appliances in the household. Three structures for industrial, business, and commercial, and residential areas have been proposed [13, 14], where the control scheme of solid-state switches used for changing network configuration and cutting off a line when abnormal voltage occurs. Further, the initial structural design of a QCC has been proposed, and the interior structures of the QCC in the scheme of FRIENDS, together with qualitative explanations for various characteristics of the proposed structures, have been provided [15–21]. Two types of QCC models that can execute unbundled power quality services to customers, along with simulation of transient behavior of the proposed model, have been investigated. Uninterruptible power supply (UPS) system, DGs, and solid-state breakers (SSB) can be incorporated into the proposed QCC models. It is expected in a QCC that electric power with various levels of quality is produced by using various types of equipment, e.g., UPS, DVR, SVC, AF (active filter), AVR (automatic voltage regulator), FCL (fault current limiter), etc., with intelligent features. A number of solid-state open/close switches, DGs, and DESSs are also important facilities for realizing many of the functions to be incorporated within a QCC. Small consumers, who are usually considered as captured consumers and cannot select their suppliers, may enjoy the multiple supplier service through the service of QCCs. This service will undoubtedly encourage competition in the electric power market. In addition, a QCC can also be considered a virtual consumer. It may choose the supplier to minimize the electricity cost.

ADPDS has been introduced as a concept of FRIENDS [22]. The major purpose in this area is to develop a method to find the realizable configuration and operation of ADPDS from a viewpoint of power quality and supply reliability under global deregulation and competition, such that the unbundled power quality services can be applied to traditional power distribution systems. In this concept, among many kinds of power qualities, power interruption or power supply reliability is the most important function of unbundled power quality services. A concept for real-time voltage regulation and a modeling method for distribution systems integrated with DGs have been recently developed [23]. It is expected that this modeling method could be widely applied to design and operation of distribution systems with many DGs.

The distribution system envisioned by FRIENDS will need to be able to operate with an array of lines, QCCs, DGs, and storage elements—all digitally controlled so that resources are allocated flexibly, yet reliably. Similarly, the transmission system must also be controlled effectively to handle the transaction increases expected. Both on the distribution and transmission sides this may be accomplished via a multilevel hierarchical control structure [24–26], which incorporates FRIENDS, FACTS, Custom Power, and wide-area communication technologies. Both a power supply and delivery technology (one that is flexible, robust, and efficient) are required to sustain the expected use of energy. FRIENDS is the key for such a successful delivery system. However, FRIENDS technology must be implemented on a value-added basis; otherwise, one risks returns on investment not well optimized with respect to the cost of capital. What is needed is a sensible roadmap of value-added innovation that fits within anticipated options available to customers. Research on other areas such as security constrained power planning and operations in competitive market [27], DSM, and computer technologies in FRIENDS [28] are under way.

19.4 The Advanced Power Electronic Technologies within QCCs

Power quality for unbundled power quality services may be considered on three levels—normal quality, high quality, and premium quality—depending on demand characteristics of customers. In the concept of FRIENDS, QCCs have transfer switches to reduce distribution losses, voltage sags, swells, flickers, interruptions, three-phase voltage unbalance, and voltage harmonics, by using advanced power electronics devices such as DSTATCOM, DVR, UPS, and SSB or SSTS.

Distribution Static Compensator

As a fast, flexible, and effective power quality mitigation tool, DTSATCOM is designed to protect the upstream system from "dirty" loads, surges, or harmonics caused by other users on the distribution system. The use of DSTATCOM devices (whose capabilities may include active filtering, reactive power supply, and flicker correction) is becoming quite attractive for customers and distributors, partly because of its direct connection to the network (just as any other load element) and mainly because the same power electronics configuration can be used, modifying the control structure, to provide the different services. The parallel-connected DSTATCOM with the load provides dynamic, subcycle voltage support and regulation of reactive power flow. It is capable of generating continuously adjustable reactive or capacitive compensation at a level up to the maximum MVA rating of the DSTATCOM inverter. Accordingly, it has better dynamic response than have conventional voltage and VAR control elements.

By varying the amplitude and phase angle of this solid-state synchronous voltage source, the dynamic subcycle response of a DSTATCOM system mitigates the terminal voltage disturbances and improves the system power factor by variable parallel-connected reactive compensation. Yet the DSTATCOM can reduce and totally eliminate voltage flicker resulting from rapid variations in load currents by dynamic nonlinear loads. By eliminating the flicker at the load, the DSTATCOM enables these loads to coexist on the same feeder as more sensitive loads, eliminating the need for separate feeders. The DSTATCOM can increase the capacity of distribution feeders by as much as 50% through power factor improvement and real-time voltage support, thereby alleviating thermal overload conditions and improving power quality for remote energy users.

Dynamic Voltage Restorer

A DVR is a custom power element for series connection into a distribution feeder. As an IGBT DC-to-AC switching inverter that injects three single-phase AC output voltages in series with a distribution line and in synchronism with the voltage of the distribution system, it provides the entire downstream load with protection from voltage sags and transients. By injecting voltages of controllable amplitude, phase angle, and frequency into the distribution feeder via a series injection transformer, the DVR can restore or supplement the voltage quality at its load-side terminals, when the quality of the source-side terminal voltage is significantly out of specification for sensitive load equipment. For example, DVR can restore the voltage quality delivered to an end-use customer when the source-side voltage deviates. During a voltage sag, the subcycle response of the DVR technology supplies what is missing in the voltage waveform, so the customer's sensitive load sees a restored nominal voltage, i.e., no sag. A DVR can supply partial power to the load from a rechargeable energy source attached to the internal DVR DC-link to keep the voltage within the requirements. During the normal line voltage conditions, the energy storage device is recharged from the AC system. Even without stored energy, the DVR can compensate for voltage fluctuations by inserting a voltage that lags or leads the line current by 90 electrical degrees, thus providing continuously variable line compensation.

A DVR (designed by Siemens, for example) is typically composed of several 2 MVA inverter modules connected in parallel at the injection transformer. Its ability to maintain maximum voltage injection depends on the kilojoules available in the energy storage subsystem (capacitors, flywheels, and SMES). The per-unit rating of a DVR can be determined by the per-unit voltage injections multiplied by the rated load current.

Solid-State Breaker

The SSB, based on GTO switch technology, is a fast-acting, subcycle breaker that can instantaneously operate to clear an electrical fault from the power system. It also can be used in combination with other power electronic devices to improve power quality performance to customers, as well as prevent excessive fault currents from developing. An SSB can be applied in a single switch, a transfer switch, a tiebreaker, or a low-level fault interrupter. An SSB may consist of two parallel-connected circuit branches: a solid-state switch composed of GTOs and another using SCRs. The capacity of the semiconductor devices used in the breaker primarily determines the operating characteristics of the SSB. Voltage and current ratings of the breaker define the number of power semiconductors required, and consequently the cost and the operating losses of the breaker. GTO breakers are used to provide rapid, subcycle current interruption and SCR breaker to provide fault current conduction for protection coordination in conventional distribution system applications.

The SSB is able to provide power quality improvements through near-instantaneous current interruption at utility distribution voltages (4 kA at 4.5 kV), an action which provides protection for sensitive loads from disturbances that conventional electromechanical breakers cannot eliminate. The SSB is designed to conduct in-rush and fault currents for several cycles, and to disconnect faulty source-side feeders in less than one half a cycle.

A high-voltage SSTS can provide nearly uninterruptible power to critical distribution-served customers who have two independent power sources. Fast-acting solid-state switches can rapidly transfer sensitive loads from a normal or preferred supply that experiences a disturbance to an alternate or backup supply such as another utility primary distribution feeder or a standby power supply operated from an integral energy storage system.

19.5 Significance of FRIENDS

FRIENDS will give consumers the ability to take advantage of multiple menu services, and utilities the ability to provide flexible, reliable, and intelligent electricity services. Since consumers are given a choice with a multiple menu service, they may have greater influence on the overall system operations. The necessary condition for the existence of FRIENDS is consumer satisfaction and improved operations for the utility. Because of deregulation and competition, it becomes important to provide reliability and tailored services and its realization will make it economically possible to create the system. From this standpoint, the reliability of the high-voltage side, such as the generation and/or transmission systems, of power networks can be less important, and the investment in it may be reduced. Therefore, a part of the investment on the high-voltage side may be diverted to the distribution side without a change in the total investment. However, implementation costs for QCCs will be high because of the SSBs in its configuration. More cost reduction is necessary to make the concept of FRIENDS realizable. FRIENDS will be developed through several items, the important of which are (1) powerful intelligent communication networks; (2) QCCs; and (3) advanced power quality enhancement systems. These items are interrelated and may be enhanced if considered together. The QCC will also have dispersed generation and storage in its configuration to achieve load leveling and to supply uninterrupted power to the loads. The QCC will change the system configuration flexibly depending on the system states, data of which are exchanged through intelligent communication networks.

As mentioned in previous sections, FRIENDS integrates a number of research concepts that have been developed individually, by adding more reliable and flexible functions. Distributed generation resources and energy storage systems, demand-side control, advanced power electronics technologies, and distributed intelligent facilities can be combined into FRIENDS systems. Generally, a FRIENDS system with QCCs will play an important role in the following ways (Fig. 19.3).

Flexibility in reconfiguration of the system. With solid-state switches in a properly used QCC, the distribution system configuration can be frequently changed according to the state and load patterns of the system to reduce distribution losses and load leveling and to avoid power interruption. QCCs may permit a reverse flow of power and have the ability to bypass selected QCCs. Power electronics switches, together with computerized relaying schemes, will be utilized to make switching operations much easier and more flexible.

Reliability in power supply. In principle, power interruption never occurs in FRIENDS. To realize an uninterrupted power supply, DESS and DGs installed at the demand side are fully used under the control of the QCC, providing flexibility in reconfiguration. A computer installed at the QCC has functions related to the control of DESS and DGs, multiquality power control, and load leveling. With consideration of power interchange between QCCs, an effective operational scheme of DESSs and DGs can be attained. It is found that there might be a trade-off between the system configuration and the capacity of DESSs and DGs.

Multimenu service for customers. A QCC produces multiquality power by using DESSs and DGs. The customers can select the supplier according to the power cost by automatically selecting the supplier and accounting for the cost of energy. For providing a multiquality power supply, the QCC can control and operate customer's switching facilities flexibly with a priority order decided by the customer's power supply reliability, such as a customer's switchboard, panel-board, and/or electric apparatus, via a powerful communication network and intelligent computers.

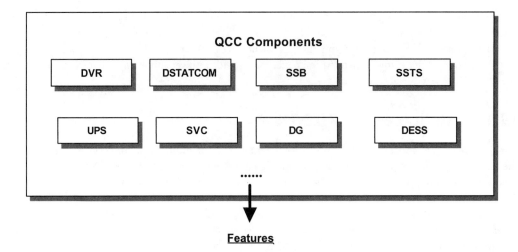

FIGURE 19.3 Features of QCCs.

Load leveling and energy conservation. Power generation in large power units can be leveled by the optimal operation of DESSs and DGs via QCC; then power plants can operate efficiently. As installation of DESSs and DGs increases, load factors at some nodes are enhanced according to the peak-cut and bottom-up approach. Further, if load factors at some nodes become 100%, it is possible to have a margin in the installation of DESSs and DGs. In FRIENDS, this margin will be effectively used to improve the total load factor by interchanging the power between QCCs.

Enhancement of information services to customers. Through the use of a powerful communication network for power system control installed along system components, much information about the power supply and other factors can be sent between a customer and a power company. For example, information of power system restoration can be sent to a customer when power interruption occurs in the system by a fault. In the view of a power supplier, this can provide a competitive edge in an open market by increasing the visibility of its company as a preferred energy provider.

Effective demand-side management. Passive and active DSM can be realized efficiently by controlling the customer's demand through the communication network and dispersed computers. The passive type of DSM uses season-of-year rates, time-of-day rates, and load control contracts. The active type of DSM uses apparatus control, voltage control, online rate system, selection of power suppliers, and so on.

Other functions. Other functions that could be accomplished by the QCC are voltage regulation, fault current limitation, prevention of momentary voltage sags, swells, and harmonics, shedding of the ordinary or low-quality loads, filter, AC-DC converters, etc. With expected progress in the study and development of FRIENDS, new ideas to further enhance the function of the QCC are expected.

In fact, FRIENDS, as a novel concept to cope with deregulation and unbundled power quality services, will impact significantly not only on the infrastructure of power delivery systems in the near future, but also on related fields such as political issues of energy policies, customer satisfaction, etc. This means that the concepts of how customers can benefit from a multiple rate service, multiple quality service, and multiple supplier service of the new systems should be well represented through multiple media.

19.6 Realization of FRIENDS

Significant progress has been achieved in several areas that form the background technologies of FRIENDS, such as information processing technologies, power electronics facilities, distributed generators and distributed energy storage systems, new communication schemes, demand-side management and deregulation, etc. Especially, ongoing development and commercial availability of high-power semiconductors will further increase opportunity and cost-effectiveness of advanced technologies as mentioned above. New system control technologies are needed to integrate safely small modular generation and storage technologies such as fuel cells, microturbines, photovoltaic, wind, batteries, flywheels, and so forth into the distributed system. The custom power/premium power pak concept is one that can integrate these state-of-the-art technologies to improve power quality, which can then be incorporated into the QCCs of FRIENDS. Related technology development in power quality monitoring, metering, and communications will also need to keep pace to support the large number of "differentiated" users that unbundling creates. Directions of development are anticipated as follows:

- Development of technologies, such as FRIENDS, premium power pak, etc., to catch up with new challenges in future electric utilities;
- Development of decision-making models to include the complexity of the present or future price structure, taking into account the technical, financial, and social issues;
- Development of new control technologies for distribution systems in a competitive power market;
- Development of technologies for revenue enhancement, instrumentation, and real-time pricing;
- Development of new rules for unbundled power quality services in energy supply systems.

To achieve the goal of realizing FRIENDS, the following technologies are being developed:

- Optimal FRIENDS configuration for reliable power supply from the viewpoints of installation and operational cost, reliability, and system operation.
- Behavior investigation and coordination of the main components in FRIENDS—Emphases will be on control coordination of fast switching actions and parameter optimization of components, determination method of optimal switching operation in normal, emergency, and restorative states, together with protection schemes to make the system reconfiguration more flexible and the power supply more reliable.
- Development of distributed generation and distributed energy storage systems—Algorithm for optimal planning and operation of DGs and DESSs, behaviors of DGs and DESSs in normal and fault states, optimal deployment, allocation, and capacity of DGs and DESS into distribution systems will be secured.
- Structure design of QCCs for unbundled power services—This task may include development of the concepts and configuration of interior structures of the QCCs, optimal selection of the power quality and power suppliers, indices for evaluating power quality quantitatively and to analyze the performance of QCC using the indices, and a hybrid system of quality control centers and intelligent communications networks.
- Efficient demand-side prospects and energy demand monitoring of FRIENDS—The efforts in this area involve efficient demand-side management systems and their evaluation methods, options of multiple-menu services for consumers using FRIENDS, and estimation of unbundled energy demands.
- FRIENDS and power system security—Control performance, involving development of new automatic generation control methodologies that take into account new industry-defined control performance standards, needs to be investigated. Further, methodologies for security-constrained generation dispatch, transfer limits evaluation and enhancement, and open transmission access in a competitive power system market should be exploited.

- Revenue enhancement and online pricing or metering scheme consisting of economic assessment of FRIENDS systems, real-time pricing methodologies, and evaluation of power quality and detection of distortion.
- Optimal frame of customer information systems for development of quality-related electricity products.
- Instrumentation to evaluate appropriate revenue metering and power quality monitoring.
- Development of other related technologies for unbundled power quality services.

These may consist of new data requirements and an integrated database for implementation of FRIENDS (static, historical, and real time), new high-level communication procedures, and new types of training simulators.

19.7 Conclusions

- Deregulation of the power industry will cause new challenges in dealing with power quality issues.
- The deregulation issues and the power quality issues cannot be dealt with in isolation.
- Deregulation is creating the opportunity for electric utilities to offer a variety of valued-added services to customers.
- Customer choice and the types of utility services provided will be the key elements in determining the success of an electric utility.
- Commercial availability of high-power semiconductors increases the opportunity and cost-effectiveness of power quality control and improvement technologies.
- The state-of-the-art, commercially available equipment allows for the development of the FRIENDS concept, and other advanced technologies provide the means for electric utilities to add value incrementally via additions to their infrastructure.
- FRIENDS provides for useful guidance and strategic implementation schemes to improve service quality, conservation, power supply planning, and possible needs in transmission and distribution line siting.
- The quality control center (QCC) is one of the facilities of FRIENDS, and may be realized by many kinds of structures according to the roles that are performed by particular types of QCCs. QCCs should not only improve electrical waveform quality, but also provide unbundled quality power to customers.

References

1. Kazibwe, W. E. and Sendaula, M. H., *Electric Power Quality Control Techniques*, Van Nostrand Reinhold, New York, 1993, chap. 1.
2. Kessinger, J. P., Advanced power electronics as an enabler of unbundled power quality services, in *Proceedings of the NSF Conference on Unbundled Power Quality Services in the Power Industry*, Key West, FL, Nov. 1996, 114.
3. Nara, K. and Hasegawa, J., An advanced flexible and reliable power delivery system, in *Proceedings of the NSF Conference on Unbundled Power Quality Services in the Power Industry*, Key West, FL, Nov. 1996, 129.
4. Nara, K. and Hasegawa, J., Future flexible power delivery system and its intelligent functions, in *Proc. of ISAP '96*, 1996, 261.
5. Nara, K. et al., Optimal allocation of dispersed energy storage and generators for power delivery system reliability enhancement, in *Proc. of the IASTED International Conference High Technology in the Power Industry*, Orlando, FL, Oct. 1997, 47.

6. Nara, K. et al., A study for composing flexible and reliable electrical energy delivery system, in *Proc. of International Conference on Electrical Engineering (ICEE'97)*, 1997, 506.
7. Nara, K. et al., Implementation of generic algorithm for distribution system loss minimum reconfiguration, *IEEE Trans. PWRS*, 7(3), 1044, 1992.
8. Nara, K., Kin, H., Suzuki, K., and Mito, A., Study for a reliable power system structure in which dispersed power generators are installed, in *IEE of Japan Power and Energy Conference*, 1996, 155, 197.
9. Nara, K., Hayashi, Y., and Mishima, Y., Optimal network planning and operation of FRIENDS, in *Proc. of the Fourth International Meeting on Systems Technologies for Unbundled Power Quality Services*, Singapore, Jan. 28, 2000, 12.
10. Domijan, A. and Heydt, G., Eds., in *Proceedings of the NSF Conference on Unbundled Power Quality Services in the Power Industry*, Key West, FL, Nov. 1996.
11. Tsuji, K., Co-generation in urban area, *Int. J. Global Energ. Issues*, 7(3/4), 137, 1995.
12. Tsuji, K., Saeki, O., and Suetsugu, J., Estimation of daily load curves by quality level based on a bottom-up simulation model, in *Proc. of the IASTED International Conference High Technology in the Power Industry*, Orlando, FL, Oct. 1997, 181.
13. Takami, M., Ise, T., and Tsuji, K., Concepts and configuration of quality control center in a new power distribution network, presented at Japan-U.S. Seminar on Intelligent Distributed Autonomous Power Systems, Hakone, 1998.
14. Tsuji, K., Current research and future research plans at Osaka University, in *Proc. of the First International Meeting on Systems Technologies for Unbundled Power Quality Services*, Sept. 7–9, Mito, Japan, 1998, 25.
15. Nishiya, K. et al., Optimal location of dispersed energy storage systems in distribution systems from a view of economical standpoint, *Trans. IEE Jpn.*, I I 4-B(12) 1257, 1994.
16. Rho, D., Kita, H., Nishiya, K., and Hasegawa, J., Evaluation of dispersed energy storage system for its installation (from a viewpoint of the secondary battery), *Trans. ME Jpn.*, 116-B(2), 187, 1996.
17. Jung, K., Kim, H., and Rho, D., Determination of the installation site and optimal capacity of the battery energy storage system for load leveling, *IEEE Trans. Energy Conversion*, I I J(1), 162, 1996.
18. Mishima, Y., Kita, H., and Hasegawa, J., Basic study on operation of quality control center for multi-quality power supply by FRIENDS, in *Proc. of the IASTED International Conference High Technology in the Power Industry*, Orlando, FL, Oct. 1997, 175.
19. Rho, D., Kim, J., Kim, E., and Hasegawa, J., Basic studies on impacts of customer voltages by the operation of FRIENDS, in *Proc. of the IASTED International Conference High Technology in the Power Industry*, Orlando, FL, Oct. 1997, 181.
20. Rho, D., Hasegawa, J. et al., Voltage regulation methods based on an extended approach and neural networks for distribution systems interconnected with dispersed storage and generation systems, *Trans. IEE Jpn.*, 17-B(3), 298, 1997.
21. Hasegawa, J., Kita, H., Tanaka, E., Mishima, Y., Wang, G., and Hara, R., Design of quality control center for unbundled power quality services, in *Proc. of the Second International Meeting on Systems Technologies for Unbundled Power Quality Services*, Gainesville, FL, Feb. 5–7, 1999, 27.
22. Kim, J. E., Park, J. K., and Kim, J. C., Research schedule and results so far in Korea: research on configuration and operation of autonomous decentralized power distribution system, in *Proc. First International Meeting on System Technologies for Unbundled Power Quality Services*, Mito, Japan, September 1998, 41.
23. Kim, J. E., Operation technologies of power distribution system with dispersed generation systems for unbundled power quality services, in *Proc. of the Second International Meeting on Systems Technologies for Unbundled Power Quality Services*, Gainesville, FL, Feb. 5–7, 1999, 64.
24. Domijan, A. and Song, Z., Incorporation of hierarchical control with FACTS technologies in large power systems, *Int. J. Power Energ. Syst.*, 20(1), 20, 2000.
25. Domijan, A. and Song, Z., Simulation on multi-machine power system with FACTS devices by hierarchical control, *Int. J. Power Energy Syst.*, 20(2), 67, 2000.

26. Domijan, A., Song, Z., Kamoto, K., and Qiu, Q., Simulation investigation on power systems with versatile models by hierarchical control and custom power elements, in *Proc. of the Third International Meeting on Systems Technologies for Unbundled Power Quality Services*, Espoo, Finland, July 4–5, 1999, 13.
27. Chang, C. S. et al., Dynamic security constrained multi-objective generation dispatch of longitudinal interconnected power systems using bicriterion global optimization, *IEEE Trans. Power Syst.*, 11(2), 1009, 1996.
28. Glamocanin, V. and Andonov, D., Data requirements for unbundled power quality services, in *Proc. of First International Meeting on System Technologies for Unbundled Power Quality Services*, Mito, Japan, September 1998, 47.

20
Unified Power Flow Controllers

Ali Feliachi
West Virginia University

Azra Hasanovic
West Virginia University

Karl Schoder
West Virginia University

20.1 Introduction 20-1
20.2 Power Flow on a Transmission Line............................ 20-2
20.3 UPFC Description and Operation............................... 20-4
   Series Converter: Four Modes of Operation • Automatic Power Control
20.4 UPFC Modeling 20-9
   UPFC Steady-State or Load Flow Model • UPFC Dynamic Model • Interfacing the UPFC with the Power Network
20.5 Control Design.. 20-14
   UPFC Basic Control Design • Power System Damping Control through UPFC Using Fuzzy Control
20.6 Case Study 20-20
   Test System • Tracking Real and Reactive Power Flows • Operation under Fault Conditions
20.7 Conclusion.. 20-24

20.1 Introduction

An electric power system is an interconnection of generating units to load centers through high-voltage electric transmission lines. It consists of generation, transmission, and distribution subsystems, which used to belong to the same electric utility in a given geographical area. But, currently, the electric power industry is in transition from large, vertically integrated utilities providing power at regulated rates to an industry that will incorporate competitive companies selling unbundled power at possibly lower rates. With this new structure, which will include separate generation, distribution, and transmission companies with an open-access policy, comes the need for tighter control strategies. The strategies must maintain the level of reliability that consumers not only have taken for granted but expect even in the event of considerable structural changes, such as a loss of a large generating unit or a transmission line, and loading conditions, such as the continuous variations in power consumption. The deregulation of electricity that is taking place now will affect all business aspects of the power industry as known today from generation, to transmission, distribution, and consumption. Transmission circuits, in particular, will be stretched to their thermal limits because existing transmission lines are loaded close to their stability limits and building of new transmission circuits is difficult, if not impossible, at least from environmental and/or political aspects. New equipment and control devices will be sought to control power flow on transmission lines and to enhance stability and reliability of the system. Flexible AC transmission systems (FACTS) and FACTS controllers, which are power electronics devices used to control the power flow and enhance stability, have become common words in the power industry, and they have started replacing many mechanical control devices. They are certainly playing an increasingly major role in the operation and control of today's power systems. This chapter describes specifically the Unified Power Flow Controller known as the UPFC. This power electronics device consists of two back-to-back converters operated from a common DC-link supplied by a capacitor. It is used to control the power flow between two nodes and also to enhance the stability of the system.

The chapter is organized as follows. First, a brief overview of the power flow on a transmission line is given. Second, the UPFC is described and its steady-state and basic operations are explained. Third, steady-state and dynamic models of the UPFC are presented. Also, a procedure to interface the UPFC with an electric power system is developed. Finally, supplementary signals through the UPFC, designed using fuzzy logic control tools, are shown to enhance the stability of the system by damping low-frequency oscillations. A two-area four-generator electric power system is used as a test system.

20.2 Power Flow on a Transmission Line

The power flow on a transmission line connecting two buses S and R (line sending and receiving buses) is briefly reviewed. The transmission line, as shown in Fig. 20.1, is modeled by a lumped series impedance, $Z = R + jX$, where R and X are the resistance and reactance of the line, respectively.

The complex power, S_S, leaving the sending bus and flowing toward the receiving bus is given by

$$S_S = \bar{V}_S \cdot \bar{I}^*_{Line} \tag{20.1}$$

where

$\bar{V}_S = V_S \angle \delta_S$ is the rms phasor voltage of the sending bus
$\bar{I}^*_{Line}$ = the complex conjugate of the phasor current on the line

The real and reactive powers are obtained from the complex power:

$$S_S = P_S + jQ_S \tag{20.2}$$

The line current, using Ohm's law, is

$$\bar{I}_{Line} = \frac{\bar{V}_S - \bar{V}_R}{Z} = (\bar{V}_S - \bar{V}_R)Y = (\bar{V}_S - \bar{V}_R)(G + jB) \tag{20.3}$$

where

$$Y = G + jB = \frac{1}{Z} = \frac{1}{R + jX} = \frac{R}{R^2 + X^2} - j\frac{X}{R^2 + X^2}$$

Therefore, the conductance and susceptance of the line are

$$G = \frac{R}{R^2 + X^2} = \frac{R}{X^2\left(1 + \left(\frac{R}{X}\right)^2\right)}$$

$$B = -\frac{X}{R^2 + X^2} = -\frac{\frac{1}{X}}{1 + \left(\frac{R}{X}\right)^2}$$

Hence, using Eqs. (20.1) and (20.3), the complex conjugate of the complex power is

$$S^*_S = P_S - jQ_S = \bar{V}^*_S \cdot \bar{I}_S = \bar{V}^*_S \cdot (\bar{V}_S - \bar{V}_R)(G + jB) = (V^2_S - \bar{V}^*_S \bar{V}_R)(G + jB) \tag{20.4}$$

Unified Power Flow Controllers

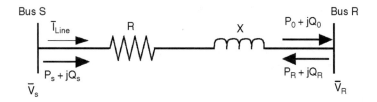

FIGURE 20.1 Transmission line.

Euler's identity, which states that $V \angle -\delta = V(\cos\delta - j\sin\delta)$, is used to write:

$$\bar{V}_S^* \bar{V}_R = (V_S \angle -\delta_S)(V_R \angle \delta_R) = V_S V_R \angle(-(\delta_S - \delta_R)) = V_S V_R (\cos(\delta_S - \delta_R) - j\sin(\delta_S - \delta_R)) \tag{20.5}$$

Substituting Eq. (20.5) into Eq. (20.4), the real and reactive powers are obtained:

$$P_S = V_S^2 G - V_S V_R G \cos(\delta_S - \delta_R) - V_S V_R B \sin(\delta_S - \delta_R) \tag{20.6}$$

$$Q_S = -V_S^2 B - V_S V_R G \sin(\delta_S - \delta_R) + V_S V_R B \cos(\delta_S - \delta_R) \tag{20.7}$$

Similarly, the real and reactive powers received at the receiving bus are

$$P_0 = -P_R = -V_R^2 G + V_S V_R G \cos(\delta_S - \delta_R) - V_S V_R B \sin(\delta_S - \delta_R) \tag{20.8}$$

$$Q_0 = -Q_R = V_R^2 B - V_S V_R G \sin(\delta_S - \delta_R) - V_S V_R B \cos(\delta_S - \delta_R) \tag{20.9}$$

In the above equations P_R and Q_R represent the powers leaving the receiving bus and flowing toward the sending bus. The power lost on the line is obtained by subtracting the power received from the power sent. Therefore, the real and reactive line losses are

$$P_L = P_S - (-P_R) = (V_S^2 + V_R^2)G - 2V_S V_R G \cos(\delta_S - \delta_R) \tag{20.10}$$

$$Q_L = Q_S - (-Q_R) = -(V_S^2 + V_R^2)B + 2V_S V_R B \cos(\delta_S - \delta_R) \tag{20.11}$$

For typical transmission lines the reactance X is a lot larger than the resistance R, i.e., the ratio R/X is very small and usually the conductance G is neglected and the susceptance B is approximated with $B = -1/X$. Using these approximations, Eqs. (20.6) and (20.8) yield the power transmitted over the line from the sending bus to the receiving bus:

$$P_S = -P_R = P_{SR} = -V_S V_R B \sin(\delta_S - \delta_R) = \frac{V_S V_R}{X} \sin(\delta_S - \delta_R) = \frac{V_S V_R}{X} \sin(\delta) = P_0(\delta) \tag{20.12}$$

where the angle $\delta = \delta_S - \delta_R$ is called the power angle.

The reactive power sent to the line from both buses is

$$Q_S = -V_S^2 B + V_S V_R B \cos(\delta_S - \delta_R) = \frac{V_S^2 - V_S V_R \cos(\delta)}{X} \quad (20.13)$$

$$Q_R = -V_R^2 B + V_S V_R B \cos(\delta_S - \delta_R) = \frac{V_R^2 - V_S V_R \cos(\delta)}{X} = -Q_0(\delta) \quad (20.14)$$

The average reactive power flow is defined as

$$Q_{SR} = \frac{Q_S - Q_R}{2} = -\frac{(V_S^2 - V_R^2)B}{2} = \frac{V_S^2 - V_R^2}{2X} \quad (20.15)$$

Equations (20.12) and (20.15) are the basis for understanding the control of power flow on a transmission line. From Eq. (20.12), it is seen that to increase the amount of real power transmitted over the line one can:

- Increase the magnitude of the voltages at either end, i.e., voltage support
- Reduce the reactance of the line, i.e., line compensation
- Increase the power angle, i.e., phase shift

One can also reverse the power flow by changing the sign of the power angle; i.e., a positive power angle will correspond to a power flow from sending to receiving, whereas a negative power angle $\delta_R > \delta_S$ will correspond to a power flow from receiving to sending.

Similarly, from Eq. (20.15), it is seen that both voltage magnitudes and line reactance will affect the reactive power. If both voltage magnitudes are the same, i.e., flat voltage profile, each bus will send half of the reactive power absorbed by the line. The power flow is from sending to receiving when $V_R < V_S$.

Hence, the four parameters that affect real and reactive power flows are V_S, V_R, X, and δ. To further understand this relationship, Eqs. (20.12) and (20.14) can be combined:

$$(P_0(\delta))^2 + \left(Q_0(\delta) + \frac{V_R^2}{X}\right)^2 = \left(\frac{V_S V_R}{X}\right)^2 \quad (20.16)$$

This equation represents a circle centered at $(0, -V_R^2/X)$, with a radius $V_S V_R/X$. It relates real and reactive powers received at bus R to the four parameters: V_S, V_R, δ, X. To see, for example, how the power angle δ affects P_0 and Q_0, assume that $V_S = V_R = V$ and $V^2/X = 1$. The P-Q locus for this case is shown in Fig. 20.2 (solid line). For a specific power angle δ, values of P_0 and Q_0 can be found, e.g., if $\delta = \pi/4$ (point A on the circle) then $P_{0A} = 0.707$ and $Q_{0A} = -0.293$. Reducing the line reactance X, say to $X' < X$, while keeping $V_S = V_R = V$, will increase the radius of the circle (dashed line). Note that the power angle δ might be constrained by stability limits.

Similarly, the relationship between the real and reactive powers sent to the line from the sending bus S can be expressed as

$$(P_S(\delta))^2 + \left(Q_S(\delta) - \frac{V_S^2}{X}\right)^2 = \left(\frac{V_S V_R}{X}\right)^2. \quad (20.17)$$

20.3 UPFC Description and Operation

The UPFC is one of the most complex FACTS devices in a power system today. It is primarily used for independent control of real and reactive power in transmission lines for a flexible, reliable, and economic operation and loading of power systems. Until recently all four parameters that affect real and reactive power flow on the line, i.e., line impedance, voltage magnitudes at the terminals of the line, and power angle, were controlled separately using either mechanical or other FACTS devices such as a static var

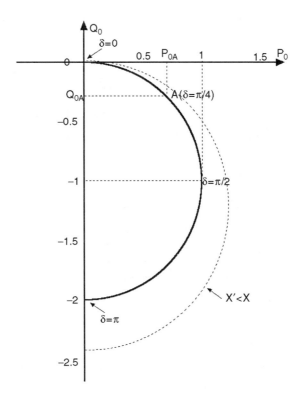

FIGURE 20.2 *P-Q* locus of the uncompensated system.

compensator (SVC), thyristor-controlled series compensation (TCSC), a phase shifter, etc. However, the UPFC allows simultaneous or independent control of all four parameters, with possible switching from one control scheme to another in real time. Also, the UPFC can be used for voltage support and transient stability improvement by damping of low-frequency power system oscillations.

The UPFC is a device placed between two buses referred to as the UPFC sending bus and the UPFC receiving bus. It consists of two voltage-sourced converters (VSCs) with a common DC-link. For the fundamental frequency model, the VSCs are replaced by two controlled voltage sources as shown in Fig. 20.3.

By applying the pulse width modulation (PWM) technique to the two VSCs, the following equations for magnitudes of shunt and series injected voltages can be obtained [2]:

$$V_{SH} = m_{SH} \frac{V_{DC}}{2\sqrt{2} n_{SH} V_B}$$
$$V_{SE} = m_{SE} \frac{V_{DC}}{2\sqrt{2} n_{SE} V_B}$$
(20.18)

where

m_{SH} = amplitude modulation index of the shunt VSC control signal
m_{SE} = amplitude modulation index of the series VSC control signal
n_{SH} = shunt transformer turn ratio
n_{SE} = series transformer turn ratio
V_B = the system side base voltage in kV
V_{DC} = DC-link voltage in kV

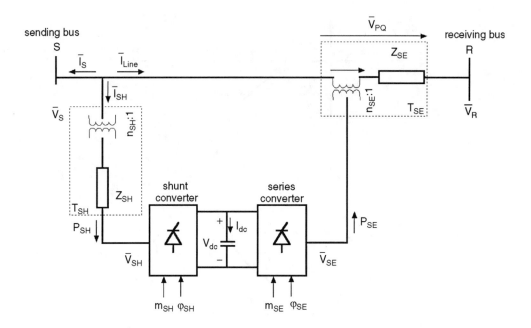

FIGURE 20.3 Fundamental frequency UPFC model.

The phase angles of $\bar{V}_{SH}$ and $\bar{V}_{SE}$ are

$$\delta_{SH} = \delta_S - \varphi_{SH}$$
$$\delta_{SE} = \delta_S - \varphi_{SE} \tag{20.19}$$

where

φ_{SH} = firing angle of the shunt VSC with respect to the phase angle of the sending bus voltage
φ_{SE} = firing angle of the series VSC with respect to the phase angle of the sending bus voltage

The voltage source at the sending bus is connected in shunt and will therefore be called the *shunt voltage source*. The second source, the *series voltage source*, is placed between the sending and the receiving buses. Both voltage sources are modeled to inject voltages of fundamental power system frequency only. The UPFC is placed on high-voltage transmission lines. This arrangement requires step-down transformers to allow the use of power electronic devices for the UPFC. The transformer impedances have been included in the model.

The series converter injects an AC voltage $\bar{V}_{SE} = V_{SE} \angle (\delta_S - \varphi_{SE})$ in series with the transmission line. The series voltage magnitude V_{SE} and its phase angle φ_{SE} with respect to the sending bus are controllable in the range of $0 \leq V_{SE} \leq V_{SE\,max}$ and $0 \leq \varphi_{SE} \leq 360°$. The shunt converter injects controllable shunt voltage such that the real component of the current in the shunt branch balances the real power demanded by the series converter. The reactive power cannot flow through the DC-link. It is absorbed or generated (exchanged) locally by each converter. The shunt converter operated to exchange the reactive power with the AC system provides the possibility of independent shunt compensation for the line. If the shunt-injected voltage is regulated to produce a shunt reactive current component that will keep the sending bus voltage at its prespecified value, then the shunt converter is operated in the *automatic voltage control mode*. The shunt converter can also be operated in the *VAr control mode*. In this case, shunt reactive current is produced to meet the desired inductive or capacitive VAr request.

Series Converter: Four Modes of Operation

As mentioned previously, the UPFC can control, independently or simultaneously, all parameters that affect power flow on a transmission line. This is illustrated in the phasor diagrams shown in Fig. 20.4 [3].

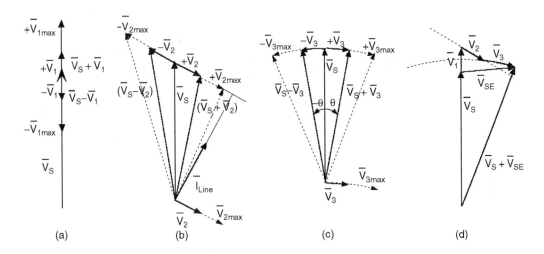

FIGURE 20.4 Phasor diagrams.

Voltage regulation is shown in Fig. 20.4a. The magnitude of the sending bus voltage $\bar{V}_S$ is increased (or decreased) by injecting a voltage $\bar{V}_1$, of maximum magnitude $V_{1\max}$, in phase (or out of phase) with $\bar{V}_S$. Similar regulation can be accomplished with a transformer tap changer.

Series reactive compensation is shown in Fig. 20.4b. It is obtained by injecting a voltage $\bar{V}_2$, of maximum magnitude $V_{2\max}$, orthogonal to the line current $\bar{I}_{\text{Line}}$. The effective voltage drop across the line impedance X is decreased (or increased) if the voltage $\bar{V}_2$ lags the current $\bar{I}_{\text{Line}}$ by 90° (or $\bar{V}_2$ leads current $\bar{I}_{\text{Line}}$ by 90°).

A desired phase shift is achieved by injecting a voltage $\bar{V}_3$, of maximum magnitude $V_{3\max}$, that shifts the phase angle of $\bar{V}_S$ by $\pm\theta$ while keeping its magnitude constant as shown in Fig. 20.4c.

Simultaneous control of terminal voltage, line impedance, and phase angle allows the UPFC to perform multifunctional power flow control. The magnitude and the phase angle of the series injected voltage $\bar{V}_{SE} = \bar{V}_1 + \bar{V}_2 + \bar{V}_3$, shown in Fig. 20.4d, are selected in a way to produce a line current that will result in the desired real and reactive power flow on the transmission line.

Therefore, the UPFC series converter can be operated in any of the following four modes:

1. Voltage regulation
2. Line compensation
3. Phase angle regulation
4. Power flow control

Automatic Power Control

The automatic power control mode cannot be accomplished with conventional compensators. To show how line power flow can be affected by the UPFC operated in the automatic power flow control mode, a UPFC is placed at the beginning of the transmission line connecting buses S and R as shown in Fig. 20.5 [3]. Line conductance is neglected. UPFC is represented by two ideal voltage sources of controllable magnitude and phase angle. Bus S and fictitious bus S_1 shown in Fig. 20.5 represent the UPFC sending and receiving buses, respectively.

In this case, the complex power received at the receiving end of the line is given by

$$S = \bar{V}_R \bar{I}^*_{\text{Line}} = \bar{V}_R \left(\frac{\bar{V}_S + \bar{V}_{SE} - \bar{V}_R}{jX} \right)^* \qquad (20.20)$$

where $\bar{V}_{SE} = V_{SE} \angle (\delta_S - \varphi_{SE})$.

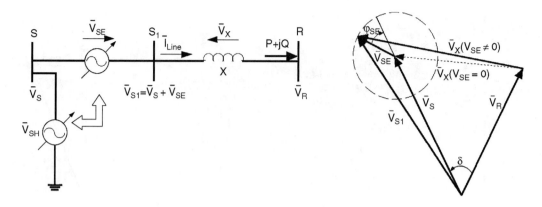

FIGURE 20.5 Transmission line with UPFC.

The complex conjugate of this complex power is

$$S^* = P - jQ = \bar{V}_R^*\left(\frac{\bar{V}_S + \bar{V}_{SE} - \bar{V}_R}{jX}\right) \quad (20.21)$$

By performing simple mathematical manipulations and separating real and imaginary parts of Eq. (20.21), the following expressions for real and reactive powers received at the receiving end of the line are

$$P = \frac{V_S V_R}{X} \sin\delta + \frac{V_R V_{SE}}{X} \sin(\delta - \varphi_{SE}) = P_0(\delta) + P_{SE}(\delta, \varphi_{SE})$$

$$Q = -\frac{V_R^2}{X} + \frac{V_S V_R}{X} \cos\delta + \frac{V_R V_{SE}}{X} \cos(\delta - \varphi_{SE}) = Q_0(\delta) + Q_{SE}(\delta, \varphi_{SE}) \quad (20.22)$$

For $V_{SE} = 0$ the above equations are identical to Eqs. (20.12) and (20.14) that represent the real and reactive powers of the uncompensated system.

It was stated previously that the UPFC series voltage magnitude can be controlled between 0 and $V_{SE\,max}$ and its phase angle can be controlled between 0 and 360° at any power angle δ. It can be seen from Eq. (20.21) that the real and reactive power received at bus R for the system, when a UPFC is installed, can be controlled between

$$P_{min}(\delta) \leq P \leq P_{max}(\delta)$$
$$Q_{min}(\delta) \leq Q \leq Q_{max}(\delta) \quad (20.23)$$

where

$$P_{min}(\delta) = P_0(\delta) - \frac{V_R V_{SE\,max}}{X}$$

$$P_{max}(\delta) = P_0(\delta) + \frac{V_R V_{SE\,max}}{X}$$

$$Q_{min}(\delta) = Q_0(\delta) - \frac{V_R V_{SE\,max}}{X}$$

$$Q_{max}(\delta) = Q_0(\delta) + \frac{V_R V_{SE\,max}}{X}$$

Unified Power Flow Controllers

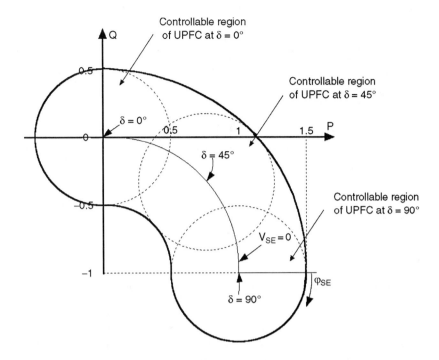

FIGURE 20.6 P-Q relationship for a simple two-bus system with a UPFC at $\delta = 0°$, 45°, and 90°.

Rotation of the series injected voltage phasor with rms value of $V_{SE\,max}$ from 0 to 360° allows the real and the reactive power flow to be controlled within the boundary circle with a radius of $V_R V_{SE\,max}/X$ and the center at $(P_0(\delta), Q_0(\delta))$. This circle is defined by the following equation:

$$(P(\delta, \varphi_{SE}) - P_0(\delta))^2 + (Q(\delta, \varphi_{SE}) - Q_0(\delta))^2 = \left(\frac{V_R V_{SE\,max}}{X}\right)^2 \qquad (20.24)$$

Figure 20.6 shows plots of the reactive power Q demanded at the receiving bus vs. the transmitted real power P as a function of the series voltage magnitude V_{SE} and phase angle φ_{SE} at three different power angles δ, i.e., $\delta = 0°$, 45°, and 90°, with $V_S = V_R = V$, $V^2/X = 1$ and $V_R V_{SE\,max}/X = 0.5$ [3]. The capability of the UPFC to control real and reactive power flow independently at any transmission angle is clearly illustrated in Fig. 20.6.

20.4 UPFC Modeling

To simulate a power system that contains a UPFC, the UPFC needs to be modeled for steady-state and dynamic operations. Also, the UPFC model needs to be interfaced with the power system model. Hence, in this section modeling and interfacing of the UPFC with the power network are described.

UPFC Steady-State or Load Flow Model

For steady-state operation the DC-link voltage remains constant at its prespecified value. In the case of a lossless DC-link the real power supplied to the shunt converter $P_{SH} = \text{Re}(\overline{V}_{SH}\overline{I}^*_{SH})$ satisfies the real power demanded by the series converter $P_{SH} = \text{Re}(\overline{V}_{SE}\overline{I}^*_{Line})$

$$P_{SH} = P_{SE} \qquad (20.25)$$

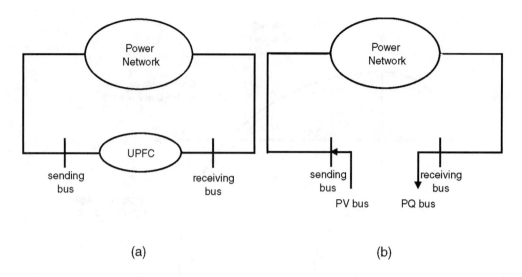

FIGURE 20.7 Power network with a UPFC included: (a) schematic; (b) LF model.

The LF model discussed here assumes that the UPFC is operated to keep (1) real and reactive power flows at the receiving bus and (2) sending bus voltage magnitude at their prespecified values [4]. In this case, the UPFC can be replaced by an "equivalent generator" at the sending bus (PV-type bus using LF terminology) and a "load" at the receiving bus (PQ-type bus) as shown in Fig. 20.7.

To obtain the LF solution of a power network that contains a UPFC, an iterative procedure is needed. Power demanded at the receiving bus is set to the desired real and reactive powers at that bus. The real power injected into a PV bus for a conventional LF algorithm is kept constant and the reactive power is adjusted to achieve the prespecified voltage magnitude. With a UPFC, the real power injected into the sending bus is not known exactly. This real power injection is initialized to the value that equals the prespecified real power flow at the receiving bus. During the iterative procedure, the real powers adjusted to cover the losses of the shunt and series impedances and to force the sum of converter interaction to become zero. The algorithm, in its graphical form, is given in Fig. 20.8.

The necessary computations are described next. The complex power injected into sending bus is

$$\bar{S}_S = \bar{V}_S \bar{I}_S^* \tag{20.26}$$

Using the voltages and currents described in Fig. 20.3

$$\bar{V}_S = \bar{V}_{SH} + \bar{V}_{Z_{SH}}$$
$$\bar{V}_{ZSH} = \bar{I}_{SH} Z_{SH} \tag{20.27}$$
$$\bar{I}_S = -\bar{I}_{SH} - \bar{I}_{Line}$$

results in

$$\begin{aligned}
\bar{S}_S &= (\bar{V}_{SH} + \bar{V}_{Z_{SH}})(-\bar{I}_{SH} - \bar{I}_{Line})^* \\
&= -\bar{V}_{SH}\bar{I}_{SH}^* - \bar{V}_{ZSH}\bar{I}_{SH}^* - \bar{V}_{SH}\bar{I}_{Line}^* - \bar{V}_{Z_{SH}}\bar{I}_{Line}^* \\
&= -\bar{V}_{SH}\bar{I}_{SH}^* - Z_{SH}\bar{I}_{SH}^2 - \bar{V}_{SH}\bar{I}_{Line}^* - Z_{SH}\bar{I}_{SH}\bar{I}_{Line}^*
\end{aligned} \tag{20.28}$$

Unified Power Flow Controllers

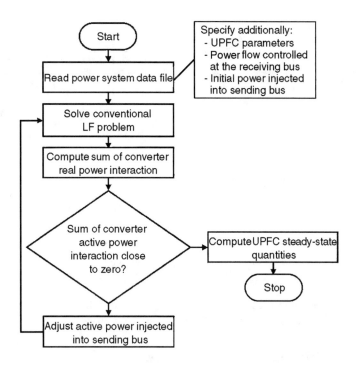

FIGURE 20.8 LF algorithm.

Computing the line current by using the bus voltages and the power flow at the receiving bus as given by the LF solution

$$\bar{I}_{Line} = -\frac{S_R^*}{\bar{V}_R^*} \tag{20.29}$$

allows one to compute the series injected voltage and the series converter interaction with the power system:

$$\bar{V}_{SE} = \bar{I}_{Line} Z_{SE} + \bar{V}_R - \bar{V}_S$$
$$P_{SE} = \text{Re}(\bar{V}_{SE}\bar{I}_{Line}^*) \tag{20.30}$$

Taking the real part of Eq. (20.28) and using Eq. (20.25), the new injected real power at the sending bus becomes

$$P_S = -P_{SE} + \text{Re}(-Z_{SH}I_{SH}^2 - \bar{V}_{SH}\bar{I}_{Line}^* - Z_{SH}\bar{I}_{SH}\bar{I}_{Line}^*) \tag{20.31}$$

It should be noted that there is no need for an iterative procedure used in Ref. 4 to compute UPFC control parameters. They can be computed directly after a conventional LF solution satisfying Eq. (20.25) is found. By neglecting transformer losses and initializing the real power injected into the sending bus to the real power flow controlled on the line, the convergence of the proposed LF algorithm is obtained within one step.

UPFC Dynamic Model

For transient stability studies, the DC-link dynamics have to be taken into account and Eq. (20.25) can no longer be applied. The DC-link capacitor will exchange energy with the system and its voltage will vary.

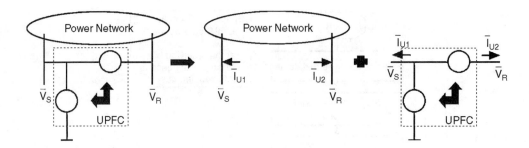

FIGURE 20.9 Interface of the UPFC with power network.

The power frequency dynamic model as given in Refs. 5 and 7 can be described by the following equation:

$$CV_{DC}\frac{dV_{DC}}{dt} = (P_{SH} - P_{SE})S_B \quad (20.32)$$

Note that in the above equation the DC variables are expressed in MKSA units, whereas the AC system variables are expressed as per-unit quantities. S_B is the system side base power.

Interfacing the UPFC with the Power Network

The interface of the UPFC with the power network is shown in Fig. 20.9 [7].

To obtain the network solution (bus voltages and currents), an iterative approach is used. The UPFC sending and receiving bus voltages $\bar{V}_S$ and $\bar{V}_R$ can be expressed as a function of generator internal voltages $\bar{E}_G$ and the UPFC injection voltages $\bar{V}_{SH}$ and $\bar{V}_{SE}$ (Eq. 20.39). Control output and Eq. (20.18) determine the UPFC injection voltage magnitudes V_{SH} and V_{SE}. However, the phase angles of the injected voltages, δ_{SH} and δ_{SE}, are unknown because they depend on the phase angle of the sending bus voltage, δ_S, which is the result of the network solution. Graphical form of the algorithm for interfacing the UPFC with the power network is shown in Fig. 20.10. Necessary computations are shown below.

Reducing the bus admittance matrix to generator internal buses and UPFC terminal buses the following equation can be written

$$\begin{bmatrix} Y_{GG} & Y_{GU} \\ Y_{UG} & Y_{UU} \end{bmatrix} \begin{bmatrix} \bar{E}_G \\ \bar{V}_U \end{bmatrix} = \begin{bmatrix} \bar{I}_G \\ \bar{I}_U \end{bmatrix} \quad (20.33)$$

where

Y_{GG} = reduced admittance matrix connecting generator currents injection to the internal generator-voltages

Y_{GU} = admittance matrix component, which gives the generator currents due to the voltages at UPFC buses

Y_{UG} = admittance matrix component, which gives UPFC currents in terms of the generator internal voltages

Y_{UU} = admittance matrix connecting UPFC currents to the voltages at UPFC buses

$\bar{E}_G$ = vector of generator internal bus voltages

$\bar{V}_U$ = vector of UPFC AC bus voltages

$\bar{I}_G$ = vector of generator current injections

$\bar{I}_U$ = vector of UPFC currents injected to the power network

Unified Power Flow Controllers

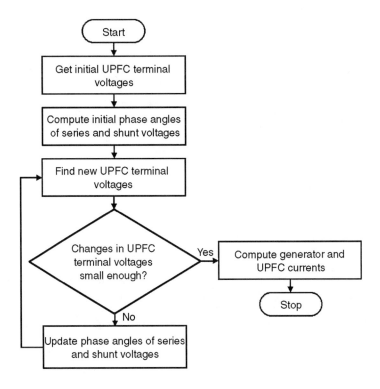

FIGURE 20.10 Algorithm for interfacing the UPFC with the power network.

The second equation of (20.33) is of the form

$$\bar{I}_U = Y_{UG}\bar{E}_G + Y_{UU}\bar{V}_U \qquad (20.34)$$

By neglecting series and shunt transformer resistances, the following equations can be written for the UPFC currents injected into the power network (see Figs. 20.3 and 20.9):

$$\bar{I}_{U1} = -\bar{I}_{SH} - \bar{I}_{Line}$$
$$\bar{I}_{U2} = \bar{I}_{Line} \qquad (20.35)$$

$$\bar{I}_{SH} = \frac{\bar{V}_S - \bar{V}_{SH}}{jx_{SH}} \qquad (20.36)$$

$$\bar{I}_{Line} = \frac{\bar{V}_{SE} + \bar{V}_S - \bar{V}_R}{jx_{SE}} \qquad (20.37)$$

Combining the above equations yields the following equation:

$$\bar{I}_U = W_U \bar{V}_U + W_C \bar{V}_C \qquad (20.38)$$

where

$$W_C = \begin{bmatrix} \dfrac{1}{jx_{SH}} & -\dfrac{1}{jx_{SE}} \\ 0 & \dfrac{1}{jx_{SE}} \end{bmatrix}$$

$$W_U = \begin{bmatrix} -\dfrac{1}{jx_{SE}} - \dfrac{1}{jx_{SH}} & \dfrac{1}{jx_{SE}} \\ \dfrac{1}{jx_{SE}} & -\dfrac{1}{jx_{SE}} \end{bmatrix}$$

$$\bar{I}_U = \begin{bmatrix} \bar{I}_{U1} \\ \bar{I}_{U2} \end{bmatrix}$$

$$\bar{V}_U = \begin{bmatrix} \bar{V}_S \\ \bar{V}_R \end{bmatrix}$$

$$\bar{V}_C = \begin{bmatrix} \bar{V}_{SH} \\ \bar{V}_{SE} \end{bmatrix}$$

By equating (20.34) with (20.38), the following equation for computation of UPFC terminal voltages can be written:

$$\bar{V}_U = (W_U - Y_{UU})^{-1} Y_{UG} \bar{E}_G - (W_U - Y_{UU})^{-1} W_C \bar{V}_C = L_G \bar{E}_G + L_C \bar{V}_C \qquad (20.39)$$

20.5 Control Design

To operate the UPFC in the automatic control mode discussed in the previous section, and also to use the UPFC to enhance power system stability and dampen low-frequency oscillations, two control designs need to be performed. A primary control design, referred to as the *UPFC basic control design*, involves simultaneous regulation of (1) real and reactive power flows on the transmission line, (2) sending bus voltage magnitude, and (3) DC voltage magnitude. A secondary control design, referred to as the *damping controller design*, is a supplementary control loop that is designed to improve transient stability of the entire electric power system. The two control designs are described in this section.

UPFC Basic Control Design

The UPFC basic control design consists of four separate control loops grouped into a *series control scheme*, whose objective is to maintain both real and reactive power flows on the transmission lines close to some prespecified values, and a *shunt control scheme*, whose objective is to control the sending bus voltage magnitude as well as the DC voltage magnitude.

Series Control Scheme

This scheme has two control loops, one for the tracking of the real power flow at the receiving bus of the line, and the second performing the same task for the reactive power flow. Specifically, the objective is to track these real and reactive power flows following step changes and to eliminate steady-state tracking errors. This is obtained by the appropriate selection of the voltage drop between the sending and the receiving buses, which is denoted $\bar{V}_{PQ}$. This voltage can be decomposed into the following two quantities, which affect the tracked powers, namely:

1. V_P = voltage component orthogonal to the sending bus voltage (it affects primarily the real power flow on the transmission line)
2. V_Q = component in phase with the sending bus voltage (it affects mainly the reactive power flow on the transmission line)

These quantities are in phasor diagram in Fig. 20.11.

Unified Power Flow Controllers

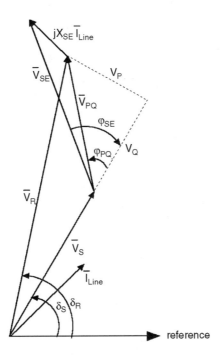

FIGURE 20.11 Phasor diagram.

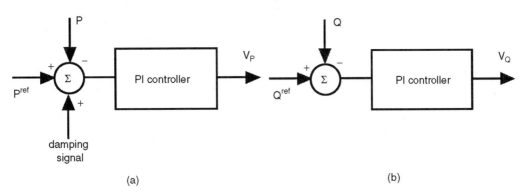

FIGURE 20.12 Series control scheme–automatic power flow mode.

Both voltages V_P and V_Q are obtained by designing classic PI (proportional-integral) controllers, as illustrated in Fig. 20.12 [7]. The integral controller will guarantee error-free steady-state control of the real and reactive line power flows.

After the V_P and V_Q components have been found, the series injected voltage are computed:

$$\begin{aligned}
V_{PQ} &= \sqrt{V_P^2 + V_Q^2} \\
\varphi_{PQ} &= tg^{-1}\frac{V_P}{V_Q} \\
\overline{V}_{PQ} &= V_{PQ} \angle(\delta_S + \varphi_{PQ}) \\
\overline{V}_{SE} &= \overline{V}_{PQ} + jX_{SE}\overline{I}_{Line}
\end{aligned} \quad (20.40)$$

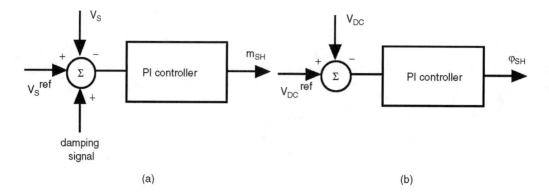

FIGURE 20.13 Shunt control scheme.

From Eqs. (20.18) and (20.19), series converter amplitude modulation index and firing angle are

$$m_{SE} = \frac{2\sqrt{2}\,n_{SE}V_{SE}V_B}{V_{DC}}$$

$$\varphi_{SE} = \delta_S - \delta_{SE} \tag{20.41}$$

Shunt Control Scheme

This control scheme also has two loops that are designed to maintain the magnitude of the sending bus voltage and the DC-link voltage at their prespecified values. The magnitude of the injected shunt voltage (Eq. 20.18) affects the reactive power flow in the shunt branch, which in turn affects the sending bus voltage magnitude. The angle between the sending bus voltage and the injected shunt voltage, φ_{SH} (Eq. 20.19), affects the real power flow in the shunt branch. It can be used to control the power flow to the DC-link and therefore the DC-link voltage. This is achieved by using two separate PI controllers as shown in Fig. 20.13 [6, 7].

Power System Damping Control through UPFC Using Fuzzy Control

Low-frequency oscillations in electric power systems occur frequently because of disturbances, such as changes in loading conditions or a loss of a transmission line or a generating unit. These oscillations need to be controlled to maintain system stability. Several control devices, such as power system stabilizers, are used to enhance power system stability. Recently [6–8], it has been shown that the UPFC can also be used to effectively control these low-frequency power system oscillations. It is done by designing a supplementary signal based on either the real power flow along the transmission line to the series converter side (Fig. 20.12a) or to the shunt converter side through the modulation of the voltage magnitude reference signal (Fig. 20.13a). The damping controllers used are of lead-lag type with transfer functions similar to the one shown in Fig. 20.14.

The authors of this chapter have designed a damping controller using fuzzy logic. This control design is presented here. But first, a brief review of fuzzy set theory and the basics of fuzzy control design is given.

Fuzzy Control Overview

Fuzzy control is based on fuzzy logic theory, but there is no systematic design procedure in fuzzy control. The important advantage of fuzzy control design is that a mathematical model of the system is not required.

Fuzzy controllers are rule-based controllers. The rules are given in the "if–then" format. The "if-side" is called condition and the "then-side" is called conclusion. The rules may use several variables both in condition and conclusion of the rules. Therefore, the fuzzy controllers can be applied to nonlinear

Unified Power Flow Controllers

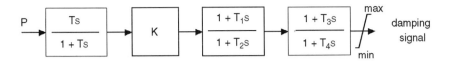

FIGURE 20.14 Lead–lag controller structure.

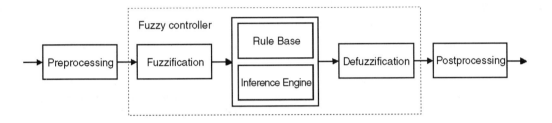

FIGURE 20.15 Fuzzy controller structure.

multiinput–multioutput (MIMO) systems such as power systems. The control rules can be found based on:

- Expert experience and control engineering knowledge of the system
- Learning (e.g., neural networks)

Fuzzy logic has its own terminology, which is reviewed below [10].

Fuzzy set: Let X be a collection of objects, then a fuzzy set A in X is defined as

$$A = \{(x, \mu_A(x)) | x \in X\} \tag{20.42}$$

$\mu_A(x)$ is called the membership function of x in A. It usually takes values in the interval [0, 1]. The numerical interval X relevant for the description of a fuzzy variable is called Universe of Discourse.

Operations on fuzzy sets: Let A and B be two fuzzy sets with membership functions $\mu_A(x)$ and $\mu_B(x)$. The *AND operator* or intersection of two fuzzy sets A and B is a fuzzy set C whose membership function $\mu_C(x)$ is defined as

$$\mu_C(x) = \min\{\mu_A(x), \mu_B(x)\}, \quad x \in X \tag{20.43}$$

The *OR operator* or union of two fuzzy sets A and B is a fuzzy set C whose membership function $\mu_C(x)$ is defined as

$$\mu_C(x) = \max\{\mu_A(x), \mu_B(x)\}, \quad x \in X \tag{20.44}$$

Next, the structure of a fuzzy controller is presented. A fuzzy controller structure is shown in Fig. 20.15 [9]. The controller is placed between preprocessing and post-processing blocks.

The *preprocessing block* conditions the inputs, crisp measurements, before they enter the controller. *First step* in fuzzy controller design is to choose appropriate input and output signals of the controller. *Second step* is to choose linguistic variables that will describe all input and output variables. *Third step* is to define membership functions for fuzzy sets. Membership functions can be of different shape, i.e., triangular, trapezoidal, Gaussian functions, etc.

Fourth step is to define fuzzy rules. For two-input one-output system each control rule R_i will be of the following form:

IF input (1) is A_{i1} AND input (2) is A_{i2} THEN output is B_i

Fifth step is to join rules by using an inference engine. The most often used inference engines are *Mamdani Max–Min and Max-Product*. The Max-Product inference procedure can be summarized as

- For the *i*th rule
  —Obtain the minimum between the input membership functions by using the AND operator
  —Re-scale the output membership function by the obtained minimum to obtain the output membership function due to the *i*th rule
- Repeat the same procedure for all rules.
- Find the maximum between output membership functions obtained from each rule by using the OR operator. This gives the final output membership function due to all rules.

Graphical technique of Mamdani (Max-Product) inference is shown in Fig. 20.16. Graphical technique of Mamdani Max-Min inference shown in Fig. 20.17 can be explained in similar matter.

Sixth step is defuzzification. The resulting fuzzy set must be converted to a number. This operation is called defuzzification. The most often used defuzzification method is the centroid method or center of area as shown in Fig. 20.18.

Post-processing. The post-processing block often contains an output gain that can be tuned.

Fuzzy Logic UPFC Damping Controller

Input signals to the controller, power flow deviation ΔP and energy deviation ΔE, are derived from the real power flow signal at the UPFC site [11]. The signals, representing the accelerating power and energy of generators at both ends of the tie line, indicate a required change in transmitted line power flow and have to be driven back to zero for a new steady state by the damping controller. The process of finding these signals requires some signal conditioning as shown in Fig. 20.19. Filters are used to remove noise and offset components [11, 12].

The inputs are described by the following linguistic variables: P (positive), NZ (near zero), and N (negative). The output is described with five linguistic variables P (positive), PS (positive small), NZ (near zero), NS (negative small), and N (negative). Gaussian functions are used as membership functions for

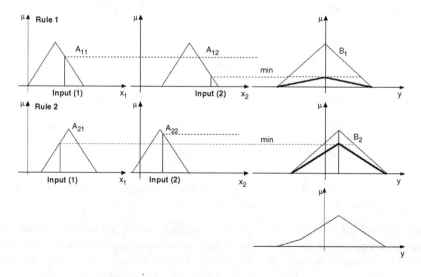

FIGURE 20.16 Mamdani Max-Product inference.

Unified Power Flow Controllers

TABLE 20.1 Fuzzy Rules

if ΔE is NZ, then damping signal is NZ
if ΔE is P, then damping signal is P
if ΔE is N, then damping signal is N
if ΔE is NZ, and ΔP is P, then damping signal is NS
if ΔE is NZ, and ΔP is N, then damping signal is PS

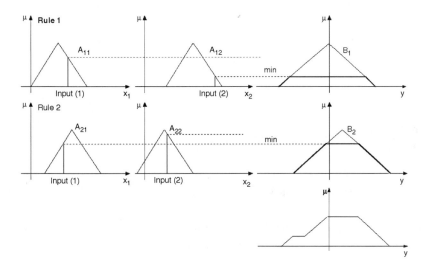

FIGURE 20.17 Mamdani (Max-Min) inference.

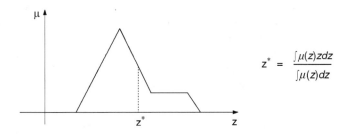

FIGURE 20.18 Centroid method.

$$z^* = \frac{\int \mu(z)z\,dz}{\int \mu(z)\,dz}$$

FIGURE 20.19 Obtaining the input signals for fuzzy controller.

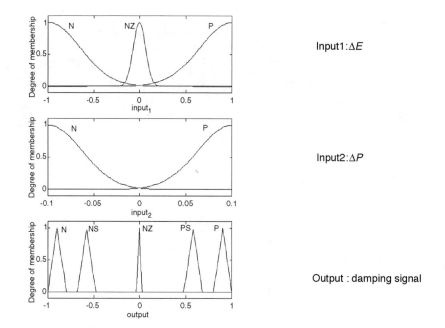

FIGURE 20.20 Fuzzy logic controller input and output variables.

both inputs, and triangular membership functions are used for output (Fig. 20.20). The damping signal is controller output. Fuzzy rules used are given in Table 20.1.

20.6 Case Study

Test System

The performance of the UPFC is tested on a two-area–four-generator system (test system) as shown in Fig. 20.21. Data for this system can be found in Ref. 1.

The two areas are identical to one another and interconnected with two parallel 230-km tie lines that carry about 400 MW from area 1 (generators 1 and 2) to area 2 (generators 3 and 4) during normal operating conditions. The UPFC is placed at the beginning of the lower parallel line between buses 101 and 13 to control the power flow through that line as well as to regulate voltage level at bus 101. Two cases are considered:

1. UPFC performance when the real and reactive power references are changed
2. UPFC performance when a fault is applied

All simulations are carried out in the Power System Toolbox (PST) [1], a commercial MATLAB-based package for nonlinear simulation of power systems that was modified by the authors to include a UPFC model.

Tracking Real and Reactive Power Flows

The objective is to keep the sending bus voltage at its prespecified value and to:

- Keep the reactive power constant while tracking the step changes in the real power (UPFC is initially operated to control the real power flow at 1.6 pu; at time 0.5 s real power flow reference

Unified Power Flow Controllers

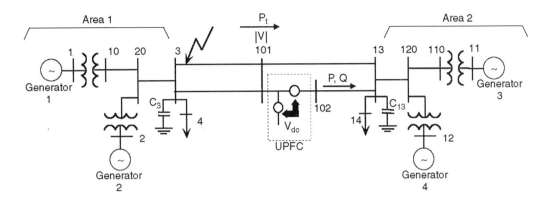

FIGURE 20.21 Two-area–four-generator test system.

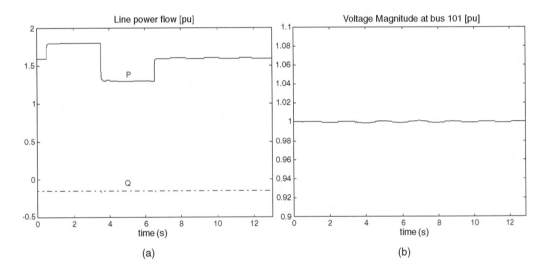

FIGURE 20.22 UPFC changing real power reference (P).

is raised to 1.8 pu and at time 3.5 s it is dropped to 1.3 pu; at time 7.5 s the system returns to the initial operating condition as shown in Fig. 20.22a.

- Keep the real power constant while tracking the step changes in the reactive power (at time 0.5 s reactive power flow reference is changed from −0.15 pu to −0.2 pu; at time 3.5 s reference is set to 0.2 pu; and at time 7.5 s the system returns to the initial operating condition as shown in Fig. 20.23a.

Results depicted in Figs. 20.22 and 20.23 show that the UPFC responds almost instantaneously to changes in real and reactive power flow reference values. For both cases the sending bus voltage is regulated at 1 pu, as shown in Figs. 20.22b and 20.23b. Plots also show that the UPFC is able to control real and reactive power flow independently.

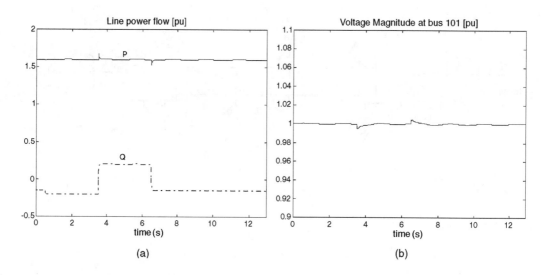

FIGURE 20.23 UPFC changing reactive power reference value (Q).

Operation under Fault Conditions

In this section test system response to a 100-ms three-phase fault applied in area 1, at bus 3, is examined. The fault is cleared by removing one line between the fault bus and bus 101. Two UPFC modes of operation are considered: (1) UPFC operated in the automatic power flow control mode, and (2) UPFC operated in the power oscillation damping control mode.

UPFC Operated in the Automatic Power Flow Control Mode

For comparison reasons, the test system without the UPFC is simulated first. Simulation results for the test system with and without UPFC are shown in Fig. 20.24.

During the steady-state operation each interconnecting tie line carries 197 MW (1.97 pu) from area 1 to area 2. It can be seen from Fig. 20.24a that the line power flow for the test system without the UPFC oscillates long after the fault is cleared, whereas the desired power flow conditions are reached quickly after the fault is cleared for the test system with the UPFC.

For the test system without the UPFC, bus 101 voltage is 0.92 pu, which is below accepted limits. Therefore, the UPFC is operated to keep bus 101 voltage at 1 pu (Fig. 20.24b).

The power angle swings for the test system with UPFC are better damped although it can be seen from Fig. 20.24c that constant power flow has negative effect on the system first swing transient, as reported in Ref. 7.

The DC capacitor voltage fluctuation during the transient is less than 1% of its 22-kV rated value (Fig. 20.24d), which is acceptable for this application.

UPFC Operated in the Power Oscillation Damping Control Mode

To improve the power oscillation damping, the UPFC is operated in the automatic power flow mode but with active damping control. To show the robustness of the proposed control scheme, based on fuzzy logic, discussed earlier, different operating conditions were simulated. The real power that each interconnecting tie-line carries during prefault operating conditions, from area 1 to area 2, is given in Table 20.2.

The fuzzy damping controller is applied to the series converter side. The inputs to the controller are derived from the total real power flow P_t at the UPFC sending bus, as shown in Fig. 20.21. The Max-Product inference engine was used and the centroid defuzzification method was applied. The results obtained with the fuzzy controller are compared with those obtained with the lead-lag controller applied

Unified Power Flow Controllers

TABLE 20.2 Operating Conditions (values in MW)

Case	Line 101–13	Line 102–13	Load bus 4	Load bus 14
(a)	232	160	976	1767
(b)	24	160	1176	1567
(c)	143	250	976	1767
(d)	−232	−160	1767	976

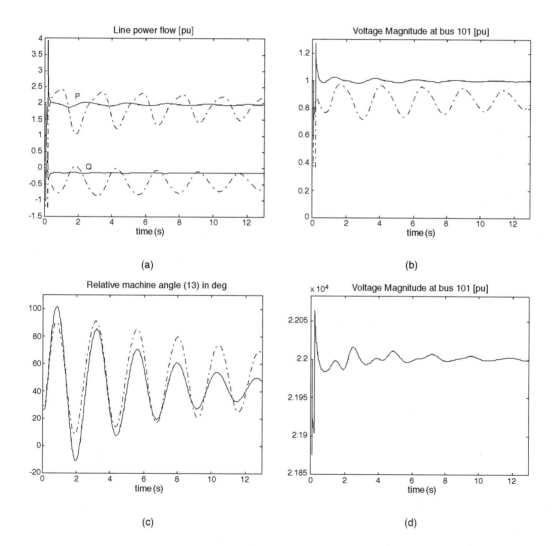

FIGURE 20.24 Simulation results: dashed lines—system without UPFC; solid lines—system with UPFC.

at the shunt converter side. Both fuzzy and lead-lag damping controllers were designed for the first operating condition, (a) in Table 20.2.

Nonlinear simulation results depicted in Fig. 20.25 show that adding a supplementary control signal greatly enhances damping of the generator angle oscillations. It can be seen that the fuzzy damping controller performs better for different operating conditions than the conventional controller.

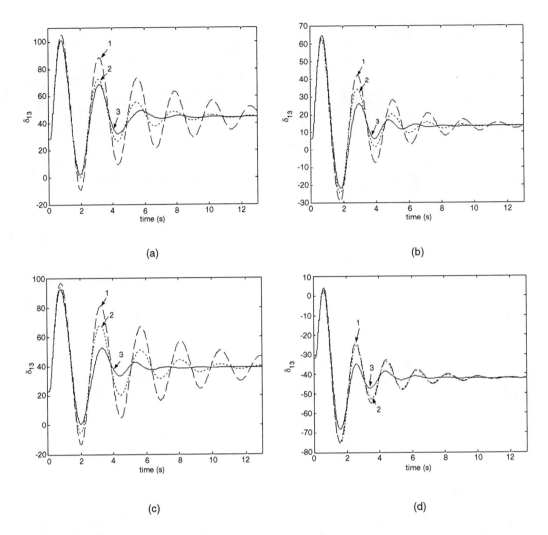

FIGURE 20.25 Relative machine angle δ_{13} in degrees for operating conditions (a to d): 1, without damping controller; 2, lead-lag damping controller; 3, fuzzy damping controller.

20.7 Conclusion

This chapter deals with the FACTS device known as the Unified Power Flow Controller (UPFC). With its unique capability to control simultaneously real and reactive power flows on a transmission line as well as to regulate voltage at the bus where it is connected, this device creates a tremendous impact on power system stability enhancement and loading of transmission lines close to their thermal limits. Thus, the device gives power system operators much needed flexibility to satisfy the demands that the deregulated power system imposes.

Specifically, in this chapter the UPFC is first described and its operation explained. Second, its steady-state and dynamic models, and an algorithm for interfacing it with the power network are presented. Third, basic and damping controller design are developed.

To operate the UPFC in the automatic control mode and to use the UPFC to damp low-frequency oscillations, two controls, basic control and the damping control, are designed. The chapter has shown that the UPFC with its basic controllers is capable of controlling independently real and reactive power flow through the transmission line, under both steady-state and dynamic conditions. Also shown is that the

UPFC can be used for voltage support and for improvement of transient stability of the entire electric power through a supplementary control loop. The proposed supplementary control, based on fuzzy control, is effective in damping power oscillations. The controller requires only a local measurement—the tie-line power flow at the UPFC location. Simulation results have shown that controller exhibits good damping characteristics for different operating conditions and performs better than conventional controllers. The performance is illustrated with a two-area–four-generator test system. Simulation is performed using the MATLAB-based Power System Toolbox package, which is modified to incorporate the UPFC model.

Acknowledgment

The National Science Foundation under Grant ECS-9870041, and the Department of Energy under a DOE/EPSCoR WV State Implementation Award, sponsored some of the research presented in this chapter.

References

1. Dynamic Tutorial and Functions, Power System Toolbox, Version 2.0, Cherry Tree Scientific Software, Ontario, Canada.
2. Mohan, N., Undeland, T. M., and Robbins, W. P., *Power Electronics: Converters, Applications and Design*, 2nd ed., John Wiley & Sons, New York, 1995.
3. Hingorani, N. G. and Gyugyi, L., *Understanding FACTS Devices*, IEEE Press, New York, 2000.
4. Nabavi-Niaki, A. and Iravani, M. R., Steady-state and dynamic models of unified power flow controller (UPFC) for power system studies, *IEEE Trans. Power Syst.*, 11(4), 1937–1943, 1996.
5. Uzunovic, E., Canizares, C. A., and Reeve, J., Fundamental frequency model of unified power flow controller, in *Proceedings of the North American Power Symposium (NAPS)*, Cleveland, OH, October 1998.
6. Uzunovic, E., Canizares, C. A., and Reeve, J., EMTP studies of UPFC power oscillation damping, in *Proceedings of the North American Power Symposium (NAPS)*, San Luis Obispo, CA, October 1999, 405–410.
7. Huang, Z., Ni, Y., Shen, C. M., Wu, F. F., Chen, S., and Zhang, B., Application of unified power flow controller in interconnected power systems—modeling, interface, control strategy and case study, presented at *IEEE Power Eng. Society Summer Meeting*, 1999.
8. Wang, H. F., Applications of modeling UPFC into multi-machine power systems, *IEE Proc. Generation Transmission Distribution*, 146(3), 306–312, 1999.
9. Jantzen, J., Design of fuzzy controllers, Technical Report No. 98-E 864 (design), Department of Automation, Technical University of Denmark, Lyngby, August 1998.
10. Hsu, Y. Y. and Cheng, C. H, Design of fuzzy power system stabilizers for multimachine power systems, *IEE Proc.*, 137C(3), 233–238, 1990.
11. Hiyama, T., Hubbi, W., and Ortmeyer, T., Fuzzy logic control scheme with variable gain for static Var compensator to enhance power system stability, *IEEE Trans. Power Syst.*, 14(1), 186–191, 1999.
12. Hiyama, T., Fuzzy logic switching of thyristor controlled braking resistor considering coordination with SVC, *IEEE Trans. Power Delivery*, 10(4), 2020–2026, 1995.

21
More-Electric Vehicles

Ali Emadi
Illinois Institute of Technology

Mehrdad Ehsani
Texas A&M University

21.1 Aircraft 21-1
Conventional Electrical Loads • Power Generation Systems • Aircraft Electrical Distribution Systems • Advanced Electrical Loads • Advanced Electrical Distribution System Architectures • Conclusions

21.2 Terrestrial Vehicles 21-6
Electrical Power Systems of Conventional Cars • Advanced Electrical Loads • Increasing the System Voltage • Advanced Distribution Systems • Electrical Power Systems of Electric and Hybrid Electric Vehicles • Automotive Electric Motor Drives • Conclusions

21.1 Aircraft

Ali Emadi and Mehrdad Ehsani

Mechanical, electrical, and centralized hydraulic and pneumatic systems are conventional power transfer systems in an aircraft. The More-Electric Aircraft (MEA) concept emphasizes utilizing electrical systems to replace more aircraft conventional power transfer systems and to facilitate new introduced electrical loads. Improving reliability, maintainability, supportability, survivability, performance, safety, emissions, and operating costs are the main motivations behind the MEA concept.

Conventional Electrical Loads

The power needed for the subsystems in an aircraft is currently derived from mechanical, electrical, hydraulic, and pneumatic sources or a combination of them [1–3]. Generally, hydraulic power transfer systems are used for most of the actuators. On the other hand, pneumatic power transfer systems are mainly employed for air-conditioning, pressurization, and ice protection systems.

Electrical and electronic systems are usually used for avionics and utility functions, such as air data instruments, communications, landing gear, lighting, navigation, and comfort of the passengers. Other conventional subsystems that are driven by the electrical sources include energy storage, engine starting, ignition, anti-skid control, and deicing and anti-icing systems [3].

Power Generation Systems

The wound-field synchronous machine has traditionally been used to generate AC electrical power with constant frequency of 400 Hz. This machine/drive system is known as a constant-speed drive (CSD) system [3–5]. Figure 21.1 shows a typical constant-speed drive system. In Fig. 21.1, synchronous generators supply AC constant-frequency voltage to the AC loads in the aircraft. Then, AC-DC rectifiers are used to convert the AC voltage with fixed frequency at the main AC bus to multilevel DC voltages at the

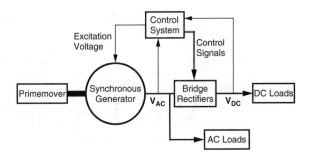

FIGURE 21.1 Typical CSD system.

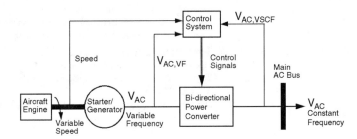

FIGURE 21.2 Typical VSCF starter/generator system.

secondary buses, which supply electrical power to the DC loads. Excitation voltage of the synchronous generator and firing angles of the bridge rectifiers are controlled via the control system of the CSD system.

Recent advances in the areas of power electronics, control electronics, electric motor drives, and electric machines have introduced a new technology of variable-speed constant-frequency (VSCF) systems. The main advantage of VSCF is that it provides better starter/generator systems. Other advantages are higher reliability, lower recurring costs, and shorter mission cycle times [5]. Figure 21.2 shows the block diagram of a typical VSCF starter/generator system. In the generating mode, an aircraft engine, which has variable speed, provides mechanical input power to the electric generator. Then, the electric generator supplies variable-frequency AC power to the bidirectional power converter, which provides AC constant-frequency voltage to the main bus. In the motoring mode, the constant-frequency AC system via the bidirectional power converter provides input electric power to the electric machine, which is a starter to the aircraft engine. Synchronous, induction, and switched reluctance machines are three candidates for VSCF starter/generator systems [3–6].

The bidirectional power electronic converter of the VSCF system is a multilevel converter, as depicted in Fig. 21.3. The input voltage is variable AC whose amplitude is not regulated. Moreover, the frequency is not constant. At the input stage of the bidirectional converter, there is an uncontrolled rectifier converting the variable AC to an unregulated DC voltage. Then, a DC voltage regulator is used to provide power for the regulated high-voltage 270-V DC system. A DC-DC converter and a DC-AC inverter connected to this system provide power for the low-voltage 28-V DC and 115/200-V, 400-Hz, three-phase AC loads, respectively. Batteries are also connected to the system via the battery charge/discharge unit.

Aircraft Electrical Distribution Systems

Because of the expansion of electrical loads and the replacement of conventional aircraft systems with the electrical counterparts, aircraft power systems are becoming more electric. As a result, in advanced aircraft, electrical distribution systems with larger capacity and more complex configuration are necessary.

More-Electric Vehicles

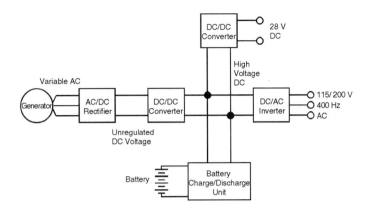

FIGURE 21.3 Multilevel conversion of the unregulated AC voltage to regulated DC and AC voltages.

Systems with constant-frequency (CF) and VSCF have 115-V AC, 400-Hz, three-phase electrical systems. They may also have a 270-V DC or higher primary power bus. The electrical system of an aircraft may have wild frequency with a variable-frequency VF generator of 115 V AC, three-phase power [7].

In the MEA electrical power systems, a number of different types of loads are used, which require power supplies different from the standard supplies provided by the main generator. Therefore, the future aircraft electrical power systems will employ multivoltage-level hybrid DC and AC systems. For example, in an advanced aircraft power system having a 270-V DC primary power supply, certain instruments and electronic equipment are employed that require 28-V DC and 115-V AC supplies for their operation. In fact, DC cannot be entirely eliminated even in aircraft that is primarily AC in concept. Furthermore, even within the items of consumer equipment themselves, certain sections of their circuits require different types of power supply and/or different levels of the same kind of the supply. It therefore becomes necessary to employ not only equipment converting electrical power from one form to another, but also equipment converting one form of supply to a higher or lower value. As a result, in a modern aircraft, different kinds of power electronic converters such as AC-DC rectifiers, DC-AC inverters, and DC-DC choppers are required. In addition, in the VSCF systems, solid-state bidirectional converters are used to condition VF power into a fixed frequency and voltage. Moreover, bidirectional DC-DC converters are used in the battery charge/discharge units.

As the AC-DC converters, conventional transformer rectifier units (TRU) are used. Each unit consists of a 12-pulse transformer and a controlled or uncontrolled rectifier. Power diodes and thyristors are used in uncontrolled and controlled rectifiers, respectively. If a constant voltage is needed, controlled rectifiers are used to regulate output voltage. And, if it is not necessary to regulate the output voltage or if there is a voltage regulator at the output side of TRU, uncontrolled rectifiers are used. However, in an advanced MEA, recent advances in the area of power electronics, such as resonant and soft switching techniques, can be used to increase the power density and improve the performance of all the power conditioning systems [8].

Advanced Electrical Loads

Performance improvements in electric actuation systems and electric motor drives are providing the impetus for the MEA concept. In fact, there is a trend toward replacement of more engine-driven mechanical, hydraulic, and pneumatic loads with electrical loads as a result of performance and reliability issues.

In an advanced aircraft, electromechanical actuators are used instead of the conventional hydraulic actuators. The expansion of this concept to braking systems results in electrically actuated braking systems [9]. Improved safety, reliability, and maintainability are the benefits that accrue through the removal of the hydraulic fluid. In addition, the efficiency is improved through better control of braking torque [9].

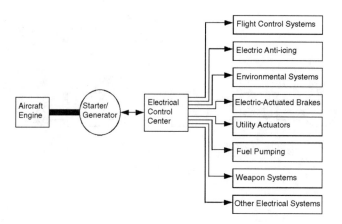

FIGURE 21.4 MEA electrical power subsystems.

Furthermore, conventional aeroengine actuators use fluid power in the form of pneumatic, hydraulic, or fueldraulics to provide the motive effort. There is also a trend toward replacing these traditional hydraulic/pneumatic/fueldraulic engine actuation systems with electromechanical actuators. The main advantages are easier interfacing, reduced maintenance costs, lighter systems, and improved reliability. The electric motor type selected is a three-phase brushless DC motor [10].

Some of the other loads considered are electromechanical and electrohydraulic flight control actuators, 270-V DC switched reluctance starter/generators, electric anti-icing systems, environmental systems, electromechanical valve controllers, air-conditioning systems, utility actuators, weapon systems, and different electric motor drives for pumps and other applications. In fact, electrical subsystems may require a lower engine power with higher efficiency. Also, they can be used only when needed. Therefore, MEA can have better fuel economy and performance. Figure 21.4 shows the main electrical power subsystems in the MEA power systems.

Advanced Electrical Distribution System Architectures

A conventional distribution network is a point-to-point topology in which all the electrical wires are distributed from the main bus to different loads through relays and switches. This kind of distribution network leads to expensive, complicated, and heavy wiring circuits. However, in an advanced aircraft, loads are controlled by intelligent remote modules. Therefore, the number and length of wires in the harness are reduced. Furthermore, by interconnection between remote modules via communication/control buses, it is possible to have a power management system (PMS). The primary function of the PMS is time-phasing of the duty cycle of loads to reduce the peak power demand [11]. Other functions of the PMS are battery management and charging strategy in a multiple-battery system, load management, management of the starter/generator system including the regulator, and provision and control of a high-integrity supply system. In addition, power management strategy can help optimize the size of the generators and batteries [11].

Figure 21.5 shows an advanced aircraft power system architecture in which there are several power electronic converters. The distribution control network of Fig. 21.5 simplifies vehicle physical design and assembly and offers additional benefits from the integration with intelligent power management control. Other advantages of this MEA technology are reduced design complexity, fewer flight test hours, reduced ground support equipment, and easier aircraft modification [7].

To power important systems in the case of an emergency, permanent magnet (PM) generators are used to generate 28-V DC voltage. Furthermore, the main distribution system can also be changed from DC to AC. The main advantage of AC distribution systems is easy conversion to different voltage levels by transformers. Also, AC machines are easy to use.

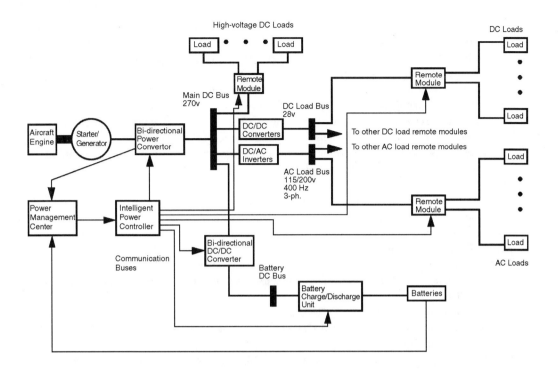

FIGURE 21.5 The concept of an advanced aircraft power system architecture of the future.

Specifications of the DC-DC converters and DC-AC inverters for MEA applications are given in Ref. 12. Two power electronic converters, which are highly compact with input nominal voltage of 270 V DC, are presented in Ref. 12. The DC-DC converter provides 5.6 kW at 29 ± 0.5 V DC with an efficiency of 90%. The DC-AC inverter provides 8 kVA of three-phase power at (115 ± 1.5)/200 V AC and 400 Hz with an efficiency of 87%. Both of these converters have high-frequency (120-kHz) resonant circuits. The reason for using the resonant circuits is that the power electronic devices are switched at zero current. This reduces the switching power losses and, in turn, increases the efficiency and switching frequency to 120 kHz [12].

Conclusions

To improve aircraft reliability, maintainability, emissions, and performance, the MEA concept emphasizes the utilization of electrical systems instead of the conventional mechanical, hydraulic, and pneumatic power transfer systems. The MEA concept facilitates high-power electric loads and requires power electronics in a solid-state rich electric environment. In fact, advanced aircraft and aerospace power systems are multiconverter power electronics–based systems. In these systems, different converters, such as AC-DC rectifiers, DC-DC choppers, and DC-AC inverters, are used to provide power at different voltage levels in both DC and AC forms. The AC system may be constant frequency, multifrequency, or wild frequency. In addition, advanced aircraft power systems employ separate buses for power and control as well as an intelligent power management center.

References

1. R. E. Quigley, More electric aircraft, in *Proc. 1993 IEEE Applied Power Electronics Conf.*, San Diego, March 1993, 609–911.
2. J. A. Weimer, Electrical power technology for the more electric aircraft, in *Proc. IEEE 12th Digital Avionics Systems Conf.*, Fort Worth, Oct. 1993, 445–450.

3. A. Emadi and M. Ehsani, Aircraft power systems: technology, state of the art, and future trends, *IEEE Aerospace Electron. Syst. Mag.*, 15(1), 28–32, 2000.
4. J. G. Vaidya, Electrical machines technology for aerospace power generators, in *Proc. 1991 Intersociety Energy Conversion Engineering Conf.*, Vol. 1, 1991, 7–12.
5. M. E. Elbuluk and M. D. Kankam, Potential starter/generator technologies for future aerospace application, *IEEE Aerospace Electron. Syst. Mag.*, 11(10), 17–24, 1996.
6. E. Richter and C. Ferreira, Performance evaluation of a 250 kW switched reluctance starter/generator, in *Proc. 1995 IEEE Industry Applications Conf.*, Orlando, Oct. 1995, 434–440.
7. M. Olaiya and N. Buchan, High power variable frequency generator for large civil aircraft, in *IEE Colloquium on Electrical Machines and Systems for the More Electric Aircraft*, London, Nov. 1999, 3/1–3/4.
8. K. W. E. Cheng, Comparative study of AC/DC converters for more electric aircraft, in *Proc. 7th International Conf. on Power Electronics and Variable Speed Drives*, Sept. 1998, 299–304.
9. FHL Division, Claverham Ltd., EABSYS: electrically actuated braking system, in *IEE Colloquium on Electrical Machines and Systems for the More Electric Aircraft*, London, Nov. 1999, 4/1–4/5.
10. R. Dixon, N. Gifford, C. Sewell, and M. C. Spalton, REACTS: reliable electrical actuation systems, in *IEE Colloquium on Electrical Machines and Systems for the More Electric Aircraft*, London, Nov. 1999, 5/1–5/16.
11. M. A. Maldonado, N. M. Shah, K. J. Cleek, and G. J. Korba, Power management and distribution system for a more electric aircraft (MADMEL)—program status, *IEEE Aerospace Electronic Syst. Mag.*, 14(12), 3–8, 1999.
12. W. G. Homeyer, E. E. Bowles, S. P. Lupan, P. S. Walia, and M. A. Maldonado, Advanced power converters for more electric aircraft applications, in *Proc. 32nd Intersociety Energy Conversion Engineering Conf.*, Aug. 1997, 591–596.

21.2 Terrestrial Vehicles

Ali Emadi and Mehrdad Ehsani

The More-Electric Vehicle (MEV) concept emphasizes the utilization of electrical systems instead of mechanical and hydraulic systems to optimize vehicle fuel economy, emissions, performance, and reliability. In addition, the need for improvement in comfort, convenience, entertainment, safety, security, and communications necessitates more electric automotive systems. As a result, an electric power distribution system with larger capacity and more complex configuration is required to facilitate increasing electrical loads.

Electrical Power Systems of Conventional Cars

The conventional electrical system in an automobile can be divided into the energy storage, charging, cranking ignition, lighting, electric motors, and instrumentation subsystems. In order for the power available at the sources to be made available at the terminals of the loads, some organized form of distribution throughout an automobile is essential. At present, most automobiles use a 14-V DC electrical system. Figure 21.6 shows the conventional electrical distribution system for automobiles. This has a single voltage level, i.e., 14-V DC, with the loads controlled by manual switches and relays [1–8]. Because of the point-to-point wiring, the wiring harness is heavy and complex.

The present average power demand in an automobile is approximately 1 kW. The voltage in a 14-V system actually varies between 9 and 16 V, depending on the alternator output current, battery age, state of charge, and other factors. This results in overrating the loads at nominal system voltage. There are several other disadvantages, which have been addressed in Refs. 1 through 4.

In addition to all the disadvantages, the present 14-V system cannot handle the future electrical loads to be introduced in the more electric environment of future cars, as it will be expensive and inefficient.

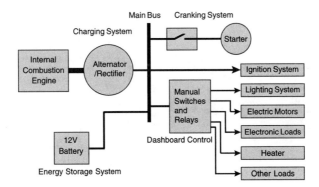

FIGURE 21.6 Conventional 14-V DC distribution system architecture.

Advanced Electrical Loads

In More-Electric Cars (MEC), there is a trend toward expanding electrical loads and replacement of more engine-driven mechanical and hydraulic systems with electrical systems. These loads include the well-known lights, pumps, fans, and electric motors for various functions. They will also include some less well known loads, such as electrically assisted power steering, electrically driven air-conditioner compressor, electromechanical valve control, electrically controlled suspension and vehicle dynamics, and electrically heated catalytic converter. In fact, electrical subsystems may require a lower engine power with higher efficiency. Furthermore, they can be used only when needed. Therefore, MEC can have optimum fuel economy and performance. There are also other loads such as antilock braking, throttle actuation, ride-height adjustment, and rear-wheel steering, which will be driven electrically in the future.

Figure 21.7 shows electrical loads in the MEC power systems. As is described in Refs. 5 through 8, most of the future electic loads require power electronic controls. In future automobiles, power electronics will be used to perform three different tasks. The first task is simple on/off switching of loads, which is performed by mechanical switches and relays in conventional cars. The second task is the control of electric machines. The third task is not only changing the system voltage to a higher or lower level, but also converting electrical power from one form to another using DC-DC, DC-AC, and AC-DC converters.

Increasing the System Voltage

Because of the increasing electrical loads, automotive systems are becoming more electric. Therefore, MEC will need highly reliable, fault-tolerant, autonomously controlled electrical power systems to deliver high-quality power from the sources to the loads. The voltage level and form in which power is distributed are important. A higher voltage such as the proposed 42 V will reduce the weight and volume of the wiring harness, among several other advantages [4, 5]. In fact, increasing the voltage of the system, which is 14 V in conventional cars, is necessary to cope with the greater loads associated with the more electric environments in future cars. The near-future average power demand is anticipated to be 3 kW and higher.

Figure 21.8 shows the concept of a dual-voltage automotive power system architecture of the future MEC. Indeed, it is a transitional two-voltage system, which can be introduced until all automotive components evolve to 42 V. Finally, the future MEC power system will most likely be a single-voltage bus (42 V DC) with provision for hybrid (DC and AC), multivoltage level distribution, and intelligent energy and load management.

Advanced Distribution Systems

The conventional automotive electrical power system is a point-to-point topology in which all the electrical wiring is distributed from the main bus to different loads through relays and switches of the dashboard control. As a result, the distribution network has expensive, complicated, and heavy wiring circuits.

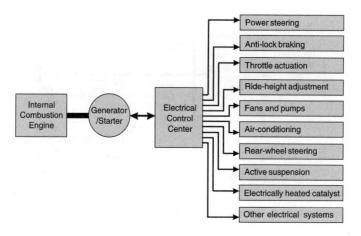

FIGURE 21.7 Electrical loads in MEC power systems.

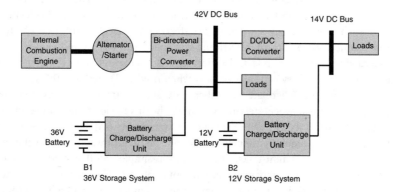

FIGURE 21.8 The concept of a dual-voltage automotive power system architecture of the future MEC.

However, in the advanced automotive electrical systems, multiplexed architectures with separate power and communication buses are used to improve the system. In a multiplexed network, loads are controlled by intelligent remote modules. Therefore, the number and length of wires in tthe harness are reduced. In addition, these systems have a power management system (PMS). The primary function of the PMS is time-phasing of the duty cycle of loads to reduce the peak power demand. Other functions of the PMS are battery management, load management, and management of the starter/generator system including the regulator. Figure 21.9 shows typical inputs and outputs of a power management center.

Figure 21.10 shows advanced multiplexed automotive power system architectures of the future with power and communication buses. The distribution control network of Fig. 21.10 simplifies vehicle physical design and assembly and offers additional benefits from the integration with intelligent power management control.

Electrical Power Systems of Electric and Hybrid Electric Vehicles

Because of environmental concerns, there is a significant impetus toward development of new propulsion systems for future cars in the form of electric and hybrid electric vehicles (EV and HEV). Electric vehicles are known as zero-emission vehicles. They use batteries as electrical energy storage devices and electric motors to propel the automobile. On the other hand, hybrid vehicles combine more than one energy source for propulsion. In heat engine/battery hybrid systems, the mechanical power available from the

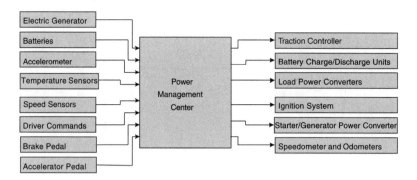

FIGURE 21.9 Power management system.

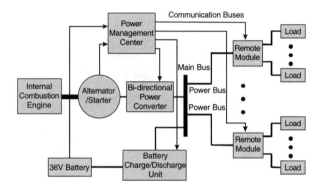

FIGURE 21.10 Advanced multiplexed automotive power system architectures of the future with power and communication buses.

heat engine is combined with the electrical energy stored in a battery to propel the vehicle. These systems also require an electric drive train to convert electrical energy into mechanical energy, as do electric vehicles.

Architectures of EV and HEV Drive Trains

Hybrid electric systems can be broadly classified as series or parallel hybrid systems [9–12]. The series and parallel hybrid architectures are shown in Figs. 21.11 and 21.12, respectively. In series hybrid systems, all the torque required to propel the vehicle is provided by an electric motor. On the other hand, in parallel hybrid systems, the torque obtained from the heat engine is mechanically coupled to the torque produced by an electric motor. In EV, the electric motor behaves exactly in the same manner as in a series hybrid. Therefore, the torque and power requirements of the electric motor are roughly equal for an EV and a series hybrid, whereas they are lower for a parallel hybrid.

Electrical Distribution System Architectures

Figure 21.13 depicts the conventional electrical power distribution system architecture for hybrid electric vehicles. It is a DC system with a main high-voltage bus, e.g., 300 or 140 V. The high-voltage storage system is connected to the main bus via the battery charge/discharge unit. This unit discharges and charges the batteries in motoring and generating modes of the electric machine operation, respectively. There are also two other charging systems, which are on-board and off-board. The off-board charger has three-phase or single-phase AC-DC rectifiers to charge the batteries when the vehicle is parked at a charging station. The on-board charger, as shown in Fig. 21.13 consists of a starter/generator and a bidirectional power converter. In the generating mode, the internal combustion engine provides mechanical input power to the electric generator. Then, the electric generator supplies electric power to the

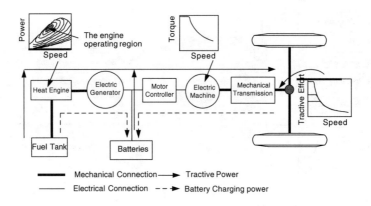

FIGURE 21.11 Series HEV architecture.

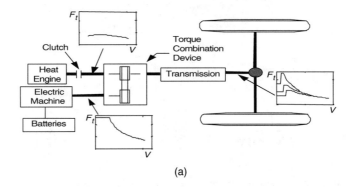

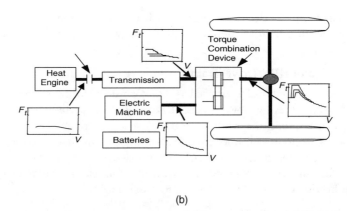

FIGURE 21.12 Parallel HEV architectures: (a) engine–motor–transmission configuration; (b) engine–transmission–motor configuration.

bidirectional power converter providing high-voltage DC to the main bus. Moreover, in the motoring mode, i.e., cranking the engine, the high-voltage DC system via the bidirectional power converter provides input electric power to the electric machine, which is a starter to the vehicle engine.

In Fig. 21.13, the electric propulsion system feeds from the main high-voltage bus. Furthermore, conventional low-power 14 and 5 V DC loads are connected to the 14-V bus. The low-voltage 14-V bus is connected to the main bus with a step-down DC-DC converter. A 12-V storage system via the battery

More-Electric Vehicles   21-11

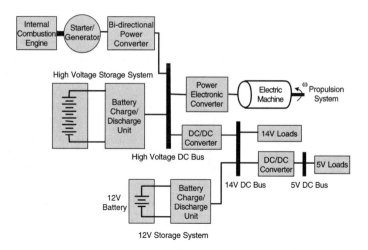

FIGURE 21.13 Conventional electrical power distribution system architecture for hybrid electric vehicles.

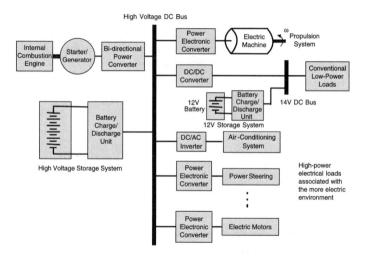

FIGURE 21.14 MEHV electrical power system architecture.

charge/discharge unit is also connected to the low-voltage bus. It should be mentioned that Fig. 21.13 without internal combustion engine, starter/generator, and bidirectional power converter shows the electrical power distribution system architecture of electric vehicles.

More-Electric Hybrid Vehicles

As described, demand for higher fuel economy, performance, and reliablility as well as reduced emissions will push the automotive industry to seek electrification of ancillaries and engine augmentations. This is the concept of MEV. Expansion of the MEV concept to HEV leads to More-Electric Hybrid Vehicles (MEHV). In the future MEV and MEHV, throttle actuation, power steering, antilock braking, rear-wheel steering, air-conditioning, ride-height adjustment, active suspension, and electrically heated catalyst will all benefit from electrical power systems.

Figure 21.14 shows the architecture of the MEHV electrical power system. It is a multivoltage hybrid (DC and AC) electrical power distribution system with a main high-voltage, e.g., 300 or 140 V, DC bus providing power for all loads. Conventional loads as well as new electrical ancillary and luxury loads associated with the more electric environment feed from the main bus via different DC-DC and DC-AC power electronic converters.

Automotive Electric Motor Drives

In a more electric car, most of the electrical loads, such as power steering and air-conditioning, require efficient, fault-tolerant, robust, simple, and compact electric motors. Different electric machines and power electronic converters will be used to facilitate conventional as well as advanced functions in future automotive systems.

DC motors with or without brushes are used for applications such as fans and pumps because of their high efficiency in addition to reduced cost, flexible control, high quality, and reliablity. Induction and variable reluctance machines are candidates for starter/generator systems. The comparison of these two machines for automotive applications is based on the electromagnetic weight, power density, efficiency, control complexity and features, complexity of design and fabrication, reliability, and thermal robustness [13, 14].

In EV and HEV, AC induction, DC commutator, and permanent magnet (PM) brushless DC motors are commonly used for propulsion systems. Recently, there has also been interest in switched reluctance motors (SRM) and synchronous reluctance motors [12, 15]. It is widely accepted that a suitable motor drive for traction applications in a vehicle should offer high efficiency, compactness, and low-cost manufacturing among other attributes. It is also important to remember that a successful candidate must provide a fault-tolerant and hazard-free operation.

A review of past works shows that considerable attention has been paid to the development of high-efficiency motor drives whereas the impact of vehicle dynamics is neglected. A field-oriented PM motor drive can have a high effciency because of the free excitation. However, a limited extended-speed constant-power region along with poor performance in the presence of short circuits and high temperature may prohibit effective use of these motor drives. Singly excited motor drives such as induction and switched reluctance machines, on the other hand, seem to have an excellent performance in the presence of partial failure and in harsh environments.

Conclusions

Advances in the areas of power electronics and electric motor drives along with fault-tolerant electrical distribution systems and control electronics enable the transforming of present automotive power systems into MEV systems. The future electrical power system for conventional cars will most likely be a single-voltage bus (42 V DC) with provision for hybrid and multivoltage-level distribution. On the other hand, the clearest direction for future HEV electrical power systems is a multivoltage system providing power for traction load and other automotive loads via high-voltage, e.g., 300 or 140 V DC, and low-voltage, e.g., 42 and/or 14 V DC, buses, respectively. The extent of these changes will certainly depend on cost-effective production of power electronics and other automotive electric and electronic components.

References

1. J. M. Miller, A. Emadi, A. V. Rajarathnam, and M. Ehsani, Current status and future trends in more electric car power systems, in *Proc. 1999 IEEE Vehicular Technology Conference*, Houston, TX, May 1999.
2. A. Emadi, M. Ehsani, and J. M. Miller, Advanced silicon rich automotive electrical power systems, presented at 18th Digital Avionics Systems Conference on Air, Space, and Ground Vehicle Electronic Systems, St. Louis, MO, Oct. 1999.
3. J. G. Kassakian, H. C. Wolf, J. M. Miller, and C. J. Hurton, Automotive electrical systems-circa 2005, *IEEE Spectrum*, 22–27, Aug. 1996.
4. J. M. Miller, D. Goel, D. Kaminski, H. P. Schoner, and T. Jahns, Making the case for a next generation automotive electrical system, presented at Convergence Transportation Electronics Association Congress, Dearborn, MI, Oct. 1998.
5. J. M. Miller and P. R. Nicastri, The next generation automotive electrical power system architecture: issues and challenges, presented at 17th Digital Avionics Systems Conference on Air, Space, and Ground Vehicle Electronic Systems, Bellevue, WA, Oct./Nov. 1998.

6. L. A. Khan, Automotive electrical systems: architecture and components, presented at 18th Digital Avionics Systems Conference on Air, Space, and Ground Vehicle Electronic Systems, St. Louis, MO, Oct. 1999.
7. S. Muller and X. Pfab, Considerations implementing a dual voltage power network, presented at Convergence Transportation Electronics Association Congress, Dearborn, MI, Oct. 1998.
8. J. G. Kassakian, The future of power electronics in advanced automotive electrical systems, in *Proc. IEEE Power Electronics Specialist Conf.*, 1996, 7–14.
9. C. C. Chan, An overview of electric vehicle technology, *Proc. IEEE*, 81(9), 1202–1213, 1993.
10. C. C. Chan and K. T. Chau, An overview of power elecetronics in electric vehicles, *IEEE Trans. Ind. Electron.* 44(1), 3–13, 1997.
11. A. Emadi, B. Fahimi, M. Ehsani, and J. M. Miller, On the suitability of low-voltage (42 V) electrical power system for traction applications in the parallel hybrid electric vehicles, *Soc. Automotive Eng. (SAE) J.*, paper 2000-01-1558, 2000.
12. M. Ehsani, K. M. Rahman, and H. A. Toliyat, Propulsion system design of electric and hybrid vehicles, *IEEE Trans. Ind. Electron.*, 44(1), 19–27, 1997.
13. J. M. Miller, P. J. McCleer, and J. H. Lang, Starter-alternator for hybrid electric vehicle: comparison of induction and variable reluctance machines and drives, in *Proc. 33rd IEEE Industry Application Society Annual Meeting*, Oct. 1998, 513–523.
14. J. M. Miller, P. R. Nicastri, and S. Zarei, The automobile power supply in the 21st century, presented at Power Systems World Conference and Exposition, Santa Clara, CA, Nov. 1998.
15. J. G. W. West, DC, induction, reluctance and PM motors for electric vehicles, *Power Eng. J.*, 77–88, April 1994.

22
Principles of Magnetics

Roman Stemprok
University of North Texas

22.1 Introduction 22-1
22.2 Nature of a Magnetic Field 22-1
22.3 Electromagnetism... 22-2
22.4 Magnetic Flux Density... 22-3
22.5 Magnetic Circuits 22-3
22.6 Magnetic Field Intensity.. 22-4
22.7 Maxwell's Equations 22-5
22.8 Inductance 22-7
22.9 Practical Considerations.. 22-7

22.1 Introduction

Although magnetism has been known since ancient times, the connection between electricity and magnetism was not discovered until the early 19th century. Electromagnetics is the study of electric and magnetic field behavior. These fields arise from charged particles both at rest and in motion, and they may exert forces on other charged particles and materials.

Hans Christian Oersted (a Danish scientist) demonstrated the relation between electricity and magnetism. In 1819, he showed that a compass needle could be deflected by a current-carrying conductor. Andrew Ampere (1775–1836) experimented with two current-carrying conductors and found that they repel or attract each other. He developed a concept to understand the electromagnetism that led to the development of transformers and electric generators.

The theory of electromagnetic fields was developed by James Clerk Maxwell (Scottish scientist) and published in 1865. His work was the culmination of a long series of experimental and theoretical research performed by a number of other scientists over the centuries. Maxwell published a set of equations that completely describe the electromagnetic field. He developed a unified theory of electromagnetism, and he also predicted the existence of radio waves. Heinrich Hertz (a German physicist) proved the existence of such waves early in the 20th century.

The electromagnetic force (emf) is one of the four known fundamental forces of nature. The emf is responsible for the functioning of a large number of devices that are important to modern civilization including radio, television, cellular telephones, computers, and electric machinery. The emf is caused by an electromagnetic field that is described by two quantities, the electric field and the magnetic field, both of which can vary in space and time.

22.2 Nature of a Magnetic Field

Magnetism can create a force between magnetic materials depending on their magnetic fields. Magnetic fields can be produced by moving charged particles in electromagnets (e.g., electrons flowing through a coil of wire connected to a battery) or in permanent magnets (spinning electrons within the atoms generate the field). Figure 22.1 shows Faraday's concept of the magnetic flux lines or lines of force on a

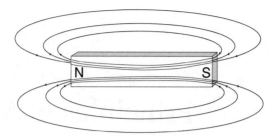

FIGURE 22.1 Magnetic field (Φ) of a bar magnet.

permanent bar magnet. The magnetic field is much stronger at the poles than anywhere else. The direction of the field lines is from the North Pole to the South Pole, and the external magnetic field lines never cross.

According to the molecular theory of magnetism, within permanent magnets there are tiny molecules or domains that can be considered micromagnets. When they line up in a row, they combine to increase the magnetic field strength. For example, in a normal piece of steel, the domains are arranged in random order having positive and negative poles scattered in all directions. When the steel is magnetized, the domains line up, allowing the whole piece of steel to act like one large magnet. By placing a magnet beneath a piece of paper and placing iron filings on top of the piece of paper, the iron filings will arrange themselves to look like the invisible magnetic force that surrounds the magnet. This invisible magnetic force, which exists in the air or space around the magnet, is known as a magnetic field and the lines are called magnetic lines of force, as shown in Fig. 22.1.

Ferromagnetic materials (e.g., iron, nickel, and cobalt) are those materials whose domains are capable of aligning to create a magnetic field. Because of this ability, they provide an easy path for external magnetic field lines. Elements and alloy substances differ in their ability to become magnetized by an external field (susceptibility). Materials can be strongly magnetized by the formation domains in which individual atoms that are weakly magnetic because of their spinning electrons align to form areas of strong magnetism. Magnetic materials lose their magnetism if heated to or above the Curie temperature. Other materials are mostly paramagnetic, that is, only weakly pulled toward a strong magnet. This is because their atoms have a low level of magnetism and do not form domains. Diamagnetic materials are the opposite of ferromagnetic materials; they are weakly repelled by a magnet since electrons within their atoms act as electromagnets and oppose the applied magnetic force. Antiferromagnetic materials have a very low susceptibility, which increases with temperature.

22.3 Electromagnetism

Electromagnetism is a magnetic effect due to electric currents. When a compass is placed in close proximity to a wire carrying an electrical current, the compass needle will turn until it is at a right angle to the conductor. The compass needle lines up in the direction of a magnetic field around the wire. It has been found that wires carrying current have the same type of magnetic field that exists around a magnet, as shown in Fig. 22.2. One can say that an electric current induces a magnetic field and the field is proportional to the current, I.

In Fig. 22.2, the "rings" represent the magnetic lines of force existing around a wire that carries an electric current, I. The magnetic field is strongest directly around the wire, and extends outward from the wire, gradually decreasing in intensity.

The direction of a magnetic field can be predicted by use of the right-hand rule. According to the right-hand rule, the right hand is placed around the wire that is carrying the current and the thumb follows the direction of current flow. Then the fingers will show the direction of the magnetic field around the conductor.

Principles of Magnetics

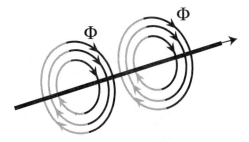

FIGURE 22.2 Magnetic field produced by current, *I*.

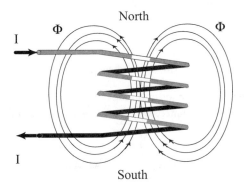

FIGURE 22.3 Magnetic field (Φ) produced by a coil.

Figure 22.3 shows a case where the wire is looped into a coil. A little magnetic field wraps around each wire and, by combining each wire turn, the coil magnetic flux (Φ) is created. It was found by experimentation that if a wire is wound in the form of a coil, the total magnetic field around the coil is magnified. This is because the magnetic fields of the turns add up to make one large flux flow, resulting in a magnetic field (Φ), shown in Fig. 22.3.

22.4 Magnetic Flux Density

Fig. 22.4 shows a ferromagnetic material where the most of flux is combined to the core. Only a small amount of the leakage flux escapes on the sides of the coil. The unit of flux (Φ) is the weber in the SI system. Flux density (*B*) is the magnetic flux per unit area. If the cross-sectional area (*A*) in SI units is m^2, then the flux density is in webers per m^2 (Wb/m^2), which is called tesla (T). Flux density (*B*) is defined by the total flux (Φ) passing perpendicularly through an area (*A*).

$$B = \frac{\Phi}{A} \quad (T) \tag{22.1}$$

22.5 Magnetic Circuits

Current (*I*) flowing through the coil shown in Figs. 22.3 and 22.4 creates the magnetomotive force ($\Im$). The greater number of turns, the greater will be the flux. The magnetomotive force, or mmf, is defined as

$$\Im = N \cdot I \quad \text{(ampere—turns, At)} \tag{22.2}$$

where *N* is the number of coil turns and *I* is the current passing through the wire. For example, a coil with 200 turns and 2 A will have an mmf of 400 At.

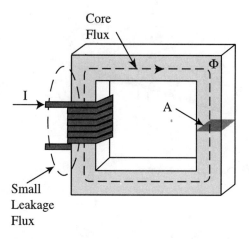

FIGURE 22.4 Magnetic flux density in a ferromagnetic material.

The magnitude of magnetic flux depends upon the opposition presented by the magnetic circuit. The opposition to the flux is called reluctance, which is similar to resistance in an electric circuit. The reluctance is defined as

$$\Re = \frac{\ell}{\mu A} \quad \text{(At/Wb)} \tag{22.3}$$

where ℓ is the length of the magnetic core and A is a cross-sectional area. The property of the material is characterized by its permeability, μ. Materials with high μ are called ferromagnetic materials, and the reluctance of those materials is low.

22.6 Magnetic Field Intensity

The magnetizing force, H, also known as magnetic field intensity, is the mmf per unit length. The magnetic field intensity is written as

$$H = \frac{\Im}{\ell} = \frac{NI}{\ell} \quad \text{(At/m)} \tag{22.4}$$

Equation (22.4) describes an ability of a coil to produce magnetic flux. If, for example, in Fig. 22.4, the coil has 1000 turns, the length of the magnetic path is 0.6 m and current through the conductor is 1 A, then the magnetic field intensity is 600 At/m.

The field intensity and the resulting flux density are related through the permeability. The flux density is

$$B = \mu H \quad \text{(T)} \tag{22.5}$$

where μ is core permeability. The core permeability is a material constant describing the level of the flux in a material. When the material constant (μ) is high, the flux density will increase. The permeability has units of webers per ampere-turn-meter in the SI system. The permeability of vacuum in free space is $\mu_o = 4\pi \times 10^{-7}$ (Wb/At-m). When magnetic flux propagates through magnetic media, other than vacuum, the flux density is

$$B = \mu_r \mu_o H \quad \text{(T)} \tag{22.6}$$

where μ_r is the relative permeability. The relative permeability is 1 for a vacuum and can reach 10,000 for ferromagnetic materials. Ferromagnetic materials have regions called domains of microscopic size.

Principles of Magnetics

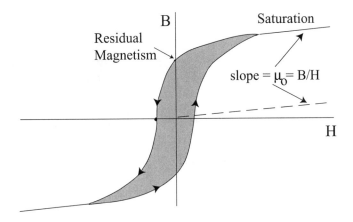

FIGURE 22.5 Hysteresis loop.

When they line up, the material is magnetized. Thus, one can increase the magnetic field until all domains are aligned, at which point the ferromagnetic material is incapable of contributing any more magnetic flux. At that point the material is saturated, as shown on the hysteresis curve in Fig. 22.5.

22.7 Maxwell's Equations

Maxwell's equations are the fundamental concept of electromagnetic (E-M) field theory. Using E-M field theory, one can calculate important quantities such as impedance, inductance, capacitance, etc. Maxwell's equations in differential form are as follows:

$$\nabla \times \bar{H} = J + \frac{\partial \bar{D}}{\partial t} \tag{22.7}$$

$$\nabla \times \bar{E} = -\frac{\partial \bar{B}}{\partial t} \tag{22.8}$$

$$\nabla \cdot \bar{D} = \rho \tag{22.9}$$

$$\nabla \cdot \bar{B} = 0 \tag{22.10}$$

where H is the magnetic field intensity (A/m), D is the electric flux density (C/m^2), B is the magnetic flux density (T), J is the current density (A/m^2), ρ is the charge density (C/m^3), and E is the electric field intensity (V/m). The other relations to Maxwell's equations are

$$\bar{D} = \varepsilon \bar{E} = \varepsilon_r \varepsilon_o \bar{E} \tag{22.11}$$

$$\bar{B} = \mu \bar{H} = \mu_r \mu_o \bar{H} \tag{22.12}$$

$$\bar{J} = \sigma \bar{E} \tag{22.13}$$

where ε_o is the permittivity of free space (8.854×10^{-12} F/m) and ε_r is the relative dielectric constant for a given material. In the second equation, μ_o is the permeability of free space ($4\pi \times 10^{-7}$) and μ_r is the relative permeability for a given material. In the third equation, σ is the electric conductivity (Siemens).

Maxwell's equations can be rewritten into integral form using mathematical tools, such as Stokes' theorem and the divergence theorem, and then they look as follows:

$$\text{Faraday's law:} \quad \oint_s \bar{H} \cdot d\bar{\ell} = I + \int_s \frac{\partial \bar{D}}{\partial t} \cdot d\bar{s} \tag{22.14}$$

$$\text{Ampere's law:} \quad \oint_s \bar{E} \cdot d\bar{\ell} = -\int_s \frac{\partial \bar{B}}{\partial t} \cdot d\bar{s} \tag{22.15}$$

$$\text{Gauss' law:} \quad \oint_s \bar{D} \cdot d\bar{s} = q_{\text{encl.}} = \int_{\text{vol}} \rho \, dv \tag{22.16}$$

$$\oint_s \bar{B} \cdot d\bar{s} = q_{\text{encl.}} = 0 \tag{22.17}$$

Michael Faraday (1791–1867) experimented with magnetic fields. He wound two coils on an iron ring. When one coil was powered from a battery, he noticed that a transient voltage developed on the second coil and, later, this led him to develop a transformer. Faraday concluded from his observations that voltage is induced in a circuit whenever the linking flux in the magnetic circuit changes. He also discovered a formula describing the change of a magnetic flux, which is proportional to the voltage change across the coil. Equation (22.14) shows the Faraday law in the integral form derived from Maxwell's equations. The first right-hand term of this equation is regular current flowing in a wire, and the second term is due to capacitive coupling called the "displacement current." In most cases the displacement current is neglected.

An example of a magnetic field density at a distance r of a long straight wire is shown in Fig. 22.6. One can assume $I = 100$ A, $r = 1$ m, and $\mu = \mu_o$. Using Eq. (22.14) one can write

$$\oint \bar{H} \cdot d\bar{\ell} = I$$

and take the integral around the wire at a constant distance r.

Integrating over a circular path,

$$\oint \bar{H} \cdot d\bar{\ell} = 2\pi r H_\Phi = I$$

and then,

$$B_\Phi = \frac{\mu I}{2\pi r} = \frac{100 \times 4\pi \times 10^{-7}}{2\pi \times 1} = 2 \times 10^{-5} \quad (\text{T})$$

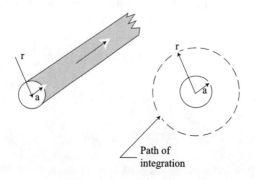

FIGURE 22.6 Long straight wire carrying a current, I.

Principles of Magnetics

In this example magnetic field lines form circles around the wire, as shown in Fig. 22.2. The magnetic field intensity decreases with distance from the wire.

By Faraday's law, the induced voltage is proportional to the rate of the magnetic flux change times the number of turns in a coil. In calculus notation, the rate of flux change times the number of turns is

$$e = N\frac{d\Phi}{dt} \quad (V) \tag{22.18}$$

where e is in volts, Φ is webers, N is the number of turns, and t is in seconds. For example, if the magnetic flux changes at a rate of 1 Wb/s in a single turn coil, the induced voltage is 1 V.

22.8 Inductance

The induced voltage in a coil is proportional to the magnetic flux change that is shown in Eq. (22.18). It is also known that the induced current is proportional to the current. Thus, one can write that the induced voltage is proportional to the current change,

$$e = L\frac{di}{dt} \quad (V) \tag{22.19}$$

where L is the inductance of the coil. The unit of inductance is the henry in the SI system. In practice, in electrical circuit calculations, the voltage across the inductance is denoted by v_L rather than e. Then, one can rewrite Eq. (22.19) as

$$v_L = L\frac{di}{dt} \quad (V) \tag{22.20}$$

22.9 Practical Considerations

Faraday's law states that a time-varying magnetic field through a surface bounded by a closed path induces a voltage inside a conductor loop. This fundamental principle has important consequences for parasitic coupling in electronic hardware, such as a noisy circuit on a printed circuit board of a computer. The problem is a significant time-varying magnetic field as a result of current levels in different conductive traces. Some of the traces can make current loops. The magnetic field generated by the loop area "looks" for other loops—the circuits where it can induce a voltage across the load. We consider the target loop area when looking at its mutual inductance, and the target loop area is determined by the closed current path in the victim circuit. The magnitude of induced voltage in the victim circuit depends on various factors:

- Current magnitude in the first circuit
- Source and load circuit impedances
- Size of the loop area in the victim circuit
- Distance between the victim and source circuits
- Orientation of the loops

This coupling can cause faulty operation of the victim circuit.

23
Computer Simulation of Power Electronics

Michael Giesselmann
Texas Tech University

23.1 Introduction 23-1
23.2 Code Qualification and Model Validation 23-2
23.3 Basic Concepts—Simulation of a Buck Converter 23-3
23.4 Advanced Techniques—Simulation
       of a Full-Bridge (H-Bridge) Converter 23-10
23.5 Conclusions 23-22

23.1 Introduction

This chapter discusses the possibilities and limitations of computer simulations for power electronics systems. Obviously, advances in raw processing power for personal computers as well as the rapid development of electronic design software have influenced the field of power electronics. In this context, electronic design software means any software used for schematic capture, circuit board layout, electrical or thermal simulation, documentation, and other applications. From the very beginning, schematic capture and circuit board design software was used for power electronics systems. Of course, by their very nature, schematic capture and layout programs had graphical user interfaces. However, long before the advent of graphical user interfaces, electronic circuits were simulated by means of computers, mainly using variations of the circuit simulation code SPICE.

SPICE, an abbreviation for **S**imulation **P**rogram with **I**ntegrated **C**ircuit **E**mphasis [14], was developed in the 1970s at the University of California at Berkeley. The initial motivation for the creation of the SPICE code was the simulation of analog electronic circuits to create integrated circuits (ICs). SPICE solves the fundamental differential equations governing electric circuits containing basic R, L, C elements and voltage (V) and current (I) sources, which can be fixed or dependent. Electronic parts, such as diodes, transistors, etc., are either implemented as native elements with equations appropriate to their nature or modeled via subcircuits containing basic and native electronic elements. Device equations are typically based on semiconductor theory and refined using semiempirical parameters.

However, the use of SPICE or similar codes for the simulation of power electronics systems proved to be difficult from the outset, because power electronics circuits typically operate in a highly discontinuous mode, with power semiconductor devices acting as almost ideal switches. The simulators typically could not follow the sudden switching transitions and would become unstable and crash. In addition, typically only transient (time domain) analyses could be performed. If the transient analysis was at all stable, it typically had to be run with very small time steps, resulting in long run times and huge output data files. Other types of analyses, such as AC (frequency domain) analysis, were not possible. For AC analysis, the circuit response is linearized around a bias point, and the small signal behavior is analyzed for a range of frequencies. Typical results are the well-known Bode plots, which have proved to be very useful for the design of feedback control loops. Of course, if a normal power electronics system has several switches, which constantly turn on and off, a single bias point cannot be found and AC analysis will fail. To make matters worse, some circuit codes will perform AC analysis anyway and give totally erroneous results.

This chapter discusses techniques to overcome these problems. With these techniques, even complex power electronics circuits can be simulated, their behavior can be studied in both the time and frequency domain, and simulation can finally be fully integrated into the electronic design process.

In the following, the possibilities and limitations of simulation tools are discussed in more detail. Computer simulation of electronic circuits in general and power electronics circuits in particular has many obvious advantages, such as:

- New topologies can be quickly tested.
- New control strategies can be studied before implementation.
- Existing topologies can be analyzed for normal and fault conditions.
- Tests can be performed safely and quickly without risk of harm for personnel or equipment.

In addition, mechanical systems such as motors and mechanical attachments can be included in the simulation of power electronics systems, thus enabling the simulation of complete mechatronics systems.

Before proceeding farther it should be acknowledged that some limitations remain, which can only be overcome in special cases and with considerable effort involving extensive fine-tuning of models from experimental data. These limitations involve the details of the switching transitions. These details include the precise transients for voltages and currents in the switching devices, including peak voltage overshoot, etc. To model these transitions precisely, which typically occur in the nanosecond time frame, not only exact models for the semiconductors are necessary, but also the parasitic circuit elements, such as the inductance of the device packages and the circuit connections, must be known and accounted for in the simulation setup. Furthermore, the precise transient traces of the control signals in the nanosecond time regime must be known and implemented.

For the above-mentioned reasons, the author does not recommend use of a simulation to verify that, for example, a certain voltage stress level on an IGBT transistor in an inverter is not exceeded. Similarly, the precise amount of switching (unlike conduction) losses is difficult to predict from a simulation. This is better left to experimental work in the laboratory. However, with the exception of a very narrow time window around individual switching transitions, the response of the circuit is very realistic. The reader should recall that circuits are typically in transition for less than 1% of total time. Therefore, the voltage and current levels in all inductors and capacitors are typically within less than 1% of the real values.

In conclusion, simulation is a great tool to study the behavior of new and existing circuits including mechanical energy conversion devices and control systems with the possible exception of a very narrow window around the switching transitions.

23.2 Code Qualification and Model Validation

Before software of any kind is used as part of a design process or in support of a comprehensive analysis of an existing system, care should be taken to ensure that the software is working correctly for the intended application. It should be pointed out that most software will work correctly for the purpose that it was designed for, but sometimes software can easily be used (or misused) in ways or for applications for which it was not intended. To make matters worse, the fact that some software should not be used for a particular problem may not be so obvious to the user. The reader should be reminded that SPICE was initially created to support the design of integrated circuits. Therefore, all basic elements are ideal and zero dimensional, meaning that a resistor has no parasitic inductance associated with it and has no propagation delay. Similarly, an inductor has no losses and no propagation delay nor any parasitic capacitance. Nevertheless, SPICE turned out to be a code that could be used for general circuit analysis and for many applications not imagined at the outset. However, every prudent engineer or engineering supervisor should always try to evaluate a computer code using a typical example with known behavior and carefully compare the simulation results with the known (measured) facts about the circuit. In this phase of code qualification, the engineer should also consult the accompanying documentation for background information about the code, its intended uses and limitations, and the internal workings of the simulation engine. This may often give important hints to the fidelity of the results of a particular application.

Close attention should also be paid to the device models that may be contained inside a particular code and their features and limitations. For example, it may be important to know if the model for a transformer uses nonlinear magnetics or not. If the code (as PSpice® and many others) allows it, custom models that have the properties needed for a given case can be added. However, in this case, the models should be carefully tested and validated before they are used, especially if critical engineering decisions are to be based on the results. The reader should also be cautioned that after an (ever more frequent) upgrade of a particular code, it is advisable to at least perform some sort of check to determine that the core of the simulation engine still behaves like before. For this purpose, the input files for a (not too simple) benchmark case should be retained along with a documentation of the output from previous versions of the code. It should also be mentioned that even "bug-fixes," "Internet-patches," or "code maintenance" can potentially cause a simulator to behave differently. (All of those things have happened to the author over the years). Sometimes the user may not even know that upgrades or the like have taken place, if software maintenance is performed by the information technology (IT) department of a company. In any event, it is always advisable to scrutinize the results of any simulation, compare them against known facts and expectations, and resolve any discrepancies.

23.3 Basic Concepts—Simulation of a Buck Converter

In the following, the simulation of a buck (step-down) converter is described to illustrate the concepts mentioned in the introduction. Good references for power electronics circuits in general are References 2, 6, and 9. The simulations have been performed using the SPICE implementation called PSpice, which was developed by MicroSim Corp., and is currently sold by Cadence, Inc [8]. As far as possible, the examples shown here can be run on the student version of the software. The examples are created using the "Schematics" editor, which provides a convenient graphical user interface.

Figure 23.1 shows the well-known topology of a switch mode, step-down (also called buck) converter. With the chosen values for the components, the converter is operating in the continuous-conduction mode (as referred to the inductor current), and the output voltage is equal to the input voltage multiplied by the duty cycle of the power MOSFET "M1."

This circuit uses only standard elements from the library of the student version. (In a real circuit a diode other than the 1N4002 would be used.) The key to a stable and quick simulation is the drive signal for the power MOSFET. The hierarchical block called "PWM-Generator" accomplishes this task. The input to this block is a voltage between 0 and 1 V, representing a duty cycle between 0 and 100%. The output

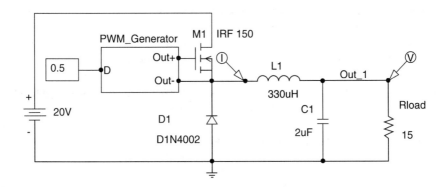

FIGURE 23.1 Schematic of a step-down converter.

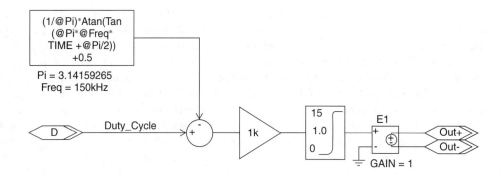

FIGURE 23.2 Schematic representing the "PWM Generator" hierarchical block of Fig. 23.1.

is a rectangular voltage, which is available between the outputs "Out+" and "Out−," which has an amplitude of 15 V, a duty cycle specified by the input and a repetition frequency, which can be freely chosen. In addition, the switching transitions of this rectangular waveform have controllable slopes with smooth edges to keep the simulator from crashing. Here is used a concept that was explained in the introduction, stating that in the interest of stable operation, short run times, and manageable output file size, it is not only permissible but recommended to replace the actual drive signal with one that is more suitable for simulation. Careful examination of the drive signals shows only very minute differences as the result of this substitution, but the advantages for stability and run times are enormous. Also as mentioned in the introduction, the total time that the circuit remains in transitions is very small. Therefore, the output voltage and the inductor current of this example are completely realistic.

For this example, the drive signals are generated entirely with so-called analog behavioral modeling (ABM) components, which have no counterpart in the real circuit. In Section 23.4 a realistic model for a real MOSFET driver circuit is presented, which also creates suitable gate drive signals.

Figure 23.2 shows the circuit that implements the PWM signals using the techniques discussed above. This circuit is an implementation of the carrier-based PWM generation method with PSpice® ABM parts. In the carrier-based PWM generation method, a voltage level, representing the duty cycle, is compared with a triangular or sawtooth-shaped carrier. A convenient way to generate such a carrier without resorting to mathematical functions with piecewise definition is to calculate the argument of a periodic trigonometric function. This is illustrated in Fig. 23.3. This figure was created using the MathCAD® [4] software package. The circuit in the upper half of Fig. 23.4 shows the implementation of a sawtooth function using basic ABM parts in PSpice in more detail.

In the lower part of Fig. 23.4, a more-compressed form with only one ABM part is shown, which generates the same output. Using the compressed form not only results in space savings on the "Schematics" page, but also reduces the total device count for the simulation. This can easily make the difference between being able to run a circuit within the limitations of the student version or not. The circuit shown in Fig. 23.2 compares the sawtooth signal with the duty cycle and amplifies the difference by a factor of 1000 (1 k) using a "Gain" device. The amplification factor controls the steepness of the transitions in the PWM signal. A soft limiter on the output of the amplifier limits the signal amplitude to the range of 0 to 15 V.

The soft limiter uses a hyperbolic tangent function to achieve its function. To illustrate this, Fig. 23.5 shows a MathCAD [4] plot of a hyperbolic tangent function for different steepness factors k. In fact, the steepness factors are just multipliers for the argument of the function. From Fig. 23.5 it is easy to see how a transition can be achieved that is steep but has rounded corners without abrupt slope changes at the same time. These signal properties are the key to a fast and stable operation of the simulator. The last element in Fig. 23.2, named "E1," is a voltage-controlled voltage source. It takes the output of the soft limiter, which is a voltage with respect to ground, and creates a voltage with a floating reference potential for driving high-side MOSFETs such as in buck converters.

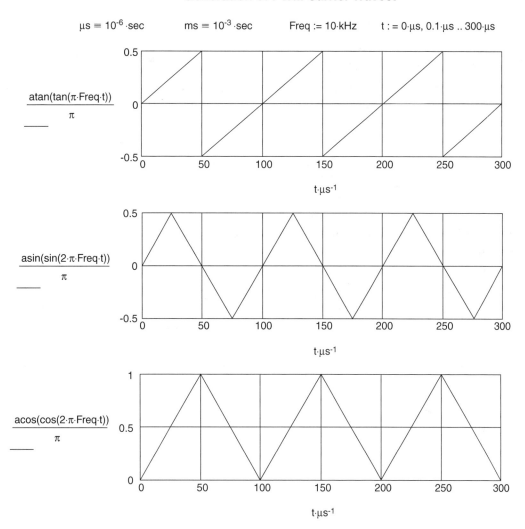

FIGURE 23.3 Illustration of the mathematical functions used for carrier wave generation.

Figure 23.6 shows the simulation results for the buck converter shown in Fig. 23.1. The simulation shows a start-up event, where a gate signal with a duty cycle of 50% is suddenly applied to MOSFET "M1" while both the inductor current as well as voltage on the output capacitor are zero. The upper half of Fig. 23.6 shows the trace of the output voltage, whereas the lower half of the graph shows the inductor current. It can be seen that both the output voltage (10 V) and the average output current (10 V/15 Ω) are represented correctly in Fig. 23.6. Since the input voltage is twice as high as the output voltage and the losses (occurring only in the MOSFET and the diode) are minimal, the average input current is half the output current. Because of the chosen gate signal generation, the simulation runs stable and fast, especially if the high switching frequency of 150 kHz is considered, which was chosen for this example.

Considering the fact that for the buck converter the ratio of the input and output voltages is proportional to the duty cycle D and the ratio of the average input and output currents is inversely proportional to D, the buck converter is acting as a transformer for DC. As in an AC transformer, the product of output voltage and the average output current is nearly identical to the product of the input voltage and the average input current. Of course, if no losses were present, the products would be precisely identical.

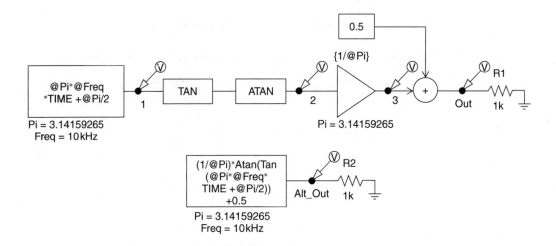

FIGURE 23.4 PSpice implementation of the sawtooth function.

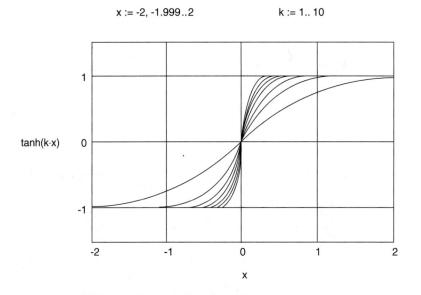

FIGURE 23.5 Hyperbolic tangent function with different steepness factors k.

This is true for both the continuous, as well as the discontinuous-conduction mode (referring to the current in the inductor), but in the latter case the dependence of the voltage and current ratio on the duty cycle D would be more complicated.

This behavior can be modeled in such a way that the switching elements in the circuit are replaced by an analog element, which is controlled by the duty cycle D. This element would create the same average voltages and currents that are present in the real circuit. However, since no actual switching takes place, the time step for the simulator can be increased dramatically, and the simulation could potentially run faster by a factor of 100 or more depending on the switching frequency of the original circuit. The reason is that, for a simulation of a circuit with switching elements, the time step (or, better, the time step ceiling, since the time step is adjusted dynamically in many simulators such as PSpice) must be small enough to ensure that the simulation can accurately follow the individual switching events. If the time step ceiling is too big, the simulator will try to finish the simulation run as fast as possible and internally select a time step that is just small enough so that the simulator remains stable. Remaining stable, however, does

Computer Simulation of Power Electronics

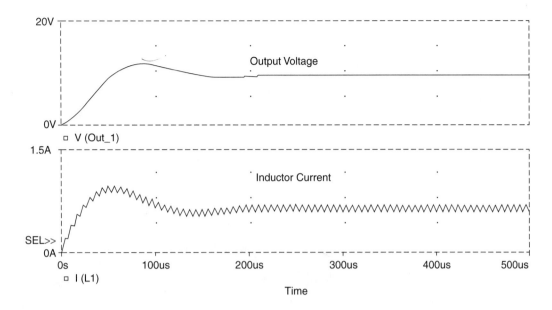

FIGURE 23.6 Simulation results for the buck converter shown in Fig. 23.1.

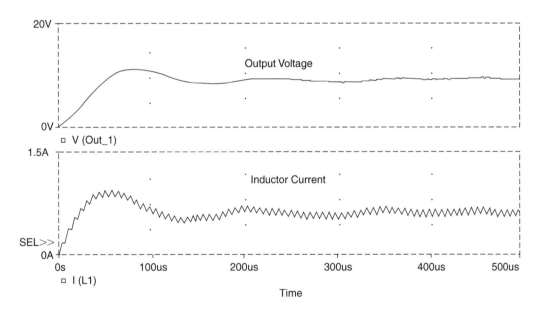

FIGURE 23.7 Incorrect simulation results for the buck converter due to improper time step settings.

not mean that the results are accurate. The size of the next time step is always predicted from the slope of the waveform just prior to the current time. If a step ceiling is set and the time step, which the simulator would choose by itself, is bigger than the time step ceiling, the time step ceiling is used instead. The choice of the proper time step ceiling requires some experience and experimentation. Figure 23.7 shows an example of a simulation that was run with a time step that is too large. It was obtained by rerunning the circuit shown in Fig. 23.1 with a different time step setting. Therefore, Fig. 23.7 can be directly compared with Fig. 23.6. It is obvious that the waveform for the inductor current in Fig. 23.7 is irregular and exhibits oscillations after the initial transient (after about 200 μs). These oscillations are caused by integration errors due to the wrong time step settings. Obviously in this example there is no reason for

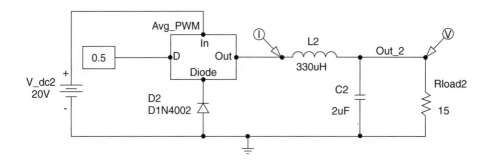

FIGURE 23.8 Buck converter with time-averaged PWM switch.

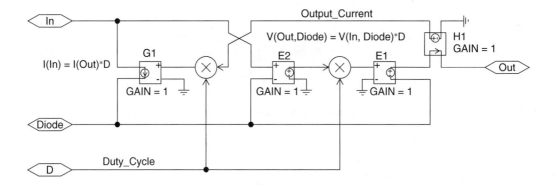

FIGURE 23.9 Subcircuit for "Avg_PWM" block.

any oscillation after the initial transient, since the circuit is run with a constant duty cycle. However, if an external feedback control system for the output voltage is present, such oscillations could occur as a result of the control action of the system, which constantly changes the duty cycle to keep the output voltage at a given value. In a case like this, considerable experience and good engineering judgment are required to avoid the wrong interpretation of the simulation results.

As mentioned above, the use of a simulation model with an analog switch replacement could be advantageous in such a case [1]. An example is shown in Fig. 23.8. A comparison with Fig. 23.1 shows that the power MOSFET "M1" has been replaced by a hierarchical block called "Avg_PWM." The associated subcircuit is shown in Fig. 23.9.

This subcircuit takes the voltage between the terminals "In" and "Diode" measured by the device "E2" and scales it with the duty cycle D. The output is provided between the terminals "Out" and "Diode." It should be noted that the "Diode" terminal is virtually at ground potential (about 0.7 V below due to the forward voltage of the diode) and the diode is not really needed for the operation of the circuit shown in Fig. 23.8.

The device "H1" measures the output current coming from the terminal "Out" and scales the value with the duty cycle D. The device "G1" will pull the scaled output current from the "In" terminal. This will implement the DC-transformer equations mentioned above. Figure 23.10 shows a comparison of the simulation output of the circuits shown in Figs. 23.1 and 23.8. It can be seen clearly, that the output of the circuit with the average PWM switch represents the "instantaneous average" (short-term average, taken over one switching cycle). In fact, if the switching frequency of the converter from Fig. 23.1 were raised high enough, the traces for both converters would be identical. This is already evident if the traces for the output voltage in Fig. 23.10 are compared since the output voltage of the switching converter has very little ripple at the chosen switching frequency of 150 kHz. Mohan [7] extends the DC-transformer approach for time-averaged modeling of H-bridge converters for motor drives.

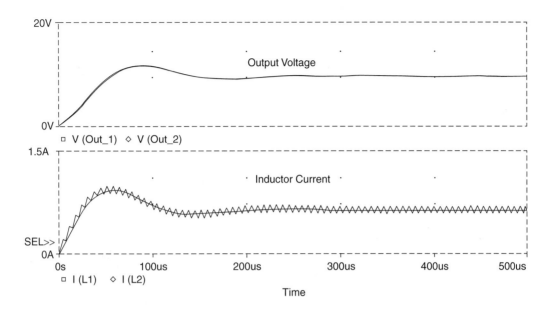

FIGURE 23.10 Combined simulation results for the buck converters from Figs. 23.1 and 23.8.

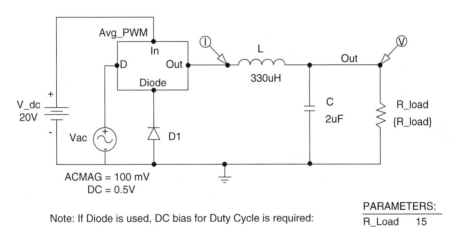

FIGURE 23.11 Simulation circuit for performing AC analysis for the buck converters from Fig. 23.8.

Besides the obvious benefit of faster simulation times, the added benefit of the buck converter with the average PWM switch is that AC or frequency domain analysis can be performed. A simulation setup for this is shown in Fig. 23.11. Here the buck converter is fed with a 50% duty cycle bias with a 10% (100 mV) AC component on top of it. The frequency of the AC component is swept from 100 mHz to 100 kHz for five different load resistors, 5W, 10W, 20W, 30W, and 40W. The result is shown in Fig. 23.12. In the upper portion of the diagram, the AC response of the output voltage is 2 V up to about 1 kHz (10% of 20 V input due to 10% AC amplitude). Above 1 kHz, the resonant peak of the LC-output filter is clearly visible for the 30-Ω load, which represents the smallest damping. Ref. 1 shows how the subcircuit in Fig. 23.9 can be used for other basic converters as well.

To model a more complex circuit, such as an H-bridge with a DC motor connected to it, in the frequency domain, each half bridge can be modeled as a DC-transformer with a transformation ratio that is controlled by the duty cycle as described in Ref. 7. As an alternative, the complete H-bridge could be modeled as a linear gain-block with the duty cycle the input and the output voltage of the H-bridge the output. This is realistic for the design of feedback control systems. In fact, H-bridge inverters for

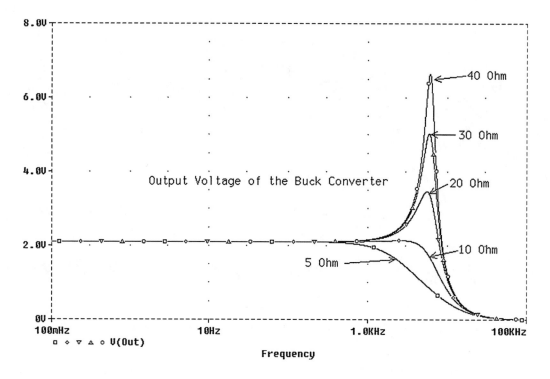

FIGURE 23.12 Simulation results for AC analysis of the buck converter from Fig. 23.11.

motor drives are often called servo-amplifiers for this reason. Some latency in the response of the amplifier could be included in the system model by adding a low-pass filter on the input.

23.4 Advanced Techniques—Simulation of a Full-Bridge (H-Bridge) Converter

In this section some advanced simulation techniques are shown using an H-bridge inverter with complementary MOSFETs. This example is part of one of the author's ongoing development projects. The goal is to build a small and efficient controller for low-voltage, high-current DC motors to be used in robotics applications. One of the design goals is the use of state-of-the-art surface-mount devices and to integrate circuit simulation into the overall design process. Figure 23.13 shows a first conceptual study for the H-bridge, which was realized entirely with parts from the PSpice student version. The upper MOSFETs ("M1" and "M3") are p-channel devices, whereas the lower ones ("M2" and "M4") are n-channel types. Therefore, and because the supply voltage is very low, no high-side drivers are needed for MOSFETs "M1" and "M3." The "PWM_Generator" is the same that was previously used, except the switching frequency is lower. Because of the well-formed signals from the PWM generator, the simulation is stable and fast. In this example bipolar switching is used, where the duty cycle controls both the polarity and the magnitude of the load current. In this mode, MOSFETs "M1" and "M4" are switched on alternating with MOSFETs "M2" and "M3."

The results of the simulation are shown in Fig. 23.14. The load, which in the final application is a DC motor with brushes, is acting as a low-pass filter for the output current. Therefore, the load current has only a relatively small amount of ripple even though the output voltage is an unfiltered PWM waveform, as shown on the upper half of Fig. 23.14. This simulation was performed to test the concept of driving both MOSFETs from the ground potential and without any special provisions for blanking time to prevent conduction overlap. The conclusion that blanking time is not needed is, however, somewhat risky, because of the previously discussed limitations for the precision of the results during switching transitions.

Computer Simulation of Power Electronics

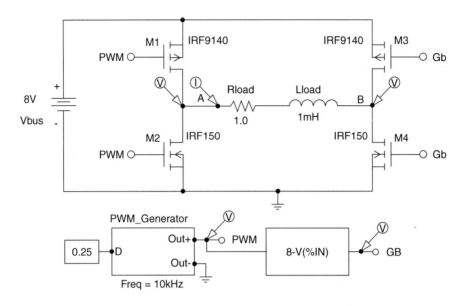

FIGURE 23.13 H-bridge with complementary MOSFETs for a low-voltage, high-current motor drive.

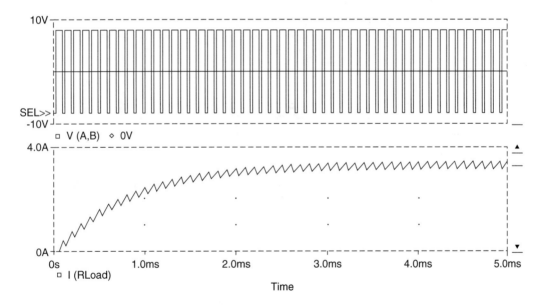

FIGURE 23.14 Diagonal voltage (upper trace) and load current (lower trace) from the circuit in Fig. 23.13.

Also, the MOSFETs used in this circuit are typically packaged in TO-220 style cases, which are not favored for the envisioned application. Therefore, to enhance the realism of the simulation study, MOSFET pairs "M1/M2" and "M3/M4" have been replaced by a custom part (NDS8858HCT from Fairchild), which represents a complementary half-bridge device packaged in a space-saving SO8 case. The new circuit is shown in Fig. 23.15. The custom part has all eight terminals of the real device and has a package definition for an SO8-type footprint, which is suitable for compact PC-board layout. In addition, the device has a "TEMPLATE" attribute, which makes it functional for simulation. During the process of "netlisting," which precedes the simulation, the "TEMPLATE" attribute generates a netlist entry. This netlist is then used as the actual input file for the simulator. Netlisting is therefore comparable with compilation of a program written in a high-level programming language. An alternative to creating a

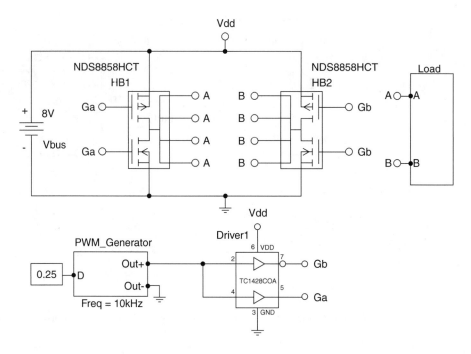

FIGURE 23.15 H-bridge circuit with custom parts for the half-bridge and the MOSFET driver.

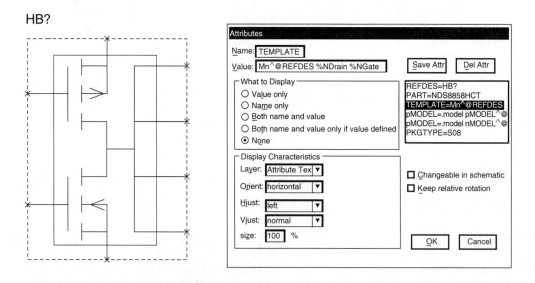

FIGURE 23.16 Symbol editor view of the NDS8858HCT complementary MOSFET half-bridge.

netlist entry via the "TEMPLATE" attribute is the creation of a subcircuit like the one shown in Figs. 23.2 or 23.9. However, if the custom part is also to be used for the generation of a PC-board, the "TEMPLATE" approach is better. Otherwise, each part in the simulation subcircuit must be listed as "SIMULATIONONLY" to prevent its inclusion on the PC-board.

Computer Simulation of Power Electronics

```
Symbol NDS8858HCT
Definition Complementary MOSFET Half Bridge
@attributes
REFDES=HB?
PART=NDS8858HCT

TEMPLATE=Mn^@REFDES %NDrain %NGate %V- %V- nMODEL^@REFDES
\n@nMODEL

\nMp^@REFDES %PDrain %Gate %V+ %V+ pMODEL^@REFDES
\n@nMODEL

\nR1^@REFDES %Vout5  %Pdrain   1u
\nR1^@REFDES %PDrain %Ndrain   1u
\nR1^@REFDES %NDrain %Vout8    1u

pMODEL=.model pMODEL^@REFDES PMOS
\n+(Level=3 W=1 L=2u Vto=-1.8 Rd=65m Cbd=1n Cgso=1n Cgdo=1n)

nMODEL=.model nMODEL^@REFDES NMOS
\n+(Level=3 W=1 L=2u Vto=1.8 Rd=35m Cbd=1n Cgso=10n Cgdo=1n)

PKGTYPE=SO8
```

FIGURE 23.17 Attributes of the NDS8858HCT complementary MOSFET half-bridge.

In PSpice custom parts can easily be created using the built-in symbol editor. It is initiated from within the graphical "Schematics" editor. The custom symbols are saved in a file with an ".slb" extension. Figure 23.16 shows the view of the NDS8858HCT complementary MOSFET half-bridge part from the symbol editor. On the right side, the window for symbol attribute entry is shown. All the attributes for this part are listed in Fig. 23.17. The "TEMPLATE" attribute creates the simulation model for the part. It inserts five parts into the PSpice netlist, a p-channel and an n-channel MOSFET (Mn^@REFDES... and Mp^@REFDES...) as well as three "dummy" resistors to connect all four output terminals together. Note that the resistors have a value of 1 $\mu\Omega$ and are essentially short circuits. However, they are needed for the syntax of the model description if all four output terminals are to be connected to the center junction of the MOSFETs. In the expression "Mn^@REFDES," "M" stands for MOSFET, "n" designates the beginning of the name for the n-channel device, and "^@REFDES" inserts the path to the part including all subcircuit names as well as the name of the "Reference Designator." This is done to create a unique name for each part. The expression "\n" inserts a new line into the netlist. "\n+" inserts a new line into the netlist and places a "+" at the beginning of the new line to create a multiline expression. For clarity, a new line is inserted in the "TEMPLATE" listings in Figs. 23.17, 23.21, 23.25, and 23.27 whenever a new line designator is encountered.

The expressions for "pMODEL" and "nMODEL" specify the characteristics for the p- and n-channel MOSFETs. The data for the Rds_{on} value (Rd=) and turn on voltage (Vto=) were taken from the data sheet of the NDS8858HCT. For more advanced requirements of model fidelity, the "PARTS" program in the PSpice group can be used to determine the model parameters if they are not otherwise (Internet) available. Today, many parts manufacturers put SPICE models on their Web sites. Figure 23.18 shows a screen from the "PARTS" program using the IRF150 MOSFET as an example. In this case, the forward conduction characteristics are studied. The parameters, which are influencing the forward conduction performance (RD), are identified by an asterisk. The parameter(s) can be adjusted to match a given behavior and the performance can be displayed graphically. An example is shown in Fig. 23.19. This screen can be called by clicking on the icon under "Plot" in Fig. 23.18.

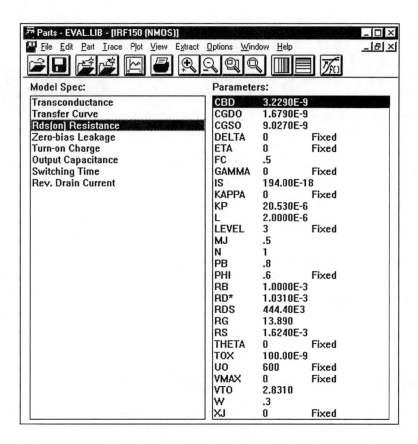

FIGURE 23.18 Screen from the "Parts" program for model parameters of the IRF150 MOSFET.

In addition to the custom part for the half-bridge, another custom part, implementing a MOSFET driver (TCI428COA from Telcom) packaged in an SO8 case, was created. For this part, a simulation model as well as a PC-board definition was included. The simulation model was created using ABM devices (basically voltage-controlled voltage sources with many options to define the control functions), to closely match the information in the data sheet. Again, the simulation model is contained in the "TEMPLATE" attribute. Figure 23.20 shows the view of the TCI428COA driver in the symbol editor. The pins 1 and 8 are marked as not included in the simulation (drawn with an interruption) since they are "not-connected" (NC) in the real part.

Voltage-controlled voltage sources, using a conditional statement—IF (Condition, Output for True, Output for False)—as their control function, have been used to implement a model for this part. The full set of attributes is shown in Fig. 23.21. The model compares the input voltage with the switching threshold (3 V) and switches between "GND" and "VDD" ±25 mV when it is crossed. A 10-Ω resistor models the output resistance for each channel. The threshold and saturation values as well as the output resistance were taken from the data sheet.

Figure 23.22 shows the simulation output of the H-bridge shown in Fig. 23.15 using the special parts described above. The output resistance of the driver in conjunction with the gate capacitance of the MOSFETs achieves smooth gate drive waveforms, which keep the simulator fast and stable and ensure realistic rise and fall times of the gate drive signals at the same time. In this circuit, the duty cycle controls both the direction and magnitude of the load current. This switching scheme is called bipolar switching. One of its disadvantages is that, without a control signal, the load (motor) is always driven at full power in one direction. Therefore, the circuit shown in Fig. 23.23 was developed. It uses a four-channel driver as

Computer Simulation of Power Electronics

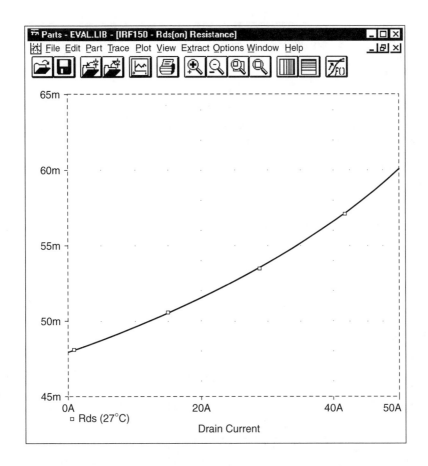

FIGURE 23.19 Graphical representation of the forward conduction of the IRF150 MOSFET.

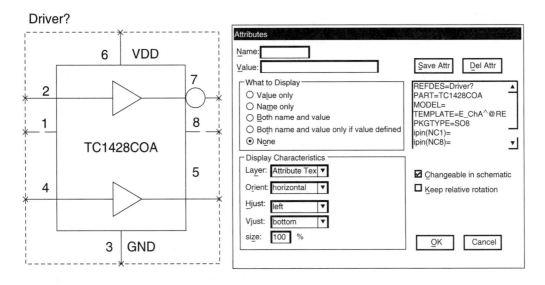

FIGURE 23.20 Symbol editor view of the TC1428COA MOSFET driver.

```
Symbol TC1428COA
Definition Complementary MOSFET Driver SO8
@attributes
REFDES=Driver?
PART=TC1428COA
MODEL=

TEMPLATE=E_ChA^@REFDES OutA^@REFDES,%GND VALUE {IF(V(%InA,%GND)>3.0V
V(%GND) + 25mV, V(%VDD,%GND)-25mV)}

\nE_ChB^@REFDES OutB^@REFDES,%GND VALUE {IF(V(%InB,%GND)>3.0V,
V(%VDD,%GND)-25mV, V(%GND)+25mV)}

\nR_OutA^@REFDES OutA^@REFDES,%OutA @Rout
\nR_OutB^@REFDES OutB^@REFDES,%OutB @Rout

PKGTYPE=SO8
ipin(NC1)=
ipin(Nc8)=
Rout=10.0
```

FIGURE 23.21 Attributes of the TC1428COA driver.

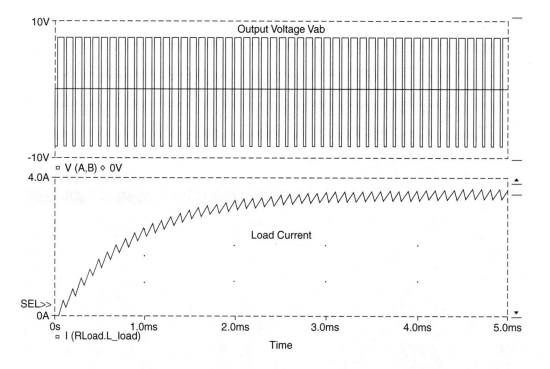

FIGURE 23.22 Simulation output for the circuit shown in Fig. 23.15.

well as an opto-coupler for isolation. In this circuit, one of the upper MOSFETs is turned on permanently for a particular direction, while the diagonally opposed (lower) MOSFET controls the speed. As an additional feature, when a lower MOSFET turns off, the MOSFET above it (which is normally not conducting for the selected direction) is turned on to shunt the current away from the body diode of the upper MOSFET and improve the efficiency. This is also referred to as synchronous rectification.

Computer Simulation of Power Electronics

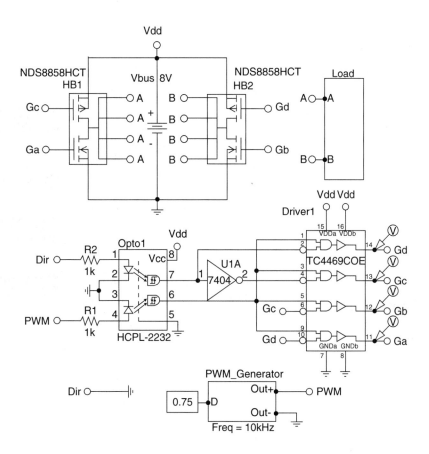

FIGURE 23.23 Advanced H-bridge circuit with synchronous rectification.

Figure 23.24 shows the view of the TC4469COE MOSFET driver from inside the symbol editor on the left side and the results of a test to check the device logic on the right side. The implementation of the simulation model is very similar to the implementation of the model for the TC1428COA driver. In fact, the "TEMPLATE" attribute from the TC1428COA was copied and then edited to implement the logical AND gates, which have one inverted and one noninverted input. The logical AND function was realized using the "&" symbol in the conditional statement inside of the IF function that controls the output voltage for each channel. Output resistors on all four outputs complete the model description. A complete list of all attributes, with the precise details of the implementation of the four-channel driver, is given in Fig. 23.25.

The correctness of the logic functions of the TC4469COE driver can be verified by inspection of the picture in the right half of Fig. 23.24. This picture was obtained by simulating a circuit containing one TC4469COE driver, a voltage source, and four termination resistors for each output. For clarity, only the TC4469COE driver is shown. In the simulation setup for this circuit, only the bias point calculation was selected. To check the logic, all possible combinations of "High" and "Low" signals were connected to the inputs. The result was made visible by selecting to display bias point voltages on the "Schematics" page. This test is another example of the model validation and verification process mentioned before.

As mentioned before, the circuit shown in Fig. 23.23 contains an opto-coupler. This particular device has an internal dielectric shield giving it a high immunity to dV/dt common mode swings on the output. Therefore, it is very suitable for use in power electronics applications. To be able to simulate the circuit containing this part, a simulation model for the opto-coupler was developed. However, in this specific case, the ability to run the simulation with the inclusion of the opto-coupler is more of a convenience than a necessity to check the performance of the circuit. In this application, the demands on the opto-coupler

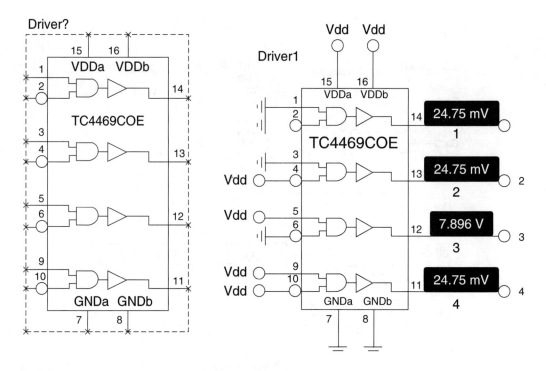

FIGURE 23.24 Symbol editor view and test output of the TC4469COE MOSFET driver.

```
Symbol TC4469COE
Definition Quad MOSFET Driver SOIC (Wide)
@attributes
REFDES=Driver?
PART=TC4469COE
MODEL=

TEMPLATE=E_Ch1^@REFDES Out1^@REFDES,%GNDa VALUE {IF(V(%1A,%GNDa)
>@Threshold & V(%1B,%GNDa)<@Threshold, V(%VDDa, %GNDa)-25mV, V(%GNDa)+25mV)}

\nE_Ch2^@REFDES Out2^@REFDES, %GNDa VALUE {IF(V(%2A,%GNDa)>@Threshold &
V(%2B,%GNDa)<@Threshold, V(%VDDa, %GNDa)-25mV, V(%GNDa)+25mV)}

\nE_Ch3^@REFDES Out3^@REFDES, %GNDa VALUE {IF(V(%3A,%GNDa)>@Threshold &
V(%3B,%GNDa)<@Threshold, V(%VDDa, %GNDa)-25mV, V(%GNDa)+25mV)}

\nE_Ch4^@REFDES Out4^@REFDES, %GNDa VALUE {IF(V(%4A,%GNDa)>@Threshold &
V(%4B,%GNDa)<@Threshold, V(%VDDa, %GNDa)-25mV, V(%GNDa)+25mV)}

\nR_Out1^@REFDES Out1^@REFDES,%1Y @Rout
\nR_Out2^@REFDES Out2^@REFDES,%2Y @Rout
\nR_Out3^@REFDES Out3^@REFDES,%3Y @Rout
\nR_Out4^@REFDES Out4^@REFDES,%4Y @Rout

PKGTYPE=SO16W
ROUT=10.0
Threshold=1.5V
```

FIGURE 23.25 Attributes of the TC4469COE driver.

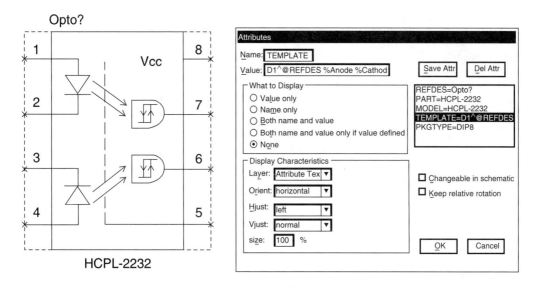

FIGURE 23.26 Symbol editor view of the TC4469COE MOSFET driver.

are quite low. The data throughput is much less than the actual bandwidth, the voltage stress is minimal, and the output does not have to drive a large load. Therefore, a rather simple model is adequate for this case. As a general rule, it is always recommended to evaluate how hard each circuit element is pushed to its limits. The harder a part is pushed, the more its behavior typically deviates from an ideal model. After this step, the choice of models becomes clearer. Complex and sophisticated models are typically needed for the parts that are pushed hard, whereas simpler models will be adequate for the parts that are operated well below their limits. It is also not recommended to use a complex model just because it is available when it is not needed. This is because the complex model typically slows the simulation down and creates stability problems.

With this said, the particular model used here for the opto-coupler can be discussed. The model comprises an input diode and a voltage-controlled voltage source for the output. Obviously, the LED current should be monitored and the output should respond to it. However, for a diode used in a simulator, the device characteristics are totally stable and voltage and current are perfectly correlated. Therefore, the model can monitor the forward diode voltage and switch the output voltage, whenever a certain level is crossed. It is also possible to include hysteresis into the model, but this is not done here. Obviously, in many other applications, in which the opto-coupler is pushed harder, a more complex model must be created. The symbol editor view of the dual channel opto-coupler is shown in Fig. 23.26. The details of the attributes including the "TEMPLATE" attribute, which defines the simulation model described above, are shown in Fig. 23.27. The easiest way to get an ASCII printout of all the attributes is to "Export" the symbol from within the symbol editor. This will create an ASCII file (extension ".sym") containing the full text of all attributes as well as textual descriptions of all the graphical elements in a format resembling postscript.

For clarity, the graphical descriptions for all custom parts have been omitted since they are better shown directly as in Fig. 23.26. The symbol file could, however, be e-mailed and be re-imported into an ".slb" library and thereby create a fully functioning part with all the graphical elements of the original.

The simulation results for the circuit shown in Fig. 23.23 are virtually the same as the results shown in Fig. 23.22 since the load is the same as in the circuit shown in Fig. 23.15. The load is a 1-Ω resistor in series with a 1-mH inductor as shown in Fig. 23.13. This simple load was replaced by a complete model for a DC motor in the circuit shown in Fig. 23.28. This circuit is suitable both for simulation and for creation of a PC-board. The reader may note a few more additions compared with the circuit shown in Fig. 23.23. One important addition is the block called "Cbuffer." This block contains a total of ten capacitors to

23-20   *The Power Electronics Handbook*

```
Symbol HCPL-2232
Definition Dual Logic Gate Optocoupler
@attributes
REFDES=Opto?
MODEL=HCPL-2232

TEMPLATE=D1^@REFDES %Anode1 %Cathode Model^@REFDES
\nD2^@REFDES %Anode2 %Cathode2 Model^@REFDES
\n.model Model^@REFDES D
\nEout1^@REFDES %Vo1, %Gnd VALUE {IF(V(%Anode1,%Cathode1)>0.65V, V(%Vcc,%Gnd), 25mV)}
\nEout2^@REFDES %Vo2, %Gnd VALUE {IF(V(%Anode2,%Cathode2)>0.65V, V(%Vcc,%Gnd), 25mV)}

PKGTYPE=DIP8
```

FIGURE 23.27 Attributes of the HCPL-2232 opto-coupler.

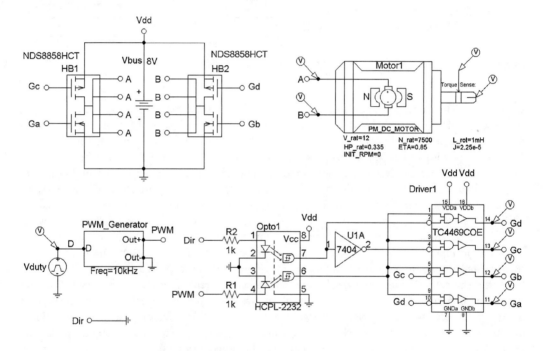

FIGURE 23.28 Advanced H-bridge circuit with motor load for simulation and PC-board layout.

buffer the DC bus. The associated subcircuit is shown in Fig. 23.29. The capacitors are not necessary for simulation, since the supply voltage "Vbus" is ideal, but needed for the real circuit.

To avoid start-up transients, all capacitors have been initialized to the bus voltage. For the PC-board layout, surface mount packages were chosen for all buffer capacitors.

In addition, parts "P1" and "P2" are screw terminals and "P6" is a 20-pin header. These parts are needed for connections to the PC-board but have no simulation model. Note also, that "Vss" is used for the ground on the primary side of the opto-coupler. Otherwise, the grounds on the PC-board would be joined and the isolation of the opto-coupler would be defeated. To be able to simulate the circuit anyway, "Vss" is tied to ground with a large resistor, which has a "SIMULATIONONLY" attribute to prevent it from being included in the circuit board layout. Also, the hidden power supply pins of the inverter "U1A" have been set to "Vdd" and "GND" so that they are connected on the PC-board.

Figure 23.30 shows the DC motor as viewed from the symbol editor. Because of the complexity of the simulation model it was not implemented using the "TEMPLATE" attribute. Instead, a subcircuit was used.

Computer Simulation of Power Electronics

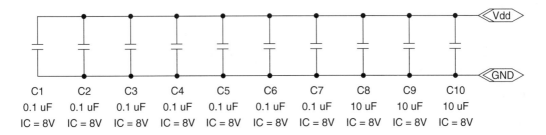

FIGURE 23.29 Subcircuit for the DC bus buffer caps.

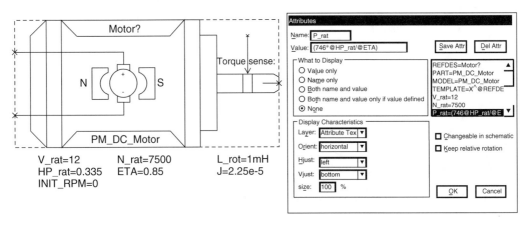

FIGURE 23.30 View of the DC motor symbol from the symbol editor.

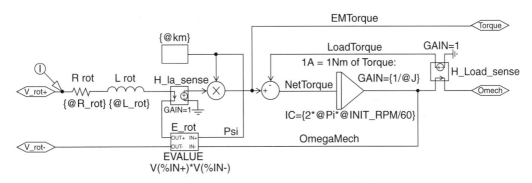

FIGURE 23.31 Schematic for the model of the DC motor.

This subcircuit is shown in Fig. 23.31. All parts are marked as "SIMULATIONONLY." The pins of the motor symbol are connected to the interface ports with the same names. This subcircuit implements both the electrical behavior of the motor and the mechanical acceleration and deceleration. The electrical circuit consists of the resistor "Rrot" and the inductor "Lrot" and the rotational voltage "E_rot". The induced voltage is equal to the motor constant "Km" multiplied with the mechanical speed. The voltage on the pin on the shaft of the machine ("Omech" terminal) is directly equivalent to the mechanical speed according to the relation: 1 V = 1 rad/s. The model also calculates the internally developed torque and makes the value available at a monitoring pin. The torque is calculated as the product of the machine constant "Km" and the rotor current, which is monitored by "H_Ia_sense". The current coming from the "Omech" terminal represents the load torque according to the relation 1 A = 1 Nm. After subtraction of the load torque from the internally generated torque, the mechanical speed is calculated by integrating the net torque weighted by the moment of inertia "J". For more details on machine theory, the reader is referred to Ref. 3.

```
Symbol PM_DC_MOTOR
DefinitionDC_Motor, Permanent Magnet
@attributes
REFDES=Motor?
PART=PM_DC_Motor
MODEL=PM_DC_Motor
TEMPLATE=X^@REFDES %V_rot+%V_rot-%Omech%Torque @MODEL
V_rat=12
N_rat=7500
P_rat=(746*@HP_rat/@ETA)
R_rot=((@V_rat-@Km*@Omega_rat)/@I_rat)
L_rot=1mH
J=2.25e-5
Pi=3.14159265
Omega_rat=(2*@Pi*@N_rat/60)
Km=(@Torque_rat/@I_rat)
HP_rat=0.335
ETA=0.85
Torque_rat=(746*@HP_rat/@Omega_rat)
INIT_RPM=0
I_rat=(@P_rat/@V_rat)
@views
DEFAULT=PM_DC_Sub.sch
```

FIGURE 23.32 Complete list of attributes for the DC motor symbol.

The subcircuit shown in Fig. 23.31 has no actual model parameters within it and is therefore truly generic. The model parameters are associated with attributes of the motor symbol itself and are passed on to the subcircuit via the "@" symbol. A complete list of all attributes of the motor symbol is shown in Fig. 23.32. The information for this figure was extracted from an exported symbol (".sym") file.

Figure 23.33 shows the results of the simulation of the most advanced H-bridge circuit shown in Fig. 23.28. Shown are the motor current and the increasing mechanical speed.

As mentioned before, the circuit shown in Fig. 23.28 can be used not only as an input for the PSpice simulator, but also as a schematic entry page for a PC-board layout to implement the circuit. This makes simulation an integrated part of the design and analysis process. Figure 23.34 shows a view of the circuit board layout program, which uses the layout netlist from the circuit shown in Fig. 23.28. It should be mentioned that the layout netlist is different from the simulation netlist (these are different files) because it was used for a different purpose. The view of Fig. 23.34 shows the footprints for all parts and their relative placement as well as the connections between them before the trace routing process. The logical connections between the parts are often referred to as "ratsnest."

23.5 Conclusions

In this chapter, the simulation of power electronics circuits has been discussed starting from principal issues to advanced techniques, including the integration of the simulation into the overall design process. The proper uses of simulation techniques, as well as limitations and potential pitfalls, have been discussed in detail. In addition, the generation of custom symbols, which can be used both for simulation as well as for PC-board layout has been shown in a number of examples. The philosophy behind the creation of simulation models for custom parts has been discussed. It should also be acknowledged, that there are many more excellent software packages for circuit simulation available. For some of the more popular

Computer Simulation of Power Electronics

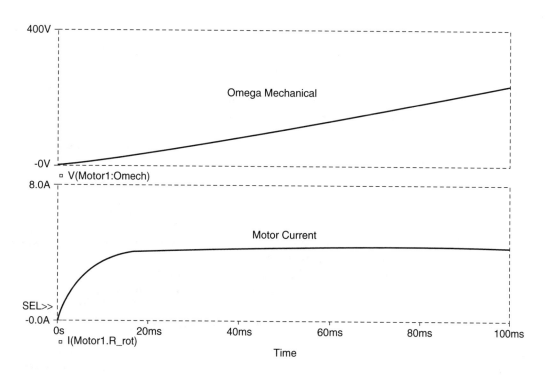

FIGURE 23.33 Simulation results for the circuit in Fig. 23.28.

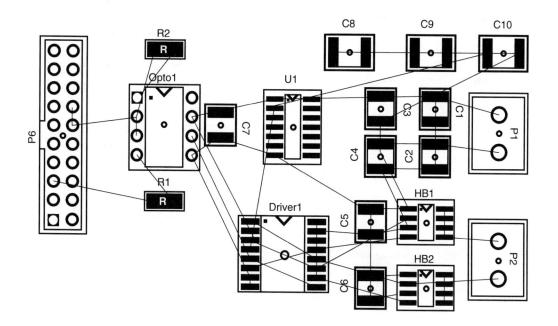

FIGURE 23.34 View of the PC-board layout for the circuit shown in Fig. 23.28.

ones, the reader is referred to Refs. 5, 12, and 13. In addition, Refs. 10, 11, and 14 give further information on simulation techniques for power electronics.

References

1. Giesselmann, M. G., Averaged and cycle by cycle switching models for buck, boost, buck-boost and cuk converters with common average switch model, in *Proceedings of the 32nd Intersociety Energy Conversion Engineering Conference,* IECEC-97, Honolulu, HI, Jul. 27–Aug. 01, 1997.
2. Kassakian, J. G., Schlecht, M. F., and Verghese, G. C., *Principles of Power Electronics,* Addison-Wesley, Reading, MA, 1991.
3. Krause, P. C., Wasynczuk, O., and Sudhoff, S. D., *Analysis of Electric Machinery,* IEEE Press, New York, 1995.
4. MathCAD® 2000 Professional, MathSoft Engineering & Education, Inc., 101 Main Street, Cambridge, MA 02142-1521, http://www.mathsoft.com.
5. MircoCap, SPECTRUM SOFTWARE, 1021 S. Wolfe Rd., Sunnyvale, CA 94086, http://www.spectrum-soft.com.
6. Mohan, N., Undeland, T., and Robbins, W., *Power Electronics: Converters, Applications, and Design,* 2nd ed., John Wiley & Sons, New York, 1995.
7. Mohan, N., *Electric Drives, An Integrative Approach,* MNPERE, Minneapolis, MN, 2000.
8. PSpice® Documentation, 555 River Oaks Parkway, San Jose, CA 95134; (408) 943 1234; http://pcb.cadence.com/.
9. Rashid, M. H., *Power Electronics: Circuits, Devices, and Applications,* Prentice-Hall, Englewood Cliffs, NJ, 1993.
10. Rashid, M. H., *SPICE for Power Electronics and Electric Power,* Prentice-Hall, Englewood Cliffs, NJ, 1993.
11. Rashid, M. H., *SPICE for Circuits and Electronics Using PSpice,* Prentice-Hall, Englewood Cliffs, NJ, 1990.
12. Saber® Mixed Circuit Simulator, Avant! Corporation, 46871 Bayside Parkway, Fremont, CA 94538; (510) 413 8000; info@avanticorp.com.
13. Simplorer®, Technical Documentation, SIMEC Corporation, 1223 Peoples Avenue, Troy, NY 12180; (518)-373-5838; Info@Simplorer.com.
14. Tuinenga, P. W., *SPICE, A Guide to Circuit Simulation & Analysis Using PSpice,* Prentice-Hall, Englewood Cliffs, NJ, 1988.

Index

A

AC-AC converters, **3**–1 to 4. *See also* Converters; Rectifiers
Acceptor impurities, **4**–5
AC generators, **16**–9 to 10
AC machines controlled as DC machines. *See* Brushless DC (BLDC) motors
ACRDCLI. *See* Actively clamped resonant DC-link inverter (ACRDCLI)
ACSL. *See* Advanced Continuous Simulation Language (ACSL)
Active filters, power quality, **17**–30 to 46
Active harmonic compensation, **4**–21
Actively clamped resonant DC-link inverter (ACRDCLI), **5**–61 to 63, **5**–65 to 66
Active vibration suppression, **12**–9
AC vs. DC, transmission, **4**–31
Adjustable speed drive (ASD), **4**–17 to 18, **4**–21
Advance angle calculation, **13**–15
Advanced Continuous Simulation Language (ACSL), **5**–76
Aerospace, **13**–19. *See also* Aircraft, more-electric vehicles (MEA)
Ahmed studies, **5**–2, **5**–7
Aircraft, more-electric vehicles (MEA)
    advanced electrical loads, **21**–3 to 4
    conventional electrical loads, **21**–1
    electrical distribution systems, **21**–2 to 3
    fundamentals, **21**–1
    power generation systems, **21**–1 to 2
    system architecture, **21**–4 to 5
Akagi, Fujita and, studies, **17**–38, **17**–39
Akagi and Fujita studies, **17**–35
Akagi-Nabae theory, **17**–32 to 34, **17**–38, **17**–41
Akagi studies, **17**–35, **17**–36, **17**–38, **17**–39
Ametek-Lamb Electric, **13**–19
Architectures, system, **21**–4 to 5, **21**–9 to 11
ARCP inverters. *See* Auxiliary resonant commutated pole (ARCP) inverters
Aredes studies, **17**–35, **17**–43
ASCR. *See* Asymmetrical silicon-controlled rectifiers (ASCR)
ASD. *See* Adjustable speed drive (ASD)
Asymmetrical silicon-controlled rectifiers (ASCR), **1**–4
Automatic power control, UPFCs, **20**–7 to 9
Automotive applications
    more-electric systems, **21**–6 to 12
    MOSFETs, **1**–59 to 60
    switched reluctance machines, **13**–19
    vehicular propulsion, **12**–9
Auxiliary resonant commutated pole (ARCP) inverters
    circuit descriptions, **5**–73 to 74, **5**–80 to 81, **5**–84 to 85
    computer study, **5**–85 to 88
    diode, low-to-high commutation, **5**–74 to 78
    fundamentals, **5**–67, **5**–88 to 89
    H-bridge analysis, **5**–80 to 83
    losses, **5**–67 to 73
    phase leg, **5**–67 to 70, **5**–73 to 80, **5**–80
    simulation phase leg, **5**–80
    switch, high current, low-to-high commutation, **5**–78 to 80
    switch, low current, low-to-high commutation, **5**–78
    three-phase analysis, **5**–83 to 87
    turn-off losses, **5**–72 to 73
    turn-on losses, **5**–70 to 72
Average forward current, **4**–4
Average output voltage, **4**–2, **4**–4
Avionics, **13**–19. *See also* Aircraft, more-electric vehicles (MEA)

B

Back-emf, **13**–5, **13**–8 to 9, **13**–15
Balbo studies, **17**–38
Barton studies, **5**–2, **5**–7
Batteries
    compared to SMES systems, **16**–9
    photovoltaic cells and systems, **18**–6 to 7
    uninterruptible power supplies (UPS), **16**–7 to 8
Bhattacharya studies, **17**–36, **17**–38, **17**–43
Biasing, bipolar junction transistors, **1**–30 to 31
Bifilar-wound hybrid motor, **14**–5, **14**–7, **14**–11 to 12
Bipolar FET (BiFET), **1**–78
Bipolar junction transistors (BJT). *See also* Insulated-gate bipolar transistors (IGBT)
    biasing, **1**–30 to 31
    characteristics, **1**–29 to 30, **1**–54 to 59
    diodes, **2**–26
    dv/dt capability, **1**–58 to 59
    fundamentals, **1**–28 to 29, **5**–21
    limitations, **1**–46
    power losses, **1**–31
    protection, **1**–31 to 44
    switches, **2**–26
    testing, **1**–31
Bird studies, **5**–2, **17**–30
BJT. *See* Bipolar junction transistors (BJT)
BLDC motors. *See* Brushless DC (BLDC) motors
Bode plots, **23**–1 to 2
Boost converter-based PFC, **17**–53 to 54

I-1

Boost (step-up) converter
    continuous-conduction mode, **2**–12 to 13
    discontinuous-conduction mode, **2**–13 to 14
    fundamentals, **2**–1, **2**–4 to 5
    ideal circuit, **2**–12
    sliding-mode control, **8**–14 to 18
Boost-type rectifier, **4**–33 to 34, **4**–34 to 38
Bose studies, **5**–2, **5**–6
Brake resistor, **5**–4 to 5
Breakover voltage, **1**–18, **1**–21
Broadband filter, **17**–23
Brumsickle studies, **5**–2
Brushless DC (BLDC) motors
    applications, **10**–15
    current source inverter, **10**–8 to 9
    disadvantages, **10**–11
    fault-tolerant configuration, **10**–8
    fundamentals, **10**–1 to 2
    machine construction, **10**–2 to 3
    modeling, **10**–6
    motor characteristics, **10**–4 to 6
    permanent magnets, **10**–2 to 3
    position sensing, **10**–9 to 11
    power converter, **10**–7 to 9
    pulsating torque components, **10**–11
    speed, torque, **10**–11 to 14
    stator windings, **10**–3
    torque, **10**–11 to 14
    unipolar excitation, **10**–7 to 8
Buck–boost converter
    capacitor value, **2**–26
    circuit analysis, **2**–18 to 24
    component selection, **2**–25 to 27
    continuous-conduction mode, **2**–18 to 22
    discontinuous-conduction mode, **2**–22 to 24
    flyback power stage, **2**–27 to 28
    fundamentals, **2**–2, **2**–17 to 18, **2**–28
    inductor value, **2**–25 to 26
    main power switch, **2**–26 to 27
    output power diode, **2**–26 to 27
    sliding-mode control, **8**–14 to 18
    small signal transfer functions, **2**–24 to 25
Buck-derived three-phase PFC, **17**–59 to 60
Buck (step-down) converter
    computer simulation, **23**–3 to 10
    continuous-conduction mode, **2**–9 to 10
    discontinuous-conduction mode, **2**–10 to 11
    fundamentals, **2**–1, **2**–4
    hysteresis feedback controllers, **7**–15 to 28
    ideal circuits, **2**–8 to 9
    sliding-mode theory application, **8**–9 to 14
    two-quadrant ZVS converter, **5**–39
"Bullet train" applications, **17**–46

C

Can-stack (permanent magnet) motor, **14**–5 to 8
Capacitance, **1**–54 to 59
Capacitors, correction, **17**–17 to 18
Capacitor value, buck–boost converter, **2**–26
Carrier devices, **1**–6
Cascaded converters
    H-bridge, **6**–11 to 14
    multilevel, **6**–15 to 16, **6**–18 to 20
    multilevel H-bridge, **6**–17, **6**–20 to 21
Cascade-5/3H inverter, **6**–20 to 21
Cascade-3/2 inverter, **6**–18 to 20
Cash and Habetler studies, **5**–1
C-dump converters, **13**–14
Cells, solar. *See* Photovoltaic (PV) cells and systems
Channel length, reduced, **1**–53
Characteristics
    bipolar junction transistors, **1**–29 to 30
    diodes, **1**–10
    MOSFETs, **1**–48 to 54
    Schottky diodes, **1**–15
    silicon-controlled rectifiers, **1**–18 to 19
Chattering, **7**–16
Choppers
    brake, **5**–5
    four-quadrant, **2**–7 to 8
    fundamentals, **2**–1, **2**–3
    more-electric aircraft, **21**–5
    one-quadrant, **2**–3 to 5
    servo drives, **15**–2
    two-quadrant, **2**–6 to 7
Circuits. *See also* Turn-off circuits
    ARCP description, **5**–73 to 74
    ARCP three-phase, **5**–84 to 85
    boost (step up) converter, **2**–12
    buck–boost converter, **2**–18 to 24
    buck (step down) converter, **2**–8 to 9
    classification, **17**–38
    controlled rectifiers, gate requirements, **4**–25
    diodes, rectifiers, **1**–13 to 14
    drive, **1**–82 to 84, **1**–94 to 95, **14**–5 to 8
    H-bridge, **4**–25 to 29, **5**–80 to 83
    magnetism, **22**–3 to 4
    parallel-resonant inverters, **5**–48 to 50
    second-order resonant circuits, **5**–25 to 30
    series-parallel-resonant inverters, **5**–51 to 52
    series-resonant inverter, **5**–42 to 45
    silicon-controlled rectifiers, **1**–19 to 20
    single-phase active PFC circuits, **17**–53 to 56
    snubber, **1**–72 to 73, **1**–76, **5**–43, **5**–58
    three-phase PFC circuits, **17**–57 to 60
Class D inverters. *See* Series-resonant inverters
Class E converter, **5**–34 to 36
Classes of equipment, **17**–50 to 51
Classification of active filters, **17**–35 to 39
Code qualification, computer simulation, **23**–2 to 3
Communication, wireless, **1**–59
Commutation, **5**–73 to 79, **17**–11
Comparators, hysteresis, **11**–13 to 17
Comparison testing
    devices used, **1**–91 to 92

forward conduction, 1–95
gate drive circuits, 1–94 to 95
loss characterization, 1–95
observations, 1–98 to 100
pulse tester, 1–89 to 91
unity gain verification, 1–92 to 94
Computer simulation
    Advanced Continuous Simulation Language (ACSL), 5–76
    ARCP phase leg, 5–85 to 88
    buck converters, 23–3 to 10
    code qualification, 23–2 to 3
    full-bridge (H-bridge) converter, 23–10 to 22
    fundamentals, 23–1 to 2, 23–22 to 24
    MathCAD, 7–1, 23–4
    MATLAB software, 20–20, 20–25
    model validation, 23–2 to 3
    PSPICE program, 2–26, 23–2 to 4, 23–22
    SABER programming code, 4–38 to 41
    SPICE program, 2–22, 23–1 to 2
Conductivity-modulated field-effect transistor (COMFET), 1–78
Conductors, power quality, 17–12
Continuous mode
    buck–boost converter, 2–18 to 22, 2–28
    fundamentals, 2–3
    step-down (buck) converter, 2–4, 2–9 to 10
    step-up (boost) converter, 2–4, 2–12 to 13
    two-quadrant choppers, 2–6 to 7
Continuous PWM control, 5–63
Control design, UPFCs, 20–14 to 20
Controlled rectifiers
    dv/dt, 4–23
    fundamentals, 4–4 to 5
    gate circuit requirements, 4–25
    gate current injection, 4–23 to 25
    H-bridge circuits, 4–25 to 29
    HVDC transmission systems, 4–30 to 31
    single-phase, 4–25 to 29
    temperature, 4–23
    three-phase, 4–29 to 30, 4–31 to 33
    thyristors, 4–25 to 29, 4–31 to 33
    voltage, 4–23, 4–30
Control strategies, SRMs, 13–14 to 16
Control strategy classification, 17–38 to 39
Converters
    AC-AC, 3–1 to 4
    BLDC motors, 10–7 to 9
    buck–boost, 2–17 to 28
    C-dump, 13–14
    choppers, 2–3 to 8
    Cúk type, 2–14 to 16
    cycloconverters, 3–1 to 3
    DC-DC, 2–1 to 28
    frequency, 3–1 to 4
    fundamentals, 2–1 to 3, 3–1
    matrix, 3–3 to 4
    PFC converters, 17–55 to 56

    six-switch PWM converters, 17–57
    step down (buck), 2–8 to 11
    step up (boost), 2–12 to 14
    switched reluctance machines (SRM), 13–11 to 14
Cúk converter
    DCM operation, 17–55 to 56
    fundamentals, 2–2, 2–14 to 16
    sliding-mode control, 8–2, 8–9, 8–18 to 20
Cúk studies, 2–19
Current-controlled drive, SRMs, 13–16
Current filamentation, gate turn-off thyristors, 1–74 to 76
Currents
    calculation, passive harmonic filters, 17–29
    control, modulation, 6–12 to 14
    diode, rectifiers, 4–2, 4–4
    limitations, sliding-mode control, 8–13 to 14
    sources, power conditioning, 17–31 to 32
    spikes, 5–43, 5–53
    transient, 1–3
Current source inverter, 10–8 to 9
Cycloconverters, 3–1 to 3

D

Damping control, UPFCs, 20–16 to 20
Darlington configurations, 1–5
Davidson studies, 13–1
DC-AC converters. *See also* Inverters
    connections, 5–9 to 11, 5–21
    filtering voltage, 5–20
    fundamentals, 5–8 to 9
    harmonics, 5–18 to 20
    output voltage, 5–11 to 20
    resonant converter techniques, 5–21 to 23
    square-wave operations, 5–9 to 11
DC-DC converters. *See also* Converters; Rectifiers
    buck–boost, 2–17 to 28
    choppers, 2–3 to 8
    Cúk type, 2–14 to 16
    fundamentals, 2–1 to 3
    parallel-loaded resonant (PLR), 5–36
    series-loaded resonant (SLR), 5–36
    sliding-mode theory application, 8–8 to 9
    step down (buck), 2–8 to 11, 5–39
    step up (boost), 2–12 to 14
    topologies, 17–54 to 55
DC generators, 16–8 to 9
DC-link choke, rectifiers, 4–19 to 20
DCM operation, unity power factor rectification, 17–55 to 56
DC motors
    AC-DC rectifier circuits, 4–25
    drives, 9–3 to 4
    fundamentals, 9–1 to 2
    SCR DC drives, 9–5 to 6
    speed control, 9–2 to 3
    transistor PWM DC drives, 9–4 to 5

DC servo drives, **15**–2 to 3
DC vs. AC, transmission, **4**–31
Dead bands, **7**–16 to 18
DeDonker and Novotny studies, **5**–6
DESSs. *See* Distributed energy storage systems (DESSs)
Detection methods, harmonics, **17**–38 to 39
Detuned capacitor bank, **17**–25
DGs. *See* Distributed generators (DGs)
DIACs, **1**–20 to 21
Di/dt capability
    gate turn-off thyristors, **1**–68
    switch requirements, **1**–62
    thyristor fundamentals, **1**–3
    voltage transfer function, **5**–53
Diode-bridge rectifiers, **17**–49 to 50
Diode-clamped multilevel converters, **6**–7 to 10, **6**–17 to 18
Diodes. *See also* Schottky diodes
    buck–boost converter, **2**–26 to 27
    characteristics, **1**–10
    conduction, rectifiers, **4**–5 to 6
    current, **4**–2, **4**–4
    forward voltage, **1**–54
    fundamentals, **1**–9, **4**–6
    germanium type, **1**–17
    low-to-high commutation, **5**–74 to 78
    pin, **1**–79
    positive pulse output waveforms, **4**–1
    protection, **1**–14
    ratings, **1**–10 to 13, **1**–14
    rectifier circuits, **1**–13 to 14, **4**–2 to 3
    reverse-biased, **1**–14
    silicon type, **1**–17
    testing, **1**–14
Direct pulse width modulation (DPWM), **5**–60. *See also* pulse width modulation (PWM)
Direct torque control (DTC) method, **11**–11 to 17
Discontinuous mode
    buck–boost converter, **2**–22 to 24, **2**–28
    fundamentals, **2**–3
    load-resonant converters, **5**–32 to 33
    step-down (buck) converter, **2**–4, **2**–10 to 11
    step-up (boost) converter, **2**–4, **2**–13 to 14
    two-quadrant choppers, **2**–6 to 7
Displacement power factor, **17**–15, **17**–52
Dissipation, MOSFETs, **1**–54
Distortion, **17**–24. *See also* Harmonics
Distributed energy storage systems (DESSs), **19**–1 to 11
Distributed generators (DGs), **19**–1 to 11
Distribution static compensator (DSTATCOM), **19**–7 to 8
Distribution systems, MECs, **21**–2 to 3, **21**–7 to 8
Divan, Lorenz and, studies, **5**–6
Divan and Skibinski studies, **5**–62
Donor impurities, **4**–5
Double-conversion, UPS, **16**–3 to 4, **16**–10

Drive circuits
    gate-commutated thyristors, **1**–82 to 84
    gate drives, **1**–94 to 95
    step motors, **14**–5 to 8
Drives, DC motors, **9**–3 to 4
Drive trains, MEVs, **21**–9
DSTATCOM. *See* Distribution static compensator (DSTATCOM)
DTC method. *See* Direct torque control (DTC) method
Duty cycles, **6**–4 to 6, **7**–5, **23**–3 to 10
Dv/dt capability
    controlled rectifiers, **4**–23
    gate turn-off thyristors, **1**–72
    MOSFETs, **1**–6, **1**–57 to 59
    reverse-conducting thyristors, **1**–4
    switches, **1**–62
DVR. *See* Dynamic voltage restorer (DVR)
Dynamic avalanche, GTOs, **1**–73 to 76, **1**–83
Dynamic characteristics, MOSFETs, **1**–54 to 59
Dynamic voltage restorer (DVR), **19**–8

E

Eddy currents, PMSM drives, **12**–3
EHPs. *See* Electron-hole pairs (EHPs)
Electric vehicles (EVs). *See* More-electric vehicles (MEV)
Electromagnetic currents, **5**–25, **22**–2 to 3
Electron-hole pairs (EHPs), **18**–2
Emitter turn-off (ETO) thyristors, **1**–85, **1**–88, **1**–91 to 100
Energy conversion, SRMs, **13**–5 to 6
Energy ratio, **13**–5
Energy storage capacity, **17**–34 to 35
Error-free start-stop rate, open loop control, **14**–16
ETO. *See* Emitter turn-off (ETO) thyristors
EVs. *See* More-electric vehicles (MEV)
Existence conditions
    buck DC-DC converters, **8**–12 to 13
    Cúk and SEPIC DC-DC converters, **8**–19
    sliding-mode theory, **8**–5, **8**–6

F

FACTS. *See* Flexible AC transmission systems (FACTS)
Faraday's law
    magnetism, **22**–1 to 2, **22**–6 to 7
    parasitic coupling, **22**–7
    PMSM drives, **12**–3 to 4
Fault-tolerant configuration, BLDC motors, **10**–8
FBSOA. *See* Forward-biased safe operation area (FBSOA)
Feedback control, hysteresis
    control circuit principles, **7**–16 to 18
    design procedure, **7**–18 to 23
    experimental results, **7**–23 to 27
    fundamentals, **7**–14 to 15, **7**–27 to 28
    simulation results, **7**–22 to 23

Feedforward controllers, **7**–15
Ferromagnetic materials, **22**–2, **22**–4
Field current regulators, **9**–3, **9**–6
Field-effect transistor (FET), **5**–21
Field-oriented control, induction machines, **11**–4, **11**–8 to 11
Filtering output voltage, **5**–20
Filters
    active, **17**–60 to 62
    passive design, **17**–23 to 29
    rectifiers, **4**–20
Firing angle, **5**–7
First quadrant, **4**–28
Fixed-frequency control, **5**–17 to 18
Flexible, Reliable, and Intelligent Electrical Energy Delivery Systems. *See* FRIENDS methodology
Flexible AC transmission systems (FACTS), **20**–1, **20**–4
Florida Solar Energy Center, **18**–9
Flux density, **22**–3
Flyback converters, **17**–55
Flyback power stage, **2**–27 to 28
Flying-capacitor multilevel converters, **6**–10 to 11
48-MVA filter, **17**–46 to 47
Forward-biased safe operation area (FBSOA), **1**–62 to 65, **1**–80
Forward blocking, **1**–4, **1**–69 to 71
Forward conduction, **1**–66 to 69, **1**–95
Forward current, average, **4**–4
Forward surge maximum (*Ifsm*), **1**–13
Forward voltage drop, **1**–62, **4**–3 to 4
Fourier series
    control strategy classification, **17**–38
    harmonics, **17**–4 to 6, **17**–9, **17**–14, **17**–20
    notching, **17**–11
    PMSM drives, **12**–4
Four-layer *pnpn* power semiconductor devices. *See* Thyristors
Four-quadrant choppers, **2**–7 to 8. *See also* Choppers
Fourth quadrant, **4**–28, **4**–29
Frequency-domain, power conditioning, **17**–38
Frequency response, passive harmonic filters, **17**–24 to 26
FRIENDS methodology. *See also* Power Quality
    advanced technologies, **19**–7 to 9
    advantages, **19**–9 to 10
    developments and implementation, **19**–5 to 7, **19**–11 to 12
    distribution static compensator, **19**–7 to 8
    dynamic voltage restorer, **19**–8
    fundamentals, **19**–1 to 4
    intelligent function, **19**–5
    quality control centers (QCCs), **19**–4 to 5, **19**–7 to 9
    solid-state breaker, **19**–8 to 9
Fujita, Akagi and, studies, **17**–35
Fujita and Akagi studies, **17**–38, **17**–39
Full-bridge inverters, **5**–9
Full-wave rectifiers, **1**–13, **4**–3 to 4, **4**–9 to 12
Fuzzy control, UPFCs, **20**–16 to 20

G

Gain-enhanced MOSFET (GEMFET), **1**–78
Gate charge, MOSFETs, **1**–56 to 57
Gate circuit requirements, **4**–25
Gate-commutated thyristors, **1**–82 to 85
Gate current injection, **4**–23 to 25
Gate drive circuits, **1**–94 to 95
Gate turn-off (GTO) thyristors
    current filamentation, **1**–74 to 76
    dynamic avalanche, **1**–73 to 74, **1**–75 to 76
    forward blocking, **1**–69 to 71
    forward conduction, **1**–66 to 69
    fundamentals, **1**–3 to 4, **1**–22, **1**–65 to 66, **1**–76, **5**–21
    hard-driven, **1**–85 to 88
    non-uniform turn-off process, **1**–74 to 76
    SCR comparison, **1**–65
    storage time difference, **1**–74 to 75
    turn-off, **1**–69 to 71, **1**–72 to 73, **1**–82 to 85
GCTs, **1**–91 to 100
Generators, **16**–8 to 10. *See also* Distributed generators (DGs)
Germanium diodes, **1**–17
Gottlieb studies, **5**–2
Grady studies, **17**–35, **17**–38
Grant and Gower studies, **1**–54
GTO thyristors. *See* Gate turn-off (GTO) thyristors
Gyugyi and Strycula studies, **17**–30

H

Habetler, Cash and, studies, **5**–1
Half-bridge inverters, **5**–9
Half-stepping phase-on excitation, **14**–13
Half-wave rectifiers, **1**–13, **4**–1 to 3, **4**–6 to 9
Hall switch, **10**–9
Hard-driven gate turn-off thyristors, **1**–85 to 88
Hard-switched inverters, **5**–67 to 73
Harmonic filter bank, **17**–25
Harmonics. *See also* FRIENDS methodology; Power quality
    broadband filter, **17**–23
    calculating currents, **17**–29
    compensation, **4**–21
    conductors, **17**–12 to 13
    correction capacitors, **17**–17 to 18
    current sources, **17**–31 to 32
    DC-AC conversion, **5**–18 to 20
    diode rectifiers, **4**–16 to 17
    effects, **17**–10 to 11, **17**–12, **17**–14 to 17
    fundamentals, **17**–3 to 7
    identified loads, **17**–30 to 31
    IEEE 519 standard, **4**–15, **4**–17 to 18, **17**–2, **17**–18 to 19, **17**–29 to 30

limits, **4**–17 to 18, **17**–50 to 51
loads, harmonic-producing, **17**–23, **17**–30 to 32
low-pass filter, **17**–23
mitigation, **4**–18 to 19
noncharacteristic, **4**–16
notching, **17**–11 to 12
origination point, **17**–8 to 10
passive filters, **4**–20, **17**–20 to 29
phase multiplication, **4**–20
power system components, **17**–12
Q value, **5**–31
sequence, **17**–7 to 8
series passive AC input reactor, **17**–22 to 23
series passive filter, **17**–21 to 22
shunt passive filter, **17**–22
SPWM signals, **7**–9 to 10
system power factor, **17**–14 to 17
system voltage, **17**–10 to 11
three-phase inverters, **5**–4
three-phase line reactors, **4**–18 to 19
three-phase neutral conductors, **17**–13
three-phase rectification, **4**–14 to 15
transformers, **17**–13 to 14
tuned harmonic filters, **17**–26 to 28
unidentified loads, **17**–30 to 31
uninterruptible power supplies (UPS), **16**–4, **16**–5
voltage sources, **17**–31 to 32
H-bridge circuits
  ARCP phase legs, **5**–67, **5**–80 to 83
  computer simulation, **23**–10 to 22
  converters, **6**–14 to 15, **6**–16 to 17
  full-wave rectifiers, **4**–10
  losses, **5**–81 to 83
  rectifiers, **4**–25 to 29
  single-switch tank, **5**–59 to 60
  three-phase semiconverter circuit, **4**–29
Hertz studies, **22**–1
HEXFET, **1**–5
High power active power line conditioners, **1**–8
Hitting condition, sliding-mode control, **8**–5, **8**–19 to 20
Holding current, **1**–20, **4**–24
Holtz, Krah and, studies, **17**–36
Holtz studies, **5**–1
HVDC transmission systems, **4**–30 to 31
Hybrid active filters, **17**–36 to 38
Hybrid electric vehicles (HEV), **21**–8 to 11. *See also* More-electric vehicles (MEV)
Hybrid step motor drives, **14**–5 to 8, **14**–11 to 12
Hysteresis
  comparators, **11**–13 to 17
  control circuit principles, **7**–16 to 18
  current controllers, **4**–37, **11**–11
  design procedure, **7**–18 to 23
  experimental results, **7**–23 to 27
  feedback control, **7**–16 to 28
  fundamentals, **7**–14 to 15, **7**–27 to 28
  simulation results, **7**–22 to 23

I

Identified loads, **17**–30 to 31
IEC standards
  555, **17**–50
  1000, **4**–15, **17**–50
IEEE standards
  519, **4**–15, **4**–17 to 18, **17**–2, **17**–18 to 19, **17**–29 to 30
  929, **18**–6, **18**–10
  1100, **17**–2, **17**–3
  C62.41, **17**–3
  utility interface issues, **17**–3
Ifsm (forward surge maximum), **1**–13
IGBT. *See* Insulated-gate bipolar transistors (IGBT)
IGCT. *See* Integrated gate commutated thyristors (IGCT)
Iizuka studies, **17**–46
Inductance, magnetism, **22**–7
Induction machines
  direct torque control, **11**–11 to 17
  field-oriented control, **11**–8 to 11
  fundamentals, **11**–1, **11**–17 to 18
  modeling, **11**–6 to 8
  scalar control, **11**–2 to 4
  space vector modulation, **11**–11 to 12
  vector control, **11**–4 to 17
Induction motor drives, **15**–3 to 6
Inductor value, buck–boost converter, **2**–25 to 26
Injector FET, **1**–78
Input time functions, **5**–30
Insulated-gate bipolar transistors (IGBT). *See also* Bipolar junction transistors (BJT)
  active harmonic compensation, **4**–21
  advantage over MOS-controlled thyristors, **1**–8
  DC-DC converters, **2**–1
  fundamentals, **1**–6 to 7, **1**–77 to 78, **5**–21
  GTO thyristor comparison, **1**–68
  operation, **1**–78 to 81
  servo drives, **15**–2
  structure, **1**–78 to 81
  testing, **1**–91 to 100
Insulated-gate FET (IGFET), **1**–78
Insulated-gate rectifier (IGR), **1**–78
Insulated-gate transistor (IGT), **1**–78
Integrated gate commutated thyristors (IGCT), **1**–85 to 86, **1**–91 to 100
Integrated series active filters, **17**–39 to 43
Intrusive (active probing) sensing methods, **13**–17, **13**–18 to 19
Inverters
  auxiliary resonant commuted pole (ARCP) circuit, **5**–67 to 89
  DC-AC conversion, **5**–8 to 24
  fundamentals, **5**–1 to 2
  issues, **5**–2
  line-commutated, **5**–7
  multilevel, **5**–6 to 7
  overview, **5**–1 to 7

Index

resonant converters, **5**–24 to 41
resonant DC-link, **5**–56 to 66
safety, **5**–55
series resonant, **5**–42 to 55
single-phase, **5**–3 to 4
three-phase, **5**–4 to 6
I^2t withstand capability
    gate turn-off thyristors, **1**–3
    power transistors, **1**–5
    thyristors, **1**–3

J

JFET. *See* Junction field-effect transistor (JFET)
von Jouanne studies, **5**–1
Junction field-effect transistor (JFET), **1**–47 to 48, **1**–51 to 52

K

Kassakian studies, **5**–2
Kawaguchi studies, **17**–38, **17**–44
Kawahira studies, **17**–38
K-factor rate transformers, **17**–13 to 14
Kirchoff's laws, **1**–30, **5**–25 to 26, **5**–75
Krah and Holtz studies, **17**–36

L

Laplace transform, **2**–24 to 25
Latching current, **1**–20, **4**–24
Lawrenson studies, **13**–1, **13**–9
Lee, Shortt and, studies, **2**–19
Lighting, DC generation applications, **16**–8
Line-commutated inverters, **5**–7
Line-interactive UPS, **16**–4 to 5
Lipo, Novotny and, studies, **5**–4
Load flow modeling, UPFCs, **20**–9 to 11
Load-resonant converters
    class E converter, **5**–34 to 36
    discontinuous mode, **5**–32 to 33
    fundamentals, **5**–21
    parallel-loaded resonant type, **5**–36
    parallel-resonant type, **5**–33 to 34
    series-loaded resonant type, **5**–36
    series-resonant type, **5**–31 to 32
    time functions, **5**–30
    types, **5**–30
Loads
    aircraft, more-electric, **17**–3 to 4
    cars, more-electric, **21**–7
    harmonic-producing, **17**–23
    power quality, **17**–30 to 32
Lorenz and Divan studies, **5**–6
Losses
    bipolar junction transistors (BJT), **1**–31
    characterization, comparison testing, **1**–95
    hard-switched inverters, **5**–67 to 73
    H-bridge analysis, **5**–81 to 83
    inverters, **5**–2
    PMSM drives, **12**–3
    resonant DC-link converters, **5**–40
    transistors, **5**–70 to 73
Low-frequency resonance, open-loop control, **14**–16
Low-pass filter, **17**–23
Low-to-high commutation, **5**–74 to 80

M

Machine construction, BLDC motors, **10**–2 to 3
Magnetism
    circuits, **22**–3 to 4
    Curie temperature, **22**–2
    electric currents, **22**–2 to 3
    field characteristics, **22**–1 to 2
    field intensity, **22**–4 to 5
    flux density, **22**–3
    fundamentals, **22**–1
    inductance, **22**–7
    Maxwell's equations, **22**–5 to 7
    parasitic coupling, **22**–7
Mamdani Inference, **20**–18
MathCAD, **7**–1, **23**–4. *See also* Computer simulation
MATLAB software, **20**–20, **20**–25
Matrix converters, **3**–3 to 4
Maximum average forward current, **1**–10
Maximum gate trigger current, **1**–20
Maximum junction temperature, **1**–13, **1**–14
Maximum surge current, **1**–13
Maxwell's equations, **22**–1, **22**–5 to 7
Maytag Neptune washing machine, **13**–19
MCT. *See* MOS-controlled thyristors (MCT)
Mean forward current, **4**–2
MegaMOS, **1**–5
MEV. *See* More-electric vehicles (MEV)
Middlebrook, Wester and, studies, **2**–19
Middlebrook studies, **2**–19
Mid-frequency resonance, open loop control, **14**–16 to 18
Miller capacitance, **1**–56
Miller's effect, **5**–43, **5**–45
Miller studies, **13**–9, **13**–13 to 14
Minimum gate trigger current, **1**–20
Minimum gate trigger voltage, **1**–20
Modeling
    brushless DC (BLDC) motors, **10**–6
    buck–boost converters, **2**–19 to 24, **2**–28
    dynamic, UPFCs, **20**–11 to 12
    induction machines, **11**–6 to 8
    load flow, UPFCs, **20**–9 to 11
    permanent-magnet synchronous machine (PMSM) drives, **12**–3 to 6
    steady state, UPFCs, **20**–9 to 11
    step motor drives, **14**–8 to 12
    unified power flow controllers, **20**–9 to 14
Model validation, computer simulation, **23**–2 to 3

Modulation techniques
    AC output, **7**–7 to 9
    current control, **6**–12 to 14
    DC average, **7**–5 to 7
    feedback control, hysteresis, **7**–14 to 28
    fundamentals, **7**–1
    generation of signals, **7**–11 to 12
    hysteresis feedback control, **7**–14 to 28
    microcontrollers, **7**–11 to 12
    multilevel converters, **6**–2 to 7
    overmodulation, **7**–5
    pulse width (PWM), **7**–2 to 9
    six-step, **7**–2
    space-vector (SVPWM), **7**–28 to 35
    third harmonic injection, **7**–9 to 10
    voltage boost, **7**–9 to 10
    voltage source-based current regulation, **6**–12 to 14
Mohan studies, **5**–2, **5**–4 to 5, **5**–7, **7**–9
Moran studies, **17**–35
More-electric vehicles (MEV), **21**–1 to 5, **21**–6 to 12
MOS-controlled thyristors (MCT), **1**–8 to 9
MOSFETs
    applications, **1**–59 to 60
    BJT comparison, **1**–45
    breakdown voltage, **1**–49 to 51
    DC-DC converters, **2**–1
    diodes, **1**–54, **2**–27
    dV/dt capability, **1**–57 to 59
    fundamentals, **1**–5 to 6, **1**–45 to 48
    gate charge, **1**–56 to 57
    gate drive circuit, **1**–5
    GTO thyristor comparison, **1**–68
    H-bridge converter, **23**–10 to 17
    IGBT comparison, **1**–80
    on-resistance, **1**–51 to 54
    portable electronics, **1**–59
    power dissipation, **1**–54
    ruggedized, **1**–6
    servo drives, **15**–2
    static characteristics, **1**–48 to 54
    switching, **1**–54 to 56, **2**–27
    threshold voltage, **1**–53 to 54
    transconductance, **1**–52 to 53
    transient response, **1**–54 to 56
    turn-off process, **1**–85
    vertical double-infused, **1**–47
    wireless communication, **1**–59
MOS turn-off (MTO) thyristors, **1**–85 to 88
Motional-emf, **13**–5
Motors, **1**–8, **10**–4 to 6, **21**–12
MTO. *See* MOS turn-off (MTO) thyristors
Multilevel converters
    diode-clamped configuration, **6**–7 to 10, **6**–17 to 18
    disadvantages, **6**–1
    examples, **6**–17 to 21
    flying-capacitor configuration, **6**–10 to 11
    fundamentals, **6**–1 to 2, **6**–21
    H-bridge configuration, **6**–11 to 14

    modulation, **6**–2 to 7
    multilevel configuration, **6**–15 to 16, **6**–18 to 20
    multilevel H-bridge configuration, **6**–14 to 15, **6**–17, **6**–20 to 21
    topologies, **6**–7 to 15
Multilevel H-bridge converters, **6**–14 to 15
Multilevel inverters, **5**–6 to 7
Multiphase stator, **12**–2

N

NDS8858HCT part, **23**–11 to 13
Negative harmonic sequence, **17**–7 to 8
Network interface, UPFCs, **20**–12 to 14
Newton's laws, **14**–9
Noncharacteristic harmonics, **4**–16
Nonintrusive sensing methods, **13**–17 to 18
Nonrepetitive peak reverse voltage, **1**–20, **4**–3, **4**–4
Non-uniform turn-off process, GTOs, **1**–74 to 76
Notching, **4**–32, **17**–11 to 12
Novotny, DeDonker and, studies, **5**–6
Novotny and Lipo studies, **5**–4
Npn transistors, **1**–28, **1**–66 to 67, **1**–69

O

Objectives classification, power conditioning, **17**–35 to 36
Oersted studies, **22**–1
Ohmmeter, **1**–14
Ohm's law, **12**–3 to 4
One phase-on excitation, **14**–13
One-quadrant choppers, **2**–3 to 5. *See also* Choppers
One-step move, open loop control, **14**–13
On-resistance, MOSFETs, **1**–51 to 54
Open-loop control, **14**–13 to 18
Output power diode, **2**–26 to 27
Overmodulation, **7**–5

P

Parallel-loaded resonant (PLR) converter, **5**–36
Parallel-resonance, **17**–25
Parallel-resonant converters (PRC), **5**–30, **5**–33 to 34
Parallel-resonant DC-link inverters (PRDCLI), **5**–22, **5**–63 to 65
Parasitic coupling, **22**–7
Passive filters, **17**–20 to 29, **17**–36 to 38
Passive harmonic filters. *See also* Harmonics
    broadband filter, **17**–23
    current calculation, **17**–29
    design, **17**–23 to 29
    distortion, **17**–24
    frequency response, **17**–24 to 26
    fundamentals, **17**–20 to 21
    IEEE standards, **17**–29 to 30
    loads, harmonic-producing, **17**–23
    low-pass filter, **17**–23

series passive AC input reactor, **17**–22 to 23
series passive filter, **17**–21 to 22
shunt passive filter, **17**–22
tuned harmonic filters, **17**–26 to 28
voltages, system, **17**–24
PCM scheme. *See* Pulse code modulation (PCM) scheme
Peak diode recovery, **1**–57
Peak forward current, **4**–2
Peak inverse voltage (PIV), **1**–10 to 11
Peak rectified forward current, **4**–4
Peak repetitive forward blocking voltage, **1**–20
Peak repetitive reverse voltage, **1**–20
Peng studies, **17**–36, **17**–38, **17**–39, **17**–43
Permanent magnet (can-stack) motor, **14**–5 to 8
Permanent-magnet synchronous machine (PMSM) drives
    BLDC motors, **10**–2 to 3
    construction, **10**–2 to 3, **12**–2 to 3
    controlling, **12**–6 to 9
    current-based drives, **12**–7 to 8
    fundamentals, **12**–1 to 2
    modeling, **12**–3 to 6
    research trends, **12**–9 to 10
    servos, **15**–4 to 6
    voltage-based drives, **12**–8 to 9
PFC converters, **17**–55 to 56
Phase-controlled rectifiers, **17**–49 to 50
Phase legs
    analysis, **5**–73 to 80
    hard-switched, **5**–67 to 70
    inverters, **5**–5, **5**–7
    simulation, **5**–80
Phase multiplication, rectifiers, **4**–20 to 21
Phase-plane, buck converters, **8**–9 to 10
Photovoltaic (PV) cells and systems
    battery backup, **18**–6 to 7
    cell performance, **18**–3 to 4
    fundamentals, **18**–1
    no-storage systems, **18**–7 to 8
    power conditioning unit (PCU), **18**–4 to 6
    simple systems, **18**–6
    solar cells, **18**–1 to 4
    stand-alone systems, **18**–7 to 10
    storage, systems with, **18**–8 to 10
    utility interactive system applications, **18**–4 to 7
Pin diodes, **1**–79
PIV. *See* Peak inverse voltage (PIV)
PLR converter. *See* Parallel-loaded resonant (PLR) converter
PMSM drives. *See* Permanent-magnet synchronous machine (PMSM) drives
PM synchronous (PM) motor drived. *See* Permanent-magnet synchronous machine (PMSM) drives
Pn-junction
    forward bias, **4**–6
    fundamentals, **1**–9

    reverse bias, **4**–6
    Schottky diodes, **1**–15
    silicon-controlled rectifiers, **1**–18
    solar cell fundamentals, **18**–2
Pnp transistors
    bipolar junction transistors, **1**–28
    gate turn-off thyristors, **1**–66 to 67, **1**–69
    insulated gate bipolar transistors, **1**–77, **1**–79
    unity turn-off gain, **1**–83
Portable electronics, MOSFETs, **1**–59
Position sensing, BLDC motors, **10**–9 to 11
Positive harmonic sequence, **17**–7 to 8
Positive pulse output waveforms, **4**–1
Power circuit classification, **17**–38
Power conditioning
    Akagi-Nabae theory, **17**–32 to 34
    applications, **17**–43 to 47
    control strategy classification, **17**–38 to 39
    current sources, **17**–31 to 32
    detection methods, harmonics, **17**–38 to 39
    energy storage capacity, **17**–34 to 35
    48-MVA filter, **17**–46 to 47
    frequency-domain, **17**–38
    future trends, **17**–43 to 44
    hybrid active filters, **17**–36 to 38
    identified loads, **17**–30 to 31
    integrated series active filters, **17**–39 to 43
    objectives classification, **17**–35 to 36
    passive filters, **17**–36 to 38
    photovoltaic system unit, **18**–4 to 6
    power circuit classification, **17**–38
    research trends, **17**–43 to 44
    series active filters, **17**–36
    shunt active filters, **17**–36, **17**–44 to 47
    system configuration classification, **17**–36 to 38
    three-phase four-wire systems, **17**–44 to 45
    time-domain, **17**–38
    unidentified loads, **17**–30 to 31
    uninterruptible power supplies, **16**–1
    voltages, **17**–31 to 32, **17**–46 to 47
Power dissipation, MOSFETs, **1**–54
Power electronics
    bipolar junction transistors, **1**–28 to 44
    diodes, **1**–9 to 17
    fundamentals, **1**–2 to 8
    gate-commuted thyristors, **1**–82 to 88
    gate turn-off thyristors, **1**–65 to 76, **1**–82 to 88
    hard-driven gate turn-off thyristors, **1**–82 to 88
    insulated gate bipolar transistors, **1**–77 to 81
    MOSFETs, **1**–45 to 60
    Schottky diodes, **1**–15 to 17
    semiconductor switches, **1**–60 to 65
    switches, **1**–60 to 65, **1**–89 to 100
    testing switches, **1**–89 to 100
    thyristors, **1**–18 to 28, **1**–65 to 76, **1**–82 to 88
    transistors, **1**–28 to 44, **1**–77 to 81
Power factors, **17**–15, **17**–52

Power quality. *See also* FRIENDS methodology; Harmonics
   active filters and techniques, **17**–30 to 46, **17**–52 to 62
   applications, **17**–43 to 47
   capacitors, correction, **17**–17 to 18
   classification, active filters, **17**–35 to 39
   conductors, **17**–12
   considerations, **17**–3 to 19
   diode bridge rectifiers, **17**–49 to 50
   filter design, passive, **17**–23 to 29
   fundamentals, **17**–1 to 3
   harmonics, **17**–30 to 32, **17**–50 to 51
   IEEE standards, **17**–2 to 3, **17**–18 to 19, **17**–29 to 30
   integrated series active filters, **17**–39 to 43
   loads, harmonic-producing, **17**–30 to 32
   notching, **17**–11 to 12
   passive filters and techniques, **17**–20 to 29, **17**–52
   phase-controlled rectifiers, **17**–49 to 50
   rectification, unity power factor, **17**–49 to 62
   series active filter, **17**–39 to 43
   theoretical approaches, **17**–32 to 35
   three-phase neutral conductors, **17**–13
   transformers, **17**–13 to 14
   unity power factor rectification, **17**–49 to 62
PRC. *See* Parallel-resonant converters (PRC)
Principle ratings, diodes, **1**–10 to 13
Proportional-integral-derivitive (PID) regulators, **9**–3
Protection, **1**–14, **1**–31 to 44
PSPICE program, **2**–26, **23**–2 to 4, **23**–22. *See also* Computer simulation
Pulsating torque components, **10**–11
Pulse code modulation (PCM) scheme, **18**–5
Pulse width modulation (PWM). *See also* Direct pulse width modulation (DPWM)
   AC output, **7**–7 to 9
   continuous, resonant DC-link inverters, **5**–63
   DC average, **7**–5 to 7
   fundamentals, **7**–2 to 5
   harmonics, **5**–18 to 20
   H-bridge, **5**–80 to 81
   microcontrollers, **7**–11 to 12
   output voltage, **5**–14, **5**–18 to 20
   third harmonic injection, **7**–9 to 10
   voltage boost, **7**–9 to 10
Punch-through phenomenon, **1**–49 to 51, **1**–80 to 81
PV cells. *See* Photovoltaic (PV) cells and systems
PWM. *See* Pulse width modulation (PWM)

Q

Quadrants, **4**–28 to 29
Quality control centers (QCCs), **19**–4 to 5, **19**–7 to 9
Quasi-resonant DC-link inverters (QRDCLI), **5**–63
Q value, **5**–31, **5**–33

R

Railroad DC generation applications, **16**–8
Rashid studies, **5**–2, **5**–4
Ratings, **1**–10 to 11, **1**–13, **1**–20
RBSOA. *See* Reverse-biased safe operation area (RBSOA)
RCT. *See* Reverse conducting thyristors (RCT)
RDCLI. *See* Resonant DC-link inverters (RDCLI)
Reach-through phenomenon, **1**–51
Reactors, three-phase line, **4**–18 to 19
Rectification, unity power factor, **17**–49 to 62
Rectifiers
   asymmetrical silicon-controlled rectifiers (ASCR), **1**–4
   boost type, **4**–33 to 38
   controlled, **4**–4 to 5, **4**–21 to 33
   DC-link choke, **4**–19 to 20
   diode-bridge, power quality, **17**–49 to 50
   diode-bridge rectifiers, **17**–49 to 50
   diode conduction, **4**–5 to 6
   dv/dt, **4**–23
   filters, **4**–20
   full-wave, **4**–3 to 4, **4**–9 to 12
   fundamentals, **4**–4 to 5, **4**–6, **4**–33
   gate circuit requirements, **4**–25
   gate current injection, **4**–23 to 25
   half-wave, **4**–1 to 3, **4**–6 to 9
   harmonics, **4**–16 to 19, **4**–21
   H-bridge circuits, **4**–25 to 29
   HVDC transmission systems, **4**–30 to 31
   indirect current control, **4**–34 to 38
   phase-controlled, power quality, **17**–49 to 50
   phase-controlled rectifiers, **17**–49 to 50
   phase multiplication, **4**–20 to 21
   silicon-controlled, **1**–18
   single-phase, **4**–1 to 4, **4**–6 to 9, **4**–25 to 29
   temperature, **4**–23
   three phase, **4**–12 to 21, **4**–31 to 33
   thyristors, **4**–25 to 29, **4**–31 to 33
   uncontrolled, **4**–4 to 21
   Vienna rectifier, **17**–59
   voltage, **4**–14, **4**–23, **4**–30
Repetitive peak reverse voltage, **4**–2 to 3, **4**–4
Resonant converters. *See also* Inverters
   application of techniques, **5**–21 to 23
   class E, **5**–34 to 36
   connections, **5**–22 to 23
   discontinuous mode, **5**–32 to 33
   fundamentals, **5**–25
   load type, **5**–30 to 36
   parallel-loaded type, **5**–36
   parallel-resonant converters (PRC), **5**–33 to 34
   resonant DC-link, **5**–40 to 41
   second-order circuits, **5**–25 to 30
   series-loaded type, **5**–36
   series-resonant converters (SRC), **5**–31 to 32
   switch converters, **5**–36 to 40
   techniques, **5**–21 to 23
   time functions, **5**–30
   two-quadrant ZVS type, **5**–39 to 40
   ZCS configuration, **5**–36 to 37, **5**–39
   ZVS configuration, **5**–38 to 41

Resonant DC-link converters, ZVS configuration, 5–40 to 41
Resonant DC-link inverters (RDCLI)
  continuous PWM control, 5–63
  disadvantages, 5–60
  fundamentals, 5–56 to 58
  inverter, 5–58 to 63
  parallel-resonant DC-link inverter (PRDCLI), 5–63 to 65
  peak voltage reduction, 5–60 to 63
  research trends, 5–65 to 66
  topology, 5–58 to 59
  ZVS achievement, 5–59
Resonant switch converters
  fundamentals, 5–21 to 22, 5–36
  two-quadrant ZVS type, 5–39 to 40
  ZCS configuration, 5–36 to 37, 5–39
  ZVS configuration, 5–38 to 39
Reverse-biased safe operation area (RBSOA)
  dynamic avalanche, 1–76
  MTO thyristors, 1–87
  semiconductor switch requirements, 1–62 to 65
  UMOS IGBT structure, 1–81
  unity gain turn-off, 1–84, 1–85
Reverse conducting thyristors (RCT), 1–4
Reverse recovery phenomenon, 4–24
RFO. *See* Rotor flux orientation (RFO)
rms forward current, diode characteristic, 4–2, 4–4
Rotary type UPS, 16–6 to 7
Rotor flux orientation (RFO), 11–8 to 11

S

SABER programming code, 4–38 to 41
Saliency ratio, 13–10
Scalar control, induction machines, 11–2 to 4
Schmitt trigger, 7–16, 15–6
Van Schoor studies, 17–38
Schottky diodes, 1–15 to 17, 5–43. *See also* Diodes
SCRs. *See* Silicon-controlled rectifiers (SCRs)
SCSs. *See* Silicon-controlled switches (SCSs)
Secondary breakdown effect, 1–6, 1–7
Second-order resonant circuits, 5–25 to 30
Second quadrant, 4–28
Semiconductor Equipment and Materials International (SEMI), 17–2
Semiconductor switches, 1–60 to 65
Sensorless control, 13–16 to 19
SEPIC converter
  CCM operation, 17–59
  DCM operation, 17–55 to 56
  sliding-mode control, 8–2, 8–9, 8–18 to 20
Sequences, harmonic, 17–7 to 8
Series active filters, 17–36, 17–39 to 43
Series control scheme, UPFCs, 20–14 to 16
Series converter, UPFCs, 20–6 to 7
Series passive AC input reactor, 17–22 to 23
Series passive filter, 17–21 to 22
Series-resonant converters (SRCs), 5–30 to 33

Series-resonant inverters
  above resonance operation, 5–45
  below resonance operation, 5–43 to 45
  circuits and waveforms, 5–42 to 45, 5–48 to 50, 5–51 to 52
  fundamentals, 5–42, 5–55
  transfer function, voltage, 5–45 to 47, 5–50 to 51, 5–52 to 55
  voltage-source parallel-resonant, 5–48 to 51
  voltage-source series-parallel-resonant, 5–51 to 55
  voltage-source series-resonant, 5–45 to 47
Servo drives, 13–16, 15–1 to 6
Shepherd and Zand studies, 5–4
Shortt and Lee studies, 2–19
Shunt active filters, 17–36, 17–43, 17–44 to 47
Shunt control scheme, UPFCs, 20–16
Shunt passive filter, 17–22
Signals, switching, 7–33 to 35
Silicon-controlled rectifiers (SCRs). *See also* Thyristors
  characteristics, 1–18 to 19
  choppers, 2–1
  drives, 9–5 to 6
  fundamentals, 1–18
  limitations, 1–65
  ratings, 1–20
  turn-off circuits, 1–19 to 20
Silicon-controlled switches (SCSs), 1–21 to 22
Simulation, computer. *See* Computer simulation
Simulation Program with Integrated Circuit Emphasis (SPICE), 23–1. *See also* Computer simulation
Single cycle surge current, 4–2, 4–4
Single-phase scheme
  active PFC circuits, 17–53 to 56
  inverters, 5–3 to 4
  rectifiers, 4–1 to 4, 4–6 to 9, 4–25 to 29
Single-stage applications, 17–56
Single-switch three-phase PFCs, 17–57 to 59
Sinusoidal current boost-type rectifier, 4–34 to 38
SIPMOS, 1–5
Six-state induction motor model, 11–6 to 7
Six-step move, open loop control, 14–13 to 16
Six-switch PWM converters, 17–57
Skibinski, Divan and, studies, 5–62
Slew rates, 1–83
Sliding line, 8–3
Sliding-mode switch-mode power supplies
  boost DC-DC converters, 8–14 to 18
  buck–boost DC-DC converters, 8–14 to 18
  buck DC-DC converters, 8–9 to 14
  control implementation, 8–20 to 22
  Cúk DC-DC converters, 8–18 to 20
  current limitation, 8–13 to 14
  DC-DC converter principles, 8–8 to 9
  equivalent control, 8–6 to 7
  existence condition, 8–6, 8–12 to 13, 8–19
  fundamentals, 8–1 to 5, 8–22
  hitting condition, 8–19 to 20
  phase-plane description, 8–9 to 10

reaching conditions, 8–6
SEPIC DC-DC converters, 8–18 to 20
sliding line selection, 8–11
stability, 8–7 to 8, 8–16 to 18, 8–20
system description, 8–6 to 7
theory, 8–5 to 8
Small-signal state space averaged model, 8–16
Small signal transfer functions, 2–24 to 25
SMPS. *See* Switch-mode power supplies (SMPS)
Snubbers, 1–72 to 73, 1–76, 5–43, 5–58
Soft-switching, 5–22
Software. *See* Computer simulation
Solar cells, 18–1 to 4. *See also* Photovoltaic (PV) cells and systems
Solid-state breaker, 19–8 to 9
Source-based current regulation, voltage, 6–12 to 14
Space vector modulation, 11–11 to 12
Space-vector pulse width modulation (SVPWM)
    fundamentals, 7–28
    implementation, 7–32 to 33
    operations, 7–28 to 32
    switching signals, 7–33 to 35
Speed control, DC motors, 9–2 to 3
SPICE program, 2–22, 23–1 to 2. *See also* Computer simulation
Spikes, currents, 5–43
Spreading velocity, 1–68
Square-wave operation, 5–9 to 11
SRCs. *See* Series-resonant converters (SRCs)
SRMs. *See* Switched reluctance machines (SRMs)
Stability
    boost and buck–boost DC-DC converters, 8–16 to 18
    Cúk and SEPIC DC-DC converters, 8–20
    sliding-mode theory, 8–5, 8–7 to 8
Standards
    555, 17–50
    C62.41, 17–3
    IEC 1000, 4–15, 17–50
    IEEE 519, 4–15, 4–17 to 18, 17–2, 17–18 to 19, 17–29 to 30
    UL 1741, 18–4, 18–6, 18–10
Standby power supplies, 16–5 to 6
Static topologies, UPS, 16–3 to 6
Static VAR compensators, 1–8
Stator windings, 10–3
Steady state modeling, UPFCs, 20–9 to 11
Step-down (buck) converters
    continuous-conduction mode, 2–9 to 10
    discontinuous-conduction mode, 2–10 to 11
    fundamentals, 2–4
    ideal circuits, 2–8 to 9
Step motor drives
    bifilar-wound hybrid motor, 14–11 to 12
    control, 14–12 to 18
    excitation, 14–13
    fundamentals, 14–1
    hybrid motor, 14–5 to 8, 14–11 to 12
    modeling, 14–8 to 12
    open-loop control, 14–13 to 18
    operation, 14–2 to 8
    permanent magnet (can-stack) motor, 14–5 to 8
    types, 14–2 to 8
    variable reluctance motors, 14–3 to 5, 14–8 to 11
Step-up (boost) converters
    continuous-conduction mode, 2–12 to 13
    discontinuous-conduction mode, 2–13 to 14
    fundamentals, 2–4 to 5
    ideal circuit, 2–12
Storage issues, 1–83, 18–7 to 10
Strycula, Gyugyi and, studies, 17–30
Sunlight conversion. *See* Photovoltaic (PV) cells and systems
Superconducting magnetic energy storage, 16–9
Surge current rating, 1–20
Survey of second-order resonant circuits, 5–25 to 30
SVPWM. *See* Space-vector pulse width modulation (SVPWM)
Switch, high current, low-to-high commutation, 5–78 to 80
Switch, low current, low-to-high commutation, 5–78
Switched reluctance machines (SRMs)
    advance angle calculation, 13–15
    advantages, 13–1 to 2
    applications, 13–19
    configuration, 13–2 to 3
    control strategies, 13–14 to 16
    converter topologies, 13–11 to 14
    current-controlled drive, 13–16
    design, 13–9 to 11
    energy conversion, 13–5 to 6
    fundamentals, 13–1 to 2
    operations, 13–4 to 8
    sensorless control, 13–16 to 19
    torque, 13–6 to 11
    voltage balance equation, 13–4 to 5
    voltage-controlled drive, 13–15
Switches
    buck–boost converter, 2–26 to 27
    Hall, 10–9
    pulse tester, 1–89 to 91
    testing, 1–89 to 100
Switching
    frequency, 1–77
    MOSFET response, 1–54 to 56
    performance, 1–95 to 98
    signals, 7–33 to 35
Switch-mode power supplies (SMPS)
    boost DC-DC converters, 8–14 to 18
    buck–boost DC-DC converters, 8–14 to 18
    buck DC-DC converters, 8–9 to 14
    control implementation, 8–20 to 22
    Cúk DC-DC converters, 8–18 to 20
    current limitation, 8–13 to 14
    DC-DC converter principles, 8–8 to 9
    equivalent control, 8–6 to 7

existence condition, **8**–6, **8**–12 to 13, **8**–19
fundamentals, **2**–1 to 2, **8**–1 to 5, **8**–22
hitting condition, **8**–19 to 20
phase-plane description, **8**–9 to 10
reaching conditions, **8**–6
SEPIC DC-DC converters, **8**–18 to 20
sliding line selection, **8**–11
stability, **8**–7 to 8, **8**–16 to 18, **8**–20
system description, **8**–6 to 7
theory, **8**–5 to 8
System configuration classification, **17**–36 to 38

T

Takeda studies, **17**–38, **17**–46
Tarter studies, **5**–2
TC4469COE driver, **23**–17
TCI428COA driver, **23**–14 to 17
Telecommunications, DC generation applications, **16**–8
Temperature, **4**–23, **22**–2
Testing
    bipolar junction transistors (BJT), **1**–31
    diodes, general, **1**–14
    Schottky diodes, **1**–15
    switches, **1**–89 to 100
Thevenin's equivalent circuit, **17**–24
Third quadrant, **4**–28
Three-phase schemes
    bridge connections, **5**–22 to 23
    four-wire systems, **17**–44 to 45
    H-bridge circuits, **4**–29
    inverters, **5**–4 to 6, **5**–83 to 87
    line reactors, **4**–18 to 19
    neutral conductors, **17**–13
    PFC circuits, **17**–57 to 60
    rectifiers, **4**–12 to 21, **4**–29 to 33
Threshold voltage, MOSFETs, **1**–53 to 54
Thyristors. *See also* specific type of thyristor
    data sheets for typical, **1**–22 to 28
    fundamentals, **1**–2 to 3, **1**–18, **4**–21, **5**–21
    rectifiers, **4**–25 to 29, **4**–31 to 33
    silicon-controlled rectifiers, **1**–18 to 20
Time-domain, **17**–38
Time functions, **5**–30
TMOS, **1**–5
Tolbert studies, **5**–2
Torque
    brushless DC (BLDC) motors, **10**–4, **10**–6, **10**–11 to 14
    controlling, **7**–12
    current-based drives, **12**–7
    DC motors, **4**–26 to 28
    PMSM drives, **12**–4
    servo drives, **15**–5
    switched reluctance machines (SRM), **13**–6 to 8, **13**–6 to 11
    voltage-based drives, **12**–8 to 9
Traction drives, **1**–4

Transconductance, MOSFETs, **1**–52 to 53
Transfer function, **5**–42 to 45, **5**–50 to 55
Transformers, **17**–13 to 14
Transient response, MOSFETs, **1**–54 to 56
Transistor PWM DC drives, **9**–4 to 5
Transistors, **1**–5, **5**–70 to 73
Transmission lines, UPFCs, **20**–2 to 4
Transmission systems, **4**–30 to 31
Trapezoidal motor, **10**–4
Triacs, **1**–2 to 3, **1**–21
True power factor, **17**–15
Trzynadlowski studies, **5**–6
Tuned harmonic filters, **17**–26 to 28
Turn-off circuits. *See also* Circuits
    advantages, **1**–84 to 85
    gate turn-off thyristors, **1**–69 to 71, **1**–72 to 73
    silicon-controlled rectifiers, **1**–4, **1**–19 to 20
    unity gain operation, **1**–82 to 85
Turn-off losses, transistors, **5**–72 to 73
Turn-on losses, transistors, **5**–70 to 72
Two phase-on excitation, **14**–13
Two-quadrant choppers, **2**–6 to 7. *See also* Choppers
Two-quadrant ZVS configuration, **5**–39 to 40

U

UL 1741, standard, **18**–4, **18**–6, **18**–10
Uncontrolled rectifiers
    DC-link choke, **4**–19 to 20
    diode conduction, **4**–5 to 6
    full-wave, **4**–3 to 4, **4**–9 to 12
    fundamentals, **4**–4 to 5
    half-wave, **4**–1 to 3, **4**–6 to 9
    harmonics, **4**–16 to 19, **4**–21
    phase multiplication, **4**–20 to 21
    single-phase, **4**–1 to 4, **4**–6 to 9
    three-phase, **4**–12 to 21
    voltage, **4**–14
Unidentified loads, **17**–30 to 31
Unified power flow controllers (UPFCs)
    automatic power control, **20**–7 to 9
    case study, **20**–20 to 24
    control design, **20**–14 to 20
    damping control, **20**–16 to 20
    dynamic model, **20**–11 to 12
    fundamentals, **20**–1 to 2, **20**–24 to 25
    fuzzy control, **20**–16 to 20
    load flow model, **20**–9 to 11
    modeling, **20**–9 to 14
    operation, **20**–4 to 9
    power network interface, **20**–12 to 14
    series control scheme, **20**–14 to 16
    series converter, **20**–6 to 7
    shunt control scheme, **20**–16
    steady state model, **20**–9 to 11
    transmission lines, **20**–2 to 4
Uninterruptible power supplies (UPS)
    AC generators, **16**–9 to 10

alternate AD and DC sources, **16**–7 to 10
basic integration, **16**–2
batteries, **16**–7 to 8
DC generators, **16**–8 to 9
double-conversion, **16**–3 to 4
enhanced integration, **16**–2
functions, **16**–1 to 3
line-interactive, **16**–4 to 5
MOS-controlled thyristors, **1**–8
network integration, **16**–2 to 3
power conditioning, **16**–1
rotary type, **16**–6 to 7
standby power supplies, **16**–5 to 6
static topologies, **16**–3 to 6
superconducting magnetic energy storage, **16**–9
system integration, **16**–2 to 3
Unipolar excitation, BLDC motors, **10**–7 to 8
Unity gain, **1**–82 to 85, **1**–92 to 94
Unity power factor rectification
    active solutions, **17**–52 to 62
    boost converter-based PFC, **17**–53 to 54
    buck-derived three-phase PFC, **17**–59 to 60
    current control, **4**–34 to 38
    DC-DC converter topologies, **17**–54 to 55
    DCM operation, **17**–55 to 56
    diode-bridge rectifiers, **17**–49 to 50
    filters, actives, **17**–60 to 62
    fundamentals, **17**–49, **17**–62
    harmonic limits, **17**–50 to 51
    passive solutions, **17**–52
    PFC converters, **17**–55 to 56
    phase-controlled rectifiers, **17**–49 to 50
    single-phase active PFC circuits, **17**–53 to 56
    single-stage applications, **17**–56
    single-switch three-phase PFCs, **17**–57 to 59
    six-switch PWM converters, **17**–57
    three-phase PFC circuits, **17**–57 to 60
    Vienna rectifier, **17**–59
UPFCs. *See* Unified power flow controllers (UPFCs)
UPS. *See* Uninterruptible power supplies (UPS)
Utility interactive systems, **18**–4 to 7. *See also* Photovoltaic (PV) cells and systems

V

Vacuum/blower SRMs applications, **13**–19
Variable frequency drive (VFD), **4**–17
Variable reluctance motors, **14**–3 to 5, **14**–8 to 11
Variable-speed constant frequency (VSCF) systems, **21**–2
Variable structure systems (VSS), **8**–1 to 3
Vector control, induction machines
    direct torque control, **11**–11 to 17
    field-oriented control, **11**–8 to 11
    formulation, **11**–4 to 6
    modeling, **11**–6 to 8
    space vector modulation, **11**–11 to 12

Vehicles. *See* Automotive applications
Very large scale integrated (VLSI) circuits, **1**–45
Vibration suppression, **12**–9
Vienna rectifier, **17**–59
VLSI circuits. *See* Very large scale integrated (VLSI) circuits
Voltage balance equation, **13**–4 to 5
Voltage-based drives, **12**–8 to 9
Voltage-controlled drive, **13**–15
Voltages
    blocking reverse, **1**–71, **4**–24
    breakdown, MOSFETs, **1**–49 to 51
    breakover, **1**–18, **1**–21
    DC-AC conversion, **5**–11 to 20
    diodes, **1**–54, **4**–2 to 3
    filtering, **5**–20
    forward, drop, **4**–3, **4**–4
    IEEE standards, **17**–3
    IGBT drop, **1**–6
    more-electric cars, **21**–7
    multilevel modulation, **6**–2 to 7
    nonrepetitive peak reverse, **4**–3, **4**–4
    output, **4**–2, **4**–4
    passive harmonic filters, **17**–24
    peak inverse voltage, **1**–10 to 11
    power conditioning, **17**–31 to 32, **17**–46 to 47
    rectifiers, **4**–14, **4**–23, **4**–30
    reduction on DC-link, **5**–60 to 63
    repetitive peak reverse, **4**–2 to 3, **4**–4
    reverse blocking, **1**–71, **4**–24
    SPWM signals, **7**–9 to 10
    threshold, MOSFETs, **1**–53 to 54
Voltage star, **11**–12
Vorpérian studies, **2**–19
VSCF systems. *See* Variable-speed constant frequency (VSCF) systems
VSS. *See* Variable structure systems (VSS)

W

Ward Leonard Motor-Generator (M-G) systems, **9**–2 to 3
Washing machine SRM applications, **13**–19
Watanabe studies, **17**–35
Waveforms
    air-gap flux density, **10**–4
    class D, **5**–42, **5**–52
    Fourier series, **17**–4 to 6
    full-wave rectifiers, **4**–9
    notching, **17**–11
    parallel-resonant inverters, **5**–48 to 50
    positive pulse output, **4**–1
    series-parallel-resonant inverters, **5**–51 to 52
    series-resonant inverter, **5**–42 to 45
    single-phase half-wave rectifier circuits, **4**–8
    six-pulse rectifier system, **4**–13
    start-up transient response, **7**–24 to 25

Index

three-phase full-wave rectifiers, **4**–12
turn-on charging current, **4**–11
Weak fields, **9**–4
Wester and Middlebrook studies, **2**–19
Wireless communication, MOSFETs, **1**–59
Van Wyk studies, **17**–38

Y

Yamanashi test line, **17**–43
Yoshida studies, **17**–44

Z

Zand, Shepherd and, studies, **5**–4
ZCS configuration
    parallel-resonant DC-link inverter, **5**–63 to 64
    resonant converter techniques, **5**–21, **5**–23
    resonant switch converters, **5**–36 to 37
    ZVS comparison, **5**–39
Zero harmonic sequence, **17**–7 to 8
Zhang studies, **17**–35
ZVS configuration
    parallel-resonant DC-link inverter, **5**–63 to 65
    resonant converter techniques, **5**–21 to 22
    resonant DC-link converters, **5**–40 to 41
    resonant DC-link inverters, **5**–59 to 60
    resonant switch converters, **5**–38 to 39
    two-quadrant type, **5**–39 to 40
    ZCS comparison, **5**–39
Van Zyl studies, **17**–38